D0074017

TARGETS IN HETEROCYCLIC SYSTEMS

Chemistry and Properties

Volume 8 (2004)

Reviews and Accounts on Heterocyclic Chemistry

http://www.sci.uniba.it

Editors

Prof. Orazio A. Attanasi

Institute of Organic Chemistry, University of Urbino
Urbino, Italy

and

Prof. Domenico Spinelli

Department of Organic Chemistry, University of Bologna
Bologna, Italy

Italian Society of Chemistry
Division of Organic Chemistry
Division of Medicinal Chemistry
Division of Mass Spectrometry

Published by:

Società Chimica Italiana

Viale Liegi, 48

00198 Roma

Italy

Volume 1 (1997) First edition 1997	ISBN 88-86208-24-3	
Second edition 1999	ISBN 88-86208-24-3	
Volume 2 (1998) First edition 1999	ISBN 88-86208-11-1	
Volume 3 (1999) First edition 2000	ISBN 88-86208-13-8	
Volume 4 (2000) First edition 2001	ISBN 88-86208-16-2	
Volume 5 (2001) First edition 2002	ISBN 88-86208-19-7	
Volume 6 (2002) First edition 2003	ISBN 88-86208-23-5	
Volume 7 (2003) First edition 2004	ISBN 88-86208-28-6	ISSN 1724-9449
Volume 8 (2004) First edition 2005	ISBN 88 86208 29 4	ISSN 1724-9449

Printed and bound in Italy by:

Arti Grafiche Editoriali s.r.l.

Via S. Donato, 148/C

61029 Urbino (Pesaro-Urbino)

Italy

May 2005

Indexed/Abstracted in: Chemical Abstracts; Current Contents; ISI/ISTP&B online database; Index to Scientific Book Contents; Chemistry Citation Index; Current Book Contents; Physical, Chemical & Earth Sciences; Methods in Organic Synthesis; The Journal of Organic Chemistry; La Chimica e l'Industria; Synthesis.

Preface

This book is dedicated to the memory of our friend Marino Novi
(Ω December 27, 2003)

Heterocyclic derivatives are important in organic chemistry as products (including natural) and/or useful tools in the construction of more complicated molecular entities. Their utilization in polymeric, medicinal and agricultural chemistry is widely documented. Both dyestuff structures and life molecules frequently involve heterocyclic rings that play an important role in several biochemical processes.

Volume 8 (2004) keeps the international standard of THS series and contains fifteen chapters, covering the synthesis, reactivity, activity (including medicinal) and mass spectrometry of different heterorings. Reviews from France, Germany, Italy, Portugal, and Spain are present in this book.

Comprehensive Reviews reporting the overall state of the art on wide fields as well as personal Accounts highlighting significative advances by research groups dealing with their specific themes have been solicited from leading Authors. The submission of articles having the above-mentioned aims and concerning highly specialistic topics is strongly urged. The publication of Chapters in THS is free of charge. Firstly a brief layout of the contribution proposed, and then the subsequent manuscript, may be forwarded either to a Member of the Editorial Board or to one of the Editors.

The Authors, who contributed most competently to the realization of this Volume, and the Referees, who cooperated unselfishly (often with great patience) spending valuable attention and time in the review of the manuscripts, are gratefully acknowledged.

The Editors thank very much Dr. Lucia De Crescentini for her precious help in the editorial revision of the book.

Orazio A. Attanasi and Domenico Spinelli

Editors

Table of Contents

(for the contents of Volumes 1-7 please visit: http://www.sci.uniba.it)

I

Five-membered heteroaryl azides and derived iminophosphoranes: useful intermediates for the construction of *b*-fused pyridine ring systems

Maria Funicello and Piero Spagnolo

Syntheses of polyhydroxylated azepanes

Soledad Pino-González, Carmen Assiego and Noé Oñas

PORPHYRINS, PHTHALOCYANINES AND RELATED COMPOUNDS AS MATERIALS FOR OPTICAL LIMITING

Danilo Dini, Guo Y. Yang and Michael Hanack[*]

University of Tübingen, Institute of Organic Chemistry, Auf der Morgenstelle 18,

Tübingen D-72076, Germany (e-mail: hanack@uni-tuebingen.de)

Abstract. *In the present work the nonlinear optical properties, namely optical limiting (OL), of metal complexes based on tetrapyrrolic macrocycles as donors are reviewed. The description of the OL properties of more than one hundred complexes among porphyrins, phthalocyanines and naphthalocyanines dispersed either in a liquid or in a self-standing matrix has been carried out. In particular, the influence of all relevant structural factors, e.g. extent of electronic conjugation, nature of the central metal, ring substituent, axial ligand and number of conjugated macrocycles per molecular unit, on the OL effect produced by this class of complexes has been systematically analyzed.*

Contents

1. Introduction

The macrocycles obtained by joining four pyrrolic units at their α-positions through carbon or nitrogen atoms constitute a broad class of compounds having a unique combination of photochemical,[1,2] redox[3,4] and

1

coordination properties.[5-9] The structures of the simplest tetrapyrrolic macrocycle, *i.e.* porphyrin (Por), and its most important derivatives porphyrazine (Pz), phthalocyanine (Pc) and naphthalocyanine (Nc) (Figure 1) are characterized by the presence of an extended network of conjugated π-electrons whose transition energies mostly fall in the range of the UV-Vis region.[10-12]

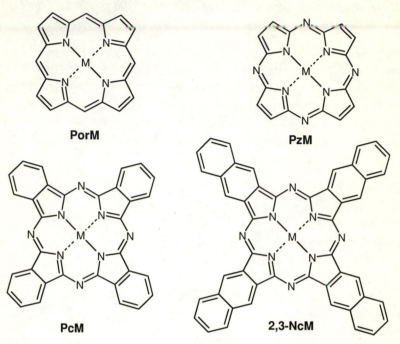

PorM

PzM

PcM

2,3-NcM

Metalloporphyrin (PorM); metalloporphyrazine (PzM); metallophthalocyanine (PcM) and
2,3-metallonaphthalocyanine (NcM).

Figure 1

This infers typical colours to these macrocyclic compounds ranging from red, green, or blue to yellow depending on the extension of the conjugated electronic network.[13] In addition, these cyclic structures present a sufficiently large central cavity with an approximate area of $7\text{-}8 * 10^{-2}$ nm^2,[14,15] which can host a great number of coordinating central atoms (Figure 2).[16]

Among various important examples of tetrapyrrolic macrocycles,[17] the magnesium-porphyrin complex *chlorophyll*, and the iron-porphyrin complex *heme*, are probably the best known for their determining roles as *in vivo* photoreceptor in the synthesis of carbohydrates and oxygen from carbon dioxide and water,[18] and as prosthetic group of numerous proteins, respectively.[19,20] On the other hand, not less relevant is the role of artificial cyclic tetrapyrrolic complexes in modern technology as demonstrated by their extensive use in, *e.g.*, xerography,[21] dying processes,[22] advanced printing and imaging technologies,[23] photovoltaic cells,[24] electrochromic devices,[25,26] organic conductors,[27a,27b] Langmuir-Blodgett films,[27c,27d] sensors,[27e,27f] and photodynamic therapy[27g,27h] among others.[28-30]

As a class of conjugated molecules, tetrapyrrolic macrocycles possess nonlinear optical (NLO) properties,[31-37] *i.e.* change their optical properties with the variation of the intensity of the radiation which interacts with them. The study of these properties accompanied by the parallel work of synthesis of materials with improved NLO characteristics is at the basis of the construction of new devices for photonics.[38-40] Among the various NLO phenomena, the effect of optical limiting (OL), consisting in the reduction of the transmitted intensity through a material or a device with the increasing intensity of the incident radiation,[33,34,41] permits the realization of smart optical devices[42] capable to protect light-sensitive elements.

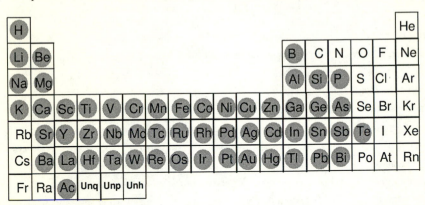

Periodic table of the elements. The shadowed boxes indicate all the elements of which tetrapyrrolic macrocycles complexes of Figure 1 are known.

Figure 2

Metal complexes of tetrapyrrolic macrocycles have proved to be one of the most effective classes of materials in the generation of OL effect in the UV-Vis region.[33,34] In fact, a great amount of results concerning the OL effect produced by conjugated tetrapyrrolic macrocycles is available following the continuous efforts from synthetic chemist and experimental physicist both involved in finding suitable and improved molecular systems for OL purposes. The present review on conjugated tetrapyrrolic macrocycle-metal complexes for OL starts with the theoretical background of the OL effect and then analyzes the implications of OL at a molecular level. In the second part the data concerning the OL effect generated by various classes of tetrapyrrolic macrocycles-metal complexes, are analyzed, discussed and ordered according to the type of conjugated macrocycle. The different classes of tetrapyrrolic macrocycles are presented with the order Pzs, Pors, Pcs and Ncs, following the criterium of increasing extension of the conjugated π-electron network on the basic skeleton.[43] The effect of the central metal M upon the OL properties of the resulting complexes[44] will be discussed within the same set of conjugated macrocycles.

2. The optical limiting (OL) effect

2.1. Intensity-dependent optical properties

Some special materials change reversibly their optical properties with the variation of the radiation intensity I. These effects are verified when I reaches a material-specific critical value I_{lim}, at which the material undergoes a reversible physical and/or chemical transition leading to the modification of the optical properties of the material itself. In general, light intensity modifies the absorptive and refractive properties of the illuminated system once $I > I_{lim}$. In the case of molecular species the extinction coefficient k can vary with the intensity I according to the relationship:[45]

$$k = k_0 \ [1/ \ 1+(\ I/I_{lim})]$$ (1)

with k_0 corresponding to the low intensity limit value of k. Equation (1) describing the optical behavior of a saturable absorber,[45] expresses the fact that the extinction coefficient k decreases with the increase of the incident intensity. In other words, a saturable absorber is a system which increases its optical transmittance upon increase of the incident light intensity. In doing so, the optical system gets more transparent at higher incident intensities, and behaves as an intensity-activated optical switch.[46,47] On the other hand, in the case of OL systems the opposite situation is verified, *i.e.* the optical system has an extinction coefficient which increases reversibly with the increase of the incident radiation intensity.[48] This phenomenon constitutes the so-called reverse saturable absorption (RSA).[49] Such an absorption-based phenomenon takes mostly place during the irradiation of certain organic dyes, donor-acceptor molecules, fullerenes and, to a less extent, inorganic semiconductors.[41] The OL effect can be generated also by means of several mechanisms other than absorption, which are based on fundamental optical processes. For example, the OL effect based on radiation refraction or scattering[45,50] does not allow the formation of a well-resolved image once the incident light rays have interacted with the limiting system. As a consequence, optical limiters based on phenomena other than absorption have the practical limitation of not being useful for the protection of those complex light-sensitive elements, *e.g.* the human eye, employed in direct viewing devices which require a clear vision of the surrounding environment. For this reason among the available systems for the limiting of intense radiations, those based upon the phenomenon of absorption are preferable.[50-52]

2.2. Models for absorbing optical limiters

The first theoretical approach leading to the definition of relationships between the net absorption coefficient α of molecular species acting like optical limiters and incident light intensity I was proposed by Hercher:[53]

$$\alpha = (\alpha_0 I_{lim} + \alpha_1 I) \ / \ (I + I_{lim})$$ (2)

where α_0 is the low intensity value of α, and α_1 is the main term of the absorption coefficient above the critical incident intensity. The OL effect occurs at the condition $\alpha_1 > \alpha_0$.

Equation (2) retains its validity if the absorption mechanism of the molecular OL system is based on the excited state absorption which involves five relevant energy levels (Figure 3).

In the five-level model the OL system absorbs at the frequency of interest ω^* through the two successive transitions [1→3] and [2→4] (or alternatively [3→5]) being $(h\omega_{13}/2\pi)=(h\omega_{24}/2\pi)=(h\omega_{35}/2\pi)$ with the absorption coefficients α_0 and α_1, respectively [Equation (2)]. ω_{ij} indicates the frequency of absorption for the transition [i→j] and h is the Planck's quantum constant. Hence, the five-level model predicts the occurrence of a two-photon absorption process due to the consecutive electronic transitions [1→3] and [2→4] (or alternatively [3→5]).

4

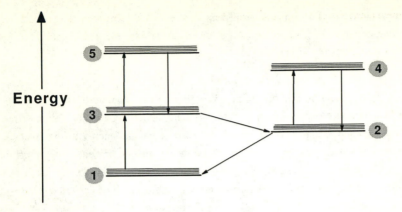

Energy diagram of a five-levels system. Full line arrows indicate absorption from ground [1→3] and excited states [2→4] and [3→5]; dashed line arrows represent intersystem crossing [3→2] and excited state fluorescence [4→2] and [5→3]; dotted line arrow indicates phosphorescence [2→1]. Thick horizontal lines indicate the ground vibrational level for every electronic level; thin horizontal lines indicate excited vibrational levels in the various electronic levels.

Figure 3

In this model the following assumptions are made:

- energy levels have no degeneracy;
- intersystem crossing (ISC) transition [3→2] is fast (it is supposed that levels 1,3 and 5 belong to singlet states);
- decay from the upper excited state [4→2] is fast (it is supposed that levels 2,4 belong to triplet states).

These assumptions are motivated by the fact that effective OL requires fast ISC in order to populate the excited state i=2. Moreover, a high value of the absorption coefficient for the transition [2→4] (or [3→5]) [α_1 in Equation (2)] is also required for the efficient absorption of the second photon. This phenomenology shows that the occurrence of OL by excited state absorbers is generally achieved within a limited range of irradiation frequencies ν by a single optical limiter depending on its specific absorption features.

The definition of the intensity-dependent net absorption coefficient α as in Equation (2), implies the modification of the Lambert-Beer law according to the new formulation:

$$- dI/dz = [(\alpha_0 I_{lim} + \alpha_1 I)/(I + I_{lim})] I \qquad (3)$$

being z the direction of light propagation.

Alternatively, the intensity-dependent net absorption coefficient can be also expressed as:[54,55]

$$\alpha' = \alpha_0 + \beta I \qquad (4)$$

with the corresponding alternative form of the modified Lambert-Beer law:

$$- dI/dz = (\alpha_0 + \beta I) I \qquad (5)$$

In Equations (4) and (5) the term β is defined as:[50,56]

$$\beta = [N_2] (\sigma_{24} - \sigma_{13}) \qquad (6)$$

being $[N_2]$ the population of the level 2 per volume unit (in m^{-3}), σ_{24} and σ_{13} the absorption cross-sections[*] for the transition $[2 \rightarrow 4]$ and $[1 \rightarrow 3]$ (in m^2), respectively (Figure 3). Equation (6) shows that effective optical limiting in the high intensity regime is achieved at two conditions: first, the population of the state which absorbs the second photon is high (from Figure 3 such a state has the energy of the level 2); second, the difference between absorption cross-section values of the second and the first electronic transition must be as large as possible. The latter condition poses some limitations to the real validity of the merit factor κ defined as the ratio σ_{24}/σ_{13},[34] and often used for evaluating the OL performance of excited state absorption (ESA) based molecular systems.[57]

For an ESA based optical limiter, the populations at the levels 1 and 2 in the steady state are given by:[56]

$$N_1(\text{Steady state}) = N_{tot}(1 + I/I_{lim})^{-1} \qquad (7)$$

$$N_2(\text{Steady state}) = N_{tot}(I/I_{lim})/(1 + I/I_{lim}) \qquad (8)$$

being $N_{tot} = N_1 + N_2$ (N_3 and N_4 or N_5 are considered negligible), and $I_{lim} = (\sigma_{13}\tau_{21})^{-1}$. The values of the transition times τ_{jk}, which actually correspond to the lifetimes of the initial states with energy level j, have important implications as far as the feasibility of the system in limiting short light pulses is concerned. Indeed, if the lifetime of the main absorbing state (in the ESA based five levels model this corresponds to the pumped state with energy level 2) is longer than the pulse duration, then the OL response will be fluence F (in $J\,cm^{-2}$) and not intensity (in $J\,cm^{-2}\,s^{-1}$) dependent.[56]

2.3. Nonlinearity of the optical limiting effect

The occurrence of the OL effect is verified when a constant output intensity I_{out} (in $J\,cm^{-2}\,s^{-1}$) is transmitted through the limiter once the incident intensity I_{in} exceeds the system-characteristic threshold value I_{lim}. This means that no matter of how many photons per unit of time will impinge the system, the flux of photons passing through the system remains constant when the irradiation of the system has levels of irradiation with $I_{in} > I_{lim}$. The optical response of an ideal optical limiter is presented in Figure 4. The transmittance T $[=d(I_{out})/d(I_{in})]$ of an ideal optical limiter is not constant within the whole regime of irradiation (nonlinear optical feature), and T is a function of I_{in} with $T \rightarrow 0$ for $I_{in} >> I_{lim}$.[58] A useful parameter for the evaluation of the OL effectiveness of different systems is the intensity threshold defined as the incident intensity value at which the transmittance of the system is equal to the 50% of the linear transmittance.

2.4. Metal complexes of conjugated macrocycles and optical limiting

NLO phenomena[58] occur in those organic compounds that have conjugated π-electrons. In fact, the presence of conjugated π-electrons infers high electronic polarizability and fast charge redistribution when

---

[*] The relationship between the absorption cross-section σ (in m^2) and the molar extinction coefficient ε (in $m^2\,mol^{-1}$) is: $\sigma = \varepsilon / (0.43 * N_A)$, N_A being the Avogadro's number.

interacting with rapidly changing intense electromagnetic fields like those of laser radiations.[59] Moreover, molecules with conjugated π-electrons can give rise to different electronic transitions at the wavelengths of irradiation in the UV-vis range, which are frequently used for the studies of nonlinear optics.[60-63]

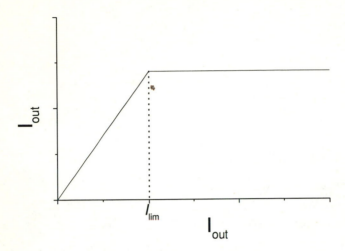

Profile of the output intensity I_{out} passing through an ideal optical limiter versus the incident input intensity I_{in}. In the figure the meaning of the threshold intensity I_{lim} at which I_{out} saturates is shown.

Figure 4

Tetrapyrrolic macrocycles show OL effect mainly through the mechanism of excited state absorption (ESA). This means the OL effect generated by those compounds is an accumulative nonlinearity because it is produced through the initial one photon absorption from the electronic ground state to an excited state and the successive absorption from this one at the frequency of the intense radiation.[50] The feature of the accumulative nonlinearity in optical limiters has practical implications for the development of smart filters[42] which become instantaneously opaque only above a system-specific threshold in order to protect the optical sensor against possible damage by high-intensity radiation.[50]

3. Porphyrins as optical limiters

The structure of a porphyrin can be varied in many ways at the different positions on the porphyrin ring without compromising its chemical stability.[64] This aspect is particularly important for the design of porphyrins and for the optimization of the NLO properties within in the spectral window of high linear optical transmission (Figure 5).[65-67]

3.1. Effect of metal coordination

In order to study the influence of the coordinating central metal M on the third-order optical nonlinearity and nonlinear absorption of PorMs, various techniques were employed for the comparison of different metal tetra-*p*-tolylporphyrins (TTPM).[68,69] The results indicated a very high nonlinear absorption for light pulses with duration in the range 10^{-12}-10^{-9} s at both 532 nm and 600 nm. The OL properties of the metallotetraphenylporphyrins (TPPM) with M=Zn[II], Cu[II], Ni[II], Co[II], V[IV]O and Fe[III]Cl were measured at 527 nm with laser pulses duration of 20 ns. Among these materials, only TPPZn and TPPVO showed OL properties comparable to those for C_{60} at the same linear transmittance at 527 nm. The NLO absorption of a series of metallooctabromotetraphenyl-porphyrins (OBTPPM) (Figure 6) were also measured at 532 nm with a laser pulse duration of 7 ns at a repetition rate of 10 Hz in order to determine the central metal effect for these β-halogenated PorMs.[70]

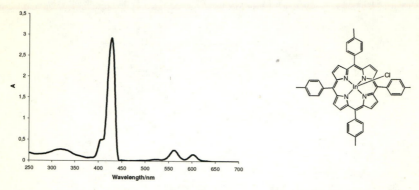

UV-vis spectrum of chloro(meso-tetra-*p*-tolylporphyrinato)indium(III).

Figure 5

The OL performance for the metal-free octabromotetraphenylporphyrin OBTPPH$_2$ and some metallated derivatives showed the following trend: OBTPPZn> OBTPPCd > OBTPPH$_2$> OBTPPPb > OBTPPCu ≈ OBTPPCo.[70,71a] This trend mostly reflected the favourable effect of the diamagnetism of the central metal in improving the OL performance of these brominated porphyrins. In particular, OBTPPZn shows a remarakable OL performance which is superior to those of C_{60} (taken as a standard)[71b] and TPPZn, and comparable to that of the state-of-the-art phthalocyanine (*t*-Bu)$_4$PcInCl.[57]

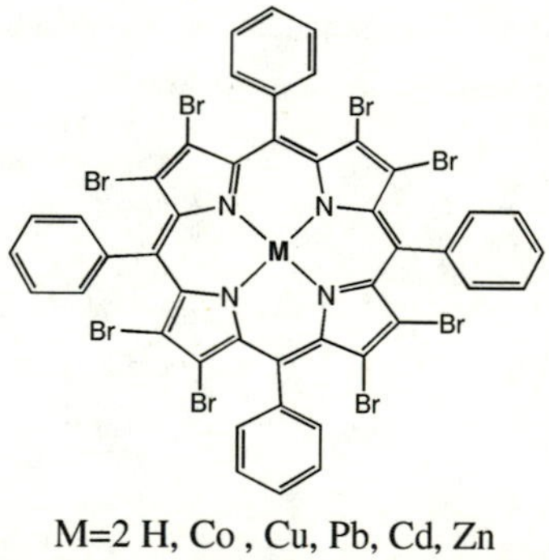

M=2 H, Co , Cu, Pb, Cd, Zn
Metallooctabromotetraphenylporphyrin (OBTPPM).
Figure 6

3.2. Effect of meso functionalization

The possibility of functionalizing a tetrapyrrolic conjugated macrocycle at the *meso* positions constitutes a considerable enrichment in terms of structural versatility of the porphyrins with respect to the tetraazaporphyrin-based molecules like phthalocyanines. The most common *meso*-substituent of a porphyrin

8

is the phenyl ring due to the relatively easiness of preparation.[72,73] One of the first demonstrations of RSA in *meso*-PorMs involved the analysis of a series of *meso*-tetraphenylmetalloporphyrins TPPM with M=Zn, Co and H$_2$ using laser pulses with 80 ps duration at 532 nm.[67]

In porphyrins the substitution of the *meso* hydrogens with conjugated groups like phenyl, alkenyl or alkynyl produces generally a red shift of the optical absorption bands thus unavoidably altering both linear and nonlinear optical properties of these molecules with respect to the unsusbstituted Pors. This approach can lead to the formation of porphyrins showing OL in spectral regions different from those of unsubstituted Pors. One indicative example is given by the substitution at the *meso* positions with the acetylene group in the series of tetra(trimethylsilylethynyl)porphyrins (TSEPMs) shown in Figure 7.[74]

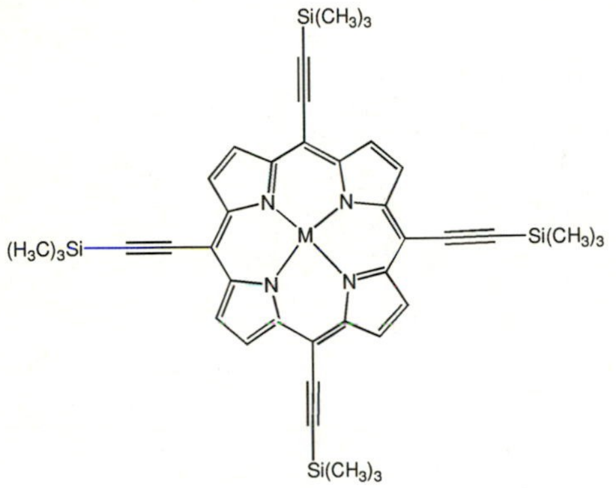

M=Zn , Pb
Metallo *meso*-tetra(trimethylsilylethynyl)porphyrin (TSEPM).

Figure 7

The results showed that by extending the conjugation length the main ground state absorptions of the modified *meso*-alkynyl porphyrins are red-shifted by several tenths of nm. The OL experiments were conducted with ps and ns lasers at 532 nm, and it was found that the heavier atom Pb enhanced the OL performance by increasing ISC rate and the excited state absorption cross-section σ_{24}.[75] Moreover, the OL performances of TSEPZn and TSEPPb were much better than that of the corresponding free base porphyrin. The introduction of a zinc atom into the TSEP chromophore leads to the reduction of the singlet state lifetime, the broadening and shifting of the excited state absorption profile together with the reduction of the ground state absorption cross-section σ_{13}. In particular, TSEPPb (Figure 7) exhibited a large merit factor $\kappa=\sigma_{24}/\sigma_{13}$ of ~45 at 531 nm. Similar results were also obtained for tetrakis(4-butylphenylethynyl)porphyrin and its Zn and Pb metal complexes,[76] with the Pb complex showing a better OL performance. Systematic work on the OL properties of *meso*-substituted porphyrins was also conducted considering the complexes obtained by insertion of atoms from the III and IV groups into tetrakis(4-triisopropylsilylethynyl)porphyrin (ISEPM),[77] with M=Al, Ga, In, Tl, Ge, Sn and Pb. The results showed that the metal complexes of the heavier atoms In[III], Tl[III] and Pb[II] offered the most sensitive response at around 530 nm with values of κ in the range 45–48. These values were the largest reported for porphyrin dyes so far. The In[III] and Tl[III] complexes displayed nonlinear absorption in the high transmission window from 480 to 620 nm and had relatively long triplet lifetimes of ~800 ns even in air saturated solution.

The determination of the OL effect of the metallo *meso*-tetra[*p*-trimethylsilylethynyl (phenyl)]porphyrin (TSEPPM) (Figure 8), of which the Zn complex has similar linear absorption to that for TPPZn but better OL performance at 532 nm,[78] indicates that the excited state absorption can be also influenced *via* fine molecular modification on the *para*-position of the *meso*-phenyl rings without severe changes of the ground state absorption.

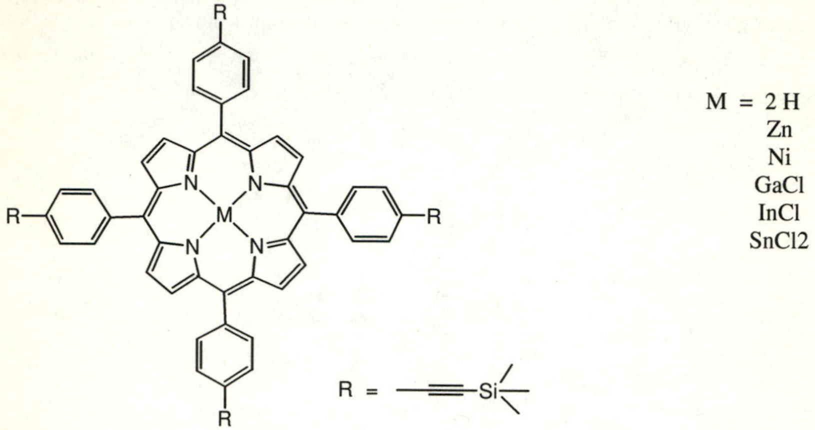
R
N
M
R
M = 2 H
Zn
Ni
GaCl
InCl
SnCl2
R =
Si

Metallo *meso*-tetra[*p*-trimethylsilylethynyl(phenyl)]porphyrin (TSEPPM).

Figure 8

The nonlinear optical response of Pors can be further enhanced *via* insertion of closed-shell metal ions. In particular, the indium complex TSEPPInCl exhibited a larger decrease of nonlinear optical transmission with respect to C_{60}, which was comparable to the one of (t-Bu_4)PcInCl (Figure 9).

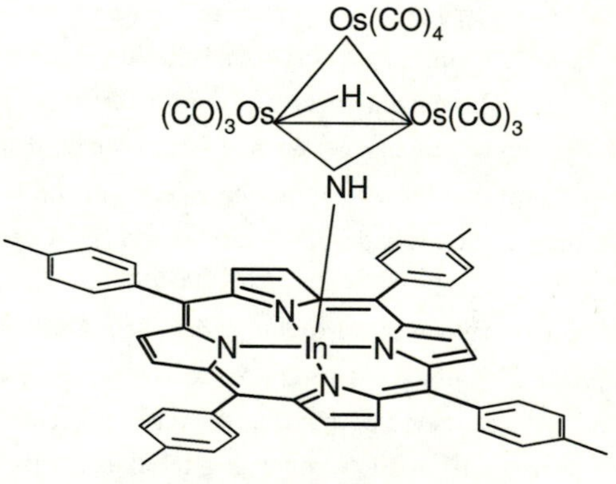
Os(CO)4
(CO)3Os
H
Os(CO)3
NH
N
In

Capped indium tetra-*p*-tolylporphyrin.

Figure 9

In a benzoporphyrin in which the phenyl group is replaced with a thiophene connected to the α-position, both the Q (0,0) and Q (1,0) bands are red-shifted respectively from 650 to 661 and from 604 nm to

611 nm, respectively. OL experiments at 532 nm demonstrated that the merit factor $\kappa = \sigma_{24}/\sigma_{13}$ for *meso*-thiophene substituted benzoporphyrin was 15.0, slightly larger than that of *meso*-phenyl benzoporphyrin ($\kappa = 14.4$).[79]

A new class of porphyrin-osmium cluster complexes (Figure 9) with an axial nitrogen bridge was recently synthesized for nonlinear optical absorption studies.[80] UV-Vis studies revealed a significant perturbation of the ring electronic distribution due to the capping moiety. In addition, the new molecule exhibited large nonlinear optical absorption at 532 nm when irradiated with ns laser pulses.

3.3. Oligomeric porphyrins

Porphyrin oligomers have been also prepared with conjugated groups at *meso* position as bridging moieties.[81] The whole delocalized π system was enlarged in this case maintaining a quasi planar and symmetric structure. The OL properties of these samples were studied at 532 nm with laser pulse duration 500 ps, and it was found that linear transmission decreased of 40%, 60% and 65% at input fluences of 0.01 J/cm^2 for the monomer, dimer and oligomer thin films, respectively. The monomer had the largest ratio $\kappa = \sigma_{24}/\sigma_{13}$ value, mainly because of the lower linear transmittance at the excitation wavelength.

3.4. Expanded porphyrins

OL studies have also been conducted on expanded porphyrins like pentaazadentate porphyrin-like metal complexes (PAPM) (Figure 10).[82-84] These compounds are characterized by the presence of a bridging group connecting the two non-pyrrolic aza groups [*cis*-dicyanoethenyl in Figure 10(a), and dibenzophenone in Figure 10(b)], which can be opportunely modified in order to obtain functional complexes with controlled optical properties. In the complexes of Figure 10 the extension of the conjugated network can be varied and the larger sizes of the central cavity allows a stable coordination environment for large cations.

Good OL performance was found for the dicyanoethenyl bridged complex [Figure 10(a)], probably due to the strong electron-withdrawing ability of the –CN groups.

OL investigations have been conducted also on pentaazadentate porphyrin-like metal complexes, in which the bridging group is a conjugated dibenzophenone [Figure 10(b)].[85] The linear absorption spectrum of this complex is characterized by a strong Q band centered at 779 nm and a B band centered at 429 nm. The molecular second-order hyperpolarizability (the microscopic parameter for third-order nonlinear effect) of this complex was relatively high (9.7×10^{-31} esu), because of the more extensive conjugation in dibenzophenone bridged compound [Figure 10(b)]. The dynamics of the NLO response showed strong fluence dependence and indicated the participation of excited states with relatively long lifetimes in the NLO process. The large third-order nonlinearity, fast time response, and effective RSA within a wider optical windows in the visible region, render dibenzophenone bridged PAPCd [Figure 10(b)] an interesting candidate for OL applications.

4. Porphyrazines for optical limiting

Tetraazaporphyrins, or porphyrazines (Pzs), are the synthetic analogous of porphyrins, in which the four *meso*-C atoms are substituted by nitrogen. The structural modification caused by azasubstitution, introduces differences between the electronic structures of porphyrazines and porphyrins as observed when comparing their absorption spectra.[86,87]

According to a proved theory,[86,87] not only the number of the observed absorption bands but also the nature of the transitions involved differ in passing from porphyrins to tetraazaporphyrins. This does not occur when the linear optical spectra of Pzs and Pcs are confronted since the absorption bands of Pcs only experience a red shift with respect to Pzs due to the extension of the conjugated π-system upon tetrabenzo-annulation, with the nature of the electronic transitions being the same for Pzs and Pcs.

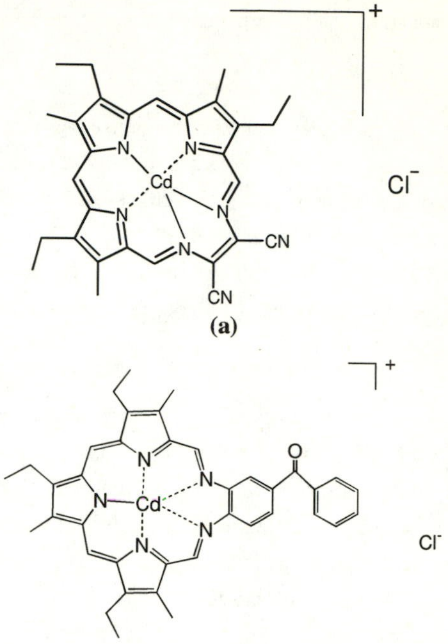

(a)

(b)

Pentaazadentate porphyrin-like complexes.

Figure 10

The ground state absorption spectra of porphyrazines and phthalocyanines, as well as of porphyrins, have been calculated[86-89] and the UV-Vis absorption energies and coefficients are tabulated for many compounds.[89] Different to porphyrins and phthalocyanines, the NLO properties of which have been extensively studied, the number of works on the NLO properties of tetraazaporphyrins is relatively low.[90-94]

The OL performance of tetraazaporphyrins in solutions at 532 nm with 3-7 ns 10 Hz laser pulses, was described for the first time for the series of axially coordinated indium octaarylporphyrazines (Ar$_8$PzInX) with the structures presented in Figure 11.[33,95] In these experiments the nature of the axial ligand was found to be not as important as the nature of peripheral groups in terms of OL effectiveness.

An improvement of the OL effect was observed when electron-withdrawing (EW) substituents were present in the peripheral aryl groups of Ar$_8$PzInX, whereas the presence electron releasing (ER) substituents was associated with lower OL effectiveness. In particular, octa(trifluoromethylphenyl) substituted In-porphyrazines showed the lowest nonlinear transmittance. The latter compounds have shown also a relatively high photostability,[95] which is an important issue to take into account for the practical realization

12

of optical limiters. In addition, there is a decrease of σ_{13} in the region 400-550 nm, which is due to the weak absorption originated by metal-to-ligand charge transfer processes in Ar_8PzInX when passing from EW to ER substituents in the peripheral aryl groups.[95] This constitutes another favourable feature of the octa(trifluoromethylphenyl) substituted complex as active OL materials since OL effectiveness depends also on the difference $(\sigma_{24} - \sigma_{13})$ [Equation (6)].

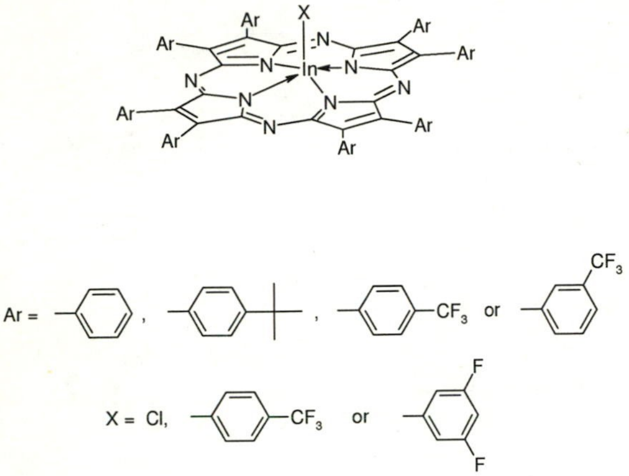

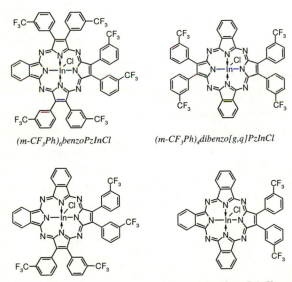

Axially substituted octa(aryl)-porphyrazinato complexes for OL.

Figure 11

$(m\text{-}CF_3Ph)_6benzoPzInCl$

$(m\text{-}CF_3Ph)_4dibenzo[g,q]PzInCl$

$(m\text{-}CF_3Ph)_4dibenzo[l,q]PzInCl$

$(m\text{-}CF_3Ph)_2tribenzoPzInCl$

Mono-, di- and tribenzo-porphyrazinato-indium chloride substituted with six, four and two
m-trifluoromethylphenyl groups, respectively.

Figure 12

13

Recently, the OL effects in the benzo-annulated derivatives of *m*-trifluoromethylphenyl (*m*-CF₃Ph) substituted chloroindium-porphyrazines (Figure 12)[96] were also investigated in our laboratory (Figure 13).[97]

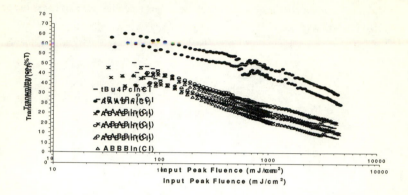

Variations of transmittance for the various *m*-trifluoromethylphenyl substituted benzoporphyrazinato-indium chloride complexes presented in Figure 12.

Figure 13

The values of the effective excited state absorption cross-sections σ_{ex}^{eff},[98] (a parameter which includes contributions from both singlet and triplet excited state absorption cross-sections), were measured at 532 nm with 5 ns pulses at the frequency 20 Hz. The value of σ_{ex}^{eff} increased with the benzo-annulation in the following order: $(m\text{-}CF_3Ph)_8PzInCl$ ($\sigma_{ex}^{eff}=2.39\times10^{-17}$ cm²) < $(m\text{-}CF_3Ph)_6benzoPzInCl$ ($\sigma_{ex}^{eff}=3.00\times10^{-17}$ cm²) < $(m\text{-}CF_3Ph)_4dibenzo[l,q]PzInCl$ ($\sigma_{ex}^{eff}=3.80\times10^{-17}$ cm²) < $(m\text{-}CF_3Ph)_4dibenzo[g,q]PzInCl$ ($\sigma_{ex}^{eff}=4.55\times10^{-17}$ cm²) < $(m\text{-}CF_3Ph)_2tribenzoPzInCl$ ($\sigma_{ex}^{eff}=5.60\times10^{-17}$ cm²). The best OL performance at 532 nm was observed for $(m\text{-}CF_3Ph)_2tribenzoPzInCl$,[97] which resulted comparable to that of $t\text{-}Bu_4PcInCl$.

The third-order NLO properties of another type of porphyrazine-based macrocycles, tetraoctanoyl-/tetraoctanyl substituted tetra-2,3-[5,6-(9,10-phenantro)pyrazino]-porphyrazinato-copper complexes, as well as of tetra-*tert*-butylpyrazinoporphyrazinato-copper, were also reported.[94] Their excited triplet-state (σ_{24} in Figure 3) and excited singlet-state (σ_{35} in Figure 3) absorption cross-sections were determined in solutions at 532 nm with 8 ns laser pulses. The values of σ_{24} for these porphyrazines were found to be an order of magnitude lower than those of copper phthalocyanines, and their RSA behavior was mainly attributed to the excited singlet-state absorption (transition [3→5] in Figure 3). It was concluded that octaazasubstitution in the annulated conjugated rings results in shortening the fluorescence life-time and decreasing the intersystem-crossing rate. The comparison of closed-aperture and open-aperture Z-scan results has also shown the stronger nonlinear refraction effects of these compounds when compared to analogues PcCu derivatives.

5. Phthalocyanines for optical limiting

Similar to porphyrins, phthalocyanines can be thoroughly modified by means of substitution at peripheral positions of the macrocycle or varying the central atom M with its possible axial ligands X.

14

Consequently, the relevant linear and NLO physical properties of these compounds can be varied within a wide range.[34,43]

Because of the larger structure of Pcs with respect to Pors and Pzs, Pcs have a stronger tendency to aggregate at high concentrations. The intermolecular interactions caused by aggregation can provoke faster decays of the excited pumped state with respect to a non aggregated molecule in NLO regimes thus decreasing the OL effectiveness for the decrease of the steady-state value of [N_2] [Equation (6)]. This effect can be present when highly concentrated solutions of Pcs or solid state configurations are analyzed for the determination of the nonlinear optical properties.[99-101] In these conditions the linear spectral properties of Pc thin films are different from those of diluted solutions for the broadening of the absorption bands and the possible shift of the absorption peaks.[102,103] Moreover, the optical susceptibility of Pcs also varies with Pc concentration and extent of aggregation in NLO regimes.[104,105] Extensive aggregation of Pcs can be suppressed by modification of the molecular shape so that the skeletons carrying the network of π-electrons are at a distance which does not allow the interference with each other.

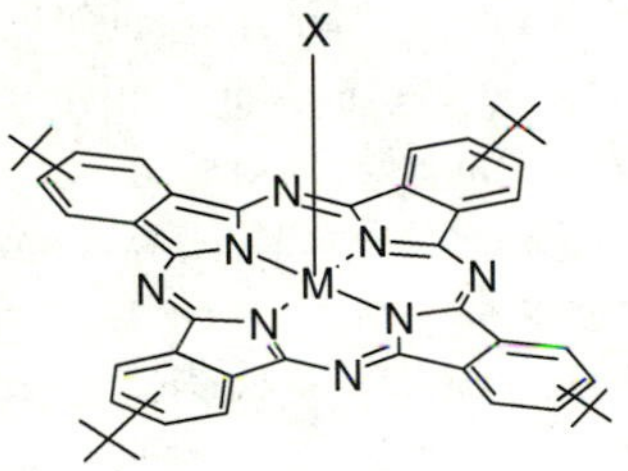

M = Ga, In; X = Cl, I, Aryl, Acetylacetonate

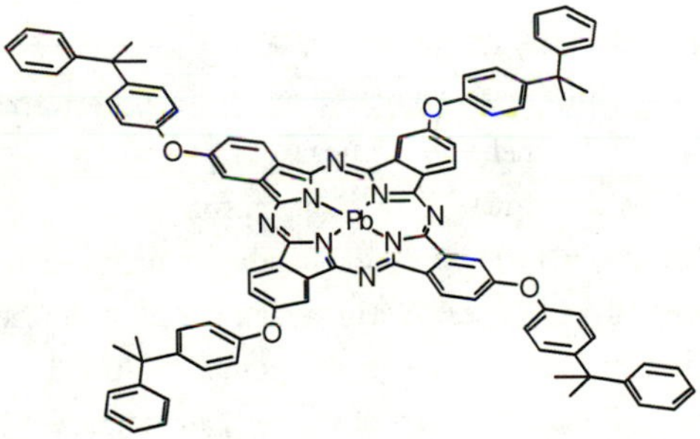

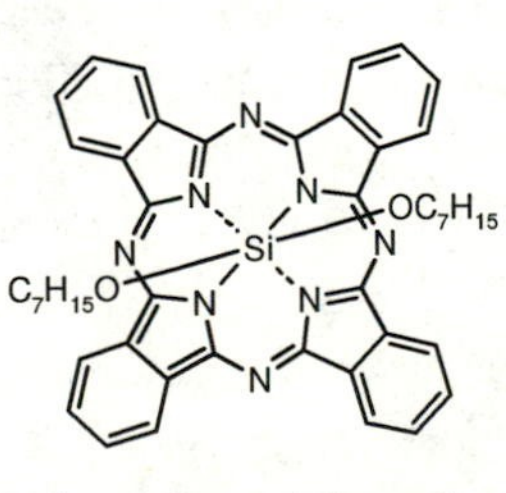

Some examples of phthalocyanine-metal complexes for OL in the visible.
Figure 14

15

Pcs display optical nonlinearities with short response times (quasi-instantaneous response),[106] and at the same time possess small dielectric constants in the UV-Vis.[34] Such general features render feasible the use of Pcs as active species (Figure 14) in OL devices based on NLO phenomena for eye and sensor protection from intense light pulses in the UV-Vis.[107,108] In the following the dependence of the OL properties on the structural features of Pcs is described with the aim of showing whereas possible the molecular features that are associated with the better OL properties.

5.1. Effect of the combination central atom-axial ligand on OL properties of Pcs

The OL properties of a large variety of Pcs differing in the nature of the central metal M, and axial ligand X have been recently reported.[109] In these studies the comparison of the linear and nonlinear optical parameteres of about fourty different metal phthalocyanines was presented. It was found the merit factor κ ($=\sigma_{24}/\sigma_{13}$) varied in the approximate range 1-30. These values were determined analyzing the various species at different values of concentration and linear optical transmittance in solution. The largest values of κ were found for the complexes with M=In, Zn, Ga, H_2, Pd and Pb, whereas nickel and cobalt phthalocyanine complexes displayed the lowest values. From the analysis of the trends of the OL response versus Pc's molecular parameters it became clear that no single molecular parameter could be used to describe quantitatively the efficacy of an optical limiter in the case of Pcs.[109] This was proved, for example, by the absence of a clear dependence of the merit coefficient κ on the atomic mass of the central atom M or the number of Pc rings per molecule. An important outcome from this compared evaluation is the idea of combining Pcs with different NLO properties in multilayered configurations. Such configurations are made by layers of Pcs having different values of saturation fluences, which can reduce the overall transmitted intensity with respect to a homogeneous single layered structure by a factor of approximately 80%. The reason of such an improvement is related to the specificity of the response of every layer to laser sources with agile features.[41,109]

From the comparison of OL responses of different Pc's, it is generally concluded that structural factors influence considerably the OL properties of phthalocyanines. In fact, the variation of the central atom can introduce significant changes in the NLO behavior as far as OL is concerned.[57,100] This is especially true in the case of PcMX's with MX=VO, TiO or InCl.[57,100,102] The reason for such variations is due to the ability of the central moieties, such as VO, TiO or InCl, of introducing dipole moments perpendicularly oriented to the Pc ring with values in the order of 10-15 Debyes,[110] which alter the electronic structure of the macrocycle in both ground and excited states, and provoke new steric effects that modify the aggregation of PcMXs.[56]

One of the first systematic studies on the effect of the combination central atom-axial ligand in PcMXs on the resulting OL effect was reported by Shirk and our group in 1998 for the series of compounds (t-Bu)$_4$PcInX [56] with X=Cl, Br, I, p-trifluoromethylphenyl (p-TMP), m-TMP, p-fluorophenyl (p-FP), phenyl, and pentafluorophenyl (PFP)[102] following the work of Perry et al. on the remarkable OL properties of (t-Bu)$_4$PcInCl.[57] Among various Pcs, the soluble and axially substituted R$_x$PcInX displayed the largest nonlinear optical absorption coefficients.[102]

A comparison between (t-Bu)$_4$PcInCl and (t-Bu)$_4$PcInI showed that these two species have similar linear optical spectra. However, in transient absorption studies, it was found that the intersystem crossing process rate of (t-Bu)$_4$PcInI was higher than for (t-Bu)$_4$PcInCl, whereas the quantum yield for the formation of the absorbing triplet excited state was approximately the same for both (t-Bu)$_4$PcInI and

(t-Bu)$_4$PcInCl.[56,102] This could be ascribed to the stronger heavy-atom effect of the heavier iodine axial ligand, which accelerates the formation of triplet excited states in the nonlinear optical regime.[111,112] On the other hand, the species (t-Bu)$_4$PcInXs having alkyl groups as axial ligands resulted photochemically unstable when exposed to both intense light and daylight.[102] These facts led to the conclusion that the In-C bond is more stable against irradiation when the C atom is sp^2 hybridized. The complexes (t-Bu)$_4$PcIn(p-TMP) and (t-Bu)$_4$PcIn(PFP) displayed higher solubility than (t-Bu)$_4$PcInCl, and exhibited higher nonlinear absorption coefficients, lower limiting thresholds and lower optical transmissions at high fluences when compared with (t-Bu)$_4$PcInCl.[56,57] For example, at 532 nm, solutions of (t-Bu)$_4$PcIn(p-TMP) and (t-Bu)$_4$PcInCl with the same linear transmittance (=80%) showed a substantially different behavior in terms of limiting threshold being the latter parameter twice lower for (t-Bu)$_4$PcIn(p-TMP) with respect to (t-Bu)$_4$PcInCl.[102] The photophysical properties of (t-Bu)$_4$PcInCl, (t-Bu)$_4$PcIn(p-TMP) and (t-Bu)$_4$PcInPFP have been studied in the nonlinear optical regime with nanosecond light pulses.[56,102] In this analysis it was found that excitation at around 500 nm produces initially excited singlet states which are converted into excited triplet states within a time lapse of about 300 ps.[56] Such a fast conversion provoked by the intersystem crossing process, indicates that the excited triplet states (with energy level 3 in Figure 3), are the main responsible for the absorption in the nonlinear optical regime. In addition to that, the lifetime of the absorbing triplet excited state was relatively longer than the duration of the light pulses in the range 1-10 ns. Consequently, these materials resulted in effective optical limiters with a photodynamic behavior which is fluence dependent.[56] In the approximate wavelength range 400-600 nm (t-Bu)$_4$PcInCl, (t-Bu)$_4$PcIn(p-TMP) and (t-Bu)$_4$PcInPFP behave as reverse saturable absorbers. The differences could be associated with the higher electronic polarizability of the axial aryl ligand with respect to chlorine atom.[102b,102c] Other facts were mostly explained in terms of reduced aggregation in the highly concentrated solutions used for the determination of OL properties, thus reducing the deleterious effects associated with molecular aggregation which mostly induces a faster decay of the excited absorbing states.[56] The diminution of molecular aggregation in phthalocyaninatoindium (III) complexes with axial aryl groups could be verified by showing the poorer dependence of the molar extinction coefficient at the different wavelengths of these molecules with respect to (t-Bu)$_4$PcInCl upon variation of the complex concentration in the linear optical regime.[56,102]

Recently, the OL response of an indium phthalocyanine with acetylacetonate as axial ligand was analyzed (Figure 14).[113] It was found that the OL effectiveness of this newly synthesized system is comparable with that of the benchmark compound (t-Bu)$_4$PcInCl.[57] Compounds with the formula PcTiX$_n$ in which the central Ti atom possesses a +4 oxidation state allows the possibility of binding different ligands, $e.g.$ Cl or O atoms belonging to thiocatechol, or O and S as single ligands, due to the bivalent nature of Ti(IV) coordinated inside the cavity of the Pc macrocycle. Our approach was the study of the NLO properties in PcTiXs as a function of the substituent in the axial aromatic ligand starting from PcTiO and peripherically substituted systems like (catecholato)-2,(3)-(tetra-$tert$-butylphthalocyaninato)titanium(IV) and analogues (Figure 15), in which the molecules with an electron-withdrawing group as substituent of the axial ligand showed the better OL response (Figure 16).[102]

The large functionalized catechol ligands introduce steric crowding and reduce the tendency to form aggregates. Electron withdrawing groups like CN, CHO, CH$_2$CN and Br in the axial catechol originate a dipole moment perpendicular to the macrocyle.

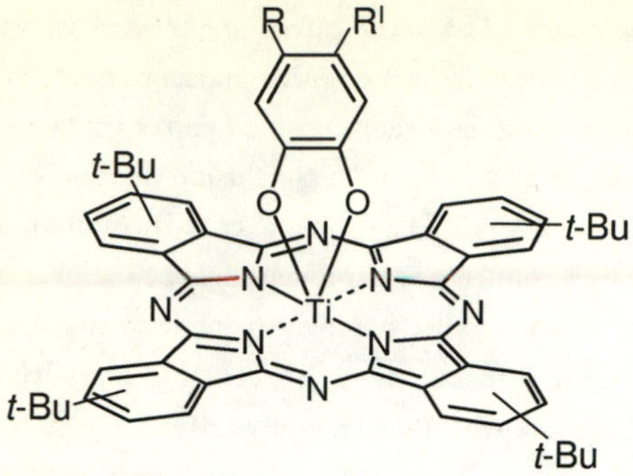

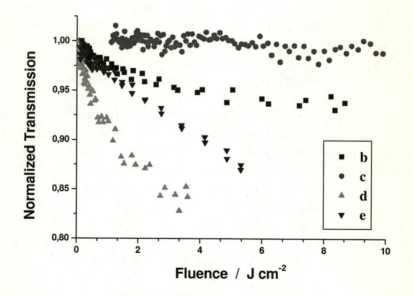

	R	R'
a	H	H
b	H	CHO
c	H	t-Bu
d	CN	CN
e	H	CH₂CN

Various (catecholato)-2,(3)-(tetra-*tert*-butylphthalocyaninato)titanium(IV) for OL in the visible.

Figure 15

Nonlinear transmission of compounds b-e of Figure 15.

Figure 16

The four complexes named above were considered as model complexes for the investigation of their OL performance. It was found the higher the dipole moment associated with the substituent on the axial catechol, the better was the OL effect. In fact, the best performance in terms of OL was achieved with the dicyano substituted material, followed by the cyano-methyl, and formyl substituted ones (Figure 15). The OL performance of these materials at 532 nm varies markedly with the different EW character of the substituents at the aromatic ligand. In ranging from compound c to compound e of Figure 15, the EW character of the axial ligand increases and a corresponding lowering of the limit transmission is observed. Such correlation is not trivial because the changes of limit transmission in PcTiXs solutions are mainly caused by the electronic effects associated with the dipole moment of the axial ligand. In these cases the steric effects associated with the different catechol based axial ligands do not play a relevant role in determining the OL performance due to the similar sizes and the similar packing properties of the different PcTiXs here considered. Such results represent the first demonstration that substituents with EW character at the axial ligand improve the OL performance of Pcs in an analogous fashion of EW substituents attached to the conjugated Pc ring.[99]

The comparison of the OL effect produced by nitrido(tetra-*tert*-butylphthalocyaninato)-rhenium (V) and the paramagnetic species nitrido(2,3-octa-*n*-pentylphthalocyaninato)-tungsten (V) shows that the tungsten complex with paramagnetic properties has a better NLO performance.[114] This is ascribed to an acceleration of the intersystem crossing process (transition [3→2] in Figure 3) which allows a faster successive absorption of the second photon (transition [2→4] in Figure 3). The presence of electronically polarizable axial ligands generally improves the NLO properties of the resulting molecule in terms of OL effectiveness.[115] The soluble species *t*-Bu$_4$PcGaCl and *t*-Bu$_4$PcGa(*p*-TMP) were also synthesized for OL studies.[116] These two materials have similar linear absorption spectra. However, the solubility of *t*-Bu$_4$PcGa(*p*-TMP) in organic solvents was enhanced compared to the chloro analogue *t*-Bu$_4$PcGaCl, showing that the usual tendency of phthalocyanines to form aggregates can be effectively suppressed by axial substitution. The NLO behaviour for *t*-Bu$_4$PcGaCl and *t*-Bu$_4$PcGa(*p*-TMP) at 532 nm in toluene indicated that they also behave as reverse saturable absorbers. The presence of *p*-TMP as axial ligand is associated with the enhancement of the OL effect at 532 nm with respect to *t*-Bu$_4$PcGaCl.

5.2. Effect of peripheral substitution in Pc

The introduction and variation of peripheral substituents in phthalocyanines can also modify the spatial arrangement of the molecule and the extent of molecular interactions between neighbouring molecules. Both the level of aggregation and the supramolecular structure of the aggregates can be strongly dependent on the nature of peripheral substituents, with the level of aggregation influencing the relevant NLO properties. A study reported the effect of the variation of the peripheral substituents in a metal free phthalocyanine R$_x$PcH$_2$ with R=*t*-Bu and x=4, in one case, and R=OCH$_2$CF$_3$ and x=8 in the other case on the NLO properties of their solutions.[117] The peripheral group OCH$_2$CF$_3$ is a good aggregation suppressor. By increasing the concentration of the phthalocyanines, the nonlinear absorption coefficient of (*t*-Bu)$_4$PcH$_2$ starts to decrease when 0.5% in weight, whereas the analogue substituted with (OCH$_2$CF$_3$) shows a nearly constant value of the nonlinear absorption coefficient within the same range of concentration.[117] This study constitutes a clear example of the effect of molecular aggregation upon NLO properties.

Recently, the OL properties of a series of several metallated and unmetallated 1,4-octaalkylsubstituted phthalocyanines (Figure 17) have been also investigated.[118] The substitution pattern of these materials infers a beneficial effect concerning the aggregation aspect, due to a partial ring distortion caused by the 1,4-substitution pattern.

Another aspect related to the presence of peripheral substituents is their influence on the NLO properties as a consequence of their electron withdrawing or releasing properties. The comparison of the OL behavior at 532 nm with 10 ns laser pulses for the series of compounds 2,(3)-tetra-*tert*-butylphthalocyaninato titanium oxide (a in Figure 18), 2,(3)-tetra-trifluoromethylphthalocyaninato titanium oxide (b in Figure 18), and tetrapyrazino-tetraazaporphyrinato titanium oxide (c in Figure 18) proved that the species possessing electron-withdrawing groups as peripheral substituent displayed a more effective OL.[119] In fact, compounds b and c in Figure 18, displayed the lowest values of optical transmittance at high levels of irradiation.[119]

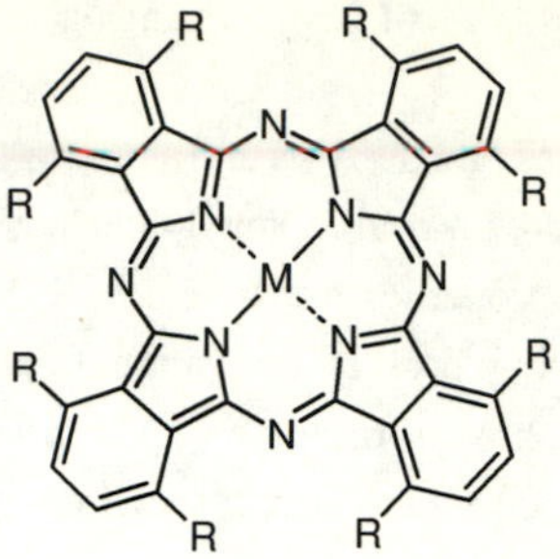

R=C$_n$H$_m$; 5 ≤ n ≤ 10; 11 ≤ m ≤ 21
M=H$_2$, Ni, Zn, Cu, Pd, Pb, InCl, Si(OH)$_2$

Octa-alkyl substituted phthalocyanines for OL.

Figure 17

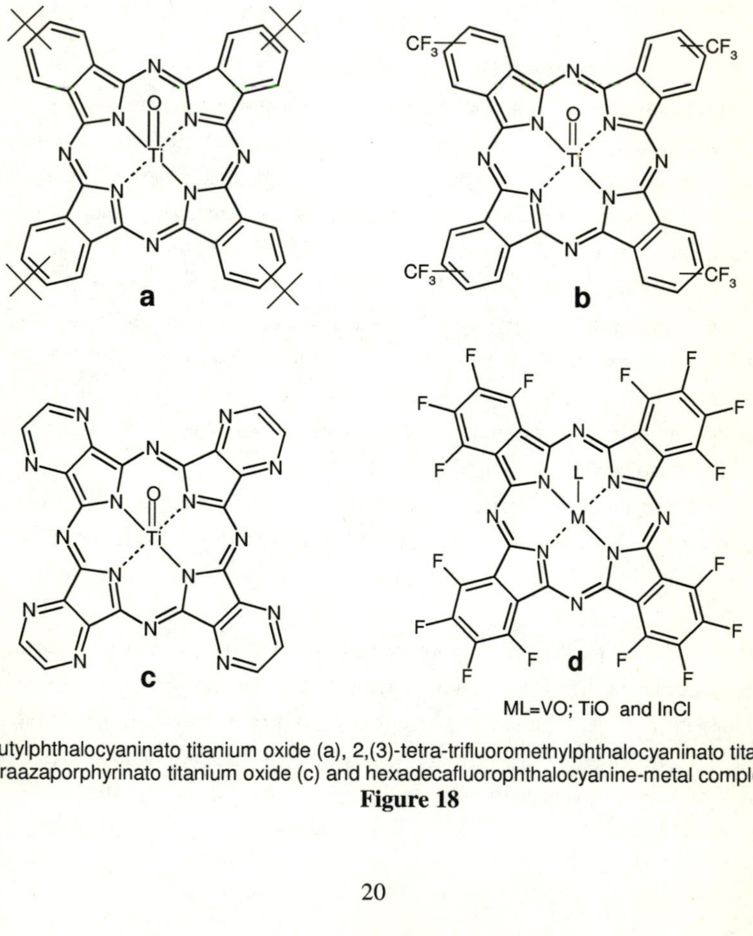

2,(3)-tetra-*tert*-butylphthalocyaninato titanium oxide (a), 2,(3)-tetra-trifluoromethylphthalocyaninato titanium oxide (b), tetrapyrazino-tetraazaporphyrinato titanium oxide (c) and hexadecafluorophthalocyanine-metal complexes (d) for OL.

Figure 18

20

These experimental facts proved that the alteration of the electronic charge distribution in the Pc ring provoked by ring substituents with strong electronic properties affects markedly the absorption properties of the complex in the excited state. It is expected that ring substituents with electron withdrawing properties produce larger variations of the transition dipole moment in correspondence of the excited state electronic transition responsible for the OL effect.[120] This would result in a net increase of the excited state absorption cross-section and an increase of the merit factor κ since the ground state absorption cross section is not as affected as the excited state cross-section by the presence of these substituents.

5.3. Effect of symmetry

The modification of the symmetry of Pcs through unsymmetrical substitution represent another approach to modulate the OL properties of phthalocyanines, since a low symmetry of the structure can change the electronic structure of the macrocycle and, consequently, the optical properties.[121] The preparation of phthalocyanines with different substituents can be accomplished by several methods,[121] but the most common one is the statistical condensation of two differently substituted phthalonitriles or diiminoisoindolines, and separation through chromatography of the statistical mixture of compounds.[122] Few reports on the NLO properties of unsymmetrical Pcs have been published, since the preparation of pure unsymmetrically substituted phthalocyanines appears to be difficult in some cases.[123]

5.4. Charge transfer complexes with Pcs

In the attempt of enhancing the OL properties of materials, the formation of charge-transfer complexes including phthalocyanines has been also considered.[124-126] A great enhancement of nonlinear optical properties has been predicted and successively verified when several components in solutions or solid films were mixed to form intermolecular charge-transfer complexes.[125] One example is given by C_{60} which is an electron deficient three-dimensional cage. In connection with an electron rich molecule, such as a phthalocyanine, intramolecular or intermolecular charge-transfer can be easily obtained. As a result, a study showed that the enhanced OL effect of $PcCu-C_{60}$ comes from a synergic interaction of PcCu and C_{60}, located in two distinct parts. This partially demonstrated that NLO properties, such as OL, can be enhanced robustly by the formation an intramolecular charge-transfer complexes.

5.5. Dimeric phthalocyanines for optical limiting

The NLO properties of dimeric Pcs formed by sandwiched bis(phthalocyaninato)lanthanides have been studied and analyzed by several groups.[127-129] It is generally found that the spectral region of effective RSA is red-shifted with respect to monomeric Pcs.[127] Moreover, modifications of the energy levels scheme involved in the NLO processes are usually taken into account due to the existence of cofacial interactions between the two Pc rings.[130] The main structural limitation associated with the use of lanthanide Pcs is the impossibility of varying the electronic properties of the sandwiched coordinating atom through, e.g. axial substitution, due to the fixed valence of the lanthanides. The OL properties of sandwich-type lanthanides diphthalocyanine were investigated at 532 and 1064 nm.[131,132] In one of these studies, it was demonstrated that $Eu[Pc(OC_5H_{11})_8]_2$ exhibited better optical limiting behavior than $Eu[Pc(C_7H_{15})_8]_2$ (Figure 19) at 532 nm.[131] A less studied group of dimeric Pcs for nonlinear optics is constituted by cofacially stacked Pc's with atoms other than lanthanides. To our knowledge the only reported example of this kind is represented by Pc_2Ti the third-order NLO properties of which were determined at 800 nm.[133] Among dimeric Pcs, the NLO properties of μ-oxo bridged Pc dimers with the general structure Pc(X)M-L-M(X)Pc [M=Fe, Ga and In; L=O, diisocyanobenzene (dib) and 2,3,5,6 tetrafluorophenylene(TFP), X=Cl and TMP][134,135] (Figure 20) have been studied at 532 nm.

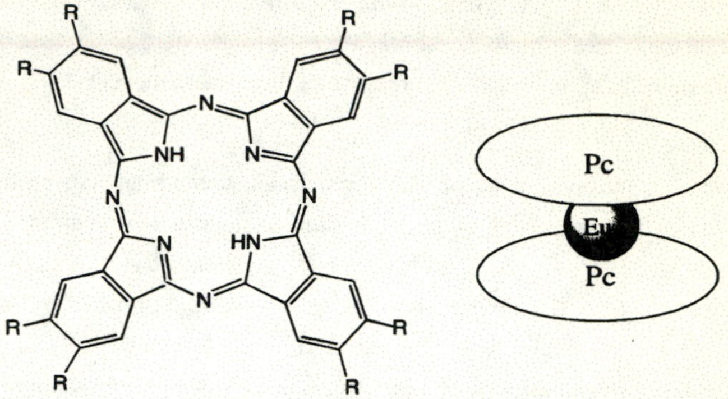

(Right) Bis-phthalocyaninato europium complex from (left) R_8PcH_2 with OL properties (R = C_7H_{15} and OC_5H_{11}).

Figure 19

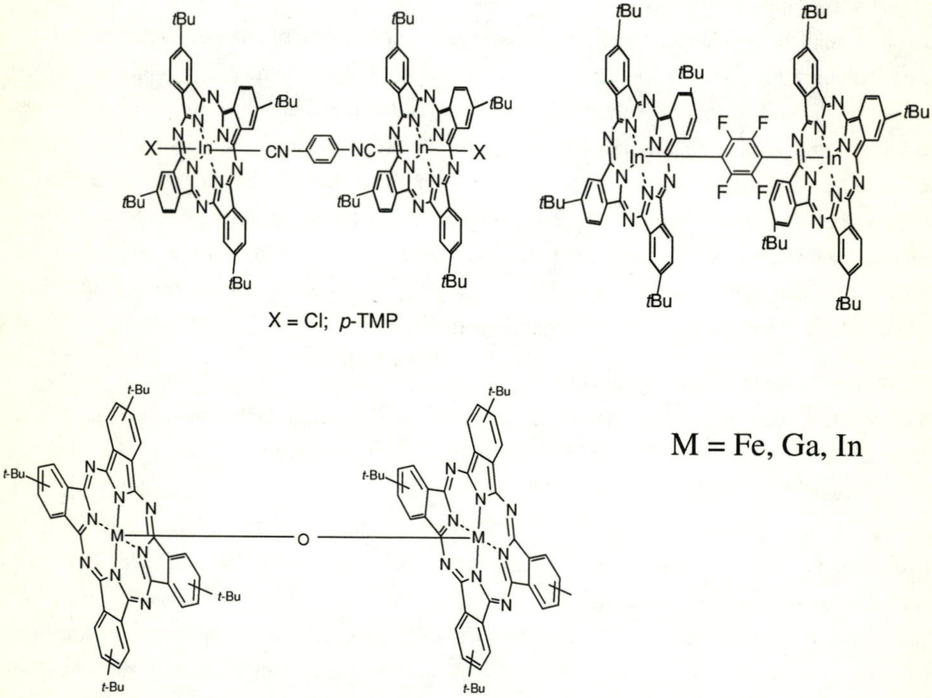

X = Cl; *p*-TMP

M = Fe, Ga, In

Axially bridged indium bisphthalocyanines and μ-oxo dimers for OL.

Figure 20

The larger OL effect generated by the dimers PcInX-dib-InPc as bridging ligand with respect to the bridged dimers PcIn-TFP-InPc, indicates the sharing of the common axial ligand TFP in PcIn-TFP-InPc reduces the excited absorption cross-section of the lower triplet excited state which is responsible for the nonlinear optical absorption.[102c] On the other hand, the similarity of the Z-scan profiles for [t-Bu$_4$PcInCl]$_2$.dib and [t-Bu$_4$PcIn(p-TMP)]$_2$.dib combined with their better OL performance, is indicative of the favourable effect associated with the presence of the additional axial electron-withdrawing groups Cl and TMP in their structures.[135b] The OL effect at 532 nm generated by the peripherally bridged dimeric species (t-Bu)$_3$PcMX(t-Bu)$_3$PcM with X=alkynyl (-C≡C-), and M=Co, Zn has been also studied.[109] The species (t-Bu)$_3$Pc(Co)-C≡C-(t-Bu)$_3$PcCo displayed the highest merit factor κ among the various peripherally alkynyl bridged Pc dimers. Moreover, the comparison between the OL performance of (t-Bu)$_3$Pc(Co)-C≡C-(t-Bu)$_3$PcCo and the monomer (t-Bu)$_3$(HC≡C)PcCo demonstrated that the dimerization leads to molecular systems with a larger merit factor.[109] Newly synthesized dimers with a direct M–M bond [t-Bu$_4$PcM]$_2$ .2L with M=Ga and In and L=dioxane and tmed (tmed=N, N, N`, N`-tetramethylenediamine) (Figure 21), display OL properties with improved features with respect to the single Pc ring coordinated by one single metal atom.[136,137]

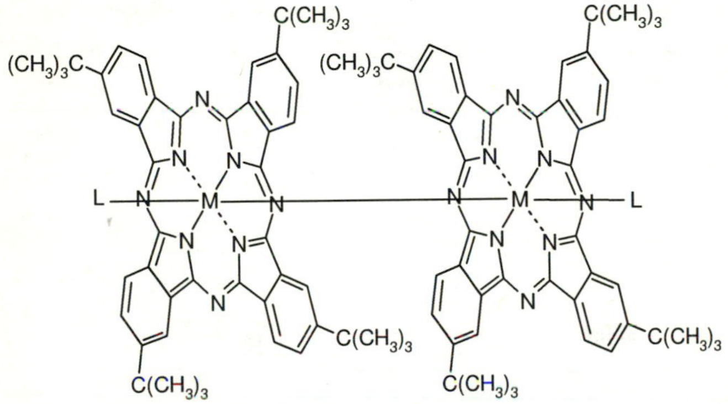

New dimeric complexes [t-Bu$_4$PcM]$_2$ .2L with direct M-M bond (M = Ga, In; L = tmed, dioxane).

Figure 21

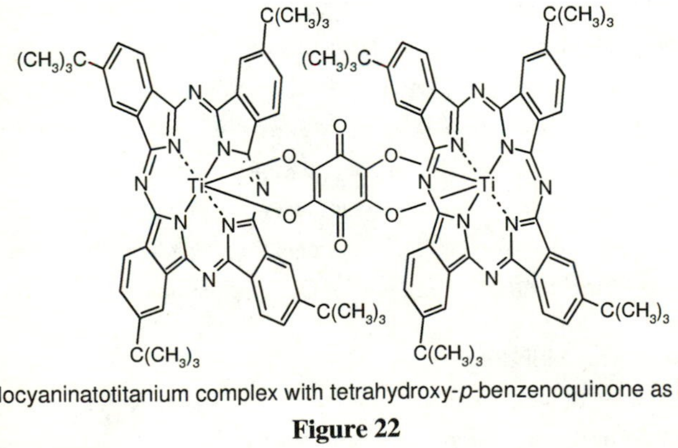

Dimeric phthalocyaninatotitanium complex with tetrahydroxy-p-benzenoquinone as bridging ligand.

Figure 22

Similarly, the Pc dimer with central Ti atoms bridged by tetrahydroxy-*p*-benzenoquinone (Figure 22) also displays an OL effect with improved characteristics if compared with the parent monomer. Such findings are explained in terms of the higher number of active Pc rings per resulting molecular unit.[102c]

Among trimeric Pc's, the linear phthalocyaninatoindium(III) acetylacetonate trimer (Figure 23) is the only example insofar the OL behavior of which has been investigated.[138] This trimer has a similar linear absorption spectrum with the chloro analogue (*t*-Bu₄)PcInCl, and shows a positive nonlinear optical absorption response at relatively low power thus indicating that trimerization can induce an enhancement of the effect of optical limiting.

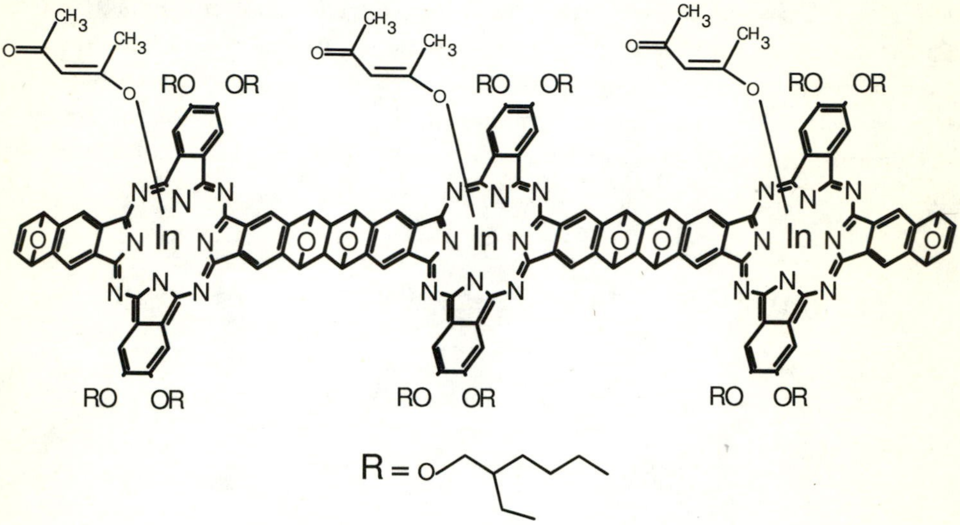

Linear trimeric phthalocyaninato indium acetylacetonate complex with OL properties.

Figure 23

5.6. Optical limiting of phthalocyanines in films

One of first studies on the OL effect generated by Pcs embedded in a sol-gel host (solid-state configuration)[139] reported the results obtained with Sn, Ge and Cu phthalocyanines with PcSnX₂ (X was not defined by the authors) showing the best OL performance.[140] An important achievement was the realization of solid-state limiters having similar properties to those shown by optical limiters based on phthalocyanine solutions. Later the OL effect generated by Pcs embedded in polymers, such as poly(methylmethacrylate) was also studied.[141] The spectral properties of PcMX thin films are different from those of diluted solutions, for the broadening of the absorption bands and the possible shifting of the absorption peaks.[56,102] As previously discussed, such results are expected as an outcome of aggregation in which intermolecular interactions are of the van der Waals type.[103,104] Systems like dendritic phthalocyanines,[142] have been considered for the reduction of molecular aggregation for OL applications.

6. Naphthalocyanines for optical limiting

As analogues of phthalocyanines, naphthalocyanines have more extended conjugated π-system due to the additional tetra-benzo-annulation (Figure 1). Such a structural modification is reflected as a large

24

bathochromic shift of the Q-band in the UV-Vis spectra of 2,3-Ncs and as the enlargement of the highly transparent window between the Q- and B-bands so important for optical limiters operating in the visible. To the contrary, UV-Vis spectra of 1,2-naphthalocyanines do not differ very much from those of Pcs: the position of the Q-band is almost not affected by 1,2-annulation indicating no strong conjugation of the annulated benzo-rings with the phthalocyanine macrocycle.[143] For this reason only 2,3-naphthalocyanines received attention as far as optical limiting properties were concerned and no OL studies on 1,2-naphthalocyanines have been reported till now. The approaches for the fine tuning of OL properties of naphthalocyanines are quite similar to those used for the modification of OL properties in phthalocyanines, *e.g.* peripheral and/or axial ligand substitution, variation of the central atom and so on. However, one should take into account that the negative effects connected with their tendency to aggregate and/or to decompose photochemically are stronger than in case of phthalocyanines.

The influence of the central atom on OL of Ncs in solutions at 532 nm was found to be not as pronounced as for Pcs. Consequently, among $NcSi(OSiHex_3)_2$, $NcSn(OSiHex_3)_2$, $(t$-$Bu)_4NcPb$ and $(t$-$Bu)_4NcInCl$ only the indium compound possesses enhanced OL. The differences between the OL performances of Si, Sn and Pb derivatives were not substantial.[144] These conclusions were drawn by comparing the values of the ratios of effective excited-state to ground-state absorption cross-sections ($\sigma_{ex}^{eff}/\sigma_g$, where σ_{ex}^{eff} is an average of the triplet and singlet excited states absorption cross-sections), and the saturation fluences F_{Sat} of the examined systems.[144] It was considered that the larger ring of Nc leads to a diminished effect of heavy central metal due to the decrease in orbital mixing with increasing size of the macrocycle, and that indium is more strongly coupled to the ring orbitals compared to the group IV-a elements.

Z-scan experiments at 640 nm were carried out for 2.4 mM solutions of $(t$-$Bu)_4NcInI$ and the results were compared with those for 2.4 mM solution of $(t$-$Bu)_4NcInCl$.[102a] Both compounds show RSA behavior, however exhibiting relatively small non-linear absorption coefficients at this concentration due to the formation of aggregates with short excited-state relaxation times. Additionally, the iodide-derivative appeared to be photochemically unstable upon extensive irradiation. It was also shown that the substitution of axial halide by bulky group, *e.g.* p-TMP, allows to increase the aggregation threshold for tetra-*tert*-butyl naphthalocyaninato indium(III) species in solutions.[102,145]

OL measurements were carried out for the first time also on axially substituted gallium tetra-*tert*-butyl naphthalocyanines $(t$-$Bu)_4NcGaCl$, $(t$-$Bu)_4NcGa(p$-$TMP)$ and μ-oxo-dimer $[t$-$Bu_4NcGa]_2O$ in chloroform solutions.[102,146] They exhibit RSA behavior at 532 nm excitation, and their transient absorption spectra recorded under 350 nm laser excitation display a positive variation of absorption coefficient in the excited state in the spectral range 500-700 nm. The substitution of axial chlorine by p-TMP group resulted in an increased value of merit coefficient κ for the studied gallium naphthalocyanines almost by a factor of 2 due to the decreased linear absorption at 532 nm. The dimeric species $[t$-$Bu_4NcGa]_2O$ exhibits the lowest value of κ (4.1 ± 0.1) but, as well, the lowest saturation density F_{Sat} among this series of gallium-naphthalocyanines. In comparison with analogous gallium-phthalocyanines all reported NcGaXs [109,146] show lower κ and F_{Sat} values.

A hexadecafluorinated gallium-2,3-naphthalocyanine $F_{16}NcGaCl$ and a μ-oxo-dimer $[F_{16}NcGa]_2O$ were also prepared and studied recently by us.[147] Although the solubility of the monomeric species appeared to be too low for NLO studies, the OL properties of the dimeric $[F_{16}NcGa]_2O$ were studied in THF and compared with those of $(t\text{-Bu})_4\text{NcGaCl}$, $[t\text{-Bu}_4\text{NcGa}]_2\text{O}$ and C_{60} at 532 nm irradiation (Figure 24). The peripherally fluorinated species exhibited enhanced OL properties as compared with non-fluorinated gallium $(t\text{-Bu})_4\text{Ncs}$ and C_{60}, as follows from their ratios of effective excited to ground state absorption cross-sections $\kappa = \sigma_{ex}^{eff}/\sigma_g$ and optical limiting thresholds. This improvement in OL performance was associated with the involvement of the strong EW effect of the fluorine substituents, which results in the increase of the difference in dipole moments between the excited states involved in second photon absorption.

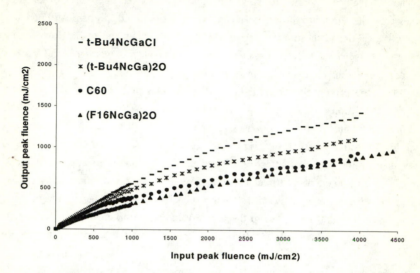

(Top) Hexadecafluorinated gallium naphthalocyanine and (bottom) OL performance of its μ-oxo-dimer in comparison to several other OL-active compounds.

Figure 24

Through the appropriate peripheral substitution, Ncs can expand the window of high optical transmission, with the red-shift of the Q-band. This is desirable for the fabrication of effective optical limiters operating in the visible-light range including NIR. A strong bathochromic shift of the Ncs Q-band can be observed upon alkoxy-substitution in the peripheral positions closer to the Ncs core (α- or 1,6-positions). This is due to the mixing of the oxygen p_z-orbitals with π-HOMO of the macrocycle and distortion of the Nc-core geometry from planar, which causes the destabilization of π-HOMO. The α-substituted 2,3-alkoxy-naphthalocyanines $(\text{BuO})_8\text{NcInCl}$ and $(\text{BuO})_8\text{NcSnF}_2$ have the maximum transmittance at 635 nm where peripherally unsubstituted $\text{NcSi}[\text{OSi}(C_6H_{13})_3]_2$ has it at 560 nm.[144] In case of $(C_5H_{11}O)_8\text{NcPb}$ the maximum linear optical absorption occurs at 910 nm and RSA takes place in the broad region 520-780 nm.[148,149] The $\sigma_{ex}^{eff}/\sigma_g$ values found for $(\text{BuO})_8\text{NcInCl}$ and $(\text{BuO})_8\text{NcSnF}_2$ in toluene solutions at 630 nm with 5 ns laser pulses are 22 and 17, respectively. These values are comparable to those exhibited by analogous Pcs at 532 nm.[144] Other substitution patterns which induce absorption in the NIR and infer high solubility to indium-naphthalocyanines as materials for OL were also realized (Figure 25).[145]

Among these octa-substituted Ncs the mixed substituent derivative $(t\text{-Bu})_4(\text{EHO})_4\text{NcInCl}$, where (EHO=2-ethylhexyloxy), and its axially substituted analogue $(t\text{-Bu})_4(\text{EHO})_4\text{NcIn}(p\text{-TMP})$ have shown high solubility and no optical evidence of aggregation at 2×10^{-3} M concentration in chloroform. The OL measurements at 580 nm on the $(t\text{-Bu})_4(\text{EHO})_4\text{NcIn}(p\text{-TMP})$ with 4×10^{-2} M concentration revealed RSA with OL threshold 10 nJ.[102a]

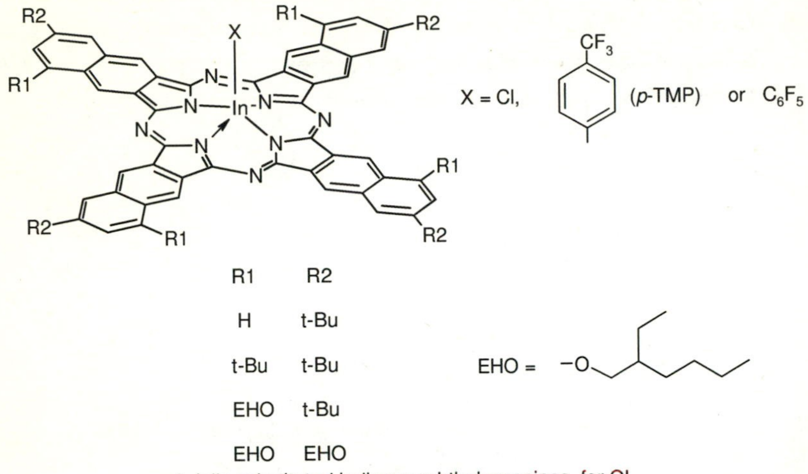

R1	R2
H	t-Bu
t-Bu	t-Bu
EHO	t-Bu
EHO	EHO

Axially substituted indium naphthalocyanines for OL.

Figure 25

At incident energies higher than 10^4 nJ, the photochemical transformation of the compound into another product was observed. Such a transformation resulted in a drastic decrease of the sample transmittance with increasing incident light intensity, and the formation of this photochemical product was found to be partially reversible.[102a] In contrast to this, the OL study on $(t\text{-Bu})_4\text{NcAlCl}$ and $(t\text{-Bu})_4\text{NcZn}$ in alcohol solutions using 5 ns pulses at 532 nm, has shown compound degradation under air and light exposure.[107] Additionally, the lower chemical stability of some Ncs compared to Pcs can result also in their facile decomposition when incorporating them into sol-gel matrixes for the manufacturing of solid optical limiters, as observed in the cases of α-octabutoxy naphthalocyaninato- copper(II) and nickel(II).[150,151]

7. Conclusions

The present review has focused on the nonlinear optical effect of optical limiting generated by metal complexes of tetrapyrrolic macrocycles like porphyrins, porphyrazines, phthalocyanines and naphthalocyanines in solution, as thin film and dispersed in solid transparent matrices. The importance of the structure of these complexes in determining the relative optical limiting effect has been evidenced in many different cases. A relatively large amount of data was made available thanks to the continuous efforts of synthetic chemists in the development and design of optimized structures for the general improvement of the optical limiting response of these complexes. The great versatility of the preparative chemistry for these complexes allows the fine tuning of the linear as well as nonlinear optical properties. Such a possibility can determine useful, although not general, structure-property relationships to model the optical limiting response of new species obtained by tetrapyrrolic macrocycles-metal complexes. This review also pointed out that further research in the field of optical limiting generated by molecular species is still necessary inorder to create practical optical limiters for the protection of the human eye. In fact, such a goal, which requires the realization of optical devices transmitting fluences below 0.1 μJ cm^{-2} for light pulses shorter than 20 nm, has not been achieved yet.

27

Acknowledgments

The authors are very grateful to Dr. James S. Shirk from Optical Sciences Division at the Naval Research Laboratory (Washington D.C., USA) for continuous and profitful discussion. Prof. Werner Blau and Dr. Sean O'Flaherty from Trinity College of Dublin are gratefully acknowledged for the realization of some of the optical limiting experiments. Financial supports from the European Community (Contract nr. HPRN-CT-2000-00020) and Deutsche Forschungsgemeinschaft (Projekt HA 280/65-1) are gratefully acknowledged.

References

1. Sima, J. *Structure and Bonding* **1995**, *84*, 135.
2. Roeder, B.; Naether, D.; Lewald, T.; Braune, M.; Nowak, C.; Freyer, W. *Biophys. Chem.* **1990**, *35*, 303.
3. Harmjanz, M.; Gill, H. S.; Scott, M. J. *J. Am. Chem. Soc.* **2000**, *122*, 10476.
4. Bernard, C.; Gisselbrecht, J. P.; Gross, M.; Jux, N.; Vogel, E. *J. Electroanal. Chem.* **1995**, *381*, 159.
5. Smith, K. M. *Heme, Chlorophyll and Bilins* **2002**, 13.
6. Buchler, J. W. *J. Porphyrins Phthalocyanines* **2000**, *4*, 337.
7. Buchler, J. W.; Dreher, C.; Kuenzel, F. M. *Structure and Bonding* **1995**, *84*, 1.
8. Gryko, D. T. *Eur. J. Org. Chem.* **2002**, 1735.
9. Hanack, M.; Heckmann, H.; Polley, R. In *Methoden der Organischen Chemie; (Houben–Weyl) 4th Ed.*; Thieme Verlag : Stuttgart, 1997; Vol. E9d.
10. Lee, L. K.; Sabelli, N. H.; Le Breton, P. R. *J. Phys. Chem.* **1982**, *86*, 3926.
11. Edwards, L.; Gouterman, M. *J. Mol. Spectr.* **1970**, *33*, 292.
12. Edwards, L.; Dolphin, D. H.; Gouterman, M.; Adler, A. D. *J. Mol. Spectr.* **1971**, *38*, 16.
13. Milgrom, L. R. *The Colors of Life: an Introduction to the Chemistry of Porphyrins and Related Compounds*; Oxford University Press: Oxford, 1997.
14. Linstead, R. P.; Robertson, J. M. *J. Chem. Soc.* **1936**, 1195.
15. Robertson, J. M.; Woodward, M. *J. Chem. Soc.* **1940**, 36.
16. Moser, F. H.; Thomas, A. L. *The Phthalocyanines-Properties*; CRC Press: Boca Raton, 1983; Vol. 1.
17. Montforts, F. P.; Glasenapp-Breiling, M. *Progr. Chem. Org. Natur. Prod.* **2002**, *84*, 1.
18. Krause, G. H.; Weis, E. *Ann. Rev. Plant Physiol. Plant Mol. Biol.* **1991**, *42*, 313.
19. Paoli, M.; Marles-Wright, J.; Smith, A. *DNA Cell Bio.* **2002**, *21*, 271.
20. Ponka, P. *Am. J. Med. Sci.* **1999**, *318*, 241.
21. Law, K. Y. *Chem. Rev.* **1993**, *93*, 449.
22. Gregory, P. *High Technology Applications of Organic Colorants*; Plenum Press: New York, 1991.
23. Gairns, R. S. *The Chemistry and Technology of Printing and Imaging Systems*; Blackie: London, 1996.
24. Yanagi, H.; Tamura, N.; Taira, S.; Furuta, H.; Douko, S.; Schnurpfeil, G.; Woehrle, D. *Mol. Cryst. Liq. Cryst. Sci. Techn. A* **1995**, *267*, 435.
25. Davidsson, A. *Chem. Phys.* **1980**, *45*, 409.
26. Nicholson, M. M. In *Phthalocyanines: Properties and Applications*; Leznoff, C. C.; Lever, A. B. P., Eds.; VCH: Cambridge, 1993; Vol. 3, p. 71.
27. (a) Hanack, M.; Lang, M. *Adv. Mat.* **1994**, *6*, 819. (b) Bouvet, M. In *Porphyrin Handbook*; Kadish, K. M.; Smith, K. M.; Guilard, R., Eds.; Elsevier Science: San Diego, 2003, Vol. 19, p. 37. (c) Cook, M. J. *Chem. Rec.* **2002**, *2*, 225. (d) Palacin, S. *Adv. Coll. Interf. Sci.* **2000**, *87*, 165. (e) Zhou, R.; Josse, F.; Gopel, W.; Oeztuerk, Z. Z.; Bekaroglu, O. *Appl. Organomet. Chem.* **1996**, *10*, 557. (f) Guillaud, G.; Simon, J.; Germain, J. P. *Coord. Chem. Rev.* **1998**, *178-180*, 1433. (g) Pandey, R. K.; Zheng, G. In *Porphyrin Handbook*; Kadish, K. M.; Smith, K. M.; Guilard, R., Eds.; Academic Press: San Diego, 2000; Vol. 6, p. 157. (h) Ben-Hur; Chan, E. W. S. In *Porphyrin Handbook*; Kadish, K. M.; Guilard, R., Eds.; Elsevier Science: San Diego, 2003, Vol. 19, p. 1.
28. Gregory, P. *J. Porphyrins Phthalocyanines* **2000**, *4*, 432.
29. Chou, J. H.; Kosal, M. E.; Nalwa, H. S.; Rakow, N. A.; Suslick, K. S. In *Porphyrin Handbook*; Kadish, K. M.; Smith, K. M.; Guilard, R., Eds.; Academic Press: San Diego, 2000, Vol. 6, p. 43.

30. Hurditch, R. *Adv. Col. Sci.Techn.* **2001**, *4*, 33.
31. Nalwa, H. S. In *Nonlinear Optics of Organic Molecules and Polymers*; Nalwa, H. S.; Miyata, S., Eds.; CRC: Boca Raton, 1997; p. 611.
32. Gubler, U.; Bosshard, C. *Adv. Polym. Sci.* **2002**, *158*, 123.
33. Hanack, M.; Dini, D.; Barthel, M.; Vagin, S. *Chem. Rec.* **2002**, *2*, 129.
34. Nalwa, H. S.; Shirk, J. S. In *Phthalocyanines: Properties and Applications*; Leznoff, C. C.; Lever, A. B. P., Eds.; VCH: Cambridge, 1996; Vol. 4, p. 89.
35. Bubeck, C. *Adv. Nonlin. Opt.* **1996**, *3*, 137.
36. Kobayashi, T. *Adv. Nonlin. Opt.* **1996**, *3*, 505.
37. Heflin, J. R.; Garito, A. F. *Opt. Eng.* **1992**, *32*, 501.
38. Zyss, J. *Nonlinear Optics: Materials, Physics and Devices*; Academic Press: Boston, 1993.
39. *Molecular Photonics: Fundamentals and Practical* Aspects; Horie, K.; Ushiki, H.; Winnik, F. M., Eds.; Wiley: New York, 2000.
40. Garnier, F.; Zyss, J. *Molecular Photonics for Optical Telecommunications: Materials, Physics & Device Technology*; Elsevier: Amsterdam, 1999.
41. Sun, Y. P.; Riggs, J. E. *Int. Rev. Phys. Chem.* **1999**, *18*, 43.
42. (a) Lampert, C. M. *Sol. En. Mat. Sol. Cells* **1998**, *52*, 207. (b) Zhao, W.; Carreira, E. M. *J. Am. Chem. Soc.* **2002**, *124*, 1582. (c) Dini, D. *Int. J. En. Env. Econ.* **2000**, *10*, 1. (d) Dini, D. *Int. J. Mol. Sci.* **2003**, *4*, 291. (e) Kato, K.; Song, P. K.; Odaka, H.; Shigesato, Y. *Jpn. J. Appl. Phys. Part 1* **2003**, *42*, 6523.
43. Nalwa, H. S.; Hanack, M.; Pawlowski, G.; Engel, M. K. *Chem. Phys.* **1999**, *245*, 17.
44. Shirk, J. S.; Lindle, J. R.; Bartoli, F. J.; Kafafi, Z. H.; Snow, A. W. In *Materials for Nonlinear Optics – Chemical Perspectives*; Marder, S. R.; Sohn, J. E.; Stucky, G. D., Eds., ACS Symposium Series 455, 1991, p. 626.
45. Boyd, R. W. *Nonlinear Optics*; Academic Press: New York, 1991.
46. Wood, O. R.; Schwarz, S. E. *Appl. Phys. Lett.* **1967**, *11*, 88.
47. Giuliano, C. R.; Hess, L. D. *Appl. Phys. Lett.* **1968**, *12*, 292.
48. Ralston, J. M.; Chang, R. K. *Appl. Phys. Lett.* **1969**, *15*, 164.
49. Harter, D. J.; Shand, M. L.; Band, Y. B. *J. Appl. Phys.* **1984**, *56*, 865.
50. Tutt, L. W.; Boggess, T. F. *Progr. Quant. Electr.* **1993**, *17*, 299.
51. Justus, B. L.; Huston, A. L.; Campillo, A. J. *U.S. Patent* **2001**, US 6297918 B1 20011002.
52. Dougherty, T. K.; Elias, W. E.; Kost, A. R.; Klein, M. B. *U.S. Patent* **1995**, US 5391329 B1 19950221.
53. Hercher, M. *Appl. Opt.* **1967**, *6*, 947.
54. Couris, S.; Koudoumas, E.; Ruth, A. A.; Leach, S. *J. Phys. B: At. Mol. Opt. Phys.* **1995**, *28*, 4537.
55. Taheri, B.; Liu, H.; Jassemnejad, B.; Appling, D.; Powell, R. C.; Song, J. J. *Appl. Phys. Lett.* **1996**, *68*, 1317.
56. Shirk, J. S.; Pong, R. G. S.; Flom, S. R.; Heckmann, H.; Hanack, M. *J. Phys. Chem. A* **2000**, *104*, 1438.
57. Perry, J. W.; Mansour, K.; Lee, I. Y. S.; Wu, X. L.; Bedworth, P. V.; Chen, C. T.; Ng, D.; Marder, S. R.; Miles, P.; Wada, T.; Tian, M.; Sasabe, H. *Science* **1996**, *273*, 1533.
58. Shen, Y. R. *The Principles of Nonlinear Optics*; John Wiley and Sons: New York, 1984.
59. Sheehy, B.; Di Mauro, L. F. *Ann. Rev. Phys. Chem.* **1996**, *47*, 463.
60. Meier, H.; Stalmach, U.; Kolshorn, H. *Acta Polym.* **1997**, *48*, 379.
61. Soltzberg, L. J. *J. Chem. Ed.* **2001**, *78*, 1432.
62. Kirtman, B.; Champagne, B. *Int. Rev.Phys. Chem.* **1997**, *16*, 389.
63. Bredas, J. L. *Eur. Mat. Res. Soc. Monogr.* **1993**, *6*, 127.
64. Montanari, F.; Banfi, S.; Pozzi, G.; Quici, S. *Rev. Heteroat. Chem.* **1992**, *6*, 94.
65. Edwards, L.; Dolphin, D. H.; Gouterman, M *J. Mol. Spectr.* **1970**, *35*, 90.
66. Gale, R.; McCaffery, A. J.; Rowe, M. D. *J. Chem. Soc., Dalton Trans. Inorg. Chem.* **1972**, 596.
67. Blau, W.; Byrne, H.; Dennis, W. M.; Kelly, J. M. *Optics Commun.* **1985**, *56*, 25.
68. Rao, S. V.; Srinivas, N. K. M. N.; Rao, D. N.; Giribabu, L.; Maiya, B. G.; Philip, R.; Kumar, G. R. *Optics Comm.* **2000**, *182*, 255.
69. Mishra, S. R.; Rawat, H. S.; Laghate, M. *Optics Comm.* **1998**, *147*, 328.
70. Su, W. J.; Cooper, T. M. *Chem. Mater.* **1998**, *10*, 1212.

71. (a) Bonnett, R.; Harriman, A.; Kozyrev, A. N. *J. Chem. Soc., Faraday Trans.* **1992**, *88*, 763. (b) Zhang, G. P.; Sun, X.; George, T. F. *Phys. Rev. B* **2003**, *68*, 165410.
72. Marsh, D. F.; Mink, L. M. *J. Chem. Ed.* **1996**, *73*, 1188.
73. Drain, C. M.; Gong, X. *Chem. Comm.* **1997**, 2117.
74. Tang, N.; Su, W.; Krein, D. M.; McLean, D. G.; Brant, M. C.; Fleitz, P. A.; Brandelik, D. M.; Sutherland, R. L.; Cooper, T. M. *Mater. Res. Soc. Symp. Proc.* **1997**, *479*, 47.
75. McEwan, K. J.; Robertson, J. M.; Wylie, A. P.; Anderson, H. L. *Mater. Res. Soc. Symp. Proc.* **1997**, *479*, 29.
76. McEwan, K. J.; Bourhill, G.; Robertson, J. M.; Anderson, H. L. *J. Nonlin. Opt. Phys. Mater.* **2000**, *9*, 451.
77. Krivokapic, A.; Anderson, H. L.; Bourhill, G.; Ives, R.; Clark, S.; McEwan, K. J. *Adv. Mater.* **2001**, *13*, 652.
78. McEwan, K. J.; Lewis, K.; Yang, G. Y.; Chng, L. L.; Lee, Y. W.; Lau, W. P.; Lai, K. S. *Adv. Funct. Mater.* **2003**, *13*.
79. Bergkamp, M. A.; Dalton, J.; Netzel, T. L. *J. Am. Chem. Soc.* **1982**, *104*, 253.
80. Yang, G. Y.; Ang, S. G.; Chang, L. L.; Lee, Y. W.; Lau, E. W. P.; Lai, K. S.; Ang, H. G. *Chem. Eur. J.*, **2003**, *9*, 900.
81. Qureshi, F. M.; Martin, S. J.; Long, X.; Bradley, D. D. C.; Henari, F. Z.; Blau, W. J.; Smith, E. C.; Wang, C. H.; Kar, A. K.; Anderson, H. L. *Chem. Phys.* **1998**, *231*, 87.
82. Wang, D.; Sun, W.; Dong, S.; Si, J.; Li, C. *Mater. Res. Soc. Symp. Proc.* **1997**, *479*, 41.
83. Si, J.; Wang, Y.; Zhao, J.; Ye, P.; Wang, D.; Sun, W.; Dong, S. *Appl. Phys. Lett.* **1995**, *67*, 1975.
84. Sun, W.; Wang, D. *Science in China (Series B)* **1996**, *39*, 509.
85. Sun, W.; Byeon, C. C.; Lawson, C. M.; Gray, G. M.; Wang, D. *Appl. Phys. Lett.* **2000**, *77*, 1759.
86. Gouterman, M. *J. Mol. Spectrosc.* **1961**, *6*, 138.
87. Gouterman, M.; Wagniere, G.; Snyder, L. C. *J. Mol. Spectrosc.* **1963**, *11*, 108.
88. Kobayashi, N.; Konami, H. In *Phthalocyanines: Properties and Applications*; Leznoff, C. C.; Lever, A. B. P., Eds.; VCH: New York, 1996; Vol. 4, p. 343.
89. Luk'yanets, E. A. *Electronic Spectra of Phthalocyanines and Related Compounds*, (catalogue in Russian); Moscow Scientific-Industrial Association (NIOPIK): Moscow, 1989.
90. Nalwa, H. S.; Engel, M. K.; Hanack, M.; Pawlowski, G. *Appl. Phys. Lett.* **1997**, *71*, 2070.
91. De la Torre, G.; Vazquez, P.; Agullo-Lopez, F.; Torres, T. *Chem. Rev.* **2004**, *104*, 3723.
92. Norwood, R. A.; Sounik, J. R. *Appl. Phys. Lett.* **1992**, *60*, 295.
93. (a) Diaz-Garcia, M. A.; Ledoux, I.; Fernandez-Lazaro, F.; Sastre, A.; Torres, T.; Agullo-Lopez, F.; Zyss, J. *Nonlinear Optics* **1995**, *10*, 101. (b) Diaz-Garcia, M. A.; Ledoux, I.; Fernandez-Lazaro, F.; Sastre, A.; Torres, T.; Agullo-Lopez, F.; Zyss, J. *J. Phys. Chem.* **1994**, *98*, 4495.
94. (a) Wen, T. C.; Chen, S. P.; Tsai, C. Y. *Synth. Met.* **1998**, *97*, 105. (b) Wen, T. C.; Tsai, C. Y. *Chem. Phys. Lett.* **1999**, *311*, 173.
95. Vagin, S.; Dini, D.; Barthel, M.; Hanack, M. *Inorg. Chem.* **2003**, *42*, 2683.
96. Vagin, S.; Hanack, M. *Eur. J. Org. Chem.* **2002**, 2859.
97. Vagin, S.; Yang, G. Y.; Lee, M. K. Y.; Hanack, M. *Opt. Comm.* **2003**, *228*, 119.
98. Perry, J. W. In *Nonlinear Optics of Organic Molecules and* Polymers; Nalwa, H. S.; Miyata, Eds.; CRC: Boca Raton, 1997, Chapter 13, p. 813-840.
99. Hosoda, M.; Wada, T.; Yamada, A.; Garito, A. F.; Sasabe, H. *Jpn. J. Appl. Phys.* **1991**, *30*, 1486.
100. Ho, Z. Z.; Ju, C. Y.; Hetherington III, W. M. *J. Appl. Phys.* **1987**, *62*, 716.
101. Snow, A. W.; Jarvis, N. L. *J. Am. Chem. Soc.* **1984**, *106*, 4706.
102. (a) Hanack, M.; Schneider, T.; Barthel, M.; Shirk, J. S.; Flom, S. R.; Pong, R. G. S. *Coord. Chem. Rev.* **2001**, *219-221*, 235. (b) Hanack, M.; Dini, D.; Barthel, M.; Vagin, S. *Chem. Rec.* **2002**, *2*, 129. (c) Dini, D.; Barthel, M; Hanack, M. *Eur. J. Org. Chem.* **2001**, 3759. (d) Calvete, M; Yang, G. Y.; Hanack, M. *Synth. Met.* **2004**, *141*, 231.
103. George, R. D.; Snow, A. W.; Shirk, J. S.; Barger, W. R. *J. Porphyrins Phthalocyanines* **1998**, *2*, 1.
104. Wu, J. W.; Heflin, J. R.; Norwood, R. A.; Wong, K. Y.; Zamani-Khamiri, O.; Garito, A. F.; Kalyanaraman, P.; Sounik, J. *J. Opt. Soc. Am. B* **1989**, *6*, 707.

105. Hosoda, M.; Wada, T.; Yamamoto, A.; Kaneko, A.; Garito, A. F.; Sasabe, H. *Jpn. J. Appl. Phys.* **1992**, *31*, 1071.
106. Orti, E.; Bredas, J. L.; Clarisse, C. *J. Chem. Phys.* **1990**, *92*, 1228.
107. Coulter, D. R.; Miskowski, A.; Perry, J. W.; Wei, T.; Van Stryland, E. W.; Hagan, D.J. *SPIE Proc.* **1989**, *1105*, 42.
108. Swalen, J. D.; Kajzar, F. *Nonlin. Opt.* **2001**, *27*, 13.
109. O'Flaherty,S.; Hold, S. V.; Cook, M. J.; Torres, T.; Chen, Y.; Hanack, M.; Blau, W. J. *Adv. Mater.* **2003**, *15*, 13.
110. Rojo, G.; Martin, G.; Agullo-Lopez, F.; Torres, T.; Heckmann, H.; Hanack, M. *J. Phys. Chem. B* **2000**, *104*, 7066.
111. Zhang, X.; Xu, H. *J. Chem. Soc. Faraday Trans.* **1993**, *89*, 3347.
112. Turro, N. J. *Modern Molecular Photochemistry*; Benjamin/Cummings: Menlo Park, 1978.
113. Youssef, T. E.; Hanack, M. *J. Porphyrins Phthalocyanines* **2002**, *6*, 571.
114. Frick, K.; Verma, S.; Sundermeyer, J.; Hanack, M. *Eur. J. Inorg. Chem.* **2000**, 1025.
115. Ottmar, M.; Ichisaka, T.; Subramanian, L. R.; Hanack, M.; Shirota, Y. *Chem. Lett.* **2001**, 788.
116. Chen, Y.; Subramanian, L. R.; Barthel, M.; Hanack, M. *Eur. J. Inorg. Chem.* **2002**, 1032.
117. Wada, T.; Yanagi, S.; Kobayashi, H.; Kumar, J.; Sasaki, K.; Sasabe, H. *SPIE Proc.* **1994**, *2143*, 172.
118. Auger, A.; Blau, W.; Burnham, P. M.; Chambrier, I.; Cook, M. J.; Isaure, B.; Nekelson, F.; O'Flaherty, S. M. *J. Mat. Chem.* **2003**, *13*, 1042.
119. Dini, D.; Barthel, M.; Schneider, T.; Ottmar, M.; Verma, S.; Hanack, M. *Sol. St. Ionics* **2003**, *165*, 289.
120. Dini, D.; Yang, G. Y.; Hanack, M. *J. Chem. Phys.* **2003**, *119*, 4857.
121. De la Torre, G.; Vásquez, P.; Agulló-López, F.; Torres, T. *J. Mat. Chem.* **1998**, *8*, 1671.
122. Schmid, G.; Sommerauer, M.; Geyer, M.; Hanack, M. In *Phthalocyanines: Properties and Applications*; Leznoff, C. C.; Lever, A. B. P., Eds.; VCH: Cambridge, 1996; Vol. 4, p. 1.
123. Liu, Y.; Xu, Y.; Zhu, D.; Zhao, X. *Thin Solid Films* **1996**, *289*, 282.
124. Perry, J. W. In *Nonlinear Optics of Organic Molecules and Polymers*; Nalwa, H. S.; Miyata, S. Eds.; CRC: Boca Raton; 1997; Chapter *13*, p. 813.
125. Brédas, J. L.; Adant, C.; Tackx, P.; Persoons, A. *Chem. Rev.* **1994**, *94*, 243.
126. Huang, W.; Wang, S.; Liang, R.; Gong, Q.; Qiu, W.; Liu, Y.; Zhu, D. *Chem. Phys. Lett.* **2000**, *324*, 354.
127. Unnikrishnan, K. P.; Thomas, J.; Nampoori, V. P. N.; Vallabhan, C. P. G. *Opt. Comm.* **2003**, *217*, 269.
128. Mendonca, C. R.; Gaffo, L.; Misoguti, L.; Moreira, W. C.; Oliveira, O. N.; Zilio, S. C. *Chem. Phys. Lett.* **2000**, *323*, 300.
129. Wen, T. C.; Lian, I. D. *Synth. Met.* **1996**, *83*, 111.
130. Manas, E. S.; Spano, F. C.; Chen, L. X. *J. Chem. Phys.* **1997**, *107*, 707.
131. Wang, X.; Liu, C. L.; Gong, Q. H.; Huanh, Y. Y.; Huang, C. H.; Jiang, J. Z. *Appl. Phys. A* **2002**, 497.
132. Shirk, J. S.; Lindle, J. R.; Bartoli, F. J.; Boyle , M. E. *J. Phys. Chem.* **1992**, *96*, 5847.
133. Fernandez-Alonso, F.; Marovino, P.; Paoletti, A. M.; Righini, M.; Rossi, G. *Chem. Phys. Lett.* **2002**, *356*, 607.
134. Philip, R.; Ravikanth, M.; Ravindra Kumar, G. *Opt. Comm.* **1999**, *165*, 91.
135. (a) Chen, Y.; Subramanian, L. R.; Fujitsuka, M.; Ito, O.; O'Flaherty, S.; Blau, W. J.; Schneider, T.; Dini, D.; Hanack, M. *Chem. Eur. J.* **2002**, *8*, 4248. (b) Chen, Y.; Hanack, M.; Dini, D.; Fujitsuka, M.; Ito, O. *J. Mat. Chem.* **2005**, *15*, publication date on the web: 24th November 2004.
136. Chen, Y.; Barthel, M.; Seiler, M.; Subramanian, L. R.; Bertagnolli, H.; Hanack, M. *Angew. Chem. Int. Ed.* **2002**, *41*, 3239.
137. Chen, Y.; O'Flaherty, S.; Fujitsuka, M.; Subramanian, L. R.; Ito, O.; Blau, W. J.; Hanack, M. *Adv. Mat.* **2003**, *15*, 899.
138. Youssef, T. E.; O'Flaherty, S.; Blau, W. J; Hanack, M. *Eur. J. Org. Chem.* **2004**, 101.
139. Fuqua, P. D.; Mansour, K.; Alvarez , D.; Marder, S. R.; Perry, J. W.; Dunn, B. *SPIE Proc.* **1992**, *1758*, 499.
140. Brant, M. C.; De Rosa, M. E.; Jiang, H.; McLean, D. G.; Sutherland, L.; Campbell, A. L. *SPIE Proc.* **1997**, *2966*, 88.
141. Mansour, K.; Fuqua, P.; Marder, S. R.; Dunn, B.; Perry, J. W. *SPIE Proc.* **1994**, *2143*, 239.

142. McKeown, N. B. *Adv. Mat.* **1999**, *11*, 67.
143. (a) Kobayashi, N. In *Phthalocyanines: Properties and Applications*, Leznoff, C. C.; Lever, A. B. P., Eds.; VCH: Cambridge, 1993; Vol. 2, p. 91. (b) Orti, E.; Piqueras, M. C.; Crespo, R.; Bredas, J. L. *Chem. Mater.* **1990**, *2*, 110.
144. (a) Perry, J. W.; Mansour, K.; Marder, S. R.; Chen, C. T.; Miles, P.; Kenney, M. E.; Kwag, G. *Mat. Res. Soc. Symp. Proc.* **1995**, volume date 1994, *374*, 257. (b) Miles, P. A. *SPIE-Proc.* **1994**, *2143*, 251.
145. Schneider, T.; Heckmann, H.; Barthel, M.; Hanack, M. *Eur. J. Org. Chem.* **2001**, 3055.
146. Chen, Y.; O'Flaherty, S.; Fujitsuka, M.; Hanack, M.; Subramanian, L. R.; Ito, O.; Blau, W. J. *Chem. Mater.* **2002,** *14*, 5163.
147. Yang, G. Y.; Hanack, M.; Lee, Y. W.; Chen, Y.; Lee, M. K. Y.; Dini, D. *Chem. Eur. J.* **2003**, 9, 2758.
148. Shirk, J. S.; Pong, R. G. S.; Flom, S. R.; Boyle, M. E.; Snow, A. W. *Mat. Res. Soc. Symp. Proc.* **1995**, volume date 1994, *374*, 201.
149. Shirk, J. S.; Flom, S. R.; Lindle, J. R.; Bartoli, F. J.; Snow, A. W.; Boyle, M. E. *Mat. Res. Soc. Symp. Proc.* **1994**, *328*, 661.
150. Acosta, A.; Sarkisov, S. S.; Wilkosz, A.; Leyderman, A.; Venkateswarlu, P. *SPIE-Proc.* **1997**, *3136*, 246.
151. De Rosa, M. E.; Su, W.; Krein, D.; Brant, M. C.; McLean, D. G. *SPIE-Proc.* **1997**, *3146*, 134.

TRIHALIDE-BASED IMIDAZOLIUM SALTS AS VERSATILE REAGENTS FOR THE HALOGENATION OF UNSATURED COMPOUNDS IN IONIC LIQUIDS

Cinzia Chiappe and Daniela Pieraccini

Dipartimento di Chimica Bioorganica e Biofarmacia, Via Bonanno 33, I-56126 Pisa, Italy

(e-mail: cinziac@farm.unipi.it)

Abstract. *This review gives a comprehensive overview of the use of trihalide-based imidazolium ionic liquids as reagent-solvents for electrophilic halogenation of carbon-carbon double and triple bonds and of aromatic compounds. Practical methods, which minimize the use of molecular solvents and avoid toxic reagents, will be described and compared with traditional routes. The stereochemical and kinetic behaviour of these reactions will be examined and compared with those characterizing the same reactions in molecular solvents, in order to put in evidence all the features that most clearly show the ability of ionic liquids to affect organic reactivity.*

Contents

1. Introduction

Room Temperature Ionic Liquids (ILs) that are air and moisture stable have been subjected to an increasing number of scientific investigations and their application as novel solvents for organic synthesis has received in the last years a great deal of attention.[1-4]

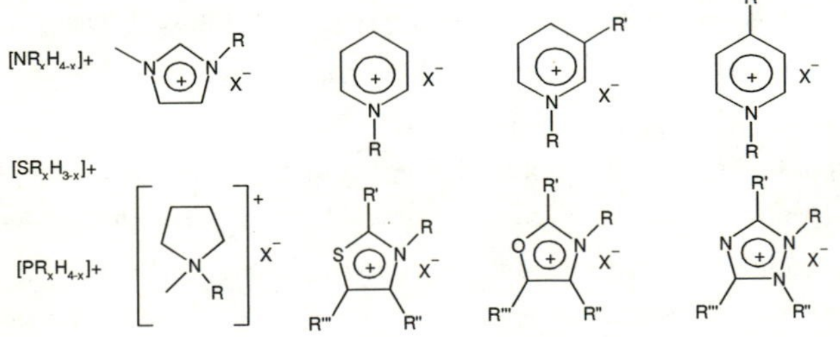

Figure 1. Some typical cations.

33

ILs are low melting salts obtained by combination of large organic cations with a variety of organic and inorganic anions. The cation is generally a bulk organic structure with a low degree of symmetry; the anion is generally a polyatomic species. Despite several cations have been described until now (ammonium, sulfonium, phosphonium, imidazolium, pyridinium, pyrrolidinium), most of the data are focussed on N,N-dialkylimidazolium salts (Figure 1).

These latter salts have attracted particular attention since they are easy to prepare, they are liquid in a large range of temperature and present the widest spectrum of physico-chemical properties. Related to the cation structure, it is often assumed that non-symmetrical N,N-dialkylimidazolium derivatives give ILs having lower melting points, although some exceptions are reported by now.[5] 1-Butyl-3-methylimidazolium ([bmim]) and 1-ethyl-3-methyimidazolium ([emim]) are the most investigated cations among the imidazolium salts, even if suitable pendant groups may be introduced on the alkyl chain. For example, ILs bearing a fluorous tail have been used to facilitate the emulsification of perfluorocarbons in ILs.[6] On the other hand, ILs having a free amine group have been tested to capture H_2S or CO_2, whereas derivatives bearing an urea or thiourea group have been used as extractants of heavy metals (Hg^{2+} and Cd^{2+}) from aqueous co-phase.[7]

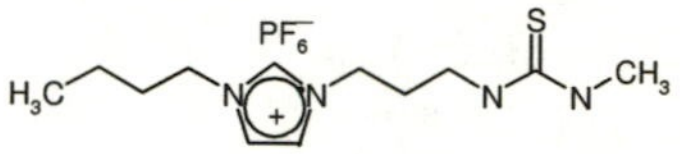

Figure 2. Task specific ionic liquid.

According to the identity of the anion, ILs may be divided into four groups:

1) systems based on $AlCl_3$ and organic salts like 1-butyl-3-methylimidazolium chloride ([bmim][Cl]), whose behaviour is strictly related to the ratio [bmim][Cl]: $AlCl_3$.[8] With a molar excess of $AlCl_3$ they behave as Lewis acids; with an excess of the organic salt they are Lewis bases, whereas equimolar amounts of [bmim][Cl] and $AlCl_3$ lead to a neutral salt. These ILs are however extremely hygroscopic and handling is possible only under dry atmosphere.

2) Systems based on anions like $[PF_6]^-$,[9] $[BF_4]^-$,[10] $[SbF_6]^-$,[11] which are nearly neutral and air stable but in presence of water may undergo slow hydrolysis leading to detectable amount of HF.

3) Systems based on anions like $[CF_3SO_3]^-$, $[(CF_3SO_2)_2N]^-$ (generally indicated as $[Tf_2N]^-$) and $[N(CN)_2]^{-1}$ which give ILs characterized by very low melting points, low viscosity and high conductivity.[12,13,14]

4) Systems based on alkylsulfate[15] and alkylsulfonate[16] anions, which are relatively cheap, easy to handle and to prepare in high purity grade under mild conditions. These latter ILs are generally characterized by a wide electrochemical window and air stability.

Dialkylimidazolium salts are usually prepared by exhaustive alkylation of 1-alkylimidazoles with alkyl halides (Scheme 1). Recently the use of microwave irradiation[17] or ultrasounds[18] has led to the expected imidazolium halides with high cleanness and good yields, reducing reaction duration. 1,3-Dialkylimidazolium chlorides, bromides or iodides can be used directly as ILs, as components for the previously described halometallate ILs or as reagents to prepare other ILs by metathesis reactions.

Therefore, ILs belonging to group 2) and 3) may contain not only traces of bases but also chloride or bromide ions arising from the preparation procedure. It is worth of note that the presence of these "impurities", even at very low concentrations, leads to variations in the physicochemical properties of the ILs[19] and may negatively affect the outcome of nucleophilic reactions as well as metal catalyzed cross-couplings.[20] Therefore, 1,3-dialkylimidazolium salts containing alkylsulfates have been proposed as alternatives to chloride and bromide salts in metathesis reactions in order to obtain ILs (hexafluorophosphates, tetrafluoroborates, bis(triflyl)imides) which are not contaminated by halides.[21]

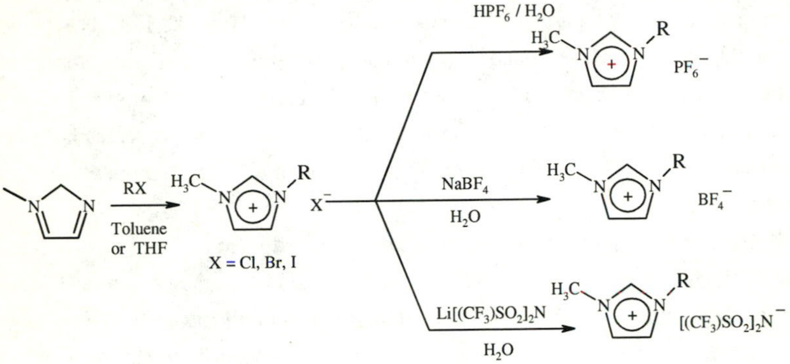

Scheme 1. General reaction scheme for ILs synthesis.

Recently, ionic liquids properly functionalized to act as catalysts have been also synthesized: for example, imidazolium salts containing anionic selenium species $[SeO_2(OCH_3)]^-$ have been used as catalysts for oxidative carbonylation of anilines,[22] whereas ILs bearing acidic counteranions (HSO_4^-, $H_2PO_4^-$) have been applied in catalysed esterifications as recyclable reaction media.[23] Finally, protonated ionic liquids have been synthesized by direct neutralization of alkylimidazoles with acids.[24-27]

Though the number of potential ionic liquids derived from the combination of all the possible anions and cations is incredibly high (evaluated to be greater than 10^{14}), all ionic liquids share some peculiar features, such as negligible vapour pressure, ability to dissolve organic as well as inorganic or polymeric materials and high thermal stability. The large possibility of synthetic variations has led to ILs being described as "Designer Solvents".[2] A large number of reactions have been successfully performed in these solvents (dimerization of alkenes,[28] Friedel-Crafts reactions,[29] Diels-Alder cycloadditions,[30] hydrogenations,[31-32] Pd-mediated cross-couplings[33-34]), showing a great potential as reaction media able to maintain and in some cases to enhance chemical reactivity as well as selectivity, with respect the molecular solvents. Moreover, ionic liquids possess a number of properties that may be of primary importance for large scale applications: in particular, their lack of measurable vapour pressure associated with the high thermal stability and the possibility of recycle make them potential green alternative solvents for synthesis as well as for catalysis. Nevertheless, the lack of a measurable vapour pressure is a two-edged weapon: the use of ILs as solvents produces no adventitious emission but introduces new challenges for products recovery and solvent purification. ILs are difficult to purify so they must be synthesized with a high purity grade. Related to products recovery, this may be problematic in particular when the compounds are exceptionally soluble in ILs: aldehydes, ketones and other dipolar organic molecules appear to be very difficult to remove either by extraction or by distillation so that new approaches have to be applied. On the other hand, products having a negligible miscibility in ILs, in particular liquid compounds, may be removed by simple decantation. Distillation can be used for highly volatile and thermally stable products, although extraction with a co-solvent poorly miscible with IL (water or organic solvent) is often used. This latter procedure, although widely applied reduces the "green" character of the reactions carried out in ILs. Therefore, more eco-friendly alternatives for product recovery have been recently developed. In particular, extraction with supercritical carbon dioxide (scCO$_2$) has received attention being this latter medium a versatile, environmentally benign solvent for a variety of applications.[35a] Moreover, the high volatility and low polarity of scCO$_2$ make this compound an interesting solvent partner with no-volatile and fairly polar ILs. The two-phase systems constituted by scCO$_2$ and ILs have found application in several areas. For example, metal-catalyzed reactions have been carried out successfully in ILs and scCO$_2$, and this procedure has been recently applied also in enzyme-catalyzed transformations. Both batch and continuous-flow processes have been developed; reactants are introduced into the IL directly or with a flow of CO$_2$ and products are recovered from decompressed CO$_2$.

Recently it has also been shown that solutes can be extracted from ionic liquids by pervaporation,[35b] a membrane process based on the preferential partitioning of the solute from a liquid feed phase into a dense, non porous membrane which is not permeable to the IL. This technique has been applied to the recovery of volatile solutes from heat sensitive reactions, such as bio-conversions, and to remove water continuously in lipase catalyzed reactions.

Finally, it must be remembered that the requirements for an ideal solvent include also a low toxicity, non-flammability, availability at low cost, high capacity and selectivity for solutes, low selectivity for carrier. At the present, ILs have some of these requirements the expectation is to address the remaining issues: cost, stability, toxicity, environmental impact. Unfortunately, today only few data have been reported about IL toxicity and on environmental impact of these salts.

2. Synthesis and characterization of trihalide based molten salts

Electrophilic halogenation of double and triple bonds is a fundamental reaction in organic chemistry,[36] providing products that are key intermediates in synthetic as well as in analytical chemistry.[37] Therefore, it is not surprising that several halogenating agents, including polymer-bound reagents were established more than 20 years ago.[38] Traditional synthetic routes to prepare dihalo derivatives involve the direct use of halogens as well as ammonium or potassium trihalides, but in particular in the case of iodohalogenating agents (ICl$_2^-$ or IBr$_2^-$) potassium salts present drawbacks associated with their ability to disproportionate.[39,40] Therefore, regio- and stereospecific iodochlorinations have been achieved mainly using ammonium or polymeric dichloroiodates in chlorinated solvents.[39,41] It is to note that these latter reagents (ammonium or polymeric trihalides) have been used successfully also for halogenation of carbonyl compounds and oxidation of alcohols.

Recently, during the investigation of new anions for the application of ILs as solvents and reagents in organic synthesis and in electrochemical devices, it became apparent that trihalide anions give hydrophobic and high stable N,N-dialkylimidazolium ILs, having lower melting points and viscosities than other common ILs. The application of these ILs for the stereoselective halogenation of unsaturated compounds has shown that: 1) these reagent-solvents maintain the same regio- and stereospecificity characterizing ammonium

36

trihalide salts in chlorinated solvents; 2) they retain the simplification of work up procedure characterizing polymeric reagents; 3) the use of toxic solvents and the formation of by-products (see below) can be avoid; 4) they are able to accelerate the electrophilic addition processes.[42]

The preparation of these functionalized ILs is extremely easy (Scheme 2). 1-Alkyl-3-methylimidazolium trihalides are generally prepared from the corresponding 1,3-dialkylimidazolium halides (Cl, Br) by addition of ICl, IBr or Br_2. All the investigated trihalide ILs are air stable and can be stored in the dark at 4 °C for several months without any degradation or loss of reactivity and selectivity. NMR measurements on the pure ILs reveal that no modification of the cationic moiety takes place during their preparation or storage.

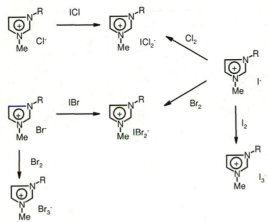

Scheme 2. General scheme for trihalide-based ILs synthesis.

ILs having ICl_2^- or IBr_2^- as counter-anion may be obtained also by addition at 0 °C of an equivalent of halogen (Cl_2 or Br_2) to a 1,3-dialkylimidazolium iodide. The redox process occurring during this reaction is in agreement with recently reported *ab initio*[43] calculations showing that the X-Y-X⁻ anions, where Y is the heaviest atom, are more stable than the isomers Y-X-X⁻ both in the gas phase and in solution. As a result, Y-X-X⁻ anions formed by Cl_2 or Br_2 oxidation of iodide ion are expected to easily isomerize to X-Y-X⁻ anions *via* cleavage of the X-X bond.

Finally, it is worth of note that trihalide-based ILs prepared by addition of I_2 at [Rmim][Br] or [Rmim][Cl] or ICl and IBr to [Rmim][I], which should have as counteranion [I_2Br]⁻ or [I_2Cl]⁻, are really complex mixtures of ILs bearing trimeric anions, in agreement with the behaviour of the corresponding ammonium salts in molecular solvents.[44, 45] ESI-MS experiments[42] carried out on trihalide based imidazolium salts reveal that the trimeric anions can be divided into two groups, according to the level of stability. In particular, derivatives like Br_3^-, I_3^-, IBr_2^- and ICl_2^- can be classified as stable species, whereas Br_2Cl^-, I_2Br^-, I_2Cl^- are rather unstable. Although the parent ions have been detected in all cases, relevant amounts of additional species were present in the case of Br_2Cl^-, I_2Br^-, I_2Cl^-. These species are presumably formed by decomposition, ion-molecule reactions, exchange or aggregation, according to Scheme 3.

Given their stability, ionic liquids based on imidazolium cation ([bmim]) and having Br_3^-, ICl_2^- and IBr_2^- as anion, have been recently characterized. Key physical properties of these salts have been determined and compared with those of other 1,3-dialkylimidazolium salts (Table 1).[46]

37

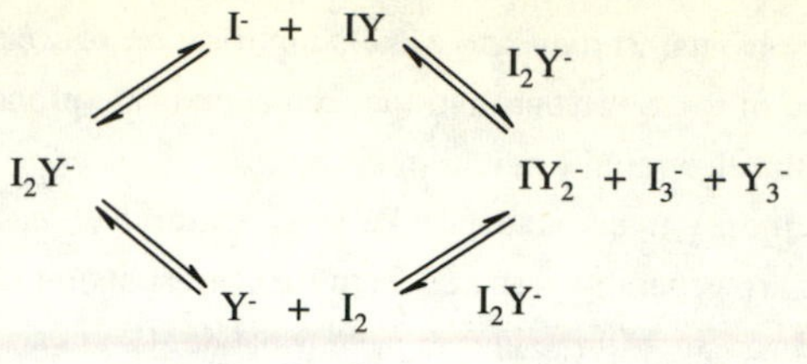

$$\text{Y} = \text{Br, Cl}$$

Scheme 3. Equilibria involving the I_2Y^- species.

Table 1. Physical properties of trihalide-based ionic liquids.

Ionic liquid	d	η	σ
	(g/ml)	(cP)	(mS/cm)
[bmim][ICl$_2$]	1.78	49.81	5.74
[bmim][IBr$_2$]	1.547	57.28	6.16
[bmim][Br$_3$]	1.702	92.53	8.93
[Hmim][HBr$_2$]	1.75[c]	17[c]	40[c]
[bmim][I]	1.44[b]	1110[b]	
[bmim][Tf$_2$N]	1.436[c]	52[a]	3.9[a]
[bmim][PF$_6$]	1.368[c]	450[b]	
[bmim][BF$_4$]	1.17[b]	219[b]	0.23[a]

[a] ref. 4; [b] ref. 47; [c] ref. 13

All the investigated trihalide ILs are miscible with solvents of medium dielectric constants (CH$_3$CN and DMSO), slightly soluble in chlorinated solvents whereas they are completely immiscible with alkanes, dioxane, aromatics and diethyl ether. Moreover, trihalide-based ILs show relatively low melting points and low viscosities, whereas density and hydrophobicity are higher than those of the corresponding chlorides, bromides or iodides.

The data reported in Table 1 reveal that the identity of the anion, probably due to a different ability of the corresponding salts to give ion pairs or ion aggregates, strongly affect viscosity and conductivity. For trihalide-based ionic liquids, the viscosity increases in the order $ICl_2^- < IBr_2^- < Br_3^-$. If only electrostatic interaction forces govern the viscosity of trihalide-based ionic liquids, then [bmim][IBr$_2$] having the largest ionic radius should be characterized by the lowest viscosity. At variance, the lowest viscosity is shown by [bmim][ICl$_2$], and the highest by [bmim][Br$_3$]. This behaviour suggests that, even in a homogenous class of ILs such as that investigated in this work (same cation, same anion nature and very similar anion structure), the combination of different effects (anion shape, polarizability, and so on) is able to affect viscosity giving often unpredictable results.

The data reported in Table 1 also reveal that [bmim][Br₃] is characterized by the highest conductivity, even though it is the most viscous among the investigated trihalide-based ionic liquids. In agreement with this behaviour, deviation from linearity can be observed by plotting ln η vs. ln σ: viscosity alone does not account for the conductive behaviour. On the other hand, the temperature dependence of conductivity for the ionic liquids under investigation exhibits a classical linear Arrhenius behaviour (Figure 3).

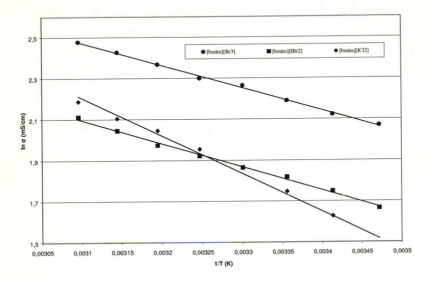

Figure 3. Temperature dependence of conductivity for trihalide-based ionic liquids.

The change in conductivity with temperature depends on the ionic liquid: while a similar behaviour has been found for the two bromide-containing species, a higher slope characterizes the straight line related to [bmim][ICl₂]. Apparently, also the temperature dependence of the conductivity is affected by the type and nature of the anionic counterpart.

3. Stereoselective bromination of alkenes and alkynes in ionic liquids

3.1. Brominating species and product distribution

Dihalo derivatives are generally prepared in chlorinated solvents (dichloromethane, chloroform, carbon tetrachloride), which are on the environmental black list. Furthermore, although CCl₄ is often used for bromination of unsaturated compounds the electrophilic addition reactions are slow in this apolar solvent and competitive radical processes, frequently leading to a different products distribution, may occur. On the other hand, the electrophilic additions in polar solvents, such as water (a green solvent) or alcohols are very fast, but they give as the main or sole products the corresponding bromohydrins and bromoethers, arising from the nucleophilic attack of the solvent on the ionic intermediate (a haloranium or β-halocarbenium ion). In this respect, the use of ILs may be advantageous, in particular considering that the polarity of these media is able to accelerate the electrophilic addition process, the non-nucleophilic nature of many ILs avoids the formation of solvent incorporating products and these reactions often occur with a high stereoselectivity.

Stereoselective brominations in ionic liquids have been easily performed by adding under stirring 1 equiv. of the unsaturated compound to a freshly prepared solution of [bmim][Br₃] in the chosen ionic medium (0.3M). The reactions of aryl and alkyl substituted alkenes and alkynes with [bmim][Br₃] in [bmim][Br] are always anti stereospecific, independently of the nature of the substituents at the double or triple bond: *erythro* (or *meso*) dibromides are obtained from *trans* olefins, while *cis* olefins give the corresponding *threo* (or *d,l*) adducts (Scheme 4 and 5).[48,49]

It is noteworthy that electrophilic addition of tribromide, generated by bromine addition to [bmim][Br], can also be carried out in other ionic liquids bearing non nucleophilic anions, such as [bmim][PF₆] or [bmim][Tf₂N]. The reactions are also in this case completely *anti* stereospecific, independently of the structure of the unsaturated compound.[50] This result suggests that, in agreement with the behaviour of tetrabutylammonium tribromide in aprotic solvents, no dissociation of the anion occurs in the investigated ILs and practically free bromine is not present in solution.

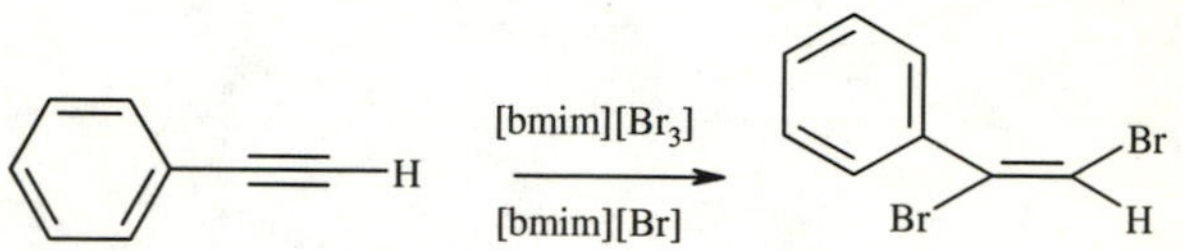

Scheme 4. Reactions of aryl alkynes with [bmim][Br₃] in [bmim][Br].

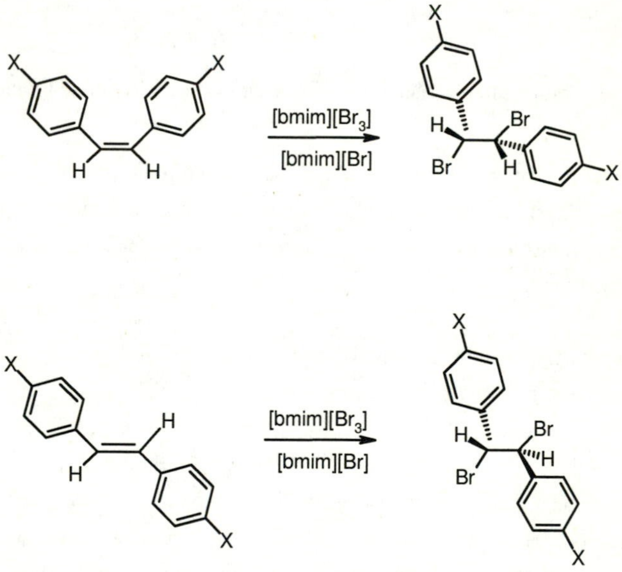

Scheme 5. Reactions of stilbenes with [bmim][Br₃] in [bmim][Br].

About the electrophilic bromination of unsaturated compounds in ILs it is however to note that also the addition of free bromine in [bmim][PF₆] and [bmim][BF₄] is an *anti* stereospecific reaction with dialkyl substituted alkenes, alkyl substituted alkynes and *trans*-stilbenes, whereas *cis*-stilbenes and aryl alkynes give under these conditions mixtures of both diastereoisomers. Bromine addition in ILs is however characterized by an increased *anti* stereoselectivity with respect to chlorinated solvents which can be attributed to a decreased lifetime of the ionic intermediates in ILs.[48]

In the case of bromine addition to diaryl substituted olefins, generally, the stereoselectivity is strictly related to the magnitude of bromine bridging of the ionic intermediate.[51] Bridged bromiranium ions give exclusively *anti* addition products, whereas open β-bromocarbenium ions give mixtures of *syn* and *anti* addition products. The nature of the substituents on the phenyl ring is responsible of the type of ionic intermediate that is formed during bromination: electron withdrawing groups favor bridged intermediates, whereas electron donating groups lead to open β-bromocarbenium ions. Open β-bromocarbenium ions can give the *syn* addition products mainly through an *anti* attack of the counter anion, after rotation around the C-C bond. Generally, the solvent is not able to affect the bridging in the ionic intermediates, it can only affect the lifetime of these transient species, *i.e.* the relative rates of nucleophilic attack and C-C rotation. (Scheme 6). Therefore, the higher *anti* stereoselectivity that characterizes the addition of free bromine to the unsubstituted *cis*-stilbene and *cis*-stilbenes bearing electrondonating groups in [bmim][PF$_6$] and [bmim][BF$_4$], with respect to chlorinated solvents, may be attributed to a reduce lifetime of the ionic intermediate (bridged bromiranium or open β-bromocarbenium) in the ionic medium determined by a faster nucleophilic trapping.

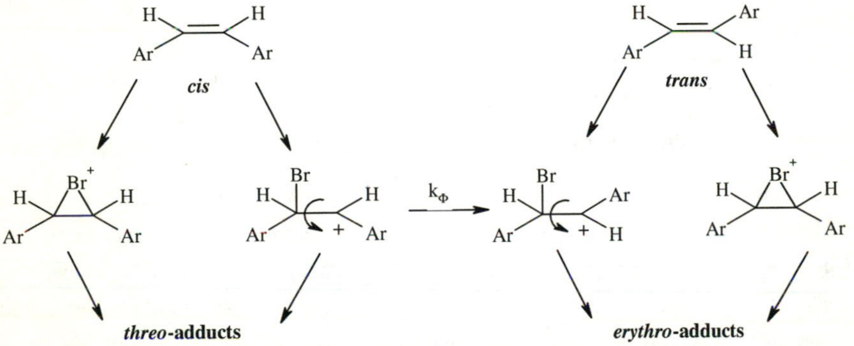

Scheme 6. Stereochemical behavior of Br$_2$ addition to stilbenes.

The capability of ionic liquids to decrease the lifetime of ionic intermediates has been also evidenced in solvolysis reactions of secondary aromatic substrates[52] as well as in nucleophilic substitution reactions by NaN$_3$ on primary and secondary substrates.[20a]

In conclusion, although also the bromine addition in ILs offers several advantages with respect the same reaction in molecular solvents (higher reaction rate, higher stereoselectivity, possibility to avoid toxic solvents) from a synthetic point of view the use of [Rmim][Br$_3$] as brominating agents in ILs may present further benefits. In particular, with respect to bromine these salts are easier to handle. With respect to the corresponding tetraalkyl ammonium or polymer bound tribromides, which must be used in chlorinated solvents, toxic media can be avoids. Furthermore, at variance with the reactions using polymeric trihalides, these additions are carried out under homogeneous conditions, avoiding problems due to the access of the substrate to the polymeric support. On the other hand, in analogy with the reactions carried out using polymeric trihalides the product recover from ILs is extremely easy and does not require any aqueous work-up. A process this latter which determines drastic reductions in the product yields in the reactions with tetraalkyl ammonium tribromides. Finally, it is to note that after the extraction of the products, the ionic liquid can be recovered and reused for further reactions. The NMR spectra of the recycled solvents generally show no evidence of degradation and no appreciable variation in reaction yields or product stereoselectivity has been observed using recovered ILs. Moreover, also the halogenating agent can be regenerated simply by adding the proper amount of free bromine.

3.2. Kinetic measurements

Related to the mechanism of bromination in ILs, it is worth of note that at least three alternative nonradical mechanistic pathways, all of which involve substrate-Br_2 π-complexes, were identified for the bromination of unsatured compounds in molecular solvents (Scheme 7).[53]

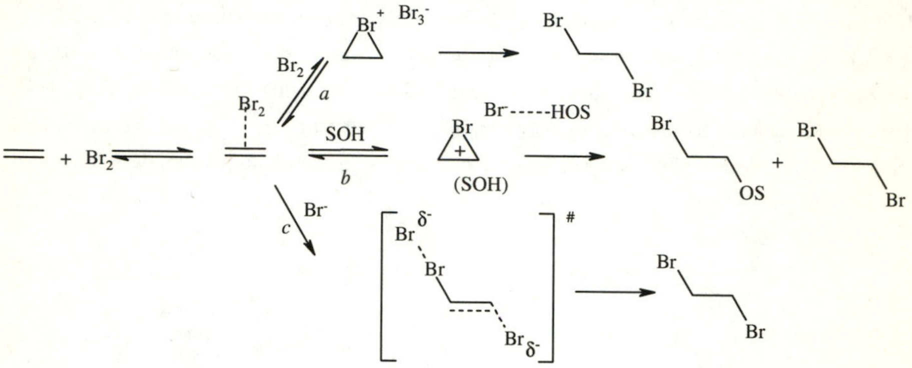

Scheme 7. Non-radical mechanistic pathways for bromine addition in molecular solvents.

The first pathway (path *a*) has been proposed for the addition of free bromine in aprotic solvents of low polarity ($4 < \varepsilon < 11$) and consists in the ionization of a 1:1 π-complex catalyzed by a second molecule of bromine to give a bromirenium or a β-bromocarbenium bromide (or tribromide) ion pair, which then collapses to give the dibromoadduct (in Scheme 7 has been represented only the bromirenium ion).[54-58] In aprotic solvents, not able to give any electrophilic assistance to the Br^- formation in the ionization step, the second molecule of bromine is believed to favor the Br-Br bond breaking by forming a charge-dispersed tribromide ion. In agreement with this mechanism, these reactions are characterized by a second-order dependence of the rate on Br_2 concentration, corresponding to the overall third order kinetic equation reported below (eq 1), where [S] is the substrate concentration.

$$-d[Br_2]/dt = k_{Br2} [Br_2]^2 [S] \quad (1)$$

In the presence of added bromide salts, which in low polarity non protic solvents bind Br_2 to give the highly stable Br_3^- species,[53] or when preformed tribromide salts are used as the brominating agents, the reaction proceeds through a rate- and product-determining nucleophilic attack of Br^- anion on the 1:1 π-complex (path *c*).

Practically no intermediate is formed in this reaction: the nucleophilic attack of Br^- and the bond breaking are indeed concerted, although not necessary syncronous.[53,57] This reaction pathway can be distinguished from path a as it is characterized by a first order dependence of the rate on the brominating agent, and is described by an overall second order kinetic equation:

$$-d[Br_2]/dt = k_{Br}\,[Br_2][Br^-]\,[S] \quad (2)$$

It is worth of note that also the addition of free bromine to olefins in protic solvents (methanol, acetic acid) follows a second order reaction law, at least when low bromine concentrations are used (eq. 3).

$$-d[Br_2]/dt = k_{Br2}\,[Br_2]\,[S] \quad (3)$$

In this case however the reaction occurs through a S_N1-like unimolecular ionization of the 1:1 π-complex to form a bromiranium, or a β-bromocarbenium, bromide ion pair.[51,59,60] Protic solvents are able to provide a specific electrophilic solvation by hydrogen bonding to the leaving bromide ion (path b), although they can give also nucleophilic assistance.[51,61] The possibility that all three pathways can contribute to the overall bromination process exists only for the reaction in protic solvents. In chlorinated solvents generally solely path a is involved in absence of added bromide ion, whereas in presence of a stoichiometric amount of Br^- or when tribromide ion is used as brominating agent, path c alone is involved.

Bromination of alkenes and alkynes using [bmim][Br$_3$] as brominating agent in [bmim][Br], [bmim][PF$_6$], [emim][Tf$_2$N], [bmim][Tf$_2$N], [hmim][Tf$_2$N], [bm$_2$im][Tf$_2$N] and [bpy][Tf$_2$N] (where emim=1-ethyl-3-methylimidazolium, hmim=1-hexyl-3-methylimidazolium, bm$_2$im=1-butyl-2,3-dimethyl imidazolium, bpy=butylpyridinium) obeyed the second order rate law reported in eq 4

$$-d[Br_3^-]/dt = k_{Br3^-}\,[Br_3^-][S] \quad (4)$$

where [S] is the substrate concentration. The second-order kinetic constants for [Br$_3$]$^-$ addition in ILs and 1,2-dichloroethane (DCE) are reported in Table 2.[50]

Table 2. Second order Rate Constants for the reaction of [Br$_3$]$^-$ with alkenes and alkynes in ionic liquids and DCE at 25 °C.[a]

Solvent	1-pentene	Styrene	ethylcinnamate	1-phenylacetylene	1-phenylpropyne
	k_{Br3^-} (M^{-1}s^{-1})	k_{Br3^-} (M^{-1}s^{-1})	k_{Br3^-} (M^{-1}s^{-1})	k_{Br3^-} (M^{-1}s^{-1})	k_{Br3^-} (M^{-1}s^{-1})
[bmim][Br]	n.d	n.d	n.d	$1.2\cdot10^{-3}$	$2.6\cdot10^{-4}$
[bmim][PF$_6$]	$3.4\cdot10^{-2}$	$6.7\cdot1^{-2}$	$2.7\cdot10^{-4}$	$9.3\cdot10^{-4}$	$1.8\cdot10^{-4}$
[emim][Tf$_2$N]	$1.2\cdot10^{-1}$	n.d[b]	$5.4\cdot10^{-4}$	$9.5\cdot10^{-4}$	$3.2\cdot10^{-4}$
[bmim][Tf$_2$N]	$1.5\cdot10^{-1}$	$1.2\cdot10^{-1}$	$1.9\cdot10^{-4}$	$6.4\cdot10^{-4}$	$2.8\cdot10^{-4}$
[hmim][Tf$_2$N]	$8.4\cdot10^{-2}$	$7.6\cdot10^{-2}$	$1.3\cdot10^{-4}$	$4.3\cdot10^{-4}$	$1.5\cdot10^{-4}$
[bm$_2$im][Tf$_2$N]	$4.2\cdot10^{-2}$	$4.1\cdot10^{-2}$	$1.2\cdot10^{-4}$	$2.7\cdot10^{-4}$	$1.3\cdot10^{-4}$
DCE	n.d[b]	n.d[b]	$1.8\cdot10^{-5}$	$5.6\cdot10^{-5}$	$2.6\cdot10^{-4}$

[a]Standard deviations in the values of k_{Br3^-} were always less than 5% and more usually between 2 and 3% of the quoted values. [b]Too fast to be measured using a UV-Vis spectrometer. n.d=not determined.

The values of the activation parameters for the reaction of ethyl *trans* cinnamate and 1-phenylacetylene obtained from fittings of the kinetic constants to the Arrhenius equation are reported in Table 3.

Table 3. Apparent activation parameters for the reaction of $[Br_3]^-$ with ethyl *trans*-cinnamate (**1**) and 1-phenylacetylene (**2**).[a]

Substrate	Solvent	k_{Br3} (298K) ($M^{-1} s^{-1}$)	$E_{a,obsd}$ (kJ mol^{-1})	$\Delta H^{\neq}$ (kJ mol^{-1})	$\Delta S^{\neq}$ (J K^{-1} mol^{-1})
1	DCE	$1.80 \cdot 10^{-5}$	69.6 (4)	67.1 (4)	-11 (2)
	[bmim][Tf$_2$N]	$1.9 \cdot 10^{-4}$	68.2 (4)	65.7 (4)	-95 (2)
	[bpy][Tf$_2$N]	$6.2 \cdot 10^{-4}$	56.1 (4)	53.6 (4)	.127 (4)
2	DCE	$5.6 \cdot 10^{-5}$	67.8 (4)	65.3 (4)	-104 (2)
	[bmim][Br]	$1.2 \cdot 10^{-3}$	21.5 (4)	19.0 (4)	-236 (6)
	[bmim][PF$_6$]	$1.8 \cdot 10^{-4}$	57.8 (3)	55.3 (3)	-118 (3)
	[emim][Tf$_2$N]	$3.2 \cdot 10^{-4}$	55.7 (4)	53.2 (4)	-124 (2)
	[bmim][Tf$_2$N]	$6.4 \cdot 10^{-4}$	60.5 (4)	58.0 (4)	-113 (2)
	[bm$_2$im][Tf$_2$N]	$2.7 \cdot 10^{-4}$	61.7 (4)	59.2 (4)	-116 (2)

[a] Standard deviations are given in parentheses.

The complete *anti* stereospecificity characterizing these reactions in ILs as in any solvent, together with the followed kinetic law lead to the conclusion that also in ionic liquids the tribromide reaction with unsaturated compounds occurs through reaction pathway *c*, *i.e.* through the reaction mechanism characterizing the addition of Br_2 in the presence of Br^- in aprotic solvents (Scheme 8).

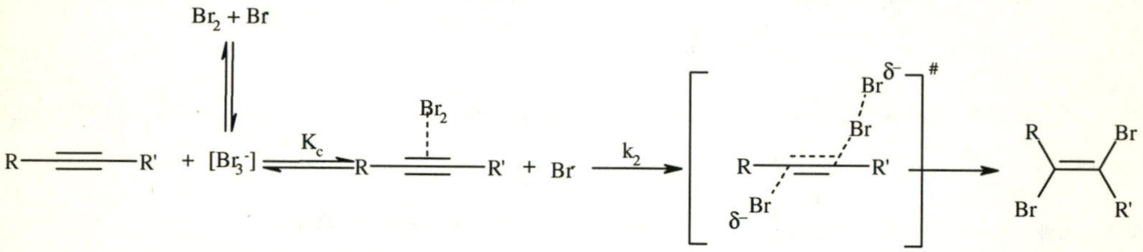

Scheme 8. Mechanistic pathway for Br_3^- addition to alkynes in ILs and molecular solvents.

The IL structure can however significantly affect reactivity. For example, the kinetic parameters characterizing the addition of Br_3^- to 1-phenylacetylene in [bmim][Br], the IL having as counter anion the nucleophile of this reaction, are quite different from those found in DCE.[49] In particular, whereas the kinetic constant for the addition to 1-phenylpropyne in this IL is of the same magnitude as that reported in DCE, the electrophilic addition of tribromide to 1-phenylacetylene is accelerated by a factor of 20. On the other hand, the data reported in Table 3 clearly show that the reaction in ILs not bearing nucleophilic anions are only slightly accelerated in comparison to DCE.[50]

The ionic solvent structure is therefore able to affect the kinetic behaviour of this reaction, but a rationalization of these effects is not easy since polarity, viscosity and hydrogen-bond donor ability of the IL[62,63] may affect the rate determining step of the reaction and/or the equilibrium constants characterizing the pre-equilibrium steps. In particular, viscosity may exert opposite effects on the reaction rates: indeed it can favour bond making and disfavour bond breaking.[64] In tribromide addition bond making and bond breaking occur in the transition state, therefore, the effect of viscosity depends on the position of the transition state with respect to the reaction coordinate. In [bmim][Br] the reaction is performed in a large excess of [Br]⁻, which is probably the true nucleophile of the addition process; under these conditions bond making may precede bond breaking, in particular when the substrate structure disfavour positive charge development.

The high viscosity of [bmim][Br], may therefore accelerate the reaction. At variance, in all the other ILs, a later transition state is more probable and the effect of solvent viscosity decreases. Considering that at 25 °C, the order of increasing viscosity is [emim][Tf$_2$N] < [bmim][Tf$_2$N] ≈ [bpy][Tf$_2$N] < [hmim][Tf$_2$N] < [bm$_2$im][Tf$_2$N] < [bmim][PF$_6$] << [bmim][Br], it is possible to note that the rate constants reported for tribromide addition in ILs do not follow the viscosity trend: in other words k$_2$ do not depend on the viscosity alone.

It is noteworthy that in chlorinated solvents tribromide addition is positively affected by the ability of the medium to give H-bonding.[53] Although, the investigated ILs have a different polarity[62-63] and are characterized by a different ability to give [65] H-bonding ([bmim][Tf$_2$N] is among the investigated ILs that having the highest H-bonding donor ability) the data reported in Table 3 show that the activation parameters are not significantly affected by this feature. This may be a consequence of the fact that in ILs, at variance with molecular solvents, both bond breaking and bond making are affected by the solvent H-bonding ability (in molecular solvents bond making is not influenced, being the entering bromide ion present as ion pair with the tetraalkylammonium cation) and when they occur simultaneously in the rate determining transition state only a very little effect, if any, may be detected.

4. Stereoselective iodochlorination and iodobromination of unsaturated compounds

4.1. Product distribution

The electrophilic properties of Br$_3$⁻ have been studied extensively in chlorinated solvents, where this species is generally produced by reaction of Br$_2$ with a tetraalkylammonium bromide. At variance, only few data about the use of ICl$_2$⁻ and IBr$_2$⁻ as electrophiles have been reported in molecular solvents.[66]

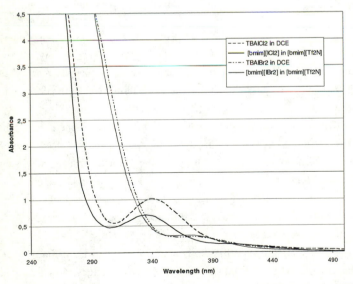

Figure 4. UV-Vis spectra of trihalide-based ionic liquids in [bmim][Tf$_2$N].

Although the stability of dichloroiodate ([ICl$_2^-$]) and dibromoiodate ([IBr$_2^-$]) anions has never been determined in ionic liquids, spectroscopic measurements as well as reactivity data suggest that also these trihalide species are highly stable in these solvents. The absorption maxima and the extinction coefficients of solutions of [bmim][ICl$_2$] and [bmim][IBr$_2$] in [bmim][Tf$_2$N] are similar to those recorded for the corresponding ammonium salts in chlorinated solvents (see Fig. 4) where the equilibrium constants are higher than 10^7 M^{-1}. At variance with protic molecular solvents, which favor trihalide dissociation, due to the large solvation energies of the small bromide and chloride anions, ionic liquids disfavor this process. The behaviour of ILs seems to be therefore similar to that of aprotic molecular solvents of medium dielectric constant.

Both [Rmim][ICl$_2$] and [Rmim][IBr$_2$] have been used as reagent-solvents or as reagents, carrying out the reactions in [bmim][PF$_6$], for the iodochlorination and iodobromination of alkenes and alkynes (Scheme 9).[42] These rather uncommon ionic liquids are excellent iodine donor systems for the stereoselective *anti* iodination of unsaturated compounds, which avoid the use of toxic reagents (I$_2$/CuCl$_2$, I$_2$/HgCl$_2$, ICN/AlCl$_3$) as well as chlorinated solvents.[67-68] Very good to almost quantitative yields of vicinal iodochloro- and iodobromo-adducts were obtained with all the examined substrates.

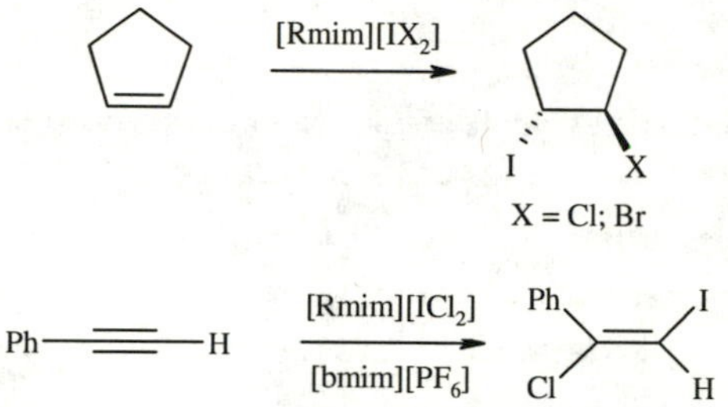

Scheme 9. Reactions of alkenes and alkynes with [Rmim][IX$_2$] in ionic liquids.

It is worth of note that, at variance with [Rmim][Br$_3$], reactions carried out with [Rmim][ICl$_2$] in [Rmim][Cl] or with [Rmim][IBr$_2$] in [Rmim][Br] showed a very low, if any, conversion into the corresponding dihalo adducts. Furthermore, it must be remarked that while the reaction of [Rmim][ICl$_2$] always proceeds with complete *anti* stereoselectivity, that of [Rmim][IBr$_2$] in the case of 1-phenylpropyne gives also the *syn* adduct.

Also the regioselectivity of the addition depends on the trihalide used and on the substrate structure. When [Rmim][ICl$_2$] have been used as halogenating agent, the Markovnikov adducts were recovered as the exclusive products from aryl-substituted double and triple bonds; on the other hand, alkyl substituted alkenes gave under the same conditions mixtures of Markovnikov and anti-Markovnikov adducts (AM), the percentage depending on the relative steric hindrance of the substituents. In particular, the increasing amount of the AM-adduct found on going from 1-pentene to 3,3-dimethyl-1-butene, clearly reveals that the addition process is affected by steric factors (Scheme 10).

The regiochemistry of the addition of [IBr$_2$]$^-$ to non crowded olefins is qualitatively similar to that of [ICl$_2$]$^-$, although in the case of aryl substituted alkynes and sterically congested olefins, higher amounts of AM-adducts are generally recovered (Scheme 11).

46

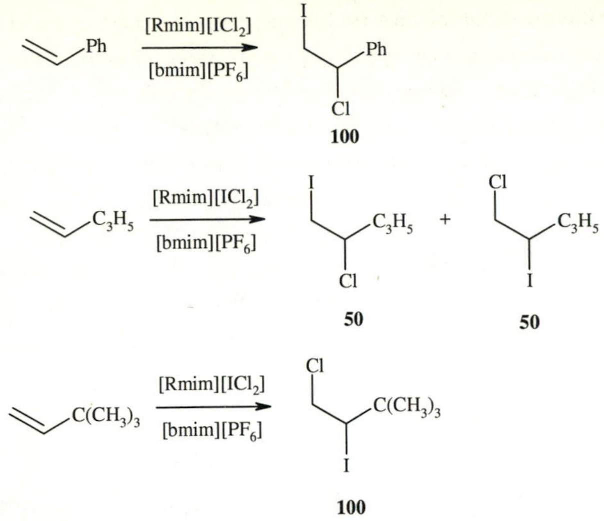

Scheme 10. Reactions of alkenes with [Rmim][ICl$_2$] in [bmim][PF$_6$].

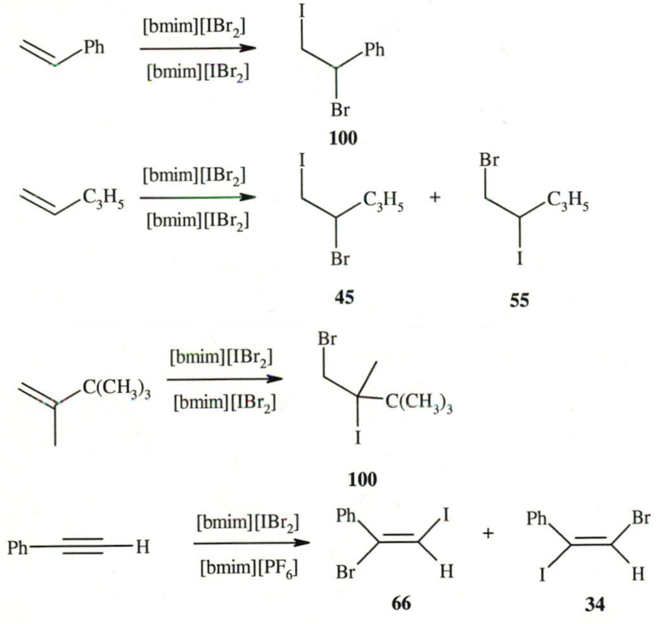

Scheme 11. Reactions of alkenes and alkynes with [bmim][IBr$_2$] in [bmim][PF$_6$].

Both reactions are however affected also by electronic effects as shown by the products arising from the addition of [IBr$_2$]⁻ and [ICl$_2$]⁻ to 2,3,3-trimethyl-1-butene and aryl substituted alkynes.

Finally, it is worth of note that the halogenations with the corresponding tetrabutylammonium salts (TBAICl$_2$ and TBAIBr$_2$) in dichloromethane[42] are characterized by a very similar chemo-, regio- and stereoselectivity. These latter reactions are however extremely slow with respect to the corresponding reactions in ionic liquids: the addition of TBAIBr$_2$ give always conversions lower than 30% even after long reaction times (more than 72 hours). This behaviour may be attributed to the decomposition of the reagent (see paragraph 4.2). In these latter reactions, therefore, ILs are not only potential green solvents offering the advantages previously evidenced for the reactions of [Rmim][Br$_3$], but they represent the best solvents to obtain mixed dihalo adducts, in high yields and with high purity degree.

4.2. Kinetic measurements

The kinetic behaviour of ICl$_2^-$ addition to several alkenes and 1-phenylacetylene has been examined both in non nucleophilic ILs and DCE.[50] In all examined solvents the reactions obey the second order rate law of eq 5.

$$-d[ICl_2^-]/dt = k_{ICl2^-} [ICl_2^-][S] \qquad (5)$$

Furthermore, at variance with the reactions performed in chlorinated solvents, where a progressive darkening of the solution allows the evaluation of only initial rate constants, the reactions carried out in ILs can be followed readily for at least two half-lives. The progressive darkening of the chlorinated solutions has been attributed to the formation of I$_3^-$, which is a consequence of the reaction of chloride anion with the addition products, according to Scheme 12. This substitution reaction, which leads to the corresponding dichloro derivatives, does not compete significantly with the addition process in ionic liquids, in particular in those having higher hydrogen bonding donor ability. In these ILs the decomposition of the ICl$_2^-$ species is practically completely repressed, in agreement with recently reported data on nucleophilic substitution reactions in ILs showing that in these media the nucleophilicity of the chloride anion is significantly reduced.[69]

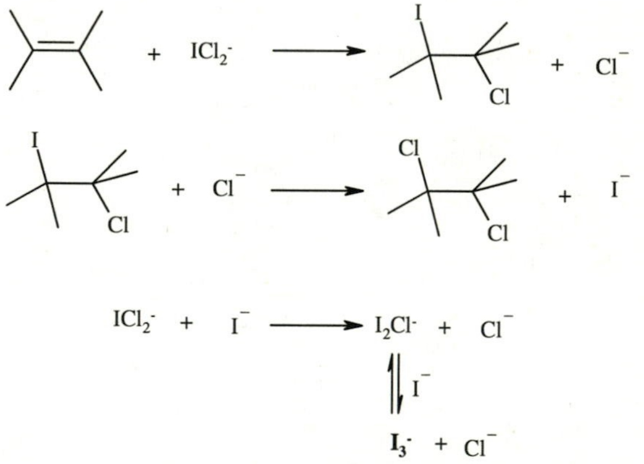

Scheme 12. Proposed mechanism for ICl$_2^-$ decomposition in chlorinated solvent.

The increase in reaction rates observed for the addition of ICl$_2^-$ on going from DCE to ILs is generally higher than that reported for the Br$_3^-$ addition in the same solvents. This behaviour suggests a different charge distribution in the two transition states. Moreover, as in the case of Br$_3^-$ addition, the activation parameters do not correlate with viscosity alone although they depend on IL structure. A possible role of the H-bonding donor ability of IL may be instead evidenced. In [bmim][Tf$_2$N], the IL having the relatively higher hydrogen bonding donor ability, the activation enthalpy is strongly reduced, while the $\Delta S^{\neq}$ value is significantly increased. At variance, in [bm$_2$im][Tf$_2$N] and [bpy][Tf$_2$N] (these ILs are characterized by cations less suitable to give hydrogen bonding) the activation enthalpies significantly increase and activation entropies decrease. It is however worth of note that the reaction performed in [bmim][PF$_6$], the IL characterized by stronger anion-cation interactions and therefore by a higher order degree, the activation enthalpy is very high whereas the activation entropy is particularly low.

The effect of IL on the addition of ICl$_2^-$ to double bonds has been therefore explained[50] considering the ability of the cation of the ionic liquid to stabilize the leaving chloride ion *via* hydrogen bonding.

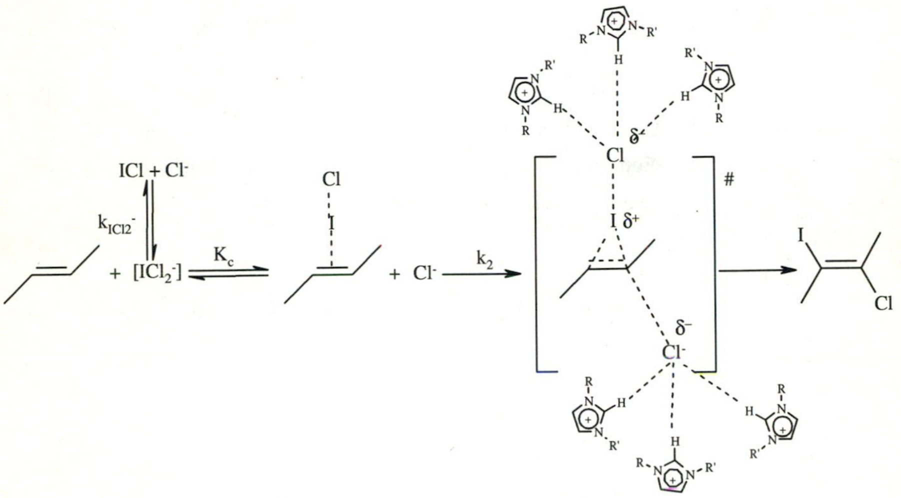

Scheme 13. Mechanistic pathway for ICl$_2^-$ addition to alkenes in ILs.

Kinetic and distribution data suggest for this reaction a concerted mechanism of the type proposed for the addition of Br$_3^-$. The nucleophilic attack of the chloride ion on the alkene-ICl π-complex should lead to the addition product(s) through a rate- and product determining transition state of the type reported above. Anyway, in the case of ICl$_2^-$ addition to olefins, at variance with the addition of Br$_3^-$, bond breaking may precede bond making, the transition state has greater iodiranium character and the attack by chloride is not itself the rate determining step. In this situation, the ability of the ionic liquid to form hydrogen bonding with the leaving chloride plays a main role, although also the IL viscosity can affect the reaction rate (Scheme 13). It is worth of note that, although activation enthalpies and entropies vary significantly passing from an ionic liquid to another one, the activation energy values are very similar in all examined ILs suggesting a compensation effect.[50] The Arrhenius plots for the reaction of *trans*-2-pentene with [bmim][ICl$_2$] in the five investigated ILs show,[50] however, two intersections points strongly suggesting[70] that not only the solvent H-bonding ability but also some other feature of these media is able to affect reactivity.

5. Halofluorination of alkenes

H. Yoshino *et al.*[71] have recently reported a successful alternative method for the synthesis of halofluoroalkanes which involves the use of 1-ethyl-3-methylimidazolium oligohydrogenfluoride ([emim][F(HF)$_{2.3}$]) as fluoride source in the presence of N-halosuccinimide (Scheme 14).

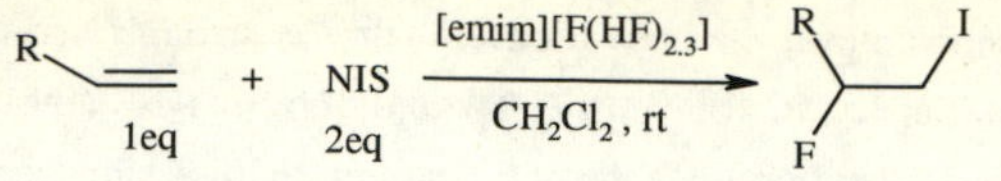

Scheme 14. Reaction of alkenes with [emim][F(HF)$_{2.3}$] in the presence of NIS.

In this procedure, the ionic liquid is used as cosolvent, together with dichloromethane. Products are recovered from reaction mixture using *n*-hexane as extractant, so that also in this case the aqueous work-up can be avoided. It is worth of note that traditional routes used to prepare vic-halofluoroalkanes are based on the reaction of olefins with halonium ions in presence of elemental fluorine,[72] hydrogen fluoride,[73] Olah's reagent,[74] ET$_3$N-3HF,[75] silver fluoride,[76] potassium fluoride-poly(hydrogen fluoride)[77] or silicon tetrafluoride,[78] reagents able to release HF after treatment with water.

The data reported by H. Yoshino clearly reveal that the nucleophilic attack of fluoride on the halonium ion is *anti*-stereospecific while the reaction occurs with a high regioselectivity. Moreover, when NIS or NBS are used to generate the halonium ion, good to excellent yields were obtained. Lower conversions characterize instead the reactions with NCS (Scheme 15).

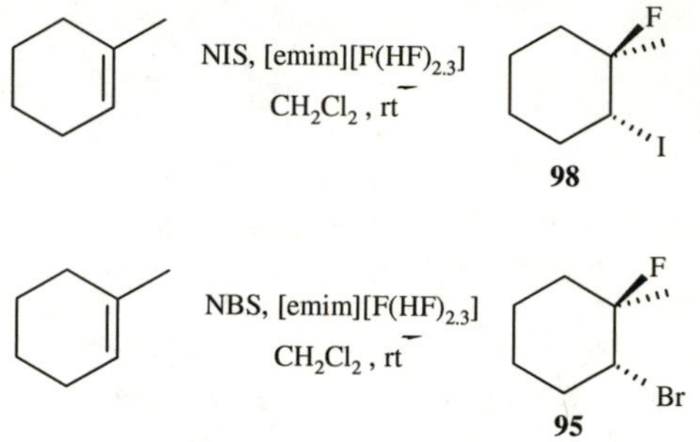

Scheme 15. Reactions of alkenes with [emim][F(HF)$_{2.3}$] in the presence of N-halosuccinimides.

Therefore, the reaction tuned by H. Yoshino *et al.* is a useful method to incorporate fluorine atoms in organic molecules and leads to products which can be further modified to obtain several organofluoro derivatives (Scheme 16).

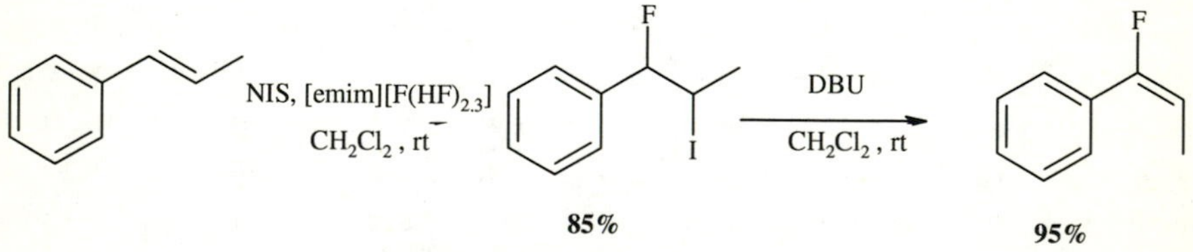

Scheme 16. Alkenes fluorination.

6. Electrophilic substitution of aromatic compounds

Halogenated aromatic compounds are key intermediates in organic synthesis as they are precursors to a number of organometallic species used for the preparation of pharmaceutically active molecules. Conventional methods applied for the direct halogenation of aromatic systems involve the use of potentially hazardous elemental halogens,[79] expensive transition metals based catalysts,[80] alkali metal halides associated with NaIO$_4$[81] or the combination of aqueous TBHP or H$_2$O$_2$ together with a hydrohalic acid.[82] Though this latter procedure is considered to be the most effective from an environmental point of view, as the sole by-product is water, it often requires an excess of hydrohalic acid to successfully perform the halogenation together with the use of chlorinated solvents, in order to achieve a complete solubilization of the substrates. Brønsted acidic ionic liquids, which have been recently applied to a variety of reactions including esterification of carboxylic acids,[83] protection of aldehydes and ketones,[84] as well as ether cleavage[25], have been used successfully as reagent-solvents in electrophilic aromatic substitutions.[85]

In particular, 3-methylimidazolium tribromide ([Hmim][Br$_3$]), whose synthesis is reported below, is more reactive than the corresponding non acidic derivative ([bmim][Br$_3$]) in aromatic brominations.

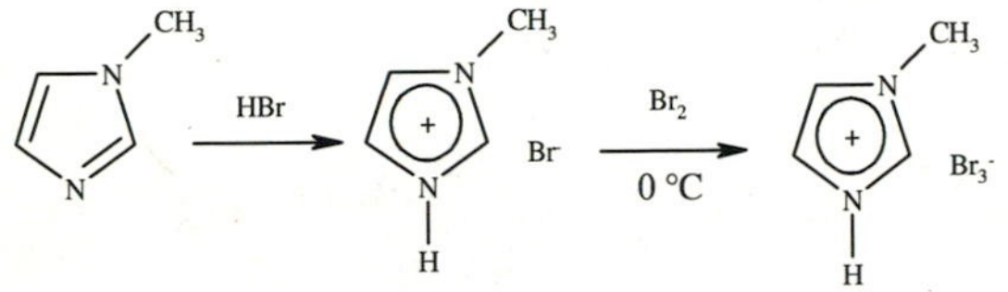

Scheme 17. General scheme for [Hmim][Br$_3$] synthesis.

Generally, these reactions can be performed in air without particular precautions and they give the mono-substitution products with a complete selectivity and in excellent yields: only in the case of N-methylaniline a polybrominated adduct has been recovered as the major product. Moreover, while using [bmim][Br$_3$] as reagent for the halogenation of *p*-xylene only α-bromination occurs, in presence of the acidic IL a 30% of *o*-bromination was achieved.

The halogenation is effective not only with activated compounds, but also with differently substituted benzaldehydes as well as with sterically hindered aromatic compounds. It is worth of note that no phenol is detected from the halogenation of anisole catalysed by [Hmim][Br$_3$], Scheme 18. Apparently, in the conditions used for the aromatic bromination, 3-methylimidazolium bromohydrogenate ([Hmim][HBr$_2$]) formed at the end of the reaction is not able to cleave the C-O bond (1:2 molar ratio was previously used[25] to cleave the ether functionality). Moreover, at variance with the reactions recently reported using pentylpyridinum perbromide, [ppy][Br$_3$],[86] bromination of anisole with [Hmim][Br$_3$] does not require any added base to prevent methoxy hydrolysis.

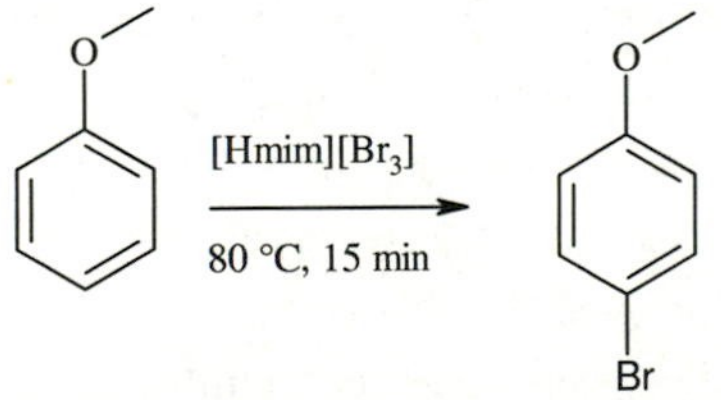

Scheme 18. Aromatic bromination.

Pentylpyridinium perbromide and [Hmim][Br$_3$] give instead similar product distributions when they are used as reagents for α-bromination of benzoketones, though the reaction of [Hmim][Br$_3$] is generally faster. Furthermore, the successful halogenation of naphtalene with the imidazolium salt ([Hmim][Br₃]), in the absence of any catalyst, is an indication of the high reactive nature of the involved halogenating specie. On the other hand, benzonitrile, bromobenzene, iodobenzene are completely non reactive towards electrophilic halogenation, either catalysed by [bmim][Br₃] or [Hmim][Br₃].

In summary, [Hmim][Br₃], [ppy][Br₃] and [bmim][Br₃] represent alternative halogenating agents for the electrophilic bromination of aromatic substrates, including some deactivated (benzaldehydes) or sterically hindered compounds, which are able to give in high yield and with high selectivity the expected products minimizing the use of organic solvents. In particular, when products are immiscible with the imidazolium salt, the recovery of the products can be achieved by simple decantation; otherwise, *n*-hexane may be used as extracting solvent. Moreover, at the end of the reaction, the halogenating agent can be readily recycled by adding the proper amount of free bromine. Alternatively, 3-methylimidazolium bromohydrogenate (the only by-product of the reaction in [Hmim][Br₃], the most efficient reagent-solvent) can be used[25] for ethers clevage, under controlled conditions (Scheme 19).

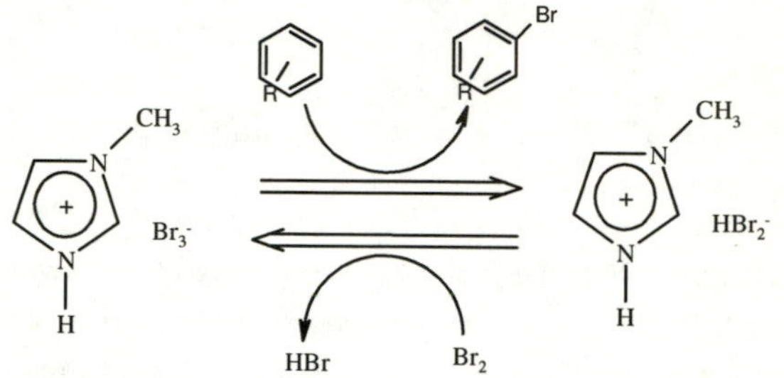

Scheme 19. Synthesis and recycle of [Hmim][Br₃].

7. Conclusions

In conclusion, trihalide-based imidazolium salts may be considered as an efficient alternative to traditional reagents for the stereoselective synthesis of halo derivatives. Their application as reagent-solvents for the halogenation of unsaturated compounds brought forth advantages in organic synthesis as it avoids the formation of solvent containing by-products, it maintains or enhances the regio- and stereoselectivity and accelerates the electrophilic addition processes; moreover it minimizes the use of molecular solvents and avoid toxic reagents. At variance with the reaction performed in molecular solvents, in which halo derivatives need to be separated from the reaction by-products (primarily ammonium salts), product recover in ILs is extremely easy and does not require any aqueous work-up. At the end of reaction, the ionic liquids as well as the halogenating reagents can be recovered and easily regenerated. Furthermore, in the case of the addition of [Rmim][ICl₂] in non nucleophilic solvents, the hydrogen bond donor ability of imidazolium ionic liquids, such as [bmim][Tf₂N], can be exploited to suppress unwanted nucleophilic substitution reactions due to the formed chloride ion on the primary reaction products .

Using [bmim][Br₃] as brominating agent either in [bmim][Br] or in non nucleophilic ionic liquids, the addition is *anti* stereospecific: *erythro* dibromides are obtained from *trans* olefins while *cis* olefins give the corresponding *threo* adducts as in any chlorinated solvents. When [Rmim][ICl₂] salts are used as reagents, the addition to alkenes and alkynes proceeds with complete *anti* stereoselectivity, while the regioselectivity depends on the substrate structure. At variance, [Rmim][IBr₂] addition to unsaturated compounds can give mixtures of *syn* and *anti* addition products at least with arylalkynes. An examination of the product distribution for the addition of IBr_2^- and ICl_2^- suggests that the two reactions are differently affected by steric and electronic effects.

Imidazolium salts having acidic halogenated anions, as [emim][F(HF)₂.₃], have been efficiently used as reagent-cosolvents to incorporate fluorine atoms in organic molecules. The nucleophilic attack of fluoride on the halonium ion is *anti*-stereospecific and occurs with high regioselectivity and good yields, when NIS or NBS are used to generate the halonium ion.

On the other hand, Brønsted acidic ionic liquids which possess perbromide species as anions can be used as reagent-solvents in electrophilic aromatic substitutions, either with non activated or sterically hindered compounds. Moreover, at the end of the reaction, 3-methylimidazolium bromohydrogenate (the only by-product of the reaction), can be recycled by adding free bromine or used for other reactions (ethers cleavage).

Kinetic investigations on tribromide as well as dichloroiodate addition to unsaturated compounds in ionic media clearly show that ionic liquids, though similar, are differently able to affect organic reactions. Moreover, ability of ionic liquids to affect organic reactivity strongly depends on the examined reaction: evidence suggests that, while the H-bonding ability of the imidazolium cation is probably the main factor able to affect the rate of the addition of ICl_2^- to olefins, this feature has no appreciable effect on the electrophilic addition of Br_3^- to the same substrates.

Understanding how chemical reactivity is influenced by the IL structure is therefore the key to avoid that these reaction media are used in a completely casual fashion, and to favour the technological improvements for a safer, more secure society and to benefit the environment and the economy.

Acknowledgments

We thank MIUR and University of Pisa for the financial support.

References

1. Rogers, R. D.; Seddon, K. R. In *Ionic liquids: Industrial Applications to Green Chemistry*; ACS Symposium Series 818, American Chemical Society: M, Washington, D. C., 2002.
2. Freemantle, M. *Chem. Eng. News* **1998**, *76*, 32.
3. (a) Holbrey, J. D.; Seddon, K. R. *Clean Products and Processes* **1999**, *1*, 223. (b) Earle, M. J.; Seddon, K. R. *Pure Appl. Chem.* **2000**, *72*, 1391. (c) Welton, T. *Chem. Rev.* **1999**, *99*, 2071. (d) Wasserscheid, P.; Keim, M. *Angew. Chem. Int. Ed.* **2000**, *39*, 3772. (e) Sheldon, R. *Chem. Commun.* **2001**; 2399. (e) Olivier-Bourbigou, H.; Magna, L. *J. Mol. Catal. A* **2002**, *182*, 419. (f) Dupont, J.; de Souza, R. F.; Suarez, P. A. Z. *Chem. Rev.* **2002**, *102*, 3667. (g) Wilkes, J. S. *J. Mol. Chem. A* **2004**, *214*, 11. (h) Poole, C. F. *J. Chrom. A* **2004**, *1037*, 49. (i) Chiappe, C.; Pieraccini, D. *J. Phys. Org. Chem.* **2005**, *18*, 275.
4. *Ionic Liquids in Synthesis* Wasserscheid, P.; Welton, T., Eds.; Wiley-VCH: Weinheim, 2003.
5. Dyzuba, S. V.; Bartsch, R. A. *Chem. Commun.* **2001**, 1466.
6. Merringan, T. L.; Bates, E. D., Dorman, S. C.; Davis, J. H. *Chem. Commun.* **2000**, 2051.
7. Visser, A. E.; Swatloski, R. P.; Reichert, W. M.; Mayton, R.; Sheff, S.; Wierzbicki, A.; Davis, J. H.; Rogers, R. D. *Environ. Sci. Technol.* **2002**, *36*, 2523.
8. Wilkes, J. S; Leviski, J. A; Zaworotko, M. J. *Inorg. Chem.* **1982**, *21*, 1236.
9. Wilkes, J. S.; Zaworotko, M. J. *J. Chem. Soc., Chem. Commun.* **1990**, 965.
10. Suarez, P. A. Z; Dullius, J. E. L; de Souza, R. F.; Dupont, J. *J. Chim. Phys. Chim. Biol* **1998**, *95*, 1626.

11. Hagiwara, R.; Ito, Y. *J. Fluor. Chem.* **2000**, *105*, 221.
12. Bonhote, P.; Dias, A. P.; Papageorgiou, N.; Kalyanasundraram, K.; Gratzel, M. *Inorg. Chem.* **1996**, *35*, 1168.
13. Dzyuba, S. V.; Bratsch, R. A. *Chem Phys Chem* **2002**, *3*, 161.
14. MacFarlane, D. R.; Forsith, S. A.; Golding J.; Deacon, G. B. *Green Chem.* **2002**, *4*, 444.
15. Holbrey, J. D.; Reichert, W. M.; Swatloski, R. P.; Broker, G. A.; Pitner, W. R.; Seddon, K. R.; Rogers, R. D. *Green Chem.* **2002**, *4*, 407.
16. Brinchi, L.; Germani, R.; Savelli, G. *Tetrahedron Lett.* **2003**, *44*, 2027.
17. Namboodiri, V. V.; Varma, R. S. *Tetrahedron Lett.* **2002**, *43*, 5381.
18. Deshmukh, R. R.; Rajagopal, R.; Srinivasan, K. V. *Chem. Commun.* **2001**, 1544.
19. Seddon, K. R.; Stark, A.; Torres, M. J. *Pure Appl. Chem.* **2000**, *72*, 2275.
20. (a) Chiappe, C.; Pieraccini, D.; Saullo, P. *J. Org. Chem.* **2003**, *68*, 6710. (b) Mathews, C. J.; Smith, P. J.; Welton, T.; White, A. J. P.; Williams, D. J. *Organometallics* **2001**, *20*, 3848.
21. Holbrey, J. D.; Reichert, W. M.; Swatloski, R. P.; Broker, G. A.; Pitner, W. R.; Seddon, K. R.; Rogers, R. D. *Green Chem.* **2002**, *4*, 407.
22. Kim, H. S.; Kim, Y. J.; Lee, H.; Park, K. Y.; Lee, C.; Chin, C. S. *Angew. Chem. Int. Ed.* **2002**, *41*, 4300.
23. Dubreuil, J. F.; Bourahla, K.; Rahmouni, M.; Bazureau, J. P.; Hamelin, J. *Cat. Commun.* **2002**, *3*, 185.
24. Ohno, H.; Yoshizawa, M. *Solid State Ionic* **2002**, *154-155*, 303.
25. Driver, G.; Johnson, K. E. *Green Chemistry* **2003**, *5*, 163.
26. (a) Susan, Md. A. B. H.; Noda, A.; Mitsushima, S.; Watanabe, M. *Chem. Commun.* **2003**, *8*, 938. (b) Noda, A.; Susan., Md. A. B. H.; Kudo, K.; Mitsushima, S.; Hayamizu, K.; Watanabe, M. *J. Phys. Chem. B.* **2003**, *107*, 4024.
27. Yoshizawa, M.; Xu, W.; Angell, C. A. *J. Am. Chem. Soc.* **2003**, *125*, 1541.
28. Dullius, J. E. L.; Suarez, P. A. Z.; Einloft, S.; de Souza, R. F.; Dupont, J.; Fisher, J.; De Cian, A. *Organometallics* **1998**, *7*, 815.
29. Boon, J. A.; Levisky, J. A.; Pflug, J. L.; Wilkes, J. S. *J. Org. Chem.* **1986**, *51*, 480.
30. (a) Abbott, A. P.; Capper, G.; Davies, D. L.; Rasheed, R. K.; Tambyrajah *Green Chem.* **2002**, *4*, 24. (b) Aggarwal, A.; Lancaster, N. L.; Sethi, A. R.; Welton, T. *Green Chem.* **2002**, *4*, 517.
31. Xu, L.; Chen, W.; Xiao, J. *Organometallics* **2000**, *19*, 1123.
32. Calò, V.; Giannoccaro, P.; Nacci, A.; Lopez, L.; Monopoli, A. *I. J. Organomet. Chem.* **2002**, *645*, 157.
33. Chiappe, C.; Imperato, G.; Napolitano, E.; Pieraccini, D. *Green Chem.* **2004**, *6*, 33.
34. McLachlan, F.; Methews, C. J.; Smith, P. J.; Welton, T. *Organometallics* **2003**, *22*, 5350.
35. (a) Dzyuba, S. V.; Bartsch, R. A. *Angew. Chem. Int. Ed.* **2003**, *42*, 148. (b) Schaefer, T.; Rodrigues, C. M.; Afonso, C. A.; Crespo, J. G. *Chem. Commun.* **2001**, *37*, 1622.
36. (a) Schmid, G. H. *The Chemistry of Double-Bonded Functional groups* Patai, S., Ed.; Wiley: New York, 1989; suppl.A, Vol. 2, Part 1, p. 699. (b) Ruasse, M. F. *Adv. Phys. Org. Chem.* **1993**, *28*, 207. (c) Brown, R. S. *Acc. Chem. Res.* **1997**, *30*, 131. (d) Melloni, G.; Modena, G.; Tonellato, U. *Acc. Chem. Res.* **1981**, *14*, 227.
37. Hardas, N. R.; Adam, R.; Uden, P. C. *J. Chromatogr. A* **1999**, *844*, 249.
38. (a) Kischning, A.; Monenschein, H.; Wittenberg, R. *Angew. Chem. Int. Ed.* **2001**, *40*, 650. (b) Sumi Mitra, S.; Sreekumar, K. *J. Polym. Sci. A: Polym. Chem.* **1997**, *35*, 1413.
39. Kajigaeshi, S; Moriwaki, M.; Fujisaki, S.; Okamoto, T. *Bull. Chem. Soc. Jpn*, **1990**, *63*, 3033.
40. Zefirov, N. S.; Sereda, S. E.; Sosonuk, N. V.; Zyk; Likhomanova, T. I. *Synthesis* **1995**, 1359.
41. Sket, B.; Zupet, P.; Zupan, M. *Tetrahedron* **1990**, *46*, 2503.
42. Bortolini, O.; Bottai, M.; Chiappe, C.; Conte, V.; Pieraccini, D. *Green Chem* **2002**, *4*, 621.
43. Ogawa, Y.; Takanashi, O.; Kikuchi, O. *TEOCHEM* **1998**, *429*, 187.
44. Eyal, E.; Treinin, A. *J. Am. Chem. Soc.* **1964**, *861*, 4287.
45. Papov, A. I.; Swensen, R. F. *J. Am. Chem. Soc.* **1955**, *77*, 3724.
46. Bagno, A.; Butts, C.; Chiappe C.; D'Amico, F.; Lord, J. C. D.; Pieraccini, D.; Rastrelli, F. *Org. Biomol. Chem.* **2005**, DOI: 10.1039/B502654G.
47. Huddleston, J. G.; Visser, A. E.; Reicjert, W. R.; Willauer, H. D.; Broker, G. A.; Rogers, R. D. *Green Chem.* **2001**, *3*, 156.

48. Chiappe, C.; Capraro, D.; Conte, V.; Pieraccini, D. *Org. Lett.* **2001**, *3*, 1061.
49. Chiappe, C.; Conte, V.; Pieraccini, D. *Eur. J. Org. Chem.* **2002**, 2831.
50. Chiappe, C.; Pieraccini, D. *J. Org. Chem.* **2004**, *69*, 6059.
51. Ruasse, M.-F.; Lo Moro, G.; Galland, B.; Bianchini, R.; Chiappe, C.; Bellocci, G *J. Am. Chem. Soc* **1997**, *119*, 12492.
52. Unpublished results from this laboratory.
53. Bellucci, G.; Chiappe, C.; Lo Moro, G. *J. Org. Chem.* **1997**, *52*, 3176.
54. Bellucci, G.; Bianchini, R.; Ambrosetti, R. *J. Am. Chem. Soc.* **1985**, *107*, 2464.
55. Bellucci, G.; Chiappe, C.; Marioni, F. *J. Am. Chem. Soc.* **1987**, *109*, 515.
56. Bellucci, G.; Bianchini, R.; Ambrosetti, R.; Ingrosso, G. *J. Org. Chem.* **1985**, *50*, 3313.
57. Modro, A.; Schmid, G. H.; Yates, K. J. *J. Org. Chem.* **1977**, *42*, 3673.
58. Bianchini, R.; Chiappe, C. *J. Org. Chem.* **1992**, *57*, 6474.
59. Dubois, J.-E.; Mouvier, G. *Bull. Soc. Chim.* **1968**, 1426.
60. McDonald, R. S.; Shapiro, S. A. *J. Org. Chem.* **1973**, *38*, 2460.
61. Modro, A.; Schmid, G. H.; Yates, K. J. *J. Org. Chem.* **1979**, *44*, 4221.
62. Charmichael, A. J.; Seddon, K. R. *J. Phys. Org. Chem.* **2000**, *13*, 591.
63. Fletcher, K. A.; Storey, I. A.; Hendricks, A. E.; Pandey, S.; Pandey, S. *Green Chem.* **2001**, *3*, 210.
64. Swiss, K. A.; Firestone, R. A. *J. Org. Chem.* **1999**, *103*, 5369.
65. (a) Crowhurst, L.; Mawdsley, P. R.; Perez-Arlandis, J. M.; Salter, P. A.; Welton, T. *Chem. Phys.* **2003**, *5*, 2790. (b) Welton, T.; Armstrong, D. W. *J. Am. Chem. Soc.* **2002**, *124*, 14247.
66. Kajigaeshi, S.; Morikavi, M.; Fujisaki, S.; Okamoto, T. *Bull. Chem. Soc. Jpn.* **1990**, *63*, 3033.
67. Sket, B.; Zupet, P.; Zupan, M. *Tetrahedron* **1990**, *46*, 2503.
68. Zefirov, N. S.; Sereda, G. A.; Sosonuk, S. E.; Zyk, N. V.; Likomanova, T. I. *Synthesis* **1995**, 1359.
69. Lancaster, N. L.; Salter, P. A.; Welton, T.; Young, G. B. *J. Org. Chem.* **2002**, *67*, 8855.
70. Liu, L.; Gou, Q.-X. *Chem. Rev.* **2001**, *101*, 673.
71. Yoshino, H.; Matsubara, S.; Oshima, K.; Matsumoto, K.; Hagiwara, R.; Ito, Y. *J. Fluor. Chem.* **2004**, *125, 455.*
72. Rozen, S.; Brand, M. *J. Org. Chem.* **1985**, *50*, 3342.
73. Bowers, A.; Denot, E.; Becerra, R. *J. Am. Chem. Soc.* **1960**, *82*, 4007.
74. Chi, D. Y.; Kilbourn, M. R.; Katznellennbogen, J. A.; Welch, M. J. *J. Org. Chem.* **1987**, *52*, 658.
75. Alvernhe, G.; Lawrent, A.; Haufe, G. *Synthesis* **1987**, 562.
76. Hall, L. D.; Manville, J. F. *Can. J. Chem.* **1973**, *51*, 2902.
77. Tamura, M.; Shibakami, Sekiya, A. *Synthesis* **1995**, 515.
78. Shimizu, M.; Nakahara, Y.; Yoshioka, H. *J. Chem. Soc., Chem. Commun.* **1989**, 1881.
79. De la Mare, P. B. *Electrophilic Halogenation*. Cambridge University Press: Cambridge, 1976; Chap.5.
80. Larock, R. C. *Comprehensive Organic Transformations*. VCH: New York, 1989; 315.
81. Dewkar, G. K.; Narina, S. V.; Sudalai, A. *Org. Lett.* **2003**, *5*, 4501.
82. Vyas, P. V.; Bhatt, A. K.; Ramachanraiah, G.; Bedekar, A. V. *Tetrahedron Lett.* **2003**, *44*, 4085.
83. Zhu, H.-P.; Yang, F.; Tang, J.; He, M.-Y. *Green Chem.* **2003**, *5*, 39.
84. Wu, H.-H.; Yang, F.; Pg, C.; Tang, J.; He, M.-Y. *Tetrahedron Lett.* **2004**, *45*, 4963.
85. Chiappe, C.; Leandri, E.; Pieraccini, D. *Chem. Commun.* **2004**, 2536.
86. Salazar, J.; Romano, D. *Synlett* **2004**, *7*, 1318.

AN EXTREMELY VERSATILE METHODOLOGY FOR THE SYNTHESIS OF ENANTIOPURE β-SUBSTITUTED PROLINE ANALOGUES WITH THE 7-AZANORBORNANE SKELETON

Ana M. Gil,[a] Elena Buñuel*[b] and Carlos Cativiela*[a]

[a]Departamento de Química Orgánica, ICMA, Universidad de Zaragoza-CSIC, E-50009 Zaragoza, Spain
(e-mail: cativiela@unizar.es)
[b]Departamento de Química Orgánica, Universidad Autónoma de Madrid, E-28049 Madrid, Spain
(e-mail: elena.bunnuel@uam.es)

Abstract. A new route for the synthesis of enantiomerically pure 2-substituted 7-azabicyclo[2.2.1]heptane-1-carboxylic acids in enantiomerically pure form is described. These compounds are new conformationally constrained β-functionalised proline analogues. Our strategy is based on the preparation of valuable azabicyclic intermediates by a key step that involves cyclisation of the derivatives obtained from the transformation of the adducts provided by the Diels–Alder reaction of C-4 unsaturated 5(4H)-oxazolones and Danishefsky's diene. The application of this procedure to the synthesis of a wide range of these enantiopure β-functionalised prolines with a 7-azanorbornane skeleton, also considered as proline–α-amino acid chimeras, illustrates the great versatility of our methodology.

Contents

1. Introduction

The discovery of new therapeutic agents based on non-proteinogenic amino acids has boosted the synthesis of conformationally constrained amino acids, particularly since their incorporation into peptides

56

was recognised as a powerful approach to generate structurally defined peptides that could be used as conformational probes and bioactive compounds.[1]

Quaternary α-amino acids represent a remarkable type of conformationally restricted α-amino acid in which the conformational rigidity is generated by the introduction of a substituent at the α-carbon. The significance of the search for new approaches for the preparation of these products has been evidenced by recent compilations of their stereoselective syntheses.[2] Among the numerous types of cyclic quaternary α-amino acids, the carbocyclic derivatives **I** have been widely studied (Figure 1). The introduction of an additional substituent in the β-position (**II**) is particularly attractive since it can turn these compounds into typical α-amino acids with a specific type of side chain restriction.

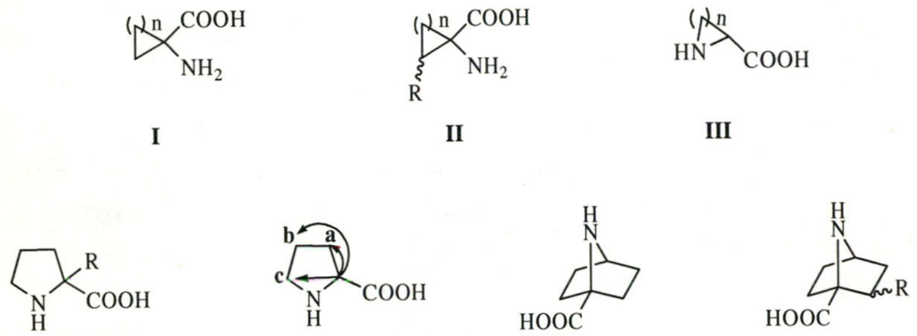

Figure 1

A remarkable way to decrease the conformational freedom of peptides involves the incorporation of residues **III** where the nitrogen forms part of the ring (Figure 1). This kind of amino acid constitutes a new family of compounds that can be considered as being related to prolines. Although there are very few references concerning the corresponding quaternary amino acids, a few α-alkylprolines (**IV**) have been studied in detail[3] and their stereoselective synthesis has also been the subject of a review.[2b] The rising interest in these amino acids is, to a large extent, caused by the nowadays extensive use of a simple proline as asymmetric catalyst in different reactions, such as aldol, Mannich, Michael, Diels–Alder reactions, and α-oxidation of aldehydes, among others.[4]

Another possibility for the conformational restriction of α-alkylprolines involves connecting the α-carbon with another proline ring carbon (Figure 1), a process that entails the creation of diverse azabicyclic structures depending on the number of atoms involved in the formation of the new cycle (**Va–c**).[5,6]

Our attention here is focused on the additional restriction to the proline ring that involves linking the α- and δ-carbons (**Vc**) through the construction of azabicyclo[n.2.1]alkane skeletons.[7,8] In particular, we are interested in the connection of the two carbons that leads to the creation of 7-azabicyclo[2.2.1]heptane rings **VI** (Figure 1).

Since the discovery of epibatidine (Figure 2) by Daly and co-workers in 1992,[9] we have witnessed a proliferation in the number of synthetic methods to construct this azabicyclic system.[10] The exceptional biological properties attributed to epibatidine and related analogues, as well as their possible clinical applications, have provided more than enough reasons to justify our interest in these systems.

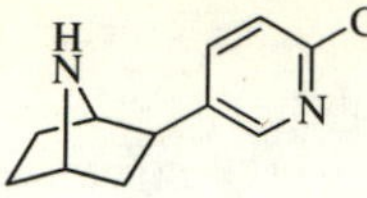

epibatidine

Figure 2

Initial results provided evidence of the benefits of proline analogues containing this structure as replacements for proline in the formation of a β-turn tripeptide mimetic[8f] and as starting materials in the synthesis of a new class of HIV-1 protease inhibitor.[8c] These findings encouraged us to intensify our synthetic efforts in this area. The introduction of a substituent in the β-position (**VII**) appears to be an interesting way to create new chimeras that combine α-amino acids and a constrained proline (Figure 1).

This paper provides a review of the development of this field since our first publication and describes new findings from our more recent research. The review is divided into two main parts: the first covers the construction of the 7-azabicyclo[2.2.1]heptane structure and the second concerns the stereocontrolled manipulations of the β-substituent in the 7-azanorbornane derivatives. Finally, we describe some of the new enantiopure proline–α-amino acid chimeras that are available by applying our methodology.

2. Building of enantiomerically pure β-substituted key intermediates with the 7-azabicyclo [2.2.1]heptane skeleton

Two main strategies have been described for building 7-azabicyclo[2.2.1]heptane skeletons (Scheme 1). One involves the cycloaddition of pyrroles by Diels–Alder reactions (a) or of azomethine ylides by [3+2] cycloadditions (b) and the other makes use of cyclisation reactions of proline (c) or cyclohexylamine derivatives (d).

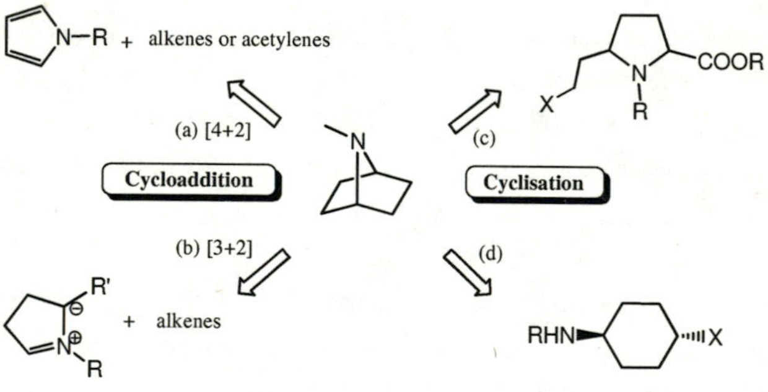

Scheme 1

An outstanding example of the asymmetric synthesis of β-substituted prolines with the 7-azanorbornane skeleton, involving the cyclisation of proline derivatives, was described by Rapoport *et al.*[7e,8d,11]

On the other hand, the synthesis of 7-azabicyclo[2.2.1]heptane-1-carboxylic acid (**VI**, Figure 1) was accomplished according to the strategy that consists of the cyclisation of cyclohexylamine derivatives. This new procedure started with the Diels–Alder reaction of methyl 2-benzamidoacrylate (Scheme 2).[12]

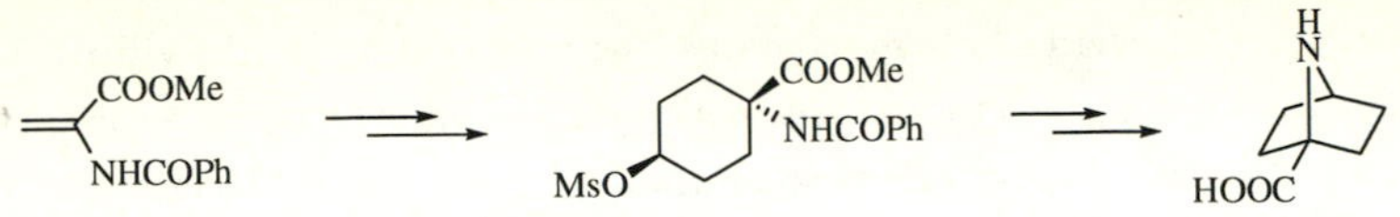

COOMe

NHCOPh

→

MsO COOMe ''NHCOPh

→

H
N

HOOC

VI

Scheme 2

Although α,β-didehydroamino acid derivatives have been widely used as useful building blocks in synthetic organic chemistry, the application of the methodology described above to the corresponding trisubstituted α,β-didehydroamino acids was not feasible due to their very low reactivity as dienophiles in Diels–Alder reactions.

Nevertheless, unsaturated C-4 oxazolones have proven to be a very valuable alternative since they show good reactivity in Diels–Alder cycloadditions.[13] Furthermore, as a part of a study on the asymmetric Diels–Alder reaction, we reported the excellent behaviour of (Z)-2-phenyl-4-[(S)-2,2-dimethyl-1,3-dioxolan-4-ylmethylene]-5(4H)-oxazolone, an unsaturated oxazolone derived from (R)-glyceraldehyde, as a dienophile towards several dienes.[14,15] These studies enabled us to prepare enantiomerically pure carbocyclic quaternary α-amino acids and encouraged us to use C-4 unsaturated 5(4H)-oxazolones as starting materials to achieve our new goal.

Our overall strategy was based on building the 7-azabicyclo[2.2.1]heptane skeleton of the key intermediates by transformation of the cycloadducts provided by the Diels–Alder reaction of C-4 unsaturated 5(4H)-oxazolones and Danishefsky's diene.

2.1. Resolution of racemic mixtures

Only one resolution procedure has been reported in the literature for the preparation of both enantiomers of a β-hydroxyproline with a 7-azanorbornane core. This strategy consists of the formation of diastereomers and subsequent separation by crystallisation, which involves the introduction of additional synthetic steps in the pathway to the final products.[16] We have developed a highly efficient and facile resolution method for an azanorbornane proline–phenylalanine chimera analogue. This process gave a valuable intermediate for the synthesis of a large variety of enantiomerically pure α-amino acid chimeras with the 7-azabicyclo[2.2.1]heptane skeleton.

The construction of the 7-azanorbornane system as a racemic mixture was achieved by the preparation of methyl exo-2-phenyl-7-azabicyclo[2.2.1]heptane-1-carboxylate, rac-5 (Scheme 3).[7c,17] The synthesis was carried out through a key step that entailed the cycloaddition between (Z)-2-phenyl-4-benzylidene-5(4H)-oxazolone (1) and Danishefsky's diene and allowed the multigram-scale synthesis of this racemic proline–phenylalanine chimera derivative.[18]

The Diels–Alder reaction of Danishefsky's diene and a C-4 unsaturated oxazolone derived from benzaldehyde, (Z)-2-phenyl-4-benzylidene-5(4H)-oxazolone (1), followed by hydrolysis of the adduct mixture and subsequent elimination of the methoxy group with oxazolone ring opening led to rac-2 in 81% overall yield (Scheme 3). Heterogeneous hydrogenation of the double bond of enone rac-2, using 20% Pd(OH)$_2$/C, afforded ketone rac-3 in 95% yield. Reduction of rac-3 with L-Selectride® (lithium tri-sec-butylborohydride) provided preferentially the *axial* alcohol (stereoselectivity ratio, 63:37). The mixture of alcohols was subsequently transformed into the corresponding methanesulfonate derivatives, *rac*-**4a** and *rac*-**4b**, in 71% overall yield from *rac*-**3** by treatment with methanesulfonyl chloride in triethylamine. In order to prepare the bicyclic product *rac*-**5**, the base-promoted internal nucleophilic displacement of the methanesulfonate group on *rac*-**4a** was achieved by treatment of the mixture of *rac*-**4a** and *rac*-**4b** with NaH in DMF, rather than potassium *tert*-butoxide in THF as previously reported.[7c,17] Finally, the separation of the product *rac*-**5** from the non-cyclisable methanesulfonate *rac*-**4b** was easily achieved by column chromatography.

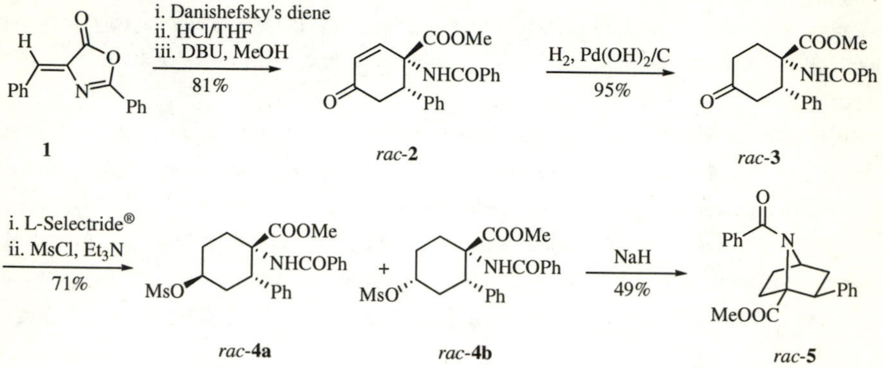

Scheme 3

The synthesis was accomplished on a multigram-scale and ca. 5 g of a racemic mixture of *rac*-**5** could be obtained from oxazolone **1** in seven steps in 27% overall yield. Once the synthesis of the racemic compound *rac*-**5** had been achieved, we attempted the isolation of this product in its enantiomerically pure form by HPLC resolution using a chiral stationary phase. We have previously shown the efficacy of this approach for resolving racemic mixtures of other constrained phenylalanine analogues.[19,20] Our aim was achieved by using a non-commercial polysaccharide-derived support consisting of mixed 10-undecenoate/3,5-dimethylphenylcarbamate of cellulose covalently attached to allylsilica gel.[21] The excellent chiral discrimination exhibited by this stationary phase towards a variety of compounds, together with its high chemical stability, make it especially suitable for resolutions on a preparative scale. The efficiency of this system has actually been demonstrated in the preparative enantioseparations of various phenylalanine surrogates.

The resolution of *rac*-**5** was initially investigated on an analytical scale (eluent, 95:5 *n*-hexane/2-propanol, k'_1=1.77, α=1.70, R_s=3.04). The optimal separation conditions were determined from these initial satisfactory results by adding a small amount of chloroform to the eluent to enhance the solubility of the compound. The extension of the analytical conditions to the preparative scale proved extremely efficient, resulting in baseline separation of the two enantiomers (see Figure 3).

Working on repetitive injection mode, 1.2 g of racemate was injected onto a 150 × 20 mm ID column with a total time of 4–5 h to complete the process. Valley-valley collection of the eluting peaks afforded the optically pure first eluted enantiomer. The second enantiomer was obtained with a 99:1 enantiomeric ratio. Recrystallisation of the last enantiomer from ethyl acetate gave an enantiopurity of >99.9:0.1. Thus, the resolution of 1.2 g of *rac*-**5** finally resulted in 1.037 g of optically pure material (ca. 517 mg and 520 mg of the first and second eluted enantiomers, respectively). The optical purity of the resolved enantiomers was assessed at the analytical level.

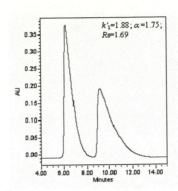

Figure 3

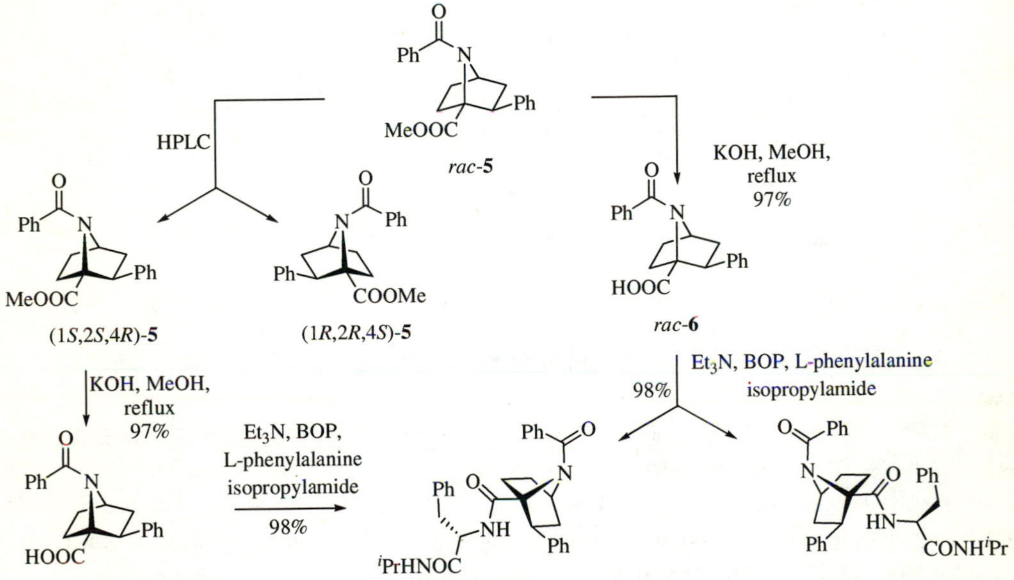

Scheme 4

The assignment of absolute configuration was performed by X-ray diffraction analysis of a single crystal, which was obtained from a solution of one of the diastereomeric dipeptides obtained by coupling the racemic mixture with an L-phenylalanine derivative. Obrecht *et al.* reported phenylalanine cyclohexylamide to be a very convenient auxiliary for the efficient assignment of the absolute configurations of α,α-disubstituted amino acids.[22] According to this approach, we coupled the racemic mixture of the 7-azanorbornane proline *rac*-**5** with this enantiomerically pure L-phenylalanine analogue. However, all attempts to obtain a suitable crystal for single crystal X-ray analysis, from either of the two resulting diastereomeric dipeptides, failed. With this result in mind, a different phenylalanine derivative was selected for coupling to the racemic 7-azanorbornane proline (Scheme 4).

The saponification of *rac*-**5** using potassium hydroxide in methanol led to the corresponding acid *rac*-**6** in 97% yield (Scheme 4). The subsequent coupling of this racemic mixture with enantiomerically pure L-phenylalanine isopropylamide using the reagent designed by Castro *et al.*,[23] (1*H*-benzotriazol-1-yloxy)tris(dimethylamino)phosphonium hexaflurophosphate (BOP), provided a mixture of two diastereomeric dipeptide derivatives **7** and **8** in 98% combined yield. The dipeptides were separated by column chromatography and both compounds were individually characterised. A crystal suitable for X-ray crystallography was finally obtained for **8**, the second chromatographic component.

The absolute configuration of the proline analogue contained in dipeptide **8** was unequivocally determined by X-ray diffraction analysis of a crystal obtained by slow evaporation from a solution in dichloromethane/*n*-hexane (Figure 4). The known (*S*) configuration for the phenylalanine residue allowed the absolute configuration of the azabicyclic residue to be assigned as (1*S*,2*S*,4*R*).

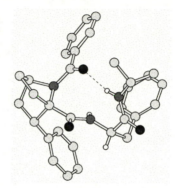

Figure 4

Independently, the first eluted enantiomer of **5** was also coupled with L-phenylalanine isopropylamide under conditions previously developed for the racemic material (Scheme 4). Comparison of the spectroscopic data of this dipeptide analogue with those obtained for compound **8**, which had provided a suitable single crystal for X-ray diffraction analysis, allowed the identification of this enantiomer as that bearing the same (1*S*,2*S*,4*R*) absolute configuration as the azabicyclic residue contained in peptide **8**.[18]

Incorporation of other constrained phenylalanine analogues, such as 1-amino-2-phenylcyclopropanecarboxylic acid (c3Phe), 1-amino-2,3-diphenylcyclopropanecarboxylic acid (c3diPhe) and 1-amino-2-phenylcyclohexanecarboxylic acid (c6Phe), into model peptides has allowed the exploration of structural properties.[24] Thus, great interest was focused on dipeptide **8** and the X-ray diffraction analysis[25] revealed that, in the solid state, this constrained peptide containing the azanorbornane proline adopted a type I β-turn whereas the analogous dipeptide sequence incorporating L-proline has been shown to accommodate a βII-turn disposition. Attractive interactions involving the central NH group and either the aromatic rings or the pyramidalised bicyclic nitrogen seem to play a role in the stabilisation of the observed βI-turn conformation. This situation is shown in the view of the molecular structure of dipeptide **8** obtained by X-ray diffraction analysis (Figure 4).

In addition, oxidative cleavage of the β-phenyl substituent on the azabicyclic ring of (1*S*,2*S*,4*R*)-**5** and (1*R*,2*R*,4*S*)-**5** was carried out. This process gave the intermediates (1*S*,2*R*,4*R*)-**9** and (1*R*,2*S*,4*S*)-**9** in 45% yield and these compounds were fully characterised (Scheme 5). Determination of absolute configuration and specific rotation of each enantiomer made these intermediates helpful reference materials for the assignment of the absolute configurations of other structural analogues containing any β-substituent capable of being transformed with stereocontrol into an *exo* carboxylic acid group.

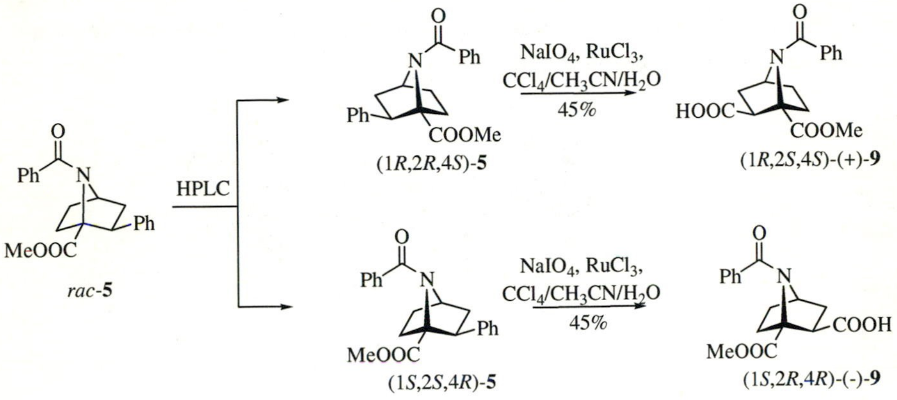

Scheme 5

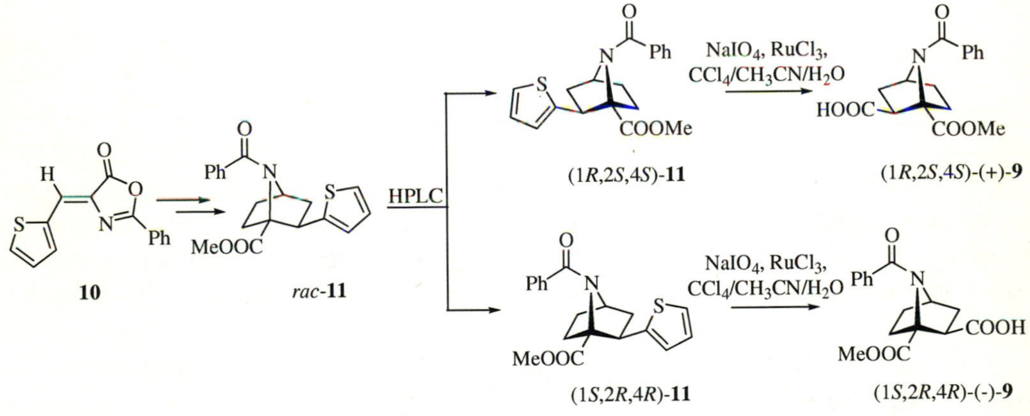

Scheme 6

As an example, our methodology was also applied to the synthesis and subsequent resolution of the 2-thienyl derivative *rac*-**11**.[26] The transformation of the Diels–Alder adducts arising from the cycloaddition of (*Z*)-2-phenyl-4-[thien-2-ylmethylene]-5(4*H*)-oxazolone (**10**) with Danishefsky's diene allowed the construction of the 7-azabicyclo[2.2.1]heptane contained in *rac*-**11** (Scheme 6). Resolution of the racemic mixture by chiral HPLC provided each enantiomer separately. Subsequent determination of absolute configurations, (1*S*,2*R*,4*R*)-**11** and (1*R*,2*S*,4*S*)-**11**, was facilitated by measurement of specific rotations on acids **9**, which were obtained by oxidative cleavage of **11** and whose absolute stereochemistry had already been established.

2.2. Asymmetric synthesis

The extension of our methodology to an asymmetric version was carried out starting from a chiral C-4 unsaturated oxazolone. We have already described the Diels–Alder reaction of Danishefsky's diene and (*Z*)-2-phenyl-4-[(*S*)-2,2-dimethyl-1,3-dioxolan-4-ylmethylene]-5(4*H*)-oxazolone (**12Z**),[14e,f,15] which is derived from 1,2-*O*-isopropylidene-(*R*)-glyceraldehyde. The thermally induced reaction afforded a 1:1 mixture of cycloadducts **13a** and **13b** (Scheme 7). Hydrolysis of this mixture and subsequent elimination of the methoxy group with ring opening provided compound **15**.

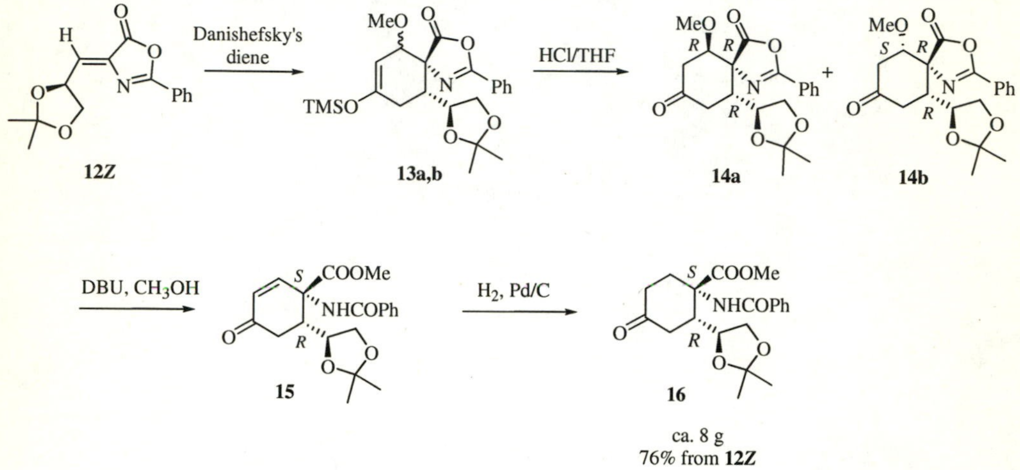

Scheme 7

The absolute stereochemistry of the enone **15** and every preceding intermediate was confirmed by single crystal X-ray diffraction analysis of compound **14b**. From this result we could confirm attack of the diene at the $C_{\alpha\text{-Re}}$ face of the dienophile **12Z** in accordance with the stereo-correlation model previously proposed for the attack of cyclopentadiene, cyclohexadiene and several open chain dienes on this oxazolone.[14,15]

Typical heterogeneous hydrogenation of the double bond of enone **15** afforded enantiomerically pure ketone **16** in nearly quantitative yield. Thus, starting from 8 g of oxazolone **12Z**, this procedure supplied 8.26 g of ketone **16** in excellent overall yield (76%). Compound **16** is not only a valuable intermediate in the synthesis of a new conformationally constrained 4-oxo-1-aminocyclohexanecarboxylic acid in enantiomerically pure form, as we reported previously,[14e,f] but is also a precursor of the aminocyclohexanol derivative that would allow us to achieve our current target.

The preparation of the desired 4-hydroxyamino derivative involved conversion of the ketone function on the cyclohexane ring into a hydroxy group (Scheme 8).[27] Moreover, in order to make the subsequent cyclisation possible, hydroxy and benzamide groups on the cyclohexane skeleton needed to be in *axial* positions, a situation that required the synthesis of the kinetically controlled alcohol **17a**.

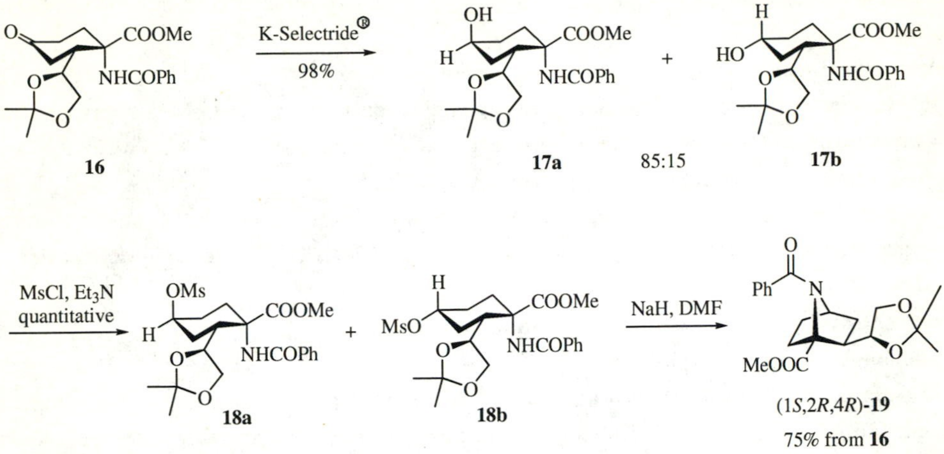

Scheme 8

Several studies have demonstrated that the use of bulky reducing agents at low temperatures produces stereoselective processes that preferentially provide the kinetically controlled product. In accordance with these reports, and taking into account the highly favourable *axial* alcohol selectivity that lithium tri-*sec*-butylborohydride (L-Selectride®) produced,[7c,8a] we decided to examine the influence of several hydride reagents on the course of the reaction.

All the reactions were performed at −78 °C and the ratio of stereoisomers was determined from the crude reaction mixture by integration of the signals in the ^{1}H NMR spectra.[27] As one might expect, L- and K-Selectride® (lithium and potassium tri-*sec*-butylborohydride) provided preferentially the *axial* alcohol **17a** with a similar stereoselectivity (85:15). However, the use of LiAlH(O^{t}Bu)$_3$ led to the thermodynamically controlled product **17b**, *i.e.*, the *equatorial* alcohol, as a single isomer. The use of KS-Selectride® (potassium trisiamylborohydride) did not lead to the desired products. Isolation of a small amount of compound **17a** allowed its characterisation and the subsequent cyclisation of its derivative enabled us to confirm the predicted stereochemistry.

From this study we selected K-Selectride® as the reducing agent of choice in terms of yield and selectivity for the preparation of alcohol **17a**. Indeed, the reduction of cyclohexanone derivative **16** with K-Selectride® in THF at −78 °C gave an 85:15 mixture of *axial* and *equatorial* alcohols **17a** and **17b** in 98% combined yield.

The resulting mixture of alcohols, **17a** and **17b**, was cleanly transformed into a mixture of the corresponding methanesulfonate derivatives, **18a** and **18b**, in nearly quantitative yield by treatment with methanesulfonyl chloride in triethylamine (Scheme 8). This mixture was purified by column chromatography and provided pure samples of both methanesulfonates, which were fully characterised.

Base-promoted internal nucleophilic displacement of the methanesulfonate group had already been achieved by treatment with potassium *tert*-butoxide in THF.[7c] However, in this case the best results for the intramolecular cyclisation of the methanesulfonate **18a** were obtained using sodium hydride (1.2 eq) with dry DMF as the solvent. These conditions provided methyl (1*S*,2*R*,4*R*)-*N*-benzoyl-2-[(*S*)-2,2-dimethyl-1,3-dioxolan-4-yl]-7-azabicyclo[2.2.1]heptane-1-carboxylate [(1*S*,2*R*,4*R*)-**19**] in good overall yield (75%) from ketone **16**.

After optimisation of this procedure we were able to scale up the process and proceeded to undertake the first asymmetric synthesis of the enantiomerically pure product (1*S*,2*R*,4*R*)-**19** on a 40 gram scale. This process involved seven fully stereocontrolled steps and gave an overall yield of 27% from an 85:15 mixture of the two geometric isomers of the chiral oxazolone, **12Z** and **12E**.

Compound **19** is a valuable building block because, as we will show below, it is an enantiomerically pure key intermediate for the synthesis of 2-substituted 7-azabicyclo[2.2.1]heptane-1-carboxylic acids.

3. Stereocontrolled manipulation of the β-substituents
3.1. Functionalisation at the γ-position

As described above, oxidative cleavage of the phenyl substituent on the azabicyclic ring of (1*S*,2*S*,4*R*)-**5** proceeded to give 45% yield and led to the intermediate (1*S*,2*R*,4*R*)-**9**.[18] Carboxylic acids constitute key intermediates for the preparation of a broad variety of other optically pure proline–α-amino acid chimeras, either directly, thanks to the versatility of the acid group, or by transformation into other synthetically flexible functions such as hydroxymethyl or formyl groups (Scheme 9).

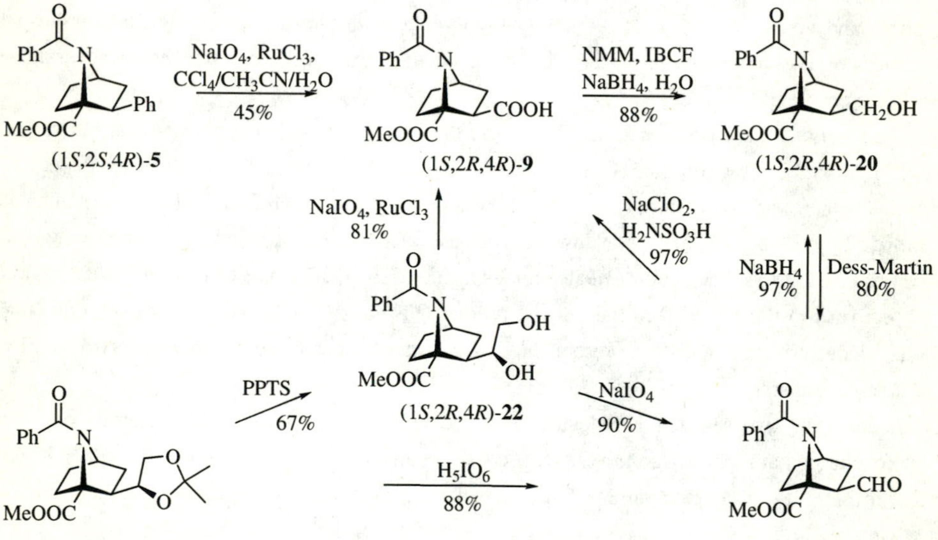

Scheme 9

While the procedure involving the acyl chloride intermediate was unsuccessful for the reduction of the acid to the alcohol, the mixed anhydride method led to excellent results (Scheme 9). Thus, treatment of (1*S*,2*R*,4*R*)-**9** with *N*-methylmorpholine (NMM) and *iso*-butylchloroformate (IBCF) and subsequent addition of $NaBH_4$ over the resulting anhydride supplied the enantiomerically pure alcohol (1S,2R,4R)-**20** in 88% yield.

The interchange between hydroxymethyl and formyl groups was also carried out. Oxidation of alcohol (1S,2R,4R)-**20** by treatment with chromium trioxide resulted in low yields (69%). On the other hand, a 70:30 mixture of the β-formyl derivative (1S,2R,4R)-**21** and its C-2 epimer was obtained in 97% combined yield when (1S,2R,4R)-**20** was treated under Swern conditions with oxalyl choride (1.5 eq), Et_3N (5 eq) and DMSO (4.5 eq) in dry dichloromethane. The best results for this oxidation reaction were definitively obtained using the Dess–Martin reagent.[28] In this case, the process took place in high yield and epimerisation was not detected in the crude reaction mixture. So, oxidation of alcohol (1S,2R,4R)-**20** using the Dess–Martin periodinane in dry dichloromethane gave the enantiopure aldehyde (1S,2R,4R)-**21** in 80% yield. On the other hand, the reduction of aldehyde (1S,2R,4R)-**21** was simply carried out by treatment with $NaBH_4$ in a methanol/water mixture and provided the primary alcohol (1S,2R,4R)-**20** (Scheme 9) in almost quantitative yield (97%).

Furthermore, the transformation of aldehyde (1S,2R,4R)-**21** into acid (1S,2R,4R)-**9** was readily performed by treatment with $NaClO_2$ and H_2NSO_3H in a water/acetone mixture. This procedure provided the carboxylic acid in excellent yield (97%).[29]

The use of the other enantiomer of the aspartic analogue (1R,2S,4S)-**9**, obtained by oxidative cleavage of the β-phenyl substituent of enantiomerically pure (1R,2R,4S)-**5**, and the application of the stereocontrolled functionalisation processes described above supplied the other enantiomeric series of derivatives (1R,2S,4S)-**9, 20, 21**.[18]

On the other hand, transformation of the 1,3-dioxolane ring in (1S,2R,4R)-**19** was also carried out in order to introduce suitable functionalisation at the γ-position of the amino acid side chain on this new enantiomerically pure building block. After creating the 7-azabicyclo[2.2.1]heptane system, we undertook the manipulation of the 1,3-dioxolane ring in compound (1S,2R,4R)-**19**. As part of this study, we also performed the stereocontrolled transformation of the dioxolane fragment of derivative (1S,2R,4R)-**19** into an acid function (Scheme 9).

To this end, hydrolysis of the acetal moiety was attempted using aqueous HCl according to the standard procedure that we described previously.[14d-f] However, this treatment did not lead to the desired diol (1S,2R,4R)-**22** due to the formation of a six-membered lactone, which was formed by cyclisation of the methyl ester with the corresponding hydroxy group.[27] The amount of by-product could be minimised by using pyridinium tosylate (PPTS). The lower acidity of PPTS, which was used in a refluxing mixture of acetone and water as the reaction solvent, afforded diol (1S,2R,4R)-**22**. This compound was purified by column chromatography (67%) and was fully characterised.

The next step involved direct oxidation of the diol moiety in (1S,2R,4R)-**22** to the carboxylic acid and this was performed under similar conditions to those reported previously.[14d,f] Treatment of (1S,2R,4R)-**22** with $NaIO_4$ in the presence of a catalytic amount of $RuCl_3$ using a two-phase solvent mixture, $CH_3CN/CCl_4/H_2O$ (1:1:3), provided the β-carboxylic acid derivative (1S,2R,4R)-**9** in 81% yield.

Given the convenient functionalisation of the side chain with a formyl group, the preparation of aldehyde (1S,2R,4R)-**21** from diol (1S,2R,4R)-**22** was also undertaken (Scheme 9). A typical oxidation procedure with $NaIO_4$ provided the carbonyl compound (1S,2R,4R)-**21** in 90% yield after isolation by column chromatography. Preparation of compound (1S,2R,4R)-**21** was achieved in 60% overall yield through a two-step process that involved hydrolysis of the acetal intermediate (1*S*,2*R*,4*R*)-**19** and oxidation of the resulting diol. The moderate yield of this procedure was mainly caused by the formation of the aforementioned undesired six-membered lactone. A re-examination of this synthetic procedure led to a highly effective synthesis of (1*S*,2*R*,4*R*)-**21** by direct transformation of the acetal moiety into the formyl group (Scheme 9).[30] Treatment of (1*S*,2*R*,4*R*)-**19** with H_5IO_6[31] gave (1*S*,2*R*,4*R*)-**21** in 88% yield through a one-step procedure, without evidence for the formation of any lactonisation product.

The stereocontrolled functionalisation of the γ-position and the facile interchanges between the acid, formyl and hydroxymethyl groups (see Scheme 9) offered extensive choice for subsequent transformations on the amino acid side chain. The already wide variety of potential modifications at the amino acid side chain was further increased as a result of the development of a procedure for the epimerisation at C-2 of aldehyde (1*S*,2*R*,4*R*)-**21** (Scheme 10).

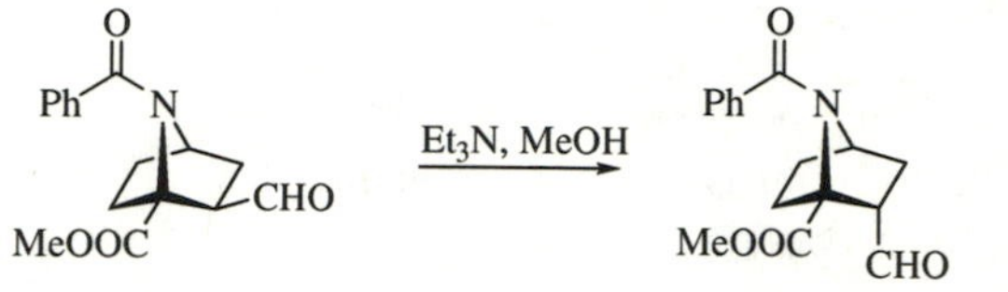

(1*S*,2*R*,4*R*)-**21**   (1*S*,2*S*,4*R*)-**21**

Scheme 10

Treatment of the formyl derivative (1*S*,2*R*,4*R*)-**21** with triethylamine in methanol allowed us to obtain the β-epimer, (1*S*,2*S*,4*R*)-**21**, which contains the β-formyl group in an *endo* disposition. The availability of the *endo* product (1*S*,2*S*,4*R*)-**21** would solve the problem of access to the derivatives that would result from the Diels–Alder reaction of Danishefsky's diene and the C-4 unsaturated *E*-oxazolones, whose preparation frequently proves to be problematic.

3.2. Reactivity studies at the γ-position

The reactivity of the functionalised derivatives was explored on the enantiomers that could be prepared in an exclusive manner from the building block (1*S*,2*R*,4*R*)-**19**, which was obtained by asymmetric synthesis as described above. However, all the procedures compiled in this section could be equally applied to the other enantiomeric series of products since the enantiomeric functionalised derivatives could be prepared by resolution of the azanorbornane proline–phenylalanine chimera analogue *rac*-**5**, as described above, or by asymmetric synthesis starting from (*Z*)-2-phenyl-4-[(*R*)-2,2-dimethyl-1,3-dioxolan-4-ylmethylene]-5(4*H*)-oxazolone, derived from (*S*)-glyceraldehyde, instead of oxazolone **12Z** arising from (*R*)-glyceraldehyde, which is described here.

3.2.1. Study of olefination reactions

The behaviour of (1*S*,2*R*,4*R*)-**21** in reactions with several triphenylphosphonium ylides was evaluated (Scheme 11).[30]

The reaction of (1*S*,2*R*,4*R*)-**21** with methylenetriphenylphosphorane involved treatment of methyltriphenylphosphonium iodide with *n*-BuLi in THF and subsequent addition of the carbonyl compound to the reaction mixture. This procedure gave the desired methyl (1*S*,2*S*,4*R*)-*N*-benzoyl-2-vinyl-7-

azabicyclo[2.2.1]heptane-1-carboxylate (**23**), which was isolated in 75% yield. In order to compare the efficacy of the Wittig procedure to obtain the vinyl product we also explored an alternative method (Scheme 12).

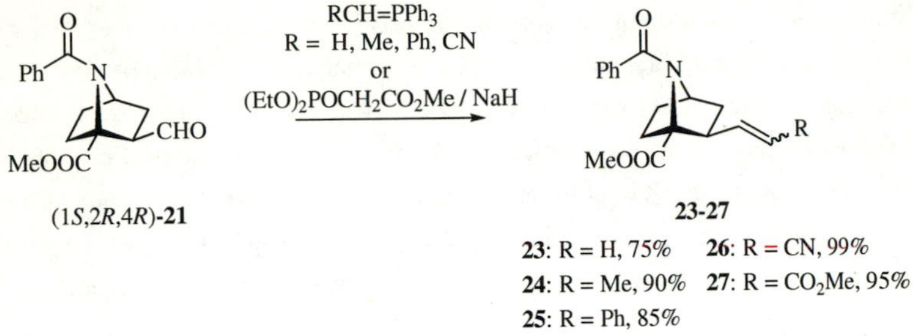

23: R = H, 75%   **26**: R = CN, 99%
24: R = Me, 90%   **27**: R = CO$_2$Me, 95%
25: R = Ph, 85%

Scheme 11

Diol (1*S*,2*R*,4*R*)-**22**, obtained by hydrolysis of the acetal in key intermediate (1*S*,2*R*,4*R*)-**19**, was converted into the vinyl azabicyclic product **23** through the two-step Corey–Winter procedure to obtain olefins from 1,2-diols *via* a thionocarbonate intermediate.[32] This method was carried out using *N,N'*-thiocarbonyldiimidazole (TCDI), as in the original protocol,[32a] instead of thiophosgene/DMAP.[32c] The procedure gave thionocarbonate **28** in good yield (83%) and treatment of this compound with trimethylphosphite afforded **23** in 72% yield. When 1,3-dimethyl-2-phenyl-1,3,2-diazaphospholidine (DMPDAP)[32c] was used instead of trimethylphosphite, the yield of **23** increased to 86%.

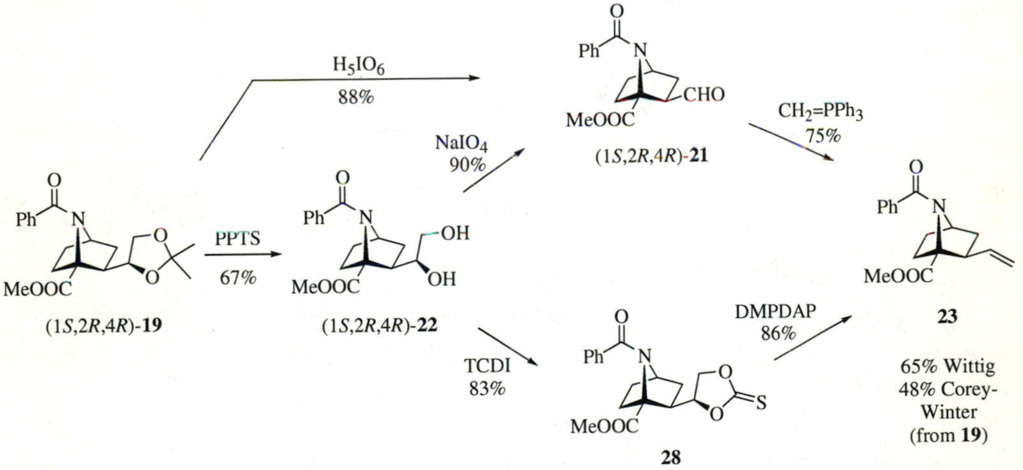

Scheme 12

In summary, this procedure gave 48% overall yield from the key intermediate (1*S*,2*R*,4*R*)-**19** through a three-step transformation. No improvement was observed in comparison to the Wittig procedure, which also provided **23** in 65% yield from (1*S*,2*R*,4*R*)-**19** through a two-step process beginning with the direct transformation of the acetal moiety into a formyl group (Scheme 12).

69

The Wittig reaction of (1S,2R,4R)-**21** with ethylidenetriphenylphosphorane (Scheme 11) provided methyl (1S,2S,4R)-N-benzoyl-2-(1-propenyl)-7-azabicyclo[2.2.1]heptane-1-carboxylate (**24**) as a 71:29 mixture of the Z and E isomers in excellent yield (90%). The determination of the Z/E ratio was carried out by integration of the appropriate signals in the ^{1}H NMR spectrum of the crude reaction mixture and was possible thanks to the partial separation of the isomers by HPLC, which afforded a pure analytical sample of the major isomer. The double bond stereochemistry of each isomer was elucidated by ^{1}H NMR experiments. Thus, irradiation of the olefinic methyl signal in the ^{1}H NMR spectrum of the pure isolated isomer of **24** gave a spectrum where the coupling constant between the olefinic protons could be clearly measured (J=11.0 Hz), allowing the assignment of the Z configuration for the double bond in the major product. A similar experiment on the mixture enriched with the minor isomer, by irradiation of the olefinic methyl signals of both isomers, revealed a coupling constant (J=15.0 Hz) consistent with an E stereochemistry for the double bond of the minor reaction product.

The reaction of (1S,2R,4R)-**21** with the semi-stabilised ylide benzylidenetriphenylphosphorane led to a 6:94 mixture of (Z/E)-methyl (1S,2S,4R)-N-benzoyl-2-(2-phenylvinyl)-7-azabicyclo[2.2.1]heptane-1-carboxylate (**25**) (Scheme 11) in 85% yield. The major product was fully characterised and analysis of the single crystal X-ray data unequivocally confirmed the stereochemistry (Figure 5). The coupling constants between the olefinic protons were determined from the ^{1}H NMR spectra obtained from benzene-d_6 solutions (J=11.4 Hz and J=15.4 Hz for the Z and E isomers, respectively).

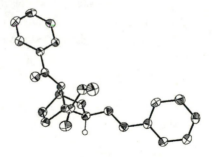

Figure 5

As far as the stabilised ylides are concerned, reaction of (1S,2R,4R)-**21** with carbomethoxymethylene-triphenylphosphorane did not give acceptable yields of the corresponding Wittig adducts under any of the reaction conditions tested. Treatment of carbomethoxymethyltriphenylphosphonium chloride with an aqueous solution of sodium hydroxide and reaction of the carbonyl compound (1S,2R,4R)-**21** with the isolated ylide gave only a very low conversion into products. The use of strong bases, such as n-BuLi or MeONa, to generate the ylide under anhydrous conditions did not improve the results. Thus, starting material **21** was completely consumed upon addition to the phosphorane generated with n-BuLi, but Wittig adducts were not detected. On using MeONa the conversion of **21** was complete at room temperature after only 20 min, but a complex mixture of products was obtained due to partial epimerisation at C-2. Formation of the ylide *in situ* using propylene oxide also led to a mixture of C-2 epimeric products.

As an alternative, we explored the reactivity of (1S,2R,4R)-**21** with cyanomethylenetriphenyl-phosphorane, which would also allow access to a carboxylic acid function by hydrolysis of the cyano group.

70

The method involving cyanomethyltriphenylphosphonium chloride and *n*-BuLi in THF proved to be unsuitable due to a lack of reproducibility in these experiments. After numerous trial reactions, generation of the ylide – from the freshly prepared phosphonium halide – with an excess of propylene oxide in the presence of the carbonyl compound (1*S*,2*R*,4*R*)-**21** cleanly led to a 52:48 *Z/E* mixture of methyl (1*S*,2*S*,4*R*)-*N*-benzoyl-2-(2-cyanovinyl)-7-azabicyclo[2.2.1]heptane-1-carboxylate (**26**). Although this process required three days for total conversion, the isomers were isolated by column chromatography in excellent yield (99%) and were separated for full characterisation. The assignment of the double bond stereochemistry for each isomer was made according to the coupling constants between the olefinic proton signals in the ^{1}H NMR spectra of the isolated products (*J*=11.0 Hz and *J*=16.2 Hz for the *Z* and *E* isomers, respectively).

The results obtained with the stabilised ylides were not completely satisfactory due to the formation of epimeric mixtures of isomers and the long reaction times required to achieve total conversion of **21**. Therefore, in an effort to improve the efficiency of this methodology, the reaction of (1*S*,2*R*,4*R*)-**21** with a phosphonate was tested (Scheme 11). The Horner–Wittig reaction, using methyl diethylphosphonoacetate and NaH in dry THF at room temperature, cleanly provided a 12:88 *Z/E* mixture of **27** in excellent yield (95%). For characterisation purposes, the isomers were separated by column chromatography and measurement of the coupling constants for the olefinic proton signals in the ^{1}H NMR spectra allowed the double bond stereochemistry to be assigned (*J*=12.0 Hz and *J*=15.0 Hz for the *Z* and *E* isomers, respectively).

Bearing in mind the possible epimerisation problem detected at C-2 and in order to verify the *exo* stereochemistry of the isolated Wittig adducts, oxidative cleavage of the double bond under Sharpless conditions[33] was carried out on **23** and the isomeric mixtures (*Z/E*)-**24**–**27**. This procedure (Scheme 13) transformed all the products into the carboxy derivative (1*S*,2*R*,4*R*)-**9**, whose stereochemistry had already been established.

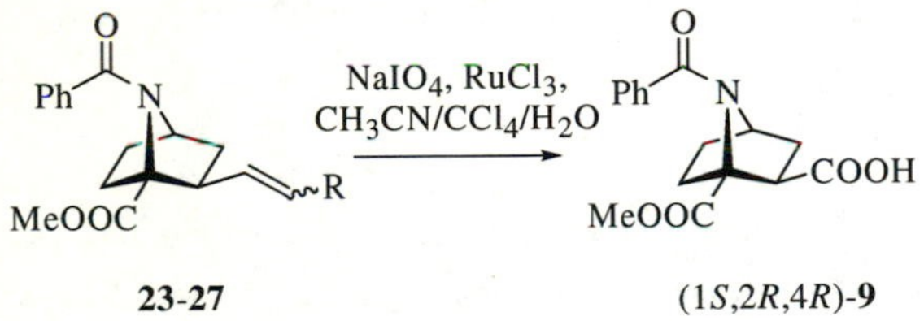

Scheme 13

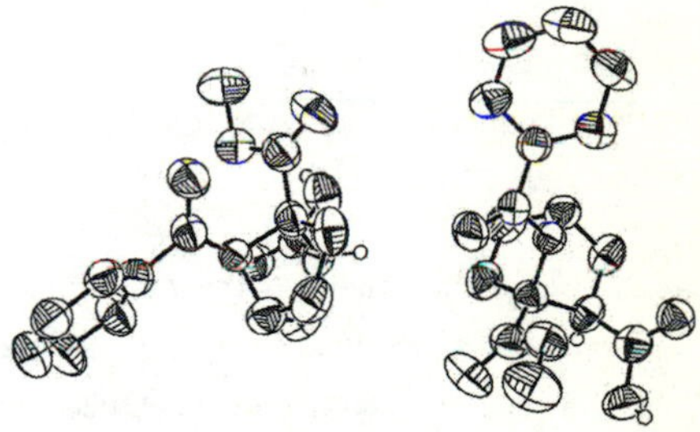

Figure 6

Moreover, an additional and rigorous proof of the assignment of the stereochemistry at C-2 was provided by single crystal X-ray analysis of (1*S*,2*R*,4*R*)-**9**, the direct precursor of the 7-azabicyclo[2.2.1]heptane L-aspartic acid analogue, as shown in Figure 6. These data unequivocally demonstrate the *exo* configuration at the β-position.

3.2.2. Study of S$_N$2 reactions

As described previously, reduction of aldehyde (1*S*,2*R*,4*R*)-**21** was carried out with NaBH$_4$ in a methanol/water mixture and this procedure provided the primary alcohol (1*S*,2*R*,4*R*)-**20** in almost

71

quantitative yield. In order to explore the versatility of the S_N2 reaction at the γ-position, the transformation of the hydroxyl group into a good leaving group was initially tested and then the displacement by different nucleophiles was attempted on the most suitable derivatives.

3.2.2.1. Introduction of leaving groups

Treatment of alcohol (1S,2R,4R)-20 with PCl₅/chloroform or SOCl₂/pyridine was tried but chloride 29 could not be obtained under any of these reaction conditions. However, the use of triphenylphosphine and dry carbon tetrachloride as the halide source gave 29 in 96% yield (Scheme 14).

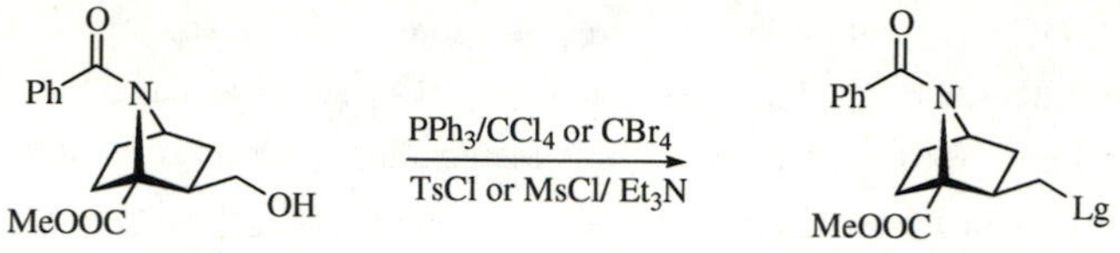

Scheme 14

The introduction of bromide as a leaving group was achieved by treatment of (1S,2R,4R)-20 with N-bromosuccinimide, which provided the bromide derivative 30 in 85% yield. A higher efficiency was obtained by treatment of the alcohol with CBr_4 and triphenylphospine (Scheme 14), which gave compound 30 in very good yield (92%).

Tosylate 31 was synthesised by treatment of the alcohol with tosyl chloride and triethylamine in dichloromethane and, in this way, derivative 31 (Scheme 14) was isolated in low yield (44%). A similar procedure, using mesyl chloride instead of tosyl chloride, readily led to isolated mesylate 32 (Scheme 14) in excellent yield (98%).

3.2.2.2. S_N2 displacements at the γ-position

Given the ready availability of mesylate 32 and its generally good behaviour as a leaving group, we initially focused our attention on a detailed study into S_N2 displacements of this compound. The exchange of the leaving group for hydride was achieved from mesylate 32 by treatment with sodium borohydride in HMPA. This provided compound 33 in good yield (79%), although it should be noted that an additional portion of the hydride had to be added to achieve complete consumption of the starting material.

Nucleophilic substitution on mesylate 32 by the thiomethoxide ion was also achieved, but this reaction was not reproducible and the isolated yield of compound 34 (Scheme 15) never surpassed 60%. The replacement of the mesylate by the benzylthiolate group was also successful and the benzylthioether 35 (Scheme 15) was isolated in 70% yield. This route provided a conveniently protected 7-azanorbornane cysteine analogue.

On the other hand, reaction of mesylate 32 with the cyanide ion proved disappointing. The reaction with sodium cyanide led to complex mixtures that could not be characterised, even on using a phase transfer catalyst [Bn(Et)$_3$NH$_4$Cl] or in the presence of sodium iodide. In an attempt to promote carbon-carbon bond formation with the cyanide ion we systematically varied the solvent and additives such as crown ether and cesium fluoride.[34] The use of KCN, DMF as solvent, CsF and 18-crown-6 ether made the process slightly more favourable and the substitution product **36** (Scheme 15) was isolated for the first time – albeit in very poor yield (<30%). The substitution product could not be detected when the solvent was changed from DMF to toluene, which has been described as the solvent of choice for other azabicyclic substrates.[34] Substitution of tosylate derivative **31** under similar reaction conditions also gave rise to very low yields (<28%).

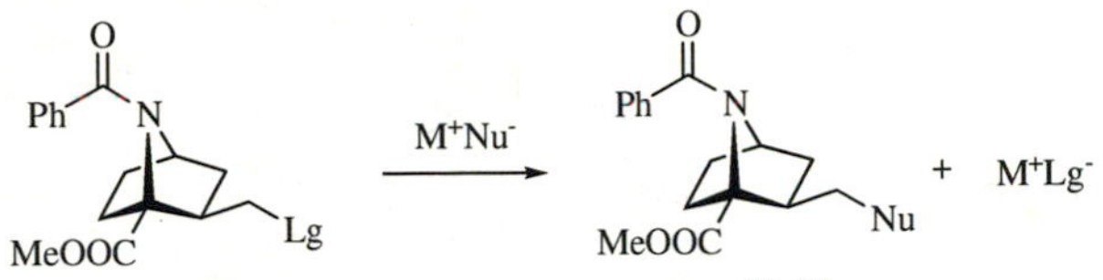

30: Lg = Br
31: Lg = OTs
32: Lg = OMs

33: Nu = H   36: Nu = CN
34: Nu = SMe 37: Nu = N$_3$
35: Nu = SBn

Scheme 15

However, introduction of the azide ion into mesylate **32** was acceptably achieved by treatment with sodium azide in DMF to give azide **37** (Scheme 15) in good yield (83%).

In addition, the mesylate leaving group was exchanged for bromide in moderate yield by treatment of **32** with lithium bromide and, in this way, bromide **30** was obtained in 75% yield.

The substitution on mesylate **32** did not produce completely satisfactory results due to low yields, long reaction times and, in some cases, lack of reproducibility. For these reasons the study was repeated on bromo derivative **30**. This compound, which contained a bromide that is considered a good leaving group, could also be obtained in very high yield from alcohol (1S,2R,4R)-**20**.

Fortunately, displacements on bromide **30** were much better in all cases than the corresponding reactions with **32**. Substitution by hydride was more easily achieved on the bromide, with a shorter reaction time and lower reagent/starting material ratio required, and the reaction gave derivative **33** in higher yield (87%) (Scheme 16). Substitutions on bromide derivative **30** also allowed the isolation of the derivatives **34** and **35** (Scheme 16) in better yields (88% and 77%, respectively) and much shorter reaction times than the same substitutions on mesylate **32**.

Displacement of the bromide group on **30** by the cyanide ion also proceeded much more efficiently and, in the absence of the two aforementioned additives, gave cyano derivate **36** in 61% yield (Scheme 16). Addition of a catalytic amount of 18-crown-6 ether led to the best result and this modification provided, from bromide **30**, cyano derivative **36** in good yield (76%).

Once again, a better result was obtained in the substitution of azide on bromide **30** and this allowed the isolation of azide **37** in excellent yield (92%) (Scheme 16) with a marked decrease in the reaction time.

The enantiomerically pure substitution products **33–37** prepared from bromide **30** are summarised in Scheme 16.

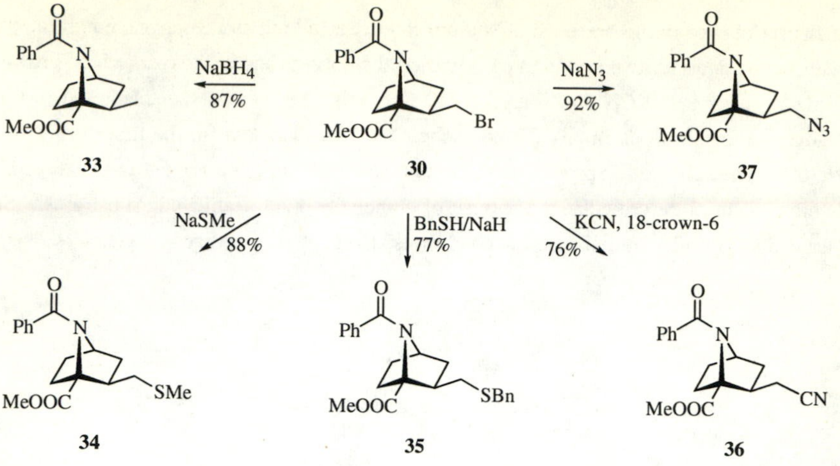

Scheme 16

3.2.3. Other transformations of the amino acid side chain

Other transformations were carried out mainly by manipulation of the side chain on derivatives obtained in the reactivity studies discussed above.

For example, hydrogenation of the double bond of compounds arising from olefination of aldehyde (1S,2R,4R)-**21**, using a catalytic amount of palladium hydroxide at room temperature and atmospheric pressure, cleanly afforded the corresponding saturated compounds **38–42** in excellent yields (≥91%) (Scheme 17).

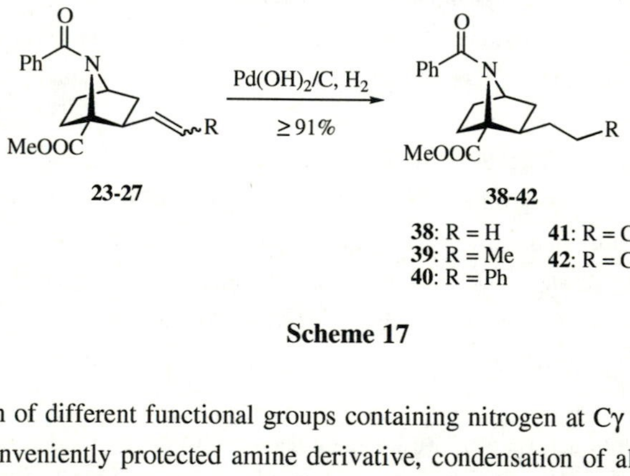

23-27   **38-42**

38: R = H   **41**: R = CN
39: R = Me   **42**: R = CO$_2$Me
40: R = Ph

Scheme 17

The introduction of different functional groups containing nitrogen at Cγ was also investigated. In an effort to prepare a conveniently protected amine derivative, condensation of aldehyde (1S,2R,4R)-**21** with benzylamine and subsequent reduction of the resulting imine derivative was carried out using NaBH$_4$ and ethanol as the solvent. This procedure led, after 3 hours, to a mixture of benzylamine derivative **43** and a small amount of compound resulting from lactamisation (**44**) (Scheme 18). Compound **43** was detected in this reaction mixture but only an analytical sample of this benzylamine could be purified and characterised by NMR spectroscopy. The extension of the reduction time with NaBH$_4$ to 3 days caused a notable increase in the proportion of lactam, which aided the complete characterisation of **44**. Reductive amination of

74

(1*S*,2*R*,4*R*)-**21** in a one-pot procedure, which simply involved mixing the aldehyde with benzylamine and sodium triacetiloxyborohydride, again led to a significant quantity of lactamisation product **44**.

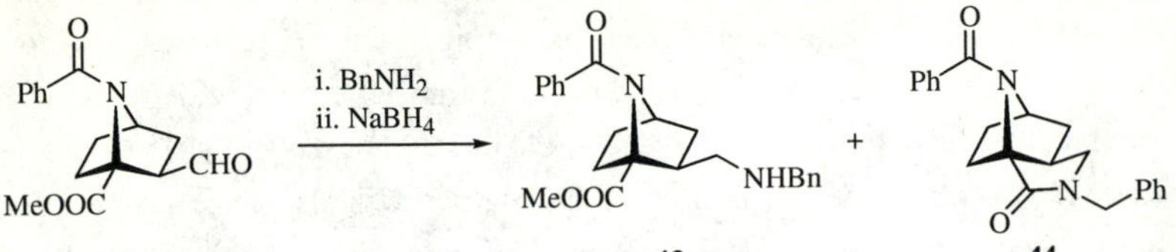
Ph
N
CHO
MeOOC
(1S,2R,4R)-21
i. BnNH2
ii. NaBH4
Ph
N
NHBn
MeOOC
43
+
Ph
N
N
Ph
44

Scheme 18

We decided to take advantage of the availability of azide **37**, obtained as an S_N2 substitution product, and explore its potential reactivity. Simple hydrogenation of **37** using Pd/C at room temperature and atmospheric pressure, cleanly yielded an unstable amine derivative **45** (Scheme 19). Despite the instability of this compound it could be well characterised by ^{1}H and ^{13}C NMR spectroscopy.

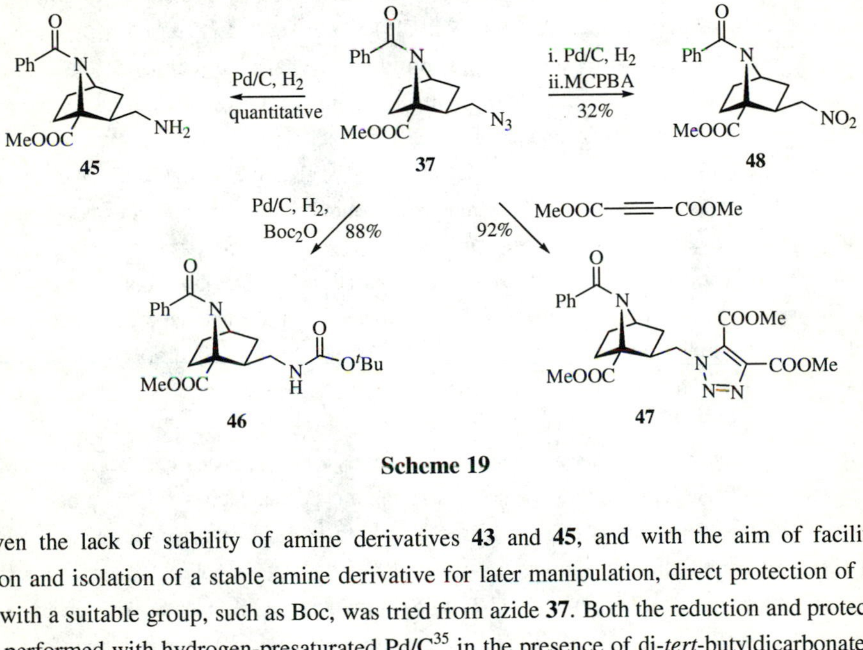
Ph
N
MeOOC
NH2
45
Pd/C, H2
quantitative
Ph
N
MeOOC
N3
37
i. Pd/C, H2
ii.MCPBA
32%
Ph
N
MeOOC
NO2
48
Pd/C, H2,
Boc2O
88%
92%
MeOOC—≡—COOMe
Ph
N
MeOOC
N
H
O'Bu
46
Ph
N
MeOOC
COOMe
N
COOMe
N=N
47

Scheme 19

Given the lack of stability of amine derivatives **43** and **45**, and with the aim of facilitating the preparation and isolation of a stable amine derivative for later manipulation, direct protection of the amine function with a suitable group, such as Boc, was tried from azide **37**. Both the reduction and protection steps could be performed with hydrogen-presaturated Pd/C[35] in the presence of di-*tert*-butyldicarbonate (Boc_2O). Under these conditions derivative **46** (Scheme 19) was isolated in good yield (88%).

In addition, other one-pot conversions of azides into Boc-protected amines were assessed. For example, treatment of azide **37** with Et_3SiH (1.5 eq), $Pd(OH)_2/C$ (10mg/mmol) and Boc_2O (1.5eq) in EtOH[36] also led to the corresponding Boc-protected amine **46**, albeit in low yield. On the other hand, formation of a Staudinger intermediate (phosphazenes from azides and tertiary phosphines) has been shown to be a convenient method for generating a nucleophilic amine from an azide.[37] In our case, according to the procedure described by Vilarrasa *et al.*,[38] the use of triphenylphosphine (1.05 eq) and 2-(*tert*-

butoxycarbonyloxyimino)-2-phenylacetonitrile (Boc-ON, 1.05 eq) enabled us to obtain directly the desired Boc-amine **46** in moderate yield from azide **37** (68%).

In order to demonstrate the synthetic versatility of the azide group in this 7-azanorbornane skeleton, we proceeded to carry out transformations into other interesting functional groups. For example, derivative **37** was treated with dimethylacetylenedicarboxylate.[39] Fortunately, the azide underwent 1,3-dipolar cycloaddition very readily to give triazol derivative **47** (Scheme 19) in excellent yield (92%).

The transformation of the azide function into a nitro group was also achieved by means of a simple two-step procedure.[40] Reduction of the azide to the amine by simple hydrogenation with Pd/C at 35 °C was followed by the addition of MCPBA to the amine solution. This simple procedure supplied the nitroderivative **48** in 32% yield (Scheme 19).

4. New enantiopure proline–α-amino acid chimeras with the 7-azabicyclo[2.2.1]heptane skeleton

Finally, the preparation of some interesting enantiopure β-functionalised prolines with the 7-azanorbornane core serves to demonstrate the versatility of our methodology.

After HPLC resolution of *rac*-**5**, the isolated enantiomers were subjected to hydrolysis. A small quantity of each enantiomer, *i.e.*, (1*S*,2*S*,4*R*) and (1*R*,2*R*,4*S*)-**5**, was separately treated with aqueous 6N HCl under reflux to provide the two optically pure proline–phenylalanine chimeras (1*S*,2*S*,4*R*)-**49** and (1*R*,2*R*,4*S*)-**49** in 95% yield (Scheme 20). The resulting amino acids, which can also be considered as (2*S*,3*S*)- and (2*R*,3*R*)-3-phenylproline analogues, were fully characterised.[18]

The neurological effects that certain amino acids, such as glutamic and aspartic acid, induce in the mammalian central nervous system are well documented. A number of these compounds act as excitatory neurotransmitters that activate the different receptors associated with a variety of physiological functions and this ability depends, in some cases, on their absolute configuration.[41,42] The introduction of conformational constraints into aspartic acid may provide useful information about conformational requirements for receptor binding and, at the same time, competitive antagonists of this type of excitatory amino acid. In addition, this approach may lead to new residues that can be considered in the design of conformationally constrained peptidomimetics with fascinating properties.

Taking into account the significance of these amino acids, we had already carried out a study into the asymmetric synthesis of L-aspartic acid analogues containing a cyclic skeleton.[14d-f,15] As a part of this study we also embarked upon the synthesis of the aspartic acid with a 7-azanorbornane skeleton. Indeed, final hydrolysis of the benzamide and ester groups of (1*S*,2*R*,4*R*)- and (1*R*,2*S*,4*S*)-**9** with 6N aqueous HCl under reflux (Scheme 20) led to (1*S*,2*R*,4*R*)- and (1*R*,2*S*,4*S*)-7-azabicyclo[2.2.1]heptane-1,2-dicarboxylic acid hydrochloride, (1*S*,2*R*,4*R*)-**50** or (1*R*,2*S*,4*S*)-**50**, in quantitative yield.[18] These compounds comprise a combination of proline and L- or D-aspartic acid and they can be also considered as (2*S*,3*R*)- or (2*R*,3*S*)-3-carboxyproline analogues, respectively, with a 7-azabicyclo[2.2.1]heptane skeleton.

More examples of new enantiopure β-substituted 7-azanorbornane prolines were obtained by hydrolysis of compounds **38–42**, which in turn were obtained by hydrogenation of the olefination products, with 6N hydrochloric acid under reflux. This procedure afforded amino acids **51–54** (Scheme 21) in very good yields (≥85%).[30] Compounds **51–54** are (2*S*,3*R*)-3-ethylproline, (2*S*,3*R*)-3-propylproline, (2*S*,3*R*)-3-(2-phenylethyl)proline and (2*S*,3*R*)-3-(2-carboxyethyl)proline analogues, respectively, and can also be considered as proline–L-norvaline (**51**), proline–L-norleucine (**52**), proline–L-α-(3-phenylpropyl)glycine (**53**), and proline–L-homoglutamic acid (**54**) chimeras with a 7-azabicyclo[2.2.1]heptane skeleton.

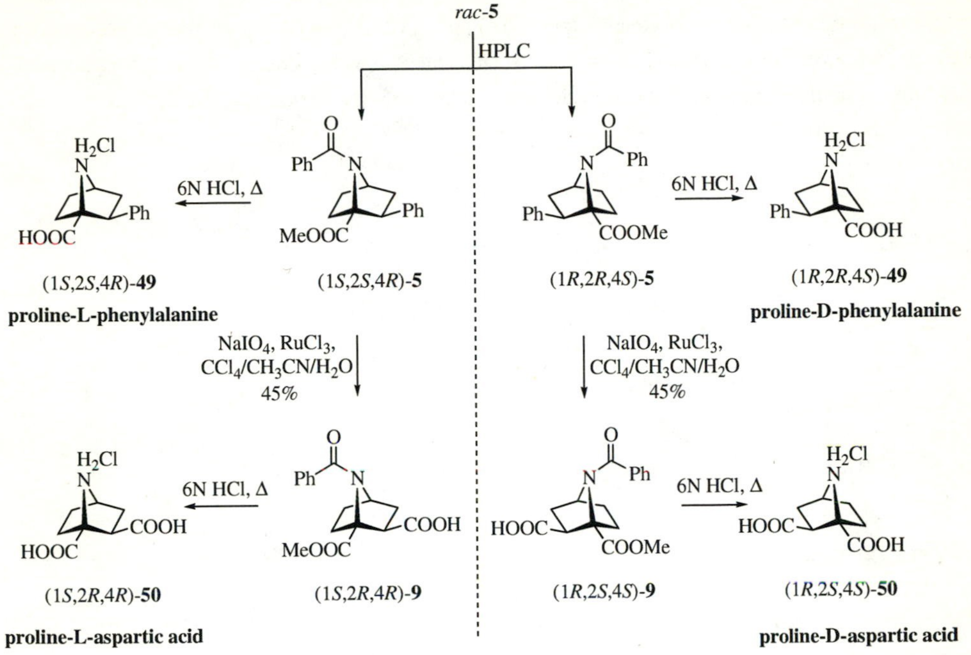

Scheme 20

In this way, olefination processes provide some very interesting β-alkyl amino acid analogues. For instance, as described above, the introduction of a propyl group on C-2 can be considered as a way to reproduce the side chain of norleucine, a non-natural amino acid whose use has been extended to diverse areas. Definite applications have already been established for this system as an active ingredient in preservatives for cut flowers or in cosmetics related to hair care. Norleucine even proved to be an efficient instrument to explore recognition sites when it was used as a residue introduced into peptides that facilitated the mapping of the S1 binding pocket of human cathepsin G.[43] Its efficacy to probe the stability of proteins generated by the introduction of non-natural amino acids, e.g. as a replacement for methionine in human recombinant annexin V,[44] is another remarkable property.

In addition, potential applications of some neutral non-natural amino acids, such as norleucine or norvaline, stem from their ability to produce cell metabolism and composition changes without modifying cell growth,[45] or to cause metabolic alterations in rats whose diets contain such compounds.[46,47] These factors must also be considered when assessing the significance of this amino acid.

Homoglutamic acid is another non-natural amino acid whose 7-azanorbornane analogue **54** can be prepared following our methodology. This compound is a constituent of novel methotrexate derivatives. These substrates have been evaluated as antirheumatic agents for humans, as suppressors of the *in vitro* cell proliferation and in an *in vivo* adjuvant arthritis model.[48] The compound has also been used for replacement probes of the chalcone isomerase,[49] aspartate aminotransferase[50] or staphylococcal nuclease mechanisms.[51]

Finally, other compounds, such as **33**, **34** and **36**, were treated with aqueous 6N HCl under reflux to provide optically pure proline–α-amino acid chimeras **55**, **56** and **57** in quantitative or nearly quantitative yields (Scheme 21). The resulting amino acids **55**–**57**, which can be also considered as (2S,3R)-3-methylproline, (2S,3R)-3-methylthiomethylproline and (2S,3S)-3-carboxymethylproline analogues, respectively, were fully characterised. These amino acids can also be considered as a combination of proline with L-valine, L-methionine and L-glutamic acid with a 7-azanorbornane skeleton.

52	**53**	**54**
proline-L-norleucine	proline-L-α-(3-phenylpropyl)glycine	proline-L-homoglutamic acid

R = Et   R = CH₂Ph   R = CH₂CN, CH₂COOMe

R = Me		R = CN
51	**33,34,36,38-42**	**57**
proline-L-norvaline		proline-L-glutamic acid

R = H   R = SMe

55	**56**
proline-L-valine	proline-L-methionine

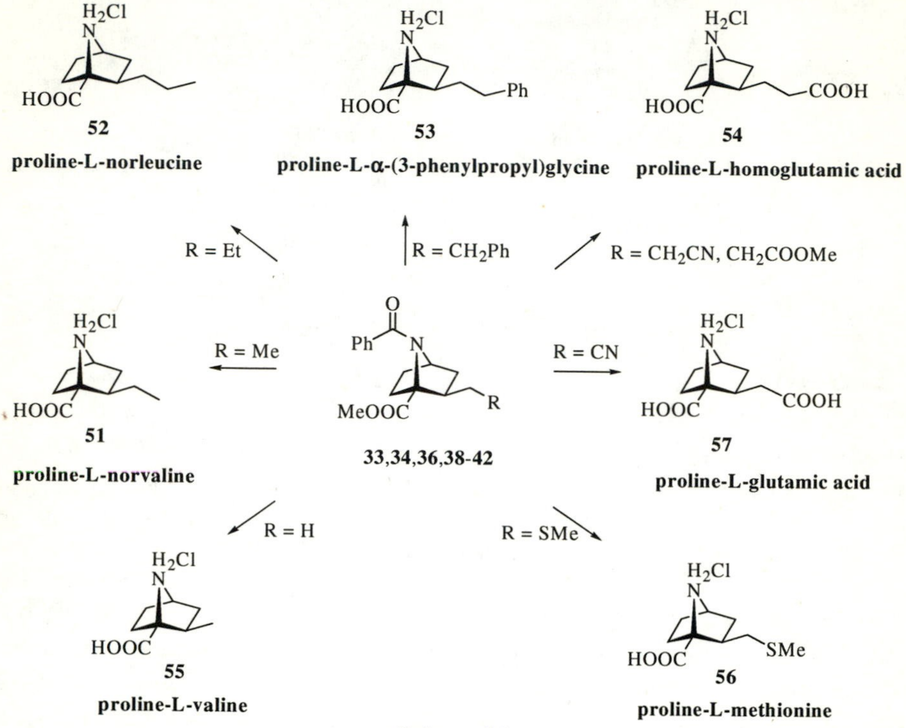

Scheme 21

5. Conclusions

A versatile and efficient methodology is reported for the synthesis of new enantiomerically pure chimeras, combinations of α-amino acids and a proline constrained by a 7-azanorbornane skeleton.

Cycloaddition of (Z)-2-phenyl-4-benzylidene-5(4H)-oxazolone and Danishefsky's diene followed by manipulation of the resulting adducts has led to the synthesis of a key intermediate, methyl N-benzoyl-2-phenyl-7-azabicyclo[2.2.1]heptane-1-carboxylate. Furthermore, the high efficacy of the HPLC resolution of this compound has provided the two enantiomerically pure enantiomers of this constrained proline–phenylalanine chimera. On the other hand, the reaction of Danishefsky's diene and (Z)-2-phenyl-4-[(S)-2,2-dimethyl-1,3-dioxolan-4-ylmethylene]-5(4H)-oxazolone, a chiral unsaturated oxazolone derived from (R)-glyceraldehyde, has allowed our methodology to be extended to the asymmetric version through the stereospecific synthesis of an enantiopure key intermediate, methyl (1S,2R,4R)-N-benzoyl-2-[(S)-2,2-dimethyl-1,3-dioxolan-4-yl]-7-azabicyclo[2.2.1]heptane-1-carboxylate.

Subsequent manipulation of the β-phenyl substituent or dioxolane moiety in the aforementioned key intermediates has made it possible to introduce β-formyl, β-carboxylic or β-hydroxymethyl groups and carry out interconversions between them. This functionalisation at the γ-position generates a huge range of possibilities for the preparation of a very special kind of amino acid where the rigidity provided by the azabicyclic skeleton is combined with the presence of a β-substituent, which mimics the α-amino acid side chain. These proline–α-amino acid chimeras are new surrogates to be incorporated into peptides whose structural and biological properties shed light upon the nature of the effects induced by this type of conformational restriction and the influence of the absolute configurations of the stereogenic centres. Besides, their use in catalysis area, in replacement for proline, could result a fascinating application field for these heavily constrained amino acids.

The excellent results obtained in this study offer clear evidence for the wide-ranging potential of our methodology.

Acknowledgments

Financial support from *DGA* (project P22/98), *Ministerio de Ciencia y Tecnología* and *FEDER* (projects PPQ2001-1834 and 2FD97-1530) is gratefully acknowledged. A. M. Gil is especially thankful to *CSIC* and *Ministerio de Educación y Ciencia* for an I3P grant.

References

1. (a) Giannis, A.; Kolter, T. *Angew. Chem. Int. Ed. Engl.* **1993**, *32*, 1244–1267. (b) Liskamp, R. M. J. *Recl. Trav. Chim. Pays-Bas* **1994**, *113*, 1–19. (c) Gante, J. *Angew. Chem. Int. Ed. Engl.* **1994**, *33*, 1699–1720. (d) Hruby, V. J.; Li, G.; Haskell-Luevano, C.; Shenderovich, M. *Biopolymers* **1997**, *43*, 219–266.

2. (a) Cativiela, C.; Díaz-de-Villegas, M. D. *Tetrahedron: Asymmetry* **1998**, *9*, 3517–3599. (b) Cativiela, C.; Díaz-de-Villegas, M. D. *Tetrahedron: Asymmetry* **2000**, *11*, 645–732. (c) Park, K.-H.; Kurth, M. J. *Tetrahedron* **2002**, *58*, 8629–8659.

3. (a) Thaisrivongs, S.; Pals, D. T.; Lawson, J. A.; Turner, S. R.; Harris, D. W. *J. Med. Chem.* **1987**, *30*, 536–541. (b) Hinds, M. G.; Welsh, J. H.; Brennand, D. M.; Fisher, J.; Glennie, M. J.; Richards, N. G. J.; Turner, D. L.; Robinson, J. A. *J. Med. Chem.* **1991**, *34*, 1777–1789. (c) Bisang, C.; Weber, C.; Inglis, J.; Schiffer, C. A.; van Gunsteren, W. F.; Jelesarov, I.; Bosshard, H. R.; Robinson, J. A. *J. Am. Chem. Soc.* **1995**, *117*, 7904–7915.

4. (a) Notz, W.; Tanaka, F.; Barbas, C. F., III *Acc. Chem. Res.* **2004**, *37*, 580–591. (b) Northrup, A. B.; MacMillan, D. W. C. *Science* **2004**, *305*, 1752–1755. (c) Northrup, A. B.; Mangion, I. K.; Hettche, F.; MacMillan, D. W. C. *Angew. Chem. Int. Ed.* **2004**, *43*, 2152–2154. (d) Storer, R. I.; MacMillan, D. W. C. *Tetrahedron* **2004**, *60*, 7705–7714. (e) Brown, S. P.; Brochu, M. P.; Sinz, C. J.; MacMillan, D. W. C. *J. Am. Chem. Soc.* **2003**, *125*, 10808–10809. (e) Alcaide, B.; Almendros, P. *Eur. J. Org. Chem.* **2002**, *10*, 1595–1601.

5. (a) Belleau, B. *Can. J. Chem.* **1957**, *35*, 663–670. (b) Armstrong, P.; Grigg, R.; Jordan, M. W.; Malone, J. F. *Tetrahedron* **1985**, *41*, 3547–3558. (c) Grigg, R.; Armstrong, P. *Tetrahedron* **1989**, *45*, 7581–7586. (d) Beak, P.; Wu, S.; Yum, E. K.; Jun, Y. M. *J. Org. Chem.* **1994**, *59*, 276–277. (e) Hercouet, A.; Bessieres, B.; Le Corre, M. *Tetrahedron: Asymmetry* **1996**, *7*, 1267–1268. (f) Ezquerra, J.; Escribano, A.; Rubio, A.; Remuinan, M. J.; Vaquero, J. J. *Tetrahedron: Asymmetry* **1996**, *7*, 2613–2626.

6. (a) Pirrung, M. C. *Tetrahedron Lett.* **1980**, *21*, 4577–4578. (b) Bell, E. A.; Qureshi, M. Y.; Pryce, R. J.; Janzen, D. H.; Lemke, P.; Clardy, J. *J. Am. Chem. Soc.* **1980**, *102*, 1409–1412. (c) Montelione, G. T.; Hughes, P.; Clardy, J.; Scheraga, H. A. *J. Am. Chem. Soc.* **1986**, *108*, 6765–6773. (d) Hughes, P.; Clardy, J. *J. Org. Chem.* **1988**, *53*, 4793–4797. (e) Griffith, D. A.; Heathcock, C. H. *Tetrahedron Lett.* **1995**, *36*, 2381–2384. (d) Esslinger, C. S.; Koch, H. P.; Kavanaugh, M. P.; Philips, D. P.; Chamberlin,

A. R.; Thompson, C. M.; Bridges, R. J. *Bioorg. Med. Chem. Lett.* **1998**, *8*, 3101–3106.

7. (a) Sato, T.; Mori, T.; Sugiyama, T.; Ishibashi, H.; Ikeda, M. *Heterocycles* **1994**, *37*, 245–248. (b) Sato, T.; Kugo, Y.; Nakaumi, E.; Ishibashi, H.; Ikeda, M. *J. Chem. Soc., Perkin Trans. 1* **1995**, 1801–1810. (c) Avenoza, A.; Busto, J. H.; Cativiela, C.; Peregrina, J. M. *Tetrahedron Lett.* **1995**, *36*, 7123–7126. (d) Ikeda, M.; Kugo, Y.; Sato, T. *J. Chem. Soc., Perkin Trans. 1* **1996**, 1819–1824. (e) Campbell, J. A.; Rapoport, H. *J. Org. Chem.* **1996**, *61*, 6313–6325. (f) Ikeda, M.; Kugo, Y.; Kondo, Y.; Yamazaki, T.; Sato, T. *J. Chem. Soc., Perkin Trans. 1* **1997**, 3339–3344. (g) Avenoza, A.; Cativiela, C.; Fernández-Recio, M.; Peregrina, J. M. *Synthesis* **1997**, 165–167.

8. (a) Avenoza, A.; Busto, J. H.; Cativiela, C.; Peregrina, J. M. *Synthesis* **1998**, 1335–1338. (b) Han, W.; Pelletier, J. C.; Hodge, C. N. *Bioorg. Med. Chem. Lett.* **1998**, *8*, 3615–3620. (c) Black, D. StC.; Craig, D. C.; Edwards, G. L.; Laaman, S. M. *Bioorg. Chem.* **1999**, *27*, 91–99. (d) Hart, B. P.; Rapoport, H. *J. Org. Chem.* **1999**, *64*, 2050–2056. (e) Avenoza, A.; Cativiela, C.; Fernández-Recio, M.; Peregrina, J. M. *Tetrahedron: Asymmetry* **1999**, *10*, 3999–4007. (f) Han, W.; Pelletier, J. C.; Mersinger, L. J.; Kettner, C. A.; Hodge, C. N. *Org. Lett.* **1999**, *1*, 1875–1877. (g) Black, D. StC.; Edwards, G. L.; Evans, R. H.; Keller, P. A.; Laaman, S. M. *Tetrahedron* **2000**, *56*, 1889–1898.

9. Spand, T. F.; Garrafo, H. M.; Edwards, M. W.; Daly, J. W. *J. Am. Chem. Soc.* **1992**, *114*, 3475–3478.

10. Chen, Z.; Trudell, M. L. *Chem. Rev.* **1996**, *96*, 1179–1193.

11. Lennox, J. R.; Turner, S. C.; Rapoport, H. *J. Org. Chem.* **2001**, *66*, 7078–7083.

12. Avenoza, A.; Cativiela, C.; Busto, J. H.; Fernández-Recio, M. A.; Peregrina, J. M.; Rodríguez, F. *Tetrahedron* **2001**, *57*, 545–548.

13. (a) Avenoza, A.; Cativiela, C.; Mayoral, J. A.; González, M.; Roy, M. A. *Synthesis* **1990**, 1114–1116. (b) Avenoza, A.; Cativiela, C.; Díaz-de-Villegas, M. D.; Mayoral, J. A.; Peregrina, J. M. *Tetrahedron* **1993**, *49*, 677–684. (c) Avenoza, A.; Cativiela, C.; Díaz-de-Villegas, M. D.; Mayoral, J. A.; Peregrina, J. M. *Tetrahedron* **1993**, *49*, 10987–10996. (d) Avenoza, A.; Cativiela, C.; Peregrina, J. M. *Tetrahedron* **1994**, 50, 10021–10028. (e) Avenoza, A.; Cativiela, C.; Peregrina, J. M. *Tetrahedron* **1994**, *50*, 12989–12998. (f) Avenoza, A.; Busto, J. H.; Cativiela, C.; Peregrina, J. M. *Synthesis* **1995**, *6*, 671–674. (g) Avenoza, A.; Busto, J. H.; Cativiela, C.; París, M.; Peregrina, J. M. *J. Heterocyclic Chem.* **1997**, *34*, 1099–1110. (h) Díaz-de-Villegas, M. D; Cativiela, C., in: *The Chemistry of Heterocyclic Compounds*; Palmer, D. C., Ed.; John Wiley and Sons, Inc., 2004; Vol. 60, p 129–330.

14. (a) Buñuel, E.; Cativiela, C.; Díaz-de-Villegas, M. D. *Tetrahedron: Asymmetry* **1994**, *5*, 157–160. (b) Buñuel, E.; Cativiela, C.; Díaz-de-Villegas, M. D.; García, J. I. *Tetrahedron: Asymmetry* **1994**, *5*, 759–766. (c) Buñuel, E.; Cativiela, C.; Díaz-de-Villegas, M. D. *Tetrahedron* **1995**, *51*, 8923–8934. (d) Buñuel, E.; Cativiela, C.; Díaz-de-Villegas, M. D. *Tetrahedron: Asymmetry* **1996**, *7*, 1521–1528. (e) Buñuel, E. Ph.D. Thesis, University of Zaragoza, 1996. (f) Buñuel, E.; Cativiela, C.; Díaz-de-Villegas, M. D. *Tetrahedron: Asymmetry* **1996**, *7*, 1431–1436.

15. Buñuel, E.; Jiménez, A. I.; Díaz-de-Villegas, M. D.; Cativiela, C., in: *Targets in Heterocyclic Systems. Chemistry and Properties*; Attanasi, O. A.; Spinelli, D., Eds; Royal Society of Chemistry: Cambridge, 2001; Vol. 5, p 79–111.

16. Avenoza, A.; Barriobero, J. I.; Busto, J. H.; Cativiela, C.; Peregrina, J. M. *Tetrahedron: Asymmetry* **2002**, *13*, 625–632.

17. Busto, J. H. Ph.D. Thesis, University of La Rioja, 1997.

18. Gil, A. M.; Buñuel, E.; López, P.; Cativiela, C. *Tetrahedron: Asymmetry* **2004**, *15*, 811–819.

19. (a) Cativiela, C.; Díaz-de-Villegas, M. D.; Jiménez, A. I.; López, P.; Marraud, M.; Oliveros, L. *Chirality*, **1999**, *11*, 583–590. (b) Alías, M.; Cativiela, C.; Jiménez, A. I.; López, P.; Oliveros, L.; Marraud, M. *Chirality*, **2001**, *13*, 48–55.

20. (a) Jiménez, A. I.; López, P.; Oliveros, L.; Cativiela, C. *Tetrahedron* **2001**, *57*, 6019–6026. (b) Royo, S.; López, P.; Jiménez, A. I.; Oliveros, L.; Cativiela, C. *Chirality*, **2002**, *14*, 39–46.

21. (a) Franco, P.; Senso, A.; Minguillón, C.; Oliveros, L. *J. Chromatogr. A* **1998**, *A 796*, 265–272. (b) Oliveros, L.; López, P.; Minguillón, C.; Franco, P. *J. Liq. Chromatogr.* **1995**, *18*, 1521–1532.

22. (a) Obrecht, D.; Bohdal, U.; Broger, C.; Bur, D.; Lehmann, C.; Ruffieux, R.; Schönholzer, P.; Spiegler, C.; Müller, K. *Helv. Chim. Acta* **1995**, *78*, 563–580. (b) Obrecht, D.; Spiegler, C.; Schönholzer, P.; Müller, K.; Lehmann, C.; Ruffieux, R. *Helv. Chim. Acta* **1995**, *78*, 1567–1587.

23. Castro, B.; Dormoy, J. R.; Evin, G.; Selve, C. *Tetrahedron Lett.* **1975**, *14*, 1219–1222.

24. (a) Jiménez, A. I.; Cativiela, C.; Aubry, A.; Marraud, M. *J. Am. Chem. Soc.* **1998**, *120*, 9452–9459. (b) Jiménez, A. I.; Cativiela, C.; París, M.; Peregrina, J. M.; Avenoza, A.; Aubry, A.; Marraud, M. *Tetrahedron Lett.* **1998**, *39*, 7841–7844. (c) Jiménez, A. I.; Vanderesse, R.; Marraud, M.; Aubry, A.; Cativiela, C. *Tetrahedron Lett.* **1997**, *38*, 7559–7562. (c) Jiménez, A. I.; Cativiela, C.; Marraud, M. *Tetrahedron Lett.* **2000**, *41*, 5353–5356. (d) Jiménez, A. I.; Cativiela, C.; Gómez-Catalán, J.; Pérez, J. J.; Aubry, A.; París, M.; Marraud, M. *J. Am. Chem. Soc.* **2000**, *122*, 5811–5821.

25. Gil, A. M.; Buñuel, E.; Jiménez, A. I.; Cativiela, C. *Tetrahedron Lett.* **2003**, *44*, 5999–6002.

26. Gil, A. M.; Buñuel, E.; Cativiela, C. unpublished results.

27. Buñuel, E.; Gil, A. M.; Díaz-de-Villegas, M. D.; Cativiela, C. *Tetrahedron* **2001**, *44*, 6417–6427.

28. (a) Dess, D. B.; Martin, J. C. *J. Org. Chem.* **1983**, *48*, 4156–4158. (b) Dess, D. B.; Martin, J. C. *J. Am. Chem. Soc.* **1991**, *113*, 7277–7287.

29. Colombo, L.; Gennari, C.; Santandrea, M.; Narisano, E.; Scolastico, C. *J. Chem. Soc., Perkin Trans. 1* **1980**, 1819–1824.

30. Gil, A. M.; Buñuel, E.; Díaz-de-Villegas, M. D.; Cativiela, C. *Tetrahedron: Asymmetry* **2003**, *14*, 1479–1488.

31. Khanapure, S. P.; Saha, G.; Powell, W. S.; Rokach, J. *Tetrahedron Lett.* **2000**, *41*, 5807–5811.

32. (a) Corey, E. J.; Winter, R. A. E. *J. Am. Chem. Soc.* **1963**, *85*, 2677–2678. (b) Corey, E. J.; Carey, F. A.; Winter, R. A. E. *J. Am. Chem. Soc.* **1965**, *87*, 934–935. (c) Corey, E. J.; Hopkins, P. B. *Tetrahedron Lett.* **1982**, *23*, 1979–1982.

33. Carlsen, H.; Katsuki, T.; Sharpless, B. *J. Org. Chem.* **1981**, *46*, 3936–3938.

34. Schrake, O.; Franz, M. H.; Wartchow, R.; Hoffmann, H. M. R. *Tetrahedron* **2000**, *56*, 4453–4465.

35. Woltering, T. J.; Weiz-Schmidt, G.; Wong, C. H. *Tetrahedron Lett.* **1996**, *37*, 9033–9036.

36. Kotsuki, H.; Ohishi, T.; Araki, T.; *Tetrahedron Lett.* **1997**, *38*, 2129–2132.

37. (a) Staudinger, H.; Meyer, J. *Helv. Chim. Acta* **1919**, *2*, 635–646. (b) Gololobov, Yu. G.; Kasukhin, L. F. *Tetrahedron* **1992**, *48*, 1353–1406.

38. Ariza, X.; Urpí, F.; Viladomat, C.; Vilarrasa, J. *Tetrahedron Lett.* **1998**, *39*, 9101–9102.

39. Abu-Orabi, S. T.; Atfah, M. A.; Sibril, I.; Mari'I, F. M.; Ali, A. A. *J. Heterocyclic Chem.* **1989**, *26*, 1461–1468.

40. Maguire, M. R.; Feldman, P. L.; Rapoport, H. *J. Org. Chem.* **1990**, *55*, 948–955.

41. (a) Watkins, J. C.; Olverman, H. J. *Trends Neurosci.* **1987**, *10*, 265–272. (b) *Excitatory Amino Acid Receptors: Design of Agonists and* Antagonists; Krogsgaard-Larsen, P.; Hansen, J. J., Eds.; Ellis Horwood, Chichester, England, 1992.

42. (a) Bigge, C. F. *Biochem. Pharmacol.* **1993**, *45*, 1547. (b) *The NMDA Receptor*; Collingidge, G. L.; Watkins, J. C., Eds.; IRL Press at Oxford University Press, Oxford, England, 1994.

43. Polanowska, J.; Krokoszynska, I.; Czapinska, H.; Watorek, W.; Dadlez, M.; Otlewski, J. *Biochim. Biophys. Acta* **1998**, *1386*, 189–198.

44. Budisa, N.; Pifat, G. *Croat. Chem. Acta* **1998**, *71*, 179–187.

45. Mayer, A. M.; Bar-Nun, N. *Phytochemistry* **1987**, *26*,1955–1958.

46. (a) Tews, J. K.; Repa, J. J.; Joyce, J.; Harper, A. E. *Pharmacol. Biochem. Behav.* **1990**, *35*, 911–921. (b) Tews, J. K.; Repa, J. J.; Joyce, J.; Harper, A. E. *J. Nutr.* **1991**, *121*, 364–378. (c) Tews, J. K.; Repa, J. J.; Joyce, J.; Harper, A. E. *J. Nutr. Biochem.* **1993**, *4*, 172–180.

47. Tews, J. K.; Harper, A. E. *J. Nutr.* **1986**, *116*, 1464–1472.

48. (a) Matsuoka, H.; Maruyama, N.; Suzuki, H.; Kuroki, T.; Tsuji, K.; Kato, N.; Ohi, N.; Mihara, M.; Takeda, Y.; Yano, K. *Chem. Pharm. Bull.* **1996**, *44*, 2287–2293. (b) Matsuoka, H.; Ohi, N.; Mihara, M.; Suzuki, H.; Miyamoto, K.; Maruyama, N.; Tsuji, K.; Kato, N.; Akimoto, T.; Takeda, Y.; Yano, K.; Kuroki, T.; *J. Med. Chem.* **1997**, *40*, 105–111.

49. Bednar, R. A.; McCaffrey, C.; Shan, K. *Bioconjugate Chem.* **1991**, *2*, 211–216.

50. Park, Y; Luo, J.; Schultz, P. G.; Kirsch, J. F. *Biochemistry* **1997**, *36*, 10517–10525.

51. Judice, J. K.; Gamble, T. R.; Murphy, E. C.; de Vos, A. M.; Schultz, P. G. *Science,* **1993**, *261*, 1578–1581.

APPLICATION OF GOLD CATALYSIS IN THE SYNTHESIS OF HETEROCYCLIC SYSTEMS

Antonio Arcadi* and Gabriele Bianchi[1]

Dipartimento di Chimica, Ingegneria Chimica e Materiali, Facoltà di Scienze, Università di L'Aquila, Via Vetoio- I-67010 Coppito(AQ)- Italy (e-mail: arcadi@univaq.it)

Abstract. Applications of gold catalysis in the synthesis of heterocyclic systems are reviewed. Gold catalysis accomplishes the functionalization of a variety of heterocyclic derivatives through gold-catalyzed C-C bond formation reactions involving C-H bond cleavage. Gold activation of alkynes towards intramolecular nucleophilic attack represents a powerful tool for annulation processes. The many personalities of gold as Lewis acid, transition-metal and dual role catalyst allow innovative synthetic approaches through domino processes.

Contents

1. Introduction

Recently, gold derivatives are growing in importance as efficient catalysts of several organic transformations.[1] Many gold(I) as well as gold(III) compounds are powerful catalysts and both oxidation states I and III have been proposed for the active species. When gold(I) compounds are used as pre-catalysts, ligands such as phosphanes[2] or phosphates[3] can be applied. The gold(III) pre-catalysts are mainly simple halides. The peculiar features of the high catalytic activity of gold salts determining the need for extremely small amount of these catalysts is of great importance on homogeneous gold catalysis. The mild conditions under which both heterogeneous and homogeneous gold catalysts operate are very attractive for the design of more environmentally friendly procedures for the synthesis of organic compounds. Activation of C-H bonds followed by C-C bond formation points out unique catalytic properties of gold derivatives. Highly effective additions of activated methylene compounds to alkenes by using Au(III) catalysts and to alkynes by means of Au(I) catalysis have been developed. It has been found that $AuCl_3$ can catalyze conjugated addition of electron-rich aromatic rings to Michael acceptors. Intermolecular gold catalyzed hydroarylation reactions of electron-deficient alkynes as well as intramolecular ones of epoxides have been discovered. Gold

---

Dedicated to the memory of Prof. Marino Novi

catalyzed addition of heteroatom nucleophiles to alkynes and activated alkenes have been explored. Alkynophilic gold(I) complexes have accomplished skeletal rearrangement and alkoxycyclization of enynes. The construction of complex molecular architectures have been carried out under very mild conditions through gold(III) chloride catalyzed domino processes. This survey would cover the literature dealing with the use of gold catalysts for the synthesis of heterocyclic systems.

2. Direct functionalization of C-H bonds by gold catalysis

The unique catalytic activity of gold toward sp^3, sp^2 and sp C-H bonds led to many applications in organic synthesis. A highly effective intermolecular addition of activated methylene compounds to alkenes was developed by using $AuCl_3/AgOTf$.[4] A tentative mechanism was proposed involving the activation of the Csp^3-H bond of the activated methylene by a Au(I) species generated *in situ*. Au(I) catalyzed Conia-ene reaction that proceeds under neutral conditions at room temperature was developed.[5] In most cases, the reaction requires low catalyst loadings, short reaction times and proceeds under "open-flask" conditions. The catalytic activation of sp^2 hybridized aryl C-H bonds by gold have been applied to the development of a variety of syntheses of heterocycles. Catalytic functionalization of aromatic C-H bonds to form C-C bonds, if realized under mild and environmentally benign conditions, will provide the most economic way to construct molecules containing arene groups. Efficient processes that work at ambient temperature in a typical organic medium or under solvent-free conditions are very useful. In 1931, Kharasch and Isbell demonstrated that anhydrous gold(III) chloride can react with neat benzene, toluene, or other aromatic compounds to form arylgold(III) complexes at room temperature.[6] Subsequent studies of this chemistry led to the isolation and characterization of several arylgold(III) species generated from the reaction.[7] This auration reaction was shown to proceed in an electrophilic manner. In one case, arylgold(III) species were shown to be stabilized by binding to 2,6-lutidine. Similarly, the reaction of *trans*-K[Au(CN)$_2$Cl$_2$] with 1,3-dimethyluracil[8] (1,3-DimeU) **1a** gives *trans*- K[Au(CN)$_2$Cl(1,3-DimeU$^-$C^5)] **2** which contains the 1.3-dimethyluracil-5-yl entity bound to Au^{III} (Scheme 1).

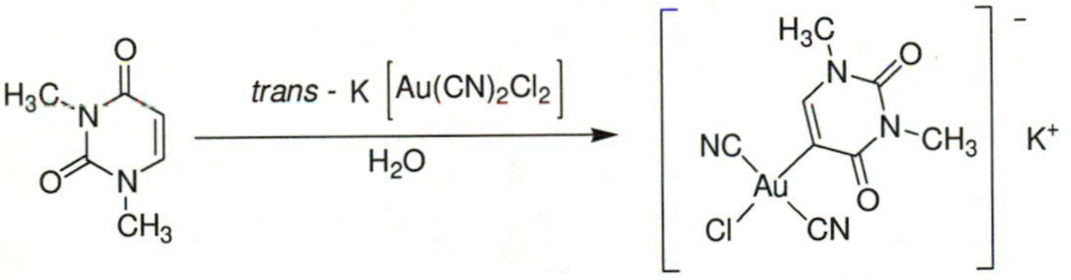

Scheme 1

Complex **2** represents the first example of an X-ray structurally characterized organogold(III) complex of a nucleobase. Interestingly, uracyl bases **1a-b** in the presence of a stoichiometric amount of NaAuCl$_4$·2H$_2$O in water and at room temperature with light excluded allowed the ready formation of C5, C5'-diuracil products **4a-b** (Scheme 2).[9]

It was assumed that the formation of the diuracyl products **4** is the consequence of a reductive elimination process of two *cis*-oriented uracil entities, each bonded to the Au^{III} centre through C5 in the intermediate **3**. Similar reactions are documented for Au^{III} alkyl compounds as well as a compound of 2-(2'-thienyl)pyridine.[10] Formation of the 5-chlorouracyl by-products **5** could occur in analogous manner.

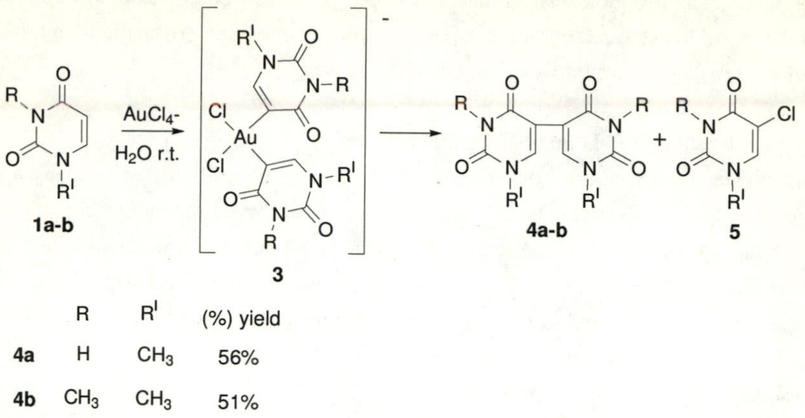
AuCl4-
H2O r.t.
1a-b
3
4a-b
5
R R' (%) yield
4a H CH3 56%
4b CH3 CH3 51%

Scheme 2

He and Shi reported an efficient hydroarylation reaction of alkynes and alkenes.[11] The reactions are catalyzed by gold(III) complexes under mild and even solvent-free conditions at ambient temperature. This method was used to construct molecules such as coumarins **7** (Scheme 3). Various coumarins were prepared from aryl alkynoates **6** in good to excellent yields. The reaction worked for aryl alkynoates that bear different substituted groups including an electron-withdrawing group, although a higher reaction temperature (70 °C) was required for some of these reactions. It should be noted that similar reactions catalyzed by palladium(II), platinum(II), platinum(IV), and ruthenium(II) have been reported, which may go through different mechanisms.[12]

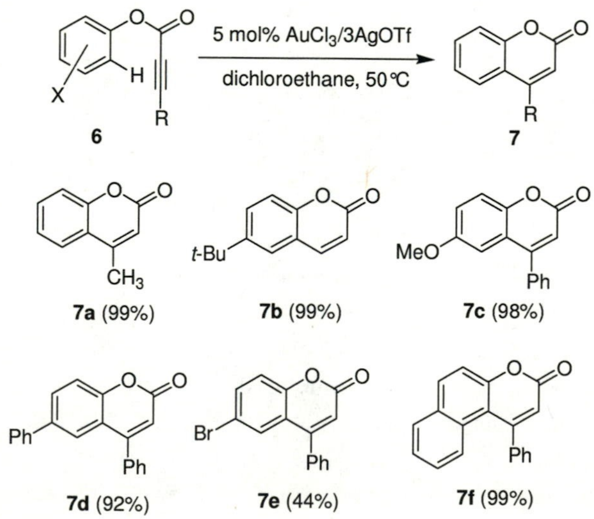
5 mol% AuCl3/3AgOTf
dichloroethane, 50 °C
6
7
7a (99%)
7b (99%)
7c (98%)
7d (92%)
7e (44%)
7f (99%)

Scheme 3

To probe the mechanism, the authors mixed stoichiometric amounts of gold(III) catalyst with pentamethylbenzene. Immediate darkening of the solution colour was observed, accompanied by a disappearance of the aromatic proton signal in the ^{1}H-NMR spectrum. This result may suggest that direct metalation of arene ring might be the first step of the reaction, which was followed by addition of the arylgold(III) to the electrondeficient alkyne substrate. To gain further insight of the mechanism, an isotope experiment was performed on the reaction of the deuterium-labeled mesitylene 8 with 9 (Scheme 4, a).

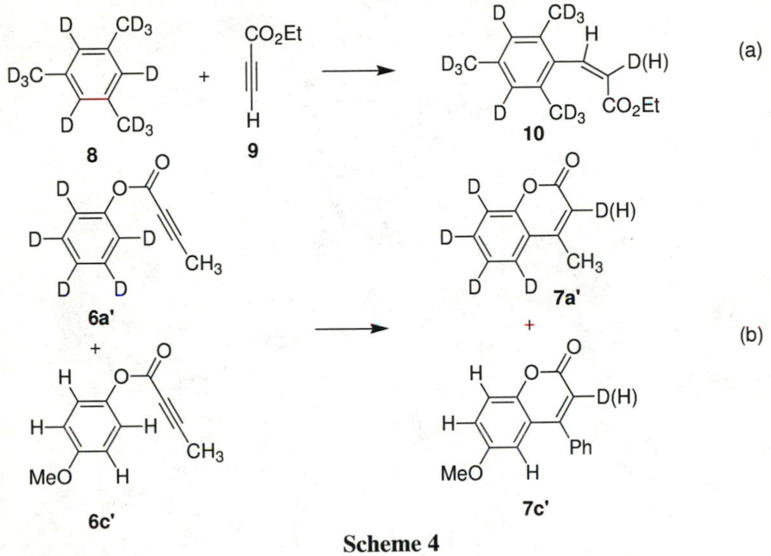

Scheme 4

The result showed that the D (H) incorporated into the vinyl part of the product 10 comes mainly from the aromatic substrate. A reaction with a mixture of two different substrates 6a' and 6c' (Scheme 4, b), one deuterium-labeled and the other proton-substituted, was also performed. Incorporation of nearly equal amounts of H and D into both products 7a' and 7c' was observed after the two independent intramolecular reactions were completed in the same solution. This result suggests that proton generated from the reaction is liberated into the solution before it is incorporated into the final product. The results observed so far seem to support the involvement of a direct metalation of electron-rich arene groups by gold(III) species to form arylgold(III) and 1 equiv of acid. Silver salt might help generate a more electrophilic gold(III) species from AuCl$_3$. AuCl$_3$ also worked but gave low yield product. The formed arylgold(III) species could add to electron-deficient alkynes and alkenes to give the final products. This step could be aided by Lewis acid activation of the alkynes and alkenes by metal ions or acid present in the reaction solution. Although the authors favour this mechanism it cannot be excluded the possibility of gold(III) simply works as Lewis acid to activate alkynes.

He's group also discovered a gold catalyzed alkylation of arenes with epoxides. Treating (phenoxymethyl)oxiranes 11 with AuCl$_3$/3AgOTf (2.5% based on gold) in dichloroethane yielded exclusively *endo* addition product 3-chromanols 12 in good yield (Scheme 5).[13]

The reaction was completed at 50 °C for more electron-rich arene substrates. For less-electron-rich substrates, higher temperature and longer reaction time were required. The reaction tolerated halide substituents, which is useful for further functionalization of the ring. Gold(III) catalyst was absolutely required for the reaction. The yields increased dramatically in the presence of 3 equiv (based on gold) of AgOTf. The silver salt might help to remove the chloride anion from $AuCl_3$ to generate more electrophilic gold(III) species. AgOTf alone did not catalyze the reaction under the same conditions. The reaction showed high diatereoselectivity with trans products obtained exclusively. Substrate **11a** was employed as a probe to examine the stereoselectivity of the reaction. Product **12a** was isolated in 82% yield (Scheme 6).

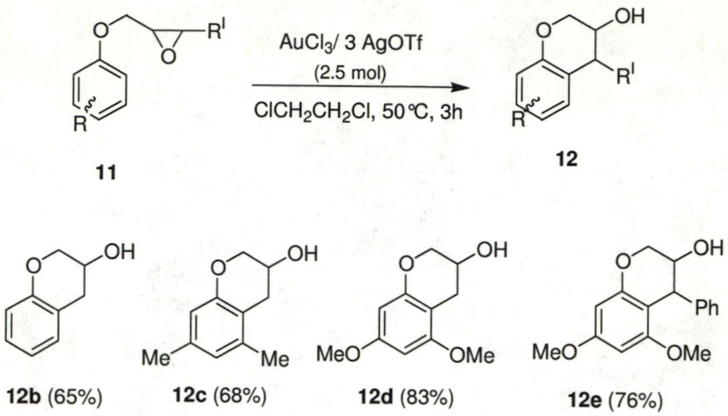

Scheme 5

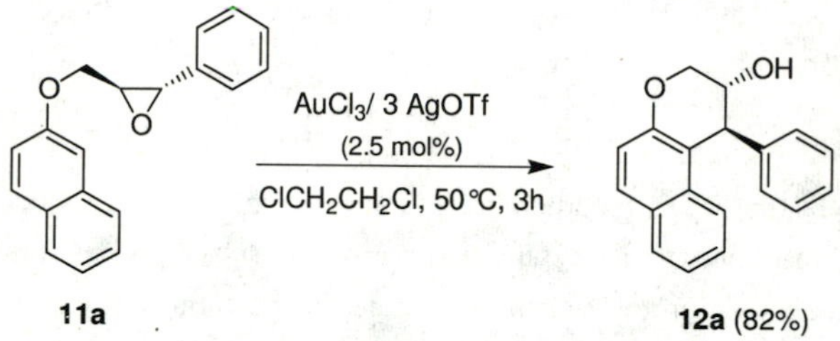

Scheme 6

Its stereochemistry, involving inversion of configuration of the carbon atom linked to the aryl ring, was confirmed by X-ray analysis. The result showed that the cyclization reaction is stereospecific. The stereochemistry of the starting epoxide substrate is relatively easy to control; thus, the method offers a good way to access 3-chromanols stereospecifically.

The intramolecular direct functionalization of arenes **13** and **14** by gold catalyzed primary alcohol sulfonate/triflate esters displacement could lead to various chromans **15** and benzopyranones **16** in good to excellent yields (Scheme 7).[14]

It should be noted that the reaction is sensitive to steric hindrance. This result further argues against a pure Lewis acid mechanism. Previously, it was shown that scandium(III) triflate can catalyze the alkylation of aromatic compounds with methanesulfonates derived from secondary alcohols through a Friedel-Crafts process, which is quite different from the reaction described here. Mechanistic studies indicate the involvement of the arylgold(III) species as the reaction intermediate. This intermediate then reacts with the sulfonate/triflate ester to give the final product.

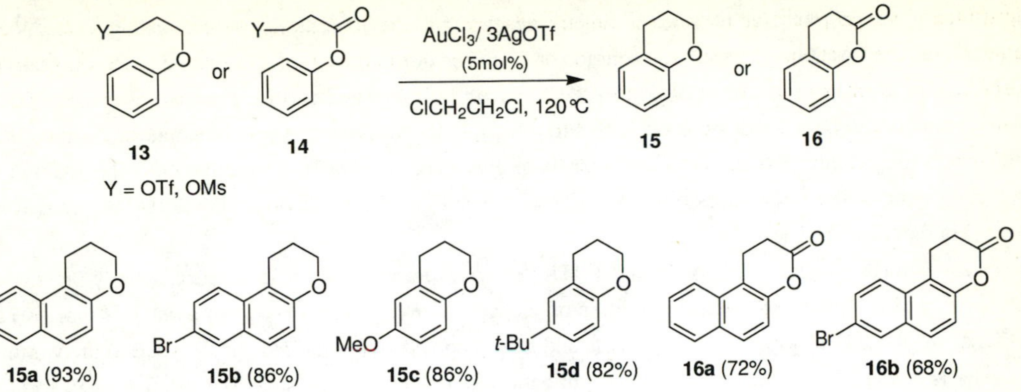
AuCl3/ 3AgOTf
(5mol%)
ClCH2CH2Cl, 120 °C
13
14
Y = OTf, OMs
15
16
15a (93%)
15b (86%)
15c (86%)
15d (82%)
16a (72%)
16b (68%)

Scheme 7

The cyclization of *N*-propargyl *N*-tosylanilines[15] **17a–c** is catalyzed by the cationic complex formed from [Au(PPh_3)Me] and HBF_4 (Scheme 8).

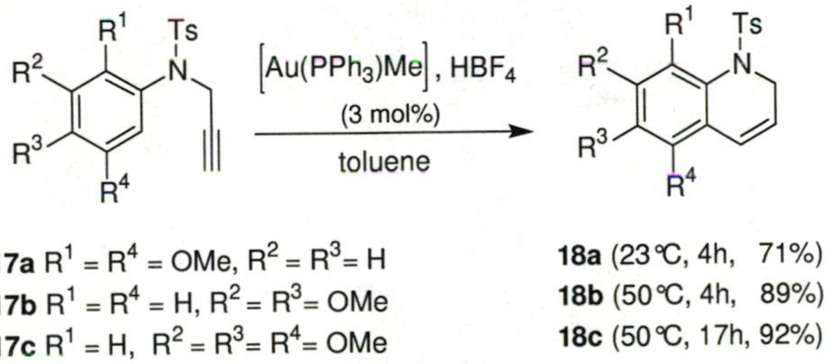
[Au(PPh3)Me], HBF4
(3 mol%)
toluene
17a R1 = R4 = OMe, R2 = R3= H
17b R1 = R4 = H, R2 = R3= OMe
17c R1 = H, R2 = R3= R4= OMe
18a (23 °C, 4h, 71%)
18b (50 °C, 4h, 89%)
18c (50 °C, 17h, 92%)

Scheme 8

This intramolecular reaction gives 1,2-dihydroquinolines **18a–c** and proceeds under milder conditions and with better yields than the cyclization catalyzed by Pt.[16]

Gold-catalyzed C/C bond forming processes under C-H activation of heterocycles are of general preparative interest. Dyker, Hashmi and co-workers[17] found that gold(III) chloride indeed nicely catalyzes the addition of 2-methyl furan **19** to α,β-enones **20**, a clean and fast reaction with a required amount of catalyst of about 1% (Scheme 9).

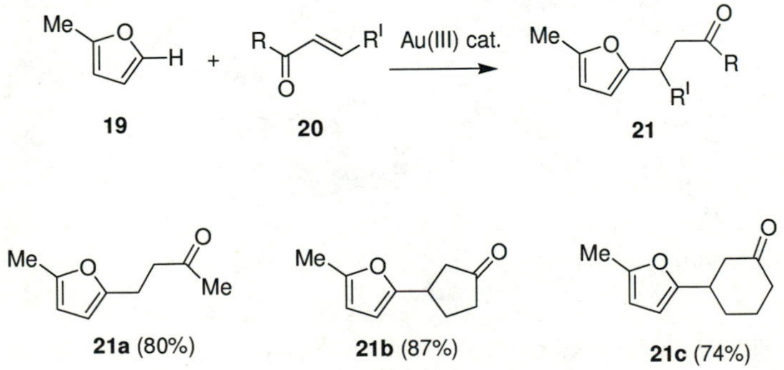
Au(III) cat.
19
20
21
21a (80%)
21b (87%)
21c (74%)

Scheme 9

NMR experiments gave hints for a direct interaction of gold(III) chloride with 2-methylfuran. When one equivalent of AuCl$_3$ was added to a solution of **19** in acetonitrile, an immediate reaction took place. The resonance of the proton at the 5-position of the furan ring disappeared and for the two resonances of the hydrogen atoms at the 3- and 4-positions only broad signals were observed. Upon standing for a longer time the spectra finally only showed very broad signals of undefined material. On the other hand the addition of one equivalent of gold(III) chloride to an acetonitrile solution of methyl,vinyl ketone (MVK) caused no change in the ^{1}H-NMR spectrum.

The gold(I) complexes PEt$_3$AuCl and THTAuCl (THT=tetrahydrothiophene) were initially inactive in the reaction of **19** with MVK, after the addition of one equivalent of AgBF$_4$ (the latter alone was inactive) an 80% conversion was observed in the first case and a 90% conversion in the second. This led to the working hypothesis that an electrophilic C-H activation might be the crucial reaction step. The auration of **19** forms a furyl-gold species **22**, which subsequently undergoes a 1,4-addition to the Michael system (Scheme 10a). Nevertheless, not only AuCl$_3$ but under appropriate conditions also Brönsted acids with a non-nucleophilic counterion can selectively catalyze the reactions between Michael acceptors and electron-rich aromatic systems by a Friedel-Crafts-type mechanism (Scheme 10b).

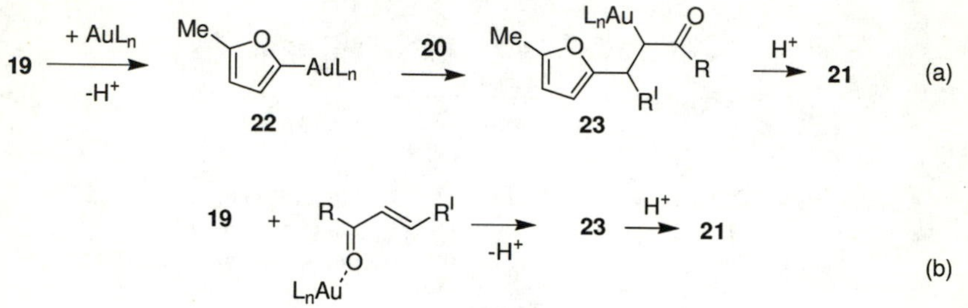

Scheme 10

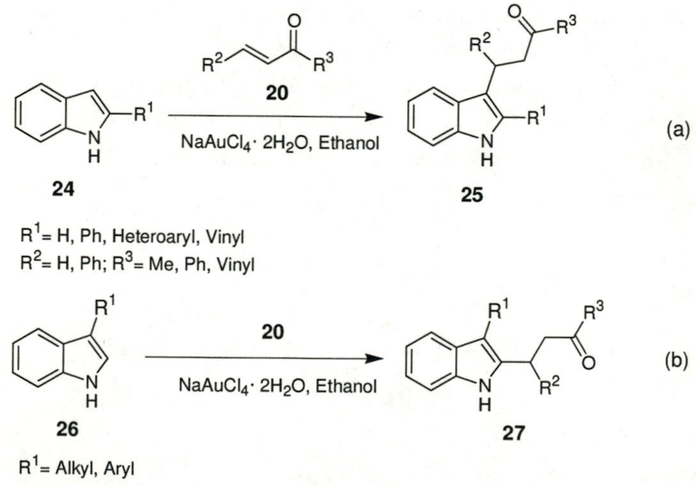

R^1= H, Ph, Heteroaryl, Vinyl
R^2= H, Ph; R^3= Me, Ph, Vinyl

R^1= Alkyl, Aryl

Scheme 11

This suggests that with the gold catalyst in the presence of water a Brönsted acid with an aurate counterion might be the active species, too. On the other hand the different selectivities observed with

Brönsted acids and gold catalysts can either be explained by specific acid catalysis or by a direct participation of the metal.

We have shown that $NaAuCl_4 \cdot 2H_2O$ is an efficient catalyst for the regioselective alkylation at the 3-position of 3-unsubstituted indoles[18] through conjugated addition type reaction with α,β-enones (Scheme 11a).

3-Substituted indoles undergo C-2 alkylation (Scheme 11b). Sequential C-3/C-2 gold catalyzed alkylation can, also, occur (Scheme 12).

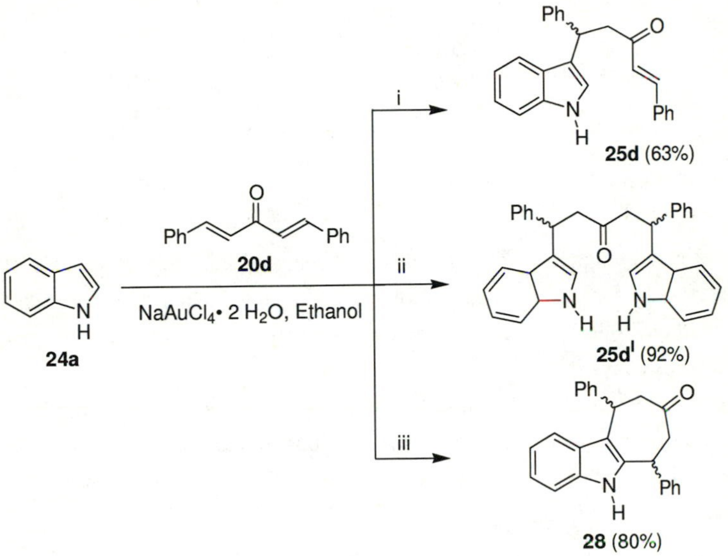

25d (63%)

25dⁱ (92%)

28 (80%)

i : **24a**:**20d**:NaAuCl$_4$· 2 H$_2$O=1:2:0.05; 30 ℃; 2.5h

ii : **24a**:**20d**:NaAuCl$_4$· 2 H$_2$O=1:0.5:0.05; 30 ℃;1.5h

iii :**24a**:**20d**:NaAuCl$_4$· 2 H$_2$O =1:2:0.05; 30 ℃ for 2h and then 60 ℃ for 3h

Scheme 12

Au^I complexes are effective in catalyzing the regioselective hydroheteroarylation of acetylencarboxylic acid ester **9** by furan **29** (Scheme 13).[19]

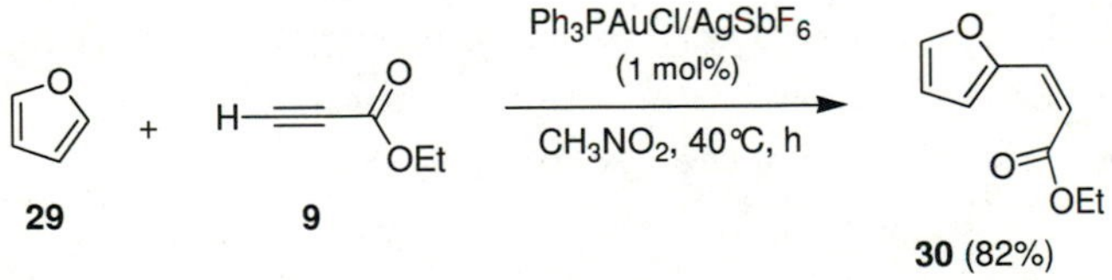

Ph$_3$PAuCl/AgSbF$_6$
(1 mol%)

CH$_3$NO$_2$, 40 ℃, h

29   **9**

30 (82%)

Scheme 13

Typically, Et$_3$PAuCl and Ph$_3$PAuCl function well, nitromethane and 1,2-dichloroethane being the best solvents. Activation by a co-catalyst such as AgSbF$_6$ or the cheaper BF$_3$·OEt$_2$ is necessary. High degrees of (Z)-selectivity are observed. Remarkably, the gold(I) complex can be prepared *in situ* simply by mixing

AuCl and equimolar amounts of a phosphorus ligand in nitromethane. In the case of this gold(I) catalyzed reactions of electron poor alkyne **9**, it was believed that the catalytic active species to be a cationic ligand-stabilized gold(I) complex L-Au$^+$ as in the previously reported additions of oxygen nucleophiles to alkynes.[20]

Gold catalysts are very soft and thus *carbophilic* rather than *oxophilic*. On the basis of this assumption a plausible mechanism for the reaction of acetylenecarboxylic acid ester **9** with furan **29** can be formulated as shown in Scheme 14.

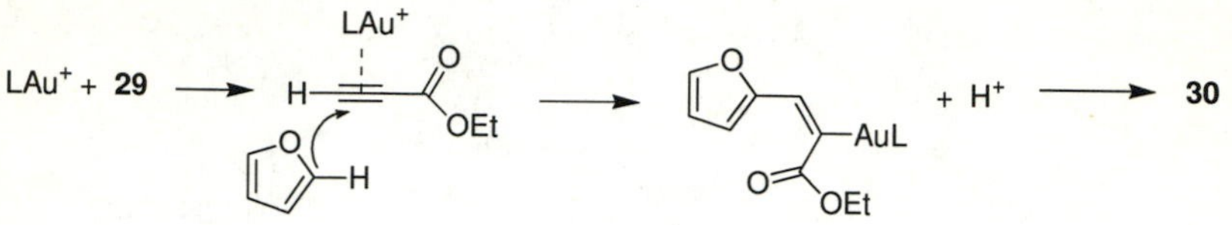

Scheme 14

The cationic gold complex coordinates to the alkyne, and nucleophilic attack of the furan from the opposite side leads to the formation of a vinylgold intermediate which is stereospecifically protonated with final formation of the (Z)-olefin **30**. Regioselectivity is dominated by electronic factors in a type of gold catalyzed Michael addition.

A highly efficient imino Friedel-Crafts type addition of heteroarenes to imines has been developed by using a combination of gold and silver catalyst.[21] Various amino acid derivatives **32** were obtained by reacting both aromatic and heteroaromatic compounds with imine **31** in the presence of a catalytic amount of AuCl$_3$/AgOTf (Scheme 15).

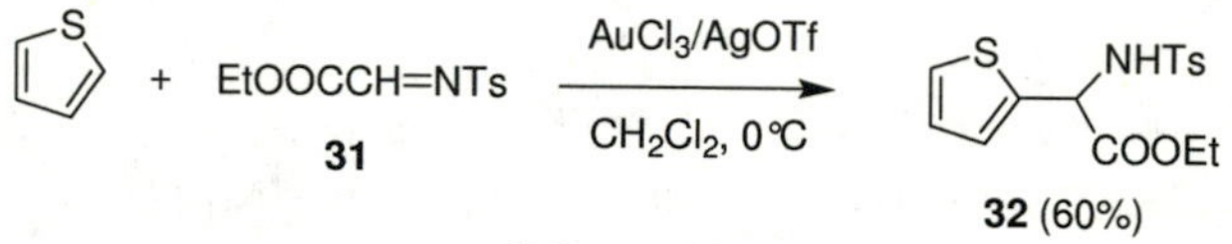

Scheme 15

No reaction was observed by using several potential catalysts such as ZnCl$_2$, AlCl$_3$, and RuCl$_3$·3H$_2$O. Even if either AgOTf or AuCl$_3$ alone can be effective catalysts an improvement of the yield of the desired product was observed when a combination of AuCl$_3$/AgOTf (2/6 mol%) was employed as catalytic system.

C$_{sp}$-H activation by gold has been achieved in water (Scheme 16).[22] A highly efficient three-component coupling of aldehyde, alkynes, and amines (A^3 Coupling) catalyzed by gold in water reaction is general and led to nearly quantitative yields of propargylamines **33** in most cases.

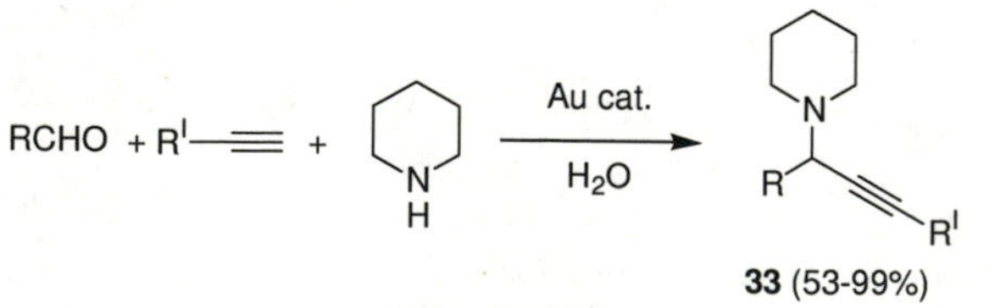

Scheme 16

90

No co-catalyst or activator is required, less than 1 mol % of catalyst is needed, and water is the only by-product for the reaction. Gold salts such as AuCl, AuI, AuBr₃, and AuCl₃ were tested, and all showed excellent catalytic activities. Au(III) salts seemed slightly better than Au(I) salts. Even when only 0.25 mol % AuBr₃ was used as the catalyst in the reaction of three-component coupling of benzaldehyde, piperidine and phenylacetylene the conversion of the reaction still was as high as 99% (the yield, 98%). Even with only 0.01 mol % AuBr₃ used as the catalyst, the reaction still led to 13% conversion during the same reaction period. As AuBr₃ is relatively cheaper among these gold salts and the most effective, it was therefore chosen as the catalyst for other substrates. Gold metal itself did not show any catalytic activity. No reaction was observed in the absence of Au(I) or Au(III) or when the reaction was carried out under an atmosphere of air. The nature of the reaction media significantly affects the reaction. Whereas reactions proceeded cleanly in water, the use of organic solvent such as THF, toluene, and DMF resulted in a low conversion and more by-products. A tentative mechanism was proposed involving the activation of the C-H bond of alkyne by a Au(I) species (for AuBr₃-catalyzed, Au(I) can be generated *in situ* from reduction of Au(III) by the alkyne) (Scheme 17).[23] The gold acetylide intermediate thus generated reacted with the immonium ion generated *in situ* from aldehydes and secondary amines to give the corresponding propargylamine **33** and regenerate the Au(I) catalyst for further reactions.

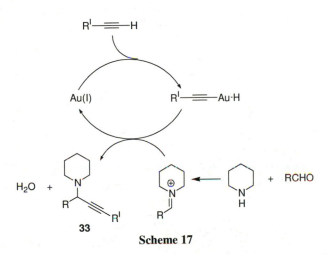

Scheme 17

3. Activation of alkynes and allenes towards nucleophilic attack

Activation of alkynes towards nucleophilic attack is one of the more interesting feature of gold salts. Both the additions of water as well as of amines to alkynes are catalyzed by different inorganic gold(III) and gold(I) complexes.[21, 24] Teles and co-workers reported the addition of alcohols to alkynes by means of cationic gold(I) complexes.[21a] Interestingly, 2-propynol **34** reacts with excess methanol to give the dioxane derivative **35** (Scheme 18).

In this cyclization, the use of extremely small amounts of the methyl(triphenylphosphane)gold(I) catalyst, is particularly impressive. H⁺ is necessary as a co-catalyst, but a Lewis acid such as boron trifluoride can also be utilized because it is rapidly hydrolyzed to trimethyl borate and HF under the reaction conditions. The turnover frequency increases in a roughly linear fashion with the concentration of H⁺.

According to Utimoto and Fukuda[25] intramolecular addition of amine to carbon-carbon triple bonds in 5-alkynylamines **36** produces 2,3,4,5-tetrahydropyridines **37** under the catalytic action of an aurate salt (Scheme 19).

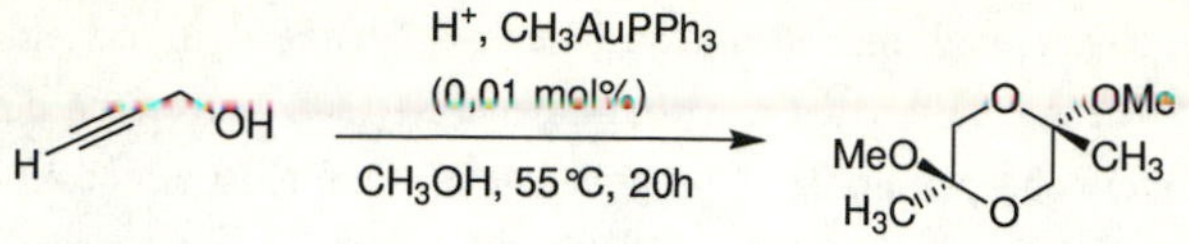

Scheme 18

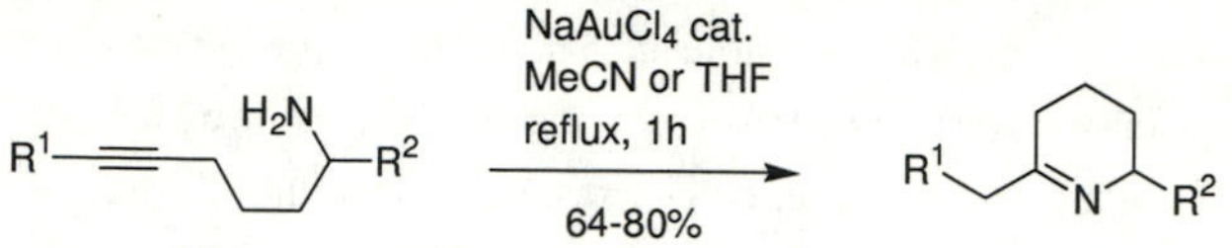

R^1 = H, Et, n-C_5H_{11}, n-C_6H_{13}, Ph
R^2 = H, Me, n-C_6H_{13}

Scheme 19

Yields of the expected products are strongly affected by the concentration of the substrates. Intramolecular reaction with a low concentrations of an alkynylamine affords a cyclized product in excellent yield, whereas the starting material is recovered in considerable amount resulting in the precipitation of metallic gold with a higher concentration of a substrate. 6-Methyl-2-undecyl-2,3,4,5-tetrahydropyridine **37a**, a component present in a venom belonging to the ant species S. Xyloni was synthesized from the corresponding alkynylamine **36a** in good yield as shown in Scheme 20. Analogously, 6-(4-pentenyl)-2,3,4,5-tetrahydropyridine **37b**, a venom alkaloid of Solenopsis (Diplorhoptrum) Puerto Rico species A, was prepared by the gold(III) catalyzed reaction of 9-decen-5-ynylamine **36b**.

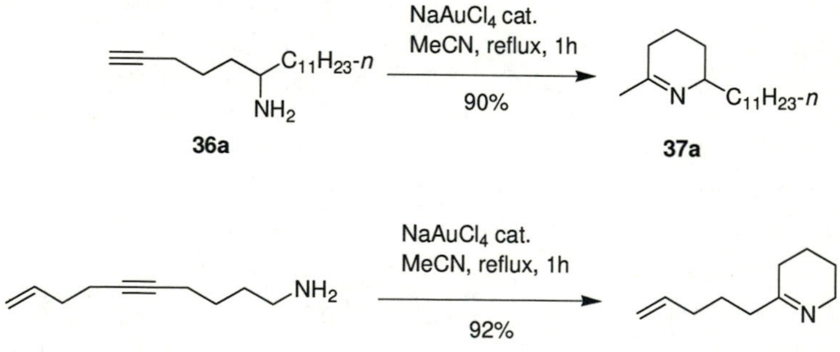

Scheme 20

The above intramolecular amination of alkynylamines **36** can be explained to proceed by the *anti*-addition of nitrogen and gold moieties in 6-*exo-dig* manner to the acetylenic bond to give a vinylaurate species, which affords the 2,3,4,5-tetrahydropyridines after protonolysis and isomerization. However 4-alkynylamines **38** cyclized exclusively in 5-*exo-dig* manner to afford dihydro-3H-pyrroles **39** as shown in the Scheme 21. The fact that any 6-*endo-dig* cyclized products could not be detected indicates that cyclization in 5-*exo-dig* manner proceeds much faster than 6-*endo-dig* type cyclization.

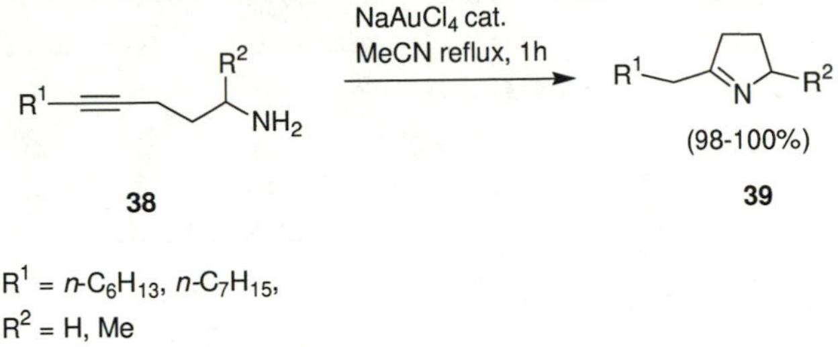
NaAuCl4 cat.
MeCN reflux, 1h
R2
R1
NH2
R1
N
R2
(98-100%)
38
39
R1 = n-C6H13, n-C7H15,
R2 = H, Me

Scheme 21

Since the substituted indole nucleus is found in many natural and synthetic products that display a wide range of interesting biological activity and is currently object of great synthetic efforts, we judged that gold(III) catalysis could improve the environmental compatibility of the preparation of these important heterocycles. One of the most versatile approach to indoles **41** is represented by the cyclization of 2-alkynylaniline or their *N*-substituted derivatives **40** (Scheme 22). Many reaction conditions have been reported. Most of them rely on transition metal or base-catalyzed processes.

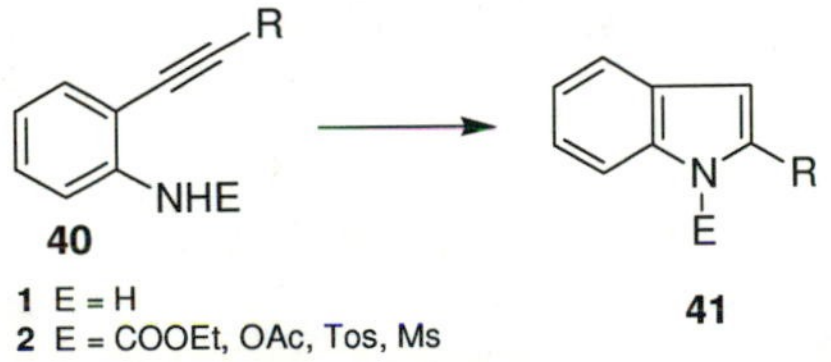
R
NHE
40
1 E = H
2 E = COOEt, OAc, Tos, Ms
N
R
E
41

Entry	Catalyst	Solvent	41a % Yield
1	$NaAuCl_4 \cdot 2H_2O$	Ethanol	83
2	"	Ethanol/Water 95/5 (v/v)	80
3	"	Ethanol/Water 66/33 (v/v)	84
4	"	Ethanol/Water 50/50 (v/v)	82
5	"	Ethanol/Water 33/66 (v/v)	62
6	$KAuCl_4$	Ethanol	76
7	AuCl	Ethanol	50
8	$PtCl_4$	Ethanol	20
9	$NaPdCl_4$	Ethanol	7
10	$PdCl_2$	Ethanol	6
11	$Pd(OAc)_2$	Ethanol	8
12	$Cu(OTf)_2$	Ethanol	10
13	$Cu(OAc)_2$	Ethanol	0

Scheme 22

Our synthetic efforts for a more environmental friendly methodology which avoids the use of toxic solvents, auxiliaries or protecting groups, leads us to select 2-(phenylethynyl)aniline **40a** as a model substrate to attempt the cyclization step at room temperature by using various catalysts (Scheme 23).

NaAuCl$_4$·2H$_2$O was the best catalyst among the gold salts tested and afforded **41a** in similar or higher yields compared with those of the reported procedures; Pd, Pt or Cu catalysts were much less effective than Au salts under the same conditions. Interestingly, the reaction can be also carried out in ethanol/water mixtures (95/5, 33/66 and 50/50 v/v) without loss of efficiency.

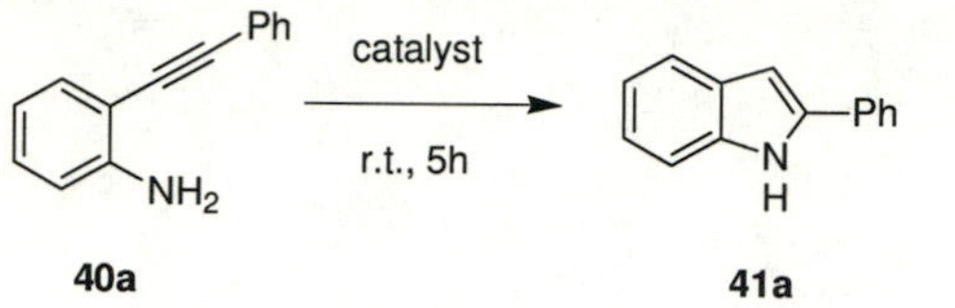

Scheme 23

Then, the process has been extended to various alkynylanilines. The results reported in the Scheme 24 show that NaAuCl$_4$·2H$_2$O in ethanol or ethanol/water mixtures afforded indoles in good to high yields under very mild conditions.[26]

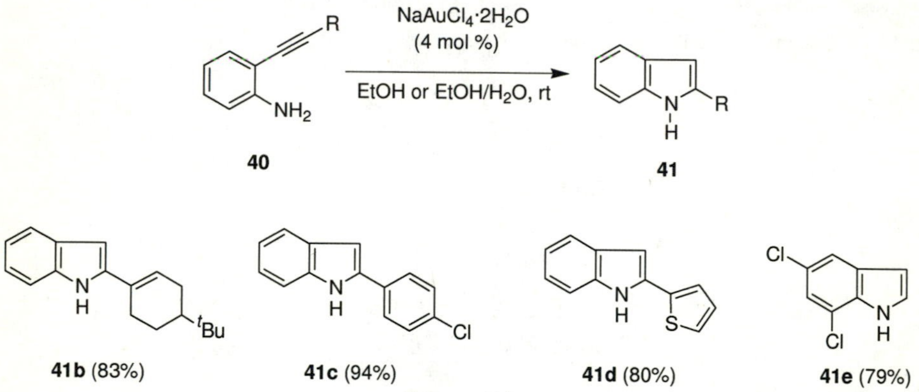

41b (83%)   **41c** (94%)   **41d** (80%)   **41e** (79%)

Scheme 24

The cyclization of *o*-(arylalkynyl)nitrobenzenes **42** was catalyzed by AuBr$_3$ to produce the corresponding isatogens **43** in good to high yields together with small amounts of anthranils **44** (Scheme 25).[27]

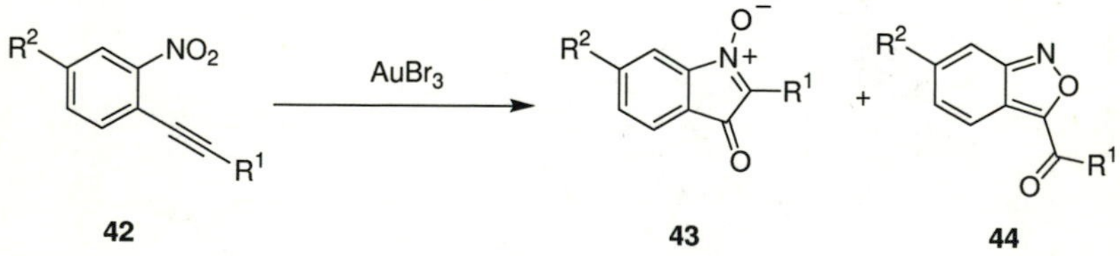

Scheme 25

94

A plausible mechanism for the AuBr$_3$-catalyzed cyclization of **42** is shown in Scheme 26. Coordination of the triple bond of **42** to AuBr$_3$ enhances the electrophilicity of the alkyne, and the subsequent nucleophilic attack of the oxygen of the nitro group to the electron-deficient alkyne forms the intermediate auric ate complex **45**. A trace amount of water in the reaction medium will generate a proton due to the presence of the AuBr$_3$. Protonolysis of **45** forms **46**, which may undergo ring opening on treatment with H$_2$O to produce the nitrosobenzene derivative **48** *via* **47**. The isatogen **43** would then be formed by cyclization/dehydration of **48** (Scheme 26, path a). On the other hand, anthranil **44** would be produced by sequential intramolecular nucleophilic addition of the enol oxygen/dehydration reaction *via* path b of Scheme 26.

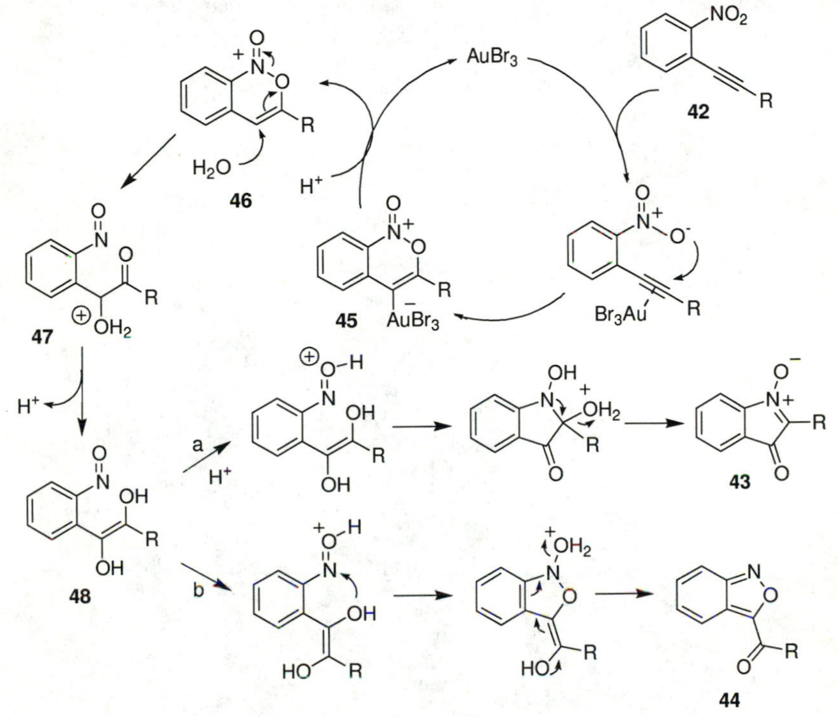

Scheme 26

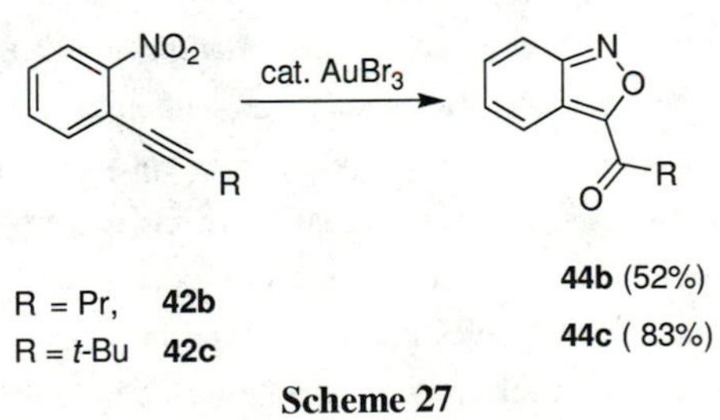

R = Pr,   **42b**
R = *t*-Bu   **42c**

44b (52%)
44c (83%)

Scheme 27

Interestingly, when the reaction was carried out using *o*-(alkylalkynyl)nitrobenzenes **42b-c**, the corresponding isatogen products **43b-c** were not obtained but the corresponding anthranil derivatives **44b-c** were obtained exclusively. Anthranil derivatives are known as useful precursors for acridones and related

95

heterocycles. The reaction of **42b** (R=Pr) afforded **44b** in 52% yield as the sole product. Moreover, the chemical yield of the anthranil increased when R became bulkier; the reaction of **42c** (R=*t*-Bu) afforded **44c** in 83% yield (Scheme 27).

Gold(III) chloride catalyzes the isomerization of alkynyl epoxides **49** to furans **50** under mild conditions (Scheme 30).[28] Compounds **49** were subjected to a catalytic amount of $AuCl_3$ in acetonitrile at room temperature. For the mechanism of the reaction it was assumed that like in many other cases the gold activates the carbon-carbon triple bond for the addition of a nucleophile by coordination as shown in **51**. In this case the epoxide oxygen would attack at the distal position of the alkyne, the gold will form a σ-bond to the other carbon atom of the alkyne and the resulting species **52** only has to loose a proton to form the aromatic furan **53** and finally the product **50** after proto-demetallation (Scheme 28).

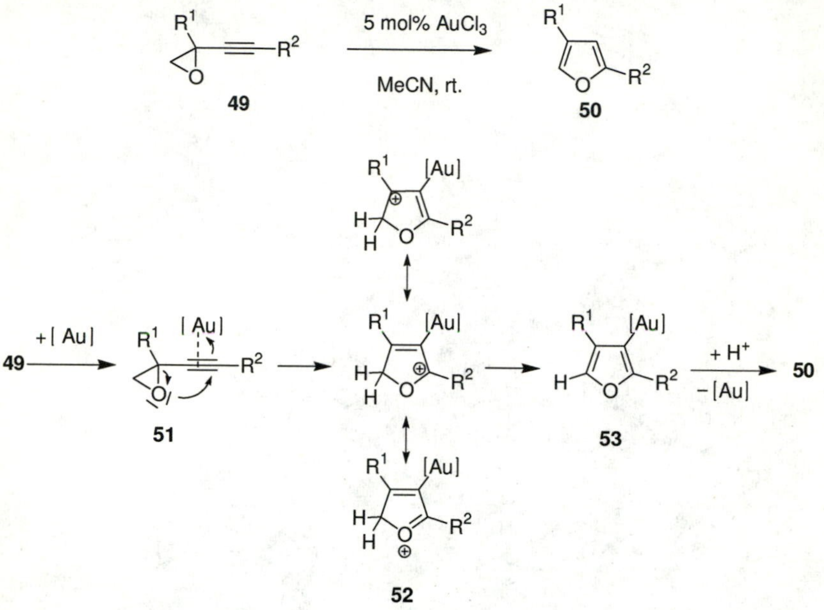

Scheme 28

Similarly, 2,5-disubstituted oxazoles **55** are synthesized from the corresponding propargylcarboxamides **54** under mild reaction conditions *via* homogeneous catalysis by $AuCl_3$ (Scheme 29).[29] The reaction can either be run in dichloromethane at ambient temperature or acetonitrile at 45°C with 5 mol % of catalyst loading to yield the products in good to excellent yields. It is noteworthy that a number of functional groups are tolerated under these conditions; however, the terminal alkyne is crucial for the reaction. A characteristic of the very mild and neutral reaction conditions is the direct observation of the intermediate methylene-3-oxazoline **58**, a species never observed in metal catalysis before with substrates that do not possess a disubstitution that blocks an isomerization to the aromatic heterocycle. This allows a deeper mechanistic insight. Using the deuterated amide it was furthermore observed only one single diastereomer **58** during the conversion showing that the activation of the alkyne **54** and the subsequent addition of the oxygen nucleophile to give **57** is strictly stereospecific. This suggests that the carbonyl-

96

oxygen stereoselectively attacks the π-coordinated alkyne from the backside and the intermediate enol-ether species is then stereospecifically proto-demetalated to **58** (Scheme 30).

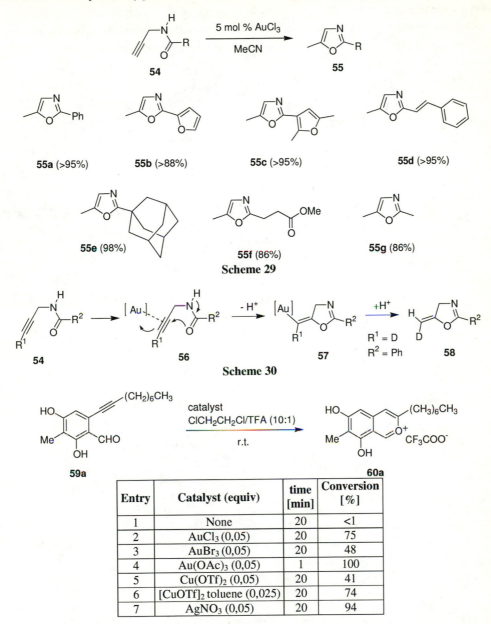

Scheme 29

Scheme 30

Entry	Catalyst (equiv)	time [min]	Conversion [%]
1	None	20	<1
2	AuCl$_3$ (0,05)	20	75
3	AuBr$_3$ (0,05)	20	48
4	Au(OAc)$_3$ (0,05)	1	100
5	Cu(OTf)$_2$ (0,05)	20	41
6	[CuOTf]$_2$ toluene (0,025)	20	74
7	AgNO$_3$ (0,05)	20	94

Scheme 31

Porco Jr. *et al.* investigated cycloisomerization reactions of *o*-alkynylbenzaldehyde **59**. It was envisaged that substrates such as **59a** could be converted directly into 2-benzopyrylium salts **60** in the presence of a catalytic amount of a *carbophilic* Lewis acid and stoichiometric amounts of a proton source. A number of Lewis acid catalysts were investigated for the cycloisomerization. Among these Lewis acids, gold(III) acetate [$Au(OAc)_3$][30] was found to be optimal and led to formation of 2-benzopyrylium salt **60a** in 1 min at room temperature with 1,2-dichloroethane/trifluoroacetic acid (10:1) as the solvent (Scheme 31).

Oxidation of **60a** with *o*-iodoxybenzoic acid (IBX) in 1,2-dichloroethane/TFA cleanly afforded the desired azaphilone **61a** in 84% yield after reductive workup (Scheme 32).[31]

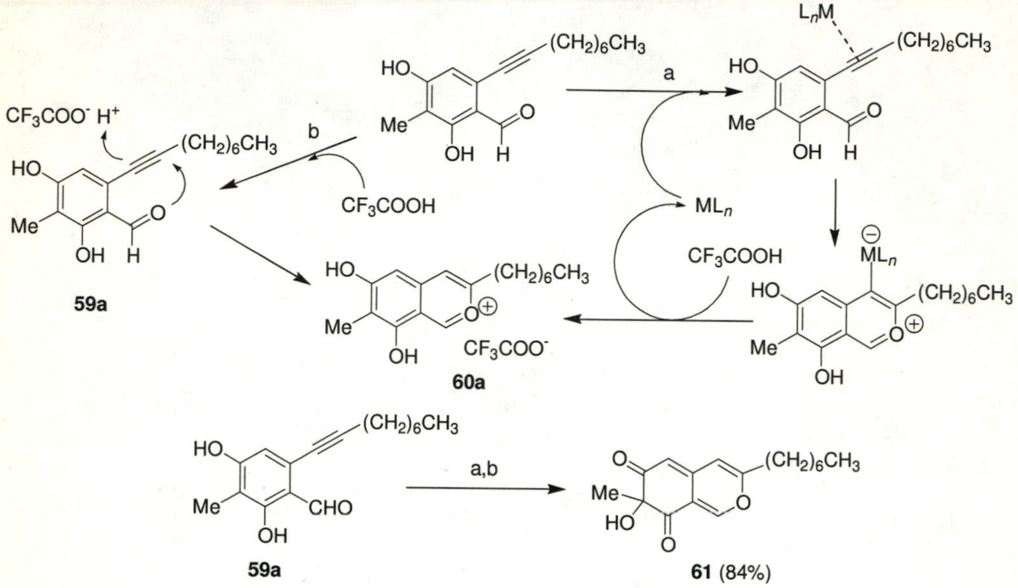

CF3COO- H+
CF3COOH
ML*n*
59a
60a
61 (84%)
a
b
a,b

a) $Au(OAc)_3$ (5 mol%), $ClCH_2CH_2Cl/CF_3COOH$ (10:1), r.t;
b) *o*-iodobenzoyc acid, tetrabutylammonium iodide (5 mol%), r.t., then sat. $Na_2S_2O_3$,

Scheme 32

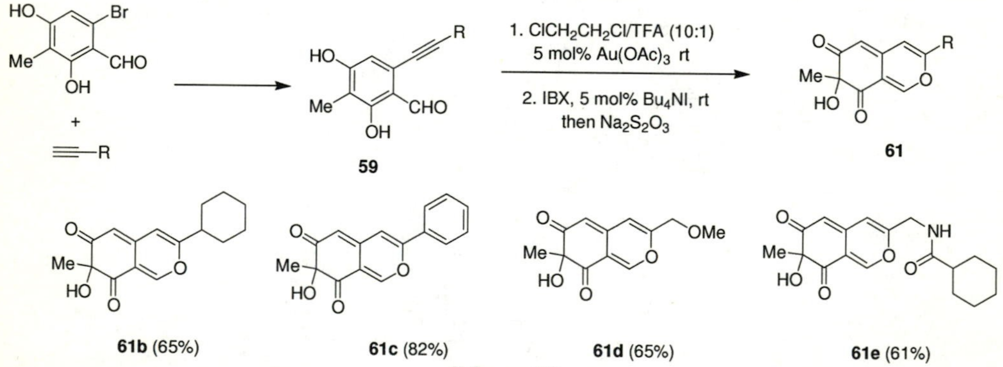

1. ClCH2CH2Cl/TFA (10:1)
5 mol% Au(OAc)3 rt
2. IBX, 5 mol% Bu4NI, rt
then Na2S2O3
59
61
61b (65%)
61c (82%)
61d (65%)
61e (61%)

Scheme 33

The cycloisomerization-oxidation sequence was next applied to the synthesis of several unnatural azaphilones. Sonogashira coupling of 2-bromobenzaldehyde with 1-alkynes in the presence of $P(t\text{-}Bu)_3$ as the ligand cleanly afforded the desired *o*-alkynylbenzaldehydes **59**. $Au(OAc)_3$-catalyzed cyclo-isomerization of the resulting o-alkynylbenzaldehydes **59** in 1,2-dichloroethane/TFA (10:1) at room temperature produced the corresponding 2-benzopyrylium salts **60**. Oxidation of the 2-benzopyrylium salts with IBX in the presence of tetrabutylammonium iodide afforded the corresponding C3-functionalized azaphilones **61b-e** (61–82%). (Scheme 33).

Hashmi *et al.* reported that propargyl ketones such as **62** and allenyl ketones such as **63** can be cyclized to furans such as **64** in reactions catalyzed by gold(III) chloride (Scheme 34).[32]

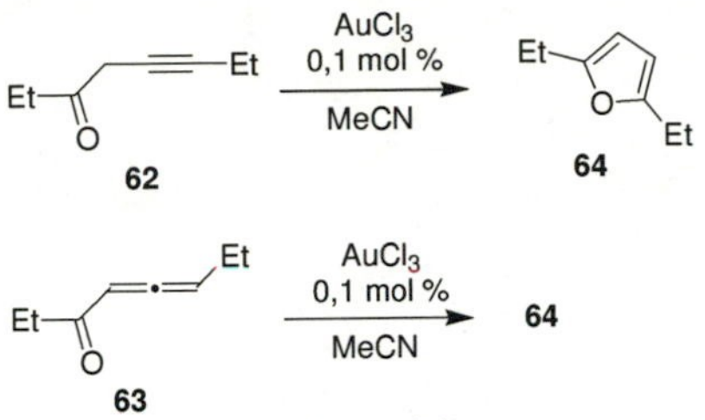

Scheme 34

2,5-Dihydrofurans and derivatives constitute pivotal structural elements of a variety of natural products with intriguing biological activities, *e.g.*, mycotoxins and polyether antibiotics. Their stereoselective preparation may occur through cyclization reaction of allenilic alcohols **65**. 2-Hydroxy-3,4-dienoates **65a-b** can be smoothly converted into the corresponding tri- and tetra-substituted 2,5-dihydrofurans **66a-b** by simple treatment with HCl gas in chloroform or more conveniently by using acidic Amberlyst 15 resin in refluxing dichloromethane. The transformation proceeds under perfect chirality transfer, furnishing the desired heterocycles in good yields and diasteroselectivities of up to 90%. However, acid-labile substrates (*i.e.* silylated hydroxyallenes) or those that would readily undergo elimination reactions to the corresponding vinyl allenes are incompatible with this method. This limitation has been overcome by the use of gold(III) chloride[33] as an electrophile (Scheme 35).

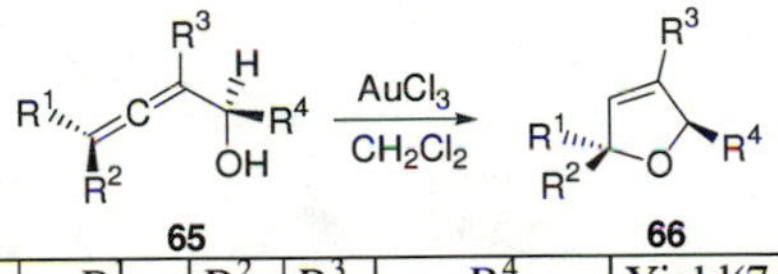

	R^1	R^2	R^3	R^4	Yield(74%)
65a	*t*-But	Me	H	CO$_2$Et	**66a** (74%)
65b	*t*-But	Me	Me	CO$_2$Et	**66b** (94%)
65c	*t*-But	H	Me	CH$_2$OTBS	**66c** (95%)
65d	*t*-But	Me	Me	CH$_2$OMe	**66d** (90%)

Scheme 35

Indeed, under the presence of a catalytic amounts of AuCl$_3$ (5-10%), the ester-substituted 2,5-dihydrofurans **65a-d** have been obtained with good to excellent yields in dry dichloromethane at r.t. As in the acid-induced cyclization, the reaction not only proceeded under complete chirality transfer, but also proved to be of wide scope with regard to the substitution pattern of the allenic entity, furnishing both, tri- and tetra-substituted 2,5-dihydrofurans. This mild and efficient method has been applied to the synthesis of natural product derivatives. In this respect, the functionalised aldehyde citreoviral **69**, which is a key intermediate in the synthesis of two structural related antiparasitic mycotoxins, citreoviridin **67** and verrucosidin **68** represented a promising target molecule (Scheme 36).

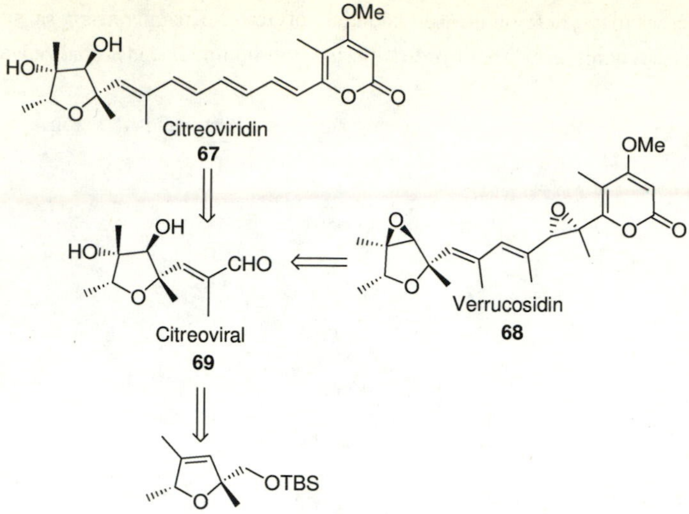

Scheme 36

One of the synthetic approach to the totals synthesis of citreoviridin **67** and its metabolite citreoviral **69** which is far less toxic, relied on the heterogeneous Ag(I) cyclization of hydroxyallene[34] **65e** to the citreoviral precursor **66e** (Scheme 37). There, almost stoichiometric amounts of silver nitrate were required to achieve a good yield of product **66e**.

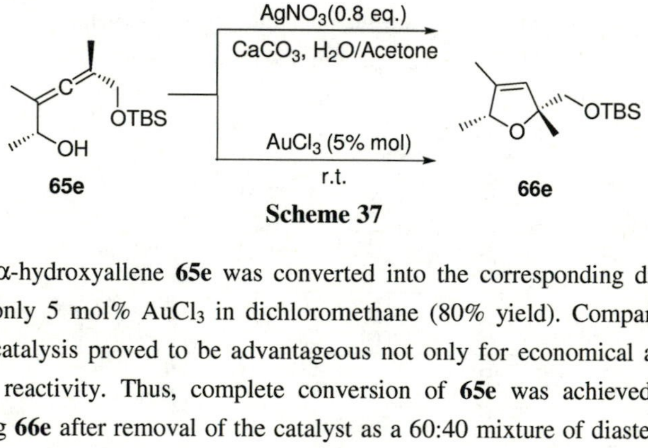

Scheme 37

The TBS-ether α-hydroxyallene **65e** was converted into the corresponding dihydrofuran derivative upon treatment with only 5 mol% $AuCl_3$ in dichloromethane (80% yield). Compared to Ag(I)-promoted cyclization, gold(III)-catalysis proved to be advantageous not only for economical and ecological reasons but also in terms of reactivity. Thus, complete conversion of **65e** was achieved in 3 hours at room temperature, furnishing **66e** after removal of the catalyst as a 60:40 mixture of diastereomers in 80% yield. No aqueous work-up is required: the desired heterocycles are obtained spectroscopically pure just after evaporation of the solvent, followed by short flash column chromatography to remove the metal catalyst.[35]

The gold(III) chloride-catalyzed cycloisomerization of various α-aminoallenes **70** gave the corresponding 3-pyrrolines **71** in good to high chemical yields (Scheme 38).[36] Treatment of the diastereomerically pure α-aminoallenes **70a-e** with 2 mol % of $AuCl_3$ in dry CH_2Cl_2 at room temperature gave the corresponding 3-pyrrolines **71a-e** in good to excellent yields. The reactions were complete after 30 min at room temperature for the *N*-protected aminoallenes **70b-e** (entries 2-7), whereas the unprotected substrate required a much longer reaction time (entry 1). A slight erosion of the stereoselectivity was observed in the cycloisomerization of the sulfonylated substrates **70b-c** which was not affected by the solvent or temperature (entries 2-5). This effect was even more pronounced for the acylated aminoallenes **70d-e** (entries 6 and 7). Only the unprotected aminoallene **70a** afforded the diastereomerically pure 3-pyrroline **71a** (entry 1).

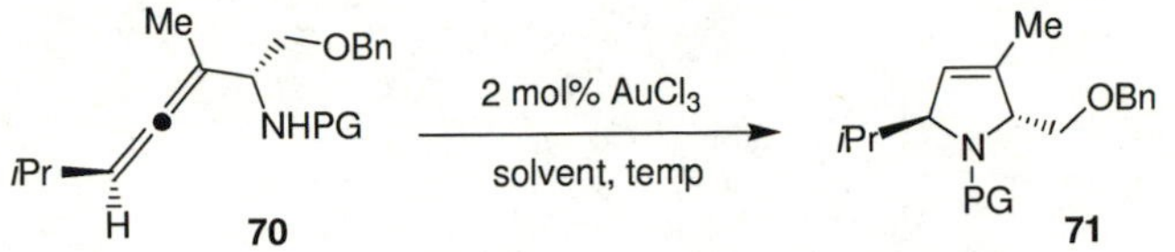

2 mol% AuCl₃, solvent, temp

70 → **71**

Entry	70	PG	solvent	T(°C)	time	71 (yield,%)	dr
1	70e	H	CH₂Cl₂	rt	5 days	71 (74%)	>99:1
2	70b	Ms	CH₂Cl₂	rt	30 min	71 (77%)	94:6
3	70c	Ts	CH₂Cl₂	rt	30 min	71 (93%)	95:5
4	70c	Ts	CH₂Cl₂	0	1h	71 (95%)	96:4
5	70c	Ts	THF	rt	1,5h	71 (95%)	93:7
6	70d	Ac	CH₂Cl₂	rt	30 min	71 (80%)	70:30
7	70e	Boc	CH₂Cl₂	rt	30 min	71 (69%)	46:54

Scheme 38

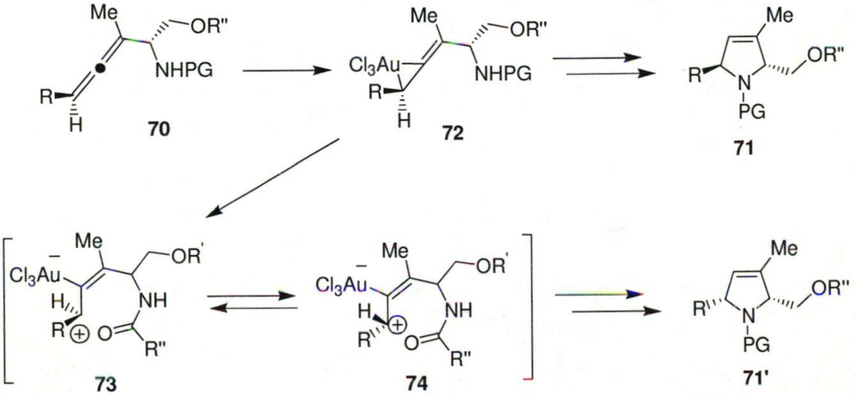

Scheme 39

A plausible mechanism of the gold(III) chloride-catalyzed cycloisomerization of α-aminoallenes is shown in Scheme 39. Thus, coordination of the *carbophilic* gold catalyst to an allenic double bond would be followed by formation of the metallacyclopropane **72**. As a consequence of the increased electrophilicity, cyclization *via* an S_N2-type transition state and subsequent proton transfer would produce the 3-pyrroline with complete axis-to-center chirality transfer. In the case of the *N*-protected aminoallenes **70b-e**, however, an oxygen atom of the protecting group could stabilize the zwitterionic complex **73** by coordination, and the cyclization would proceed with partial isomerization (*via* single bond rotation) to complex **74**, leading to a diminished diastereoselectivity.

A group of synthetically useful transformations of α,ω-enynes are catalyzed by electrophilic transition-metal complexes or halides MX_n to give a variety of heterocycles.[37] Coordination of MX_n to the alkyne forms a (η^2-alkyne)metal complex **75**, which evolves to form the metal cyclopropyl carbene complexes **76**

101

(5-*exo-dig*) or **79** (6-*endo-dig*) (Scheme 40). Then skeletal rearrangement of α,ω-enynes may proceed *via* intermediates **76** (best envisioned *via* canonical form **76'**) to form conjugated dienes **77** or **78**. β-Hydrogen elimination from **79** can give derivative **80** or **81**.

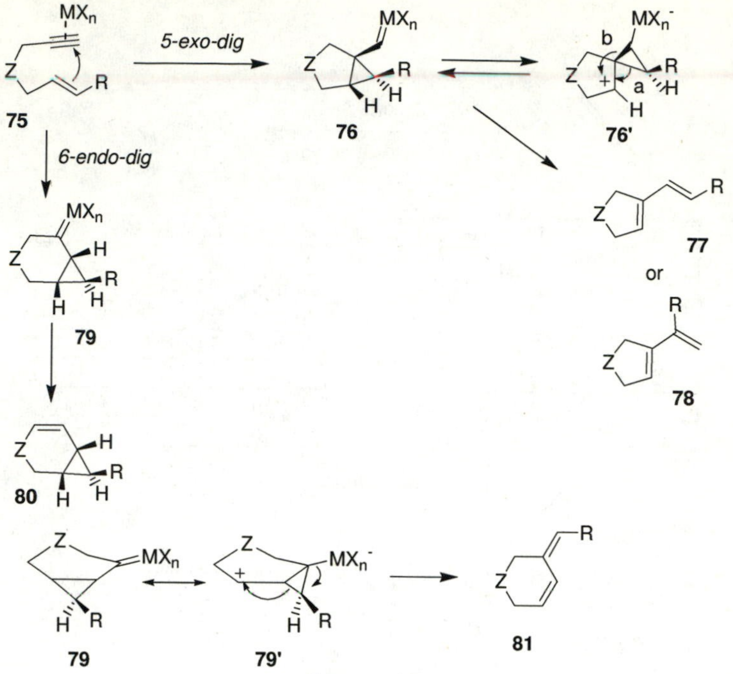

Scheme 40

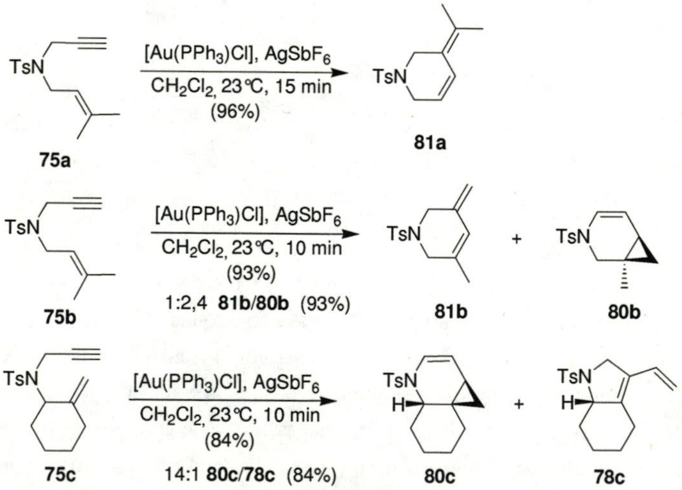

Scheme 41

Interestingly, the endocyclic rearrangement depicted in Scheme 40 was evidenced in the clean gold-catalyzed transformation of enyne **75a** into diene **81a**. Similarly, enyne **75b** gives rearranged **81b**, along with **80b**. In the case of **75c**, the major endocyclic intermediate evolves to give **80c**, along with exocyclic rearranged derivative **78c** (Scheme 41).

Gold(I)-catalyzed 5-*endo-dig* carbocyclization of the indole derivative **82** afforded the tricyclic product **83** (Scheme 42).[38]

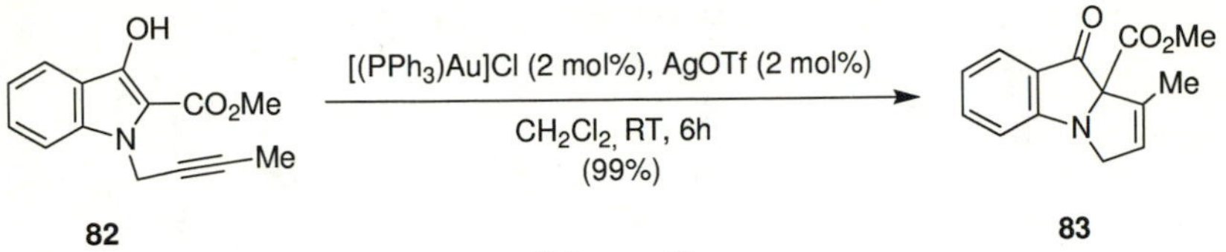

Scheme 42

4. Gold derivatives as Lewis acids

Two different types of molecular catalysts, Lewis acids and transitions metals are becoming important in modern organic reactions.[39] A typical role of Lewis acid MX_n is the formation of a complex with lone pairs. A first landmark in gold catalysis was made by Ito and Hayashi when they reported an elegant synthesis of oxazolines **87** utilizing a gold(I)-catalyzed aldol reaction in the presence of chiral ferrocenylamine ligands that possess both planar and central chirality (Scheme 43).[40]

Mechanistic aspects of this gold(I)-catalyzed aldol reaction were investigated. It was proposed that the terminal amino group of pendant side chain of ferrocenylphosphine ligand **86** participates in the formation of enolate of the isocyanate coordinated with gold and the participation permits a favourable arrangements of enolate and aldehyde on the gold at the stereodifferentiating transition state to bring about high stereoselectivity (Scheme 43, Figure A).[41]

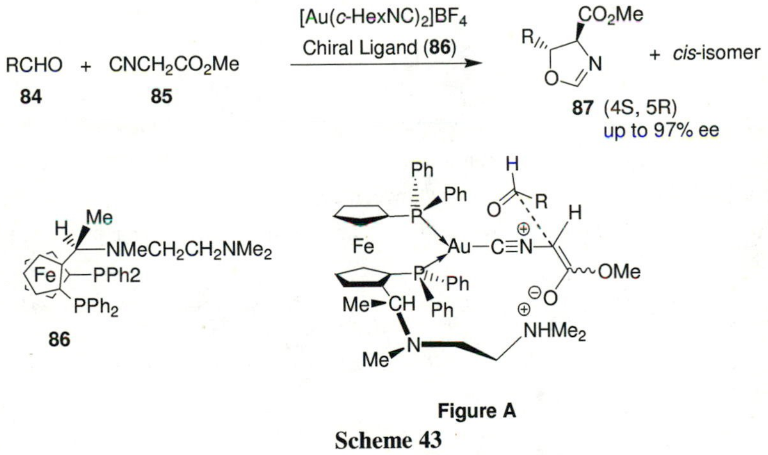

Figure A

Scheme 43

On the basis of the proposed mechanism, appropriate modification of the terminal amino group of the ligand increased the stereoselectivity. The ferrocenylphosphine ligand containing morpholino or piperidino group at the end of ferrocene side chain gave the best results.[42] Aldol reaction of methyl α-isocyanocarboxylates **88** with paraformaldehyde in the presence of 1% of a chiral (aminoalkyl)-ferrocenylphosphine-gold(I) complex **89** gave optically active 4-alkyl-2-oxazoline-4-carboxylates (up to 83% ee) **90a-e** wich were readily hydrolized to α-alkylserines (Scheme 44).[43]

103

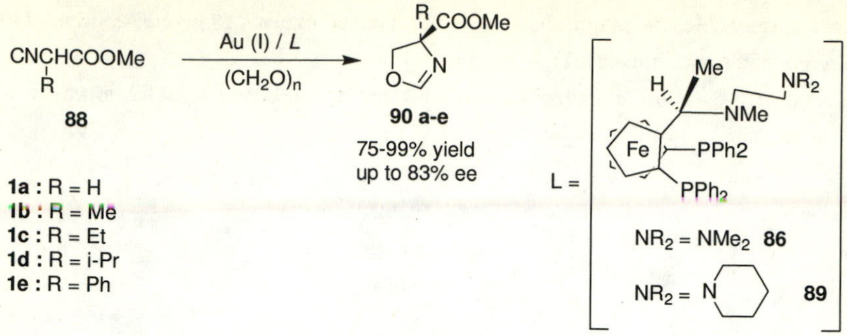

Scheme 44

Asymmetric aldol reaction of methyl α-isocyanoacetate **85** with (E)-2-hexandecenal **91** in the presence of 1 mol% of a chiral (aminoalkyl)ferrocenylphosphine-gold(I) complex **92** afforded optically active trans-4-(methoxycarbonyl)-5-((E)-1-pentadecenyl)-2-oxazoline **93** (93% ee) which was readily converted into D-threo- and erythro-sphingosines (Scheme 45).[44]

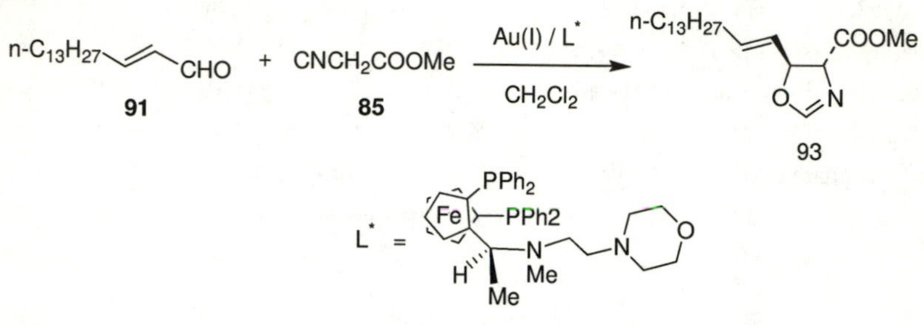

Scheme 45

Another typical role of Lewis acid MX_n is enhancing the reactivity of a substrate by the formation of a complex with lone pairs of C=Y (Y=O, N) multiple bond, facilitating the nucleophilic attack of Nu-H to the carbon bearing a positive charge. In this context, we reported the gold catalyzed condensation of 1,3-dicarbonyl compounds with ammonia, primary and secondary amines to give β-enaminones **94** according to the Scheme 46.[45]

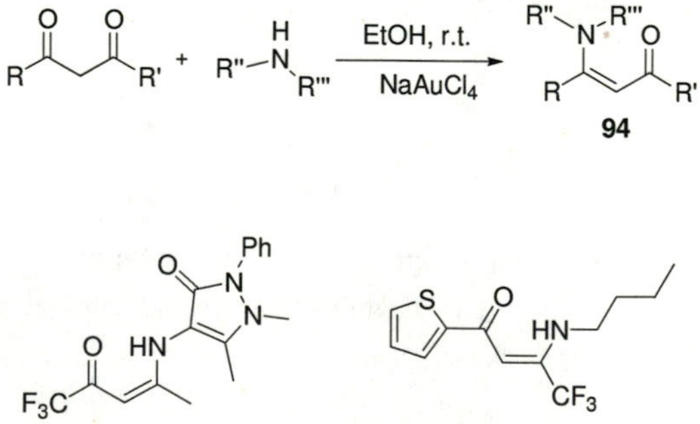

94a 96%   **94b** 95%

Scheme 46

104

Since we were interested in the synthesis of functionalized quinolines, we were driven to investigate the extension of the gold(III) catalysis to the development of a mild version of the Friedländer synthesis of quinolines.[46] Indeed, by choosing as the model system the reaction of the ethyl acetoacetate **95a** with 2-aminobenzophenone **96a** (Scheme 47), we failed to obtain the corresponding quinoline derivative **97a** without adding any catalyst; the starting 2-aminobenzophenone **96a** was recovered (90%) after reacting at 40 °C in ethanol for 6h. By contrast, in the presence of a catalytic amount of NaAuCl$_4$·2H$_2$O the quinoline **97a** was isolated in 83% yield under the same reaction conditions. Furthermore, the reaction can be run very efficiently at r.t. A screening of the efficiency of other transition metal salts revealed that palladium(II)-, copper(I)-, and silver(I) salts were ineffective as catalysts; ZnCl$_2$ can lead to the formation of **97a** (43% yield), even if its efficiency is lower than that of gold(III).

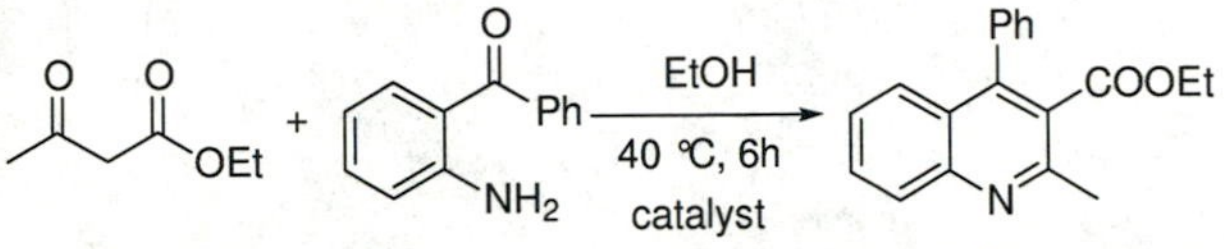

95a + **96a** → **97a**

catalyst	Yield (%)
/	/
NaAuCl$_4$	83%
Na$_2$PdCl$_4$	/
CuI	/
ZnCl$_2$	43
AgNO$_3$	/

Scheme 47

Typically, *o*-amino-substituted aromatic ketones **96** were taken along with a carbonyl compounds **95** and a catalytic quantity of NaAuCl$_4$·2H$_2$O (2.5 mol%) in ethanol at r.t. to give quinoline derivatives **97**. (Scheme 48). When cyclic ketones were used as reagents the reaction was carried out at 60 °C. Then, gold(III) catalysis accomplished a "green" approach to the Friedländer synthesis of quinolines that requires neither harsh conditions nor using hazardous acids or bases.

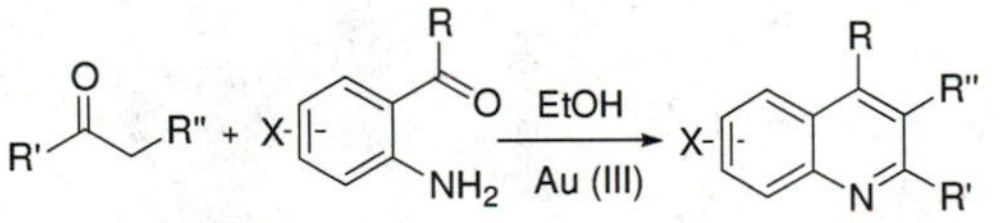

95 + **96** → **97**

R = Ph, Me
X = H, Cl

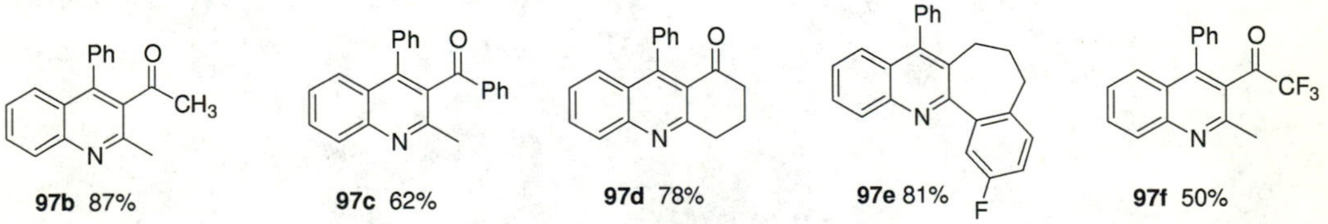

97b 87%   **97c** 62%   **97d** 78%   **97e** 81%   **97f** 50%

Scheme 48

105

5. Dual role Au(III) catalysis

From all possible retrosynthetic schemes of six-membered heterocycles, the best of them require one C-C bond and one C-hetero atom disconnection. It was plausible to suppose that condensation reaction of **98** with carbonyl derivatives **99** could give pyridines **101** if after the formation of a C-N linkage giving *N*-propargylenamine derivatives **100,** the required C-C bond could be formed by a sequential regioselective *6-endo-dig* annulation/aromatization reaction (Scheme 49).

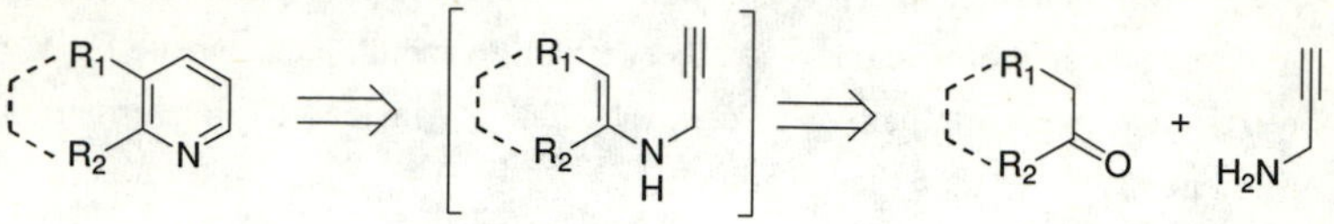

R_1 = H, alkyl, aryl
R_2 = H, alkyl, aryl, heteroaryl

Scheme 49

The *endo* mode of annulation remains relatively unexplored compared to the exo cyclizations. A variety of methods of selective *5-endo-dig* cyclization of 5-en-1-ynes have been developed, but selective *6-endo-dig* cyclizations are scarcely known. We have found that reaction of carbonyl compounds with **98** gives rise to the formation of pyridines **101** through gold-catalyzed sequential amination/*6-endo-dig* annulation/aromatization reaction (Scheme 50).[47]

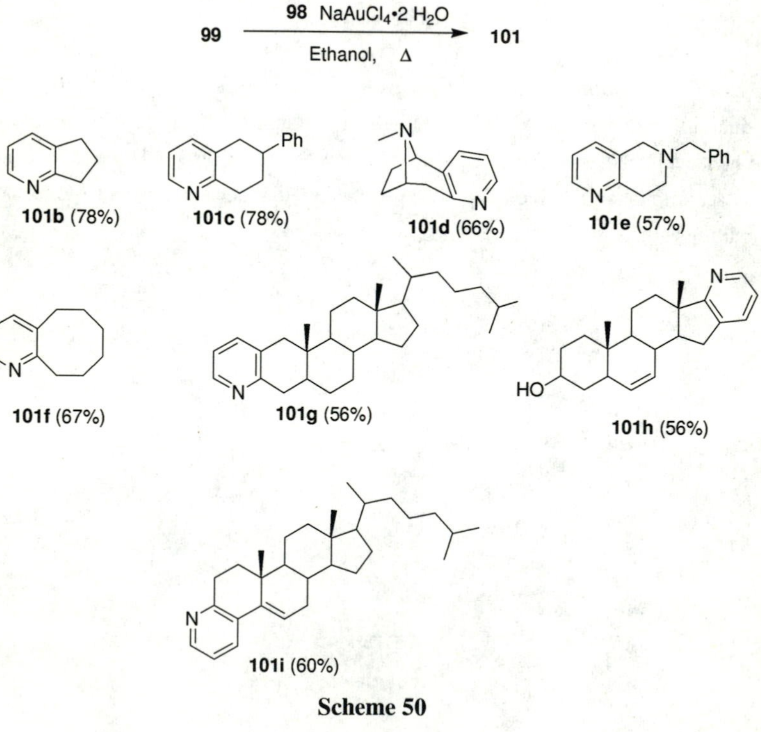

Scheme 50

106

Gold(III) catalysts were very effective and from all the catalyst screened, NaAuCl$_4$·2H$_2$O was the most efficient. Copper salts, were also, effective catalysts. Ir(I) and Rh(I) complexes are known to catalyze the addition of amines to acrylic acid derivatives, but in our cases these complexes were not active. ZnCl$_2$, AlCl$_3$, TiCl$_4$ and other Lewis acids have been reported to be efficient catalysts and water scavengers in the condensation of ketones with amines, but in our hands they showed low activity. A plausible mechanism consists of the followings steps, (1) formation of the imino intermediate **102** through the metal salt catalyzed condensation reaction of the ketone **99** with **98**, (2) imine-enamine isomerization to give **100**, (3) regio-selective *6-endo-dig* intramolecular nucleophilic attack of the carbon of the enamino moiety to the activated carbon-carbon triple bond to give an organometallic intermediate **103**, (5) protonolysis of the C$_{sp2}$-M bond to give the dehydropyridine **104**, and (6) dehydrogenation reactions to afford the pyridine derivative **101** (Scheme 51).

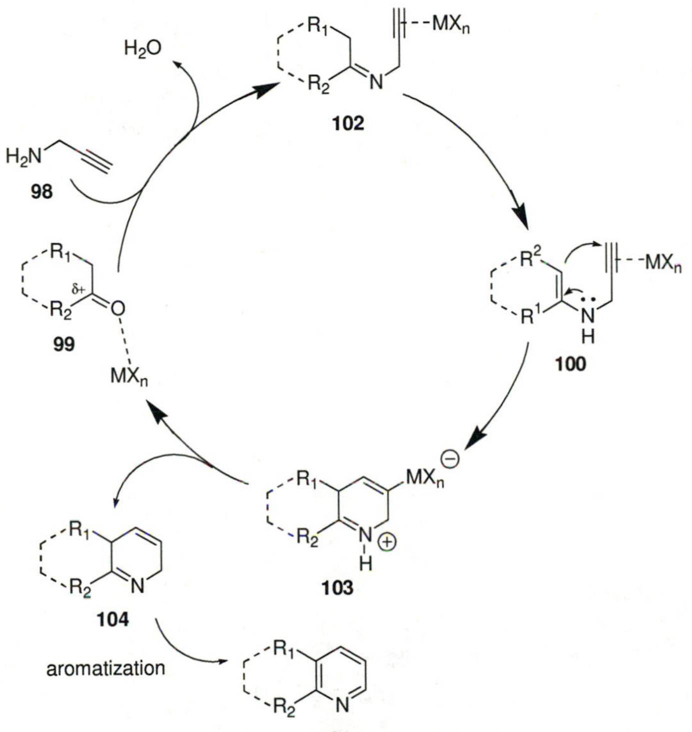

Scheme 51

An effective catalyst of this sequential amination/cyclization reaction should act simultaneously as a Lewis acid and as a transition-metal. MX$_n$ should catalyze the nucleophilic attack of the amino group of **98** to the carbonyl carbon through the formation of a complex with the carbonyl group. MX$_n$ should act as transition-metal catalyst by forming a complex with the alkyne group that facilitates the nucleophilic attack of enamine to the electron-deficient carbon.

In previous studies we observed the high efficiency of gold(III) both as Lewis acid/transition metal catalyst in sequential amination/annulation reactions of α-propargyl dicarbonyls **105** with amines, aminoalcohols and α-aminoesters[48] **106**. We prepared functionalised pyrroles **107** from 2-propargyl-1,3-dicarbonyls **105** using a simpler, and more efficient one-pot procedure compared to the classical stepwise methodology (Scheme 52).

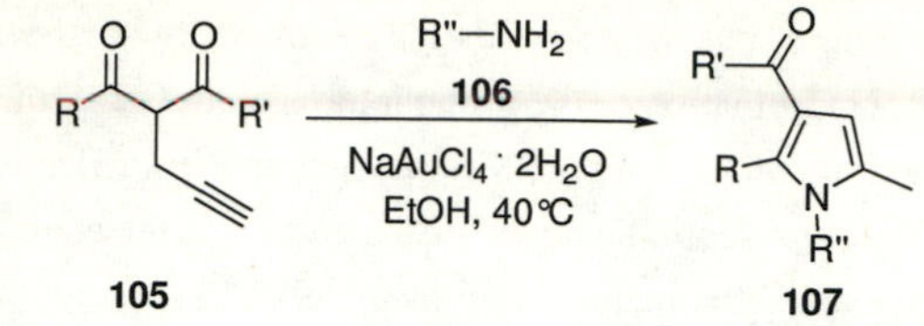

Scheme 52

The reaction of the 2-acetyl-5-hexyn-2-one **105a** with the benzylamine **106a** in dry toluene or ethanol under reflux led to the formation of the enaminone derivative **108a** which was unable of undergoing annulation reaction under the reaction conditions (Scheme 53).

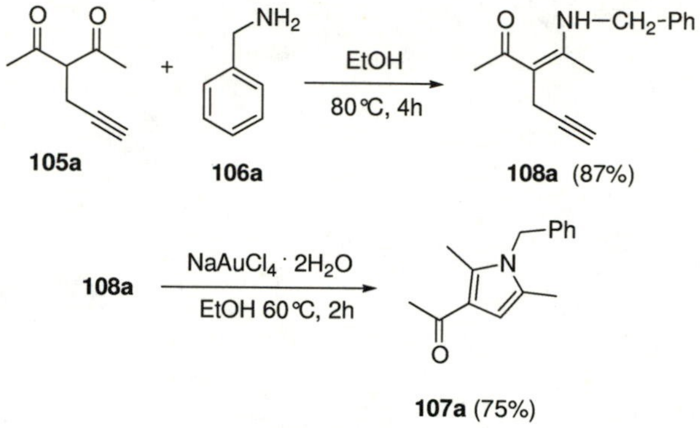

Scheme 53

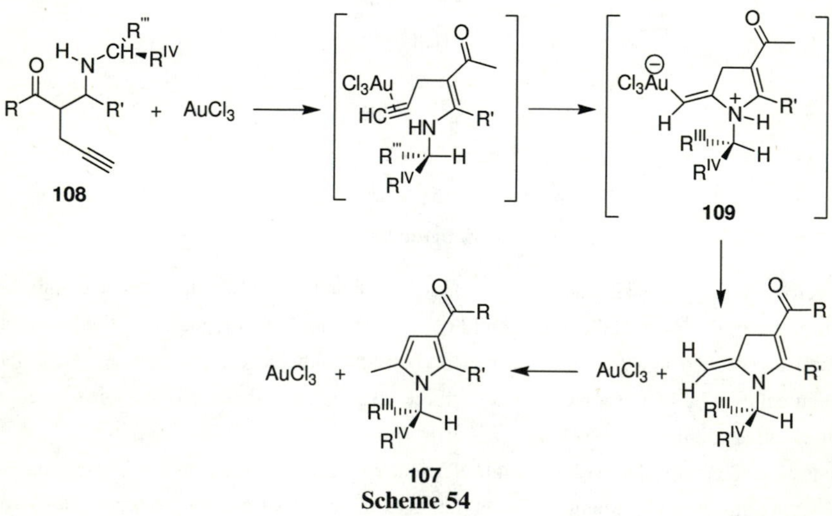

107
Scheme 54

108

Very likely, the *5-exo-dig* annulation reaction of **108a** as well as of a variety of alkynes containing nucleophiles near the alkyne moiety is dependent on intriguing combination of electronic, coordinating, and medium factors. Among them, the strength of the nucleophile appear to play a prominent role. The presence of the electron-withdrawing acetyl group in **108a** implies that the electron density of the NH-Ph group is lower than it would be if there were no –M resonance effects of the β-substituent of the enamine. Consequently, the intramolecular nucleophilic attack of the NH-Ph on the carbon-carbon triple bond was hampered by the weakness of the nucleophile. The regioselective intramolecular addition of enamine to carbon-carbon triple bond in **108a** to give the target **107a** can be easily accomplished under the catalytic action of NaAuCl$_4$·2H$_2$O. The above intramolecular amination of **108a** can be explained to proceed by the *anti*-addition of nitrogen and gold moieties in *5-exo-dig* manner to the acetylenic bond to give a vinylaurate species **109**, which affords the pyrrole **107** after protonolysis and isomerization (Scheme 54).

Moreover, the functionalised pyrrole **107** could be obtained in moderate to quantitative yields in the presence of phosphane-free metal catalysts through a sequential amination/annulation/protonolysis isomerization reaction of 2-propynyl-1,3-dicarbonyl compounds.

When the system was extended to other 2- propargyl-1,3-dicarbonyls and primary amino group the gold(III) catalysed sequential amination/annulation/protonolysis isomerization reaction to give pyrroles **107** turned out to be quite general. Subsequent extension of this procedure to enantiomerically pure amines, β-amino alcohols and α-aminoesters led to the preparation of the target 1,2,5-trisubstituted pyrroles **107** in homochiral form (Scheme 55).[49]

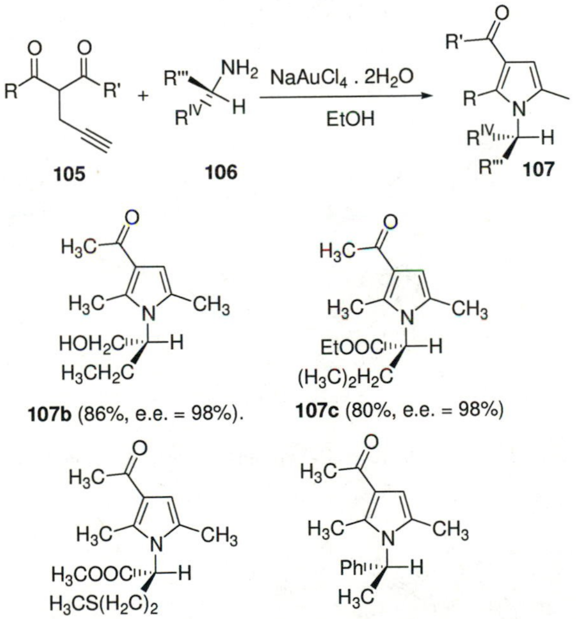

107b (86%, e.e. = 98%).

107c (80%, e.e. = 98%)

107d (74%, e.e. = 99%)

107e (96%, e.e. = 97%)

Scheme 55

Based on the assumption that the stereogenic centre in the starting amine is not affected during this transformation, the absolute stereochemistry of **107** can be assigned. Moreover, it should be pointed out that the enantiomeric excess of the resulting products **107** was always identical with that of the corresponding starting amine derivatives, thus demonstrating the complete stereospecificity of the reaction sequence.

AuCl₃ exhibited again a dual role catalyst for the synthesis of *N*-(alkoxycarbonyl)indoles **111** from 2-alkynyl)phenylisocianates **110** and alcohols (Scheme 56).[50]

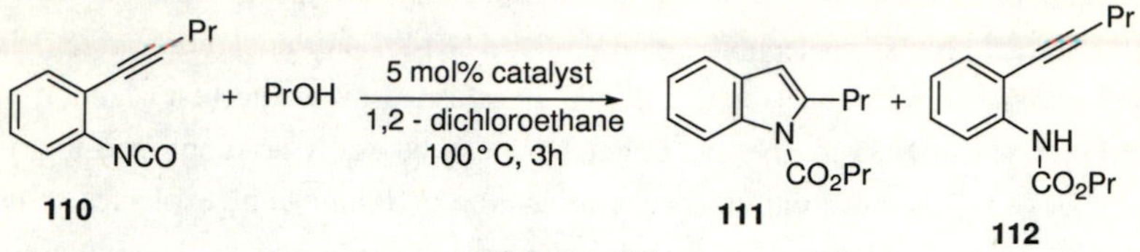

110   **111**   **112**

Scheme 56

Pd(II) or Pt(II) catalyst exhibited also similar behavior: as Lewis acids they accelerate the addition of alcohols to isocyanates and as typical transition-metal catalysts they activated the subsequent cyclization according to the Scheme 57.

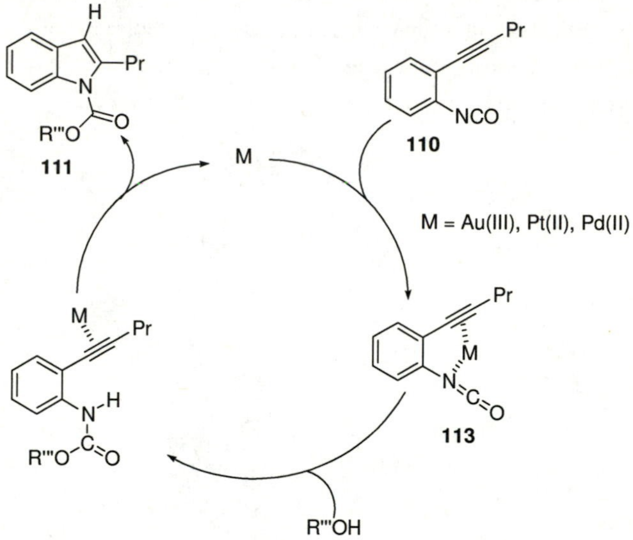

M = Au(III), Pt(II), Pd(II)

111   **114**   **113**

R'''OH

Scheme 57

Larock *et al.* envisioned that 2-(1-alkynyl)-2-alken-1-ones **115** might undergo a gold-catalyzed cyclization to highly substituted furans **116** (Scheme 58).[51]

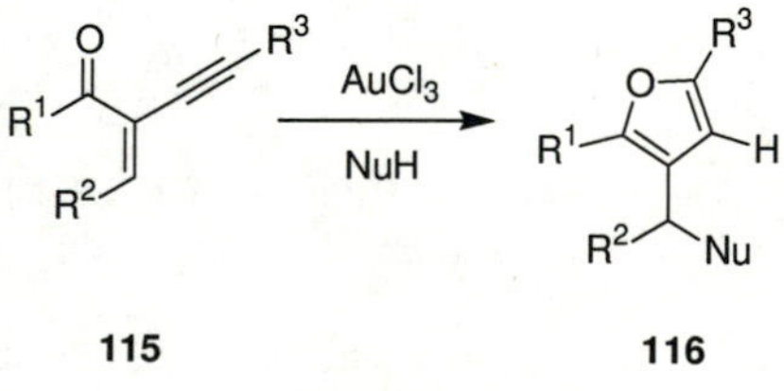

115   **116**

Scheme 58

Overall, ready access to a wide variety of polysubstituted furans **116** is achieved upon reaction of various 2-(1-alkynyl)-2-alken-1-ones **115** with an unprecedented set of nucleophiles under very mild reaction conditions. Alcohols and 1,3-diketones, as well as various electron-rich aromatics, serve as efficient nucleophiles in this process. At least two conceivable mechanisms have been proposed for this gold-catalyzed cyclization (Scheme 59).

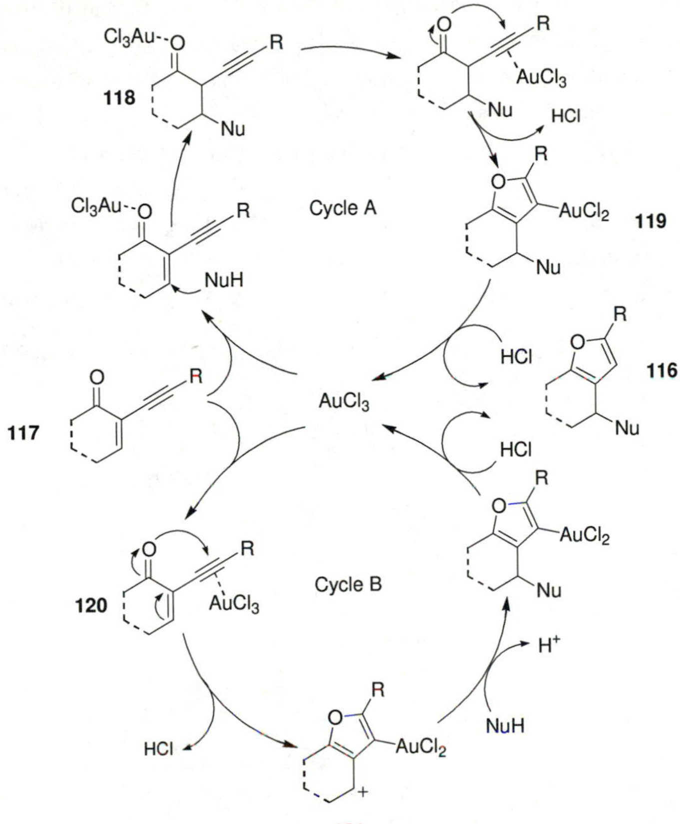

Scheme 59

In one (Scheme 59, cycle A), gold functions as both a Lewis acid and a transition metal. AuCl₃ first acts as a Lewis acid, forming the complex **117**. This facilitates 1,4-addition of the nucleophile to the carbon-carbon double bond to produce **118**. Subsequent coordination of the alkynyl moiety of the alkenynone **118** to AuCl₃ induces a cyclization of the carbonyl oxygen onto the triple bond, followed by elimination of a proton and protonation of the resulting organogold intermediate **119** to afford furan **116** and simultaneously regenerate the catalyst AuCl₃. An alternative mechanism in which AuCl₃ functions simply as a transition metal is also possible (Scheme 59, cycle B).The coordination of the triple bond to AuCl₃ in **120** enhances the electrophilicity of the triple bond, and subsequent nucleophilic attack of the carbonyl oxygen on the electron-deficient triple bond generates carbocation **121**. Intermolecular nucleophilic attack on the carbocation and subsequent protonation of the carbon-gold bond afford furan **116** and regenerate the catalyst AuCl₃.

111

6. Gold-catalyzed domino reactions

The recent trends in organic reaction development are oriented toward synthetic efficiency to alleviate their environmental impact. Homogeneous gold catalysis represents a methodology that perfectly meets the requirements of the design of environmentally friendly procedures for the synthesis of organic compounds. A challenging goal in organic chemistry is to link together in one reaction flask three or more components *via* sequential bond-forming processes. Indeed, it would be much more efficient if one could form several bonds in one sequence without isolating the intermediates, changing the reaction conditions, or adding reagents. It is obvious that this type of reaction would allow the minimisation of waste since compared to stepwise reactions the amount of solvents, reagents and energy would be dramatically decreased. Success would provide rapid and efficient means for transforming simpler molecules into more complex, useful compounds. Gasparrini *et al.* described a one-flask multicomponent gold-catalyzed reaction of (2+2+1) type leading to the formation of 3,5-disubstituted isoxazoles **122**.[52] They find that terminal alkynes react smoothly with nitric acid under biphasic conditions (nitromethane/water, 1:1 v/v) and in the presence of a catalytic amount of tetrabutylammonium tetrachloroaurate (TBA$^+$Cl$^-$) to give **122** (Scheme 60).

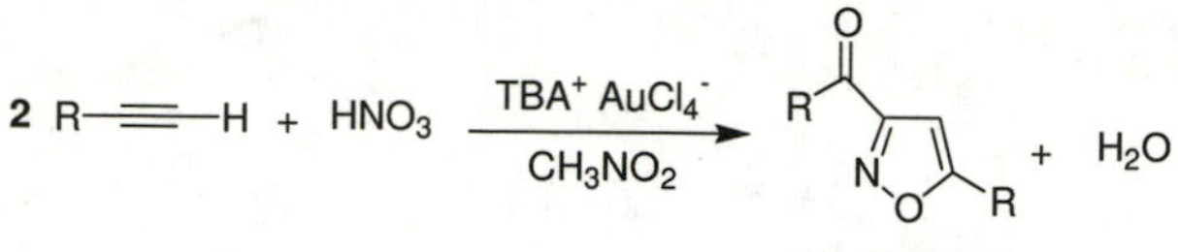

Scheme 60

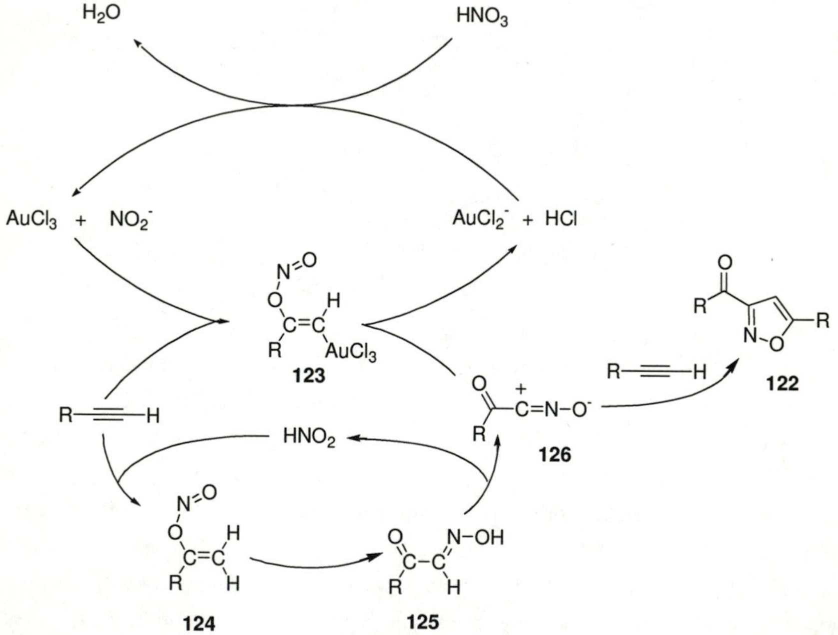

Scheme 61

The yield of the isolated 3,5-disubstituted isoxazoles **122** was in the range 35-50% and there are no apparent limitations to the nature of R. During the reaction, most of the gold is present in the reduced form,

AuCl$_2^-$, indicating that the gold(III) takes part in the redox process, and the rate of reduction by the alkyne is faster than is reoxidation by nitric acid. Most probably, the simultaneous attack on the triple bond of an electrophile (AuCl$_3$ and H$^+$) and of a nucleophile (NO$_2^-$) causes the formation of a vinyl nitrite **123** and **124**, respectively (Scheme 61).

The intermediate **123** can rearrange to the nitrile oxide **126** *via* AuCl$_2^-$ and HCl elimination, whereas **124** can first isomerize to the acyloxime **125** and then be oxidized to the nitrile oxide **126** by nitric acid. The corresponding isoxazole **122** results from 1,3-dipolar cycloaddition of the alkyne to **126**.

Regio- and chemoselective construction of polysubstituted aromatic compounds has been a challenging problem in organic synthesis. In particular the synthesis of highly substituted arenes have been accomplished through gold catalyzed domino processes. As shown for **127**, simultaneous intramolecular C-O and C-C bonds occur to provide the derivative **129** (Scheme 62).[33]

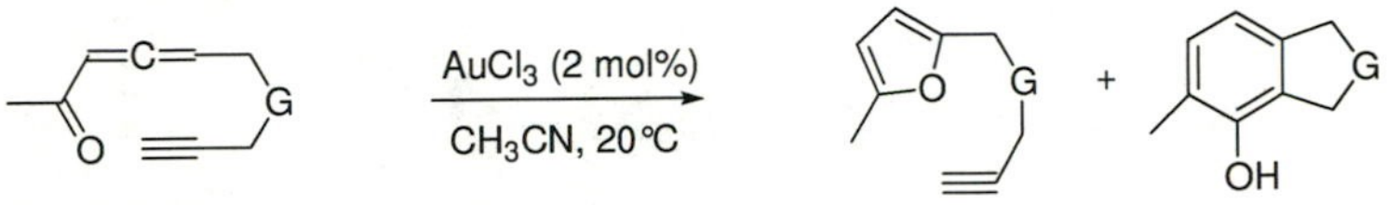

127	G	128(%)	129(%)
127a	CH$_2$	24	35
127b	O	5	52
127c	NTs	55	36

Scheme 62

Further experiments proved that AuCl$_3$ transform cleanly the furan derivative **128** to **129**. Due to their good accessibility, the use of furans derivatives **128** instead of the ketones **129** resulted synthetically more appealing. For a successful reaction a substituent in the 5-position of the furan ring was crucial together with the terminal alkyne moiety. No reaction was observed with AgNO$_3$, Hg(ClO$_4$)$_2$, Pd$_2$dba$_3$·CHCl$_3$, Pd(PPh$_3$)$_4$, Rh(PPh$_3$)$_4$, Rh(PPh$_3$)$_3$Cl, ZnI$_2$, AlCl$_3$, FeCl$_3$, InCl$_3$ and p-TsOH (all in MeCN), heating up to 170 °C in mesitylene for several hours. Gold(I) compounds showed a very low activity. Exclusion of oxygen or humidity was not necessary in these conversions. For gold(III) acetonitrile is the best solvent, in acetone the reaction is slightly faster but less selective, and in DMF even after long reactions times apart from starting material only traces of product could be detected (Scheme 63).[53]

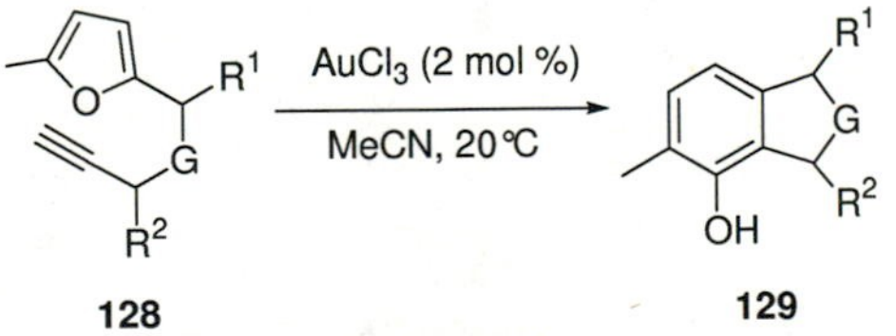

Scheme 63

For the mechanism of the reaction, it was proposed the gold coordination of **128** increase the electrophilicity of the triple bond. Then, an intramolecular Diels-Alder reaction could provide **130**. Usually furans are unreactive diene partners in Diels-Alder reactions.

113

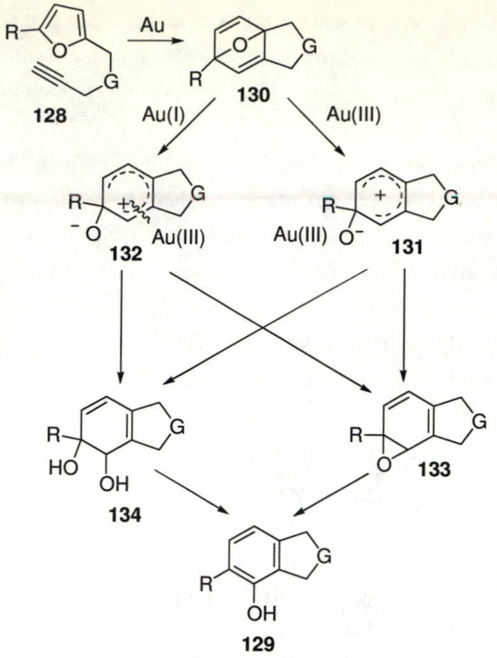

Scheme 64

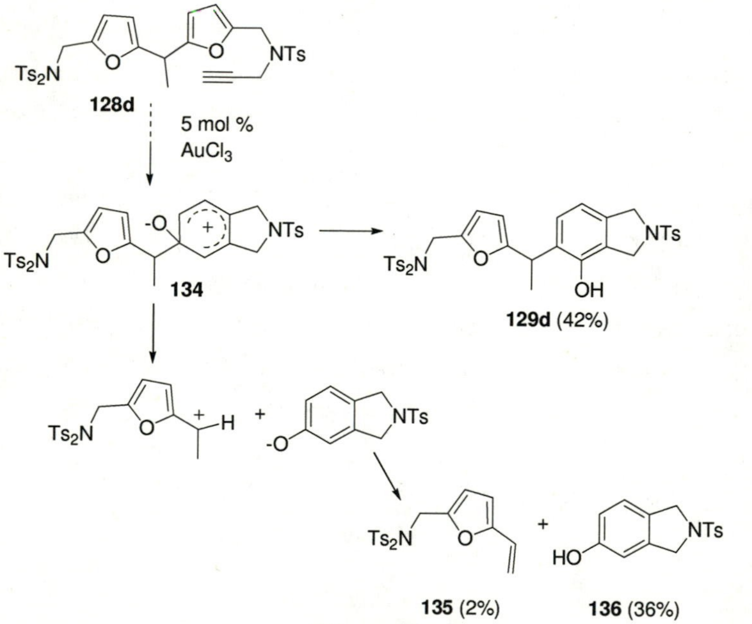

Scheme 65

In the next step the oxygen must be broken, and this could either be induced by gold (III) acting as a Lewis acid (forming **131**) or by Au(I) which forms a Au(III)-stabilized pentadienyl cations **132**. Then either the nucleophilic oxygen atom or external water may serve as nucleophiles and possible intermediates are **133** and **134**. They finally lead to the product **129** (Scheme 64).

The reaction of **128** in the presence of $H_2^{18}O$ revealed not only that the chemoselectivity of the reaction was high (no addition of water to alkyne was observed) but also that oxygen transfer was mainly intramolecular (by MS analysis of the isotope pattern of the molecular ion, less than 1% of incorporation of ^{18}O was detected).[54] When the reaction was run in methanol as solvent, also no incorporation of this stronger nucleophile was observed. This is in accordance with the formation of the epoxide intermediate **133** which gives the final product through a NIH-shift reaction *via* cationic intermediates. Further support for this came from substrate as **128d**, which in addition of the target **129d** led to the highly sensitive **135** (polymerizes readily) and to **136** after a C-C bond cleavage. During this C-C bond cleavage the arenium ion dissociates into the stable arene and a well stabilized furylmethyl cation (Scheme 65).

Turning up to sp^3 and sp 2,5-disubstituted furan derivative **128e** the AuCl$_3$ catalyzed reaction in MeCN gave the alkynyl phenol **129e** and the benzofuran **137**; now even five bonds are broken and five new bonds are formed. The gold catalyzed conversion of **129e** to **137** proved how **137** was formed (Scheme 66).

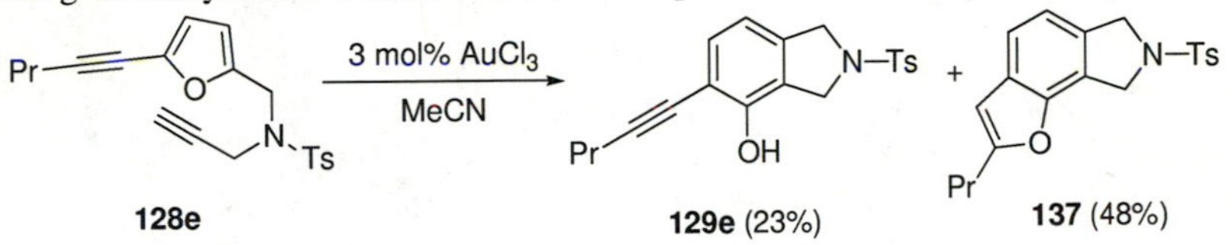

Scheme 66

The reactions of 5-aryl-substituted substrates **128f-g** show that substituents that are conjugated to the π-system of the furan do not cause difficulties and biphenyls **129f-g** can be isolated in good yields (Scheme 67).[55]

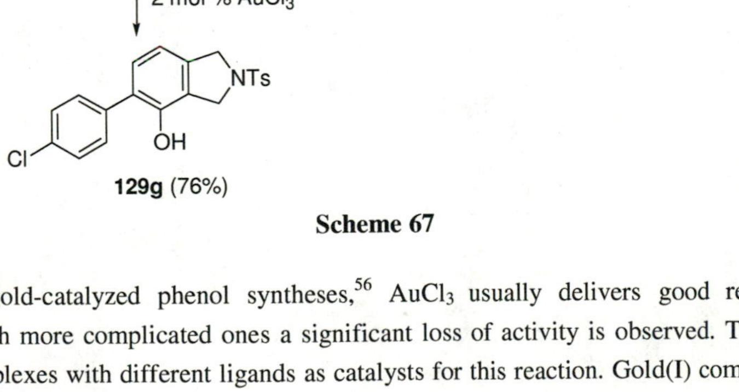

Scheme 67

For these gold-catalyzed phenol syntheses,[56] AuCl$_3$ usually delivers good results with simple substrates, but with more complicated ones a significant loss of activity is observed. Then several gold(I) and gold(III) complexes with different ligands as catalysts for this reaction. Gold(I) complexes showed low selectivity and led to several side products. Satisfactory results in terms of activity, long-term stability and product selectivity were obtained only with gold(III) complexes with pyridine derivatives, some of which contained chelating oxygen functionalities. The major benefit in preparative terms of using these pre-catalysts in this reaction is that they enable the preparation of benzoannelated heterocycles with a phenolic hydroxyl group in a position which is difficult to achieve by employing classic synthetic methodology. The most significant targets are antitumor antibiotics from the tetrahydroisoquinoline family, which with this substitution pattern are not readily accessible by the Pictet–Spengler or related reactions. As reported previously, with 5 mol% or less AuCl$_3$ no complete conversion could be achieved, but **139** accomplished a quantitative conversion of **138** at room temperature.[57] Cationic gold(I) complexes failed to achieve the product selectivity reported here (Scheme 68).

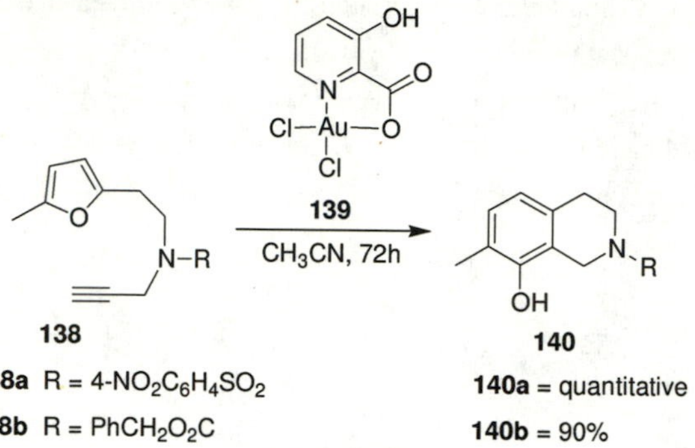

Scheme 68

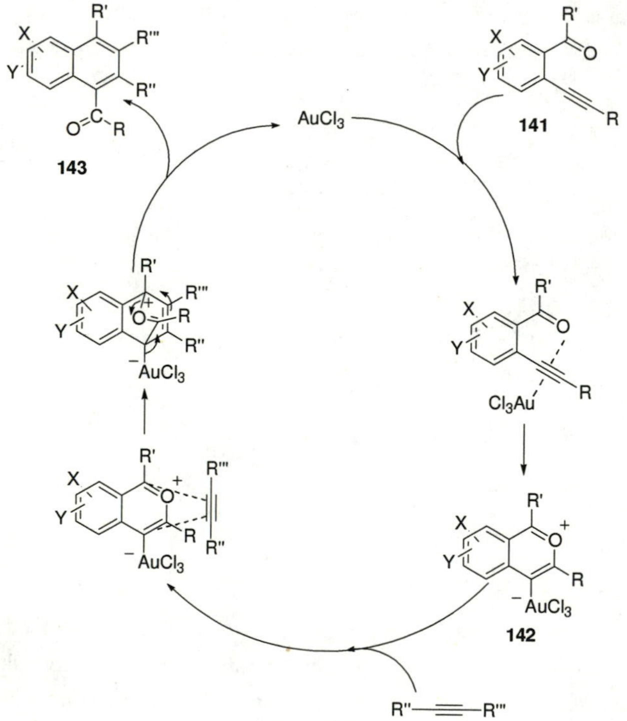

138a R = 4-NO$_2$C$_6$H$_4$SO$_2$
138b R = PhCH$_2$O$_2$C

140a = quantitative
140b = 90%

Scheme 69

116

As a result of its synthetic potential the electrophilic cyclization of phenylacetylenes **141** that have a carbonyl group in the ortho position is currently a popular research topic. These reactions evidently profit energetically from the formation of heteroaromatic isobenzopyrylium salts. Recently, Yamamoto and co-workers[58] found that gold(III) chloride can catalyze a domino process in which the formation of isobenzopyrylium derivative **142** was followed by an intermolecular Diels–Alder reaction with an alkyne functionality to give highly substituted naphthalenes **143**. The mechanism proposed by the Yamamoto group has been confirmed by quantum-chemical calculations (Scheme 69).[59]

Under the assumption that the final aromatization step is not crucial for the efficiency of the domino process, Dyker *et al.* tested electron-rich heteroarenes as especially interesting reaction partners of the *in situ* generated isobenzopyrylium derivative **142**.[60] Their reactions led, through the construction of stereogenic centers, to a wide range of products with remarkable structural diversity. 2,5-Dimethylfuran **144** and benzo[b]furan **146**, both known as electron-rich dienophiles for Diels-Alder reactions with inverse electron demand reacted with 2-phenylethynyl benzaldehyde in the presence of AuCl$_3$ and a small amount of water to afford the fascinating double acetal **145** in the former case and the secondary alcohol **147** in the latter. Benzimidazole **148** trapped the pyrylium cation intermediate to give the N,O-acetal **149** (Scheme 70).

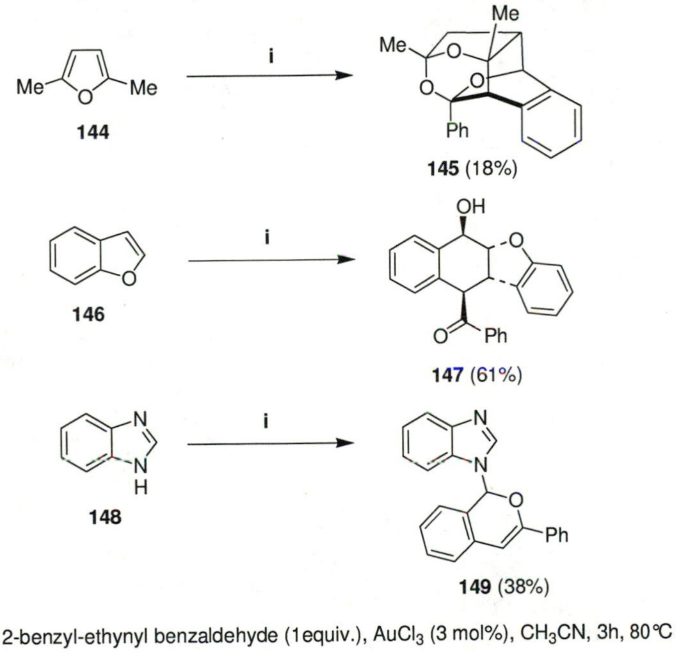

i = 2-benzyl-ethynyl benzaldehyde (1 equiv.), AuCl$_3$ (3 mol%), CH$_3$CN, 3h, 80 °C

Scheme 70

7. Conclusions

Recently, the peculiar features of the high catalytic activity of gold derivatives determining the need for extremely small amount of these catalyst has spurred on growing interest in developing new synthetic methods. Thought to be noble, gold received only little attention in the past, but actually has a rich chemistry. Fascinating in its complexity, this chemistry has found applications in materials science, medicine and, in both heterogeneous and homogeneous catalysis. The mild conditions under which gold catalysts operate have determined the design of more environmentally friendly procedures for the synthesis of organic compounds. Applications of gold catalysis in the synthesis of heterocyclic systems have deserved this review.

Acknowledgments

Work carried out in the framework of the National Project " La Catalisi dei Metalli di Transizione nello Sviluppo di Strategie Sintetiche Innovative di Eterocicli " supported by the Ministero dell'Università e della Ricerca Scientifica e Tecnologica, Rome, and by the University of L'Aquila.

References

1. For reviews of Au-catalyzed reactions see: (a) Hoffmann-Röder, A.; Krause, N. *Org. Biol. Chem.* **2005**, *3*, 387. (b) Arcadi, A.; Di Giuseppe S. *Curr. Org. Chem.* **2004**, *8*, 795. (c) Hashmi, A. S. K. *Gold Bull.* **2003**, *23*, 3. (d) Thompson, D. T: *Appl. Catal. A: General* **2003**, *243*, 201. (e) Dyker, G. *Angew. Chem. Int. Ed.* **2000**, *39*, 4237.

2. (a) Baker, R. T.; Nguyen, P.; Marder, T. B.; Westcott, S. A. *Angew. Chem.* **1995**, *107*, 1451; *Angew. Chem. Int. Ed. Engl.* **1995**, *34*, 1336. (b) Baker, R. T.; Calabrese, J. C.; Westcott, S. A. *J. Organomet. Chem.* **1995**, *498*, 109.

3. Sundermeyer, J.; Jost, C. DE 10041510, 1999 [Chem. Abs. 2001, 134, 280723].

4. Yao, X.; Li, C.-J. *J. Am. Chem. Soc.* **2004**, *126*, 6884.

5. Kennedy-Smith, J. J.; Stanben, S. T.; Toste, F. D. *J. Am. Chem. Soc.* **2004**, *126*, 4526.

6. Kharasch, M. S.; Isbell, H. S. *J. Am. Chem. Soc.* **1931**, *53*, 3053.

7. (a) Liddle, K. S.; Parkin, C. *J. Chem. Soc., Chem. Commun.* **1972**, 26. (b) de Graaf, P. W. J.; Boersma, J.; van der Kerk, G. J. M. *J. Organomet. Chem.* **1976**, *105*, 399. (c) Fuchita, Y.; Utsunomiya, Y.; Yasutake, M. *J. Chem. Soc., Dalton Trans.* **2001**, 2330.

8. Zamora, F.; Zangrando, E.; Furlan, M.; Randaccio. L.; Lippert, B. *J. Organomet. Chem.* **1998**, *552*, 127.

9. Zamora, F.; Amo-Ochoa, P.; Fischer, B.; Schimanski, A.; Lippert, B. *Angew. Chem. Int. Ed.* **1999**, *38*, 2274.

10. (a) Komiya, S.; Albright, T. A.; Hoffmann, R.; Kocki, J. K. *J. Am. Chem. Soc.* **1976**, *98*, 7255. (b) Constable, E. C.; Henney, R. P. G.; Tocher, D. A. *J. Chem. Soc., Dalton Trans.* **1992**, 2467.

11. Shi, Z.; He, C. *J. Org. Chem.* **2004**, *69*, 3669.

12. (a) Jia, C. G.; Piao, D. G.; Oyamada, J. Z.; Lu, W. J.; Kitamura, T.; Fujiwara, Y. *Science* **2000**, *287*, 1992. (b) Fürstner, A.; Mamane, V. *J. Org. Chem.* **2002**, *67*, 6264. (c) Pastine, S. J.; Youn, S. W.; Sames, D. *Org. Lett.* **2003**, *5*, 1055. (d) Youn, S. W.; Pastine, S. J.; Sames, D. *Org. Lett.* **2004**, *6*, 581 and references therein.

13. Shi, Z.; He, C. *J. Am. Chem. Soc.* **2004**, *126*, 5964.

14. Shi, Z.; He, C. *J. Am. Chem. Soc.* **2004**, *126*, 13596.

15. Nieto-Oberhurer, C.; Muñoz, M. P.; Buñuel, E.; Nevado, C.; Cárdenas, D. J.; Echavarren, A. M. *Angew. Chem. Int. Ed.* **2004**, *43*, 2402.

16. (a) MartOn-Matute, B.; Nevado, C.; Cárdenas, D. J.; Echavarren, A. M. *J. Am. Chem. Soc.* **2003**, *125*, 5757. (b) Pastine, S. J.; Youn, S. W.; Sames, D. *Org. Lett.* **2003**, *5*, 1055.

17. Diker, G.; Muth, E.; Hashmi, A. S. K.; Ding, L. *Adv. Synth. Catal.* **2003**, *345*, 1247.

18. Arcadi, A.; Bianchi, G.; Chiarini, M.; D'Anniballe, G.; Marinelli, F. *Synlett* **2004**, 944.

19. Reetz, M. T.; Sommer, K. *Eur. J. Org. Chem.* **2003**, 3485.

20. Gold(I)-catalyzed addition of oxygen nucleophiles to alkynes: (a) Teles, J. H.; Brode, S.; Chabanas, M. *Angew. Chem.* **1998**, *110*, 1475; *Angew. Chem. Int. Ed.* **1998**, *37*, 1415. (b) Tales, J. H.; Schulz, M. BASF AG, WO-A 97/21648, **1997** [*Chem. Abstr.* **1997**, *127*, 121499u]. (c) Mizushima, E.; Sato, K.; Hayashi, T.; Tanaka, M. *Angew. Chem.* **2002**, *114*, 4745; *Angew. Chem. Int. Ed.* **2002**, *41*, 4563.

21. Luo, Y.; Li, C.-J. *J. Chem. Soc. Chem. Commun.* **2004**, 1930.

22. Wei, C.; Li, C.-J. *J. Am. Chem. Soc.* **2003**, 125, 9584.

23. For synthesizing alkynyl gold complexes, see: Vicente, J.; Chicote, M. T.; Abrisqueta, M. D. *J. Chem. Soc., Dalton Trans.* **1995**, 497.

24. (a) Fukuda, Y.; Utimoto, K. *J. Org. Chem.* **1991**, *56*, 3729. (b) Arcadi, A.; Cerichelli, G.; Chiarini, M.; Di Giuseppe, S.; Marinelli, F. *Tetrahedron Lett.* **2000**, *41*, 9195. (c) Raubenheimer, H. G.; Cronje, S *J. Organomet. Chem.* **2001**, *617-618*, 170. (d) Casado, R.; Contel, M.; Laguna, M.; Romero, P.; Sanz, S. *J. Am. Chem. Soc.* **2003**, *125*, 11925. (e) Mizushima, E.; Hayashi, T.; Tanako, M. *Org. Lett.* **2003**, *5*, 3349. (f) Roembke, P., Schmidbaur, H.; Cronje, S.; Raubenheimer, H. *J. Mol. Cat. A: Chemical* **2004**, *212*, 35.

25. Fukuda, Y.; Utimoto, K. *Synthesis* **1991**, 975.

26. Arcadi, A.; Bianchi, G.; Marinelli, F. *Synthesis* **2004**, 610.

27. Asao, N., Sato, K., Yamamoto, Y. *Tetrahedron Lett.* **2003**, *44*, 5675.

28. Hashmi, A. S. K.; Sinha, P. *Adv. Synth. Catal.* **2004**, *346*, 432.

29. Hashmi, A. S. K.; Weyrauch, J. P.; Frey, W.; Bats, J. W. *Org. Lett.* **2004**, *6*, 4391.

30. Seker, E.; Gulari, E. *Appl. Catal. A.* **2002**, *232*, 203.

31. Zu, J.; Germain, A. R.; Porco Jr. J. A. *Angew. Chem. Int. Ed.* **2004**, *43*, 1239.

32. Hashmi, A. S. K.; Schwarz, L.; Choi, J.-H.; Frost, T. A. *Angew. Chem. Int. Ed.* **2000**, *39*, 2285.

33. Hoffmann-Röder, A.; Krause, N. *Org. Lett.* **2001**, *3*, 2537.

34. Marshall, J. A.; Pinney, K. G. *J. Org. Chem.* **1993**, *58*, 7180.

35. Krause, N.; Hoffmann-Röder, A.; Canisius, J. *Synthesis* **2003**, 1759.

36. Morita, N.; Krause, N. *Org. Lett.* **2004**, *6*, 4121.

37. (a) Lloyd-Jones, G. C. *Org. Biomol. Chem.* **2003**, *1*, 215. (b) Aubert, C.; Buisine, O.; Malacria, M. *Chem. Rev.* **2002**, *102*, 813.

38. Staben, S. T.; Kennedy-Smith, J.; Toste, F. D. *Angew. Chem. Int. Ed.* **2004**, *43*, 5350.

39. Asao, N.; Nogami, T.; Takahashi, K.; Yamamoto, Y. *J. Am. Chem. Soc.* **2002**, *124*, 764.

40. Ito, Y.; Sawamura, M.; Hayashi, T. *J. Am. Chem. Soc.* **1986**, *108*, 6405.

41. (a) Sawamura, M.; Ito, Y.; *Tetrahedron Lett.* **1990**, *31*, 2723. (b) Togni, A.; Pastor, S. D. *J. Org. Chem.* **1990**, *55*, 1649.

42. Ito, Y.; Sawamura, M.; Hayashi, T. *Tetrahedron Lett.* **1987**, *28*, 6215.

43. Ito, Y.; Sawamura, M.; Shirakawa, E.; Hayashizaki, K.; Hayashi, T. *Tetrahedron Lett.* **1988**, *29*, 235.

44. Ito, Y.; Sawamura, M.; Hayashi, T. *Tetrahedron Lett.* **1988**, *29*, 239.

45. Arcadi, A.; Bianchi; G., Di Giuseppe, S.; Marinelli, F. *Green Chemistry*, **2003**, 5, 64.

46. Arcadi, A.; Chiarini, M.; Di Giuseppe, S.; Marinelli, F. *Synlett*, **2003**, 203.

47. Abbiati, G.; Arcadi, A.; Bianchi, G.; Di Giuseppe, S.; Marinelli, F.; Rossi, E. *J. Org. Chem.* **2003**, *68*, 6959.

48. Arcadi, A.; Di Giuseppe, S.; Marinelli, F.; Rossi, E. *Adv. Synth. Catal.* **2001**, *343*, 443.

49. Arcadi, A.; Di Giuseppe, S.; Marinelli, F.; Rossi, E. *Tetrahedron: Asymmetry* **2001**, *12*, 2715.

50. Kamijo, S.; Yamamoto, Y. *J. Org. Chem.* **2003**, *68*, 4764.

51. Yao, T.; Zhang, X.; Larock, R. C. *J. Am. Chem. Soc.* **2004**, *126*, 11164.

52. Gasparrini, F.; Giovannoli, M.; Misiti, D.; Natile, G.; Calmieri, G.; Maresca, L. *J. Am. Chem. Soc.* **1993**, *115*, 4401.

53. Hashmi, A. S. K.; Frost, T. M.; Bats, J. W. *J. Am. Chem. Soc.* **2000**, *122*, 11553.

54. Hashmi, A. S. K.; Frost, T. M.; Bats, J. W. *Org. Lett.* **2001**, *3*, 3769.

55. Hashmi, A. S. K.; Frost, T. M.; Bats, J. W. *Catalysis Today* **2002**, *72*, 19.

56. Hashmi, A. S. K.; Ding, L.; Bats, J. W.; Fischer, P., Frey, W. *Chem. Eur. J.* **2003**, *9*, 4339.

57. Hashmi, A. S. K.; Weyrauch, J. P.; Rudolph, M.; Kurpejović, E. *Angew. Chem. Int. Ed.* **2004**, *43*, 6545.

58. Asao, N.; Takahashi, K.; Lee, S.; Kasahara, T.; Yamamoto, Y. *J. Am. Chem. Soc.* **2002**, *124*, 12650.

59. Straub, B. F. *J. Chem. Soc. Chem. Commun.* **2004**, 1726.

60. Dyker, G.; Hildebrandt, D.; Liu, J.; Merz, K. *Angew. Chem. Int. Ed.* **2004**, *43*, 4399.

PYRAZOLINO[60]FULLERENES: SYNTHESIS AND PROPERTIES

Fernando Langa and Frédéric Oswald

Departamento de Química Orgánica, Universidad de Castilla-La Mancha

Campus de la Fábrica de Armas, E-45071-Toledo, Spain (e-mail: Fernando.LPuente@uclm.es)

Abstract. *We describe our contribution to the application of heterocycle formation, by means of 1,3-dipolar cycloaddition reactions, in the preparation of pyrazolino[60]fullerenes. Several advantages that are inherent in the use of nitrile imines, with respect to other procedures for the derivatization of C_{60}, will be discussed. Electrochemical and photochemical properties of this new family of fullerene derivatives suggest that they are good candidates for incorporation into new Donor-Acceptor assemblies.*

Contents

1. Introduction
2. Synthesis
3. Electrochemical properties
   3.1. Influence of the nature of the heterocycle
   3.2. Substitution on the *C*-side of the pyrazoline
   3.3. Substitution on the *N*-side of the pyrazoline
4. Photophysical properties
5. Concluding remarks
Acknowledgments
References

1. Introduction

The dramatic increase in research into fullerenes, particularly C_{60}, since the first multigram preparation in 1990[1] continues today.[2] Several years have been spent developing new procedures for the functionalization of these materials and intense effort is currently focused on the design and synthesis of new fullerene derivatives, particularly in the fields of medicinal chemistry[3,4] and material science.[5-7]

Among the many reactions available for the preparation of fullerene derivatives, cycloaddition reactions – principally 1,3-dipolar cycloadditions – represent a powerful tool due to the fact that C_{60} behaves as an electron-deficient olefin.

The addition of a 1,3-dipole to an alkene to give a five-membered ring is a classical organic reaction. Indeed, 1,3-dipolar cycloaddition reactions are useful for the formation of carbon-carbon bonds and for the preparation of heterocyclic compounds. Molecules that undergo this kind of reaction are called 1,3-dipoles and can be classified into two types depending on their electronic structure. The first type of 1,3-dipole is isoelectronic with the allyl anion. These compounds are characterized by a system of four π-electrons in three parallel atomic pz orbitals. The second type of 1,3-dipole, called the propargyl-allenyl type, contains an additional π orbital orthogonal to the allyl anion-type molecular orbital. In one canonical form, this type of dipole contains a triple bond. Allyl-type dipoles are bent whereas propargyl-allenyl types are linear.

The transition state of the concerted 1,3-dipolar cycloaddition is controlled by the frontier molecular orbitals of the substrates. A classification of the 1,3-dipolar cycloaddition has been performed on the basis of the relative frontier molecular orbital energies between the alkene and the 1,3-dipole (Figure 1). The type I cycloaddition is ruled by an interaction between the HOMO of the dipole and the LUMO of the alkene. In type II additions, the energy level of the frontier orbitals of both dipole and alkene are similar, which makes both interactions between HOMO$_{dipole}$-LUMO$_{alkene}$ and HOMO$_{alkene}$-LUMO$_{dipole}$ important. Type III reactions are based on the interaction between the LUMO of the dipole and the HOMO of the alkene. The rate of the reaction depends on the gap between the controlling orbitals, with smaller gaps associated with faster cycloadditions. Substitution by electron-donating or electron-withdrawing groups on the dipole and/or the alkene may change the relative frontier orbital energies and result in a change in the rate and reaction type.

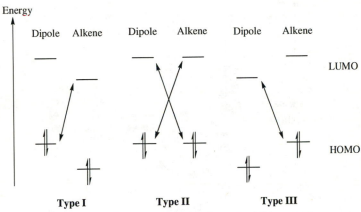

Figure 1. Classification of 1,3-dipolar cycloaddition reactions.

The low value of the LUMO level of C$_{60}$ (–2.88 eV) suggests that type I interactions should be the most favoured approach and, consequently, those dipoles with higher HOMO levels should react more easily with C$_{60}$.

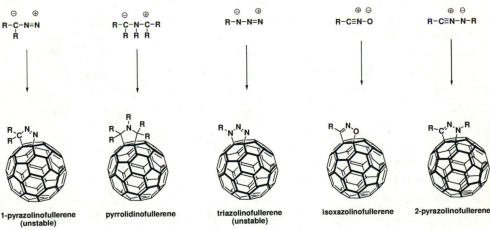

In this way, diazomethanes,[8] azides,[9,10] azomethine ylides[11-16] and nitrile oxides[16-22] have been used for the preparation of fullerene derivatives in attempts to increase the solubility of C$_{60}$ in common solvents, to

121

diminish the aggregation to make materials that are easier to handle or to prepare tailor-made novel fullerene derivatives for specific applications in medicinal or material chemistry.

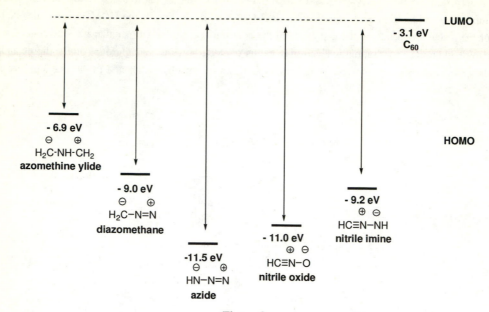

Figure 2

The formation of fulleropyrrolidines using 1,3-dipolar cycloadditions between fullerene and azomethine ylides, which was developed by Prato, has been widely used.[23] Azomethine ylides can be generated *in situ* from a wide variety of readily available starting materials and the HOMO-LUMO gap between the LUMO of C_{60} and the HOMO of the dipole (Figure 2) is appropriate for a facile reaction. For example, *N*-methyl fulleropyrrolidine can be obtained by cycloaddition to C_{60} of the azomethine ylide resulting from the decarboxylation of *N*-methylglycine in the presence of paraformaldehyde in refluxing toluene. The simplicity of this procedure and the good yields achieved explain the popularity of this approach in fullerene chemistry. Fulleropyrrolidines have been widely used in different areas of chemistry. For example, water soluble or protein-modified fulleropyrrolidines have been prepared for medical applications.[24-27] In 1993, C_{60} was proposed as an inhibitor of HIV protease[24] and, more recently, fulleropyrrolidine derivatives bearing two ammonium groups on the surface of the fullerene sphere showed significant anti-HIV activity.[25]

The design of covalently linked donor-fullerene systems that are capable of undergoing photoinduced electron transfer has been extensively studied in recent years as a result of the unique photophysical[28,29] and electrochemical[30] properties exhibited by fullerenes. Cyclic voltammetry experiments have shown that C_{60} is an excellent electron acceptor.[31] One drawback, however, is that most of the fullerene adducts reported to date show worse electron-accepting capacity than the parent C_{60}[32] as a consequence of the saturation of one double bond and the loss of conjugation. In order to increase the electron affinity of C_{60}, Wudl *et al.* undertook the synthesis of a methanofullerene bearing two cyano groups.[33] This compound showed a first reduction potential that was 156 mV more positive than the parent C_{60}. A number of heterofullerenes[34] and some multifluorinated derivatives[35,36] exhibit better electron affinity than pristine fullerene. Another example concerns a series of pyrrolidinium salts that show better electron-accepting properties than the corresponding pyrrolidines. This behaviour was attributed to the inductive effect that the heteroatoms within these structures have on the fullerene sphere.[37]

One solution in the search for donor-C_{60} systems with a better – or at least equivalent – electron affinity than the parent C_{60} would be to use electronegative or electrodeficient atoms directly connected to the fullerene cage. Such a goal could be achieved by the addition of nitrile oxides to form isoxazolinofullerene, or the addition of nitrile imines to give pyrazolinofullerene. Unlike other 1,3-dipolar cycloadditions, the reaction of C_{60} with 1,3-nitrile imines to give pyrazolino[60]fullerene has not been widely exploited.

In this contribution we have aimed to provide an introduction to the synthesis of pyrazolino[60]fullerene systems and to present interesting electrochemical and photophysical properties. Certain applications of these materials in photovoltaic devices is also covered in an effort to stimulate new work with these compounds in the field of material science.

2. Synthesis

The first 1,3-dipolar cycloadditions to C_{60} were described in the early days of fullerene chemistry. However, the first example of reaction between a nitrile imine and C_{60} was not described until 1994, when Mathews et al.[38] prepared a mixture of bisadducts **1** by reaction of C_{60} and 1,3-diphenylnitrile imine, generated in situ from benzhydrazidoyl chloride, in the presence of triethylamine during ten days. This C_{60} bisadduct was characterized by FAB mass spectrometry and IR spectroscopy as well as by proton and carbon NMR spectroscopy. The monoadduct was not isolated on following this procedure (Scheme 1).

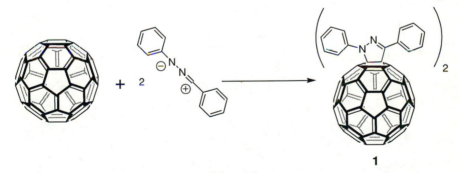

Scheme 1

In 1995, Yoshida et al.[39] described the preparation of a 1,3-diphenyl-2-pyrazolinofullerene. Furthermore, the synthesis and electrochemical and photochemical properties of some 2-pyrazoline ring-fused C_{60} derivatives were described in 1996[40] by the same authors. These adducts were prepared by cycloaddition of C_{60} to nitrile imines generated in situ from the corresponding N-chlorobenzylidene derivatives in the presence of triethylamine in refluxing benzene for two hours (Scheme 2). The benzylidene derivatives can be obtained in two steps from the corresponding acid chloride and hydrazine, followed by reaction with phosphorous pentachloride.

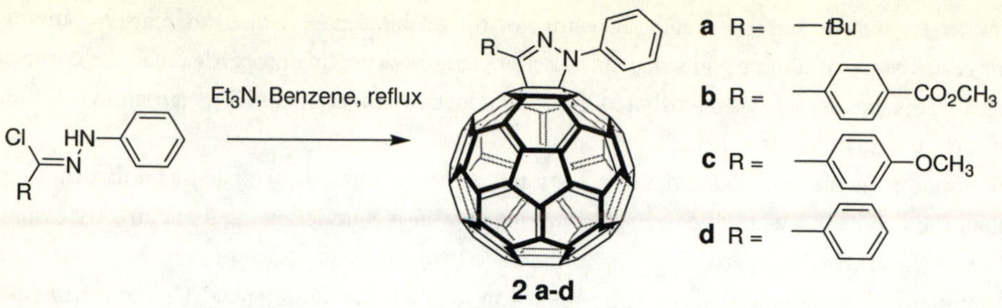

2 a-d

Scheme 2

Another approach to 2-pyrazolines involves the use of azomethine imines. Hydrazones give a thermal 1,2-hydrogen shift to afford azomethine imine intermediates that can undergo a 1,3-dipolar cycloaddition.[41,42] Bipyrazoles were obtained from pyrazolylhydrazones under microwave irradiation. The synthesis involved [3+2] cycloadditions with alkenes after an efficient hydrazone-azomethine imine tautomerization (Scheme 3). It should be noted that this reaction does not proceed under classical heating conditions.[43]

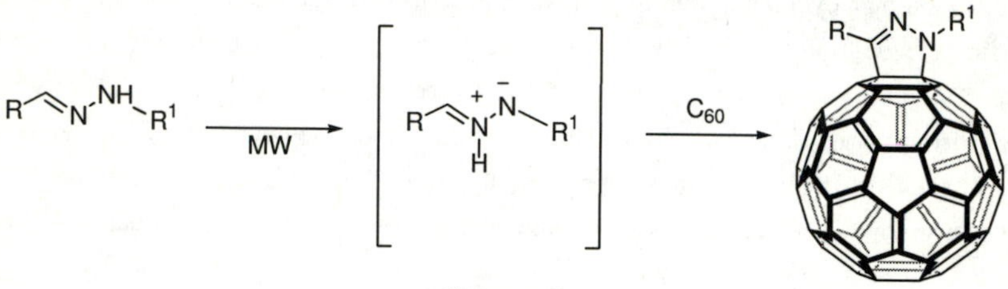

Scheme 3

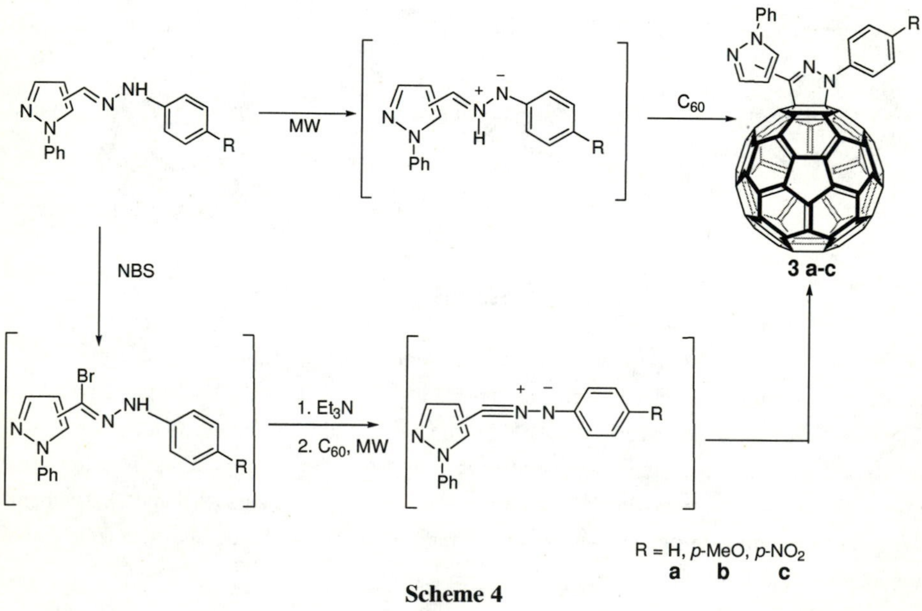

3 a-c

R = H, *p*-MeO, *p*-NO$_2$
   a   b   c

Scheme 4

This procedure was used to react different pyrazolylhydrazones (which are easily available from aldehydes in almost quantitative yields) with C_{60} under microwave irradiation to form the corresponding fullerene cycloadducts.[44] However, the yields obtained were less than 6%. An alternative route to the pyrazolino[60]fullerenes indicated in Scheme 3 involved the use of nitrile imines as dipoles, but in this case the procedure was modified in that the dipoles were prepared in a one-pot procedure starting from hydrazones. These 1,3-dipoles were generated *in situ* directly from the hydrazones and *N*-bromosuccinimide in the presence of triethylamine and then reacted with C_{60} under microwave irradiation.[44] The corresponding adducts (Scheme 4) were obtained in 20–38% yield.

A modification of this procedure was used for the preparation of various pyrazolinofullerenes[45] bearing different functional groups as substituents. Such a modification avoids the use of microwave irradiation and the reaction can be carried out at room temperature. The starting hydrazone was treated with *N*-chlorosuccinimide in the presence of pyridine; the resulting chloro-derivative was directly treated with triethylamine and C_{60} at room temperature to give the corresponding cycloadduct in good yield (Scheme 5).

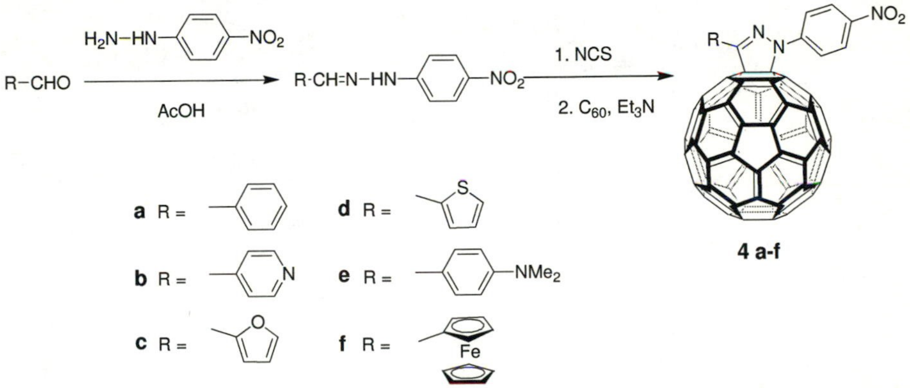

Scheme 5

This synthetic strategy was used in conjunction with a wide variety of hydrazones to prepare numerous different pyrazolino[60]fullerenes. For example, an oligophenylenevinylene (OPV) trimeric subunit was attached to C_{60} through a pyrazoline ring[46] (compound **5**) and a number of new phenylenevinylene (PV) fullerodendrimers have been synthesized[47] (compounds **6a–b**).

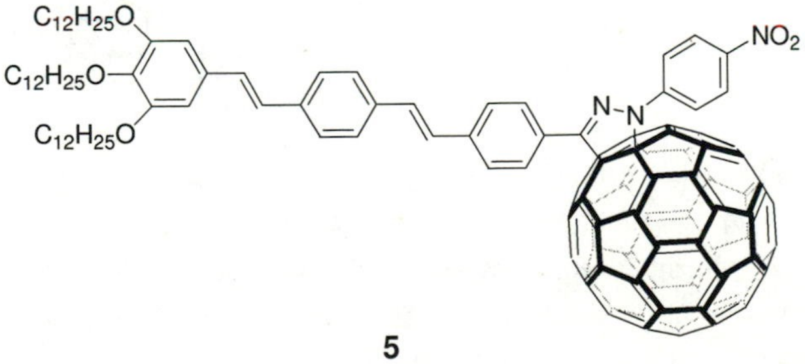

5

In 2002, symmetrically substituted oligophenylenevinylene derivatives bearing terminal *p*-nitrophenyl-hydrazone groups were prepared and used for the synthesis of dumb-bell-shaped bis-

(pyrazolino[60]fullerene)-oligophenylenevinylene systems[48] (compound **7**). Unlike other functionalization methods, the formation of stereoisomers does not occur and this method allows pure cycloadducts to be obtained in good yields.

One drawback of this functionalization method is that, although the reaction is general for any kind of aldehyde (*C*-side of the final pyrazolino[60]fullerene), in terms of the *N*-side the route is limited by the availability of the starting hydrazine to form the hydrazone. As indicated above, substituents that raise the energy of the dipole LUMO should make the cycloaddition easier and, consequently, frontier orbital theory predicts that *N-p*-nitrophenyl hydrazones should react more slowly than *N*-phenyl or *N-p*-methoxyphenyl derivatives. Despite the fact that *N*-phenyl and *N-p*-methoxyphenyl pyrazolino[60]fullerenes have been prepared, the best results have been obtained from *p*-nitrophenyl hydrazones.[49] Nevertheless, the frontier orbital theory (Scheme 6) can not be applied in this case as the nitrile imine intermediate is stabilized by the *p*-nitrophenyl group[50] (Scheme 7).

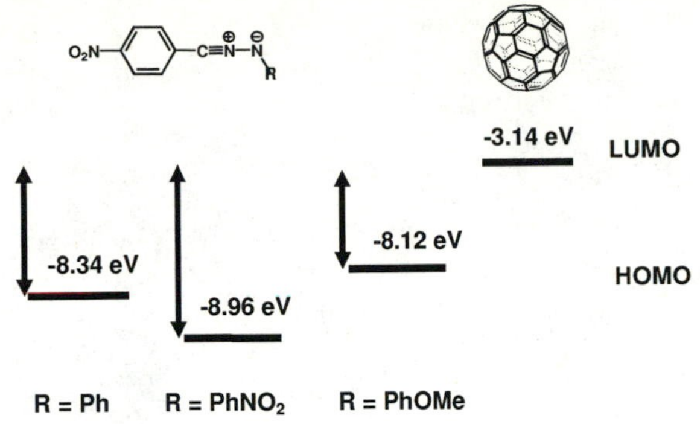

Scheme 6

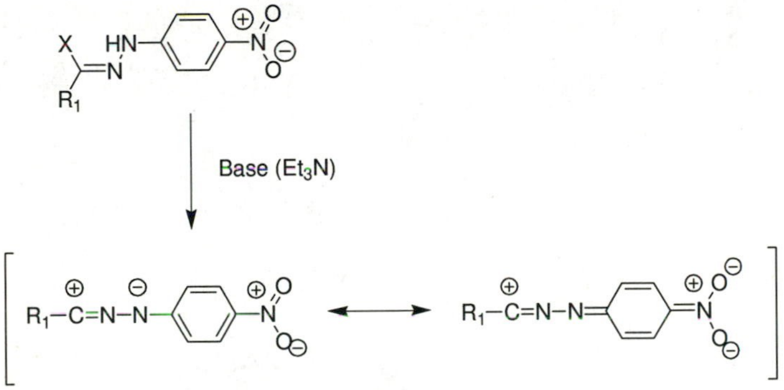

Base (Et₃N)

Scheme 7

A simple procedure that allows the preparation of pyrazolino[60]fullerene with substitution on the *N*-side of the pyrazoline ring was developed[51] and involves reduction with tin of the *p*-nitrophenyl moiety (Scheme 8). This method enabled the introduction of a range of desirable moieties on the *N*-side of the pyrazolino[60]fullerene system by functionalization of the amino group. The resulting *N*-anilinopyrazolino[60]fullerenes are convenient synthons for the preparation of other fullerene derivatives with the appropriate functionalities for a given application (Scheme 9).

This methodology can be used as well for the functionalization of single-walled carbon nanotubes (SWNTs): two soluble, photoactive single-walled carbon nanotubes containing *n*-pentyl esters at the tips and 2,5-diarylpyrazoline units at the walls of tubes were synthesized by 1,3-dipolar cycloaddition of nitrile imines to the sidewalls of SWNTs[52] (Scheme 10). The success of the synthesis was demonstrated by solution ¹H- and ¹⁹F-NMR as well as by UV-Vis and FT-IR spectroscopy. These two compounds were selected to have soluble single-wall carbon nanotubes substituted either with an electron-poor (3,5-bistrifluoromethylphenyl, **12**) or electron-rich (4-*N,N*-dimethylaminophenyl, **13**) unit on the C atom of the pyrazoline ring. Laser flash photolysis in dynamic flow of these carbon nanotubes in acetonitrile revealed the formation of several transient species decaying in the microsecond time scale. All of the available data –

including the photochemistry of model compounds, quenching experiments with electron donors and acceptors and the influence of acids – are compatible with the occurrence of electron transfer from the electron-rich substituents of the pyrazoline unit to the electron-acceptor terminal that incorporates electron-poor aryl rings and nanotube walls.

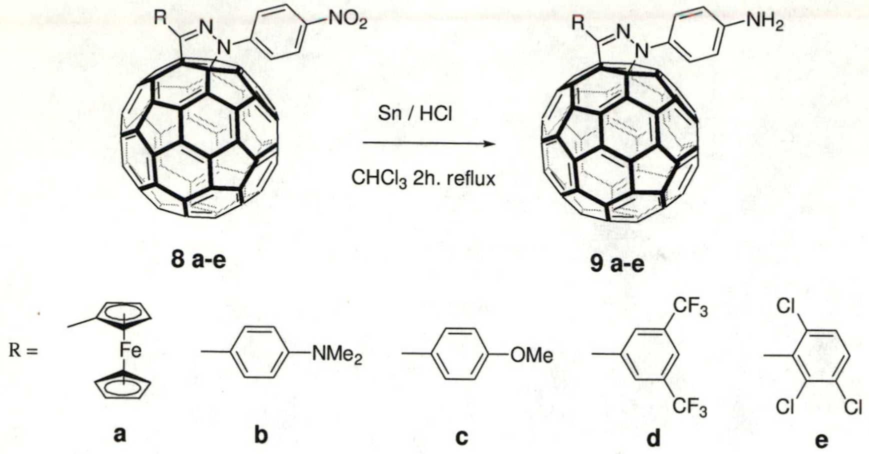

Scheme 8

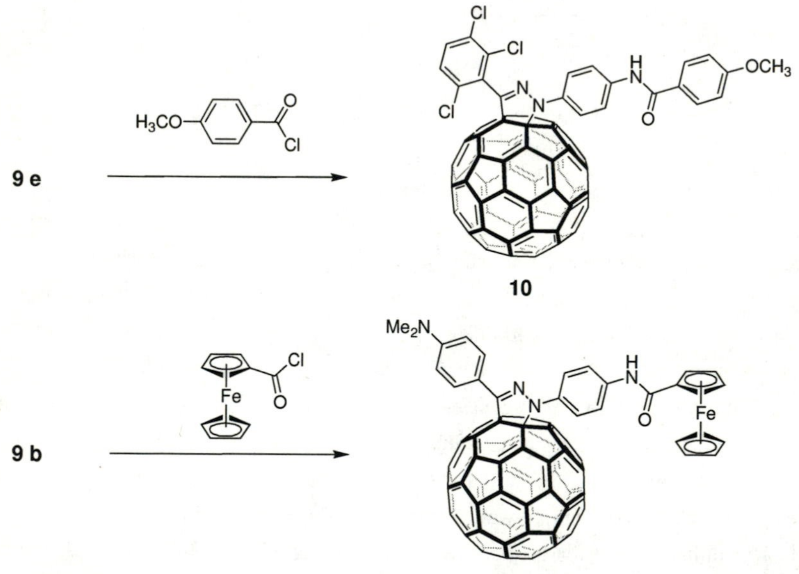

Scheme 9

It has been demonstrated that this method is a general and versatile procedure for the functionalization of fullerenes and carbon nanotubes in a one-pot reaction starting from the corresponding hydrazone. The hydrazones are stable compounds and are readily available from aldehydes in almost quantitative yield.

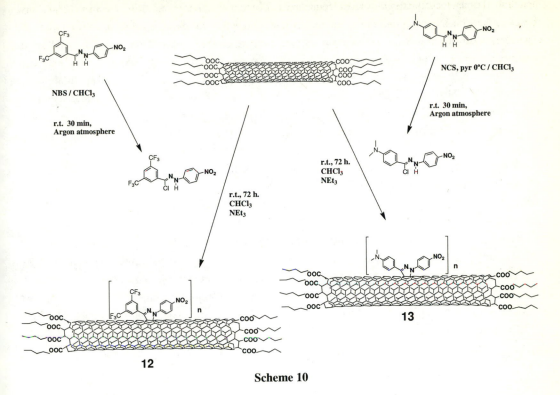

Scheme 10

3. Electrochemical properties

The study of the electrochemical properties of fullerenes[53] has been a field of intense research since these materials were first reported in 1990.[54] As a consequence of its LUMO triple degeneracy, C_{60} can be reduced up to the hexaanion[55] by a stepwise reduction. The reductions were found to be reversible and the different potentials appeared to be almost equidistant (Figure 3). The low value of the first reduction potential of C_{60}, which is similar to that of quinones, makes it an interesting partner in the preparation of donor-acceptor (D-A) systems due to the interesting electronic and optical properties that these materials can display. As a result, several groups have focussed their attention on the preparation of C_{60}-donor dyads[29,56-62] and the study of the existence of inter- or intramolecular electronic interactions in the ground state.

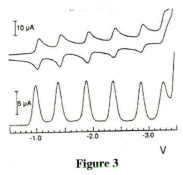

Figure 3

The characteristic features of the electrochemical properties of C_{60} are also seen in monoadducts. However, the reduction potentials of the monoadducts are about 100 mV more negative than those of C_{60}[30,63] and a great deal of effort has been spent on attempts to increase the electron affinity of C_{60}. Wudl *et al.* described the synthesis of a methanofullerene bearing two cyano groups[33] and this showed a first reduction potential 156 mV more positive than the parent C_{60}. Several heterofullerenes[34] and multifluorinated derivatives[35,36] exhibit better electron affinity than pristine fullerene. Another example is represented by a series of pyrrolidinium salts that showed better electron-accepting properties than the corresponding pyrrolidines. This behaviour was attributed to the inductive effect of the heteroatoms on the fullerene sphere.[37]

Consequently, one possible way to prepare functionalized fullerenes with a better, or at least equivalent, electron affinity to the parent C_{60} would be to use electronegative or electrodeficient atoms directly connected to the fullerene cage – as in the addition of nitrile oxides to form isoxazolinofullerene[16-20,64,65] or the addition of nitrile imines to prepare pyrazolinofullerenes.

The most simple pyrazolino[60]fullerene prepared to date, methyl derivative **14**, was synthesized and its electrochemical behaviour studied[66] by Osteryoung square wave (OSWV) and cyclic voltammetry (CV). This compound showed four reversible reduction waves and one irreversible reduction wave attributed, respectively, to the fullerene core and the *p*-nitrophenyl moiety. Compared to the parent C_{60}, this adduct showed similar values for the first and the second fullerene-based reduction waves in an *o*-dichlorobenzene (ODCB)/acetonitrile (4:1) mixture (E^1_{red}=-0.63 V and E^2_{red}=-1.04 V for **14** compared to respectively -0.64V and -1.05V for C_{60}).

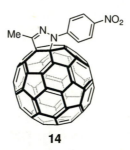

14

3.1. Influence of the nature of the heterocycle

In order to compare the electrochemical properties of different fullerene derivatives with a heterocyclic five-membered ring linked to the C_{60} cage, the synthesis, spectroscopic characterization, electrochemistry and photochemical measurement of three 2-methoxyphenylfullerene derivatives with pyrrolidine, isoxazolone and pyrazoline rings were undertaken and their properties compared.[67] Cyclic and Osteryoung square wave voltammetry studies indicated that the pyrazolino[60]fullerene **17** showed improved electron affinity in comparison to the parent fullerene (Table 1).

Cyclic and Osteryoung square wave voltammetry studies of pyrrolidino[60]fullerene **15** showed four reversible reduction peaks and confirmed that these derivatives retain the electrochemical pattern of C_{60}. However, as previously described for other fullerene derivatives, the reduction potentials are shifted to more negative values compared to those of C_{60}. A different trend was found for isoxazolo[60]fullerene **16**, for

130

which reduction potentials were slightly shifted to more positive values than the parent C_{60}. This behaviour is similar to that described previously for this class of fullerene derivative.[68-71] Finally, pyrazolino[60]fullerene presented four reversible reduction potentials corresponding to the fullerene core as well as a fifth reduction potential located between the second and third potential of C_{60}. This compound showed an anodic shift relative to both the corresponding isoxazolino[60]fullerene and C_{60}. This shift is thought to result from the inductive effect of the nitrogen atom directly connected to C_{60}.

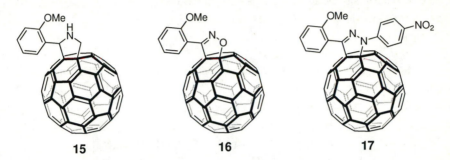

15   **16**   **17**

Table 1. Redox potential (OSWV) of organofullerenes **15–17** and C_{60} measured at room temperature in *o*-dichlorobenzene (ODCB)/acetonitrile (4:1).[a]

Compounds	E^1_{red}	E^2_{red}	E^3_{red}	E^4_{red}
C_{60}	–0.99	–1.42	–1.87	–2.41
15	–1.10	–1.51	–2.04	–2.36
16	–0.97	–1.37	–1.88	–2.36
17	–0.95	–1.38	–1.80	–2.30

[a] Experimental conditions: V *vs* Ag/AgNO₃; GCE as working electrode; 0.1 M TBAP; scan rate: 100 mV/s.

From this study, it can be concluded that in those cases, when high electron affinity of the C_{60} cage is a desired property, the preparation of pyrazolino[60]fullerenes shows advantages with respect to other functionalization procedures.

3.2. Substitution on the *C*-side of the pyrazoline

The influence that the substituent on the *C*-side of the pyrazoline ring has on the electrochemical properties of *p*-nitrophenyl-substituted pyrazolino[60]fullerenes has been studied.[45] The cyclic voltammograms of compounds **4a–f** were recorded (Table 2) and showed four quasi-reversible reduction waves for the fullerene moiety, a situation similar to that found for the parent C_{60}. In addition, another non-reversible reduction wave, between the second and the third fullerene-based reduction waves, was observed and was assigned to the *p*-nitrophenyl moiety. The influence of strong electron donors (*e.g.*, ferrocene or *N,N*-dimethylaniline) and an electron acceptor (*e.g.*, pyridine) was studied. The introduction of an electron-acceptor moiety, like 4-pyridinyl, in compound **4b** resulted in the most positive first reduction potential in this series of compounds. Indeed, the electron affinity of this compound was better than the parent C_{60}. When electron donors like *N,N*-dimethylaniline (**4e**) or furan (**4c**) were introduced on the C atom of the pyrazolino ring, the most negative values in this series were found, with first reduction potentials similar to the parent C_{60}. The influence that the nature of the substituent on the C-side of the pyrazoline has on the electron affinity of C_{60} can be explained in terms of through-bond or through-space interactions between the substituent and the fullerene cage. Comparison of the reduction potentials of the p-nitrophenyl moiety of these compounds showed a shift to more negative values as the electron-donor ability of the substituent on the other side of the molecule was increased. This reduction potential was shifted by 60 mV when a strong electron-donor was linked to the pyrazoline ring. This fact demonstrates the electronic communication between the C-substituent and the sp^3 nitrogen atom of the pyrazoline ring. As a consequence, the electron-transfer process from this nitrogen atom should be favoured when an electron-donor is incorporated into the C-side of the pyrazoline ring.

Table 2. Redox potential (OSWV) of organofullerenes **4a–f** and C_{60} measured at room temperature in o-dichlorobenzene (ODCB)/acetonitrile (4:1).[a]

Compounds	E^1_{red}	E^2_{red}	E^3_{red}	E^4_{red}	E^5_{red}
C_{60}	−0.95	−1.36	-	−1.83	−2.29
4a	−0.93	−1.33	−1.70	−1.87	−2.24
4b	−0.91	−1.32	−1.66	−1.82	−2.22
4c	−0.95	−1.36	−1.70	−1.84	−2.33
4d	−0.92	−1.32	−1.70	−1.85	−2.22
4e	−0.95	−1.35	−1.72	−1.91	−2.28
4f	−0.94	−1.35	−1.72	−1.92	−2.30

[a] Experimental conditions: V vs Ag/AgNO$_3$; GCE as working electrode; 0.1 M TBAP; scan rate: 100 mV/s.

3.3. Substitution on the N-side of the pyrazoline

The electrochemical behaviour of a series of pyrazolylpyrazolino[60]fullerenes (**3a–c**) was investigated by cyclic voltammetry.[49] The first reduction potential, which is attributed to the fullerene moiety, showed an anodic shift of up to 80 mV compared to the parent C_{60} (Table 3). This difference became less pronounced as the N-phenyl group attached to the nitrogen atom of the pyrazoline ring became more electron rich. The inductive effect of the substituents is one of the most important factors in the determination of the redox properties of organofullerenes.[72] Fulleropyrrolidine **18** was synthesized and the electrochemical properties were compared to the corresponding pyrazolino[60]fullerene.

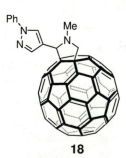

18

It was found that the fulleropyrrolidine presents a markedly more negative value for the first reduction potential than the pyrazolino[60]fullerene. The formation of the pyrrolidine ring usually leads to a displacement of the reduction potentials to more negative values, which results from a decrease in the electron affinity of the fullerene cage.[73]

Table 3. Redox potential (OSWV) of organofullerenes **3a–c** and C_{60} measured at room temperature in *o*-dichlorobenzene (ODCB)/acetonitrile (4:1).[a]

3 a-c

R	E^1_{red}	E^2_{red}	E^3_{red}	E^4_{red}
C_{60}	−0.59	−0.98	−1.48	−2.95
3a H	−0.55	−0.97	−1.46 (−1.71)	−2.22
3b *p*-MeO	−0.57	−0.99	−1.44 (−1.74)	−2.23
3c *p*-NO$_2$	−0.51	−0.91	−1.19 (−1.73)	−2.21

[a] Experimental conditions: V *vs* Ag/AgNO$_3$; GCE as working electrode; 0.1 M TBAP; scan rate: 100 mV/s.

Interestingly, amino pyrazolino[60]fullerene showed different electrochemical behaviour in comparison to C_{60}[51] and the difference depended on the solvent used for the studies (Table 4). In the less polar mixture (ODCB/dichloromethane), compounds **9a–e** showed reduction potentials that were shifted to more positive values than the parent C_{60}. The materials exhibited an improved electron affinity compared to the parent fullerene. However, in the most polar solvent mixture (ODCB/acetonitrile), the reduction potentials were shifted to slightly more negative values compared to C_{60}. This phenomenon was explained by possible electronic interactions between the fullerene cage and the aniline substituent in the ground state, which should be stabilized in the more polar solvent.

Table 4. $E_{1/2}$ values measured as the average of the anodic and cathodic peak potentials (CV) of compounds **9a–e** and C_{60} at room temperature.[a]

Solvent	ODCB/Acetonitrile 4/1					ODCB/CH$_2$Cl$_2$ 4/1
Compounds	E^1_{red}	E^2_{red}	E^3_{red}	E^1_{ox}	E^2_{ox}	E^1_{red}
C_{60}	−0.94	−1.36	−1.84	-	-	−0.94
9a	−1.01	−1.41	−1.91	0.13[c]	0.39[b]	−0.91
9b	−1.00	−1.40	−1.90	0.16	0.32[b]	−0.88
9c	−1.00	−1.39	−1.90	0.37[b]	0.57[b]	−0.89
9d	−0.98	−1.39	−1.91	0.30[b]	-	−0.89
9e	−1.00	−1.40	−1.89	0.32[b]	-	−0.89

[a] Experimental conditions: V *vs* Ag/AgNO$_3$; GCE as working electrode; 0.1 M TBAP; scan rate: 100 mV/s. [b] irreversible. [c] reversible, corresponding to the ferrocene addend.

As shown above, when compared to other C_{60} derivatives, pyrazolino[60]fullerenes must be considered as interesting electroactive compounds. Unlike other fullerene adducts, most of these compounds showed similar or better electron-accepting properties than the parent fullerene, thus making them good candidates for the preparation of new electroactive fullerene assemblies.

4. Photophysical properties

Fullerenes are characterized by excited levels that are relatively low in energy. The singlet and triplet excited states of C_{60} are localized, respectively, at 1.7^{74} and 1.5^{75} eV above the ground state. In addition, the position of these low-lying electronic states in functionalized fullerenes changes very little compared to the parent C_{60}. The singlet excited state is partially deactivated to the ground state through a radiative process that leads to the emission of fluorescence at 700 nm.[74,76,77] The main deactivation process for the fullerene singlet excited state is the cross intersystem pathway that leads to the formation of the triplet excited state[75] (Figure 4).

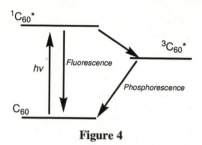

Figure 4

The fact that fullerenes are good acceptors with low-energy electronic levels is an important aspect for the photochemistry of fullerene derivatives: the excitation of a molecule in the presence of fullerene (whether it is covalently attached or not) can be followed by two processes. The first involves an electronic transfer (ET) and the second an energy transfer (EN) (Figure 5). These two processes are in competition with one another.

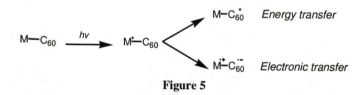

Figure 5

The fluorescence spectra of fullerene derivatives **15**, **16**, **17** and the model *N*-methylpyrrolidino[60]fullerene were recorded at room temperature in toluene and benzonitrile with excitation at 430 nm (Figure 6). Solutions with the same absorbance were employed so that the fluorescence intensity and quantum yield could be correlated.

When toluene was used as the solvent the model compound presented a maximum at 717 nm while **15** showed a maximum at 718 nm – both values are in the expected region for this system.[78] The emission of **15** was of a similar intensity to that of the reference *N*-methylpyrrolidino[60]fullerene, indicating that **15** did not undergo photoinduced charge transfer under these conditions. The fluorescence intensities were found to be similar and this shows that the quantum yields are almost equal. A similar pattern was observed for **16** but the intensity of the emission was slightly lower and the maximum appeared at 706 nm. Pyrazolino[60]fullerene **17** showed a maximum at 709 nm and a marked reduction in the fluorescence emission was observed in comparison to the reference.

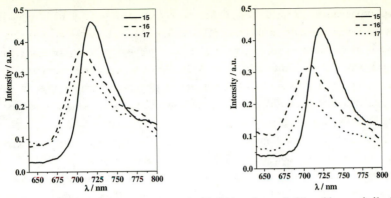

Figure 6. Fluorescence spectra of compounds **15–17** in toluene (left) and benzonitrile (right).

When benzonitrile was used as the solvent, the maximum of **15** was red shifted to 722 nm and the emission intensity remained virtually unchanged with respect to that found in toluene (considered as 100%). The reference compound showed the same behaviour.

A similar red shift and a slight decrease in the fluorescence (85% with respect to that found in toluene) emission were found in the case of **16**. In contrast, the maximum for pyrazolino[60]fullerene **17** was blue shifted to 705 nm. In addition, a substantially lower fluorescence intensity (60% with respect to that found in toluene) was observed. The quenching of the fluorescence in **17** as the polarity of the solvent increased suggests the occurrence of an electron-transfer process from the lone pair of the sp^3 nitrogen atom to the C_{60} cage in benzonitrile.

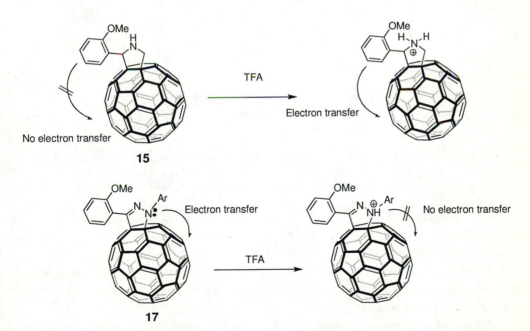

Figure 7. Proposed mechanism for the quenching of the electron transfer.

135

The addition of trifluoroacetic acid (TFA) to benzonitrile solutions of **15** and **17** produced the opposite effect. In the case of compound **15** the addition of TFA resulted in a weak quenching of the fluorescence and the maximum was blue shifted. This behaviour can be explained in terms of the formation of the pyrrolidinium salt and the possibility that a weak electron-transfer process can occur from the 2-methoxyphenyl moiety. The addition of TFA to a solution of **17** in benzonitrile resulted in a slight increase in the fluorescence. Such an increase suggests quenching of the electron-transfer process because the nitrogen lone pair is no longer available (Figure 7).

Compound **19** exhibited a faint fluorescence band in CH_2Cl_2 at room temperature (λ_{max}=706 nm) and the emission quantum yield was about two orders of magnitude lower than that of **20**. The fluorescence intensity was dependent on solvent polarity, with lower polarity solvents giving rise to stronger emission (Figure 8a). Fluorescence enhancement was obtained on addition of TFA and the addition of about 30% TFA (v/v) led to an intensity plateau (Figure 8b). The luminescence signal reverted to its initial intensity after extraction of the CH_2Cl_2/TFA solution with saturated aqueous NaOH. These results all suggest that the lowest fullerene singlet excited state was quenched by electron transfer from the lone pair of the pyrazoline sp^3 nitrogen, in analogy with similar substituted fullerenes.[69,79-83] Accordingly, fulleropyrazoline **19** was described as a redox dyad in which the electron donor and acceptor units were linked through an sp^3 carbon as a very simple spacer, as suggested by Sun *et al.* for aminofullerene derivatives.[80]

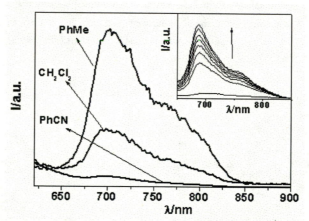

Figure 8. (a) Fluorescence spectra of **19** in various solvents under the same experimental conditions; λ_{exc}=430 nm, A=0.400. (b) Inset: Fluorescence spectrum of **19** in CH_2Cl_2 upon addition of increasing amount of acid up to 30% (v/v); λ_{exc}=424 nm.

The fluorescence spectra of **4a–f** [45] were measured in both toluene and benzonitrile upon excitation at 430 nm at room temperature. The solutions used had the same absorbance so that the fluorescence intensity and quantum yields could be correlated.

The fluorescence spectrum of pyrazolino[60]fullerene **4a** in toluene was similar to those of other fullerene derivatives, with an emission maximum at 705 nm and a shoulder at 772 nm due to the intersystem crossing into the excited triplet state.

The intensities of the emission maxima in **4b**, **4c** and **4d** were lower and the maxima were red shifted with respect to those of **4a**. This trend indicates that the excited state had been partially quenched. Evidence for emission was not found when the strong donors *N*,*N*-dimethylaniline (**4e**) or ferrocene (**4f**) were linked to the pyrazolino[60]fullerene system.

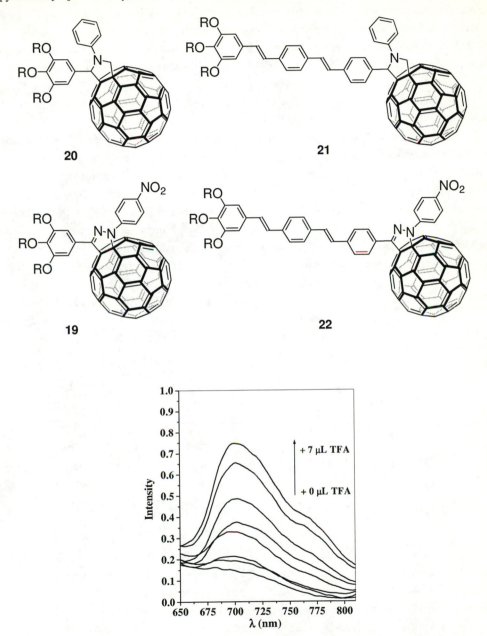

Figure 9. Fluorescence spectra of compound **4e** upon addition of increasing amounts of trifluoroacetic acid.

This quenching was explained in terms of a photoinduced electron-transfer process (PET), which is not common in non-polar solvents like toluene. On the other hand, the fluorescence intensity in benzonitrile was lower and was also associated with solvatochromic red shifts of the maxima in all cases. This solvatochromic effect in benzonitrile when compared to toluene was assigned to a weak CT character. The quenching of the fluorescence was ascribed to a weak PET interaction that occurs on increasing the polarity of the solvent and this would be expected to lead to increased stability of the separated charge state.

Addition of TFA to a benzonitrile solution of **4e** (R=*N*,*N*-dimethylaniline) caused an enhancement in the fluorescence, suggesting quenching of the PET process as the lone electron pairs of the two nitrogen atoms (*N*,*N*-dimethylaniline group and pyrazoline ring) become unavailable (Figure 9). One (or both) of the electron pairs should be responsible for the PET process; the addition of base (pyridine) caused quenching of the fluorescence, indicating the reversibility of the process.

The photochemical behaviour of pyrazolino[60]fullerenes **3a–c** was investigated.[49] These compounds share common absorption bands but the peak at 425 nm, which is characteristic of [6,6]-bridged monoadducts, only appeared as a shoulder in the nitro adduct **3c** due to the absorption of the nitrophenyl group. A band at around 687 nm, attributed to the 0-0* transition, was also observed. A weak, broad band was observed in the 450–650 nm region. This band was stronger in dyads containing an electron-donating substituent on the nitrogen atom of the pyrazoline ring (*e.g.*, **3b**). Similar bands have been reported for intramolecular charge transfer in some C_{60}-based dyads[84-86] or intermolecular charge transfer interactions with other donors.[87] These broad absorptions were red-shifted on increasing the polarizability of the solvent. This fact suggests that weak charge-transfer complexes are formed in the ground state. The intensity of this band was found to increase with the polarity of the solvent, supporting the formation of a charge-transfer complex in more polar solvents. This band was not observed in **3c** because the electron-withdrawing moiety on the nitrogen atom of the pyrazoline ring inhibited the formation of a charge-transfer complex in this derivative.

a R = H

b R = OMe

c R = NO$_2$

3 a-c

18

The effect of solvent on the fluorescence has been studied in fullerene derivatives. Previous studies on the photochemical properties of *N*-methylpyrrolidino[60]fullerenes indicated that the nitrogen atom of the pyrrolidine ring is not involved in charge transfer.[80,84] The fluorescence spectrum of fulleropyrrolidine **18** in cyclohexane was independent of the excitation wavelength and similar to the spectrum of the reference unsubstitued *N*-methylfulleropyrrolidine. The fluorescence study showed that this compound does not undergo photoinduced charge transfer in non-polar solvents.

An increase in the solvent polarity caused quenching of the fluorescence. This quenching, as well as the red shift observed in the more polar solvent benzonitrile, was ascribed to a weak charge transfer interaction in the excited state between the C_{60} and the pyrazole substituent.

The fluorescence spectra of the pyrazolino[60]fullerenes **3a–c** were recorded in different solvents. These compounds all showed a weak fluorescence compared to the fulleropyrrolidine discussed above. The fluorescence experiments demonstrated efficient deactivation of the fullerene singlet excited state in pyrazolinofullerene derivatives when compared to fulleropyrrolidine derivatives. The fluorescence quenching of these compounds was attributed to an intramolecular electron-transfer process from the pyrazole substituent to the C_{60} cage.

A series of isoindazole-C_{60} dyads based on pyrazolino[60]fullerene has been prepared[66] (Scheme 11).

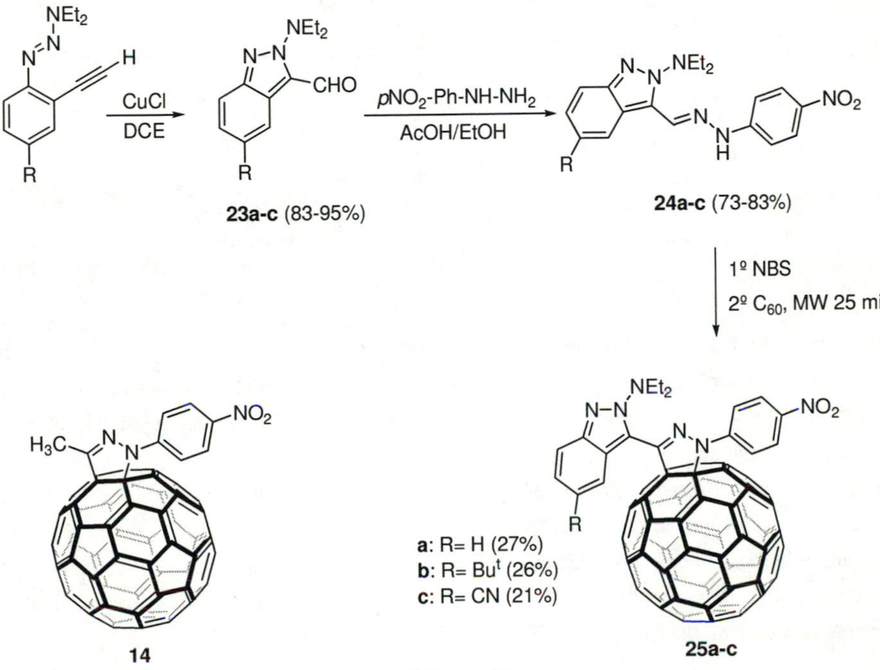

a: R= H (27%)
b: R= Bu^t (26%)
c: R= CN (21%)

Scheme 11

The fluorescence spectra of the dyads **25a–c** and the model compound *N*-methylpyrazolino[60]fullerene **14** were measured in toluene and benzonitrile (Figure 11). The toluene solutions gave peaks at 697 nm due to the fullerene moiety. Fluorescence was not observed in the 500–600 nm region for **25a–c**, indicating that energy transfer takes place efficiently from the excited singlet state of the isoindazole (or hydrazone) moiety to the fullerene moiety. Appreciable substituent effects were not observed in compounds **25a–c**. In contrast, in the more polar solvent (benzonitrile) the fluorescence intensities of the fullerene moieties at 697 nm changed considerably from **14** (reference) and **25c** to **25a** and **25b**. As the electron-donating ability of the substituents of the isoindazole moiety increased, the fluorescence intensity decreased significantly.

This trend was confirmed by the shortening of fluorescence lifetimes, which allowed intramolecular charge-separation rates and efficiencies *via* the excited singlet states of the fullerene moiety to be evaluated. The yields of the triplet states in the polar solvent decreased with electron-donating ability. This observation is consistent with the competitive formation of the charge-separated state by intersystem crossing from the excited states. Thus, isoindazole[60]fullerene **25b** can be considered as a molecular switch with an AND logic gate (Figure 10).

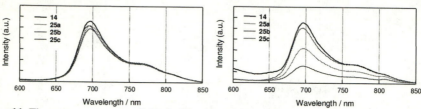

I₁	I₂	O
0	0	0
0	1	0
1	0	0
1	1	1

Figure 10. AND logic gate and its corresponding truth table for 4b (I_1: Irradiation at 430 nm; I_2: addition of TFA; O: fluorescence; 1=yes, O=not.

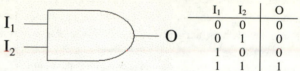

Figure 11. Fluorescence spectra of dyads **25a–c** and **14** in toluene (on the left) and in benzonitrile (on the right); λ_{exc}=400 nm.

The fluorescence emission behaviour of two pyrazolino[60]fullerene phenylenevinylene dendrimers (**6a** and **6b**) has been studied[47] (Figure 12). The fluorescence emission was strongly quenched (around 60% for **6a** and 50% for **6b**) compared to the corresponding starting hydrazones, indicating significant interaction between the singlet excited state of the phenylenevinylene moiety and the fullerene ground state. Moreover, different behaviour was observed for the generation one (**6a**) and generation two (**6b**) dendrimers. Dendrimer **6a** showed emission at 694, 707 (ascribed to the fullerene cage as a consequence of an energy transfer), 965 and 983 nm. These bands were not observed for **6b**, suggesting that an electron transfer might occur in this case. The fluorescence spectra were taken at different concentrations and were independent of the concentration – indicating that the process involved was intramolecular.

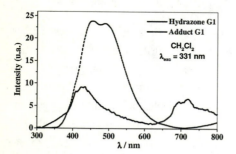
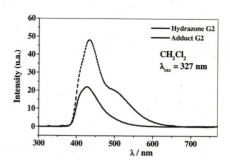

Figure 12. Fluorescence spectra of phenylenevinylene dendrimers **6a** (left) and **6b** (right) in benzonitrile. The spectra of the precursor hydrazones are shown for comparison.

The pyrazolino[60]fullerene phenylenevinylene dendrimer **6a** has been studied in solid molecular films for the first time.[88] Good-quality Langmuir–Blodgett (LB) films and cast dropped films have been prepared in order to compare their photophysical (Figure 13), electrochemical and structural properties. These studies showed how the assembly of these donor-acceptor dyads in thin films seemed to favour the photoinduced electron transfer with respect to that observed in solution. This difference in behaviour was related to intermolecular effects, as suggested by the steady-state photophysical outcomes of LB and cast dropped films, as well as the electrochemical behaviour of the LB films, in which the electron affinities of the fullerene redox states were markedly improved. However, a competitive energy transfer between the two chromophores was also assumed and this was optimized in case of the LB film due to the favourable intermolecular alignment. Finally, in an effort to further understand the structural aspects related to the properties outlined above, the structural organisation of the fullerodendrimer in the films has been elucidated by means of Reflection-Absorption Infrared Spectroscopy. This investigation showed that the dyad achieved a preferential intermolecular order in the LB film due to the anchoring of the –NO$_2$ group on the substrate. In contrast, cast films of **6a** consisted of molecules that were randomly orientated but more efficiently packed. Atomic Force Microscopy topographic images show that the LB films present a growing aggregation on increasing the number of monolayers, but homogeneity was attained throughout the film. In contrast, the cast films consisted of dispersed and aggregated crystallites.

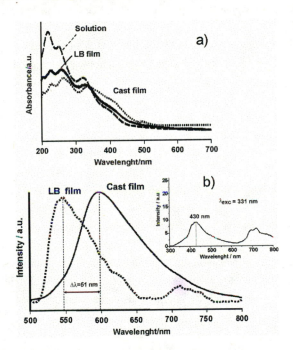

Figure 13. a) Steady-state UV-Vis absorption; b) fluorescence spectra of **6a** in molecular films at λ_{ex}=488 nm (spectra corrected to arbitrary units). Inset: fluorescence spectrum of a ca. 10^{-5} M CH$_2$Cl$_2$ solution excited at 331 nm.

A new pyrazolino[60]fullerene (**22**) in which an OPV unit (3PV) is linked to a fulleropyrazoline (**19**) moiety has been prepared.[46] The corresponding fulleropyrrolidines **21** and **20** were synthesized for comparison purposes.

The photophysical properties of **19**, **21** and **22** were studied in three different solvents of increasing polarity, as well as at two different temperatures (298 and 77 K). In the case of **19**, a photoinduced pyrazoline-to-C_{60} electron transfer was observed at room temperature in all solvents. This transfer could be blocked by the addition of acid or on conversion to a rigid matrix at 77 K. In **21** an OPV-to-C_{60} energy transfer was found in all solvents and this was followed by charge separation in PhCN to a significative extent. Compound **22** was arranged in such a way that the C_{60} unit could act either as an energy (for the OPV moiety) or electron (for the pyrazoline moiety) acceptor following excitation with light. Excitation of the OPV moiety or the fullerene chromophore triggered distinct photo processes. In addition to the choice of excitation wavelength, control over electron transfer was achieved by varying several parameters (or combinations of parameters) such as solvent polarity, acidity and temperature.

The incorporation of both of these features within a C_{60} unit was reported for the first time in **22**. These changes introduced a very interesting pattern in terms of the photophysical properties in solution, making **22** a fullerene-based molecular switch in which the switchable parameters are photoinduced processes. The incorporation of **22** into photovoltaic devices resulted in a lower light-to-current efficiency than in **21**. This lower efficiency is due to the fact that charge separation involving the fullerene moiety and the pyrazoline nitrogen atom was unable to contribute to the photocurrent and, as such, the pyrazoline unit can act as an electron trap.

From a photochemical point of view, the behaviour of systems incorporating a pyrazolino[60]fullerene moiety is complex. As discussed above, the pyrazoline ring itself should be considered as a photoactive unit and should be taken in account during the study of photoinduced processes.

5. Concluding remarks

The cycloaddition of nitrile imines to [60]fullerene to afford pyrazolino[60]fullerene is a general and versatile procedure for the functionalization of C_{60}. This method has several advantages over other derivatization procedures: (1) Hydrazones are easily available in one step from aldehydes in almost quantitative yields, (2) the cycloadducts are obtained in good yields, (3) the formation of stereoisomers does not occur – in contrast to other functionalization methods – and, most importantly, (4) unlike other fullerene derivatives, pyrazolino[60]fullerenes have similar reduction potentials to the parent C_{60}, which has proven to be an excellent three-dimensional electron acceptor. The electrochemical and photophysical properties of this new family of fullerene derivatives indicates that they are excellent candidates for the preparation of tailor-made donor-acceptor dyads or donor-donor-acceptor triads for photoinduced electron-transfer processes of interest in artificial photosynthetic systems and optoelectronic devices.

Acknowledgments

Financial support for this work was provided by grants from the Ministerio de Educación y Ciencia of Spain, FEDER funds (Project CTQ2004-00364/BQU) and Junta de Comunidades de Castilla-La Mancha (Project PAI-02-023). One of us (F.O.) acknowledges a grant from Research Training Networks: FAMOUS (HPRN-CT-2002-00171)

References
1. Kräschmer, W.; Lamb, L. D.; Fotstiropoulos, K.; Huffman, D. R. *Nature* **1990**, 354.
2. Braun, T.; Schubert, A. P.; Kostoff, R. N. *Chem. Rev.* **2000**, *100*, 23.

3.   Tagmatarchis, N.; Shinohara, H. *Mini Reviews in Medicinal Chemistry* **2001**, *1*, 339.
4.   Ros, T. D.; Spalluto, G.; Prato, M. *Croatica Chemica Acta* **2001**, *74*, 743.
5.   Prato, M. *J. Mater. Chem.* **1997**, 7, 1097.
6.   Diederich, F.; Gómez-López, M. *Chem. Soc. Rev.* **1999**, *28*, 263.
7.   Nierengarten, J. F. *Chem. Eur. J.* **2000**, *6*, 3667.
8.   Wudl, F.; Hirsch, A.; Khemani, K. C.; Suzuki, T.; Allemand P.M.; Koch A.; Ecket H.; Srdanov, G.; Webb H. In *ACS Symposium Series 468*; Hammond, G. S., Kuck, V. S., Eds. Washington D. C., 1992, p. 161.
9.   Cases, M.; Durán, M.; Mestres, J.; Martín, N.; Sola, M. *J. Org. Chem.* **2001**, 433.
10. Hirsch, A. *Acc. Chem. Res.* **1999**, 795.
11. Maggini, M.; Menna, E. In *Fullerenes: From Synthesis to Optoelectronic Properties*; Guldi, D. M., Martin, N., Eds. 2002, p. 1–50.
12. Novello, F.; Prato, M.; Ros, T. D.; Amici, M. D.; Bianco, A.; Toniolo, C.; Maggini, M. *Chem. Commun.* **1996**, 903.
13. Bianco, A.; Maggini, M.; Scorrano, G.; Toniolo, C.; Marconi, G.; Villani, C.; Prato, M. *J. Am. Chem. Soc.* **1996**, 4072.
14. Tan, X.; Schester, D. I.; Wilson, S. R. *Tetrahedron Lett.* **1999**, 4187.
15. Illescas, B.; Rifé, J.; Ortuño, R. M.; Martín, N. *J. Org. Chem.* **2000**, 6246.
16. Ros, T. D.; Prato, M.; Luchini, V. *J. Org. Chem.* **2000**, 4289.
17   Meier, M. S.; Poplawska, M. *J. Org. Chem.* **1993**, 4524.
18. Meier, M. S.; Poplawska, M. *Tetrahedron* **1996**, 5043.
19. Ros, T. D.; Prato, M.; Novello, F.; Maggini, M.; Amici, M. D.; Micheli, C. D. *Chem. Commun.* **1997**, 60.
20. Irngartinger, H.; Weber, A.; Escher, T.; Fettel, P. W.; Gassner, F. *Eur. J. Org. Chem.* **1999**, 2087.
21. Irngartinger, H.; Weber, A.; Escher, T. *Eur. J. Org. Chem.* **2000**, 1647.
22. Irngartinger, H.; Fettel, P. W.; Escher, T.; Tinnefeld, P.; Nord, S.; Sauer, M. *Eur. J. Org. Chem.* **2000**, 455.
23. Prato, M.; Maggini, M. *Acc. Chem. Res.* **1998**, *31*, 519.
24. Wudl, F.; Friedman, S. H.; Decamp, D. L.; Sijbesma, R. P.; Srdanov, G.; Kenyon, G. L. *J. Am. Chem. Soc.* **1993**, 6506.
25. Ros, T. D.; Prato, M. *Chem. Commun.* **1999**, 663.
26. Marcorin, G. L.; Ros, T. D.; Castellano, S.; Stefancich, G.; Bonin, I.; Miertus, S.; Prato, M. *Org. Lett.* **2000**, 3955.
27. Kordatos, K.; Ros, T. D.; Bosi, S.; Vázquez, E.; Bergamin, M.; Cusan, C.; Pellarini, F.; Tomberli, V.; Baiti, B.; Pantarotto, D.; Georgakilas, V.; Spalluto, G.; Prato, M. *J. Org. Chem.* **2001**, 4915.
28. Guldi, D. M. *Chem. Commun.* **2000**, 321.
29. Gust, D.; Moore, T. A.; Moore, A. L. *Acc. Chem. Res.* **2001**, *34*, 40.
30. Echegoyen, L.; Echegoyen, L. E. *Acc. Chem. Res.* **1998**, *31*, 593.
31. Xie, Q.; Arias, F.; Echegoyen, L. *J. Am. Chem. Soc.* **1993**, *115*, 9818.
32. Suzuki, T.; Li, Q.; Khemani, K. C.; Wudl, F.; Almarson, Ö. *Science* **1991**.
33. Kesharavz, M.; Knight, B.; Haddon, R. C.; Wudl, F. *Tetrahedron* **1996**, 5149.
34. Hummelen, J. C.; Knight, B.; Pavlovich, J.; González, R.; Wudl, F. *Science* **1995**, 1554.
35. Zhou, F.; Berkel, G. J. V.; Donovan, B. T. *J. Am. Chem. Soc.* **1994**, 5485.
36. Liu, N.; Touhara, H.; Morio, Y.; Komichi, D.; Okino, F.; Kawasaki, S. *J. Electrochem. Soc.* **1996**, 214.
37. Prato, M.; Ros, T. D.; Carano, M.; Ceroni, P.; Paolucci, F.; Bofia, S. *J. Am. Chem. Soc.* **1998**, 11645.
38. Mathews, C. K.; Rao, P. R. V.; Ragunathan, R.; Maruthamuthu, P.; Muthu, S. *Tetrahedron Lett.* **1994**, *35*, 1763.
39. Matsubara, Y.; Tada, H.; Nagase, S.; Yoshida, Z. *J. Org. Chem.* **1995**, *60*, 5372.
40. Matsubara, Y.; Muraoka, H.; Tada, H.; Yoshida, Z.. *Chem. Lett.* **1996**, 373.
41. Grigg, R.; Kemp, J.; Thompson, N. *Tetrahedron Lett.* **1978**, 2827.
42. Grigg, R. *Chem. Soc. Rev.* **1987**, 89.
43. Arrieta, A.; Carrillo, J. R.; Cossío, F. P.; Díaz-Ortiz, A.; Gómez-Escalonilla, M. J.; Hoz, A. d. l.; Langa, F.; Moreno, A. *Tetrahedron* **1998**, *54*, 13167.

44. Cruz, P. d. l.; Díaz-Ortiz, A.; García, J. J.; Gómez-Escalonilla, M. J.; Hoz, A. d. l.; Langa, F. *Tetrahedron Lett.* **1999**, *40*, 1587.

45. Espíldora, E.; Delgado, J. L.; Cruz, P. d. l.; Hoz, A. d. l.; López-Arza, V.; Langa, F. *Tetrahedron* **2002**, *58*, 5821.

46. Armaroli, N.; Accorsi, G.; Gisselbrecht, J.-P.; Gross, M.; Krasnikov, V.; Tsamouras, D.; Hadziioannou, G.; Gómez-Escalonilla, M. J.; Langa, F.; Eckert, J.-F.; Nierengarten, J.-F. *J. Mater. Chem.* **2002**, 12.

47. Langa, F.; Gómez-Escalonilla, M. J.; Díez-Barra, E.; García-Martínez, J. C.; Hoz, A. d. l.; Rodríguez-Lopéz, J.; González-Cortés, A.; López-Arza, V. *Tetrahedron Lett.* **2001**, *42*, 3435.

48. Gómez-Escalonilla, M. J.; Langa, F.; Rueff, J.-M.; Oswald, L.; Nierengarten, J.-F. *Tetrahedron Lett.* **2002**, *43*, 7507.

49. Langa, F.; Cruz, P. d. l.; Espíldora, E.; Hoz, A. d. l.; Bourdelande, J. L.; Sánchez, L.; Martín, N. *J. Org. Chem.* **2001**, *66*, 5033.

50. Kabada, P. *Synthesis* **1973**, 71.

51. Delgado, J. L.; Cruz, P. d. l.; López-Arza, V.; Langa, F. *Tetrahedron Lett.* **2004**, *45*, 1651.

52. Alvaro, M.; Atienzar, P.; Cruz, P. d. l.; Delgado, J. L.; Garcia, H.; Langa, F. *J. Phys. Chem. B* **2004**, *108*, 12691.

53. Chlistunoff, J.; Clieff, D.; Bard, A. J. In *Handbook of Organic Conductive Molecules and Polymers*; Nalwa, H. S., Ed.; John Wiley and Sons, Ltd.: New York, 1997; Vol. 1, p 333–412.

54. Haufler, R. E.; Conceicao, J.; Chibante, L. P. F.; Chai, Y.; Byrne, N. E.; Flanagan, S.; Haley, M. M.; O'Brien, S. C.; Pan, C. *J. Phys. Chem.* **1990**, *94*, 8634.

55. Xie, Q.; Perez-Cordero, E.; Echegoyen, L. *J. Am. Chem. Soc.* **1992**, 3978.

56. Guldi, D. M. *Chem. Soc. Rev.* **2002**, *31*, 22.

57. Liddell, P. A.; Kodis, G.; Moore, A. L.; Moore, T. A.; Gust, D. *J. Am. Chem. Soc.* **2002**, *124*, 7668.

58. Pérez, I.; Liu, S.; Martín, N.; Echegoyen, L. *J. Org. Chem.* **2000**, *65*, 3796.

59. Segura, J. L.; Priego, E. M.; Martín, N. *Tetrahedron Lett.* **2000**, *41*, 7737.

60. Zandler, M. E.; Smith, P. M.; Fujitsuka, M.; Ito, O.; D'Souza, F. *J. Org. Chem.* **2002**, *67*, 9122.

61. Fujitsuka, M.; Tsuboya, N.; Hamasaki, R.; Ito, M.; Onodera, S.; Ito, O.; Yamamoto, Y. *J. Phys. Chem. A* **2003**, *107*, 1452.

62. Guldi, D. M.; Swartz, A.; Luo, C.; Gómez, R.; Segura, J. L.; Martín, N. *J. Am. Chem. Soc.* **2002**, *124*, 10875.

63. Echegoyen, L.; Herranz, M. A. In *Fullerenes: From Synthesis to Optoelectronic Properties*; Guldi, D. M., Martín, N., Eds. 2002, p. 267.

64. Irngartinger, H.; Weber, A.; Oeser, T. *Angew. Chem. Int. Ed. Engl.* **1999**, *38*, 1279.

65. Yashiro, A.; Nishida, Y.; Kobayasi, K.; Ohno, M. *Synlett* **2000**, 361.

66. Delgado, J. L.; Cruz, P. d. l.; López-Arza, V.; Langa, F.; Kimball, D. B.; Haley, M. M.; Araki, Y.; Ito, O. *J. Org. Chem.* **2004**, 2661.

67. Langa, F.; Cruz, P. d. l.; Delgado, J. L.; Gómez-Escalonilla, M. J.; González-Cortés, A.; Hoz, A. d. l.; López-Arza, V. *New J. Chem.* **2002**, 76.

68. Cruz, P. d. l.; Espíldora, E.; García, J. J.; Hoz, A. d. l.; Langa, F.; Martín, N.; Sánchez, L. *Tetrahedron Lett.* **1999**, 4889.

69. Langa, F.; Cruz, P. d. l.; Espíldora, E.; González-Cortés, A.; Hoz, A. d. l.; López-Arza, V. *J. Org. Chem.* **2000**, 8675.

70. Irgantingen, H.; Fettel, P. W.; Escher, T.; Tinnefeld, P.; Nord, S.; Sauer, M. *Eur. J. Org. Chem.* **2000**, 455.

71. Illescas, B. M.; Martin, N. *J. Org. Chem.* **2000**, 5986.

72. Suzuki, T.; Maruyama, Y.; Akasaka, T.; Ando, W.; Kobayashi, K.; Nagase, S. *J. Am. Chem. Soc.* **1994**, *116*, 1359.

73. Maggini, M.; Karlsson, A.; Scorrano, G.; Sandonà, G.; Farnia, G.; Prato, M. *J. Chem. Soc., Chem. Commun.* **1994**, 589.

74. Sibley, S. P.; Argentine, S. M.; Francais, A. H. *Chem. Phys. Lett.* **1992**, 187.

75. Hung, R. R.; Grabonsi, J. J. *J. Phys. Chem.* **1991**, 6073.

76. Kim, D. H.; Lee, M. Y.; Suth, Y. D.; Kim, S. K. *J. Am. Chem. Soc.* **1992**, 4429.

77. Sun, Y. P.; Wang, P.; Hamilton, N. B. *J. Am. Chem. Soc.* **1993**, 6378.
78. Guldi, D. M.; Prato, M. *Acc. Chem. Res.* **2000**, *33*, 695.
79. Williams, R. M.; Zwier, J. M.; Verhoeven, J. W. *J. Am. Chem. Soc.* **1995**, *117*, 4093.
80. Sun, Y. P.; Bunker, C. E. *J. Phys. Chem.* **1998**, 7580.
81. Thomas, K. G.; Biju, V.; George, M. V.; Guldi, D. M.; Kamat, P. V. *J. Phys. Chem. A* **1998**, *102*, 5341.
82. Lawson, G. E.; Kitaygorodskiy, A.; Sun, Y. P. *J. Org. Chem.* **1999**, *64*, 5913.
83. Komamine, S.; Fujitsuka, M.; Ito, O.; Moriwaki, K.; Miyata, T.; Ohno, T. *J. Phys. Chem. A* **2000**, *104*, 11497.
84. Guldi, D. M.; Maggini, M.; Scorrano, G.; Prato, M. *J. Am. Chem. Soc.* **1997**, 974.
85. Llacay, J.; Veciana, J.; Vidal-Gancedo, J.; Bourdelande, J. L.; González-Moreno, R.; Rovira, C. *J. Org. Chem.* **1998**, 5201.
86. Nakamura, Y.; Minowa, T.; Tobita, S.; Shizuka, H.; Nishimura, J. *J. Chem. Soc., Perkin Trans. 2* **1995**, 2351.
87. Rath, M. C.; Pal, H.; Mukherjee, T. *J. Phys. Chem. A* **1999**, 4993.
88. Parra, V.; Caño, T. d.; Gómez-Escalonilla, M. J.; Langa, F.; Rodríguez-Méndez, M. L.; Saja, J. A. D. *Synth. Met.* **2005**, *148*, 47.

THIOLYSIS OF 1,2-EPOXIDES
UNDER ENVIRONMENTALLY FRIENDLY CONDITIONS

Francesco Fringuelli,* Ferdinando Pizzo, Simone Tortoioli and Luigi Vaccaro*

Centro di Eccellenza Materiali Innovativi Nanostrutturati – CEMIN

Dipartimento di Chimica, Università di Perugia, Via Elce di Sotto 8, I-06123, Perugia

(e-mail: frifra@unipg.it; luigi@unipg.it)

Abstract. *Available catalysts and new protocols for the thiolysis of 1,2-epoxides in water or under solvent-free condition have been illustrated.*

Contents

1. Introduction

1,2-Epoxides are among the most versatile and useful class of intermediates in organic synthesis. Due to their reactivity toward various nucleophiles, they play a pivotal role in the synthesis of a range of important commercial products,[1] pharmaceuticals,[2] and in general, in the stereoselective preparation of 1,2-difunctionalized molecules. There is an extensive literature covering the numerous protocols for the preparation[3] and for the nucleophilic ring-opening reactions of 1,2-epoxides.[4]

Thiolysis of 1,2-epoxides consists in the oxirane ring-opening by thiols, and is a direct access route to β-hydroxy sulfides. Besides its synthetic utility, the invaluable importance of the thiolysis of 1,2-epoxides resides in the fact that oxidation-thiolysis sequence is used by Nature for metabolic detoxification of olefinic xenobiotics and is also an adjuvant mechanism for cancer chemotherapy, enhancing cytotoxicity of alkylating drugs (such as olefins and 1,2-epoxides) in cancer cell lines.[5] Most of the known, highly biologically active metabolites containing β-hydroxy sulfide moieties are formed through this pathway.[6] Leukotrienes (**1**),[6a] glutathione conjugate of cryptophycin 52 (**2**),[6c] and diltiazem (**3**)[6d,f] are representative examples of biologically active β-hydroxy sulfides (Figure 1). The potent HIV-1 PR protease inhibitors (**4**),

146

are an example of a recently prepared, totally-synthetic C_2-symmetric drug possessing a β-hydroxy sulfide motif (Figure 1).[7]

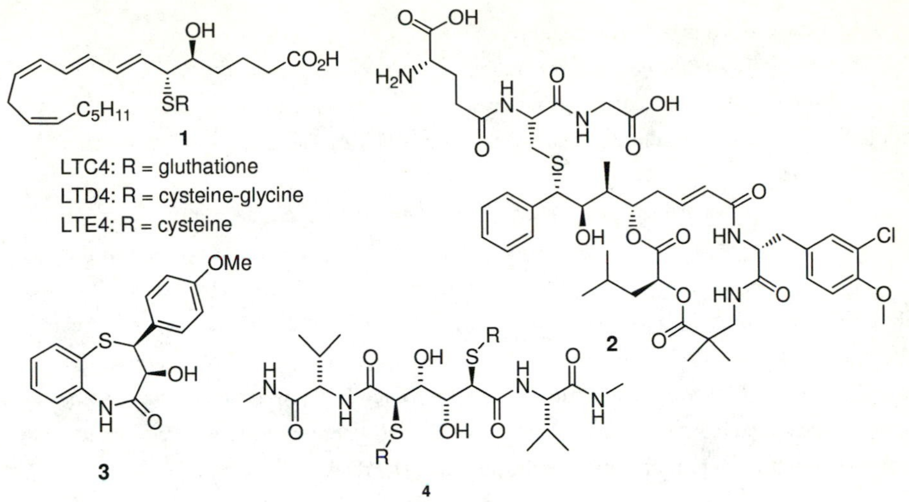

LTC4: R = gluthathione
LTD4: R = cysteine-glycine
LTE4: R = cysteine

R = Ph, Bn, 2-Py, allyl, 2-thienyl

Figure 1

In the physiological environment, deprotonation of thiols (*i.e.*, glutathione or cysteine)[8] is induced by lowering the pKa of the thiol functionality (typically by means of a tyrosine or a serine residue of an appropriate enzyme); the corresponding thiolate then reacts with the oxirane ring.[8]

Organic chemists have usually performed thiolysis in organic solvents (THF, CH_2Cl_2, MeOH, MeCN) generating reactive thiolate under anhydrous conditions.[9] The reactions occur with good yields in short times but reaction conditions are often harsh and only functional groups which tolerate basic conditions can be used.[9a] When thiols are used as nucleophiles, milder reaction conditions are usually required but an activating agent is necessary. The promoting agents include onium ions,[10] polyethylene glycols,[11] common Lewis acids,[12] solid neutral or acidic catalysts,[13] and optically active Lewis-acid complexes.[14] In these cases, the reaction times are generally long and the yields are not always good.

The search for new protocols and new catalysts to improve the efficiency of this process is still topical. In particular, greener synthetic methodologies are required to face the urgent environmental issues, especially for such a widely studied transformation as thiolysis of 1,2-epoxides. It is therefore necessary to define protocols for carrying out this process in alternative reaction media, using recoverable and reusable catalysts and reaction media. Considering the high interest in the study of thiolysis in aqueous medium in order to better understand this process under physiological conditions, particular attention has been given to water as reaction medium.[15]

Some years ago we started a project aimed at re-examining the thiolysis of 1,2-epoxides according to modern efficiency's parameter of green chemistry. Our research group has been working in this field for many years and has contributed to the revaluation of water as reaction medium for the synthesis of target molecules,[16] to the development of Lewis-acid catalysis in aqueous medium,[17] and, more recently, to the use of Solvent-Free Condition (SFC).[18]

147

In this review we will present the results reported in the literature in the last few years about the environmentally-benign protocols for the thiolysis of 1,2-epoxides.

2. Thiolysis of 1,2-epoxides in water as reaction medium

While Nature uses water as medium for all the fundamental chemical processes, organic chemists have been developing organic transformations using organic solvents as reaction media, believing that a high solubility of the reagents is essential to ensure a high reactivity of the process and that many Lewis acids "decompose" in water.

In the last decades, water has been receiving growing attention as reaction medium, as the need for environmentally responsible organic processes has become more urgent and the concept of a modern, green chemistry has been generally accepted. Certainly, the paper published in 1980 by Breslow[19] is considered a milestone in the use of water as reaction medium, because it showed that organic chemists can exploit the unique chemical properties of water (hydrogen bonding, pH control, high internal pressure,...) to reach some results that are impossible when organic media are used. Since Breslow's paper, much attention has been given and numerous organic transformations are now efficiently performed in water,[20] and in many cases it has been pointed out that water is essential for the process to be realized.[16,17,20]

2.1. Thiolysis under basic conditions

Nucleophilic ring-opening of 1,2-epoxides in water has been listed among those transformations which meet Click Chemistry criteria.[21] It has been pointed out that the use of water as reaction medium facilitates the nucleophilic addition to 1,2-epoxides because i) it responds to the demand for hydrogen bonding of these substrates, ii) allows the pH to be changed and controlled together with the concentration of nucleophiles and aqua ion, when the process is catalyzed by a Lewis acid.[17]

2.1.1. Thiolysis of alkyl- and aryl 1,2-epoxides

Both aryl- (pKa 6-8)[22] and alkylthiols (pKa 10-11)[22] can be deprotonated in water by hydroxide ions, to form the corresponding highly nucleophilic thiolates.

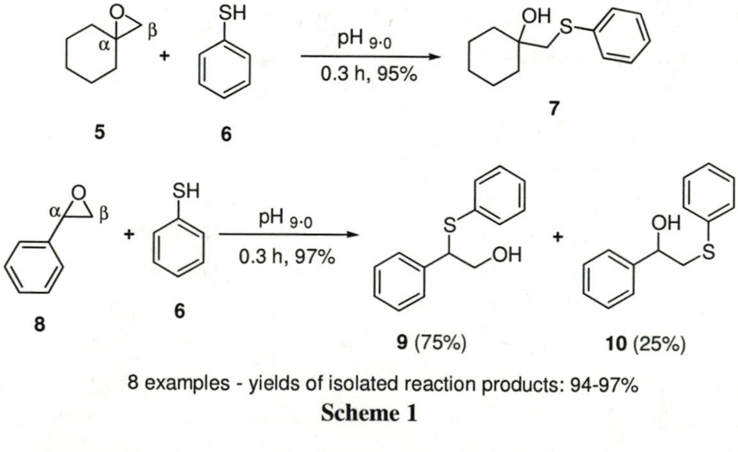

8 examples - yields of isolated reaction products: 94-97%
Scheme 1

In the first account of this topic, we performed thiolysis of a variety of alkyl- and aryl substituted 1,2-epoxides by phenylthiol (**6**) (Scheme 1) in water at pH 9.0 and at 30 °C.[23] The reactions were performed by adding phenylthiol (pKa 6.5)[24] in water and preparing *in situ* the corresponding phenylthiolate by adding an aqueous solution of NaOH and fixing the pH of the reaction medium at 9.0. Under these conditions thiolysis proceeded *via* an S_N2 mechanism mainly or exclusively, at the less substituted β-carbon of all the 1,2-epoxides with the exception of styrene oxide, in which the attack of the nucleophile is driven predominantly, as expected, at the benzylic α-carbon by electronic effects (Scheme 1). Yields of the isolated products were always very good.

Consequent extension[25] to a variety of substituted arylthiols showed that thiolyses under basic conditions in water were fast and in 0.08-4.0 h a complete conversion was reached at 30 °C with the prevalent formation of the β-products (> 95%) coming from the totally *anti* nucleophilic attack at the less substituted carbon of the oxirane ring. A little α-addition (3-5%) was sometimes observed. The thiolysis of 1-methylcyclohexene oxide (**11**) with *p*-acetamido phenylthiol (**12**) to give corresponding β-hydroxy sulfide **13** is illustrated as an example in Scheme 2. The higher nucleophilicity of ArS$^\ominus$, with respect to $N_3^\ominus$,[22] makes the thiolysis of alkyl oxiranes in basic aqueous medium much more β-regioselective than the azidolysis one.[26] Formation of diol products due to the competition of nucleophilic oxygen species (OH$^\ominus$, H_2O) with ArS$^\ominus$ was rarely observed[14,27] and this by-product was never an obstacle to the purification of sulfide because the latter is poorly soluble in aqueous medium and it can be easily separated as a solid crystalline by filtration. Also, in the case of highly sterically hindered thiols or 1,2-epoxides such as *ortho*-methyl-phenylthiol or 2-methyl-2,3-heptene oxide, the reactions were complete after a reasonable time (4.8-22 h). In all cases the yields of the isolated β-hydroxy sulfides are very satisfactory (≥ 85%).

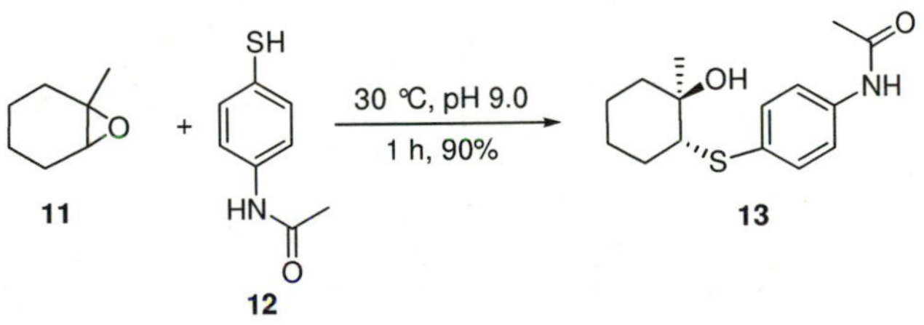

18 examples - yields of isolated reaction products: 80-95%

Scheme 2

The basic aqueous reaction medium was recovered after filtering the solid reaction products and reused without adding any additional base. Water not only allowed a chemically efficient protocol but it has proved to be essential for reducing the environmental impact of the process.

Hou *et al.* reported that tributylphosphine (10 mol%) is an efficient catalyst for the ring-opening reaction of 1,2-epoxides and aziridines by various nucleophiles.[28] Thiolysis by phenylthiol (**6**) was investigated and it was proved that the best results were obtained by using only water as reaction medium, while tributylphosphine was not active when the reactions were performed in MeCN. In the case of cycloaliphatic 1,2-epoxides, *e.g.* cyclohexene oxide (**14**), only *trans* adduct was observed, confirming that tributylphosphine did not attack the oxirane ring to form the corresponding betaine (Scheme 3).[22b] It has been proposed that tributylphosphine reacts with the oxygen of the oxirane ring forming a salt which acts as a base and producing the thiolate. By reacting with 1,2-epoxide, this species furnishes the product which is responsible for further deprotonation of other molecules of nucleophile completing the catalytic cycle.

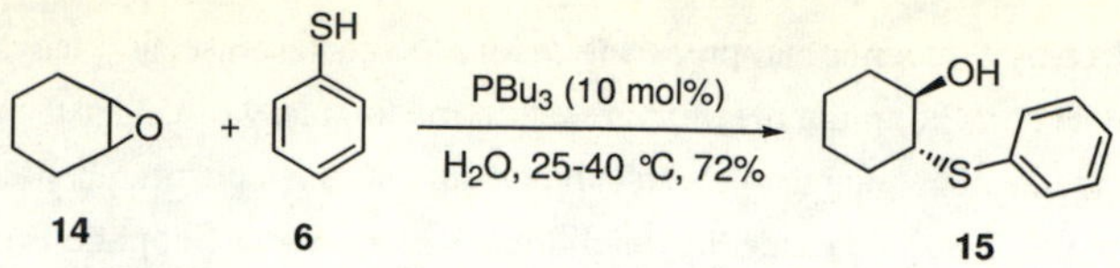

6 examples - yields of isolated reaction products 72-88%

Scheme 3

2.1.2. Thiolysis of α,β-epoxyketones and α,β-epoxycarboxylic acids

Thiolysis of α,β-epoxyketones in water has been used as a key step for the one-pot multisteps synthesis of α-carbonyl vinylsulfoxides starting from the corresponding α,β-unsaturated ketones.[29] Generally the thiolysis of such a class of 1,2-epoxides is considered to be neither regio- nor stereoselective at the C-α position, especially in the case of acyclic substrates.[30]

The basicity of the reaction medium plays a crucial role in the reaction of the representative 3,4-epoxyheptan-2-one (**16**) (Scheme 4) depending on the nature of the thiol employed in the process. β-carbonyl-β-hydroxy sulfides are very base-sensitive and easily give epimerization reaction at C-3, retroaldol condensation and dehydration reactions to give complex reaction mixtures.

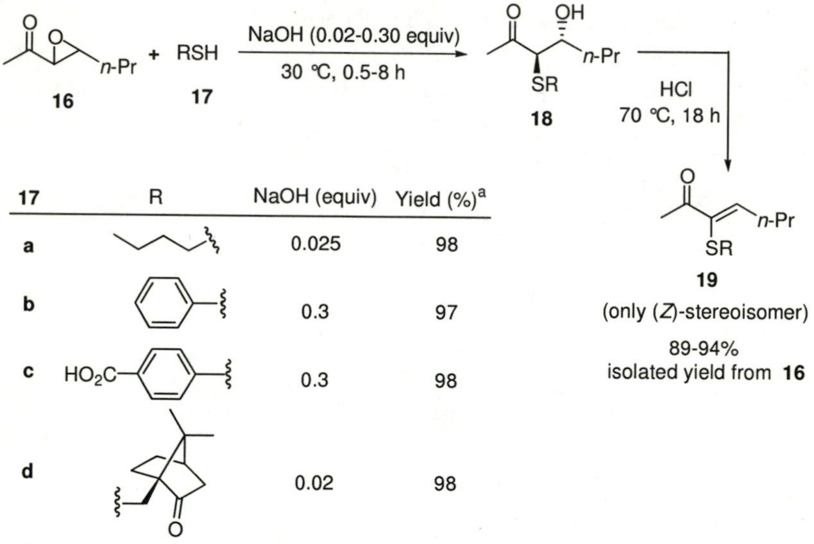

17	R	NaOH (equiv)	Yield (%)[a]
a		0.025	98
b		0.3	97
c	HO$_2$C-	0.3	98
d		0.02	98

[a] Yield of the isolated crude product **18**

Scheme 4

Thiolysis of 1,2-epoxides proceeded with the formation of an alkoxide ion which is responsible for further basification of the reaction medium, therefore the amount of base must be precisely chosen. In an accurate study on the influence of pH on this transformation, we found that a catalytic amount of NaOH (0.02-0.3 molar equiv) was sufficient to complete the thiolysis of **16** in water at 30 °C with thiols **17a-d**. The process is completely α-regio- and *anti*-stereoselective with formation of only *anti*-β-hydroxy sulfides **18a-d** with excellent yields (97-98%). The one-pot synthesis of the corresponding vinyl sulfides was accomplished by coupling the thiolysis process with a stereoselective dehydration achieved by treating compounds **18a-d** with HCl at 70 °C for 18 h (Scheme 4).

150

α,β-Epoxycycloalkenones gave the corresponding α-carbonyl vinyl sulfide directly upon treatment of thiol with an appropriate amount of NaOH (0.025-0.5 equiv) (Scheme 5).[29] The possibility of including thiolysis step in a one-pot multisteps procedure makes these results interesting and gives an innovative approach to α-carbonyl vinyl sulfide starting from α,β-unsaturated ketones.

An important class of 1,2-epoxides is constituted by α,β-epoxycarboxylic acids. Thiolysis of these substrates is a key synthetic step in the preparation of calcium channel blocker Diltiazem (3).[6d,f] We investigated the thiolysis by phenylthiol (6) of a series of α,β-epoxycarboxylic acids in sole water.[31]

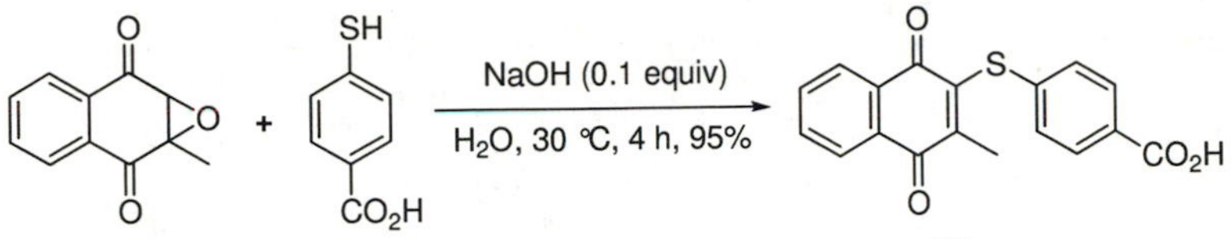

8 examples - yields of isolated reaction products: 91-96%

Scheme 5

The reactions were very slow under acidic conditions (pH 4.0) and sometimes occurred with very low conversions, while they became very fast at pH 9.0 and occurred quantitatively (Table 1). As generally observed,[16b,e,17] C-α regioselectivity increased under basic conditions, where phenylthiolate predominantly attacked the more electrophilic C-α carbon, except in the case of β-phenyl-substituted α,β-epoxypropanoic acid (24) and when an alkyl substituent was present at C-α position, as in the case of 25 and 26.

2.2. Thiolysis under acidic conditions

It has been reported that in water, moving from a basic to acidic pH, it is possible to significantly influence the regio- and stereoselectivity of azidolysis, iodolysis and bromolysis of alkyl- and aryl-substituted 1,2-epoxides and α,β-epoxycarboxylic acids.[17]

In the case of halogenolysis and azidolysis (pKa (HN₃)=4.70) under acidic conditions, highly nucleophilic iodine and azido ions are present at significant concentrations while, in the thiolysis process at a pH below 5, nucleophilic attack is formally accomplished by thiol, considering that thiolate species is present at a very low concentration. Thiol is much less nucleophilic than the corresponding thiolate and therefore, under these conditions, the thiolysis reaction is much slower and the hydrolysis process is competitive.[23] Therefore activation of the oxygen ring is necessary, essentially through the coordination of the oxygen atom by a Lewis acid (*e.g.* a metal cation or a proton).

2.2.1. In(III) and Zn(II): the best catalysts for the thiolysis of 1,2-epoxides

We have demonstrated that the best catalytic efficiency for a metal ion in water is expected at a pH value lower than its $pK_{1,1}$ hydrolysis constant, at which the maximum concentration of the aqua ion is present (the species expected to possess the highest Lewis acidity).[17b-d]

Taking into account these results, we have started a study on the thiolysis of cyclohexene oxide (14) with phenylthiol (6) by comparing the catalytic efficiency of metal salts possessing a $pK_{1,1}$ hydrolysis constant smaller than 5, such as InCl₃, Sc(OTf)₃, and AlCl₃, and bigger than 7, such as ZnCl₂, CoCl₂,

Cu(NO$_3$)$_2$, Ni(NO$_3$)$_2$, at pH 4.0 and 30 °C (Table 2). InCl$_3$ and ZnCl$_2$ were by far the most efficient catalysts allowing to achieve the complete conversion to β-hydroxy sulfide **15** after only 10 min. These two main classes of catalysts have been investigated separately.

Table 1. Thiolysis of α,β-epoxy acids **22-26** by phenylthiol (**6**) in water.

Entry	Epoxy acid	pH	T	time	C	C-α/C-β
			(°C)	(h)	(%)[a]	
1	**22**	4.0	55	48	19	80/20
2		9.0	55	3	99	96/4
3	**23**	4.0	55	48	84	93/7
4		9.0	55	1	99	95/5
5	**24**	4.0	30	10	99	15/85
6		9.0	30	0.5	99	21/79
7	**25**	4.0	30	192	16	1/99
8		9.0	30	72	99	57/43
9	**26**	4.0	30	72	32	69/31
10		9.0	30	8	99	51/49

[a] Reaction conversion.

The catalytic efficiency of InCl$_3$ significantly decreased by performing the reaction of **14** with **6** at pH values above pH 4.0, in agreement with the pK$_{1,1}$ hydrolysis constant value of InCl$_3$, (~ 4.0).[32] InCl$_3$-catalyzed thiolysis conditions were extended to various 1,2-epoxides. At pH 4.0, in the absence of InCl$_3$, the α-regioselectivity of the reaction increased and, in some cases, the C-β hydroxyl C-α-phenyl sulfide was the prevalent product, but long reaction times (2-36 h) were generally required and in some cases very low conversions were reached. The presence of 10 mol% of InCl$_3$ strongly accelerated the reactions that occurred in 3-10 min and likewise greatly changed the α/β ratios of regioisomer products with respect to the reactions carried out at pH 9.0 (see previous section). Thus, the α-regioselectivity of thiolysis of 1-oxaspiro[2.5]octane (**5**), 1-methyl-cyclohexene oxide (**11**), and 2-methyl-hept-2-ene oxide (**27**) increased by 733, 40, and 186 times, respectively, and that of the other 1,2-epoxides by 3-12 times (Scheme 6).

The strong catalytic effect of InCl$_3$ and the increased α-regioselectivity has been justified considering that the aqua ion species In(H$_2$O)$_6^{+3}$, present in high concentration at pH 4.0, coordinates the oxirane oxygen producing a considerable positive charge on the more substituted α-carbon and favoring the nucleophilic attack according to a S$_N$2 -*borderline* mechanism. Phenylthiol probably participates in the formation of a complex with the 1,2-epoxide and the metal ion and this complex is probably the real reason why reactivity of the oxirane ring is higher than that produced by proton-complexation, while the regioselectivities of the proton- and indium-catalyzed reactions are roughly the same, indicating that the charge density ratio on C-α and C-β of protonated epoxide is very similar to that of indium-complexed epoxide.

Table 2. Thiolysis of 1,2-cyclohexene oxide (**14**) by phenylthiol (**6**) with various catalysts in water.

Entry	Metal Salt (10 mol%)	t (min)	C (%)[a]	15/30
1	-	1440	90	94/6
2	Ni(NO₃)₂	1440	73	29/71
3	Sc(OTf)₃	420	33	73/27
4	CoCl₂	420	95	85/15
5	Cu(NO₃)₂	240	99	32/68
6	AlCl₃	240	37	75/25
7	ZnCl₂	10	> 99	> 99/< 1
8	InCl₃	10	> 99	> 99/< 1

[a] Reaction conversion.

8 examples - yields of the isolated reaction products: 94-97%
Scheme 6

Within a project aimed at defining a one-pot protocol starting from 1,2-epoxides that would allowed the selective preparation of β-hydroxy sulfoxides or β-hydroxy sulfones, we were searching for greener metal catalysts and the attention was directed toward *classical* Lewis acids such as Zn(II), Cu(II), Ni(II), and Co(II) salts. As mentioned above, Zn(II) salts were found to be truly active (Table 2).[33] At pH 4.0 and 30 °C, ZnCl₂ (10 mol%) proved to be a very efficient catalyst in the reaction of phenylthiol (**6**) and cyclohexene oxide (**14**) giving the *trans*-β-phenylthiocyclohexanol (**15**) solely in only 10 min similar to InCl₃.[23] At same pH, Co(II), Ni(II), and Cu(II) salts gave slower reactions with concomitant formation of the byproduct *trans*-1,2-cyclohexandiol (**30**) (15, 71, and 68% respectively). Excellent results were also obtained reducing ZnCl₂ to 5 mol%. This protocol was extended to a variety of 1,2-epoxides.[33] Most of the reactions were complete in less than one hour and gave excellent yields (85-96%). An interesting result was obtained[33] in the case of *trans*-2,3-epoxyhexanol *O*-tosyl derivative (**31**), which, in organic solvent (sodium thiolate, THF, 30 min, r.t.), gave only substitution at C-1 carbon, resulting in the formation of **32** and leaving the oxirane ring intact

153

(Scheme 7). In water at pH 4.0 and in the absence of catalyst, the reaction was very slow and after 168 h, gave only 6% of product **33**, which was the result of the attack at the C-1 and C-3 carbons. In water at pH 4.0 and in the presence of ZnCl$_2$ (5 mol%), the main product was **34**, coming from the C-3 oxirane ring-opening and was obtained in 90% yield after 24 h.

The recovery and reuse of reaction medium, excess of thiol and catalyst was accomplished by varying the pH of the final reaction mixture to dissolve Zn(II) as tetrahydroxyzincate and phenylthiol as phenylthiolate in water. Extraction of the organic layer gave the desired product leaving them in water. The one-pot chemoselective preparation of the sulfoxide and a sulfone starting from cyclohexene oxide (**14**) was also realized.

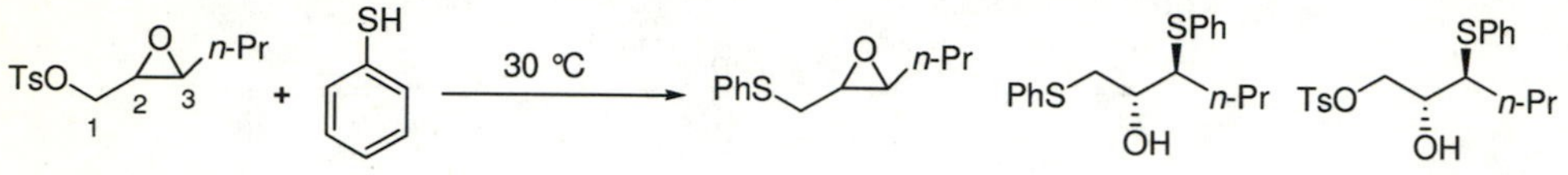

Reaction conditions	Conversion (%)			
		32	**33**	**34**
PhSNa, THF, 0.5 h	100	100	0	0
pH 4.0, H$_2$O, 168 h	6	0	100	0
pH 4.0, H$_2$O, ZnCl$_2$ (5 mol%)	100	0	6	94

Scheme 7

2.3. Zn (II)-Catalyzed thiolysis of 1,2-epoxides under biomimetic conditions

The research was then directed toward identification of a suitable metal catalyst for performing thiolysis of 1,2-epoxides in water under physiological (*i.e.* biomimetic) conditions.

Metal-thiol and metal-thiolate interactions are extremely important in biochemical processes.[34] *In vivo*, Nature regulates the catalytic efficiency of biocatalysts with ions such as Co^{2+}, Ni^{2+}, Cu^{2+}, and Zn^{2+} exploiting their interactions with sulfur-containing functionalities.[35] This biological behavior suggests that these cations could be efficient catalysts in water even under neutral conditions. Their pK$_{1,1}$ hydrolysis constants (9.65, 9.86, 7.97, and 8.96, respectively)[32] support the hypothesis that they could be efficient catalysts at pH 7.0.[17b-d]

Therefore, *classical* Lewis acids such as Co(II), Ni(II), Cu(II), and Zn(II) salts offer the possibility of performing the thiolysis of 1,2-epoxides in aqueous medium under virtually bio-mimetic conditions (water as medium, pH 7.0, room temperature), especially for acid- and base-sensitive substrates.

At pH 7.0,[36] ZnCl$_2$ (10 mol%) gave total conversion of **14** to **15** in only 5 min confirming that this metal salt can be used as an efficient catalyst for the thiolysis process in water under neutral condition. At this pH value, Cu(NO$_3$)$_2$ and Ni(NO$_3$)$_2$ also showed a catalytic efficiency but only CoCl$_2$ gave results comparable to ZnCl$_2$ with a complete conversion after only 10 min. All the Zn(II) salts gave the same catalytic efficiency. The reaction was slightly slower when the amount of ZnCl$_2$ was reduced to 2 mol% (Table 3).

The use of ZnCl$_2$ under neutral conditions was then extended[33] to the thiolysis of a variety of 1,2-epoxides. In all cases, excellent yields (94-97%) and short reaction times were obtained (5-300 min) except in the case of slowly reactive *trans*-2,3-epoxyhexanol *O*-tosyl derivative (**31**) which was converted into thiolysis products after 1200 min.

Table 3. Thiolysis of 1,2-cyclohexene oxide (**14**) by phenylthiol (**6**) with various catalysts in water.

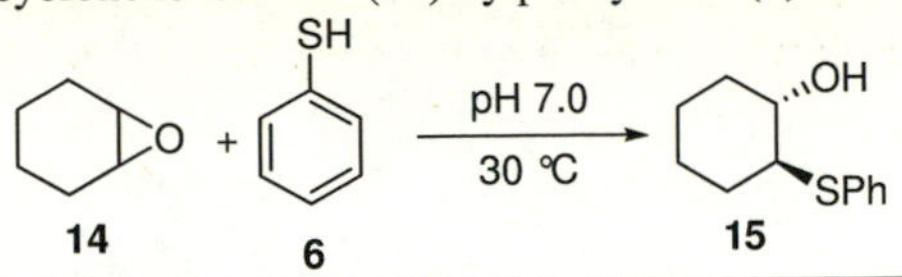

Entry	Metal Salt	t (min)	C (%)^a
1	-	480	95
2	Cu(NO$_3$)$_2$ (10 mol%)	130	95
3	Ni(NO$_3$)$_2$ (10 mol%)	85	97
4	CoCl$_2$ (10 mol%)	10	100
5	Zn(OTf)$_2$ (10 mol%)	5	100
6	Zn(NO$_3$)$_2$ (10 mol%)	5	100
7	Zn(ClO$_4$)$_2$ (10 mol%)	5	100
8	ZnCl$_2$ (10 mol%)	5	100
9	ZnCl$_2$ (5 mol%)	5	97
10	ZnCl$_2$ (2 mol%)	15	97

^{a}Reaction conversion.

The thiolysis by a variety of substituted arylthiols was also investigated.[37] An example is illustrated in Scheme 8. No catalytic effect was observed in the case of o- and p-NH$_2$, and o- and p-CO$_2$H-subtituted phenylthiols, supposedly due to the formation of a stable complex with Zn^{2+} and its consequent deactivation as oxirane ring-opening catalyst. The efficiency of ZnCl$_2$ as catalyst was regained in the case of o-Me-, p-NHAc, and o-CO$_2$Me phenylthiols, that is, when the thiol carries functionalities with reduced binding properties.

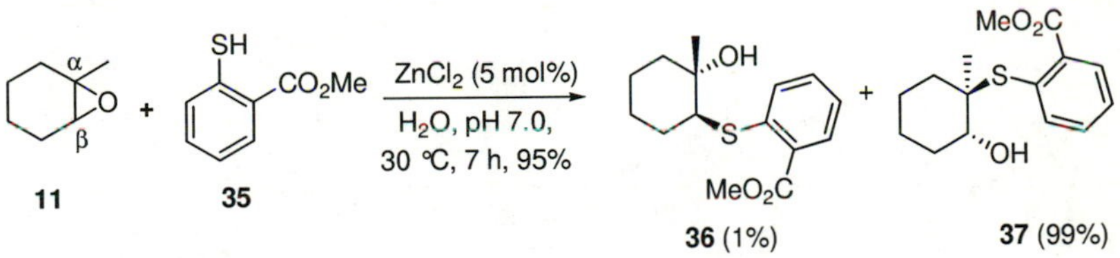

20 examples - yields of the isolated reaction products: 94-97%

Scheme 8

ZnCl$_2$ has proved to be a more efficient catalyst than InCl$_3$ in water toward alkyl- and aryl-substituted 1,2-epoxides, showing a catalytic efficiency under various pH conditions. On the contrary, in the case of strong biaptic 1,2-epoxides such as α,β-epoxycarboxylic acids, ZnCl$_2$ was a less efficient catalyst. InCl$_3$ (10 mol%) proved to be an efficient catalyst for the β-regio- and anti-stereoselective thiolysis of α,β-epoxycarboxylic acids in water at pH 4.0, also for α,β-epoxyhexanoic acid (**22**) and α,β-epoxybutanoic acid (**23**), which have under uncatalyzed conditions, a great tendency to be attacked at more electrophilic C-α position (Scheme 9).

155

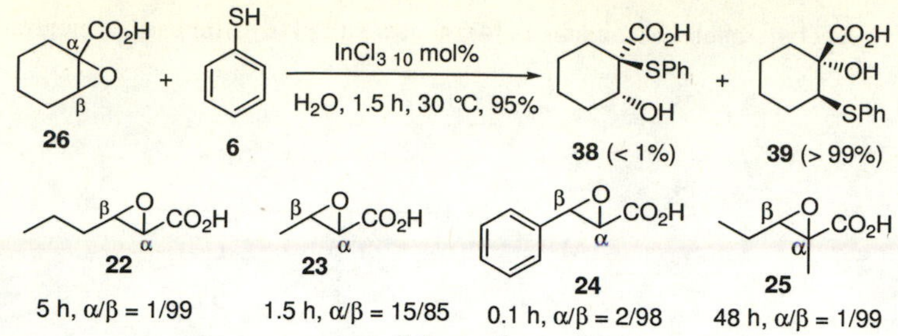

Scheme 9

Compounds 38 (< 1%) and 39 (> 99%)

- **22** 5 h, α/β = 1/99
- **23** 1.5 h, α/β = 15/85
- **24** 0.1 h, α/β = 2/98
- **25** 48 h, α/β = 1/99

3. Thiolysis of 1,2-epoxides under solvent-free condition

For historical and cultural reasons, organic reactions have generally been developed in solution while, a higher chemical efficiency would be expected if there was no medium interposed between the reactants. To minimize the amount of harmful organic solvents used in chemical processes, attention has recently been given to the possibility of performing chemical processes in the absence of solvent (solvent-free condition, SFC).[37,38] The examples reported demonstrate that no-solvent reactions are generally faster, give higher selectivities and yields, and usually require easier work-up procedures and simpler equipment.[18,38,39]

3.1. Base-catalyzed thiolysis

Thiols when treated with an organic or inorganic base under SFC can be deprotonated or can form an intimate ionic couple where a partial negative charge is formally allocated on the sulfur atom. In this manner, nucleophilic properties of thiols can be significantly marked. Following this idea, K_2CO_3, as a Brönsted base, and n-Bu$_3$P, as a Lewis base, were used in thiolysis of a variety of 1,2-alkyl-and aryl-substituted 1,2-epoxides with phenylthiol (**6**) under SFC.[40]

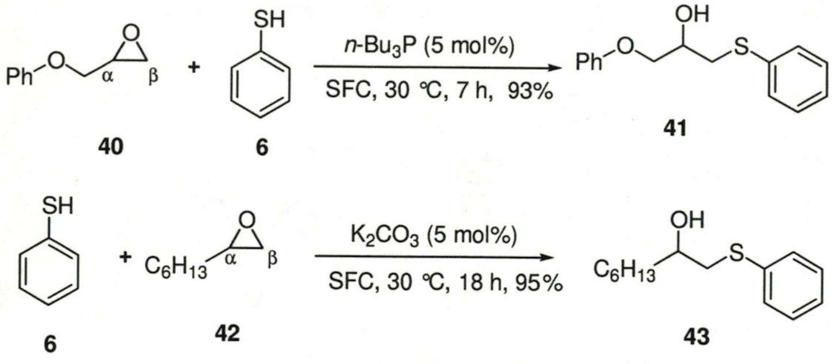

8 examples - yields of the isolated reaction products: 72-95%

Scheme 10

While n-Bu$_3$P in acetonitrile or dichloromethane was not effective at all, it showed to be an efficient and mild catalyst under SFC and only 5 mol% of this phosphine was sufficient to complete the reactions of a variety of 1,2-epoxides with phenylthiol (**6**) in 7-500 h. The Brönsted base K_2CO_3 was much more efficient and, when used at 5-10 mol%, it efficiently catalyzed the thiolysis process giving the corresponding β-hydroxy sulfides after 1-300 h in excellent yields. The reactions were predominantly β-regioselective, even in the case of styrene oxide which usually gives a major percentage of α-product. Two representative examples are illustrated in Scheme 10.

156

Penso *et al.* reported[10b] that tetrabutylammonium fluoride (TBAF) efficiently promoted the reaction of a variety of aryl- and alkylthiols with variously substituted 1,2-epoxides under SFC (Scheme 11). Excellent yields were obtained (88-100%) in all cases. All the processes were completely β-regioselective with the exception of styrene oxide (**8**) and 1-phenylcyclohexene oxide which also gave the α-products, as expected. Same authors have also reported that TBAF was much less efficient when used in acetonitrile as reaction medium. Other ammonium salts such as bromide, hydrogen sulfate and dyhydrogen trifluoride, were less efficient and required higher temperatures and longer reaction times to complete the process.

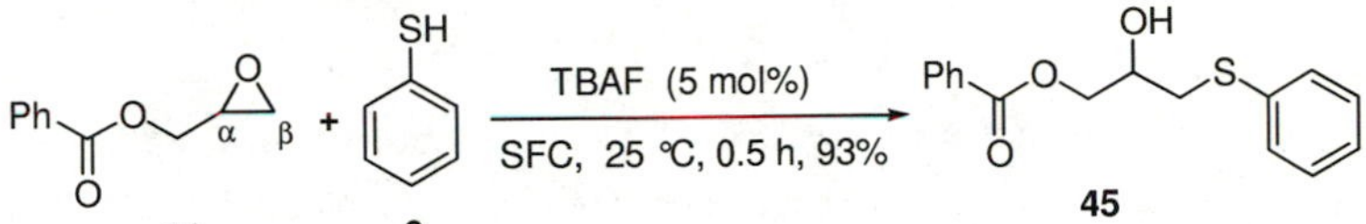

14 examples - yields of the isolated reaction products: 88-100%

Scheme 11

Polystyryl-supported 1,5,7-triazabicyclo[4.4.0]dec-5-ene (PSTBD) was recently used for the first time as a catalyst under SFC in a variety of organic transformations[39e] including the thiolysis of 1,2-epoxides by allylthiol (**46**). SFC allowed to overcome problems related to the swelling of the polymer support. In fact 5 mol% of PSTBD promoted the complete conversion of styrene oxide (**8**) to the corresponding β-hydroxy sulfides in only 0.5 h at 30 °C under SFC, while very low conversion was reached in acetonitrile (Scheme 12). Although styrene oxide (**8**) is usually attacked at the more electrophilic α-position, in this case steric factors prevailed and a β-regioselectivity was observed (α/β 20/80).

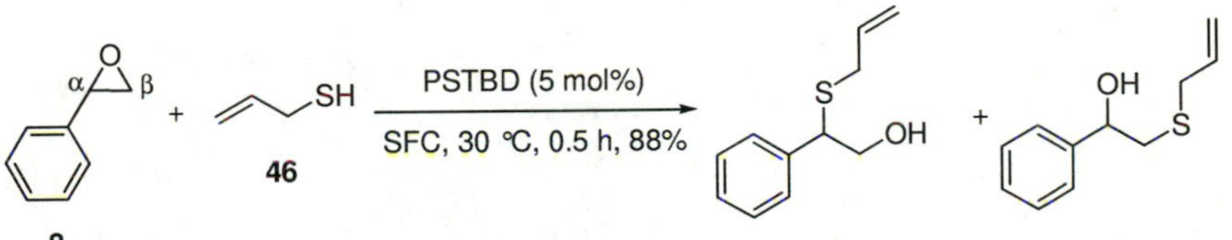

Scheme 12

3.2. Acid-catalyzed thiolysis

InCl$_3$ was chosen to investigate the Lewis-acid catalyzed thiolysis of 1,2-epoxides under SFC[40] since this salt showed to be a very good catalyst in aqueous medium. InCl$_3$ (5 mol%) was found to be extremely active giving the complete conversion of 1,2-epoxides to the corresponding β-hydroxy phenyl sulfides in only 1-10 min and with isolation yields ranging from 77 to 95%. One example is illustrated in Scheme 13. Only in the case of phenyl glycidol (**40**), the reaction with phenylthiol (**6**) stopped at 51% conversion and 10 mol% of InCl$_3$ were needed to reach a conversion of 84%. In the case of styrene oxide (**8**) the process was too fast and the by-product 1,2-bis(phenylthio)-ethylbenzene was formed. By reducing the catalyst loading to 1 mol%, the by-product formation was significantly reduced (8%).

The influence of a Brönsted acid as catalyst, such as *p*-TsOH was also investigated. Poor catalytic effects were generally observed. The only good result was achieved in the thiolysis of cyclohexene oxide

(**14**) that gave the corresponding *trans*-β-hydroxy sulfide (**15**) in quantitative yields after 48 h at 30 °C. The same process, when performed in acetonitrile, gave only 2% conversion after 48 h at 30 °C.

8 examples - yields of the isolated reaction products: 77-95%
Scheme 13

We recently published[41] the first one-pot protocol for the preparation of benzo[*e*]1,4-oxathiepin-5-ones based on the β-regio- and *anti*-stereoselective self-promoted addition of thiosalycilic acid (**49**) on 1,2-epoxides as illustrated in Scheme 14. To achieve this goal we initially investigated the thiolysis under SFC in the presence of various solvents and in the absence and presence of Lewis and Brönsted catalysts. The best results were obtained by avoiding the use of any external catalyst and the best yields (85-99%) and reaction conditions (50-60 °C, 2-8 h) were obtained by simply mixing thiosalycilic acid (**49**) with an appropriate 1,2-epoxide under SFC. An example is illustrated in Scheme 14. This self-promoted thiolysis could be the consequence of a favorable hydrogen bond between the oxirane ring and the thiosalicylic acid (**49**), which favors the approach of the reagents, particularly in aprotic solvents and under SFC.

9 examples - yields of the isolated reaction products: 85/94%
Scheme 14

4. Conclusions

Thiolysis of 1,2-epoxides is a very useful organic process and can completely exploit its synthetic utility through the use of alternative reaction media such as water or SFC.

These reactions in water under basic conditions give high yields and are generally fast. At acidic pH, Lewis acids are necessarily used to catalyze and to control the regio- and stereoselectivity of the process. In(III) and Zn(II) salts have showed the highest catalytic efficiency at pH 4.0, while under neutral condition Zn(II) and Co(II) are the only effective metal catalysts found. SFC strongly increases the reactivity of the oxirane ring-opening allowing generally faster reactions and smaller catalyst loadings. The results illustrated in this review show that the use of water or SFC can play an important role for the greener preparation of target molecules *via* thiolysis of 1,2-epoxides.

Aknowledgments

The Ministero dell'Istruzione dell'Università e della Ricerca (MIUR) and the Università degli studi di Perugia (within the funding projects: COFIN, COFINLAB (CEMIN) and FIRB 2001) and Fondazione Cassa di Risparmio di Perugia are thanked for financial support.

References
1. Bauer, K.; Garbe, D.; Surburg, H. In *Common Fragrance and Flavor Materials*; Wiley-VCH: New York/Weinheim, 1997.
2. Some, among many, notable examples: (a) Itazaki, H.; Nagashima, K.; Sugita, K.; Yoshida, H.; Kawamura, Y.; Yasuda, Y.; Matsumoto, K.; Ishii, K.; Uotani, N.; Nakai, H.; Terui, A.; Yoshimatsu, S. *J. Antibiot.* **1990**, *43*, 1524. (b) Bollag, D. M.; McQueney, P. A.; Zhu, J.; Hensens, O.; Koupal, L.; Liesch, J.; Goetz, M.; Lazarides, E.; Woods, C. M. *Cancer Res.* **1995**, *55*, 2325-2333. (c) Nakajima, H.; Takase, S.; Terano, H.; Tanaka, H. *J. Antibiot.* **1997**, *50*, 96. (d) Askin, D. *Curr. Opin. Drug. Disc. Devel.* **1998**, *1*, 338-348. (e) Liu, P.; Panek, J. S. *J. Am. Chem. Soc.* **2000**, *122*, 1235-1236. (f) Lindsay, K. B.; Pyne, S. G. *J. Org. Chem.* **2002**, *67*, 7774-7780. (g) Furutani, T.; Imashiro, R.; Htsuda, M.; Seki, M. *J. Org. Chem.* **2002**, *67*, 4599-4601. (h) Fringuelli, F.; Pizzo, F.; Rucci, M.; Vaccaro, L. *J. Org. Chem.* **2003**, 68, 7041-7045. (i) Raheem, I. T.; Goodman, S. N.; Jacobsen, E. N. *J. Am. Chem. Soc.* **2004**, *126*, 706-707. (j) Marco-Contelles, J.; Molina, M. T.; Anjum, S. *Chem. Rev.* **2004**, *104*, 2857-2900.
3. For reviews, see: (a) Katsuki, T. In *Comprehensive Asymmetric Catalysis*; Jacobsen, E. N.; Pfaltz, A.; Yamamoto, H., Eds.; Springer: New York, 1999; Vol. 2, Chap. 18.1. (b) Jacobsen, E. N.; Wu, M. H. ibid., Chap. 18.2. (c) Frohn, M.; Shi, Y. *Synlett* **2000**, 1979-2000. (d) Hofstetter, K.; Lutz, J.; Lang, I.; Witholt, B.; Schmid, A. *Angew. Chem. Int. Ed.* **2004**, *43*, 2163-2166.
4. (a) Parker, R. E.; Isaacs, N. S. *Chem. Rev.* **1959**, *59*, 737-799. (b) Rao, A. S.; Paknikar, S. K.; Kirtane, J. G. *Tetrahedron* **1983**, *39*, 2323-2367. (c) Smith, J. G. *Synthesis* **1984**, 629-656. (d) Bartók, M.; Láng, K. L. *Small Ring Heterocycles*. In *The Chemistry of Heterocyclic Compounds*, Hassner, A., Ed.; Wiley: New York, 1985, Vol. 42, Part 3, Chapter 1. (e) Zwanenburg, B. *Pure Appl. Chem.* **1999**, *71*, 423-430. (f) Righi, G.; Bonini, C. *Targets in Heterocyclic Chemistry*, **2000**, *4*, 139-165. (g) Amantini, D.; Fringuelli, F.; Piermatti, O.; Tortoioli, S.; Vaccaro, L. *Arkivoc* **2002**, (xi), 293-311.
5. Magdalou, J.; Fournel-Gigleux, S.; Testa, B.; Ouzzine, M. In *The Practice of Medical Chemistry*; Wermuth, C. G., Ed.; Academic Press: San Diego, CA; 2nd edition, 2004, p. 541.
6. (a) Corey, E. J., Clark, D. A.; Goto, G.; Marfat, A.; Mioskowski, C.; Samuelsson, B; Hammarström, S. *J. Am. Chem. Soc.* **1980**, *102*, 1436-1439. (b) Corey, E. J.; Clark, D. A.; Marfat, A.; Goto, G. *Tetrahedron Lett.* **1980**, *21*, 3143-3146. (c) Sharma, S.; Mesic, T. M.; Matin, R. A. *Tetrahedron* **1994**, *50*, 9223-9228. (d) Sauders, J. in *Top Drugs, Top Synthetic Routes*, Oxford University Press: Oxford, UK, 2000, and refereces cited therein. (e) Martinelli, M. J.; Vaidyanathan R.; Van Khau, V.; Staszak, M. A. *Tetrahedron Lett.* **2002**, *43*, 3365-3367. (f) Furutani, T.; Imashiro, R.; Hatsuda, M. *J. Org. Chem.* **2003**, *67*, 4599-4601.
7. Mühlman, A.; Classon, B.; Anders, H.; Samuelsson, B. *J. Med. Chem.* **2001**, *44*, 3402-3406.
8. Armostrong, R. N. *Curr. Opin. Chem. Biol.* **1998**, *2*, 618-623.
9. (a) Abul-Hajj, Y. J. *J. Med. Chem.* **1986**, *29*, 582-584 [Na, neat]. (b) Behrens, C. H.; Sharpless, K. B. *J. Org. Chem.* **1985**, *50*, 5696-5704 [NaH, THF]. (c) Yamada, O.; Ogasawara, K. *Synlett* **1995**, 427-428 [NaH, THF]. (d) Justo De Pomar, J. C.; Soderquist, A. *Tetrahedron Lett.* **1998**, *29*, 4409-4412 [DBU]. (e) Adams, H.; Bell, R.; Cheung, Y.–Y.; Jones, N. D.; Tomkinson, N. C. O. *Tetrahedron: Asymmetry*, **1999**, *10*, 4129-4142 [Na].
10. (a) Iizawa, T.; Goto, A.; Nishikubo, T. *Bull. Chem. Soc. Jpn.* **1989**, *62*, 597-598 [Bu_4NX]. (b) Albanese, D.; Landini, D.; Penso, M. *Synthesis* 1994, 34-36 [Bu_4NF].
11. (b) Maiti, A. K.; Bhattacharyya, P. *Tetrahedron* **1994**, *50*, 10483-10490 [PEG 4000]. (b) Younes, M. R.; Chaabouni, M. M.; Baklouti, A. *Tetrahedron Lett.* **2001**, *42*, 3167- 3169 [Triton-B].
12. (a) Chong, J. M.; Sharpless, K. B. *J. Org. Chem.* **1985**, *50*, 1560-1563 [$Ti(O\text{-}i\text{-}Pr)_4$]. (b) Vougioukas, A. E.; Kagan, H. B. *Tetrahedron Lett.* **1987**, *28*, 6065-6068 [Lanthanides complexes]. (c) Bortolini, O.; Di Furia, F.; Licini, G., Modena, G. *Phosphorous Sulfur* **1988**, *37*, 171-174 [$LnCl_3$]. (d) Guivisdalsky, P. N.; Bittman, R. *J. Am. Chem. Soc.* **1989**, *111*, 3077-3079 [$BF_3 \cdot Et_2O$]. (e) Iqbal, J.; Pandey, A.; Shukla, A.; Srivastava, R. R.; Tripathi, S. *Tetrahedron* **1990**, *46*, 6423-6432 [$CoCl_3$]. (f) Chini, M.; Crotti, P.; Giovani, E.; Macchia, F.; Pineschi, M. *Synlett* **1992**, 303-305 [$LiClO_4$]. (g) Lin, G.; Shi, Z.; Zeng, C. *Tetrahedron: Asymmetry* **1993**, *4*, 1533-1536 [$Ti(O\text{-}i\text{-}Pr)_4$]. (h) Still, I. W. J.; Martin, L. P. J. *Synthetic Commun.* **1998**, *28*, 913-923 [SmI_2]. (i) Yadav, J. S.; Reddy, B. V. S.; Baisha, G. *Chem Lett.* **2002**, 906-907 [$InCl_3$]. (j) Chanrasekar, S.; Reddy, Ch. R.; Babu, B. N.; Chandrashekar G. *Tetrahedron*

Lett. **2002**, *43*, 3801-3803 [B(C₆F₅)₃]. (k) Devan, N.; Sridhar, P. R.; Prabhu, K. R.; Chandrasekaran, S. *J. Org. Chem.* **2002**, *67*, 9417-9420 [tetrathiomolybdate]. (l) Sasaki, M.; Tanino, K.; Hirai, A.; Miyashita, M. *Org. Lett.* **2003**, *5*, 1789-1791 [(CH₃O)₃B].

13.   (a) Posner, G. H.; Rogers, D. Z. *J. Am. Chem. Soc.* **1977**, *79*, 8208-8213 [Al₂O₃]. (b) Nishikubo, T.; Iizawa, T.; Shimojo, M.; Kato, T.; Shiina, A. *J. Org. Chem.* **1990**, *55*, 2536-2542 [polymer-onium ions]. (c) Maiti, A. K.; Biswas, G. K.; Bhattacharyya, P. *J. Chem. Res.: Synop.* **1993**, 325 [Montmorillonite]. (d) Takeuchi, H.; Kitajima, K.; Yamamoto, Y.; Mizuno, K. *J. Chem. Soc, Perkin Trans. 2* **1993**, 199-203 [X-type zeolite]. (e) Raubo, P.; Wicha, J. *Pol. J. Chem.* **1995**, *69*, 78- [SiO₂]. (f) Choi, J.; Yoon, N. M. *Synthesis* **1995**, 373-375 [hydrosulfide exchange resin]. (g) Bandini, M.; Fagioli, M.; Melloni, A.; Umani-Ronchi, A. *Adv. Synth. Catal.* **2004**, *246*, 573-578 [Amberlist-In]. (h) Polshettiwar, V.; Kaushik, M. P. *Catal. Commun.* **2004**, *5*, 515-518 [CsF-Celite].

14.   (a) Yamashita, H. *Bull. Chem. Soc. Jpn.* **1998**, *61*, 1213-1220 [zinc tartrates]. (b) Iida, T.; Yamamoto, N.; Sasai, H.; Shibasaki, M. *J. Am. Chem. Soc.* **1997**, *119*, 4783-4784 [gallium complex]. (c) Fukuzawa, S.; Kato, H.; Ohtaguchi, M.; Hayashi, Y.; Yamazaki, H. *J. Chem. Soc., Perkin Trans. 1* **1997**, 3059-3064 [carbonyl chromium complex]. (d) Wu, M. H.; Jacobsen, E. N. *J. Org. Chem.* **1998**, *63*, 5252-5254 [(salen) chromium complex]. (e) Wu, J.; Hou, X.-L.; Dai, L.-X.; Xia, L.-J.; Tang, M.-H. *Tetrahedron: Asymmerty* **1998**, *9*, 3431-3436 [titanium complex]. (f) Leong, P.; Lautens, M. *J. Org. Chem.* **2004**, *69*, 2194-2196 [rhodium complex].

15.   (a) Meister, A.; Andersen, M. E. *Ann. Rev. Biochem.* **1983**, *52*, 711-760. (b) Conchillo, A.; Camps, F.; Messeguer, A. *J. Org. Chem.* **1990**, *55*, 1728-1735. (c) Wipf, P.; Jeger, P.; Kim, Y. *Bioorg. Med. Chem. Lett.* **1998**, *8*, 351-356.

16.   (a) Fringuelli, F.; Matteucci, M.; Piermatti, O.; Pizzo, F.; Burla, M. C. *J. Org. Chem.* **2001**, *66*, 4661-4666. (b) Amantini, D.; Fringuelli, F.; Pizzo, F.; Vaccaro, L. *J. Org. Chem.* **2001**, *66*, 4463-4467. (c) Amantini, D.; Fringuelli, F.; Piermatti, O.; Pizzo, F.; Vaccaro, L. *Green Chem.* **2001**, *3*, 229-232. (d) Amantini, D.; Fringuelli, F.; Pizzo, F. *J. Org. Chem.* **2002**, *67*, 7238-7243. (e) Fringuelli, F.; Pizzo, F.; Rucci, M.; Vaccaro, L. *J. Org. Chem.* **2003**, *68*, 7041-7045. (f) Amantini, D.; Fringuelli, F.; Piermatti, O.; Pizzo, F.; Vaccaro, L. *J. Org. Chem.* **2003**, *68*, 9263-9268.

17.   (a) Fringuelli, F.; Pizzo, F.; Vaccaro, L. *Synthesis* **2000**, 646-650. (b) Fringuelli, F.; Pizzo, F.; Vaccaro, L. *J. Org. Chem.* **2001**, *66*, 3544-3548. (c) Fringuelli, F.; Pizzo, F.; Vaccaro, L. *Tetrahedron Lett.* **2001**, *42*, 1131-1134. (d) Fringuelli, F.; Pizzo, F.; Vaccaro, L. *J. Org. Chem.* **2001**, *66*, 4719-4722. (e) Fioroni, G.; Fringuelli, F.; Pizzo, F.; Vaccaro, L. *Green Chem.* **2003**, *5*, 425-428.

18.   (a) Amantini, D.; Beleggia, R.; Fringuelli, F.; Pizzo, F.; Vaccaro, L. *J. Org. Chem.* **2004**, *69*, 2896-2898. (b) Fringuelli, F.; Pizzo, F.; Tortoioli, S.; Vaccaro, L. *J. Org. Chem.* **2004**, *69*, 7745-7747.

19.   Rideout, D. C.; Breslow, R. *J. Am. Chem. Soc.* **1980**, *102*, 7816-7817.

20.   (a) Li, C. J.; Chang, T. H. *Organic Reactions in Aqueous Media*; Wiley: New York, 1997. (b) Fringuelli, F.; Piermatti, O.; Pizzo, F. In *Organic Synthesis in Water*; Grieco P. A., Ed.; Blackie Academic and Professional: London, 1998; pp. 223-261.

21.   Kolb, H. C.; Sharpless, K. B. *Angew. Chem., Int. Ed.* **2001**, *40*, 2004-2021.

22.   (a) Streitwieser, A. *Chem. Rev.* **1956**, *56*, 571, cfr. page 582. (b) Smith, M. B.; March, M. B. In *March's Advanced Organic Chemistry* fifth edition; Wiley-Interscience Publication, 2001, p. 329-332.

23.   Fringuelli, F.; Pizzo, F.; Tortoioli, S.; Vaccaro, L. *Adv. Synth. Catal.* **2002**, *344*, 379-384.

24.   Kreevoy, M. M.; Eichinger, B. R.; Stary, F. E.; Katz, E. A.; Sellstedt, J. M. *J. Org. Chem.* **1964**, *29*, 1641-1645.

25.   Fringuelli, F.; Pizzo, F.; Tortoioli, S.;Vaccaro, L. *Green Chem.* **2003**, *5*, 436-440.

26.   Fringuelli, F.; Piermatti, O.; Pizzo, F.; Vaccaro, L. *J. Org. Chem.* **1999**, *64*, 6094-6097.

27.   This result was expected considering that nucleophilic ability of phenylthiolate is ca. 50.000 times that of water, while under our reaction conditions there are ca. 75 mmol of water per 1 mmol of phenylthiolate.[21] Hydroxide ion and phenylthiolate have comparable nucleophilic ability[14] but under our reaction conditions (pH 9.0) the concentration of $OH^{\ominus}$ is ca. 7.5 x 10⁻⁶ smaller than that of the thiolate.

28.   Fan, R.-H; Hou, X.-L. *J. Org. Chem.* **2003**, *68*, 726-730.

29.   Fringuelli, F.; Pizzo, F.; Vaccaro, L. *J. Org. Chem.* **2004**, *69*, 2315-2321.

30. (a) McElroy, A. B.; Warren, S. *Tetrahedron Lett.* **1985**, *26*, 5709-5712. (b) Lauret, C. *Tetrahedron: Asymmetry* Report N. 52 **2001**, *12*, 2359-2383.

31. Fringuelli, F.; Pizzo, F.; Tortoioli, S.; Vaccaro, L. *Unpublished results.*

32. (a) Baes Jr., C. F.; Mesmer, R. E. *The Hydrolysis of Cations*; Wiley: New York, 1976. (b) Richens, D. T. *The Chemistry of Aqua Ions*; Wiley: New York, 1997.

33. Amantini, D.; Fringuelli, F.; Pizzo, F.; Tortoioli, S.; Vaccaro, L. *Synlett* **2003**, 2292-2296.

34. Srogl, J.; Liu, W.; Marshall, D.; Liebeskind, L. S. *J. Am. Chem. Soc.* **1999**, *121*, 9449-9450.

35. *Handbook of Metal-Ligand Interactions in Biological Fluids*; Berthon, G., Ed.; Marcel Dekker: New York-Basel-Hong Kong, 1995.

36. Fringuelli, F.; Pizzo, F.; Tortoioli, S.; Vaccaro, L. *J. Org. Chem.* **2003**, *68*, 8248-8251.

37. a) Tundo, P.; Ananstas, P. T. In *Green Chemistry: Theory and Practice*; Oxford University Press: Oxford, 1998. (b) De Simone, J. M.; *Science*, **2002**, *297*, 799-803.

38. (a) Metzger, J. O. *Angew. Chem. Int. Ed.* **1998**, *37*, 2975-2978. (b) Varma, R. S. *Green Chem.* **1999**, 43-55. (c) Tanaka, K; Toda, F. *Chem Rev.* **2000**, *100*, 1025-1074. (d) Varma, R. S. *Pure Appl. Chem.* **2001**, *73*, 193-198. (e) Cave, G. W. V.; Raston, C. L.; Scott, J. L. *Chem. Commun.* **2001**, 2159-2169. (f) Tanaka, K.; Toda, F. In *Solventy-free Organic Synthesis*; Wiley-VCH, 2003.

39. (a) Loh, T.-P.; Huang, J.-M.; Goh, S.-H.; Vittal, J. J. *Org. Lett.* **2000**, *2*, 1291-1294. (b) Hajipour, A. R.; Arbabian, M.; Ruoho, A. E. *J. Org. Chem.* **2002**, *67*, 8622-8624. (c); Yadav, L. D. S.; Singh, S. *Synthesis* **2003**, *1*, 63-66. (d) Lee, J. C.; Bae, Y. H. *Synlett* **2003**, *4*, 507-508. (e) Xu, Z.-B-; Lu, Y.; Guo, Z.-R. *Synlett* **2003**, *4*, 564-566. (e) Fringuelli, F.; Pizzo, F.; Vittoriani, C.; Vaccaro, L. *Chem. Commun.* **2004**, 2756-2757.

40. Fringuelli, F.; Pizzo, F.; Tortoioli, S.; Vaccaro, L. *Tetrahedron Lett.* **2003**, *44*, 6785-6787.

41. Fringuelli, F.; Pizzo, F.; Tortoioli, S.; Vaccaro, L. *J. Org. Chem.* **2004**, *69*, 8780-8785.

SYNTHESIS OF HETEROCYCLIC COMPOUNDS BY [2+2] CYCLOADDITION TO OXA- AND AZA-NORBORNENIC SYSTEMS (INCLUDING CARBOMETALLATION REACTIONS)

Rocío Medel and Joaquín Plumet*

Departamento de Química Orgánica I, Facultad de Química, Universidad Complutense de Madrid, E-28040 Madrid, Spain (e-mail: plumety@quim.ucm.es)

Abstract. *Heteroatom substituted norbornene derivatives -mainly oxa- and azanorbornene derivatives- are useful synthetic intermediates in the synthesis of heterocyclic compounds. For this purpose the cycloaddition reactions achieved on the endocyclic carbon-carbon double bond constitute a versatile tool. Among these the [2+2] cycloaddition reactions have been scarcely studied. In this account the state-of-art of this methodology, including the metathesis reactions (ROM-CM) of these compounds are reviewed.*

Contents

1. Introduction

Reactions which form in one single step rings and/or stereocenters are useful methods for the convenient assembly of molecular structures. In this context, cycloadditions reactions hold a very important place in the arsenal of synthetic methods currently available to organic chemist. Even more, when these cycloadditions reactions are achieved on chiral, rigid structures, the chiral information stored into the rigid skeleton may be transferred to the obtained cycloadducts. That is the case of the bicyclic compounds formally derived from norbornene by introduction of one or more heteroatoms - mainly oxygen or nitrogen - into the parent structure. In these compounds the functionalization of the endocyclic double bond constitutes the first step in the synthesis of complex molecules. In this way some nucleophilic and electrophilic additions achieved with or without concomitant bridge opening have been extensively considered. To a lesser extent, the cycloaddition reactions to the endocyclic double bond were also used both with synthetic and theoretical purposes. In particular the [2+2] cycloaddition reactions on these double bonds give rise to fused four-membered rings but their value in synthesis goes beyond the straightforward formation of fused cyclobutane or cyclobutene derivatives because, in some cases, subsequent ring-expansion, ring-contraction or rearrangement may provide access to otherwise difficulty accessible compounds. The present report will concentrate on this less known aspect of the chemistry of some norbornene derivatives, in particular, 7-oxanorbornenes **1**, 7-azanorbornenes **2** and 2-azanorbornenes **3** (Figure 1).

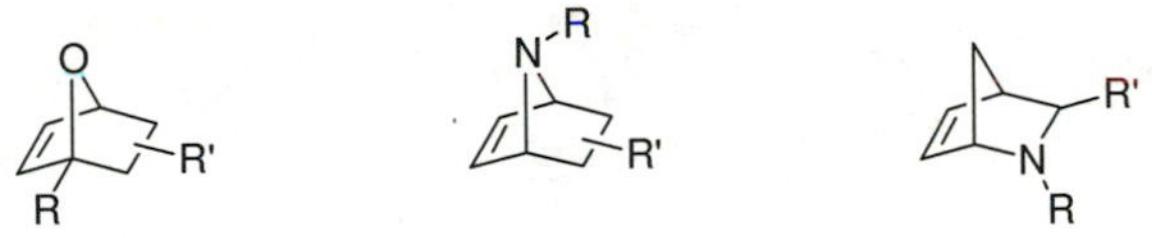

1   **2**   **3**

Figure 1

The chemistry of 7-oxa,[1,2,3,4] 7-aza[5] and 2-aza[6] norbornene derivatives has already been reviewed several times during the last ten years. In some of these reviews fragmentary aspects related with these cycloaddition reactions have been described. These will be conveniently referenced at the right moment but they will not be considered in detail.

This review covers the literature on the topic between 1990 and the end of September 2004 although some previous and specially significant reports have been included.

2. The [2+2] cycloaddition reactions. An overview

The [2+2] cycloaddition of alkenes and/or alkynes represents an important strategy for the synthesis of cyclobutan or cyclobuten derivatives.[7] The reaction is thermally forbidden by the Woodward-Hoffmann rules[8] because it must be antarafacial on one component. However, both geometrical and orbital constraints imposed by this approach, ensure that they are possible only in special circumstances. By these reasons thermal [2+2] cycloadditions of alkenes take place by a stepwise pathway involving diradical[9] or zwitterionic[10] intermediates. On the other hand, these reactions may be achieved photochemically,[11] or under thermal conditions using Lewis acid catalyst[12] or transition metal catalyst.[13]

On the other hand, [2+2] cycloaddition reactions of heterocumulenes (ketenes, ketene-iminium salts and isocyanates) with olefinic and other multiple bonds may occur by a concerted thermal pathway because the twisted double bond allows easier antarafacial addition of one component. However, for these compounds a range of mechanism of various degrees of concertedness have been proposed.[14] In any case, the resulting products specially oxetanes,[15] β-lactones[16] and β-lactams[17] are important synthetic objectives.

163

3. Thermal [2+2] cycloaddition reactions to oxa- and azanorbornene derivatives

To the best of our knowledge, the thermal, uncatalyzed [2+2] cycloaddition reactions of oxa- and azanorbornene derivatives have never been described. For instance, the reluctance of the endocyclic double bond to react with the *in situ* generated ketenes has been observed in the case of 7-oxanorbornenone **4**[18] and the related imine **5**.[19] In both cases β-lactone **6** and β-lactams **7** were obtained without detection of products arising from the [2+2] cycloaddition reaction to the carbon-carbon double bond (Scheme 1).[20]

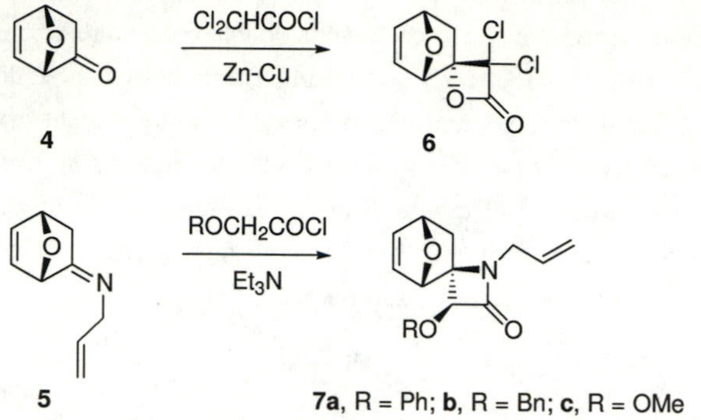

7a, R = Ph; **b**, R = Bn; **c**, R = OMe

Scheme 1

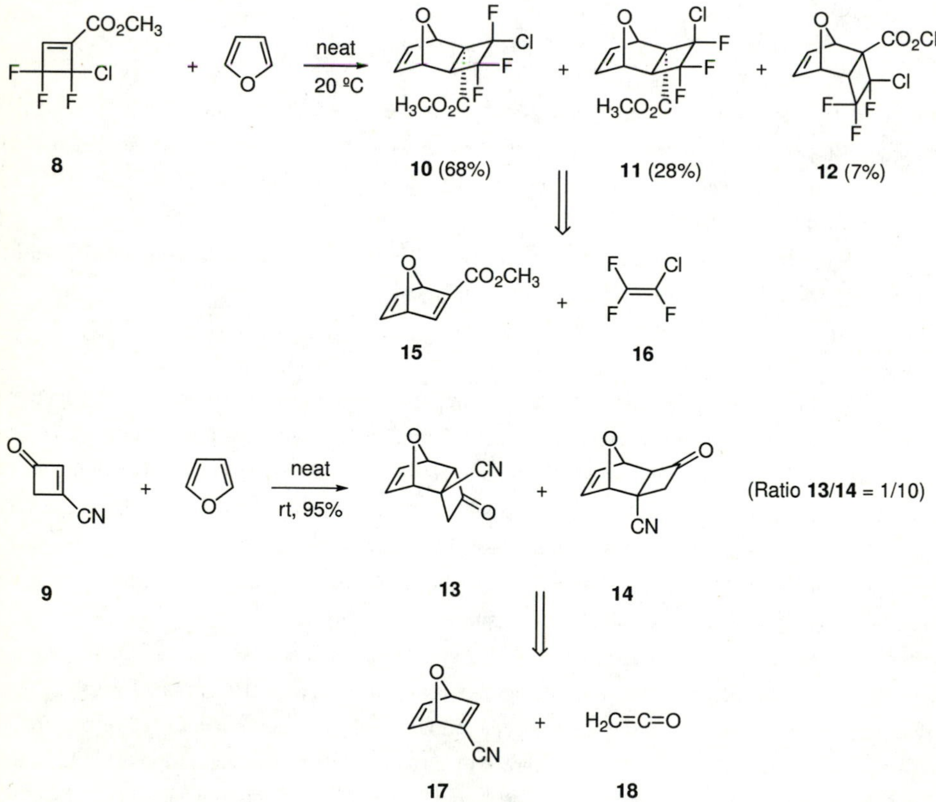

Scheme 2

164

A formal [2+2] cycloaddition reaction of oxanorbornadienes may be achieved by Diels-Alder reaction of furan with some cyclobutene derivatives. For instance, reaction of cyclobutenes **8**[21] and **9**[22] with furan as solvent afforded cycloadducts **10-12** and **13-14** respectively (Scheme 2). Compounds **10-12** are the formal [2+2] cycloadducts of oxanorbornadiene **15** with trifluorchloroethylene **16** and compounds **13** and **14** are the formal cycloadducts of oxanorbornadiene **17** and ketene **18**.

4. Photochemically induced [2+2] cycloaddition reactions of oxanorbornene derivatives

The photochemically induced [2+2] cycloaddition reactions of 7-oxanorbornene derivatives and alkenes has only one isolated precedent which also shows an interesting case of chemoselectivity.[23] The UV irradiation of *exo,endo*-2,7-bis-(methoxycarbonyl)-11,12-dioxatetracyclo[6.2.1.1^{3,6}.0^{4,7}]dodeca-4,9-diene **19** in the presence of *p*-benzoquinone in CDCl$_3$ afforded photoadduct **20** (oxetane) in 78% yield (Scheme 3). Under the same reaction conditions, the *exo,exo* isomer **21** was photochemically inactive. In the same way, the UV irradiation of **19** with 1,4-naphtoquinone gave a single product **22** in 44% yield.[24]

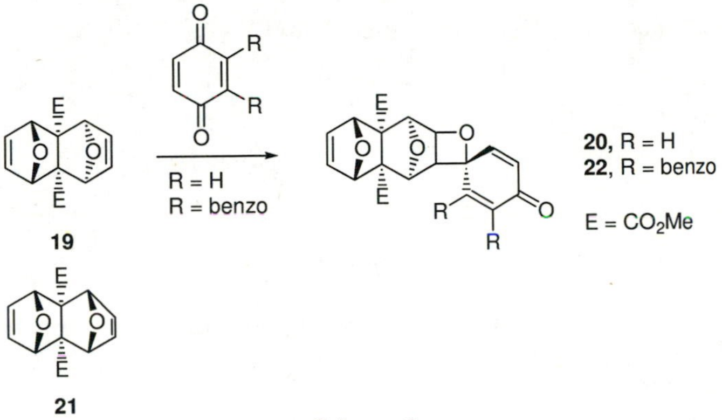

Scheme 3

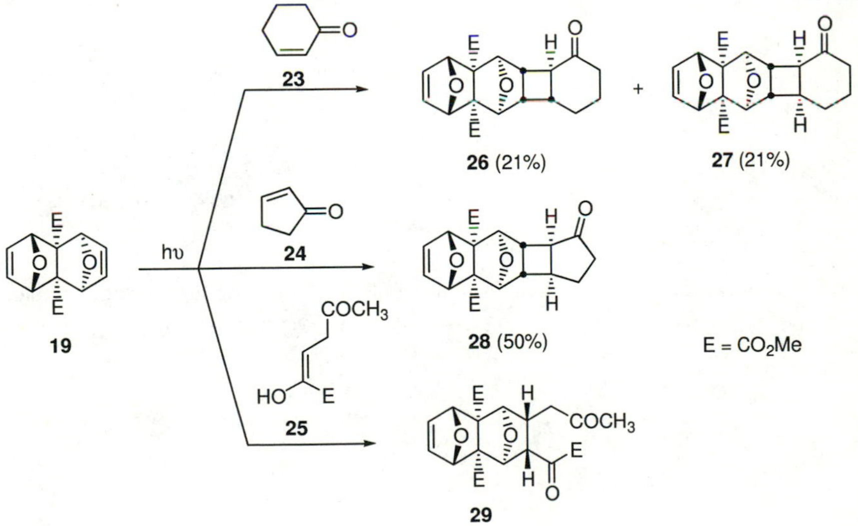

Scheme 4

165

In sharp contrast, cyclic enones **23** and **24** and the α,β-unsaturated ketone **25** by reaction with **19**, gave the intermolecular [2+2] cycloaddition across the carbon-carbon double bond affording cycloadducts **26-29** (Scheme 4). In the case of the reaction with enone **23** others minor diastereoisomers (overall yield 1.5%) have also been obtained.

It should be indicated that, at difference of **19**, the monocyclic 7-oxabicyclo[2.2.1]heptadiene **30**, by irradiation with **25** solely gave the photoisomer 7-oxaquadricyclane derivative **31** without observation of the intermolecular photoadduct (Scheme 5).[25]

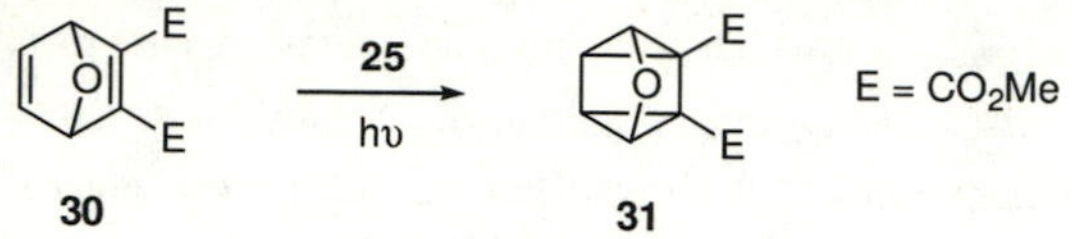
E
E
25
hυ
E
E
E = CO_2Me
30
31

Scheme 5

The thermal cycloreversion reactions of these photocycloadducts were also described. For instance, compound **28** was transformed into **32** by reflux in xilene at 140 °C (Scheme 6).

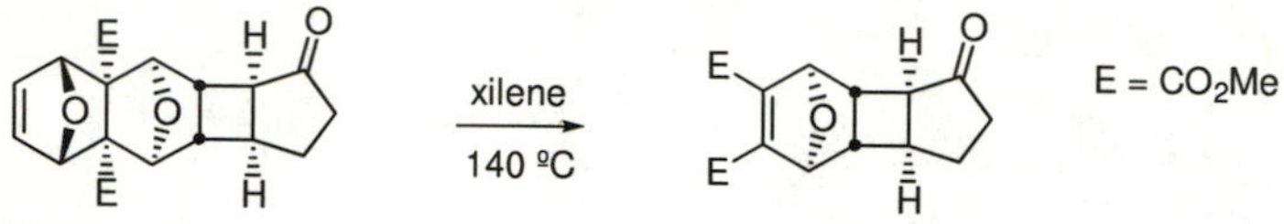
E
H
O
E
H
xilene
140 ºC
E
H
O
E
H
E = CO_2Me
28
32 (60%)

Scheme 6

5. High-pressure induced [2+2] cycloaddition reactions of oxanorbornene derivatives

Cyanoacetylene **33** behaves as a moderate dienophile in Diels-Alder cycloadditions.[26] However, when the oxanorbornenic cyano derivative **34** was treated with **33** in a high-pressure experiment (40 °C, 12 Kbar), the [2+2] cycloadduct **35** and the homo Diels-Alder product **36** were obtained in ratio 86:14 (GC analysis). The reaction probably occurs *via* a stepwise pathway (Scheme 7).[27]

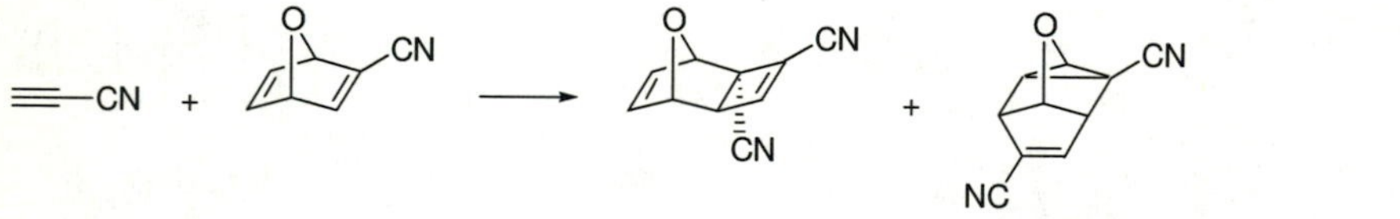
CN
+
O
CN
O
CN
CN
+
O
CN
NC
33
34
35
36
(35 : 36 = 86 : 14)

Scheme 7

Compound **34** was synthesized by flash-vacuum-pyrolysis (FVP) of a mixture of compounds **37** and **38**, previously obtained in the cycloaddition of **33** with furan at 140 °C, 1 bar (Scheme 8).

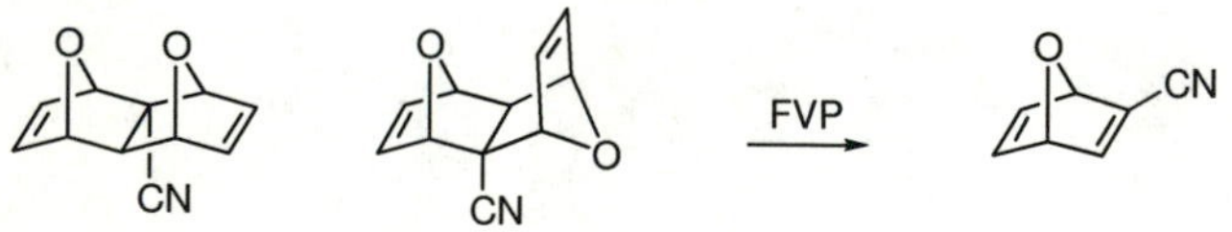
O
O
CN
O
O
CN
FVP
O
CN
37
38
34

Scheme 8

6. Transition metal mediated [2+2] cycloaddition reactions of oxa- and azanorbornene derivatives

Transition metal mediated cycloaddition reactions is today a standard method in synthetic organic chemistry and new advances on this methodology are very often described.[13,28] Cycloaddition reaction such as [2+2+2], [3+2], [4+2], [5+2] and [5+3] catalyzed by transition metal complexes are well documented. However, only few reports concerns the metal-promoted [2+2] cycloaddition reactions and, among them, the reactions of strained alkenes have received the most attention. However, and in contrast with norbornene derivatives, the related oxa- and aza- analogues have been only scarcely studied. Metal complexes derived from Co, Ru and Ni have been effectively utilized in these reactions and the most significant results will be describe in the next paragraphs. Metathesis reactions which imply an initial Ru-mediated [2+2] cyclometallation reaction will be describe independently (see below).

6.1. Co-catalyzed [2+2] cycloaddition reactions of oxa- and azanorbornene derivatives and alkynes

Oxa- and azabenzonorbornadiene derivatives **39** react with mono- and disubstituted alkynes to give cyclobutene derivatives **40** and **41** using $[Co(PPh_3)_2I_2]$ as catalyst and in the presence of Zn powder and PPh_3 (Scheme 9).[29]

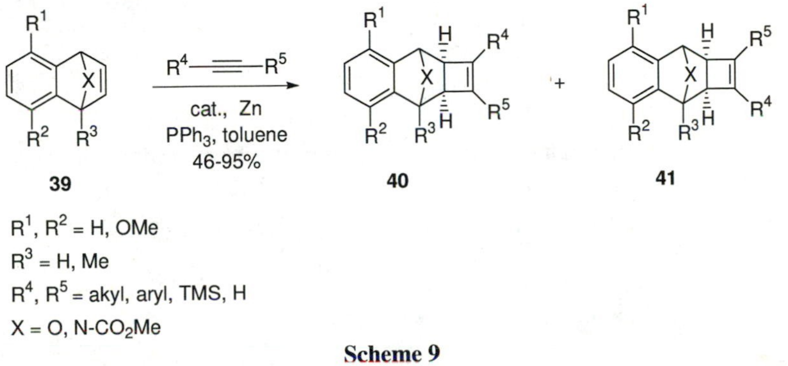

R^1, R^2 = H, OMe
R^3 = H, Me
R^4, R^5 = akyl, aryl, TMS, H
X = O, N-CO$_2$Me

Scheme 9

In the case of unsymmetrical oxa- and azabenzonorbornadienes and unsymmetrical alkynes very low regioselectivities were obtained reflecting the fact that the substituents R^1, R^2 are far away from the reaction centers.[30] In addition to oxa- and azabenzonorbornadienes, oxanorbornene such as **42** also underwent facile cross [2+2] cycloaddition with diphenylacetylene under standard conditions to give cyclobutene **43** in 90% yield (Scheme 10).

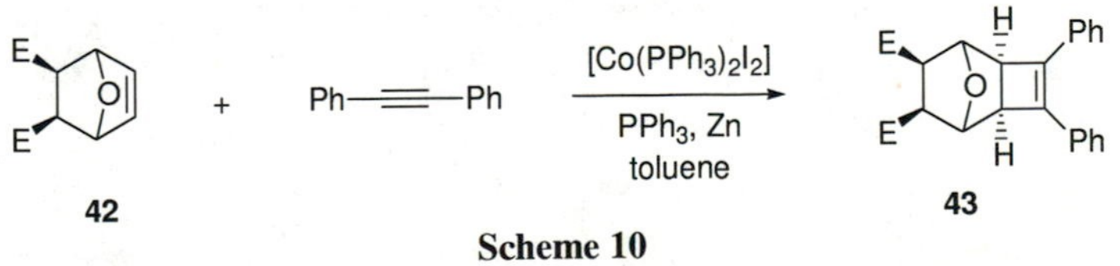

Scheme 10

Products arising from the homodimerization of the starting material **39** were observed in some cases. The amounts of these side products were minimized reducing the initial concentration of **39** or by addition of THF to the initial solution.

The proposed mechanism for this cycloaddition is depicted in the Scheme 11. Reduction of $Co(PPh_3)_2X_2$ to a Co(I) specie by the action of Zn metal initiates the catalytic cycle. Coordination of the alkyne and the starting alkene to the cobalt center followed by oxidative cyclometallation affords cobaltcyclopentane intermediate. Subsequent reductive elimination gives the cyclobutene and regenerates the Co(I) species.

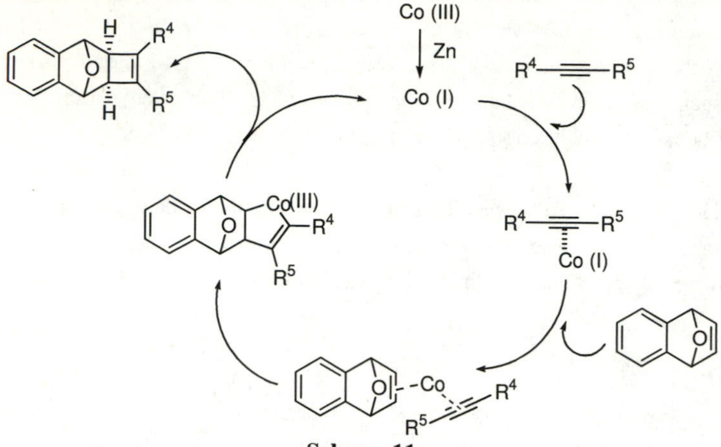

Scheme 11

6.2. Ni-catalyzed [2+2] cycloaddition reactions of oxa- and azanorbornene derivatives and alkynes[31]

The Ni-catalyzed [2+2] cycloadditions of oxa- and azabenzonorbornadiene derivatives **44** with alkynes to give the cyclobutenes **45** has been described by Cheng and co. (Scheme 12).[32]

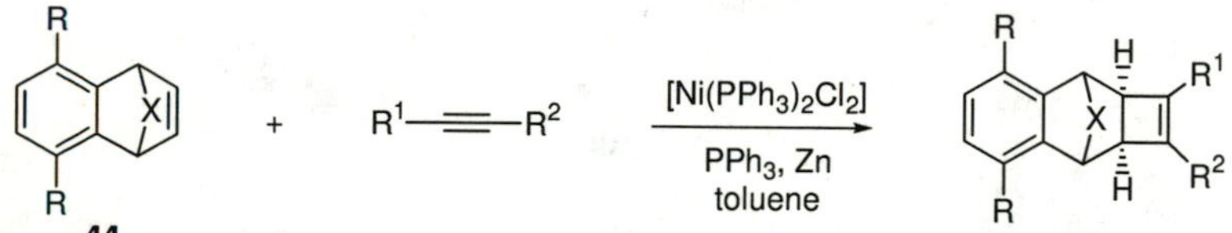

X = O, R = H
X = O, R = Me
X = N-CO$_2$Me, R = H

Scheme 12

In this reaction mono- and disubstituted alkynes afforded the corresponding *exo* cyclobutene derivatives in fair to excellent yields. In this context, two aspects should be commented: first, the reaction of 7-oxanorbornadiene **46** with alkyne **47** gave the formylderivative **48** in 90% yield without isolation of the corresponding diethylacetalderivative (Scheme 13).

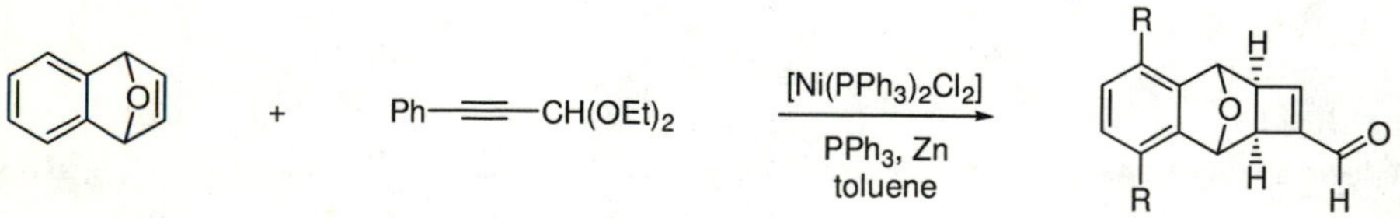

Scheme 13

168

Second, the reaction of diethylacetylenedicarboxylate with **46** depends on the solvent. Thus, the reaction was not observed in toluene but when the catalytic reaction was achieved in acetonitrile the [2+2] cycloadduct was isolated in 83% yield. Finally, it should be pointed out that the same Ni catalyst was inactive for the [2+2] cycloaddition of diethylacetylene and *bis*-(trimethylsilyl)acetylene.

On the other hand, when methoxyderivative **49** was treated with methylphenylacetylene, two regioisomers **50** and **51** were obtained in ratio 1:1 (Scheme 14).

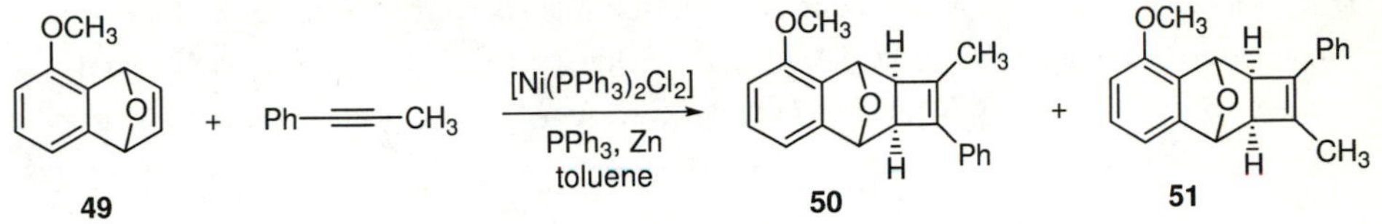

Scheme 14

In this reactions competition between [2+2] and [2+2+2][33] reactions have been observed. This competition can be clearly seen in the reaction of 1-heptyne and oxabenzonorbornadiene **46**. At ambient temperature (Scheme 15) the reaction gave the [2+2+2] cycloaddition products **52** in 68% yield whereas when the reaction temperature increases to 80 °C the [2+2] cycloaddition giving **53** dominates.

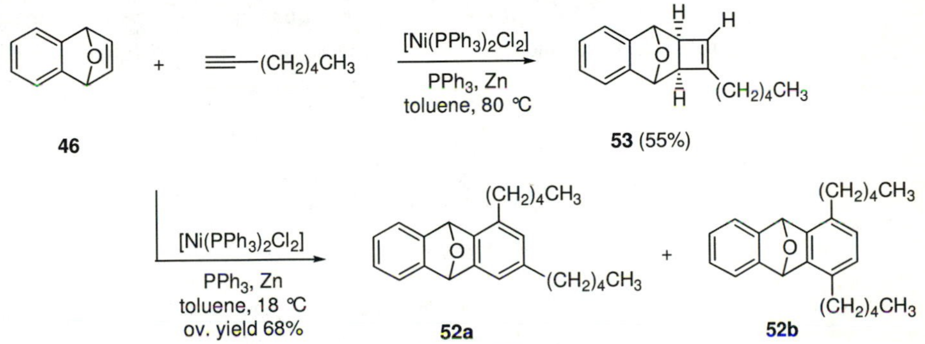

Scheme 15

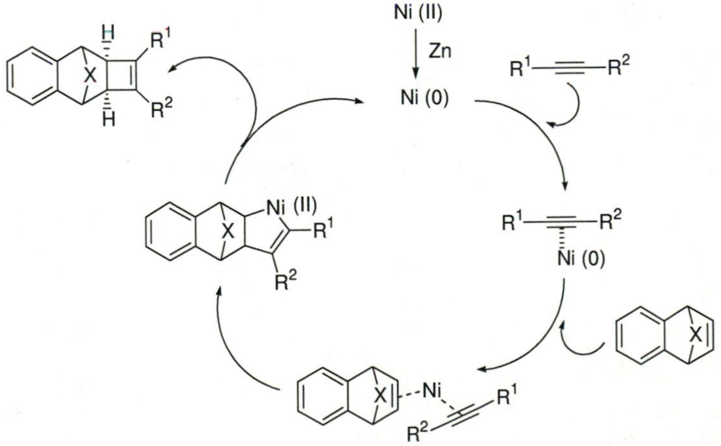

Scheme 16

169

The proposed mechanism for this [2+2] cycloaddition reaction (very similar to those proposed for the Co-catalyzed [2+2] cycloadditions, see Scheme 11) is depicted in Scheme 16. The reduction of [Ni(PPh$_3$)$_2$Cl$_2$] to Ni (0) species by Zn metal initiates the catalytic reaction. Coordination of an alkyne and 7-oxabenzonorbornadiene to the metallic center followed by oxydative cyclometallation affords a nickelcyclopentane intermediate. Subsequent reductive elimination of the metallocycle gives the cyclobutene and regenerates the Ni (0) species.

The cyclobutene derivatives **45** are quite stable at room temperature. However the FVP at 500 °C of these compounds readily affords dienes **54** in high yields (Scheme 17). When X=O and R^1=R^2=Ph, deoxygenation using TiCl$_4$ and Zn in THF affords the corresponding cyclooctatriene derivative **55** in 89% yield.

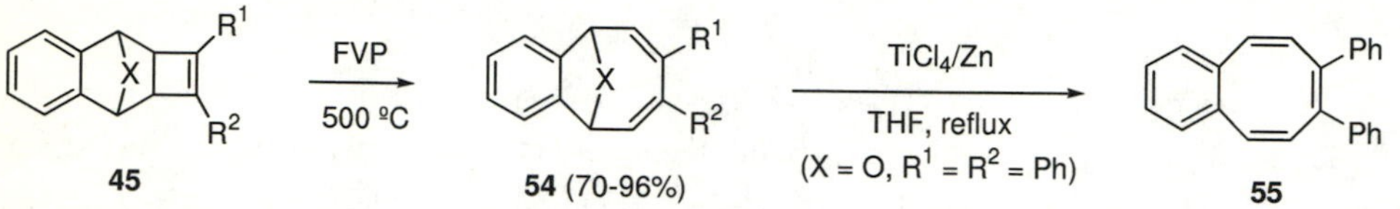

Scheme 17

The proposed mechanism for the reaction **45-54** involves a scission of the carbon-carbon single bond between the two fused carbons of the cyclobutene moiety (Scheme 18) followed by disrotatory process and bond rearrangement.

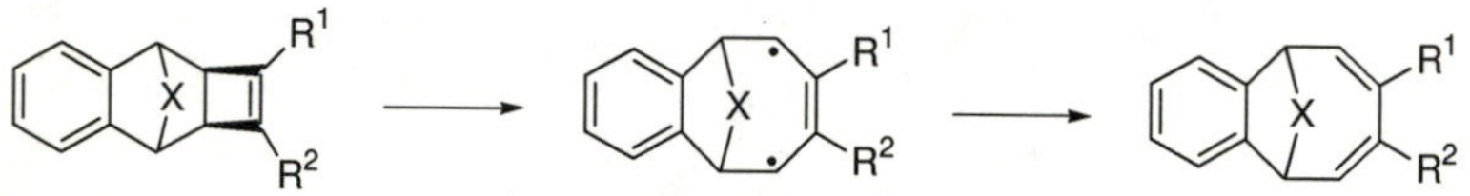

Scheme 18

6.3. Ru-catalyzed [2+2] cycloaddition reactions of oxa- and azanorbornene derivatives and alkynes

Although the Ru-catalyzed [2+2] cycloadditions of norbornene derivatives and alkynes have attracted considerably attention[29,34,35] there are only few reports concerning the analogues reactions achieved on the related 7-oxa- and 7-aza analogues.

The first report on the [2+2] cycloaddition of a 7-oxanorbornadiene and methylacetylenedicarboxylate was published by Mitsudo and co. in 1979.[36] Thus, reaction of compound **46** with methylacetylene-dicarboxylate in the presence of RuH$_2$(CO)$_n$(PAr$_3$)$_{4-n}$ (n=0, 1) or (cyclooctatriene)(cyclooctadiene) Ru (0) catalysts afforded [2+2] cycloadduct **56** in moderate yield (Scheme 19).

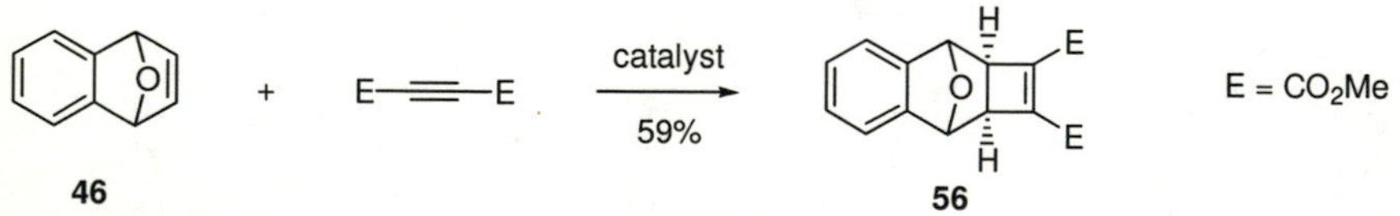

Scheme 19

Interestingly, this reaction may be formally achieved on the generated *in situ* oxanorbornene derivatives **57**. Thus, when a 1:2 mixture of furan or 2-methylfuran react with methylacetylenedicarboxy-late in the presence of Ru catalysts, cycloadducts **58** and **59** were obtained albeit in poor yields (Scheme 20).

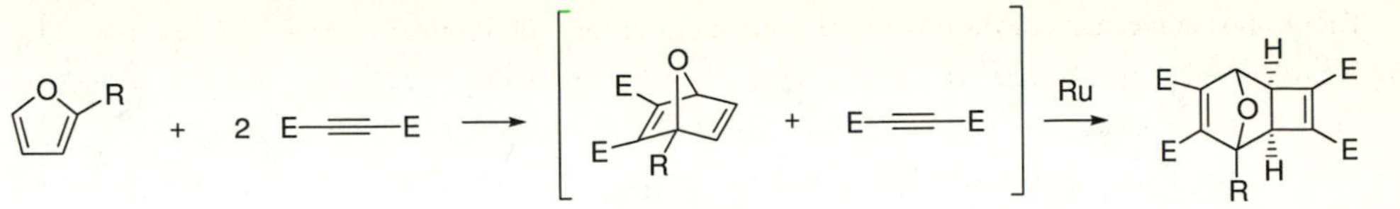

R = H, Me
E = CO$_2$Me

57
a, R = H
b, R = Me

58, R = H (21%)
59, R = Me (10%)

Scheme 20

Cycloadducts such as **56**, the azanalogues **60**[37] and the *bis* adducts **61**[38] and **62**[39] (Figure 2), prepared using the same Ru catalyzed [2+2] cycloaddition have been used by Warrener and co., among others, in the synthesis of semi-rigid U-shaped molecules designed as host to provide enzyme-like pockets.[40]

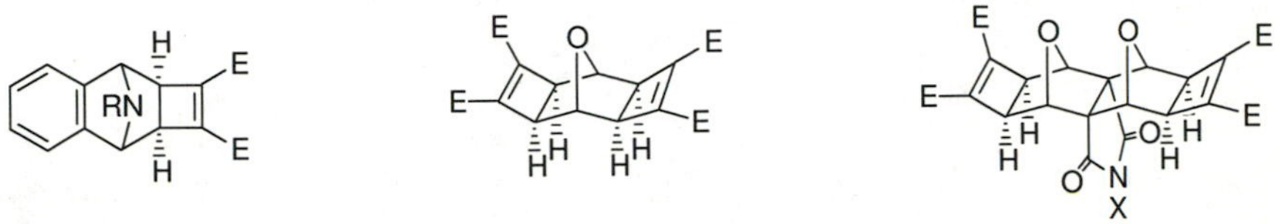

60   **61**   **62**

E = CO$_2$Me
X = Troc, *n*-Pr, *i*-Pr

Figure 2

In the same way, unsymmetrical adducts of 1,4-*bis*-benzyne such as **81-83** have been synthesized by the addition of dienes to monobenzynes generated sequentially from 1,2,4,5-*tetrakis*(trimethylsilyl)benzene **63** (Scheme 21).[41] Thus, reaction of **63** with iodobenzene diacetate (PhI(OAc)$_2$) and trifluoromethanesulfonic acid (TfOH) in the presence of diisopropylamine afforded phenyl-[2,4,5-*tris*(trimethylsilyl)phenyl]iodonium triflate **64** which, by reaction with TBAF, generates benzyne **65**. This was trapped *in situ* with *N*-Cbz-pyrrole, furan or cyclopentadiene to give cycloadducts **66-68**. The Ru-catalyzed [2+2] cycloaddition of **66-68** with dimethylacetylenedicarboxilate gave rise compounds **69-71** in high yields. These news aryne precursors react in turn with cyclopentadiene, *N*-Cbz-pyrrole and furan giving adducts **72-80** as a mixture of *syn* and *anti* isomers in yields in the range 69-98%. The steroisomers of **72, 78** and **79** were separated using centrifugal radial chromatography and the *syn* isomers subjected to a second Ru-catalyzed addition with dimethylacetylenedicarboxilate affording compounds **81-83**.

7. Metathesis reactions in oxa- and azanorbornene derivatives

The conversion of oxa- and azanorbornenic compounds into the related cyclohexan- or cyclohexen derivatives have been carried out under different conditions (Scheme 22, a)[42] and, in all cases, these reactions were synthetically exploited. However, the alternative ring opening of these compounds to substituted five-membered ring heterocycles have been less considered besides their synthetic potential (Scheme 22, b). For these purposes only the oxidative cleavage of the endocyclic double bond[43] as well as the related *cis*-diols[44] may be considered as a general method. In some cases cycloreversion reactions[45] and Grob-type fragmentations[46] were also successfully used.

171

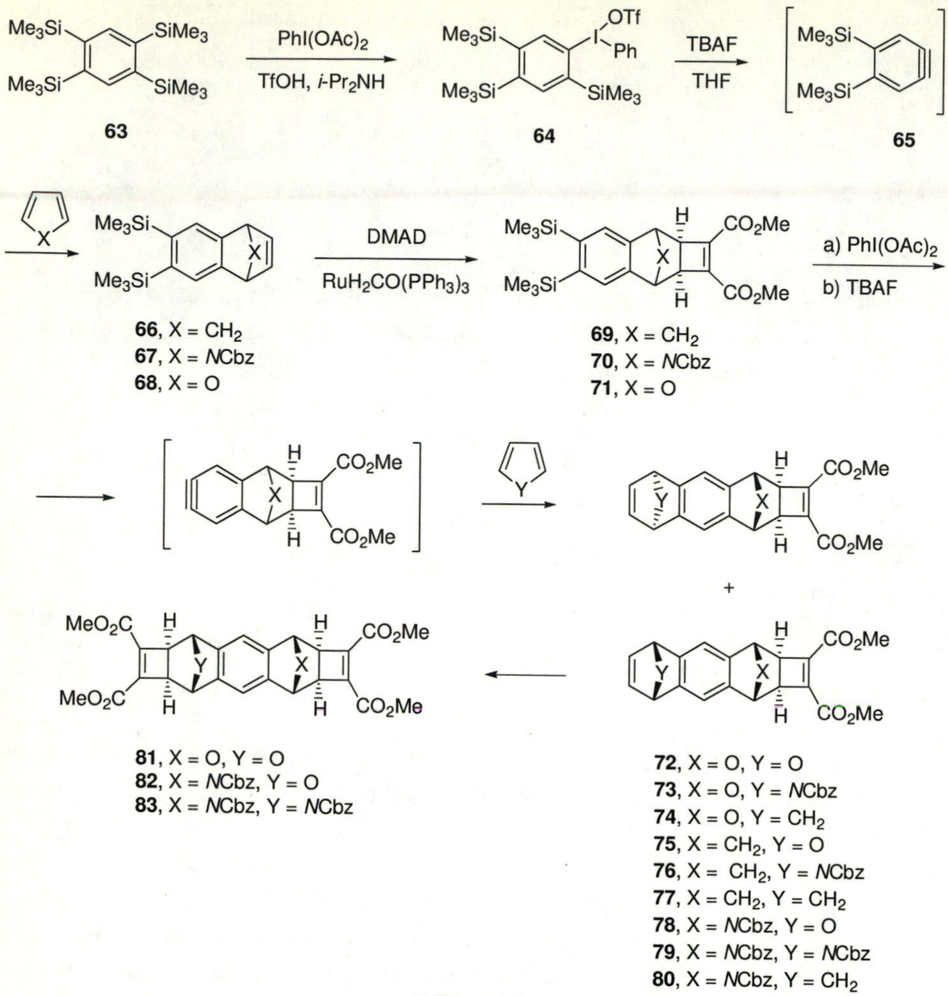

Scheme 21

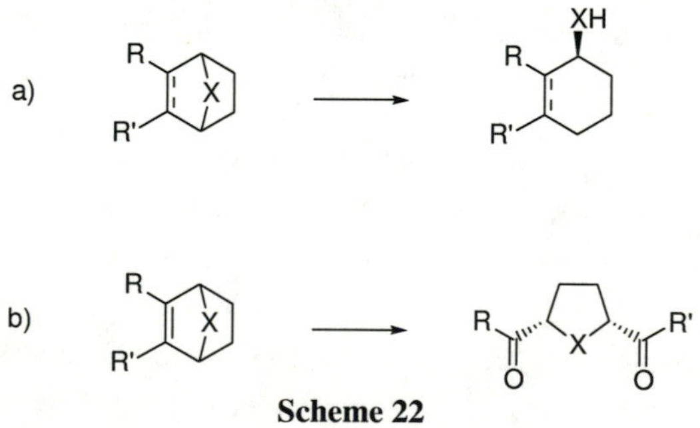

Scheme 22

An alternative method in order to achieve this transformation is the use of the sequence ROM (ring opening metathesis)-CM (cross-metathesis) reactions. In the next paragraphs we will disclose the more recent advances of this methodology which has been successfully incorporated to the arsenal of the synthetic methods in heterocyclic chemistry.

7.1. The metathesis reaction. An overview

Alkene metathesis is a reaction by which the carbon-carbon double bond of an alkene is broken and reformed by exchange of the alkylidene groups (Scheme 23). The reaction, catalyzed by a metal carbene complex, is today clearly established as a valuable synthetic tool in organic chemistry.[47]

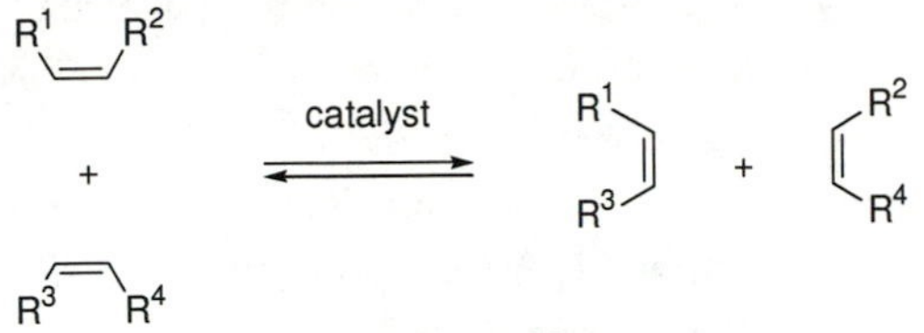

R^1 R^2

catalyst

+

R^3 R^4

R^1 R^2

+

R^3 R^4

Scheme 23

The first report on double bond scrambling reactions was reported in 1956[48] and during almost three decades this reaction (refereed to as alkene metathesis) was restricted to the field of polymer chemistry.

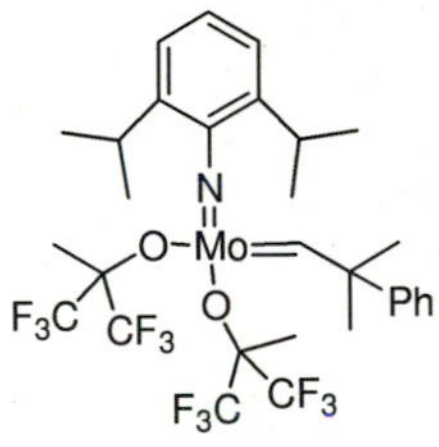

84

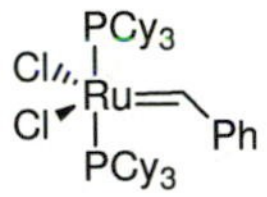

85

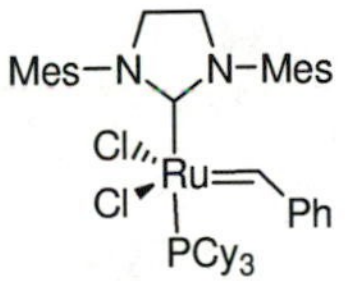

86

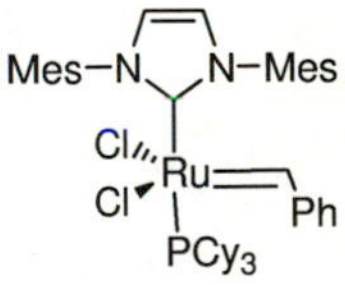

87

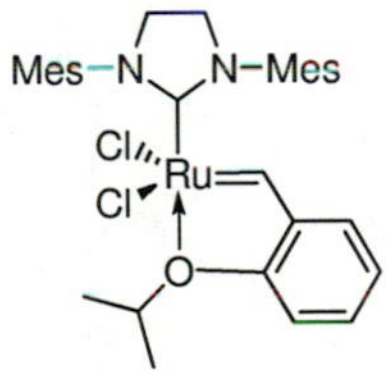

88

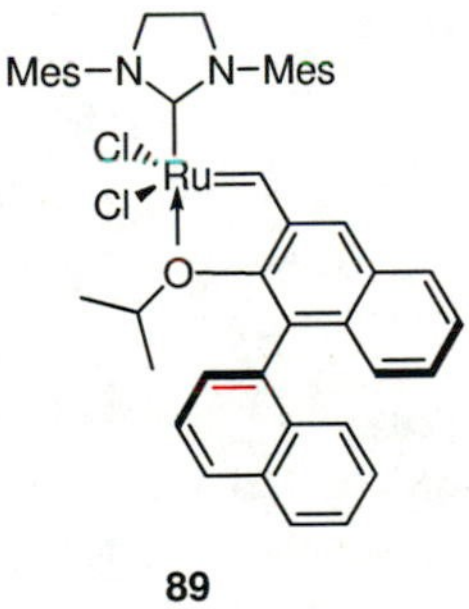

89

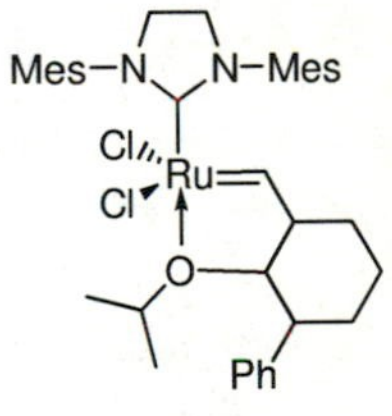

90

Cy = cyclohexyl
Mes = 2,4,6-trimethylphenyl

Figure 3

However, the discovery of well defined, high-performing, reasonably stable and -most important- functional-group tolerant alkylidene metal-complexes catalyst (or precatalyst that transforms into metal alkylidenes *in situ*) have allowed their widespread use in organic synthesis.[49,50,]

To the data, and summarizing briefly the current scenario, three generations of catalyst may be consider. The Mo catalyst **84** (Schrock's catalyst, Figure 3), commercially available, is very active although highly sensitive to oxygen and humidity and usually it is necessary to use Schlenck techniques for their manipulation. The second generation (Ru catalyst **85-87**, Figure 3) combine high activity with an excellent functional group tolerance. Catalyst **85** and **86** also are commercially available. Finally, a third generation of robust Ru catalyst has been developed on the basis of the coordinative effects between an isopropoxy group attached to the aromatic ring and the metallic center. Examples of these catalyst are compounds **88-90** (Figure 3), being catalyst **88** commercially available. It should be pointed out that this third generation of metathesis catalyst show an interesting combination between high activity, functional group tolerance and exceptional stability.

Metathesis reactions have a variety of applications. These include the more classical ring-opening metathesis polymerization (ROMP), (Scheme 24, a) ring closing metathesis (RCM) (Scheme 24, b), acyclic diene metathesis polymerization (ADMET) (Scheme 24, c), ring opening metathesis (ROM) (Scheme 24, d) and cross metathesis (CM) (Scheme 24, e). Some of the most important achievements include the use of ROMP (for the synthesis of functionalized polymers), RCM (for the preparation of small to large carbocyclic and heterocyclic compounds) and CM (for the synthesis of alkenes with different pendant functional groups).

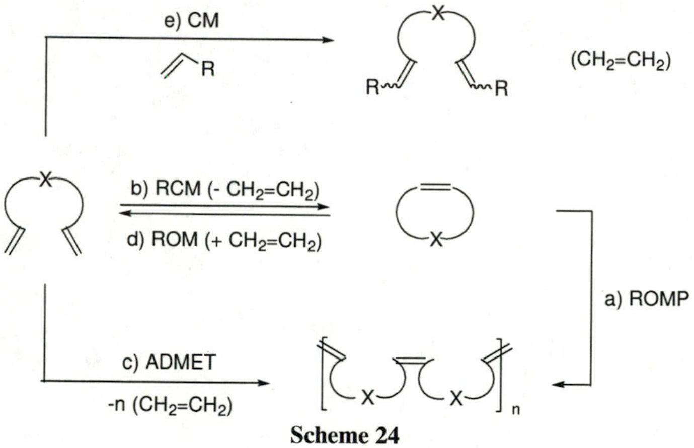

Scheme 24

On the basis of different experimental studies,[49d] mainly kinetic, the general accepted mechanism can be formulated as follows (Scheme 25)[51] using as model the catalyst **85**.

Thus, after coordination of the catalyst with the alkene, elimination of a phosphine ligand and cyclometallation (*cis* or *trans* regarding the "external" double bond) a cycloreversion step releases the new alkylidene carbene A. This is the first catalyst turnover. This carbene A undergoes a new [2+2] cycloaddition reaction (inter or intramolecular) affording the different products which should be expected from the different variants included in Scheme 24.

Metathesis reaction is a reversible process and in, consequence, it is necessary to shift the equilibrium in order to make the metathesis reaction useful in synthetic terms. In some cases the forward process is entropically driven because one of the reaction products is volatile (ethylene, Scheme 24).

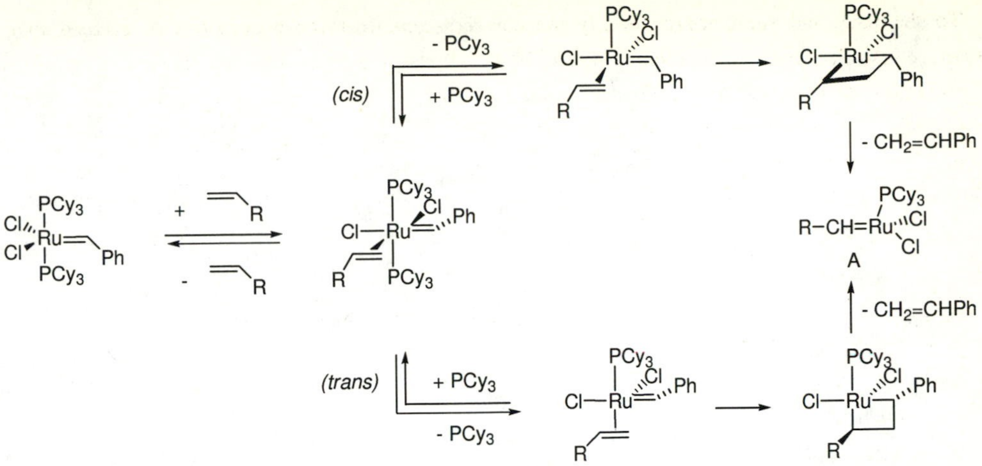

Scheme 25

7.2. The metathesis reactions of oxa- and azanorbornene derivatives

The consideration of bicyclic alkenes in the context of the metathesis reactions deserves special attention because the strain release associate with the transformation of the reactants to products constitutes and appropriate driving force for the ROM and ROMP reactions. In the next paragraph we will disclose the more recent aspects of the metathesis reactions of these bicyclic compounds.

7.2.1. ROMP of oxa- and azanorbornene derivatives

The ROM reactions applied to bicyclic alkenes constitutes the basis of the ROMP of these compounds.[52] For instance, substituted 7-oxanorbornadiene **91** undergo ROMP generating a variety of functional polymers such as **92** (Scheme 26).[53]

Scheme 26

Interestingly the reaction has also been applied to the preparation of biomimetic polymers[54] and to the synthesis of chiral catalytic active polymers useful in enantioselective reactions.[55]

The ROMP reaction of oxanorbornene derivatives has been applied to synthetic organic chemistry for the preparation of amines and hydrazines using a "capture-ROMP-release" process.[56] The sequence starts by the Mitsonobu reaction of the oxabicyclic derivative **93** and a variety of alcohols giving the *N*-substituted oxabicyclic compounds **94** (Scheme 27). The reaction crude was submited to ROMP using **86** as catalyst (0.5-1 mol %; 0.2 M concentration in CH_2Cl_2). After quenching with excess of ethylvinylether, polymer **95** was purified by precipitation in MeOH. Excess of Mitsonobu reagents and byproducts were next removed by filtration and **95** was obtained in good to excellent yields and with a high degree of purity. Finally, amine or hydrazine **96** were liberated by reaction **95** with hydrazine. The overall yields of the sequence lie in the range 71-99 %. Optically active alcohols were used in some cases affording amines or hydrazines with e.e. in the range 92-96 %.

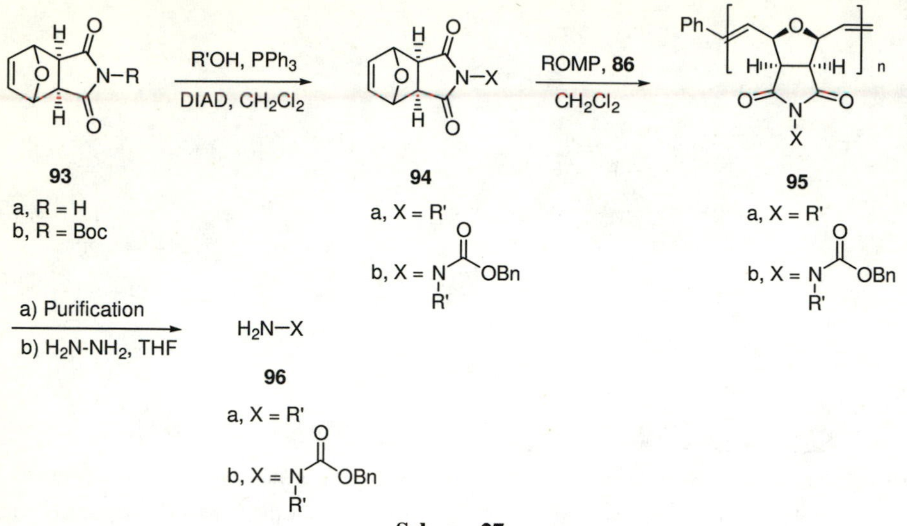

Scheme 27

The ROMP of azabicyclic systems has also been considered using Mo-catalyst.[57]

7.2.2. The sequence ROM-CM reactions in oxa- and azanorbornene derivatives. Synthesis of five-membered ring heterocycles

The ROMP reaction can be truncated by addition of a second "external" alkene. In this case, and after the first sequence carbometallation-metallacyclobutane ring opening, an alternative reaction may occurs by reaction of the intermediate carbene with the "external" alkene giving a new metallacyclobutane intermediate (Scheme 28).

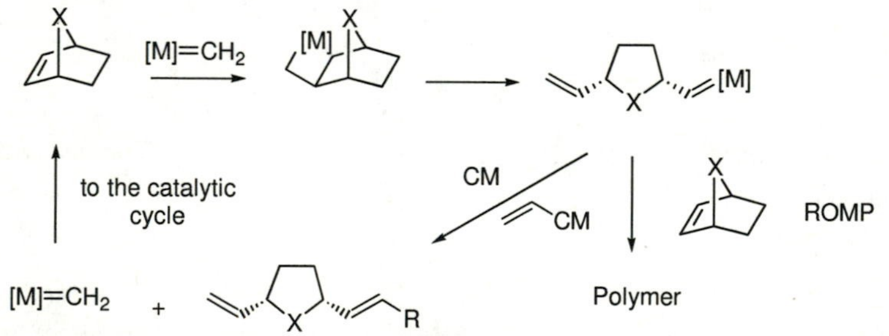

Scheme 28

This losses the catalytic agent, which reenters into the cycle, affording the ROM-CM product. The method constitutes another useful variation of olefin metathesis. Some reviews on this metathesis domino[58] have been published in the last years and only the more significant aspects of this methodology together with the more recent achievements will be discussed in this account.

7.2.2.1. The concurrence between ROMP and ROM-CM

Two important problems of the protocol ROM-CM despicted in Scheme 28 are: a) How interrupt the polymerization reaction specially in the case of reactive alkenes such as norbornene derivatives? and b) how avoid the homodimerization of the CM partner ("external" alkene)?

Regarding the first question, the relative weight of different experimental parameters on the concurrence ROM-CM sequence versus ROMD (ring opening metathesis dimerization)-CM sequence have been considered. For instance, the reaction of solutions of optically pure compound **97** in CH$_2$Cl$_2$ at 25 °C with alkene **98** in the presence of Grubb's ruthenium catalyst **85** (6% mol) afforded, after catalytic hydrogenation, a mixture of the tetrahydrofuran derivatives **99** and **100**, arising from the sequence ROM-CM, and dimeric products **101** and **102**[59] (aspects concerning the regioselectivities of these reactions will be discussed below) together with unidentified polymeric material (Scheme 29).

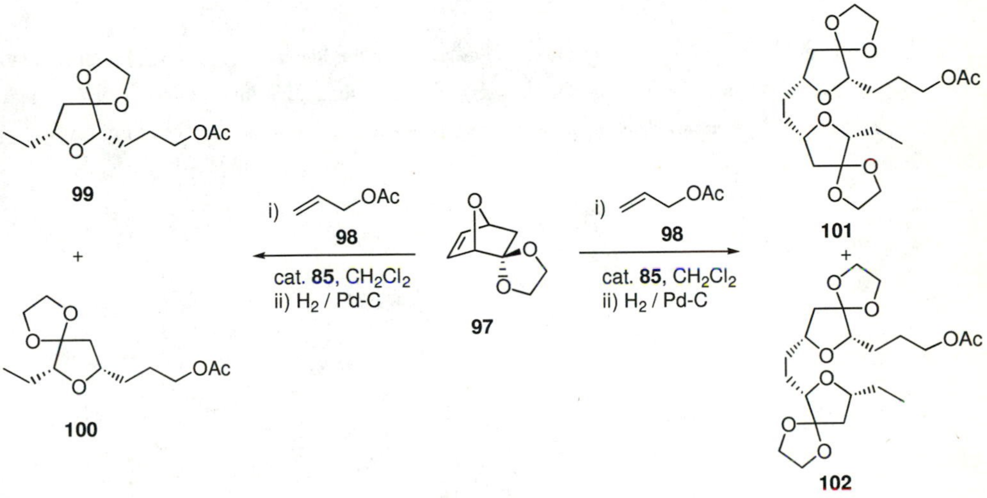

Scheme 29

Different reaction conditions were tested in order to enhance the yields of monomeric (**99** and **100**) or dimeric (**101** and **102**) products. The results are quoted in Table 1.

The inspection of these data put forward that, together with an increase in concentration (entry 2), an increase in the reaction temperature also favored the dimerization process (entry 3). Extensive polymerization was observed when the solvent was changed to toluene (entry 4) or THF (entry 5) and when the reaction was carried out with 0.5 equivalents of alkene **98**. Regarding the control of the ROM-CM sequence *versus* ROMD-CM sequence the best results were obtained controlling both the amount of catalyst **85** and the dilution of the solution in CH$_2$Cl$_2$ (entries 1 and 7).

The modification of the catalyst appears does not play an important role in the concurrence between both reactions. Thus, experiments achieved with catalyst **85**, **86** and **88** gave almost the same results regarding the reactions considered in Scheme 29.[60]

Regarding the second question (the homodimerization of the CM partner) it should be pointed out that a categorization of olefin reactivity towards the CM reaction has been empirically proposed for three commercially available catalyst **84**, **85** and **86**.[61]

Table 1. Reaction of **97** with acetate **98** in presence of catalyst **85**.

Entry	Molar ratio 97:98	Solvent (M)	[85]	[99+100] Ratio [101+102]	Overall yield[a] (%)
1	1:1	CH$_2$Cl$_2$ (0.017)	0.06	8.0:1	90
2	1:1	CH$_2$Cl$_2$ (0.034)	0.06	0.7:1	80
3[b]	1:1	CH$_2$Cl$_2$ (0.034)	0.06	0.6:1	65
4	1:1	Toluene (0.034)	0.06	1.2:1	26
5	1:1	THF (0.034)	0.06	1.7:1	55
6	2:1	CH$_2$Cl$_2$ (0.017)	0.06	2.0:1	30
7	1:1	CH$_2$Cl$_2$ (0.034)	0.12	0.4:1	85

[a] The remains material were unidentified polymeric products. No traces of starting **97** were observed.
[b] Reaction performed at 40 °C

This analysis describes olefins on a gradient scale on their tendency to undergo homodimerization (CM). Although bicyclic alkenes have not been included, the experience is that these compounds[49b] (and also cyclobutenes[62]) undergo CM reaction with terminal alkenes faster than the homodimerization of those.[63]

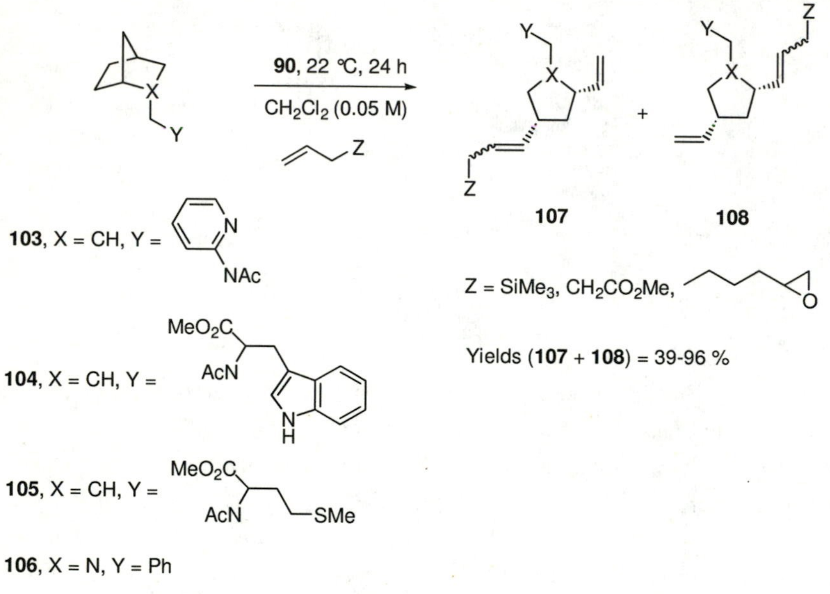

103, X = CH, Y =

104, X = CH, Y =

105, X = CH, Y =

106, X = N, Y = Ph

Z = SiMe$_3$, CH$_2$CO$_2$Me,

Yields (**107** + **108**) = 39-96 %

Scheme 30

7.2.2.2. The compatibility of the catalyst with functional groups attached to the reactive partners

It is known that several functional groups such as amines and sulfides are poor substrates in metathesis reactions probably due to their ability to coordinate to the metallic center. The consequence is the transformation of the catalyst in a unreactive form. The problem was circumvented by the use of catalyst such as **88-90** containing a 2-isopropoxystyrene moiety or the related biphenyl derivative. Thus, the ROM-CM reactions of norbornene and azanorbornene derivatives bearing unprotected Lewis-basic nitrogen (indol, pyridine) or sulfur (unhindered sulfides) functionalities proceeded in highly efficient fashion in the presence of allylic alkenes (Scheme 30).[64]

7.2.2.3. The question of the *E-Z* diasteroselectivity

Obviously this aspect concerns the CM step and it is often difficult to control or predict. Changes of the diastereoselectivity are probably associated with the structure of the substrate, the nature of the catalyst and probably, with coordination effects involving the three species.[65] For instance, in the case of the sequence ROM-CM of several 7-oxanorbornene derivatives, the diastereomeric ratios change significatively by simple modification of the extra-bicyclic moiety and/or the nature of the CM partner (Scheme 31).[66,67,68]

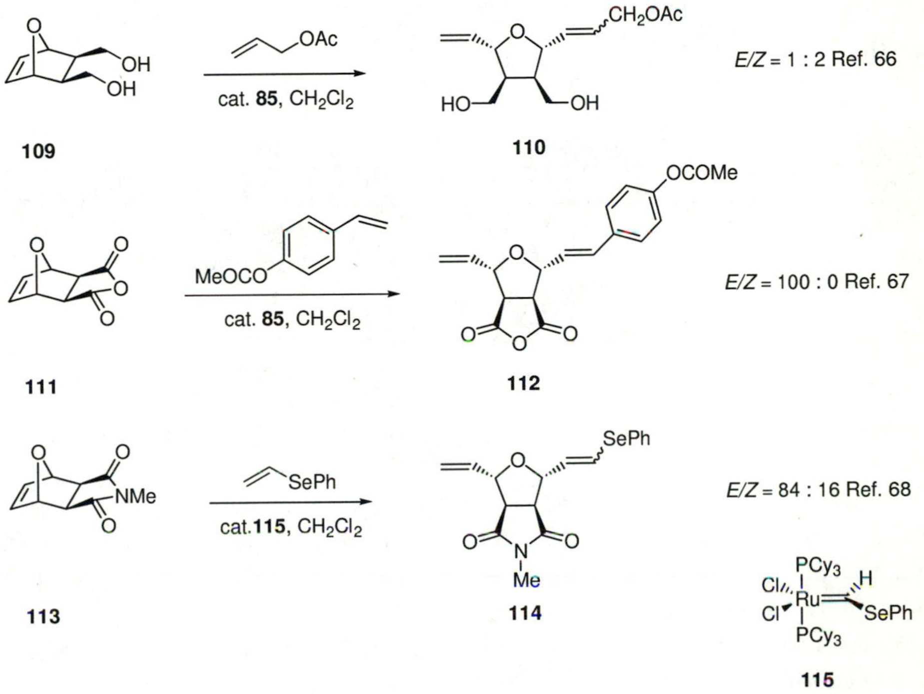

OH
OH
OAc
cat. 85, CH_2Cl_2
CH_2OAc
HO
OH
E/Z = 1 : 2 Ref. 66
109
110
OCOMe
MeOCO
cat. 85, CH_2Cl_2
E/Z = 100 : 0 Ref. 67
111
112
NMe
SePh
cat.115, CH_2Cl_2
SePh
N
Me
E/Z = 84 : 16 Ref. 68
PCy_3
H
Cl
Ru
Cl
SePh
PCy_3
113
114
115

Scheme 31

7.2.2.4. The question of the regioselectivity

The sequence ROM-CM of a symmetrically substituted bicyclic alkene using an external olefin as cross-partner obviously produces only one regioisomer.

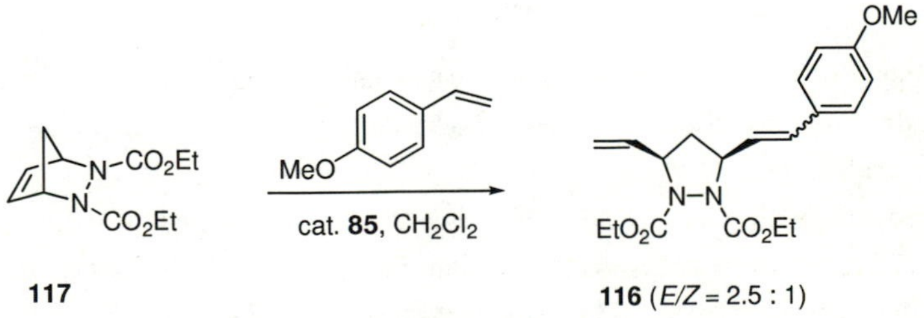

N-CO_2Et
N
CO_2Et
MeO
cat. 85, CH_2Cl_2
OMe
N-N
EtO_2C
CO_2Et
117
116 (E/Z = 2.5 : 1)

Scheme 32

That is also the case when the cross-partner is ethylene (see Scheme 31). The synthetic utility of the reaction is evident. For instance, cyclic hydrazines **116** have been prepared by the sequence ROM-CM reactions of the cycloaddict of cyclopentadiene and diethylazodicarboxilate **117** (Scheme 32).[69]

However, two regioisomeric products should be expected for the reaction of an unsymmetrically substituted bicyclic alkene with a CM partner different to ethylene. We have studied the regioselectivity of the sequence ROM-CM in 2-substituted 7-oxanorbornene derivatives with different results depending on the nature of the substituent at C-2[70,71] (Scheme 33, Table 2).[72]

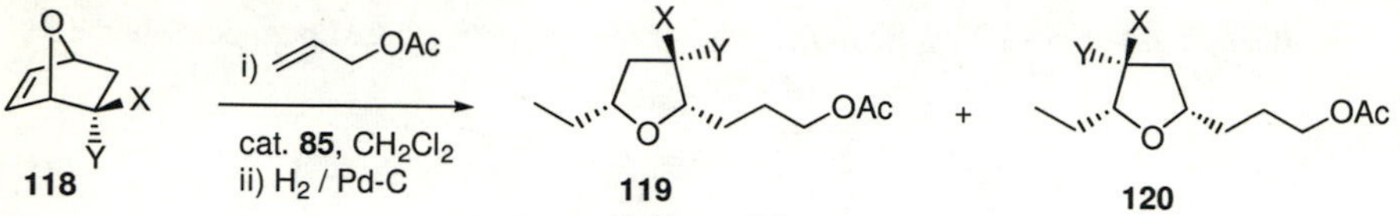

Scheme 33

Table 2. ROM-CM reactions of 7-oxanorbornenes **118** with allylacetate.

Entry		X, Y	Regioisomeric ratio	
1	**118a**	X=Y=CO	**119a/120a**	1:1
2	**118b**	X=H, Y=OH	**119b/120b**	1:1
3	**118c**	X=H, Y=OAc	**119c/120c**	0.2:1
4	**118d**	X=H, Y=OCOCH=CH$_2$	**119d/120d**	0.3:1
5	**97**	X=Y=OCH$_2$CH$_2$O	**99/100**	1:4

In the case of 7-oxanorbornenone **118a** an equimolecular mixture of both regioisomer was obtained. This result contrast with those obtained in the case of the acetate and the vinyl ester. Both compounds gave rise to the formation of the regioisomeric mixtures in which the major products shows the larger alkenyl group located on the less hindered side of the tetrahydrofuran ring. The regioselectivity of the reactions appears to be independent of the structure of the external alkene because the same results were obtained using allyl benzyl ether instead of allyl acetate.

These results should be compared with those obtained by Tam and co. using the related norbornene derivatives as bicyclic alkene (Scheme 34, Table 3).[73]

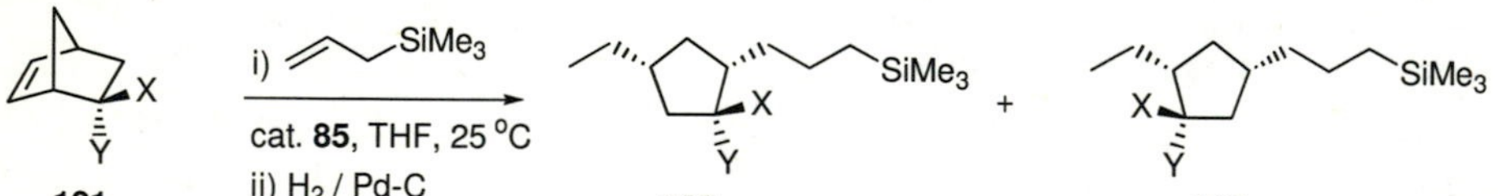

Scheme 34

From the comparison of the results quoted in Tables 2 and 3 may be deduced that the modification of the substituent ant the position 7 in the bicyclic systems seems not play an important role in determining the regioselectivity of the reaction.

In sharp contrast with this results, the *endo*-2-*p*-toluenesulfonyl derivative of 7-oxa and 7-azanorbornenes **124** show a total regioselectivity being the only isolated regioisomer the opposite of that observed in the case of the norbornenes **121** and 7-oxanorbornenes **118** (Scheme 35, Table 4). Noteworthy,

180

when the *p*-toluenesulfonyl group was placed at the *exo* position a 1 : 1 mixture of regioisomeric products was observed.[74]

Table 3. ROM-CM reactions of norbornenes **121** with allyltrimethylsilane.

Entry		X, Y	Regioisomeric ratio	
1	**121a**	X=Y=CO	**122a/123a**	1:1
2	**121b**	X=H, Y=OH	**122b/123b**	0.6:1
3	**121c**	X=H, Y=OAc	**122c/123c**	0.7:1
4	**121d**	X=OH, Y=H	**122d/123d**	0.6:1
5	**121e**	X=OAc, Y=H	**122e/123e**	0.4:1

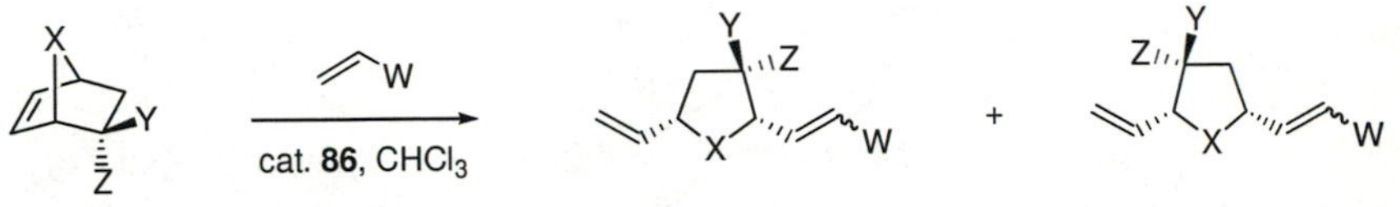

124 → cat. **86**, CHCl₃ → **125** + **126**

Scheme 35

Table 4. ROM-CM reactions of 2-*p*-toluenesulfonyl substituted oxa- and azanorbornenes.

Entry		X	Y	Z	Regioisomeric ratio	
1	**124a**[a]	O	H	Ts	**125a/126a**	1:0
2	**124b**[a]	NBoc	H	Ts	**125b/126b**	1:0
3	**124c**[b]	NBoc	Ts	H	**125c/126c**	1:1

[a] Vinyl acetate (among others) was used as CM partner.
[b] 1-octene and methyl pent-4-enoate were used as CM partners.

Also cycloadduct **127** prepared by the periodate oxidation of benzohydroxamic acid in the presence of 1,3-cyclopentadiene provides an opportunity to examine the ROM-CM sequence of a non-symmetric substrate containing multiple heteroatoms.[69,75] In this case, a nearly 1:1 mixture of regioisomers **128** and **129** was obtained (Scheme 36).

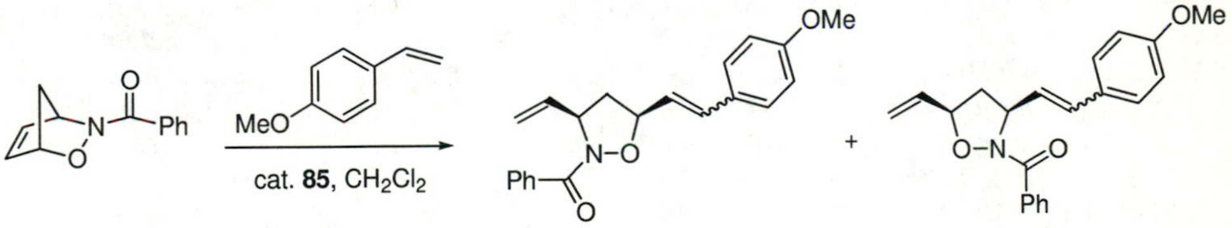

Scheme 36

The question of the regioselectivity has also been studied in the case of the *N*-substituted 2-azabicyclo[2.2.1]hept-5-en-3-one **130**. The first example of a Ru-catalyzed metathesis reaction of this compound was reported by Blechert's group.[66] These authors observe that the ROM-CM of the *N*-Boc protected compound proceeds in a totally regioselective fashion giving only a regioisomer **131** as a *E/Z* mixture (Scheme 37).

181

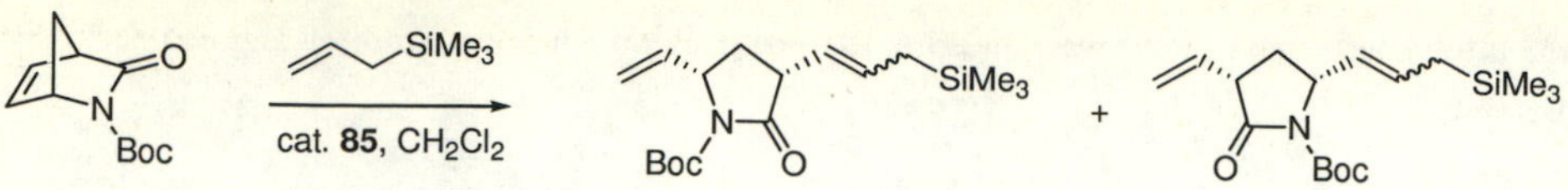

Scheme 37

Moreover the same reaction has been revisited by Ishikura and co.[76] who found that a mixture of regioisomeric products was obtained, determined after hydrogenation of the reaction mixture, being **131** the major regioisomer. Independently of the quantitative disparity of these results appears to be clear the tendency of this reaction to give the less encumbered product as major regioisomer.

8. Conclusions

The [2+2] cycloaddition reactions on the endocyclic carbon-carbon double bond of oxa- and azanorbornene derivatives constitute a valuable tool for the synthesis of fused bicyclic cyclobutan or cyclobutene heterocyclic systems. These reactions may be induced using different methodologies and the synthetic opportunities offered by the final products have been only scarcely explored. On the mechanistically point of view the sequence ROM-CM may be included in this category of reaction being a useful tool for the synthesis of five membered ring heterocycles.

Acknowledgments

Financial support of Ministerio de Ciencia y Tecnología of Spain (Project BQU2003-04967) are gratefully acknowledged.

References and Notes

1. **General:** Vogel, P.; Cossy, J.; Plumet, J.; Arjona, O. *Tetrahedron* **1999**, *55*, 13521.
2. **Synthesis:** (a) Shipman, M. *Contemporary Org. Synth.* **1995**, *2*, 1. (b) Kappe, O. C.; Murphree, S.; Padwa, A. *Tetrahedron* **1997**, *53*, 14179.
3. **Ring opening reactions:** (a) Chiu, P.; Lautens, M. *Topics in Current Chem.* **1997**, *190*, 1. (b) Woo, S.; Keay. B. A. *Synthesis* **1999**, 669. (c) Arjona, O.; Plumet, J. *Recent Res. Develop. Organic Chem.* **1999**, *3*, 265. (d) Arjona, O.; Plumet, J. *Curr. Org. Chem.* **2002**, *6*, 571.
4. **Synthetic uses:** 4-1) *Carbohydrates and analogues:* (a) Hudlicky, T.; Entwistle, D. A. Pitzer, K. K.; Thorpe, A. J. *Chem. Rev.* **1996**, *96*, 1195. (b) Vogel, P.; Ferrito, R.; Kraehenbuehl, K.; Baudat, A. In *Carbohydrate Mimics*; Chapleur, Y. Ed.; 1998; p. 14. (c) Vogel, P. *Curr. Org. Chem.* **2000**, *4*, 455. (d) Vogel, P. *Chimia* **2001**, *55*, 359 (see p. 361-365). 4-2) *Shikimic acids and analogues:* (e) Jiang, J.; Singh, G. *Tetrahedron* **1998**, *54*, 4697. 4-3) *Zaragozic acids and analogues:* (f) Jotterand, N.; Vogel, P. *Curr. Org. Chem.* **2001**, *5*, 637. 4-4) *Cyclitols, conduritols and related compounds*: (g) Gultekin, M. S.; Celik, M.; Balci, M. *Curr. Org. Chem.* **2004**, *8*, 1159.
5. (a) Chen, Z.; Trudell, M. L. *Chem. Rev.* **1996**, *96*, 1179. The chemisry of 7-azanorbornene derivatives has also been reviewed in the context of the synthesis of epibatidine. See: (b) Olivo, H. F.; Hemenway, S. *Org. Prep. Proc. Int.* **2002**, *34*, 1.
6. Brandt, P.; Andersson, P. G. *Synthesis* **2000**, 1092.
7. See, for instance: (a) *Comprehensive Organic Synthesis;* Trost, B. M.; Fleming, I.; Paquette, L. A., Eds.; Pergamon: Oxford, 1991; Vol. 5, chps 1-9. (b) *Advances in Cycloadditions;* JAI Press: Greenwich, 1988-1999; Vols. 1-6. For more specific aspects, see: (c) Yamazaki, S. *Reviews on Heteroatom Chem.* **1999**, *21*, 43. (d) Yamazaki, S. *J. Synth. Org. Chem. Jpn.* **2000**, *58*, 50. For the use of cyclobutanes in organic synthesis, see: (e) Namyslo, J. C.; Kauffmann, D. E. *Chem. Rev.* **2003**, *103*, 1485.

8. (a) *The Conservation of Orbital Symmetry*; Woodward, R. B.; Hoffmann, R., Eds.; Academic Press: New York, 1970. (b) Carruthers, W. In *Cycloaddition Reactions in Organic Synthesis*; Tetrahedron Organic Chemistry Series, Pergamon, 1990; Vol. 8. For a seminal work on the mechanism of the [2+2] cycloadditions, see: (c) Bartlett, P. D. *Science* **1968**, *159*, 833.

9. Baldwin, J. E. In *Comprehensive Organic Synthesis*; Trost, B. M.; Fleming, I.; Paquette, L. A., Eds.; Pergamon: Oxford, 1991; Vol. 5, p. 63.

10. For two excellent reviews concerning zwitterionic [2+2] cycloaddition reactions, see: (a) Huisgen, R. *Acc. Chem. Res.* **1977**, *10*, 117. (b) Huisgen, R. *Acc. Chem. Res.* **1977**, *10*, 199. See also: (c) Huisgen, R. *Pure & Appl. Chem.* **1981**, *53*, 171.

11. (a) Crimmins, M. T. In *Comprehensive Organic Synthesis*; Trost, B. M.; Fleming, I.; Paquette, L. A., Eds.; Pergamon: Oxford, 1991; Vol. 5, p. 123. For more specific aspects, see: (b) Wagner, P. J. *Acc. Chem. Res.* **2001**, *34*, 1. (c) Jones, G. *Photochromism: Molecules and Systems* **2003**, 514.

12. For some selected references, see: (a) Engler, T. A.; Letavic, M. A.; Reddy, J. P. *J. Am. Chem. Soc.* **1991**, *113*, 5068. (b) Mitani, M.; Sudoh, T.; Koyama, K. *Bull. Chem. Soc. Jpn.* **1995**, *68*, 1683. (c) Knolker, H. J.; Baum, E.; Schmitt, O. *Tetrahedron Lett.* **1998**, *39*, 7705.

13. For an excellent review, see: (a) Lautens, M.; Klute, W.; Tam, W. *Chem. Rev.* **1996**, *96*, 49. For reviews of more limited scope, see: (b) Montgomery, J. *Acc. Chem. Res.* **2000**, *33*, 467. (c) Barluenga, J. *Pure & Appl. Chem.* **2000**, *74*, 1317.

14. Whereas for the [2+2] cycloaddition of ketene to alkenes a concerted mechanism has been proposed, the cycloaddition ketene to imine appears to be zwitterionic. See: *mechanism ketene-alkene*: (a) Wang, X.; Houk, K. N. *J. Am. Chem. Soc.* **1990**, *112*, 2106. (b) Arrieta, A.; Cossio, F. P. *J. Org. Chem.* **2000**, *65*, 8458. *Mechanism ketene-imine*: (c) Sordo, J. A.; González, J.; Sordo, T. L. *J. Am. Chem. Soc.* **1992**, *114*, 6249. (d) Palomo, C.; Cossio, F. P.; Cuevas, C.; Lecea, B.; Mielgo, A.; Román, P.; Luque, A.; Martínez-Ripoll, M. *J. Am. Chem. Soc.* **1992**, *114*, 9360. (e) López, R.; Sordo, T. L.; Sordo, J. A.; González, J. *J. Org. Chem.* **1993**, *58*, 7036. (f) Arrieta, A.; Lecea, B.; Cossio, F. P. *J. Org. Chem.* **1998**, *63*, 5869. For an experimental evidence, see: g) Bandini, E.; Favi, G.; Martelli, G.; Panunzio, M.; Piersanti, G. *Org. Lett.* **2000**, *2*, 1077. Regarding the [2+2] cycloaddition of chlorosulfonylisocyanate to olefins the reaction appears to proceed *via* a concerted mechanism rather than a stepwise process, see: (h) Furman, B.; Borsuk, K.; Kaluza, Z.; Lysek, R.; Chmielewski, M. *Curr. Org. Chem.* **2004**, *8*, 463.

15. See, for instance: (a) Porco, J. A.; Schreiber, S. L. In *Comprehensive Organic Synthesis*; Trost, B. M.; Fleming, I.; Paquette, L. A., Eds.; Pergamon: Oxford, 1991; Vol. 5, p. 151. (b) Mattay, J.; Conrads, R.; Hoffmann, R In *Methoden der Organische Chemie* (Houben-Weyl); Helmchen, G.; Hoffmann, R. W.; Mulzer, J.; Schaumann, E., Eds.; Thieme: Stuttgart, 1995; Vol. IV. E21c., p. 3133.

16. See, for instance: (a) Zhang, W. *Curr. Org. Chem.* **2002**, *6*, 1015. (b) Nemoto, H.; Fukumoto, K. *Synlett* **1997**, 863.

17. See, for instance: (a) Palomo, C.; Aizpurua, J. M.; Ganboa, I.; Oiarbide, M. *Eur. J. Org. Chem.* **1999**, 3223. (b) Alcaide, B; Almendros, P. *Curr. Org. Chem.* **2002**, *6*, 245. (c) France, S.; Weatherwax, A.; Taggi, A. E.; Lectka, T. *Acc. Chem. Res.* **2004**, *37*, 592. (d) Palomo, C.; Aizpurua, J. M.; Ganboa, I.; Oiarbide, M. *Curr. Med. Chem.* **2004**, *11*, 1837. (e) Deshmukh, A. R. A. S.; Bhawal, B. M.; Krishnaswamy, D. *Curr. Med. Chem.* **2004**, *11*, 1889.

18. (a) Arjona, O.; de la Pradilla, R. F.; Pérez, S.; Plumet, J.; Carrupt, P. A.; Vogel, P. *Tetrahedron Lett.* **1986**, *27*, 5505. (b) Arjona, O.; de la Pradilla, R. F.; Pérez, S.; Plumet, J. *Tetrahedron* **1988**, *44*, 1235.

19. Arjona, O.; Csaky, A. G.; Murcia, M. C.; Plumet, J. *Tetrahedron Lett.* **2002**, *43*, 6405.

20. Inter- and intramolecular cycloaddition of ketenes to the endocyclic bond of norbornene have been reported. See, for instance: (a) Krepski, L. R.; Hassner, A. *J. Org. Chem.* **1978**, *43*, 2879. (b) Bak, D. A.; Brady, W. T. *J. Org. Chem.* **1979**, *44*, 107. (c) Sauers, R. R.; Kelly, K. W. *J. Org. Chem.* **1970**, *35*, 3286.

21. Plancquaert, M. A.; François, P.; Merenyi, R.; Viehe, H. G. *Tetrahedron Lett.* **1991**, *32*, 7265.

22. Bienfait, B.; Coppe-Motte, G.; Merenyi, R.; Viehe, H. G.; Sicking, W.; Sustmann, R. *Tetrahedron* **1991**, *47*, 8167.

23. Tian, G. R.; Mori, A.; Kato, H.; Takeshita, H. *Bull. Chem. Soc. Jpn.* **1989**, *62*, 506.

183

24. The exclusive oxetane formation in photoreactions of *p*-quinones is a well-known reaction mode of these compounds. See: (a) Bryce-Smith, D.; Gilbert, A.; Johnson, M. G. *J. Chem. Soc. (C)* **1967**, 383. (b) Mori, A.; Takeshita, H. *Chem. Lett.* **1975**, 599. (c) Mori, A.; Takeshita, H. *Bull. Chem. Soc. Jpn.* **1985**, *58,* 1581.

25. Prinzbach, H.; Argüelles, M.; Druckey, E. *Angew. Chem., Int. Ed.* **1966**, *5*, 1039.

26. See, for instance: (a) Hopf, H.; Witulski, B. *Pure & Appl. Chem.* **1993**, *65,* 47. See, also: (b) Breitkopf, V.; Hopf, H.; Klärner, F. G.; Witulski, B.; Zimny, B. *Liebigs Ann. Chem.* **1995**, 613.

27. Breitkopf, V.; Bubenitscheck, P.; Hopf, H.; Jones, P. G.; Klärner, F. G.; Schomburg, D.; Witulski, B.; Zimny, B. *Liebigs Ann. Chem.* **1997**, 127.

28. (a) Hegedus, L. S. *Coord. Chem. Rev.* **1997**, *161*, 129 M. (b) Lautens, M.; Tam, W. In *Advances in Metal Organic Chemistry*; Liebeskind, L. S., Ed.; JAI Press: Greenwich, 1998; Vol. 6, p. 49.

29. Chao, K. C.; Rayabarapu, D. K.; Wang, C. C.; Cheng, C. H. *J. Org. Chem.* **2001**, *66*, 8804.

30. The role of the "proxime" substituent R^3 has not been determined. Only one example of cycloaddition of oxanorbornadiene with a R^3 substituent different to H (R^3=Me) was described. In this case the reaction was achieved with diphenylacetylene and hence not regioselectivity is possible.

31. Other synthetically useful reactions of oxa- and azabenzonorbornadienes derivatives catalyzed by Ni complexes have been recently reported. See, for instance: (a) Rayabarapu, D. K.; Cheng, C. H. *Pure & Appl. Chem.* **2002**, *74*, 69. (b) Rayabarapu, D. K.; Cheng, C. H. *Chem. Eur. J.* **2003**, *9*, 3164. (c) Rayabarapu, D. K.; Shukla, P.; Cheng, C. H. *Org. Lett.* **2003**, *5*, 4903.

32. Huang, D. J.; Rayabarapu, D. K.; Li, L. P.; Sambaiah, T.; Cheng, C. H. *Chem. Eur. J.* **2000**, *6*, 3706.

33. For recent reports on Ni-catalyzed [2+2+2] cyclotrimerizations of oxa-and azanorbornadienes, see: (a) Sambaiah, T.; Li, L. P.; Huang, D. J.; Rayabarapu, D. K.; Cheng, C. H. *J. Org. Chem.* **1999**, *64*, 3663. (b) Sambaiah, T.; Huang, D. J.; Cheng, C. H. *J. Chem. Soc., Perkin Trans 1* **2000**, 195.

34. For recent reports on the Ru-catalyzed [2+2] cycloaddition reactions to norbornene derivatives, see: (a) Jordan, R. W.; Tam, W. *Org. Lett.* **2000**, *2*, 3031. (b) Jordan, R. W.; Tam, W. *Org. Lett.* **2001**, *3*, 2367. (c) Jordan, R. W.; Tam, W. *Tetrahedron Lett.* **2002**, *43*, 6051. (d) Villeneuve, K.; Jordan, R. W.; Tam, W. *Synlett* **2003**, 2123. (e) Tenaglia, A.; Giordano, L. *Synlett* **2003**, 2333. (f) Villeneuve, K.; Tam, W. *Angew. Chem., Int. Ed.* **2004**, *43*, 610.

35. A report of the Ru-catalyzed [2+2] cycloaddition (cyclopropanation) of 7-oxanorbornene **42** and propargyl alcohol **133** catalyzed by the complex **134** has been reported (Scheme 38). The reaction affords a mixture of cyclopropanes **135** and **136** in ratio **135 : 136**=6 : 1. See: (a) Matsushima, Y.; Kikuchi, H.; Uno, M.; Takahashi, S. *Bull. Chem. Soc. Jpn.* **1999**, *72*, 2475. See also: (b) Kikuchi, H.; Uno, M.; Takahashi, S. *Chem. Lett.* **1997**, 1273.

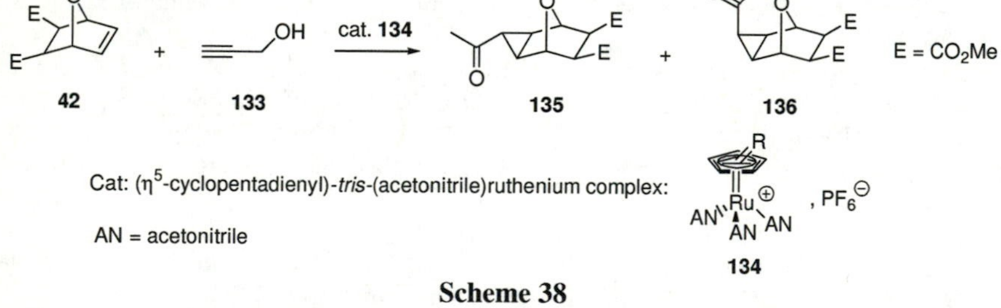

Cat: (η^5-cyclopentadienyl)-*tris*-(acetonitrile)ruthenium complex:

AN = acetonitrile

Scheme 38

36. (a) Mitsudo, T.; Kokuryo, K.; Shinsugi, T.; Nakagawa, Y.; Watanabe, Y.; Takegami, Y. *J. Org. Chem.* **1979**, *44*, 4492. See also: (b) Mitsudo, T.; Kokuryo, K.; Takegami, Y. *J. Chem. Soc., Chem. Commun.* **1976**, 722.

37. Warrener, R. N.; Margetic, D.; Butler, D. N.; Sun, G. *Synlett* **2001**, 202 and references therein.

38. Warrener, R. N.; Margetic, D.; Foley, P. J.; Butler, D. N.; Winling, A.; Beales, K. A.; Russell, R. A. *Tetrahedron* **2001**, *57*, 571.

39. Head, N. J.; Oliver, A. M.; Look, K.; Lokan, N. R.; Jones, G. A.; Paddon-Row, M. N. *Angew. Chem., Int. Ed.* **1999**, *38*, 3219.

40. Reviews: (a) Warrener, R. N.; Butler, D. N. *Aldrichimica Acta* **1997**, *30*, 119. (b) Warrener, R. N.; Butler, D. N. Russell, R. A. *Synlett* **1998**, 566. (c) Warrener, R. N. *Eur. J. Org. Chem.* **2000**, 3363. (d) Warrener, R. N. In *Progress in Heterocyclic Chemistry*; Pergamon-Elsevier: Oxford, 2001; Vol. 13, p. 25.

41. Winling, A.; Russell, R. A. *J. Chem. Soc., Perkin Trans. 1* **1998**, 3921.

42. For reviews, see ref. 3. For other more recent accounts mainly using organometallic reagents, see, among others: (a) Rayabarapu, D. K.; Cheng, C. H. *Chem. Eur. J.* **2003**, *9*, 3164 and references therein. (b) Li, L. P.; Rayabarapu, D. K.; Nandi, M.; Cheng, C. H. *Org. Lett.* **2003**, *5*, 1621. (c) Nakamura, M.; Matsuo, K.; Inoue, T.; Nakamura, E. *Org. Lett.* **2003**, *5*, 1373. (d) Arrayás, R. G.; Cabrera, S.; Carretero, J. C. *Org. Lett.* **2003**, *5*, 1333. (e) Chen, C. L.; Martin, S. F. *Org. Lett.* **2004**, *6*, 3581. (f) Lautens, M.; Mancuso, J. *J. Org. Chem.* **2004**, *69*, 3478.

43. See, for instance: (a) Kobayashi, S.; Sato, M.; Eguchi, Y.; Ohno, M. *Tetrahedron Lett.* **1992**, *33*, 1081. (b) Tochtermann, W.; Kraft, P. *Synlett* **1996**, 1029. (c) Li, C. C.; Wu, H. J. *Synthesis* **1996**, 715.

44. See, for instance: Shizuri, Y.; Nishiyama, S.; Shigemori, H.; Yamamura, S. *J. Chem. Soc., Chem. Commun.* **1985**, 292.

45. See, for instance: (a) Bloch, R.; Perffeti, M. T. *Tetrahedron Lett.* **1990**, *31*, 2577. (b) Bloch, R.; Brillet, C. *Tetrahedron: Asymm.* **1991**, *2*, 797. (c) Brown, P. S.; Greeves, N.; McElroy, A. B.; Warren, S. *J. Chem. Soc., Perkin Trans. 1* **1991**, 1485. (d) Bloch, R.; Bortolussi, M.; Girard, C.; Seck, M. *Tetrahedron* **1992**, *48*, 453. (e) Song, Z. Z.; Ho, M. S.; Wong, H. N. C.; *J. Org. Chem.* **1994**, *59*, 3917. (f) Aitken, R. A.; Cadogen, J. I. G.; Gosney, I.; Newlands, S. F. *J. Chem. Soc., Perkin Trans. 1* **1994**, 2301. (g) Chambers, R. D.; Roche, A. J.; Rock, M. H. *J. Chem. Soc., Perkin Trans. 1* **1996**, 1095. (h) Bortolussi, M.; Cinquin, C.; Bloch, R. *Tetrahedron Lett.* **1996**, *37*, 8729. (i) Mandville, G.; Girard, C.; Bloch, R. *Tetrahedron: Asymm.* **1997**, *8*, 3665.

46. See, for instance: (a) Cinquin, C.; Sharper, I.; Mandville, G.; Bloch, R. *Synlett* **1995**, 339. (b) Koshimizu, H.; Baba, T.; Yoshimitsu, T.; Nagaoka, H. *Tetrahedron Lett.* **1999**, *40*, 2777. See also: (c) Arjona, O.; Csaky, A. G.; Murcia, M. C.; Plumet, J. *Helv. Chem. Acta* **2001**, *84*, 3667.

47. For a recent monograph, see: *Handbook of Metathesis*; Grubbs, R. H., Ed.; Wiley-VCH: Weinheim, 2003; Vol. 13.

48. (a) Anderson, A. W.; Merckling, N. G.; US Patent 2, 721,189,1955. *Chem. Abst.* **1956**, *50*, 3008i. See also: (b) Calderón, N. *Acc. Chem. Res.* **1972**, *5*, 127.

49. The enormous impact of the metathesis reactions on chemical research as well as on chemical industry is well documented in many reviews articles. For a short selection of the more recent, see: *General:* (a) Fürstner, A. *Angew. Chem., Int. Ed.* **2000**, *39*, 3012. (b) Connon, S. J.; Blechert, S. *Angew. Chem., Int. Ed.* **2003**, *42*, 1900. (c) Grubbs, R. H. *Tetrahedron* **2004**, *60*, 7117. *Catalysts:* (d) Trnka, T. M.; Grubbs, R. H. *Acc. Chem. Res.* **2001**, *34*, 18. *Enyne metathesis:* (e) Diver, S. T.; Giessert, A. J. *Chem. Rev.* **2004**, *104*, 1317. (f) *Enyne metathesis: a catalytic, cross selective diene synthesis*; Poulsen, C. S.; Modien, R. *Synthesis* **2003**, 1. *Catalytic asymmetric olefin metathesis:* (g) Hoveyda, A. R.; Schrock, R. R. *Chem. Eur. J.* **2001**, *7*, 945.

50. For some selected, recent reviews on application of metathesis reactions in heterocyclic synthesis, see: (a) *Recent advances in the synthesis of heterocycles by RCM*; Walters, M. A. In *Progress in Heterocyclic Chem.* 2003; Vol. 15, p. 1. (b) *Cross metathesis of nitrogen containing systems;* Vernall, A. J.; Abell, D. A. In *Aldrichimica Acta* **2003**, *36*, 93. (c) Nakamura, I.; Yamamoto, Y. *Chem. Rev.* **2004**, *104*, 2127. (d) Deiters, A.; Martin, S. F. *Chem. Rev.* **2004**, *104*, 2199. (e) McReynolds, M. D.; Dougherty, J. M.; Hanson, P. R. *Chem. Rev.* **2004**, *104*, 2239.

51. This mechanism is usually referred as "Chauvin mechanism" and it is the currently accepted. See: Herisson, J. L.; Chauvin, Y. *Makromol. Chem.* **1970**, *141*, 161.

52. For a recent review see: Slugovc, S. *Macromol. Rapid Commun.* **2004**, *25*, 1283.

53. See, for instance: Madan, R.; Srinastava, A.; Arnand, R. C.; Varma, I. K. *Progress in Polymer Science* **1998**, *23*, 621.

54. For a seminal work on the synthesis of a carbohydrate-bearing polymer that can block protein-initiated cell agglutination starting from an oxanorbornene derivative, see: Martell, K. H.; Gingras, M.; Kiesling, L. L. *J. Am. Chem. Soc.* **1994**, *116*, 12053.

55. Bolm, C.; Dinter, C. L.; Seger, A.; Höcker, H.; Brozio, J. *J. Org. Chem.* **1999**, *64*, 5730.

56. Mukherjee, S.; Poon, K. W. C.; Flynn, D. L.; Hanson, P. R. *Tetrahedron Lett.* **2003**, *44*, 7187.

57. Prehihuber-Pflugl, P.; Buchacher, P.; Eder, E.; Schitter, R. M.; Stelzer, F. *J. Mol. Cat.* **1998**, *133*, 151 and references therein.

58. (a) Blechert, S. *Pure & Appl. Chem.* **1999**, *71*, 1393. (b) Arjona, O.; Csaky, A. G.; Plumet, J. *Synthesis* **2000**, 857. (c) Arjona, O.; Csaky, A. G.; Plumet, J. *Eur. J. Org. Chem.* **2003**, 611.

59. Arjona, O.; Csaky, A. G.; Mula, M. B.; Murcia, M. C.; Plumet, J. *J. Organomet. Chem.* **2001**, *627*, 105.

60. Monterde, M. I.; Plumet, J., unpublished.

61. Chaterjee, A. K.; Choi, T. L.; Sanders, D. P.; Grubbs, R. H. *J. Am. Chem. Soc.* **2003**, *125*, 11360.

62. See, for instance: (a) Snapper, M. L.; Tallarico, J. A.; Randall, M. *J. Am. Chem. Soc.* **1997**, *119*, 1478. (b) Tallarico, J. A.; Randall, M.; Snapper, M. L. *Tetrahedron* **1997**, *53*, 16511.

63. It should be pointed out that CM dimerization reactions of terminal olefins have been successfully exploited in synthesis of macrocyclic systems. For some selected accounts, see: (a) Smith III, A. B.; Adams, C. M.; Kozmin, S. A.; Paone, D.V. *J. Am. Chem. Soc.* **2001**, *123*, 5925 and references therein. (b) Fürstner, A.; Grabowski, J.; Lehmann, C.W.; Kataoka, T.; Nagai, K. *Chem. Biochem.* **2001**, *2*, 60. (c) Iwamoto, H.; Itoh, K.; Nagamiya, H.; Fukazawa, Y. *Tetrahedron Lett.* **2003**, *44*, 5773. (d) Tsae, J.; Yang, Y. K. *Org. Lett.* **2003**, *5*, 741.

64. Dunne, A. M.; Mix, S.; Blechert, S. *Tetrahedron Lett.* **2003**, *44*, 2733 and references therein.

65. For a report concerning the importance of the chelation effects on the *E/Z* diastereoselectivity in CM reactions, see: (a) Engelhardt, F. C.; Schmitt, M. J.; Taylor, R. E. *Org. Lett.* **2001**, *3*, 2209. (b) Engelhardt, F. C.; Schmitt, M. J.; Taylor, R. E. *J. Am. Chem. Soc.* **2001**, *123*, 2964.

66. Schneider, M. F.; Lucas, N.; Velder, J.; Blechert, S. *Angew. Chem., Int. Ed.* **1997**, *36*, 257.

67. Cuny, G. D.; Cao, J.; Sidhu, A.; Hauske, J. R. *Tetrahedron* **1999**, *55*, 8169.

68. Catalyst **115** has been used in reaction with phenyl vinyl selenide as CM reagent without notable amount of polymeric products; see: (a) Katayama, H.; Urushima, H.; Nishioka, T.; Wada, C.; Nagao, M.; Ozawa, F. *Angew. Chem., Int. Ed.* **2000**, *39*, 4513. For other recently introduced Ru catalyst with excellent results in the sequence ROM-CM, see: (b) Krause, J. O.; Nuyken, O.; Wurst, K.; Buchmeiser, M. R. *Chem.Eur. J.* **2004**, *10*, 777.

69. Ellis, J. M.; King, S. B. *Tetrahedron Lett.* **2002**, *43*, 5833.

70. Generally speaking, the control of the regioselectivity in the additions or cycloaddition reactions to the endocyclic double bond of a remote-substituted bicyclic system is an old question not yet well explained. For instance, some electrophilic additions show total regioselectivity depending on the nature of the remote substituent. In contrast, cycloaddition reactions such as 1,3-dipolar, Diels-Alder and Pauson-Khand have different levels of regioselectivity depending on both the remote substituent and the nature of the reagent involved in the reaction. These questions will be discussed in a next article of this serie.

71. Arjona, O.; Csaky, A. G.; Murcia, M. C.; Plumet, J. *J. Org. Chem.* **1999**, *64*, 9739.

72. All regioisomeric products were obtained as a diastereomeric *E/Z* mixture. In order to avoid complications in the determination of the regioselectivity the purified reaction mixtures were submitted to catalytic hydrogenation. As this last reaction was held in all cases with yields between 93-100 % the determination of the ratio of hydrogenated isomers may be consider as a convenient measurement of the regioselectivity of the reaction.

73. Mayo, P.; Tam, W. *Tetrahedron* **2002**, *58*, 9513.

74. Weeresakare, G. M.; Liu, Z.; Rainier, J. D. *Org. Lett.* **2004**, *6*, 1625.

75. The influence of the remote substituents in the sequence ROM-CM reactions has also been considered in the case of 8-oxabicyclo[3.2.1]octane derivatives. See: (a) Whright, D. L.; Usher, L. C.; Jimenez, M. E. *Org. Lett.* **2001**, *3*, 4275. (b) Usher, L. C.; Jimenez, M. E.; Ghiviriga, I.; Whright, D. L. *Angew. Chem., Int. Ed.* **2002**, *41*, 4560.

76. (a) Ishikura, M.; Saijo, M:, Hino, A. *Heterocycles* **2002**, *57*, 241. (b) Ishikura, M.; Saijo, M:, Hino, A. *Heterocycles* **2003**, *59*, 573. (c) Ishikura, M.; Hasunuma, M.; Saijo, M., *Heterocycles* **2004**, *63*, 5.

PROPERTIES, SYNTHESIS AND FUNCTIONALIZATIONS OF β-LACTAMS

Luigino Troisi,* Catia Granito and Emanuela Pindinelli

Dipartimento di Scienze e Tecnologie Biologiche ed Ambientali, University of Lecce,

Via Prov.le Lecce-Monteroni, I-73100 Lecce, ITALY (e-mail: luigino.troisi@unile.it)

Abstract. A novel and efficient synthetic method for preparing in high stereoselectivity β-lactams poly-functionalized is reported. A palladium-catalyzed [2+2] carbonylative cycloaddition of simple allyl halides and imines affords 2-azetidinones having alkyl or aryl substituents at the heterocyclic nitrogen atom, aryl or heteroaryl moiety linked at the C-4 carbon, and alkenyl groups beared to the C-3 carbon. Moreover, through the formation of a stable carbanion at the C-3 and the C-4, it is possible to further functionalize these latter positions with various electrophiles inserting, thus, more akyl groups or alcoholic and epoxidic functions in the azetidinone ring.

Contents

1. Introduction

After the discovery of penicillins and cephalosporins as β-lactam antibiotics and clinically useful active agents, the past few decades have witnessed remarkable growth in the field of β-lactam chemistry.[1] The need for potent effective β-lactam antibiotics as well as more effective β-lactamase inhibitors has motivated synthetic organic and medicinal chemists to design new functionalized 2-azetidinones. Apart from their clinical use as antibacterial agents, these compounds have been used as synthons in the preparation of various heterocyclic compounds of biological significance.[2] The potential use of some β-lactams as therapeutic agents for lowering plasma cholesterol levels has been documented as well.[3] Extensive studies of human leukocyte elastase (HLE) inhibitory mechanisms and biological activity of this class of compounds have also been published.[4]

1.1. Inhibitors for β-lactamase

The emergence of pathogenic microorganisms resistant to multiple classes of antibiotics is a serious clinical challenge.[5] Among these classes of antibacterial, β-lactam antibiotics are still the most commonly used over 50 years after their initial introduction. The most common mechanism for resistance to β-lactam antibiotics is the ability of bacteria to produce β-lactamases.[6,7] These enzymes hydrolyze the β-lactam moiety in these drugs, a reaction that inactivates the antibiotic. There are four distinct classes of β-lactamases, of which, class A enzymes are the most common.[7]

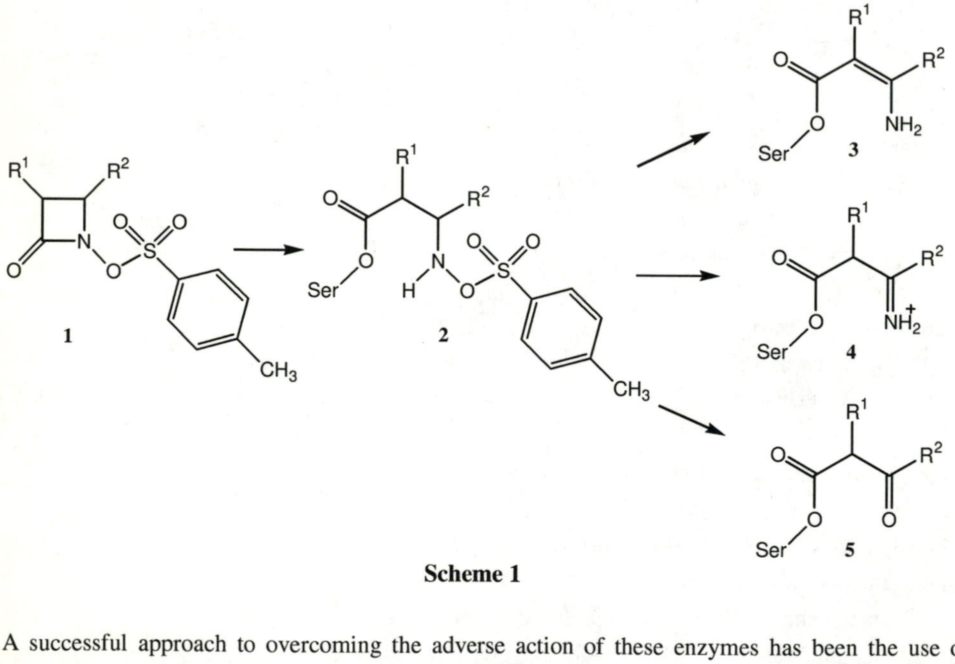

Scheme 1

A successful approach to overcoming the adverse action of these enzymes has been the use of β-lactamase inhibitors together with the typical β-lactam antibiotics, such as pennicillins.[7,8] Unfortunately, this approache has been compromised as well by the discovery of the new variants of β-lactamases, resistant to inhibition by known inhibitors.[9] Therefore, development of novel β-lactam inhibitors that operate by entirely different mechanisms[10] was highly desirable. The mechanism of action of a novel class of monobactam inhibitors for the class A β-lactamases is showed in Scheme 1.[11,12] As exemplified by compound 1, the inhibitor acylates the active site serine of β-lactamases, a process that is usually rapid. On acylation of the active site, the tosylate is released from species 2. The acyl-enzyme species undergoes fragmentation, resulting in enzyme inhibition by three distinct products, depending on the nature of the functionalities that have been incorporated into the inhibitor (species 3, 4, or 5).

The presence of the phenyl group at C-4 position of the azetidinone ring favored a specific hydrophobic interaction with the active site of class A β-lactamases. Furthermore, the stereochemistry at C-4 position appeared not to be important for inhibition.[11a] Studies reported recently for the structure-function analyses of the sulfonate moiety argue for the requirement of a hydrophobic functionality, but its size does not appear to be limiting. The absence of any hydrophobic functionality at this position impairs the ability of the molecules to inhibit β-lactamases.[13]

1.2. Inhibition of Human Leukocyte Elastase (HLE)

Time-dependent inhibitors of enzyme human leukocyte elastase based on the cephem nucleus have been developed. A series of cephalosporin tert-butyl esters has been examined, and the activity of these compounds has been found to be very sensitive to C-7 substituents, with small, α-oriented, electron-withdrawing groups showing greatest activity. Additionally, the oxidation state of the sulphur atom has been found to play a role in potency, with sulfones showing considerably greater activity than the corresponding sulfides or β-sulfoxides. The α-sulfoxides resulted inactive.[14]

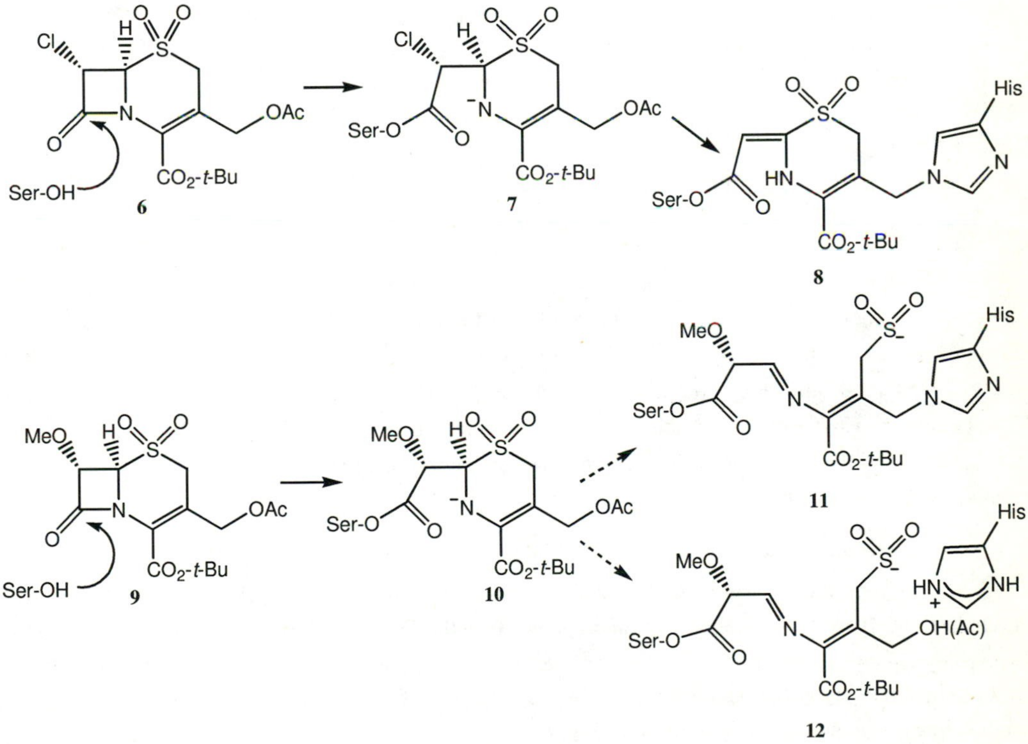

Scheme 2

189

Scheme 2 depicts, as reported in literature,[14] the mechanisms thought to be at work when molecules of this type inhibit HLE. It appears that at least two types of inhibited complexes can be formed with HLE. The initial event in all cases is considered to be β-lactam ring opening by the OH of Ser-195 of the enzyme catalytic triad (Ser-195, His-57, Asp-102) to form complex **7**. In case where the C-7 substituent is chloro, expulsion of the 3'-acetate, loss of HCl, and Michael addition of N-1 of the imidazole ring of His-57 lead to formation of inhibited complex **8**. However, in the case where the C-7 substituent is a poorer leaving group such as methoxy (Scheme 2, structure **9**), crystallographic data suggested that this group is not lost and the thiazoline ring appears to open to generate an inhibited species. At this point, it is not possible to discern whether there is a second covalent linkage between the inhibitor and the enzyme (Scheme 2, structure **11**) or alternatively that a critical salt bridge between the liberate sulfinate and the imidazole of His-57 is formed which slows hydrolysis of intermediate **12** presumably because the imidazole moiety is not aligned properly to deliver water to the serine ester carbonyl.

1.3. Cysteine protease inhibitors

The cysteine proteases cathepsin B, L, K and S are involved in diseases[15] such as osteoporosis, cancer metastasis, rheumatoid arthritis, and infections diseases. Thus, these proteases are important targets for the development of inhibitors as therapeutic agents.[16] Recently a series of 4-substituted-3-Cbz-Ph-β-lactam compounds, like the azetidinone **13** (Figure 1), were identified as a novel class of cysteine proteases inhibitors.[17]

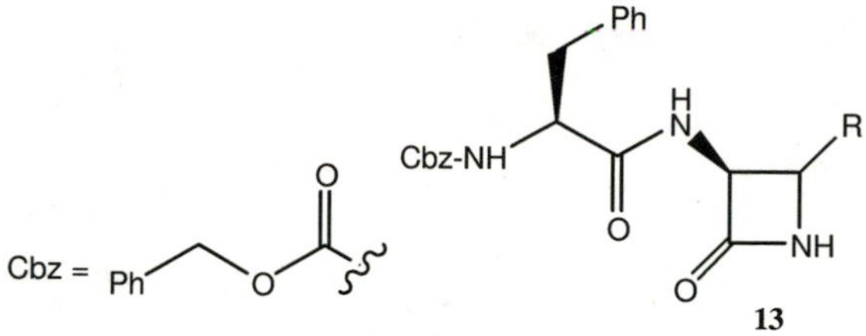

Figure 1

Several studies suggested the importance of the (3S) stereo configuration of the C-3 carbon, which is equal to the configuration of natural L-amino acids, on the interactions with the enzymes. The different substitution at C-4 showed a significant effect on inhibitory potency. C-4 substituents such as –OR', -OCOR' (R'=generic aliphatic group), -OPh, -SPh and –S(O)$_2$Ph appear to be necessary for a good inhibitory activity. However, the C-4 substituent may not function only as a leaving group; it is also involved in interactions with the S' subsite of the enzyme. The stereo configuration requirement of C-4 depends upon the nature of the substituents.[17]

1.4. Hypocholesterolemic activity

Atherosclerotic coronary artery disease (CAD) is a major cause of death. Although reducing dietary fat and cholesterol is still considered the appropriate first-line therapy, the advent of more effective pharmacological agents has led to an increased use of drug therapy to control serum cholesterol.[18] Serum cholesterol can be reduced by inhibiting endogenous cholesterol biosynthesis, promoting hepatic cholesterol clearance from the plasma, and inhibiting the absorption of dietary and biliary cholesterol from the intestines.[19]

190

2-Azetidinones such as **14a** (SCH 48461 – Figure 2) have been reported as effective inhibitors of cholesterol absorption in a cholesterol-fed hamster model.[20] Subsequently, the compound **14a** has been shown to reduce serum cholesterol in human clinical trials.[21] Although this class of compounds was initially designed as acyl coenzyme A cholesterol transferases (ACAT) inhibitors, early structure-activity studies demonstrated a striking divergence of *in vitro* ACAT inhibition and *in vivo* activity in the cholesterol-fed hamster. A detailed examination of the hypocholesterolemic activity of **14a** indicates that it acts at intestinal wall to inhibit cholesterol absorption through a novel mechanism.[22]

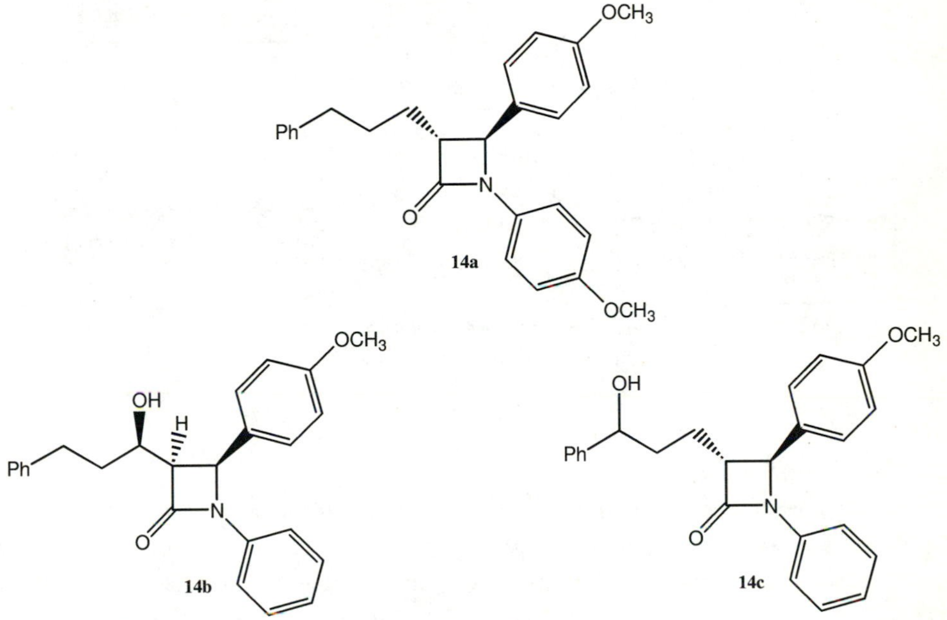

Figure 2

Reported studies on **14a** and compounds of this class, have shown that the azetidinone nucleus is a critical element for *in vivo* activity.[23] On the basis of these finding, an investigation of structure-activity relationships around the 2-azetidinone nucleus has been done. These studies have revealed clear structure-activity relationships (SAR) for cholesterol absorption inhibition which are distinct from the modest ACAT inhibitory activity shown by these compounds. Thus, these compounds appear to be acting *via* a novel mechanism which may be fundamentally important in the intestinal absorption of cholesterol.[24] Moreover, structurally related 2-azetidinones **14b** and **14c** (Figure 2), containing hydroxyl groups on the phenylalkyl substituent at C-3, have been reported as cholesterol absorption inhibitors more active than their deshydroxy analogues.[20,25] β-Lactams having at C-3 a substituent containing a ketonic moiety have been also reported as good inhibitors for cholesterol absorption.[26]

A summary of the basic structure-activity relationships in the 2-azetidinone compounds is showed in Scheme 3. The presence of a 4-methoxyphenyl or similar hydrogen bonding moiety and proper absolute stereochemistry at the C-4 carbon are both critical determinants of activity. The phenylalkyl group at C-3 and monosubstitution at C-3 and C-4 is less critical. An N-aryl group is required, but there is considerable tolerance for substitution on the phenyl ring. The azetidinone ring is required, but there is no evidence that it acts as anything more than a scaffold to correctly position the pharmacophore groups.[23]

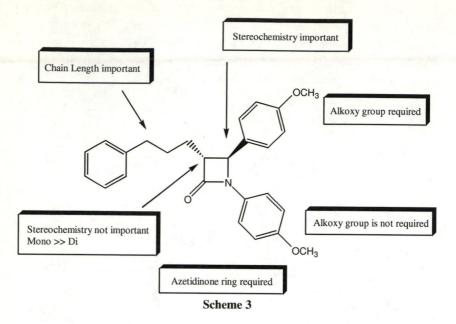

Scheme 3

1.5. Anti-hyperglycemic activity

Monocyclic β-lactam compounds showing similar structure of **14a** have been reported for antidiabetic activity, as these were able to control diabetic hypercholesterolemia.[27] Induction of diabetes was confirmed by a significant rise in serum glucose and depression in hepatic glycogen contents, which control cholesterol metabolism. Tests performed on the above monocyclic β-lactams significantly lowered the serum glucose levels in rats, indicating their anti-hyperglycemic activity. Moreover, an increase in the activity of glycogen synthetase enzyme was noticed, as evidenced by augmented liver glycogen contents in tested rats. As far as the order of anti-diabetic activity of tested 2-azetidinones is concerned, few important structure-activity relationships were reported. Phthalimido subtitution at the C-3 carbon showed the best activity followed by phenyl and phenoxy substitution. 1,3-Butadienyl substitution resulted in total loss of activity. N-1 substitution by cyclohexyl and isopropyl groups favored the activity more than phenyl and para-methoxy phenyl substitution. Styryl and para-methoxy phenyl substitution at C-4 was more favorable than phenyl.

1.6. Anticancer activity

Recently discovered antitumor monocyclic and bicyclic β-lactam systems[28] in general are in good agreement with the phenomenon of azetidin-2-one pharmacophore of inexhaustible pharmacological potential due to specific ability of its numerous derivatives to inhibit not only bacterial transpeptidase but also mammalian serin and cystein proteases.[14,29] As a measure of cytotoxicity, some compounds have been assayed against nine human cancer cell lines. Structure-activity studies have revealed that N-chrysenyl- and N-phenantrenyl 3-acetoxy-4-aryl-2-azetidinones **15** and **16**, respectively, have potent anticancer activity (Figure 3). The comparable N-anthracenyl, N-pyrenyl and N-naphthalenyl derivatives resulted inactive. It is evident that the minimal structural requirement of the aromatic moiety for cytotoxicity is at least three aromatic rings in an angular configuration. The presence of the acetoxy group at the C-3 position of the β-lactams has proved to be obligatory for their antitumor activity.[28d]

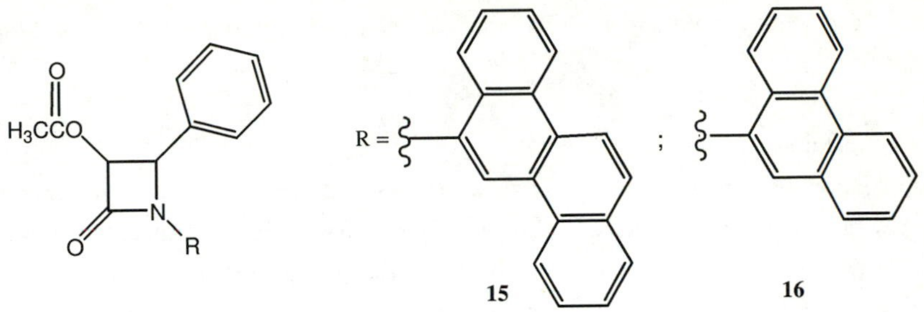

Figure 3

Moreover, potent inhibiting properties exhibited by 7-alkylidene substituted cephalosporanate sulfones against tumor strains both *in vitro* and *in vivo*[28c] motivated the researchers to subject penicillanate sulfones, togheter with 4-heteroaryldithio- and 4-methylsulfonyl azetidin-2-ones, containing alkylidene side chain at the C-6 and the C-3 positions, to similar biological investigations.[30] Cytotoxic properties of these compounds were tested *in vitro*. Their analysis evidenced that the incorporation of *tert*-butoxycarbonylmethylene, benzylidene and 4-nitrobenzylidene structures at the C-6 position of penicillanate sulfoxides and sulfones, the same as at C-3 position of 4-heteroaryldithio and 4-methylsulfonyl azetidin-2-ones, in many cases provided antitumor effect.

1.7. Antiviral activity

Human cytomegalovirus HCMV, a β-herpesvirus, is an opportunistic pathogen in immunocompromised individuals such as AIDS patients and organ transplant recipients.[31] Thus the need for effective, safe therapeutic agents for HCMV disease continues to exist. HCMV protease has become a viable target for antiviral chemotherapy because of its critical role in capsid assembly and viral maturation.[32] Several monobactams incorporating a benzyl side chain at the C-4 carbon, such as compound **17**, have been reported to be selective inhibitors of HCMV protease (Figure 4).[33]

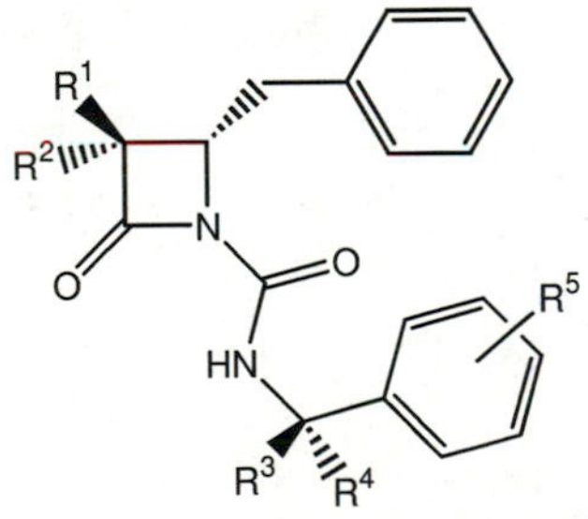

17

Figure 4

Substitution at the C-3 position was tolerated and gave small increases in stability and enzymatic activity. These compounds were much less selective, however, than the corresponding inhibitors which were unsubstituted at C-3. Substitution of the urea moiety suggested that benzyl groups were the best choice at this position. Both tri- and tetra substituted ureas were effective with tetra substitution giving a slight stability advantage. Modification of the benzyl function indicated that strong electron withdrawing groups at the *para* position had the best activity. Moreover, mechanistic investigations indicate that these compounds are reversible and competitive inhibitors of HCMV protease and that inhibition involves the formation of an acyl-enzyme species.[33]

2. Synthesis of β-lactams

As a consequence of the large pharmacological potency and use of the β-lactam systems, intensive research has generated numerous methods for synthetizing this skeleton, and the topic has been amply documented and reviewed several times.[34] Moreover, as documented in the previous paragraphs, the chemical reactivity of the β-lactam ring depends strongly on the substitution at the C-3 and C-4 positions.

2.1. Previous synthetic methodologies

The Staudinger reaction,[35] which is a stereo-controlled ketene-imine cycloaddition, has been used extensively for the synthesis of diversely substituted 2-azetidinones. The stereochemistry varies, including *trans*, *cis* and a *trans-cis* mixture, depending on the substituents present in the ketene and the imine, and the conditions of the reaction. Bose and Manhas[36,2a,28d] devised a direct and stereoselective synthesis of α-vinyl-β-lactams **20** by the reaction of an α,β-unsaturated acid chloride **18** with Schiff bases **19** (Scheme 4).

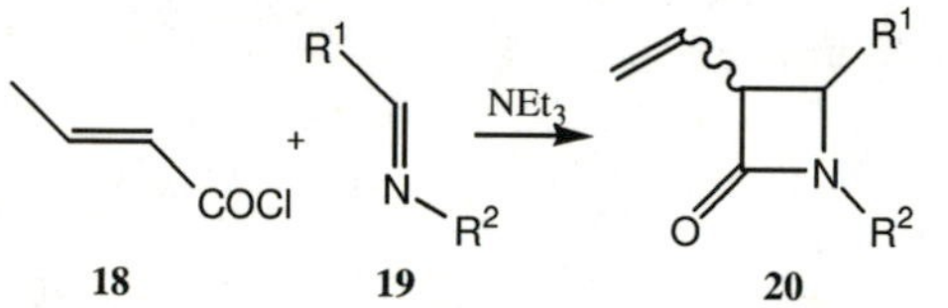

R^1 and R^2 = alkyl and aryl groups
Scheme 4

Zamboni and Just[37] used this reaction, for preparing a number of *trans* α-vinyl-β-lactams as potential synthons for β-lactams antibiotics. By a similar protocol *trans* β-lactams **23** were prepared by Georg *et al.*,[38] using crotonic acid **22** in place of α,β-unsaturated chloride, and Mukaiyama's reagent[39] (Scheme 5).

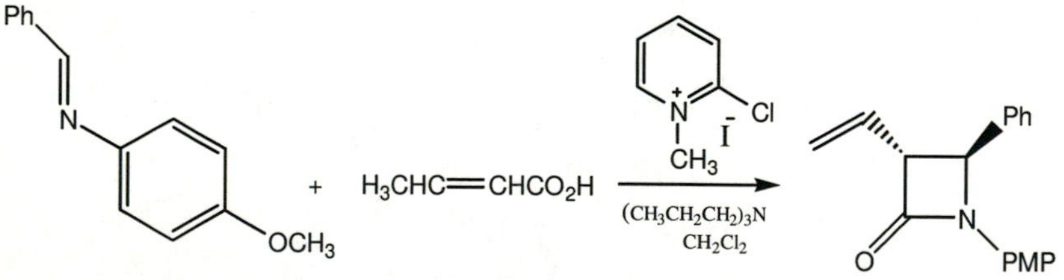

Scheme 5

The ester enolate-imine condensation[40] (the Gilman-Speeter reaction) is another of the most popular entries to β-lactams such as compounds **26** (Scheme 6). This reaction has been constantly refined to become a powerful and reliable synthetic tool.[41]

194

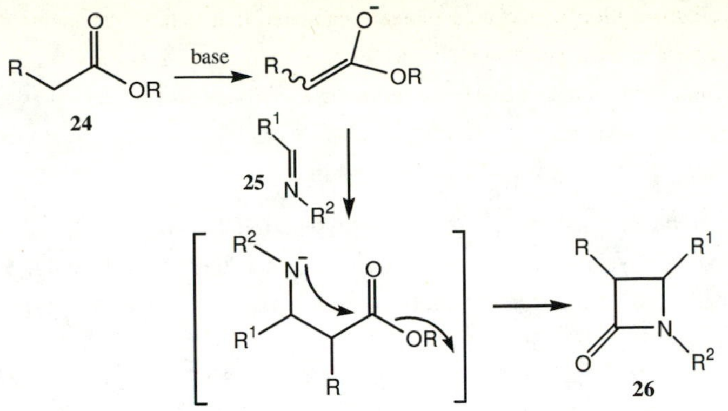

Scheme 6

A recent application of this reaction is reported by Benaglia *et al.*:[42] the S-thioester metal enolate-imine condensation represents a mild, efficient and straightforward route to the preparation of β-lactams in a stereoselective way.

Other notable methods are sometimes reported for preparing the β-lactam nucleus, including the isocyanate-alkene cycloaddition,[43] the chromium carbene-imine reaction,[44] the hydroxamate cyclization,[45] the photocyclization reactions[46] and the radical cyclizations.[47]

Another versatile method extremely used for the construction of substituted 2-azetidinones, is the transition metal-catalyzed carbonylation reaction.[48] Alper *et al.* prepared a wide range of β-lactams **28** (Scheme 7) in a stereospecific manner by the carbonylative ring expansion of aziridines **27** catalyzed by rhodium (*e.g.* [Rh(CO)₂Cl]₂),[49] palladium (e.g. Pd(PPh₃)₄, Pd(OAc)₂/PPh₃ and Pd(dba)₂/CH₂Cl₂),[50] and cobalt (*e.g.* Co₂(CO)₈)[51] complexes.

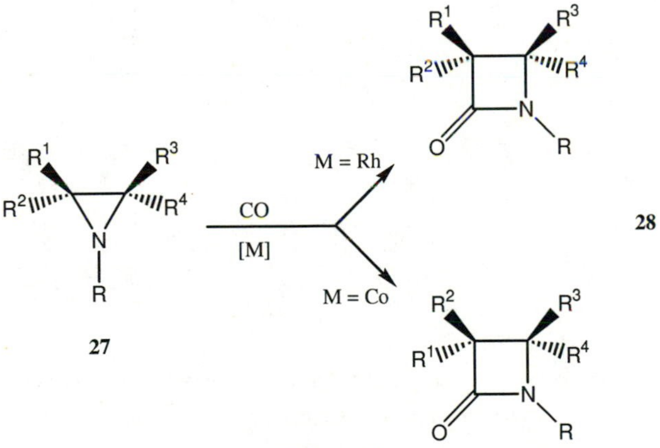

Scheme 7

The reaction stereochemistry is strongly dependent on the catalyst used, with retention of configuration occuring at the aziridine ring substituents when rhodium complexes were employed, and inversion of configuration found when Co₂(CO)₈ was used as catalyst.

Mori *et al.*[52] reported the palladium-catalyzed carbonylation of the aminovinyl bromide **29** to synthetize vinyl β-lactams **30** in a convenient way (Scheme 8).

195

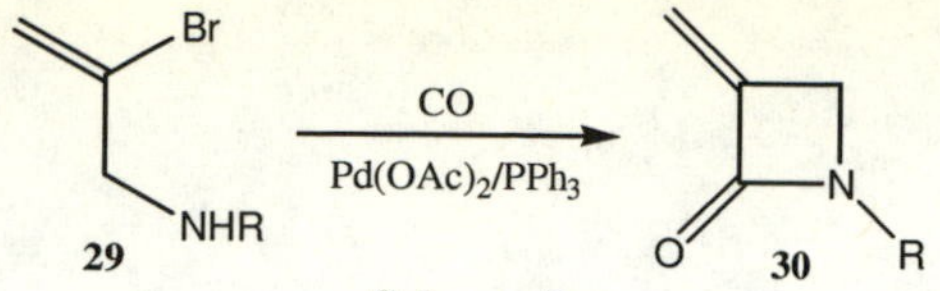

Scheme 8

The synthesis of biciclic β-lactams **32** was also reported by Alper *et al.* (Scheme 9) using the carbonylative coupling and cyclization reaction of 2-aryl-1,3-thiazines **31** with triallyl phosphate, catalyzed by bis(benzonitrile)palladium dichloride. Several rhodium complexes are also effective catalysts for this process.[53]

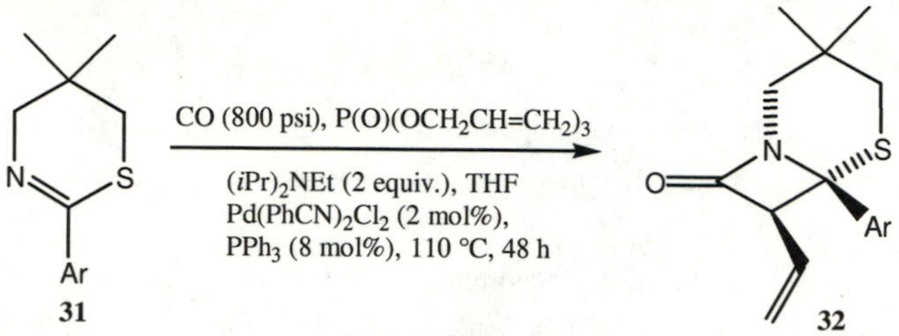

Scheme 9

Torii *et al.*[54] reported the stereoselective synthesis of *trans* and *cis*-β-lactams **35** and **37** by palladium-catalyzed carbonylation of allyl phosphate **33** in the presence of imines, under CO pressure.

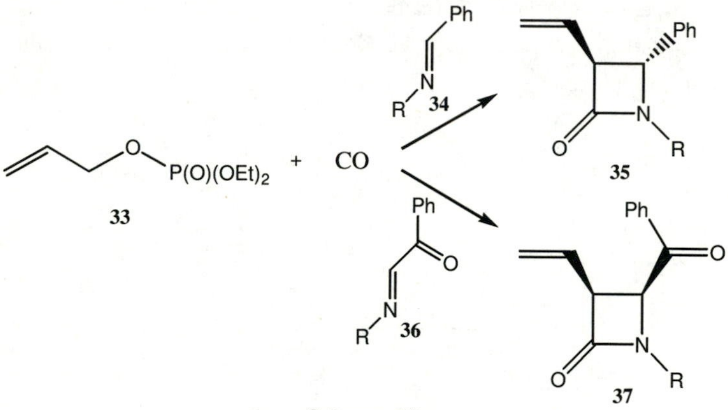

Scheme 10

The reaction stereochemistry depends on the imine employed for the coupling. The imine **36** conjugated with a carbonyl group gave the *cis*-β-lactam, whereas the unconjugated imine **34** afforded the *trans* isomer. No reaction products, or just traces of β-lactams, were reportedly formed when allyl bromide,[55] allyl acetate,[56] allyl phenyl ether,[57] allyl carbonate[58] or allyl sulfone[59] were used under similar reaction conditions.

2.2. Novel synthetic methodology

In contrast with the observations reported by Torii *et al.*, we found that simple allyl halides of various structure react with imines, under slightly different reaction conditions, leading to β-lactams in a regio- and stereoselective way.[60] The imines **38a-d**, used as substrates (Scheme 11), were prepared starting from the benzaldehyde and the corresponding amines.[61]

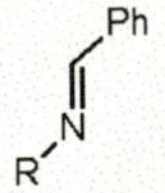

Imine	R
38a:	Ph
38b:	CH$_2$Ph
38c:	CH(CH$_3$)Ph
38d:	CH(Ph)CH$_2$OCH$_3$

Scheme 11

The commercial allyl halides of different structure **39a-l** used are detailed on Scheme 12.

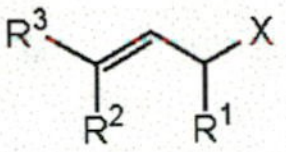

Allyl halide	X	R^1	R^2	R^3
39a:	Br	H	H	H
39b:	Cl	H	H	H
39c:	Cl	H	CH$_3$	CH$_3$
39d:	Br	H	H	CH$_3$
39e:	Cl	H	H	Ph
39f:	Br	H	H	(CH$_2$)$_2$CH$_3$
39g:	Br	-CH$_2$-CH$_2$-CH$_2$-		H
39h:	Cl	CH$_3$	H	H
39i:	Br	CH$_3$	H	H
39l:	Br	(CH$_2$)$_2$CH$_3$	H	H

Scheme 12

The [2+2] cycloaddition of allyl halides **39a-l**, with simple imines **38a-d**, under CO pressure, in the presence of Et$_3$N, triphenylphosphine and a catalytic amount of Pd(OAc)$_2$, led to β-lactams **40a-o**, **41a-o**, **42a-o**, **43a-o** in good yields and high diastereoselectivity (entries 1-14, Scheme 13). Slightly different reaction conditions than those reported in the literature, such as the use of Et$_3$N instead of i-Pr$_2$NEt, allyl halides as substrates instead of less easily accessible allyl-phosphates, and a lower pressure of CO, make this methodology much simpler than those reported in the literature.[60]

The *trans* relative configuration was found to predominate both in the β-lactam ring and in the C-C double bond of the products. The stereochemistry of the C-C double bond in the products was assigned on the basis of the $^3J_{H,H}$ coupling constants across the double bond. For the β-lactam moiety itself, the stereochemistry was assigned based on the coupling constants between the protons beared to C-3 and C-4, according to the literature report that for small rings $J_{cis} > J_{trans}$[62] The diastereoisomers **40n,40n'** and**40o,40o'**

197

showed a *trans*-type coupling constants $J_{H,H}$ for the β-lactam ring, similar to those measured for the other compounds. The protons of the R substituent on the nitrogen atom showed a difference in chemical shift (δ) between compounds **40n** and **40n'** and between **40o** and **40o'**. This suggests two possible stable configurations for the β-lactamic nitrogen.[60]

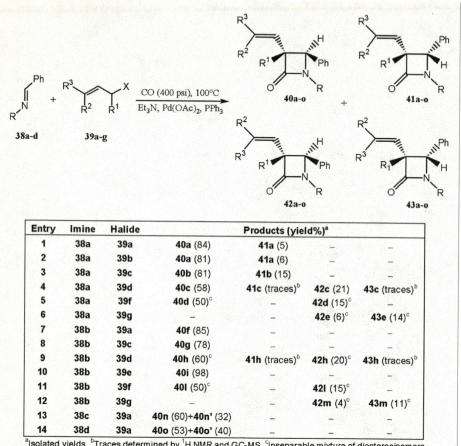

Entry	Imine	Halide	Products (yield%)[a]			
1	38a	39a	40a (84)	41a (5)	–	–
2	38a	39b	40a (81)	41a (6)	–	–
3	38a	39c	40b (81)	41b (15)	–	–
4	38a	39d	40c (58)	41c (traces)[b]	42c (21)	43c (traces)[b]
5	38a	39f	40d (50)[c]	–	42d (15)[c]	–
6	38a	39g	–	–	42e (6)[c]	43e (14)[c]
7	38b	39a	40f (85)	–	–	–
8	38b	39c	40g (78)	–	–	–
9	38b	39d	40h (60)[c]	41h (traces)[b]	42h (20)[c]	43h (traces)[b]
10	38b	39e	40i (98)	–	–	–
11	38b	39f	40l (50)[c]	–	42l (15)[c]	–
12	38b	39g	–	–	42m (4)[c]	43m (11)[c]
13	38c	39a	40n (60)+40n' (32)	–	–	–
14	38d	39a	40o (53)+40o' (40)	–	–	–

[a]Isolated yields. [b]Traces determined by ^{1}H NMR and GC-MS. [c]Inseparable mixture of diastereoisomers (ratio measured by GC and ^{1}H NMR).

Experimental details: **38a-d** (1.0 mmol), **39a-l** (1.5 mmol), PPh₃ (0.08 mmol), Pd(OAc)₂ (0.02 mmol), Et₃N (2 mmol), were dissolved in THF (10 ml) and placed in a 45 ml autoclave. The autoclave was purged, pressurized (400 psi CO), and heated to 100 °C for 18 h. The reaction mixture was then cooled to room temperature and worked-up with H₂O (5 ml), extracted with Et₂O (3 x 5 ml) and purified by column chromatography (silica gel, petroleum ether/Et₂O, 7:3).

Scheme 13

According to the mechanism proposed by Torii[54a] (Scheme 14), the reaction starts by the formation of the π-allyl palladium complex **I**. In the acyl complex **II**, derived from the insertion of CO, the protons alpha to the carbonyl group show appreciable acidity.

Consequently, deprotonation by Et₃N could lead to the formation of a carbanion, or even a ketene,[63] which, through an addition reaction with the imine, could form a four membered ring. Moreover, we

198

presume that the catalytic complex involved is $(Ph_3)_4Pd(0)$ formed according to Scheme 15; indeed, using the commercial *tetrakis*(triphenylphosphine)-palladium(0) $[Pd(PPh_3)_4]$, the reaction showed equally high yields. The metallic Pd(0) and the Ph_3PO were observed at the end of the reaction.[64]

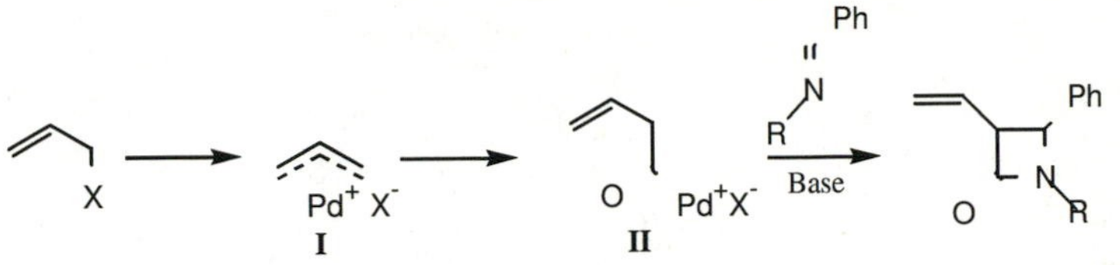

Scheme 14

$$Pd(AcO)_2 \ + \ Ph_3P \ + \ H_2O \ \longrightarrow \ Ph_3PO \ + \ Pd(0) \ + \ 2\ AcOH$$

Scheme 15

According to previously reported data,[54] as well as the data of Scheme 13, the CO insertion would always occur between the C-3 carbon directly linked to the halogen and the palladium (structure **II**). When the halogen was not in the terminal position of the allylic chain, we found, instead, that the insertion occurred at the less hindered C-1 carbon atom of the allyl halide structure.

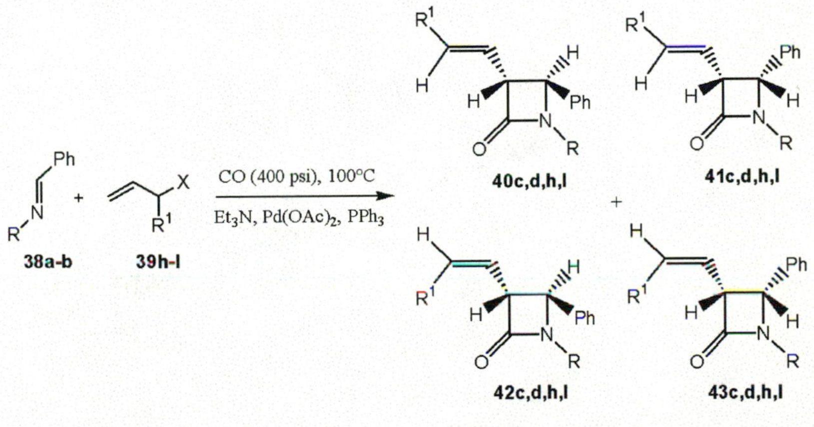

Entry	Imine	Halide	Products (yield%)[a]			
1	38a	39h	40c (60)	41c (traces)[b]	42c (20)	43c (traces)[b]
2	38b	39h	40h (60)[c]	41h (traces)[b]	42h (25)[c]	43h (traces)[b]
3	38a	39i	40c (55)	41c (traces)[b]	42c (23)	43c (traces)[b]
4	38b	39i	40h (65)[c]	41h (traces)[b]	42h (23)[c]	43h (traces)[b]
5	38a	39l	40d (52)[c]	–	42d (14)[c]	–
6	38b	39l	40l (53)[c]	–	42l (15)[c]	–

[a] Isolated yields. [b] Inseparable mixture of diastereoisomers (ratio measured by GC and ^{1}H NMR). [c] Traces determined by ^{1}H NMR and GC-MS.

Scheme 16

In particular, when the halo-substrate was 3-chloro-1-butene (**39h**) or 3-bromo-1-butene (**39i**) ($R^2=R^3=H$, $R^1=CH_3$), or 3-bromo-1-hexene (**39l**) $[R^2=R^3=H, R^1=(CH_2)_2CH_3]$, carbonylation occurred at the C-1 position, and not at the C-3 (Scheme 16). In fact, in these cases, we isolated the same reaction products as were obtained when crotyl bromide (**39d**) or 1-bromo-2-hexene (**39f**) were used to form the intermediate π-allyl palladium complex (entries 4, 5, 9 and 11, Scheme 13). The low yield obtained with 3-bromocyclohexene (**39g**) (entries 6 and 12, Scheme 13) could be caused by steric effects. The formation of the π-allyl palladium complex, and the subsequent terminal double-bond shift, could not proceed easily for steric reasons, resulting in a considerably lower yield for this reaction.[60]

2.3. Synthesis of chiral β-lactams

To further investigate the applications of our methodology we considered the possibility of inducing stereoselectivity on the two new stereocenters (C-3 and C-4), formed on the β-lactamic ring, through a cyclocarbonylation on chiral optically pure imines **38c** and **38d**.[64] The enantiopure (S)-(–)-benzilidene(2-methoxy-1-phenylethyl)amine (–)-**38d** was reacted with allyl bromide **39a**, under CO pressure, with a catalytic amount of palladium (II) complexed by PPh$_3$, according to the synthetic protocol described in Scheme 13. Optically pure *trans* β-lactams (–)-**40o** and (+)-**40o''** (in a diasteromeric ratio of 53/43) together with a small amount of *cis* β-lactam (–)-**41o** were isolated with an overall yield of 98% (Scheme 17). The three diastereomers were obtained in pure form after column chromatography (silica gel, petroleum ether/Et$_2$O, 7:3) and the optical rotation values were measured.

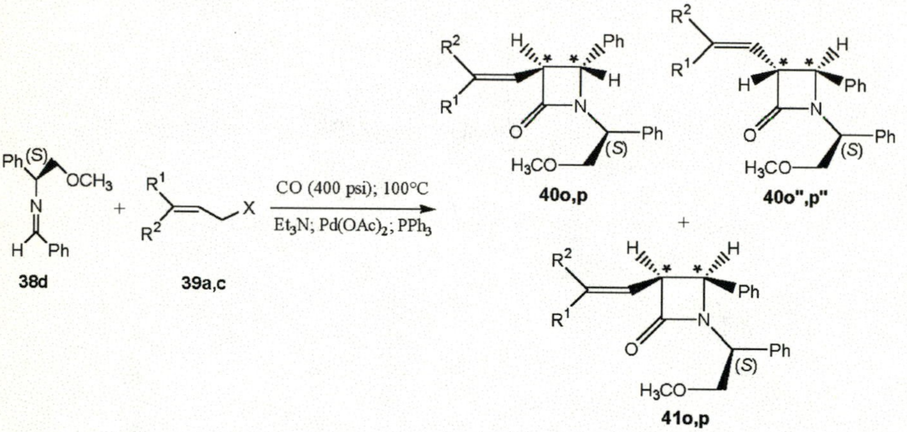

Imine absol. configuration	R^1, R^2	X	Yield (%)[a]	dr[b] [α]$_D^{22c}$		
(S)	R^1 =R^2 =H	Br	98	(–)-**40o** (53) [–44.7]	(+)-**40o''** (43) [+57.2]	(–)-**41o** (4) [–37.2]
(R)[d]	R^1 =R^2 =H	Br	95	(+)-**40o** (53) [+43.4]	(–)-**40o''** (43) [–55.4]	(+)-**41o** (4) [+35.0]
(S)	R^1 =R^2 =CH$_3$	Cl	54	(–)-**40p** (53) [–40.0]	(+)-**40p''** (47) [+24.0]	**41p** (traces)[e]
(R)[d]	R^1 =R^2 =CH$_3$	Cl	40	(+)-**40p** (53) [+40.8]	(–)-**40p''** (47) [–21.5]	**41p** (traces)[e]

[a]Isolated yields. [b]Diasteromeric ratios measured by GC and ^{1}H NMR of the crude product. [c]c 0.01–0.03, CHCl$_3$ [d]Using the enantiopure imine (R)-**38d**, the products **40o,p**, **40o'',p''** and **41o** are the enantiomers of those obtained with the enantiopure imine (S)-**38d**. [e]Traces determined by ^{1}H NMR and GC-MS.

Scheme 17

The *trans* and *cis* configurations, in this and in the following cases, have been again assigned from the ^{1}H NMR spectra through the $^3J_{H,H}$ coupling constants between the two protons on the C-3 and the C-4 carbons, $(J_{cis} > J_{trans})$.[62] X-ray crystal structure analysis[64] of (−)-**40o** confirmed the ^{1}H NMR assignment and showed that the absolute configuration of the new two centers was 3S and 4R. In the other *trans* structure (+)-**40o''** the new two centers, C-3 and C-4, were obviously of opposite configuration, 3R and 4S. The configuration of structure (−)-**41o** was assigned in comparison with the previous two structures: in this case a stronger $^3J_{H,H}$ coupling constant between the two protons in the β-lactamic ring was consistent with a *cis* structure (3R*, 4R*). Moreover, when the imine **38d** was used in the optically pure form (R)-(+), a similar procedure led to the enantiomers (+)-**40o**, (−)-**40o''**, and (+)-**41o** (Scheme 17). These three new compounds showed, in fact, identical ^{1}H NMR and ^{13}C NMR spectra but optical rotation of opposite sign and equal absolute value.[64] Scheme 17 collects also the results of the reaction of imine **38d**, in both enantiomeric forms, with 1-chloro-3-methyl-2-butene **39c**.

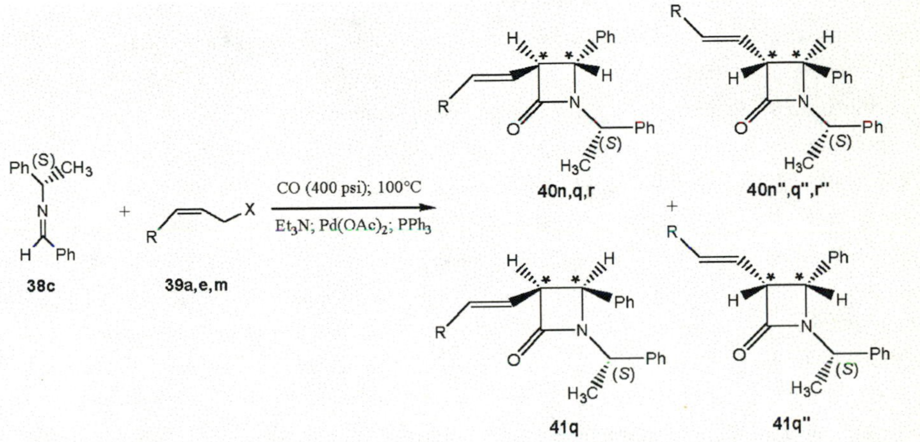

Scheme 18

Entry	Imine absol. configuration	R	X	Halide	Yield (%)[a]	dr[b] [α]$_D$^{22c}		
1	(S)	H	Br	39a	93	(+)-**40n** (66) [+34.0] (−)-**40n''** (34) [−35.0]		−
2	(R)[d]	H	Br	39a	92	(−)-**40n** (66) [−39.6] (+)-**40n''** (34) [+34.1]		−
3	(S)	Ph	Cl	39e	63	(+)-**40q**(74) [+199.7] (−)-**40q''**(20) [−194.3]	**41q +41q''** (6)[e]	
4	(R)[d]	Ph	Cl	39e	90	(−)-**40q** (74) [−226.7] (+)-**40q''**(20) [+210.3]	**41q +41q''** (6)[e]	
5	(S)	CH$_3$	Cl	39m	85	(+)-**40r** (60) [+62.9] (−)-**40r''** (40) [−41.0]		−
6	(R)[d]	CH$_3$	Cl	39m	78	(−)-**40r** (60) [−65.5] (+)-**40r''** (40) [+40.1]		−

[a]Isolated yields. [b]Diasteromeric ratios measured by GC and ^{1}H NMR of the crude product. [c]c 0.01–0.05, CHCl$_3$ [d]Using the enantiopure imine (R)-**38c**, the products 40n,q,r, 40n'',q'',r'' and 41q, q'' are the enantiomers of those obtained with the enantiopure imine (S)-**38c**. [e]Traces determined by ^{1}H NMR and GC-MS.

This latter gave, with smaller yields, approximately the same amount of *trans* diastereomers (−)- and (+)-**40p**, (+)- and (−)-**40p''** and traces of the **41p** isomer. The absolute configurations of (−)-**40p** and (+)-**40p''**, have been assigned as (3S,4R) and (3R,4S), from the comparison of the chemical shifts with (−)-**40o**

201

and (+)-**40o''**, respectively. Also the relative yields of **40p** and **40p''** followed a similar trend to **40o** and **40o''**compounds.[64] The HPLC analysis of the crude mixture, obtained by the reaction of the racemic imine **38d** with **39c** showed two pairs of peaks of nearly equal intensity, corresponding to the diastereoisomers (±)-**40p** and (±)-**40p''**, respectively. When the same reaction was carried out with the enantiopure *S* imine **38d** (Scheme 17), the HPLC analysis showed only two peaks. One peak corresponded to the first peak of the first pair, and the other to the second peak of the second pair: they were related to the isomers (−)-**40p** and (+)-**40p''**, respectively. The remaining two peaks, one for each pair, were observed on the HPLC analysis of the reaction carried out with the enantiopure *R* imine **38d** (Scheme 17). These latter two peaks were ascribed to the isomers (+)-**40p** and (−)-**40p''**, respectively.[64] Analogous results were obtained performing a similar reaction with a second imine, benzylidene(1-phenyl-ethyl)amine **38c**, used in both the enantiomerically pure forms (S)-(+) and (R)-(−). The cyclocarbonylation of **38c** with different allyl halides led, in similar reaction conditions, each time, to two optically active *trans* diastereomers with only traces of the *cis* form. The results of these cyclocarbonylations are collected in Scheme 18.

An attempt was made to assign the absolute configurations of the newly induced chiral centers, comparing the chemical shifts and the $^3J_{H,H}$ coupling constants of the C-3 and C-4 protons of these latter structures with those of Scheme 17. We assigned the configurations (3S,4R) and (3R,4S) to the structures **40n,q,r**, depending whether the starting imine had configuration S or R, respectively. *Vice versa* we assigned the configurations (3R,4S) and (3S,4R) to the structures **40n'',q'',r''**, depending whether the starting imine had configuration S or R, respectively. *Cis* type diastereomers, as inseparable mixture (dr=7/3, yields 6%) have been also isolated in the carbonylation with cinnamyl chloride **39e** (entries 3 and 4, Scheme 18). The configuration of the vinylic moiety of the compounds **40q,q''**, **41q,q''**, and **40r,r''**, ($^3J_{H,H}$ =16.0 Hz for vinylic protons) was found to be always *trans*. Finally, when the 3-chloro-1-butene was used in the carbonylation of **38c** (in the enantiomerically pure (R)-(−) configuration) the isomers (−)-**40r** and (+)-**40r''** were isolated with the same relative yields of entry 6. As we reported previously,[60] for similar reactions performed with non chiral imines, an isomerization occurs during the catalytic cycle with the insertion of CO at the C-1 carbon of the alkene.[64]

2.4. Synthesis of β-lactams 4-heteroaryl substituted

In our opinion, the presence of an heterocycle bonded to the azetidinone ring could induce an important increase of solubility to these structures in polar medium. Moreover, the resulting greater susceptibility to further synthetic elaborations should favour an increased biological activity. In particular, heterocycles like the benzothiazole and the thiazole can undergo easily to ring opening reaction freeing the masked carbonylic function.[65] Furthemore, being the heteroaryl moiety beared at the C-4 carbon, their electron withdrawing effect could increase the acidity of the C-4 proton, adding more opportunity of functionalizing these structures. N-unsubstituted β-lactams bearing 2-furyl substituent at the C-4 carbon are reported in the literature.[66] A novel series of β-lactams 4-heteroaryl substituted[67] have been, then, synthetized following a synthetic procedure analogous to that reported in Scheme 13. The heteroaryl imines used, for this purpose, were prepared, in good yields, by coupling reactions of aniline with the appropriate aldehydes (Scheme 19).[61] The imines **38e-l** were reacted with allyl bromide **39a** by a [2+2] cycloaddition, according to the procedure described on Scheme 13, for 30-35 h, affording the 4-heteroaryl substituted β-lactams **44a-f**, **45a-f**, **46b**, and **47b** (Scheme 20).

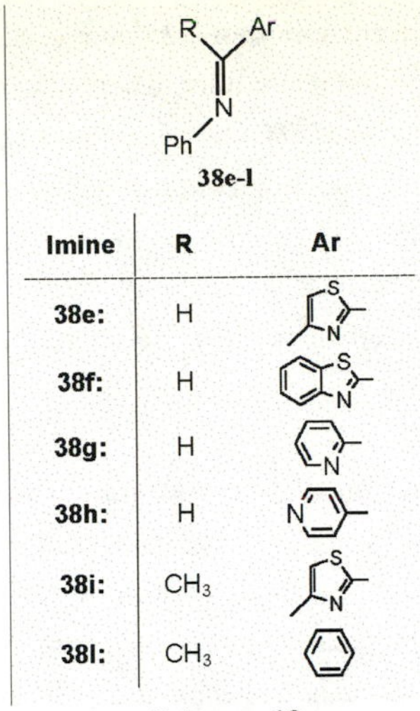

38e-l

Imine	R	Ar
38e:	H	(4-methylthiazol-2-yl)
38f:	H	(benzothiazol-2-yl)
38g:	H	(pyridin-2-yl)
38h:	H	(pyridin-4-yl)
38i:	CH₃	(4-methylthiazol-2-yl)
38l:	CH₃	(phenyl)

Scheme 19

$$38e\text{-}l + 39a \xrightarrow[\text{Et}_3\text{N, Pd(OAc)}_2,\ \text{PPh}_3]{\text{CO (400 psi), 100°C}} 44a\text{-}f + 45a\text{-}f$$

(products **44a-f**, **45a-f**, **46b**, **47b**)

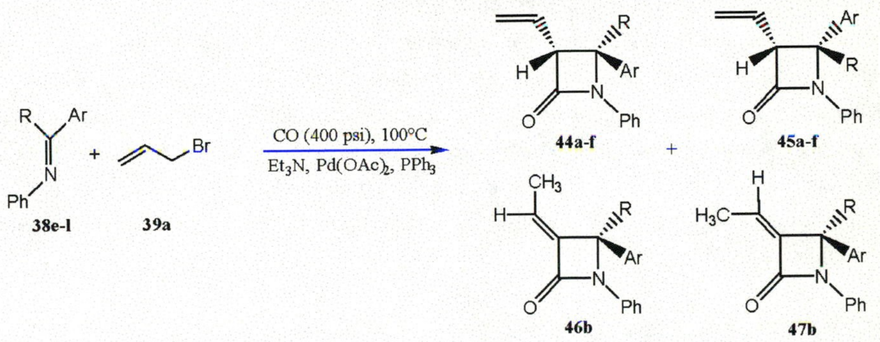

Entry	Imine	Total Yield (%)[a]	Product distributions[b]			
1	38e	90	44a (86)	45a (14)	–	–
2	38f	60	44b (78)	45b (12)	46b (5)	47b (5)
3	38g	75	44c (86)	45c (14)	–	–
4	38h	50	44d (85)	45d (15)	–	–
5	38i	40	44e (47)	45e (53)	–	–
6	38l	80	44f (65)	45f (35)	–	–

[a]Isolated yields. [b]Diastereomeric ratios evaluated by GC and ^{1}H NMR spectroscopy.

Scheme 20

The β-lactams 4-heteroaryl substituted were isolated with high stereoselectivity: the *trans/cis* ratios resulting always high excepted entries 5 and 6 (Scheme 20), where the presence of two groups (methyl and aryl) on the starting imines reduced the selectivity of the cycloaddition reaction. Moreover, when the heterocycle was benzothiazole (entry 2, Scheme 20), two new products were observed (**46b** and **47b**) together with the expected compounds (**44b** and **45b**). The **46b** and **47b** fractions were small, but they could increase for longer reaction times. Their formation should be due to the isomerization of **44b** and **45b** to the more stable α-β-unsaturated carbonylic structures. In fact, when **44b** and **45b** were warmed up with Et$_3$N in THF, an analogous transformation was noticed.[67] The *trans* and *cis* configurations, have been again assigned from the ^{1}H NMR spectra through the $^3J_{H,H}$ coupling constants between the two protons at C-3 and C-4, (J_{cis} > J_{trans}).[62] For compounds showing a methyl group linked at the C-4, the relative configuration was assigned from the coupled ^{13}C NMR spectra. A very small or negligible $^3J_{CH3-H}$ coupling constant corresponded to a *trans* configuration, while a larger $^3J_{CH3-H}$ corresponded to a *cis* configuration.[68]

3. Functionalization of β-lactams

As amply discussed in the introduction, numerous examples in literature suggest that the biological activity of β-lactams is largely affected by the kind of substituent connected to the ring. We therefore considered the opportunity of functionalizing the 2-azetidinone systems, previously synthetized, as a natural extension of our previous works. This can be achieved through the formation of a stable carbanion on the ring followed by a coupling reaction with various electrophiles.

3.1. Functionalization of β-lactams 4-phenyl substituted

Treatment of **40a** and **41a** with a base like lithiumdiisopropylamide (LDA) afforded a stable azetidinyl anion which, trapped by various electrophiles, gave β-lactams functionalized exclusively at the C-3.[69] This behaviour may be explained by the greater acidity of the hydrogen bonded to the C-3, with respect to that bonded to the C-4. When the C-3 is deprotonated, in fact, the anion conjugation (Scheme 21), produced by the neighbouring carbonylic and vinylic groups, is larger than that produced at the deprotonated C-4, by the phenyl group and the inductive effect of the nitrogen. Similar results were observed also using stronger bases such as *sec*-BuLi or *n*-BuLi.

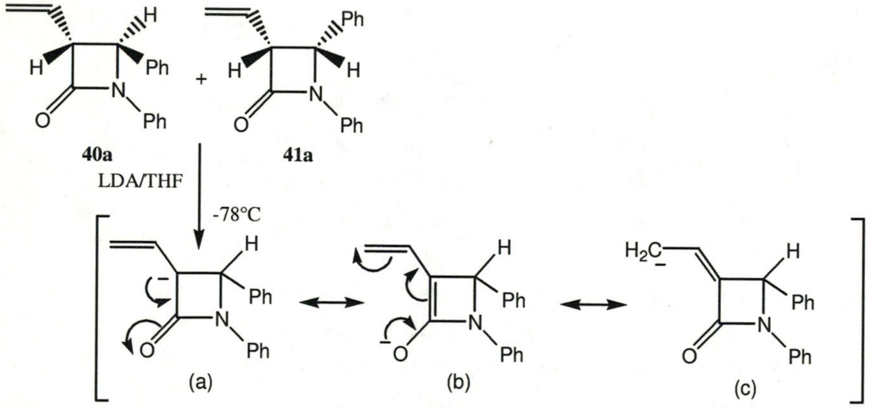

1 mmol of **40a** or **41a** was dissolved in THF (10 ml) and treated with LDA (1.1 mmol) at –78 °C, under N$_2$.

Scheme 21

In particular, treating the solution of azetidinyl anion, generated indifferently by **40a** or **41a**, with small dimension electrophiles such as H$^+$ or D$^+$, it was observed a diastereomeric mixture of products **40a,41a**, and **48a,49a** in equal ratios of *trans* and *cis* forms (entries 1-4, Scheme 22).[69]

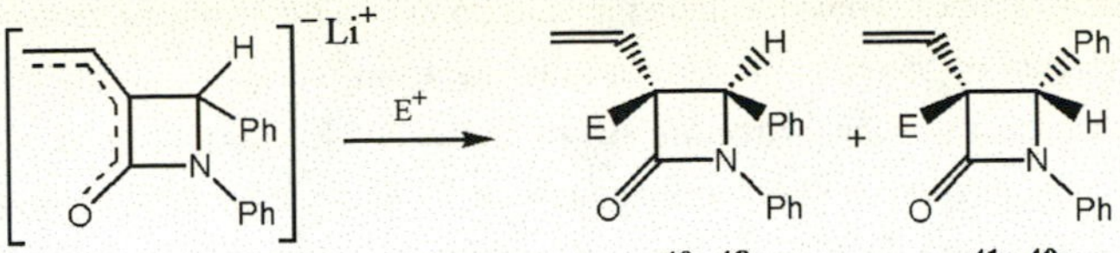

Entry	β-Lactam	E	Total Yield (%)[a]	Product distributions[b]	
1	40a	D_2O	87	48a (45)	49a (55)
2	41a	D_2O	86	48a (46)	49a (54)
3	40a	H_2O	88	40a (44)	41a (56)
4	41a	H_2O	89	40a (45)	41a (55)

[a]Isolated yields. [b]Diastereomeric ratios evaluated by GC and ^1H NMR spectroscopy.

The azetidinyl anion, obtained by treatment with LDA, was captured at –78 °C, under N_2, by the corresponding electrophile in a ratio of 1:1.

Scheme 22

Probably, the formed carbanions have always the same planar structure (Scheme 21) that makes possible for the smaller electrophyles (H^+ or D^+) to bind indifferently both sides of the molecule, leading to almost equal diasteromeric ratios. The quenching reaction became stereoselective when the electrophiles had larger dimensions as the alkyl halides.[69] The results of this investigation are collected on Scheme 23.

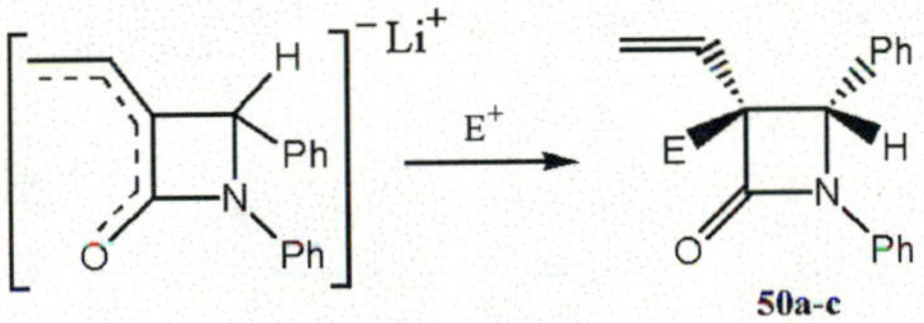

Entry	β-Lactam	E	Total Yield (%)[a]	Product distributions[b]		
1	40a	CH_3I	97	40a+41a (20)	50a (80)	–
2	41a	CH_3I	98	40a+41a (22)	50a (78)	–
3	40a	⟋⟍Br	95	40a+41a (15)	50b (85)	–
4	40a	$PhCH_2Cl$	80	40a+41a (78)	50c (15)	47g (7)

[a]Isolated yields. [b]Diastereomeric ratios evaluated by GC and ^1H NMR spectroscopy.

The azetidinyl anion, obtained by treatment with LDA, was captured at –78 °C, under N_2, by the corresponding electrophile in a ratio of 1:1.

Scheme 23

The most hindered electrophile binds exclusively the C-3 (**50a-c**), from the less hindered side, *i.e.* in *anti* with respect to the phenyl connected to the C-4. The transformation yields were always high, with the

205

exception of the benzyl chloride (entry 4, Scheme 23); in this case, we also isolated the isomerization product **47g**. The formation of this latter compound derives from the quenching with H^+ of the azetidinyl anion, corresponding to the resonance structure (c) of Scheme 21, confirming the presence of a partial negative charge delocalized also on the terminal carbon of the vinylic chain. From all reactions, we recovered a mixture of products **40a** and **41a** in an almost unitary ratio. These could be formed back from the quenching of the anion, not captured by the electrophile, by the water proton as in entries 3 and 4 of Scheme 22.[69] We assigned the *cis* configuration to compound **50a** on the basis of the large $^3J_{CH3-H}$ coupling constant between the methyl on C-3 and the hydrogen on C-4 ($J_{cis} > J_{trans}$).[68] Moreover, the ^{1}H NMR signals of the vinylic system move to higher fields when the group is located on the same side of the phenyl for an evident case of diamagnetic anisotropy produced by the aromatic ring and felt by the vinylic hydrogens.

When planar electrophiles, such as alifatic aldehydes or ketones, were employed in the quenching of the azetidinyl anion, a different reactivity and stereoselectivity was found. Together with the initial substrates (**40a** and **41a**, found in equal ratios), there were isolated alcohols **51a-c** and **52a-c**, deriving from both attack of *anti* and *syn* type with respect to the phenyl bonded to the C-4, with a strong dominance of the former (*% tran > % cis*).[69] Scheme 24 collects the results of such investigation.

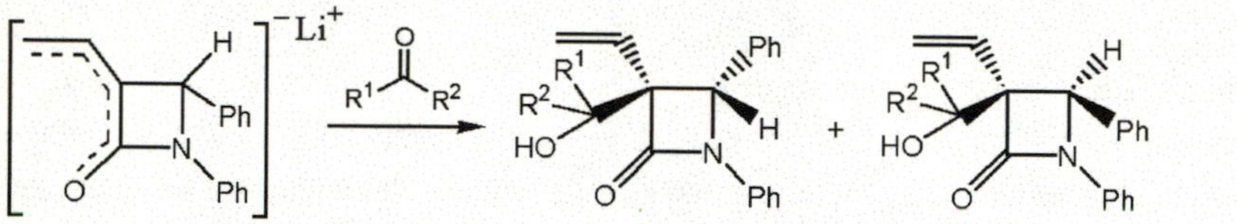

Entry	β-Lactam	R^1	R^2	Total Yield (%)^a	Products distributions^b		
1	40a,41a	H	CH$_3$	95	40a+41a (36)	51a (56)	52a (8)
2	40a	CH$_3$	CH$_3$	82	40a+41a (56)	51b (43)	52b (traces)
3	40a,41a	CH$_3$	CH$_2$Cl	90	40a+41a (48)	51c (39)	52c (13)

^{a}Isolated yields. ^{b}Diastereomeric ratios evaluated by GC and ^{1}H NMR spectroscopy.

The azetidinyl anion, obtained by treatment with LDA, was captured at –78 °C, under N$_2$, by the corresponding electrophile in a ratio of 1:1.

Scheme 24

When the capture of the azetidinyl anion took place with aromatic aldehydes, two new products were isolated. Together with the expected alcohols **53a,b** and **54a,b**, deriving from the quenching at the C-3 on the α position, there were isolated in greater yields alcohols **55a,b** and **56a,b**, deriving from the electrophilic quenching at the terminal carbon atom of the vinylic chain, on the γ position (Scheme 25). Probably, the tridentate nature of the reagent influences the regioselectivity of the electrophilic attack which may be directed to the α or γ position, under kinetic or thermodynamic control, respectively.[70]

We then showed how it is possible to functionalize a β-lactam ring with an oxiranyl group. Employing α-chloro-ketones as electrophiles we already obtained a chloridrine moiety in the chain linked at the C-3

(51c) (entry 3, Scheme 24). The treatment of such compound with NaOH in isopropanol produced a nucleophilic internal substitution with the formation of an epoxidic function in good yield (57) (Scheme 26).[69]

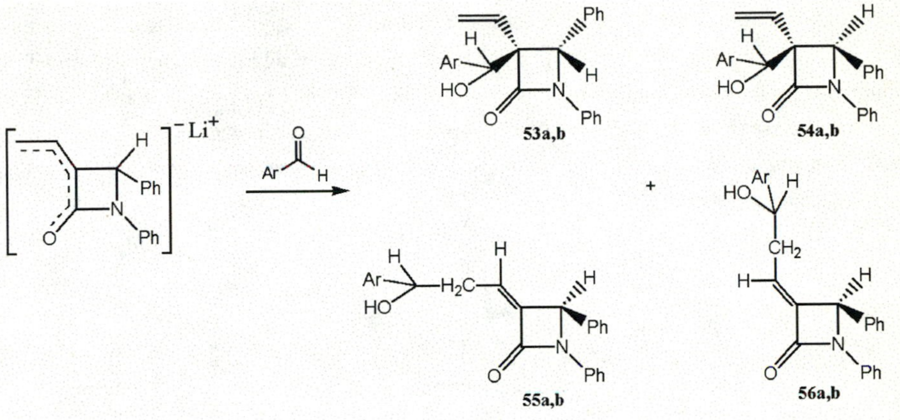

β-Lactam	Ar	Total Yield (%)[a]	Product distributions[b]				
40a,41a	Ph	93	40a+41a (72)	53a (traces)[c]	54a (traces)[c]	55a (3)	56a (24)
40a,41a	p-CF₃Ph	97	40a+41a (29)	53b (23)	54b (7)	55b (5)	56b (36)

[a]Isolated yields. [b]Diastereomeric ratios evaluated by GC and ¹H NMR spectroscopy. [c]Traces determined by ¹H NMR and GC-MS.

The azetidinyl anion, obtained by treatment with LDA, was captured at –78 °C, under N₂, by the corresponding electrophile in a ratio of 1:1.

Scheme 25

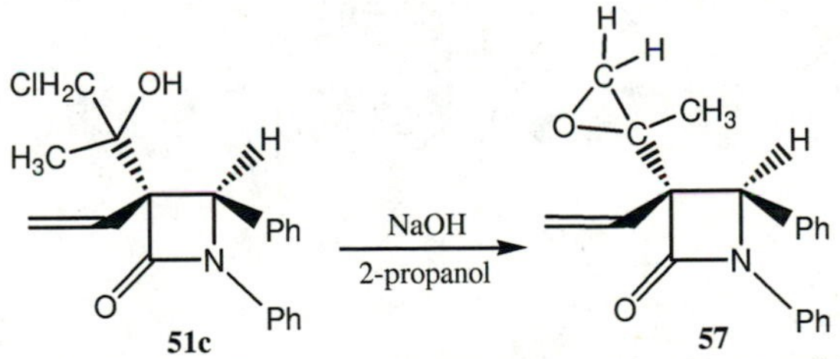

51c (0.2 mmol), NaOH 1 N (1 mmol), 2-propanol (10 mL), r.t., yield > 99%.

Scheme 26

3.2. Functionalization of β-lactams 4-heteroaryl substituted

None of the reactions reported in the previous paragraph afforded compounds deriving from a deprotonation at the C-4 carbon. Even deprotonating substrates having no hydrogens at the C-3, like compound **50a**, no carbanions at the C-4 position, enough stable to be captured by an electrophile, were formed.[69] At this point, we thought that the presence of an heteroaryl moiety linked at the C-4, instead of a phenyl group, should increase the acidity at this position, and, then, could generate a carbanion stable at the C-4. For these reasons, β-lactams *trans* **44a-c**, and *cis* **45a-c** were treated with LDA and quenched with various electrophiles.[67] Similarly to β-lactams 4-phenyl substituted, these compounds can have two kinds of acid protons, linked to the C-3 and the C-4 carbon atoms, respectively. The deprotonation of either of them led, in any case, to the formation of an azetidinyl anion stabilized by a large conjugation: by structures A, B, and C in the deprotonation of the C-3, by structures D, E and by an additional inductive effect of the β-lactamic nitrogen, in the deprotonation of C-4 (Scheme 27). However we noticed that the deprotonation, again, occurred exclusively at the C-3 allylic carbon. In fact, adding an electrophile (**E⁺**), the carbanion was trapped affording four different quenching products, resulting from A and\or C structure, according to the suggested mechanism of Scheme 27.[67]

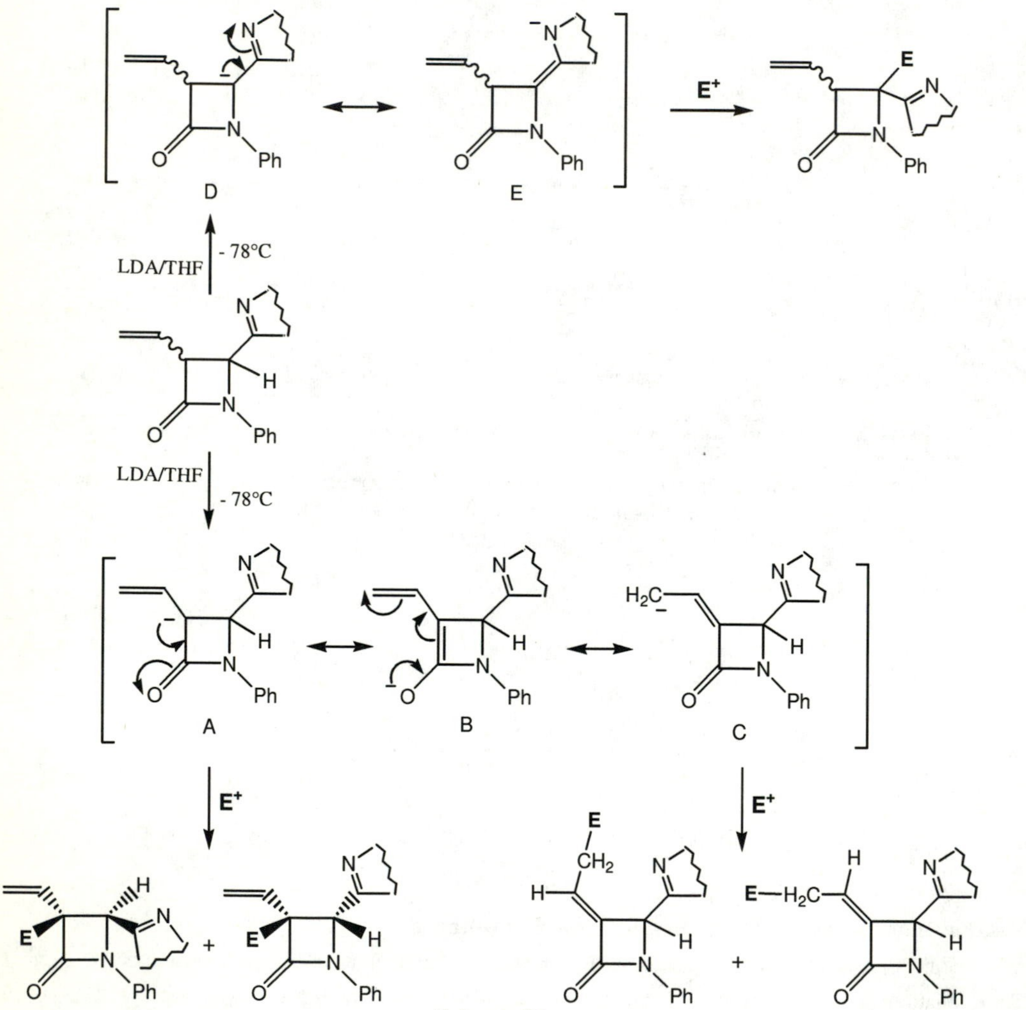

Scheme 27

208

The deprotonation of **44a,c** and the relative treatment of the azetidinyl anion with various electrophiles afforded functionalized β-lactams, as resumed in Scheme 28.

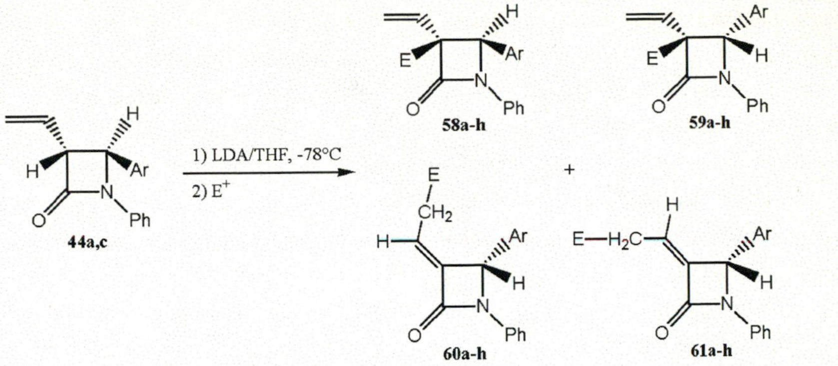

Entry	β-lactam	Ar	E	Total Yield (%)[a]	Product distributions[b]				
1	44a	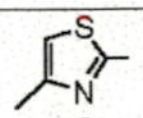 (thiazolyl)	H_2O	90	44a,45a (50,50)	–	–	–	–
2	44a	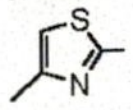 (thiazolyl)	D_2O	90	–	58a (42)	59a (58)	–	–
3	44a	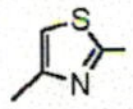 (thiazolyl)	CH_3I	99	44a+45a (24)	–	59b (76)	–	–
4	44a	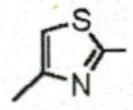 (thiazolyl)	$PhCH_2Cl$	85	44a–47a (74)	–	59c (26)	–	–
5	44a	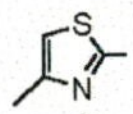 (thiazolyl)	 (allyl)Br	95	44a+45a (19)	–	59d (81)	–	–
6	44a	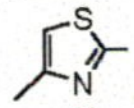 (thiazolyl)	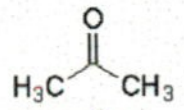 H_3C–CO–CH_3	87	44a+45a (76)	–	–	60e (12)	61e (12)
7	44a	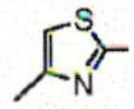 (thiazolyl)	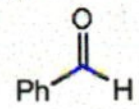 Ph–CO–H	86	44a+45a (75)	–	–	60f (22)	61f (3)
8	44c	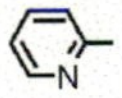 (pyridyl)	D_2O	85	–	58g (50)	59g (50)	–	–
9	44c	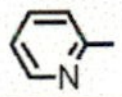 (pyridyl)	CH_3I	90	–	58h (10)	59h (90)	–	–

[a]Isolated yields. [b]Diastereomeric ratios evaluated by GC and ^{1}H NMR spectroscopy.

1 mmol of **44a** or **44c** was dissolved in THF (10ml) and treated with LDA (1.1mmol) at –78 °C, under N_2, then added of 1 mmol of the corresponding electrophile.

Scheme 28

All the reactions performed showed high yields, referred to the reacted product. The functionalization followed the similar behaviour described in the previous paragraph for 4-phenyl substituted β-lactams. In particular, using small electrophiles like H^+ or D^+, equimolecular mixtures of diastereomers *trans* and *cis* were observed (entries 1 and 2, Scheme 28). When alkyl halides were used, the reaction became highly diastereoselective, the *cis* isomer being the major reaction product (entries 3 and 5, Scheme 28). When the electrophile was the benzyl chloride (entry 4, Scheme 28), in addition to the expected *cis* product, also two more isomers were generated, **46a+47a**, which had an unsaturation at the C-3 carbon. Those latter α-β-unsaturated β-lactams should result from the quenching with H_2O of the resonance structure C (Scheme 27).

209

The equimolecular mixtures of **44a** and **45a** isolated in almost every reaction performed with **44a** (entries 1-7, Scheme 28) could be formed back from the quenching with H_2O of the carbanion not captured by the electrophile. The results confirmed a planar structure, or a tetrahedral configurationally unstable structure, for the carbanion generated by the deprotonation at the C-3. While a small electrophile such as H^+ or D^+, can bind indifferently from both sides of the molecule, bulkier electrophiles, like alkyl halides, prefer an *anti* type attack with respect to the heterocycle bonded at the C-4, leading stereoselectively to the *cis* 2-azetidinones. With carbonyl compounds as electrophiles, like acetone or benzaldehyde, the quenching occured at the terminal carbon atom of the vinylic chain on the γ position (structure C, Scheme 27); the thermodynamic products **60e,f** and **61e,f** (entries 6 and 7, Scheme 28) were almost exclusively isolated, instead of the kinetic products arising from the electrophilic attack on the α position.[70] Finally, **44c** deprotonated with LDA, at –78 °C in THF, produced the anion after losing the proton on C-3. The subsequent quenching with D_2O led to an equimolecular mixture of *trans* and *cis* isomers (entry 8, Scheme 28). A large diastereoselectivity was again found when the carbanion was quenched with CH_3I, having isolated a *cis/trans* mixture in the 9/1 ratio (entry 9, Scheme 28).[67]

Still, deprotonating **44b** with LDA, at –78 °C in THF, togheter with the products of functionalization at the C-3 carbon **58i** and **59i,l**, also a surprising partial functionalization at the C-4 was noticed (**62i,l**) (entries 1 and 2, Scheme 29).

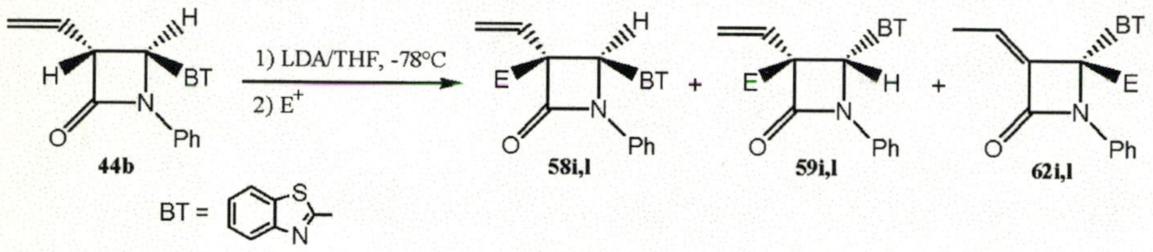

Entry	β-lactam	E	Total Yield (%)[a]	Product distributions[b]		
1	44b	D_2O	85	58i (40)	59i (42)	62i (18)
2	44b	CH_3I	90	–	59l (45)	62l (55)

[a]Isolated yields. [b]Diastereomeric ratios evaluated by GC and ^{1}H NMR spectroscopy.

1 mmol of **44b** was dissolved in THF (10 ml) and treated with LDA (1.1 mmol) at –78 °C, under N_2, then added of 1 mmol of the corresponding electrophile.

Scheme 29

The formation of the product functionalized at the C-4 may be explained by the deprotonation of the isomeric forms **46b** and **47b**, which are generated in a basic medium only when the heterocyclic moiety is the benzothiazole, as already described (entry 2, Scheme 20). Evidently, the stronger electron withdrawing effect of this latter heterocycle, compared to the others, can increase the acidity of the C-4 hydrogen.[67]

Finally, the same isomeric ratios and transformation yields were obtained deprotonating β-lactams of *cis* configuration **45a-c** and trapping them with D_2O or CH_3I. This behaviour provides a further support to the hypothesis of a planar structure of the carbanion stabilized by the resonance structures A, B, and C of Scheme 27.[67]

Excluding compounds **44b** and **45b**, none of the reactions carried out with the above substrates showed the formation of products deriving from the deprotonation of the C-4 carbon, neither using stronger bases like *sec*-BuLi or *n*-BuLi, denoting, therefore, a strong difference of acidity between the hydrogens linked to the C-3 and the C-4. These results prove that, even bearing an heterocyclic moiety like thiazole or 2-pyridine to the C-4, the increased acidity of the C-4 hydrogen is still weaker than that of the C-3 hydrogen.

However, a complete functionalization of the C-4 carbon atom was achievable only with β-lactams doubly functionalized at the C-3 and 4-heteroaryl substituted, as some of those isolated from the previous reactions, and using a strong base like *n*-BuLi. In fact, when **59h** was treated with *n*-BuLi, in THF at −78 °C, the carbanion was formed at the C-4 with planar or unstable tetrahedral structure (structures D and E, Scheme 27), since the subsequent quenching with electrophiles, like D$_2$O or CH$_3$I, led to products exclusively functionalized at the C-4 (entries 1 and 2, Scheme 30). No relevant diastereoselectivity was noticed, probably because of the two groups linked at the close C-3, which did not allow the electrophile to distinguish between the two sides of the carbanion.[67]

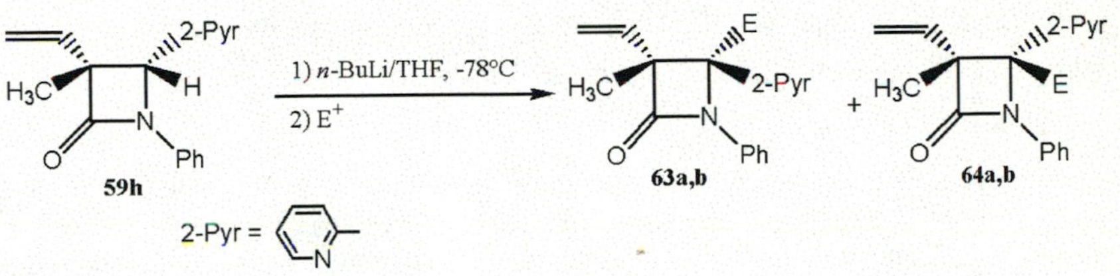

Entry	β-lactam	E	Total Yield (%)[a]	Products distributions[b]	
1	59h	D$_2$O	90	63a (49)	64a (51)
2	59h	CH$_3$I	85	63b (48)	64b (52)

[a]Isolated yields. [b]Diastereomeric ratios evaluated by GC and ^{1}H NMR spectroscopy.

1 mmol of **59h** was dissolved in THF (10 ml) and treated with *n*-BuLi (1.1 mmol) at −78 °C, under N$_2$, then added of 1 mmol of the corresponding electrophile.

Scheme 30

The relative configurations of functionalized β-lactams 4-heteroaryl substituted was assigned on the basis of NMR spectroscopic data, as described in the previous paragraphs.

4. Conclusions

A large number of β-lactams 4-aryl substituted have been synthetized by a novel palladium-catalyzed [2+2] carbonylative cycloaddition of imines with allyl halides. The novelty of our protocol is essentially based on the use of Et$_3$N as base, and simple allyl halides as substrates instead of less easily accessible allyl phosphates. A new aspect of the suggested mechanism has been shown: the CO insertion can occur on either of the carbon atoms of the intermediate π-allyl palladium complex, depending on the steric environment. Several syntheses have been performed in a stereoselective manner, leading to optically pure enantiomers. A simple deprotonation of the obtained β-lactams 4-aryl substituted, followed by trapping of the formed carbanion with various electrophiles allowed the insertion, in the lactamic skeleton, of new functionalities like alkylic, alcoholic and epoxidic groups. The functionalization reaction was stereoselective as the electrophile bound preferentially the least sterically hindered side of the azetidinyl anion formed by deprotonation. Moreover, the functionalization took place preferably at the C-3 carbon atom which tourned out to have the greater acidity compared to the C-4, also with an heteroaryl moiety like thiazole and 2-pyridine linked at the C-4. Only the 4-benzothiazolyl β-lactam showed a partial functionalization of the C-4 together with the products of functionalization of the C-3 carbon. A complete functionalization of the C-4 was, however, observed only with β-lactams 4-heteroaryl substituted and doubly functionalizated at the C-3. The presence on the β-lactamic ring of various functionalities susceptible to further synthetic elaborations, such as heterocycles, unsaturated fragments, and alcoholic groups, make this class of compounds particularly interesting for their potential biological and pharmacological activities.

Acknowledgments

This work was carried out within the framework of the National Project "Stereoselezione in Sintesi Organica: Metodologie ed Applicazioni" financially supported by the Ministero dell'Istruzione, dell'Università e della Ricerca (MIUR, Rome) and by the University of Lecce.

References

1. For a recent review see: (a) Bose, A. K.; Manhas, M. S.; Banik, B. K.; Srirajan, V. In *The Amide Linkage: Selected Structural Aspects in Chemistry, Biochemistry, and Material Science* Greenberg, A.; Breneman, C. M.; Liebman, J. F., Eds; Wiley- Interscience: New York, 2000, Chapter 7, p. 157-214. (b) Singh, G. S. "β-Lactams in the new millenium. Part. I: Monobactams and Carbapenems", *Mini Review in Medicinal Chemistry*, **2004**, *4*, 69.

2. (a) Manhas, M. S.; Banik, B. K.; Mathur, A.; Vincent, J. E.; Bose, A. K. *Tetrahedron* **2000**, *56*, 5587. (b) Bose, A . K.; Mathur, C.; Wagle, D. R.; Manhas, M. S. *Tetrahedron* **2000**, *56*, 5603. (c) Ojima, I. *Acc. Chem Res.* **1995**, *28*, 383. (d) Banik, B. K.; Manhas, M. S.; Bose, A. K. *J. Org. Chem.* **1994**, *59*, 4714. (e) Banik, B. K.; Manhas, M. S.; Bose, A. K. *J. Org. Chem.* **1993**, *58*, 307.

3. (a) Clader, J. W.; Burnett, D. A.; Caplen, M. A.; Domalski, M. S.; Dugar, S.; Vaccaro, W.; Sher, R.; Browne, M. E.; Zhao, H.; Burrier, R. E.; Salisbury, B.; Davis, H. R. *J. Med. Chem.* **1996**, *39*, 3684. (b) Burnett, D. A.; Caplen, M. A.; Darris, H. R.; Burrier, R. E.; Clader, J. W. *J. Med. Chem.* **1994**, *37*, 1733.

4. Finke, P. E.; Shan, S. K.; Fletcher, D. S.; Ashe, B. M.; Brause, K. A.; Chandler, G. O.; Dellea, P. S.; Hand, K. M.; Maycock, A. L.; Osinga, D. G.; Underwood, D. J.; Weston, H.; Davies, P.; Doherty, J. B. *J. Med. Chem.* **1995**, *38*, 2449.

5. (a) Cohn, M. L. *Science* **1992**, *257*, 1050. (b) Neu, H. C. *Science* **1992**, *257*, 1064. (c) Davies, J. *Science* **1994**, *264*, 375. (d) Shlaes, D.; Levy, S.; Archer, G. *ASM News* **1991**, *57*, 455. (e) *The World Health Report 1996* World Health Organization: Geneva, Switzerland, 1996. (f) *Emerging infections: Microbial Threats to Health in the United States* Lederberg, J.; Shope, R. E.; Eds., National Academy Press: Washington, DC, 1992.

6. (a) Bush, K.; Jacoby, G. A.; Medeiros, A. A. *Antimicrob. Agents Chemother.* **1995**, *39*, 1211. (b) Massova, I.; Mobashery, S. *Acc. Chem. Res.* **1997**, *30*, 162. (c) Kuzin, A. P.; Nukaga, M.; Nukaga, Y.; Hujer, A.; Bomono, R. A.; Knox, J. R. *Biochemistry* **2001**, *40*, 1861. (d) Yang, Y.; Rasmussen, B. A.; Shales, D. M. *Pharmacol. Ther.* **1999**, *83*, 141.

7. Bush, K.; Mobashery, S. in *Resolving the Antibiotic Paradox: Progress in Understanding Drug Resistance and Development of New Antibiotics.* Rosen, B. P., Mobashery, S. Eds., Plenum Press: NY, 1998; pp. 71-98.

8. (a) Bush, K. *Clin. Microbiol. Rev.* **1988**, *1*, 109. (b) Bush, K. in *Antibiotic and Chemotherapy* Churchill Livingstone, Ed.: Edinburgh, UK, **1997**; pp 306-327.

9.   (a) Vedel, G.; Belllouaj, A.; Gilly, L.; Labia, R.; Phillippon, A.; Nevot, P.; Paul, G. J. *Antimicrob. Chemother.* **1992**, *30*, 449. (b) Sirot, D.; Chanal, C.; Henquell, C.; Labia, R.; Sirot, J.; Cluzel, R. J. *Antimicrob. Chemother.* **1994**, *33*, 1117. (c) Belaaouaj, A.; Lapoumeroulie, C.; Canica, M. M.; Vedel, G.; Nevot, P.; Krishnamoorthy, R.; Paul, G. *FEMS Microbiol. Lett.* **1994**, *120*, 75. (d) Brun, T.; Péduzzi, J.; Canica, M. M.; Paul, G.; Nevot, P.; Barthélémy, M.; Labia, R. *FEMS Microbiol. Lett.* **1994**, *120*, 111. (e) Stemmer, W. P. C. *Nature* **1994**, *370*, 389. (f) Vakulenko, S. B.; Geryk, B.; Kotra, L. P.; Mobashery, S.; Lerner, S. A. *Antimicrob. Agents Chemother.* **1998**, *43*, 1542, and references cited therein.

10.   Carland, M. V.; Martin, R. L.; Schiesser, C. H. *Org. Biomol. Chem.* **2004**, *2*, 2612.

11.   (a) Bulychev, A.; O'Brien, M. E.; Massova, I.; Teng, M.; Gibson, T. A.; Miller, M. J.; Mobashery, S. *J. Am. Chem. Soc.* **1995**, *117*, 5938. (b) Swarén, P.; Massova, I.; Bellettini, J. R.; Bulychev, A.; Maveyraud, L.; Kotra, L. P.; Miller, M. J.; Mobashery, S.; Samama, J.-P. *J. Am. Chem. Soc.* **1999**, *121*, 5353.

12.   Mourey, L.; Kotra, L. P.; Bellettini, J.; Bulychev, A.; O'Brien, M. E.; Miller, M. J.; Mobashery, S.; Samama, J. P. *J. Biol. Chem.* **1999**, *274*, 25260. Teng, M.; Nicas, T. L.; Grissom-Arnold, J.; Cooper, R. D. G.; Miller, M. *J. Bioorg. Med. Chem.* **1993**, *1*, 151.

13.   Bulychev, A.; Bellettini, J. R.; O'Brien, M. E.; Croker, P.; Samama, J.-P.; Miller, M. J.; Mobashery, S. *Tetrahedron* **2000**, *56*, 5719.

14.   Doherty J. B.; Ashe, B. M.; Barker, P. L.; Blacklock, T. J.; Butcher, J. W.; Chandler, G. O.; Dahlgren, M. E.; Davies, P.; Dorn, C. P.; Finke, P. E.; Firestone, R. A.; Hagmann, W. K.; Halgren, T.; Knight, W. B.; Maycock, A. L.; Navia, M. A.; O'Grady, L.; Pisano, J. M.; Shah, S. K.; Thompson, K. R.; Weston, H.; Zimmerman, M. *J. Med. Chem.* **1990**, *33*, 2513.

15.   (a) Delaisse, J. M.; Ledent, P.; Vales, G. *Biochem. J.* **1991**, *279*, 167. (b) Maciewicz, R. A. ; Van der Stapper, J. W. J.; Paraskewa, C.; Williams, A. C.; Hague, A. *Biochem. Soc. Trans.* **1991**, *19*, 362. (c) Esser, R. E.; Angelo, R. A.; Murphey, M. B.; Watts, L. M.; Thornburg, L. P.; Palmer, J. T.; Talhouk, J. W.; Smith, R. E. *Arthritis Rheum.* **1994**, *37*, 236. (d) Thomson, S. K.; Halbert, S. M.; Bossard, M. J. *Proc. Natl. Acad. Sci. U.S.A.* **1997**, *94*, 14249. (e) Semenov, A.; Olson, J. E.; Rosental, P. J. *Antimicrobiol. Agents Chemother.* **1998**, *42*, 2254. (f) Engel, C.; Doyle, P.; Hsieh, I.; McKerrow, J. H. *J. Exp. Med.* **1998**, *188*, 725.

16.   (a) Chapman, H. A.; Riese, R. J.; Shi, G. P. *Annu. Rev. Physiol.* **1997**, *59*, 63. (b) Smith, W. W.; Abdel-Meguid, S. S. *Exp. Opin. Ther. Pat.* **1999**, *9*, 683. (c) Henkin, J. *Annu. Rep. Med. Chem.* **1993**, *28*, 151. (d) Michaud, S.; Gour, B. J. *Exp. Opin. Ther. Pat.* **1998**, *8*, 645. (e) Elliott, E.; Sloane, B. F. *Exp. Opin. Ther. Pat.* **1996**, *6*, 12. (f) Veber, D. F.; Thomson, S. K. *Curr. Opin. Drug Disc.Dev.* **2000**, *3*, 362.

17.   Zhou, N. E.; Guo, D; Thomas, G; Reddy, A. V. N.; Kaleta, J; Purisima, E; Menard, R; Micetich, R. G.; Sing, R. *Bioorg. Med. Chem. Lett.* **2003**, *13*, 139.

18.   (a) The Nutrition Committee of the American Heart Association in, *Dietary guidelines for healthy American Adults: a statement for physicians and health professionals,* Circulation **1988**, 77, 721A. (b) The Expert Panel, *Summary of the Second Report of the National Cholesterol Education Program (NCEP) Expert Panel on Detection, Evalution, and Treatment of High Blood Cholesterol in Adults (Adult Treatment Panel II). J. Am .Med. Assoc.* **1993**, *269*, 3015.

19.   For a review of pharmacological approaches to the treatment of atherosclerosis, see: Krause, B. R.; Sliskovic, D. R.; Bocan, T. M. A. *Exp. Opin. Invest. Drugs* **1995**, *4*, 353.

20.   Burnett, D. A.; Caplen, M. A.; Davis, H. R. Jr.; Burrier, R. E.; Clader, J. W. *J. Med. Chem.* **1994**, *37(12)*, 1733.

21.   Bergam, M.; Morales, H.; Mellars, L.; Kosoglou, T.; Burrier, R.; Davis, H. R.; Sybertz, E. J. ; Pollare, T. *12th International Symposium on Drugs Affecting Metabolism* Houston, TX, Nov. 7-10, **1995.**

22.   Salisbury, B. G.; Davis, H. R.; Burrier, R. E.; Burnett, D. A.; Boykow, G.; Caplen, M. A.; Clemmons, A. L.; Compton, D. S; Hoos, L. M; McGregor, D. G.; Schnitzer-Polokoff, R.; Smith, A. A.; Weig, B. C.; Zilli, D. L.; Clader, J. W; Sybertz, E. J. *Atherosclerosis* **1995**, *115*, 45.

23.   For a discussion of the activity of other lactams, see: Dugar, S.; Kirkup, M. P.; Clader, J. W; Lin, S. I.; Rizvi, R.; Snow, M. E.; Davis, H. R.; McColombie, S.W. *Bioorg. Med. Chem. Lett.* **1995**, *5*, 2947.

24. Clader, J. W.; Burnett, D. A.; Caplen, M. A.; Dolmalski, M. S.; Dugar, S.; Vaccaro, W.; Sher, R.; Browne, M. E.; Zhao, H.; Burrier, R. E.; Salisbury, B.; Davis, H. R. *J. Med. Chem.* **1996**, *39 (19)*, 3684.

25. McKittrick, B. A.; Ma, K.; Huie, K.; Yumibe, N.; Davis, H. Jr.; Clader, J. W.; Czarniecki, M.; McPhail, A. T. *J. Med. Chem.* **1998**, *41*, 752.

26. Altmann, S. W.; Davis, H. R. Jr.; Zhu, L.; Yao, X.; Hoos, L. M.; Tetzloff, G.; Iyer, S. P. N.; Maguire, M.; Golovko, A.; Zeng, M.; Wang, L.; Murgolo, N.; Graziano, M. P. *Science* **2004**, *Feb 20, 303,* 5661, 1201.

27. Goel, R. K.; Mahajan, M. P.; Kulkarni, S. K. *J. Pharm. Pharmaceut. Sci* **2004**, *7 (1),* 80.

28. (a) Otani, T.; Oie, S.; Matsumoto, H.; Tempest, M.; Micetich, R. G.; Singh, R.; Yamashita, T. WO 01,109, 1994 *Chem. Abstr.* **1994**, *121*, 57497. (b) Alpegiani, M.; Bissolino, P.; Perrone, E.; Pesenti, E. WO 02,603 1995 *Chem. Abstr.* **1995**, *122*, 205179. (c) Veinberg, G.; Vorona, M.; Shestakova, I.; Kanepe, I.; Zharkova, O.; Mezapuke, R.; Turovskis, I.; Kalvinsh, I.; Lukevics, E. *Bioorg. Med. Chem.* **2000**, *8*, 1033. (d) Banik, I.; Becker, F. F.; Banik, B. B. *J. Med. Chem.* **2003**, *46*, 12.

29. (a) Finke, P. E.; Dahlgren, M. E.; Weston, H.; Maycock, A. L.; Doherty, J. B. *Bioorg. Med. Chem. Lett.* **1993**, *3*, 2277. (b) Zhou, N. E.; Kaleta, J.; Prisima, E.; Menard, R.; Micetich, R. G.; Singh, R. *Bioorg. Med. Chem. Lett.* **2002**, *12*, 3417. (c) Zhou, N. E.; Guo, D.; Thomas, G.; Reddy, A. V. N.; Kaleta, J.; Prisima, E.; Menard, R.; Micetich, R. G.; Singh, R. *Bioorg. Med. Chem. Lett.* **2003**, *13*, 139.

30. Veinberg, G; Shestakova, I; Vorona, M; Kanepe, I.; Lukevics, E. *Bioorg. Med. Chem. Lett.* **2004**, *14*, 147.

31. Alford, C. A.; Britt, W. J. In *The Human Herpesviruses* Roizman, B.; Whitley, R. J.; Lopez, C., Eds.; Raven Press: New York, 1993, p. 227-255.

32. (a) Welch, A. R.; McNally, L. M.; Gibson, W. *J. Virol.* **1991**, *6*, 4091. (b) Welch, A. R.; Woods, A. S.; McNally, L. M.; Cotter, R. J.; Gibson, W. *Proc. Natl. Acad. Sci. U.S.A.* **1991**, *88*, 10792.

33. Yoakim, C.; Ogilvie, W. W.; Cameron, D. R.; Chabot, C.; Guse, I.; Haché, B.; Naud, J.; O'Meara, J. A.; Plante, R.; Déziel, R. *J. Med. Chem.* **1998**, *41*, 2882.

34. For comprehensive general reviews see: (a) Koppel, G. A., in *Small Ring Heterocycles* Hassner, A. Ed., Wyley, New York, **1983**, *vol. 42,* p. 219. (b) De Kimpe, N., In *Comprehensive Heterocyclic Chemistry II*; Katritzky, A. R.; Rees, C. W.; Scriven E. F. V.; Padwa A., Eds.; Pergamon: Oxford, 1996, Vol. 1B, p. 507.

35. Staudinger, H. *Justus Liebigs Ann. Chem.* **1907**, *356*, 51.

36. Bose, A. K.; Spiegelman, G.; Manhas, M. S. *Tetrahedron Lett.* **1971**, 3167.

37. Zamboni, R.; Just, G. *Can. J. Chem.* **1979**, *57*, 1945.

38. Georg, G. I.; Mashava, P. M.; Guan, X. *Tetrahedron* **1991**, *32*, 581.

39. Mukaiyama, T. *Angew. Chem. Int. Ed. Engl.* **1979**, *18*, 707.

40. Gilman, H.; Speeter M. *J. Am. Chem. Soc.* **1943**, *65*, 2255.

41. For reviews see: (a) Hart, D. J.; Ha D.-C. *Chem. Rev.* **1989**, *89*, 1447. (b) Brown, M. J. *Heterocycles* **1989**, *29*, 2225. (c) Cainelli, G.; Panunzio, M. *Il Farmaco* **1991**, *46*, 177. (d) Fujisawa, T.; Shimizu, M. in *Reviews on Heterocyclic Chemistry*; Oae, S. Ed.; Myu: Tokyo, 1996; Vol. 15, pp. 203-225.

42. Benaglia, M.; Cinquini, M.; Cozzi, F. *Eur. J. Org. Chem.* **2000**, 563.

43. Chmielewsky, M.; Kaluza, Z.; Furman, B. *Chem. Commun.* **1996**, 2689.

44. Hegedus, L. S. *Acc. Chem. Res.* **1995**, *28*, 299.

45. Miller, M. J. *Acc. Chem. Res.* **1986**, *19*, 49.

46. Ikeda, M.; Uchino, T.; Yamano, M.; Watanabe, Y.; Ishibashi, H.; Kido, M. *Chem. Pharm. Bull.* **1986**, *34*, 4997.

47. Ishibashi, H.; Kameoka, C.; Kodama, K.; Ikeda, M. *Tetrahedron* **1996**, *52,* 489.

48. Colquhoun, H. M.; Thompson, D. G.; Twigg, M. V. In *Carbonylation*; Plenum Press: New York, 1991.

49. (a) Alper, H.; Urso, F.; Smith, D. J. H. *J. Am. Chem. Soc.* **1983**, *105*, 6737. (b) Calet, S.; Urso, F.; Alper, H. *J. Am. Chem. Soc.* **1989**, *111*, 931.

50. (a) Alper, H.; Hamel, N. *Tetrahedron Lett.* **1987**, *28*, 3237. (b) Alper, H.; Delledonne, D.; Kameyama, M.; Roberto, D. *Organometallics* **1990**, *9*, 762.

51. Piotti, M. E.; Alper, H. *J. Am. Chem. Soc.* **1996**, *118*, 111.

52. (a) Mori, M.; Chiba, K.; Okita, M.; Ban, Y. *J. Chem. Soc., Chem. Commun.* **1979**, 698. (b) Chiba, K.; Mori, M.; Ban, Y. *Tetrahedron* **1985**, *41*, 387. (c) Mori, M.; Chiba, K.; Okita, M.; Kayo, I.; Ban, Y. *Tetrahedron* **1985**, *41*, 375.

53. Zhou, Z.; Alper, H. *J. Org. Chem.* **1996**, *61*, 1256.

54. (a) Torii, S.; Okumoto, H.; Sadakane, M.; Hai, A. K. M. A.; Tanaka, H. *Tetrahedron Lett.* **1993**, *34*, 6553. (b) Tanaka, H.; Hai, A. K. M. A.; Sadakane, M.; Okumoto, H.; Torii, S. *J. Org. Chem.* **1994**, *59*, 3040.

55. (a) Tsuji, J.; Kiji, J.; Imamura, S.; Morikawa, M. *J. Am. Chem. Soc.* **1964**, *86*, 4350. (b) Dent, W. T.; Long, R.; Whitfield, G. H. *J. Chem. Soc.* **1964**, 1588. (c) Merrifield, J. H.; Godschalx, J. P.; Stille. J. K. *Organometallics* **1984**, *3*, 1108.

56. (a) Murahashi, S.; Imada, Y.; Taniguchi, Y.; Higashiura, S. *Tetrahedron Lett.* **1988**, *29*, 4945. (b) Murahashi, S.; Imada, Y.; Taniguchi, Y.; Higashiura, S. *J. Org. Chem.* **1993**, *58*, 1538

57. Tsuji, J.; Kobayashi, Y.; Kataoka, H.; Takahashi, T. *Tetrahedron Lett.* **1980**, *21*, 1475.

58. (a) Tsuji, J.; Sato, K.; Okumoto, H. *J. Org. Chem.* **1984**, *49*, 1341. (b) Tsuji, J.; Sato, K.; Okumoto, H. *Tetrahedron Lett.* **1982**, *23*, 5189.

59. Trost, B. M.; Schmuff, N. R.; Miller, M. J. *J. Am. Chem. Soc.* **1980**, *102*, 5979.

60. Troisi, L.; De Vitis, L.; Granito, C.; Epifani, E. *Eur. J. Org. Chem.* **2004**, 1357.

61. (a) Taguchi, K.; Westheimer, F. H. *J. Org. Chem.* **1971**, *36*, 1570. (b) De Vitis, L.; Florio, S.; Granito, C.; Ronzini. L.; Troisi, L.; Capriati, V.; Luisi, R.; Pilati, T. *Tetrahedron* **2004**, *60*, 1175.

62. (a) Wartski, L. *Bull. Soc. Chim. France* **1975**, 1663. (b) Sharma, A. K.; Mazumdar, S. N.; Mahajan, M. P. *J. Org. Chem.* **1996**, *61*, 5506. (c) Florio, S.; Capriati, V.; Di Martino, S.; Abbotto, A. *Eur. J. Org. Chem.* **1999**, 409.

63. Guangzhong, W.; Shimoyama, I.; Negishi, E. *J. Org. Chem.* **1991**, *56*, 6506.

64. Troisi, L.; De Vitis, L.; Granito, C.; Pilati, T.; Pindinelli, E. *Tetrahedron* **2004**, *60*, 6895, and references therein cited.

65. Dondoni, A.; Perrone, D. *Aldrichimica Acta* **1997**, *30*, 35.

66. Barbaro, G.; Battaglia, A.; Guerrini, A.; Bertucci, C. *J. Org. Chem.* **1999**, *64*, 4643.

67. Troisi, L.; Granito, C.; De Vitis, L.; Pindinelli, E.; Ronzini, L. data submitted for pubblication, and available contacting: luigino.troisi@unile.it.

68. Kingsbury, C. A.; Durham D. L.; Hutton, R. *J. Org. Chem.* **1978**, *43*, 4696.

69. Troisi, L.; Pindinelli, E.; De Vitis, L.; Granito, C.; data submitted for pubblication, and available contacting: luigino.troisi@unile.it.

70. Cainelli, G.; Cardillo, G. *Acc. Chem. Res.* **1981**, *14*, 89. Henryck, C. A.; Willy, W. E.; McKean, D. R.; Baggiolini, E.; Siddal, B. J. *J. Org. Chem.* **1975**, *40*, 8. Casinos, I.; Mestres, R. *J. Chem. Soc., Perkin Trans.* **1978**, *1*, 165.

SUBSTITUTED PROLINES:
SYNTHESES AND APPLICATIONS IN STRUCTURE-ACTIVITY RELATIONSHIP STUDIES OF BIOLOGICALLY ACTIVE PEPTIDES

Philippe Karoyan,* Sandrine Sagan, Olivier Lequin, Jean Quancard,
Solange Lavielle and Gérard Chassaing

Université Pierre et Marie Curie, UMR-7613-CNRS, Synthèse, Structure et Fonction de Molécules Bioactives, 4 place jussieu F-75252 Paris cedex 05, France (e-mail: philippe.karoyan@upmc.fr)

Abstract. *The control of the threedimensional structure of peptides and proteins by chemical modification of proteinogenic amino acids has been the keystone of peptide chemistry for the past decades. The major difficulty remains in inducing the correct folding of the peptide while retaining the critical recognition elements involved in the interaction. With this aim, substituted prolines represent useful tools. Indeed, these cyclic amino acids constrain the Φ-value of the peptide backbone around -65° while insertion of the side chain of natural amino acids on the pyrrolidine ring might give data both on its conformation and the importance of the information it carries. This review will focus on the chemical syntheses of substituted prolines, their effects on peptide conformation and structure and on the biological activities of the resulting analogues of several peptide families.*

Contents

1. Introduction

  Since the pioneering work of du Vigneaud *et al.*[1,2] first reporting the synthesis of deaminooxytocin, interest in analogues of natural amino acids has increased for several reasons. They can be valuable starting material for the synthesis of other products[3] or simply be metabolic antagonists of natural amino acids.[4]

However, the main interest is probably linked to their use in structure-activity relationship (SAR) studies of biologically active peptides. Indeed, the control of the threedimensional structure of peptides and proteins by chemical modification of proteinogenic amino acids has been the keystone of peptide chemistry for the past decades. The aim of these studies is the development of compounds with improved selectivity, bioavailability, stability and permeability.[5-8] The major difficulty remains in inducing the correct folding of the peptide while retaining the critical recognition elements involved in the interaction.[9] Among the various approaches (Disulfide bridging, lactam cyclization, N-methylation...) described in the *de novo* design of peptides with high propensity to fold with predetermined secondary structure, the replacement of a native residue by proline chimeras has been widely used.[10-13] Indeed, these cyclic amino acids constrain the Φ-value of the peptide backbone around -65° while insertion of the side chain of natural amino acids on the pyrrolidine ring might give data both on its conformation and the importance of the information it carries. It has been demonstrated that the insertion of a proline in a heterochiral peptide sequence remains the simplest strategy to induce a β-turn and incorporation of substituted prolines is a valuable approach to mimic natural β-turns found in proteins with retention of side chain functionality in the turn.[14] These secondary structures are known to play several roles (as stabilizing tertiary structure, initiating folding and facilitating intermolecular recognition).[13,15] Recently polyproline II (PPII) helices have received much attention. They are believed to be the dominant conformation of proline-rich sequences[16] and are often involved in mediating protein-protein interactions.[17] The introduction of chemical diversity into polyproline II helices through proline chimeras may lead to compounds able to inhibit protein-protein interaction by creating new stabilizing interactions as suggested by Chmielewski.[18] All these possible applications led to the rise of many strategies of variously substituted proline syntheses.

One review by Mauger and Wiktop[19] on the syntheses of substituted prolines has been published covering the period up until 1966. The present review will focus on the chemical syntheses of exclusively mono-substituted (in position 2, 3, 4 or 5) prolines including contributions to the literature from 1966 to the end of 2004. The effects of the incorporation of these modified prolines on peptide conformation and structure and on the biological activities of the resulting analogues will be discussed.

2. Syntheses of substituted prolines

The synthesis of mono-2,3,4, or 5-substituted prolines can be considered through several pathways leading to racemates or optically pure products and involving either direct functionalization of proline itself or proline derivatives, or by inter- or intramolecular cyclizations *via* C-C or C-N bond formation, using anionic, cationic or radicalar processes.

2.1. 2-Substituted prolines syntheses
2.1.1. 2-Substituted prolines syntheses starting from proline or proline derivatives

The direct alkylation of *N*-protected proline esters seems to be the most convenient method for the synthesis of 2-substituted prolines and several approaches have been reported.

In 1983, Seebach and co-workers[20] have proposed the actually most employed methodology for the synthesis of 2-substituted prolines in peptidomimetics and SAR studies (Scheme 1).

Proline is condensed with pivalaldehyde to give a single stereoisomer of 2-*tert*-butyl-1-aza-3-oxabicyclo[3.3.0]octan-4-one. Deprotonation with LDA followed by alkylation of the chiral enolate with various electrophiles yielded α-alkylated proline derivatives. The overall process is an electrophilic substitution of the α-proton of proline with retention of configuration and was called self-reproduction of chirality by the authors. The main difficulty remains the cleavage of the protecting group which needs heating in HBr from 15% to 48% for several hours, depending on the nature of R. Cleavage of the oxazolidinone has also been reported using thionyl chloride in refluxing methanol leading to α-methyl proline methyl ester in 58% yield.[21]

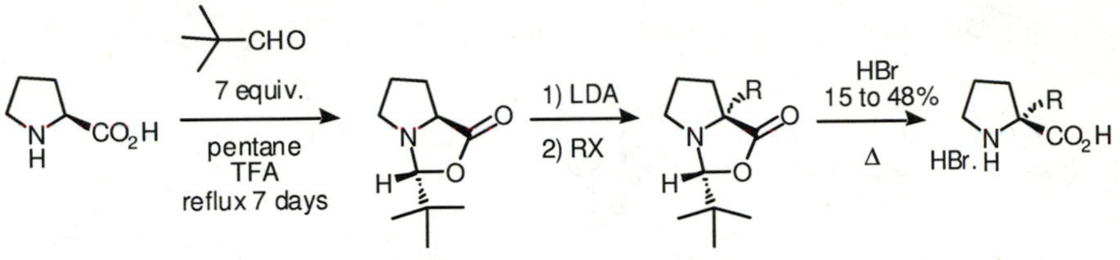

R = -CH$_3$, -CH$_2$-CH=CH$_2$, -CH$_2$Ph, -CH$_2$N(CH$_3$)$_2$,-CH$_2$CO$_2$CH$_3$, -CH$_2$-CON(CH$_3$)$_2$...

Scheme 1

More recently, Wang *et al.*[22] have reported the use of 4-alkyl-2-trichloromethyloxazolidin-5-ones (instead of Seebach's *tert*-butyloxazolidinone) as precursors to enantiomerically pure *C*- and *N*-protected α-alkyl prolines (Scheme 2).

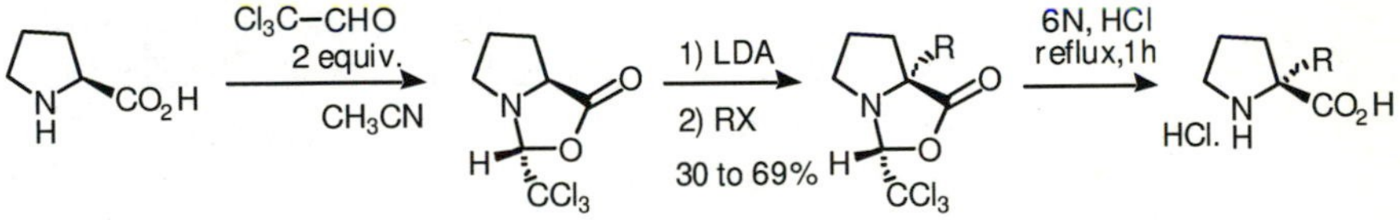

R = -CH$_3$, -CH$_2$-CH=CH$_2$, -CH$_2$Ph, -CH$_2$CO$_2$CH$_3$

Scheme 2

The results are similar to those reported by Seebach in term of stereochemical course but the authors argue that the use of trichloromethyloxazolidinone may be preferable due to its greater stability and its lower cost production. Matsumura and co-workers[23] have proposed the asymmetric synthesis of α-methylproline by alkylation of 5-phenylthio-*N*-methoxycarbonyl-L-proline methyl ester (Scheme 3). The chiral centre in position 5 of the pyrrolidine ring here governs the asymmetric alkylation :

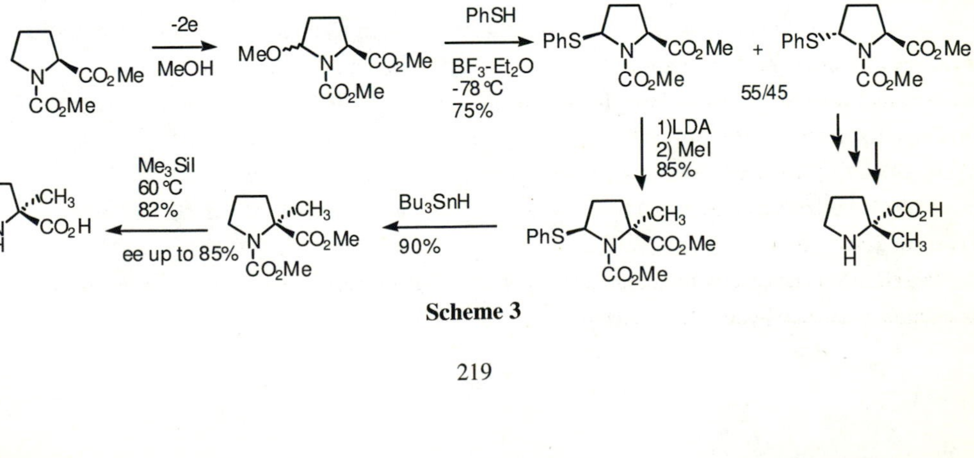

Scheme 3

219

Electrochemical oxidation in methanol followed by substitution of the 5-methoxy group by a phenylthio group in the presence of Lewis acid led to 5-phenylthioproline methyl ester as a separable mixture of diastereoisomers in a ratio of 55:45 in favour of the *cis*-isomer. The α-methylation was performed after deprotonation with LDA. Both enantiomers of α-methylproline could be obtained starting from *cis*-or *trans*-5-thiophenylproline derivatives. The *ee* were determined on the fully deprotected compounds obtained by reaction with tributyltin hydride followed by treatment with iodotrimethylsilane.

Starting from *N*-benzyl proline ethyl ester, Arboré and co-workers[24] have proposed to access to 2-allyl-proline by [2,3]-Stevens rearrangement (Scheme 4).

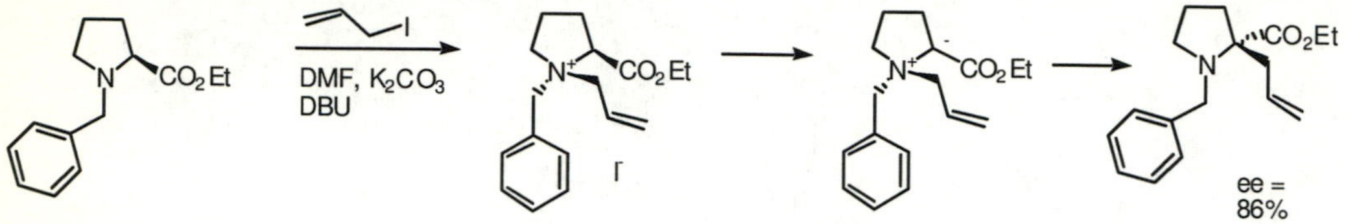

Scheme 4

Allylation of *N*-benzyl proline methyl ester gave the chiral ammonium salt by transfer of chirality from carbon to nitrogen. Loss of the original chiral centre by proton abstraction gives the ylide and the chiral centre is regenerated by [2,3]-Stevens rearrangement with 86% *ee* in the best case, reported herein.

Several racemic alkylation approaches starting from proline have been reported, simply differing by the protecting groups of the nitrogen and the carboxylic acid. Thus, in 1988, Confalone and co-workers[25] have reported a version using *N*-Boc protecting group and methyl ester (Scheme 5).

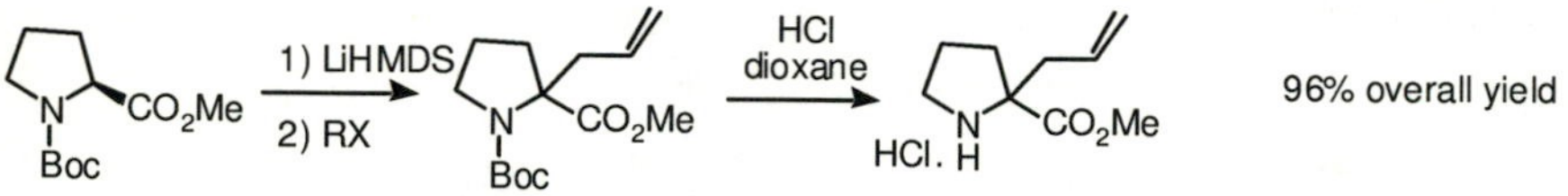

Scheme 5

Witter obtained similar results[26] with the benzyl ester. The racemate approach was also used by Ward *et al.*[27] using the *tert*-butyl ester and *N*-benzyloxycarbonyl as protecting groups, by Chan,[28] Pfeifer[29] and Cox[30] starting from the methyl ester. In the case of Pfeifer, the reaction was used in order to introduce a *tert*-butyl acetyl ester function in α-position. The α-methylation of the *N*-benzoyl-L-proline methyl ester has been reported by Culbertson[31] and then by Sato[32]. More recently, Papillon *et al.*,[33] interested in the synthesis of the octane ring system found in Oxazolomycin, have reported alkylation of variously protected proline derivatives (*N*-Boc, *N*-Z, *N*-benzyl and pentafluorophenyl-, Phenyl-, Succinyl-, Me- and Butyl- esters). The introduction of chiral auxiliary by Sato[34][(+)-menthol] on the ester function of *N*-Boc protected proline led to poor diastereoselectivity (66/34).

In 1970, Gallina and co-workers[35] have reported the synthesis of *N, N*-diacetyl-2-amino-D,L-proline and 5-amino-D,L-proline (see below) by addition of acetamide to *N*-acetyl-Δ^2-pyrrolidine-2-carboxylic acid upon heating.

The title compound was not deprotected and was characterized through α-keto-δ-acetylaminovaleric acid obtained after acid hydrolysis.

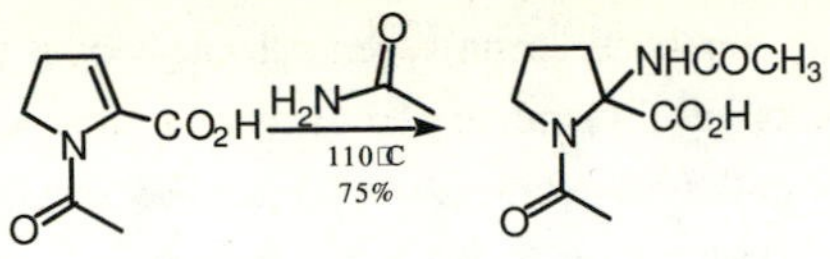

Scheme 6

2.1.2. 2-Substituted prolines syntheses by intramolecular C-N bond forming cyclization

In 1986, Bajgrowicz and co-workers[36] have reported a new method for the synthesis of cyclic imino acids (Scheme 7).

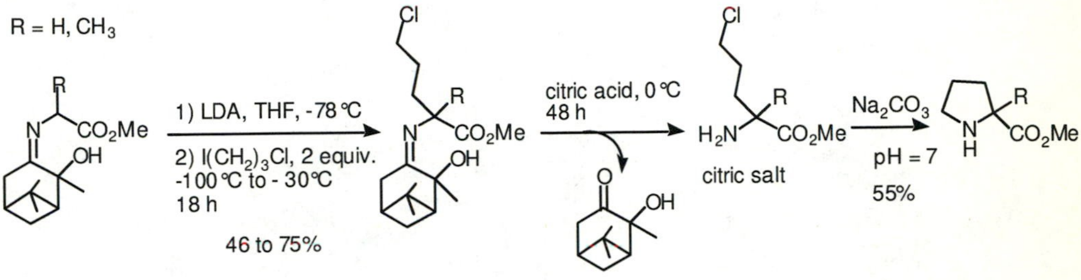

Scheme 7

The diastereoselective alkylation of Schiff bases (derived from α-amino esters and (+) or (-) 2-hydroxypinan-3-one) with dihalogeno alkane, followed by smooth hydrolytic cleavage and cyclization afforded the imino esters with *ee* up to 95%. Five or six membered rings were obtained by this method albeit with poor *ee* in the case of hindered R for six membered cyclic imino esters (*ee*=5% for R=*n*-propyl). In 2002, Scott[37] have reported the alkylation of resin-bound Schiff bases leading to methyl proline in the same manner but as a racemate.

In 1987, Schöllkopf and co-workers[38] have reported the asymmetric synthesis of α-methylproline using the bislactim ether method. The cyclic amino acid is constructed by diastereoselective alkylation (Scheme 8).

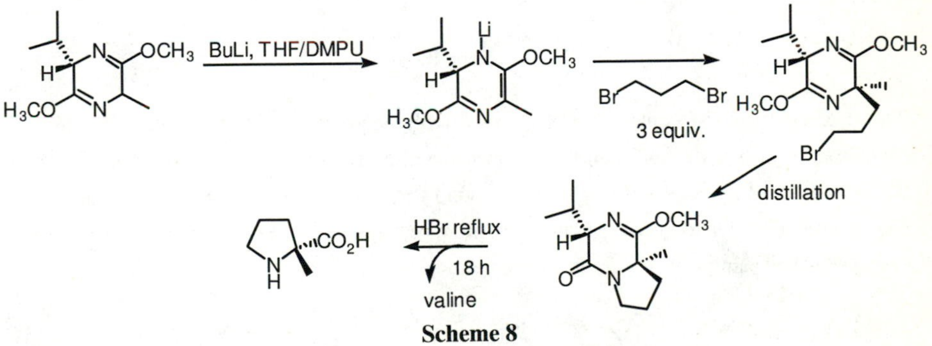

Scheme 8

After alkylation with the dihalide, cyclization of the bromo-intermediate occurs upon distillation. The 2-methylproline is obtained by hydrolytic cleavage with refluxing HBr for 18 h. The main interest as

221

claimed by the authors in comparison of Seebach's general method is the possible application of the bislactim approach to the synthesis of 6-membered rings. A similar approach has been reported by Chinchilla[39] starting from 3,6-dihydro-2H-1,4-oxazin-2-one instead of Schöllkopf's bislactim.

The synthesis of substituted prolines by C-N bond formation has also been reported earlier from substituted hydantoin leading however to racemate compounds. Smissman et al.[40] have thus reported the synthesis of 2-phenyl proline as outlined in Scheme 9.

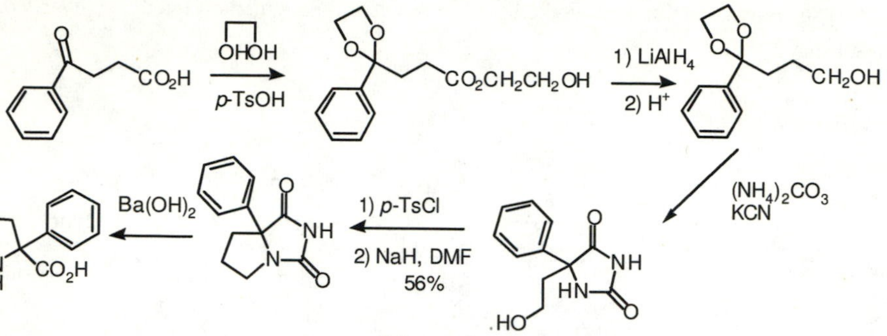

Scheme 9

After conversion of 3-benzylpropionic acid to 4-hydroxybutyrophenone by lithium aluminium hydride reduction of the corresponding ethylene ketal monoethylene glycol ester, the hydantoin was obtained by condensation with ammonium carbonate and potassium cyanide. The cyclization using NaH in DMF was performed after tosylation of the hydroxyl function, yielding to the phenylpyrrolidino [1,2-d]-1,3-(2H)-imidazolidinone. 2-Phenyl proline was obtained in 61% yield after base hydrolysis with ammonium hydroxide-hydrogen sulfide solution or in barium-hydroxyde solution. The same methodology was reported by Ellington et al.,[41] interested in the synthesis of α-methyl amino acids. The synthesis of 2-methyl proline from substituted hydantoin was reported as outlined in Scheme 10.

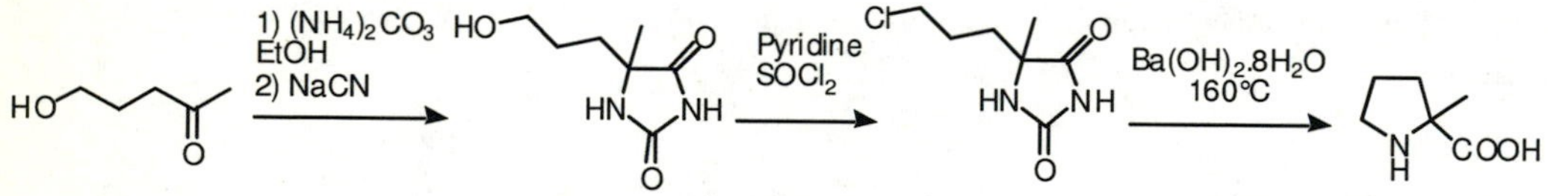

Scheme 10

The cyclisation of 5-(3-chloroxypropyl)-5-methylhydantoin (obtained by reduction of 2-methylfuran and conversion of the 5-hydroxy-2-pentanone) occurred after hydrolysis leading to 2-methyl proline.

The synthesis of pyrrolidines rings by C-N bond forming reaction has also been reported by Shono et al.[42] The intramolecular anodic C-N bond formation between a tosylamino group and an active methyne group was promoted using iodide as mediator (Scheme 11).

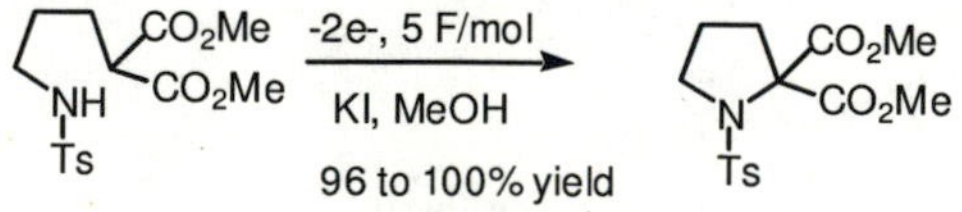

Scheme 11

222

Although as claimed by the authors this reaction seems worthwhile from a synthetic point of view, it is limited by the nature of the substituents introduced in position 2 of the pyrrolidine ring.

There are several contributions from Mark Holladay and co-workers concerning the synthesis of variously substituted prolines. In 1991,[5] they reported the synthesis of constrained phenylalanine analogs starting from L-phenylalanine (Scheme 12).

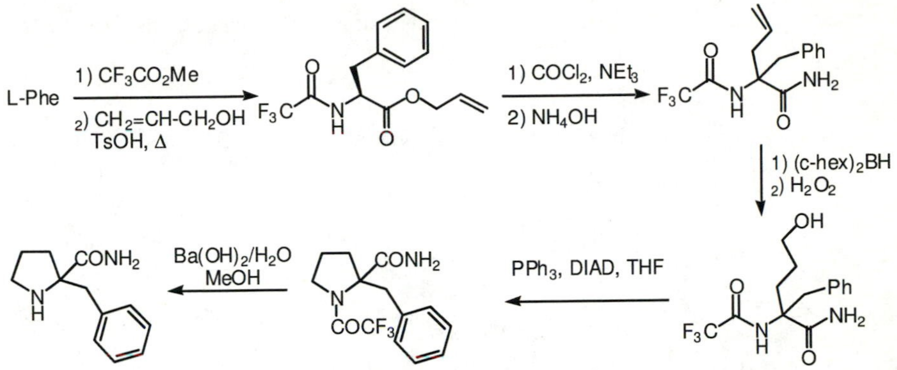

Scheme 12

The allyl-substituted phenylalanine derivative was obtained through a tandem dehydration/aza-Cope rearrangement reaction. The hydroboration of the allyl function with dicyclohexylborane in oxidative conditions followed by dehydrating cyclization under Mitsunobu conditions smoothly produced the substituted proline in 67% yield as a racemate. The N-deprotection was realized by aqueous Ba(OH)$_2$ in MeOH. In 1997, Betsbrugge and co-workers[43] have reported a similar approach for the cyclization process through Mitsunobu's reaction starting from optically pure product obtained by enzymatic resolution (Scheme 13).

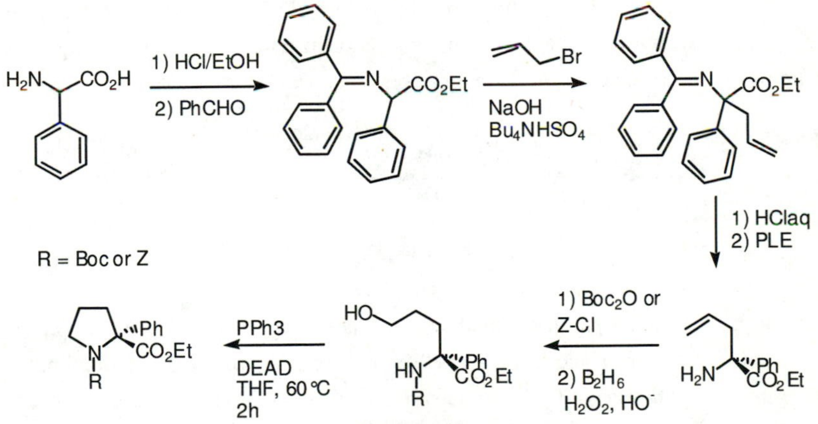

Scheme 13

Starting from the racemic (R,S)-phenylglycine, N-benzylidene phenylglycine ethyl ester was prepared in 89% yield. An allyl group was introduced under phase-transfer catalysis conditions. N-benzylidene-α-

223

allylphenylglycine ethyl ester was then hydrolyzed and the resulting α-allylphenylglycine ethyl ester was enzymatically resolved. After *N*-protection (Boc- or Z-) and oxidative hydroboration, the cyclization was performed *via* Mitsunobu reaction.

2.1.3. 2-Substituted prolines syntheses by C-C bond forming cyclization

Only one example using this approach has been reported by Kawabata in 2003[44] who describes the cyclic amino acids with quaternary stereocenter as a new class of unnatural amino acids (Scheme 14).

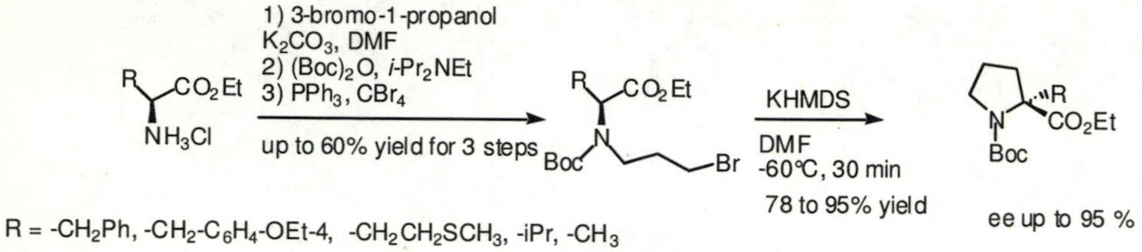

R = -CH$_2$Ph, -CH$_2$-C$_6$H$_4$-OEt-4, -CH$_2$CH$_2$SCH$_3$, -iPr, -CH$_3$

Scheme 14

The substrates were obtained from *S*-amino acids through *N*-alkylation with 3-bromo-1-propanol, Boc protection and conversion of the hydroxy group into bromine without loss of enantiomeric purity. The Boc group is essential for the generation of chiral nonracemic enolate intermediate, which was obtained by deprotonation with KHMDS in DMF. The cyclization by C-C bond forming occurred without loss of the chirality of the parent amino acid during enolate formation. This method seems to be the easiest one to obtain in four steps, with very good yield and *ee*, directly *N*-Boc protected 2-substituted prolines starting from natural amino acids.

2.1.4. 2-Substituted prolines syntheses by alternative methods

In 1982, Duhamel[45] has reported ring contraction of heterocyclic enamines allowing the synthesis of substituted pyrrolidines. Indeed, upon treatment by bromine followed by water-triethylamine, the cyclic enamine underwent ring contraction, yielding the heterocyclic aldehyde, which was subsequently engaged in Wittig olefination and hydrogenation over platinum.

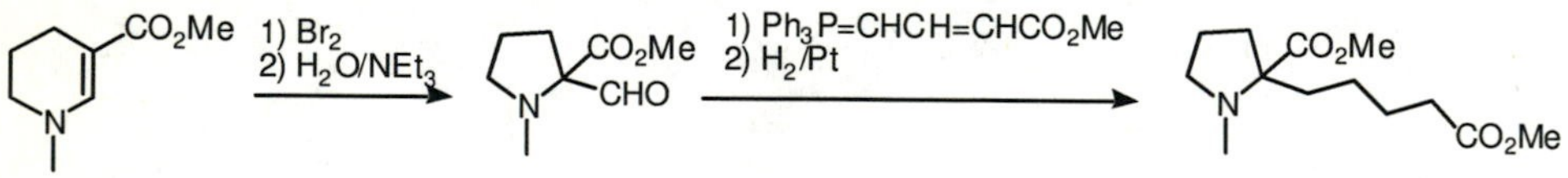

Scheme 15

Although elegant, the main problem remaining in this strategy was the nitrogen deprotection. This methodology was used by Donati and co-workers[46] with *N*-benzyl protected α-bromoiminium bromide. Treatment of the iminium with bases afforded various ring contraction products (Scheme 16) as racemates.

In 2000, Trancard and co-workers[47] have reported a study on the diastereoselectivity of the ring-contraction process. The best results were obtained by introducing the chirality on the nitrogen (85/15 *d.e* ratio with α-methyl naphtyl). Shatzmiller and co-workers[48] have reported the asymmetric synthesis of α-methyl proline using chloromethyl-menthyl ether (Scheme 17).

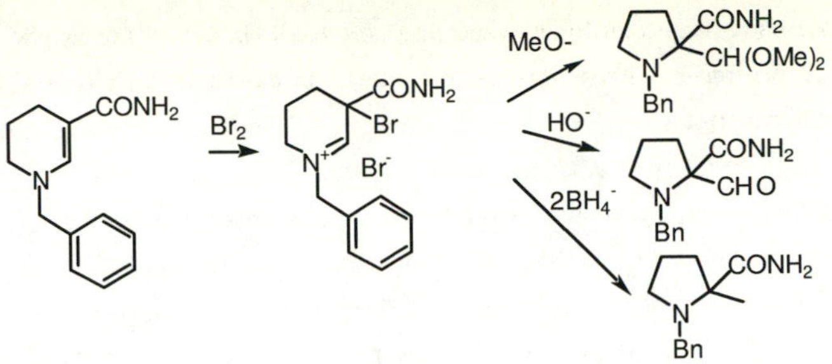

Scheme 16

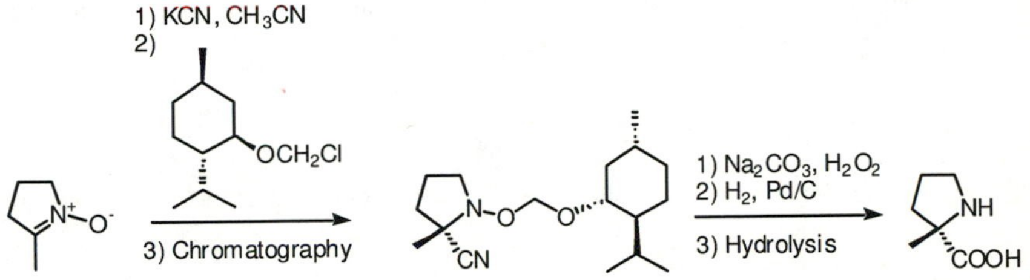

Scheme 17

The treatment of 3,4-dihydro-5-methyl-2H-pyrrole 1-oxide with 3 equivalents of KCN in acetonitrile at 65 °C in an ultrasonic bath for 48h followed by addition of chloromethyl(-)-menthyl ether gives a 1:1 mixture of the diastereomeric amino nitriles in 75% yield. The isomers were separated by chromatography. Hydrolysis of the nitrile and hydrogenolysis of the N-O bond yielded both enantiomers of α-methyl proline.

2.2. 3-Substituted prolines syntheses
2.2.1. 3-Substituted prolines syntheses starting from proline or proline derivatives

Häusler[49,50] has reported the synthesis of 3-substituted prolines through nucleophilic substitution (NS) of bromine intermediate (Scheme 18).

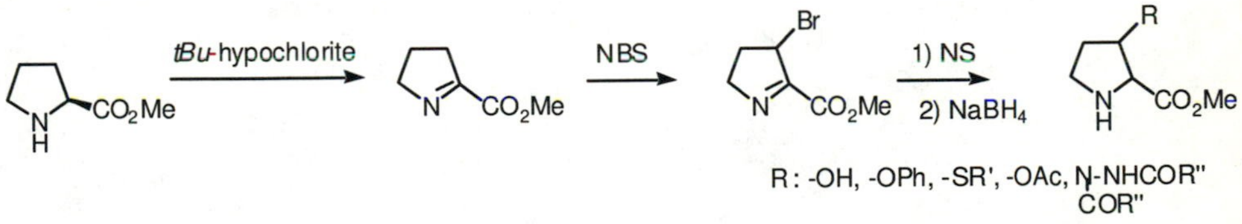

R: -OH, -OPh, -SR', -OAc, N-NHCOR"
COR"

Scheme 18

The oxidation of proline methyl ester was followed by bromination of the imine intermediate. Substitutions were performed with a few nucleophiles. The compounds were obtained as racemate mixtures.

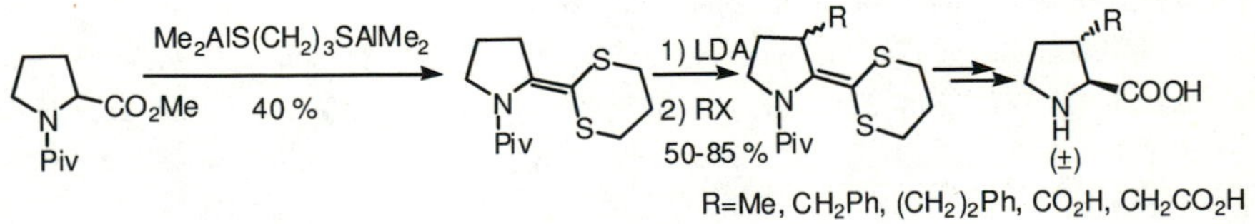

R=Me, CH$_2$Ph, (CH$_2$)$_2$Ph, CO$_2$H, CH$_2$CO$_2$H

Scheme 19

225

The synthesis of 3-substituted prolines by regioselective alkylation of the allylic anion of the ketene S,S-acetal derived from proline has been reported by Moss and co-workers.[51] Several functionalities were introduced using this approach, leading to racemates (Scheme 19). After regeneration of the carboxylic acid function, the *trans* derivative was isolated as the major product.

3-Hydroxyproline has rarely been used to synthesize 3-substituted prolines by SN_2 displacement of the tosylate or mesylate, probably because of the competing elimination reaction due to Hα acidity. However, one group described the synthesis of *cis*- and *trans*-3-fluoroprolines from *trans*- and *cis*-3-hydroxyproline,[52] but the yield of the fluorination is significantly lower compared to the corresponding SN_2 from 4-hydroxyprolines (see below, Scheme 34).

Recently, the introduction of several substituents by palladium coupling on an enol triflate derived from 3-hydroxyproline has been described by Kamenecka[53] (Scheme 20).

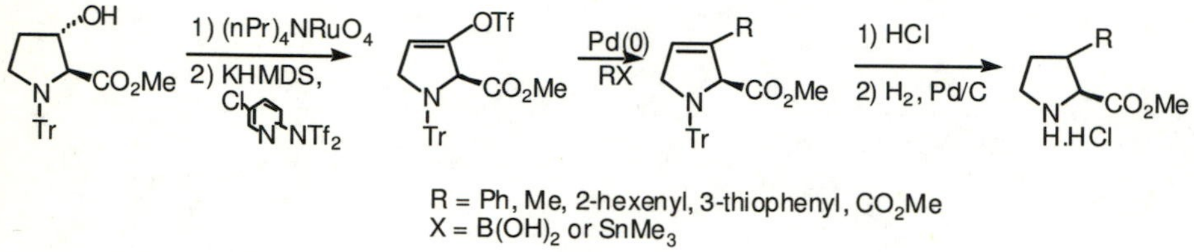

R = Ph, Me, 2-hexenyl, 3-thiophenyl, CO₂Me
X = B(OH)₂ or SnMe₃

Scheme 20

In contrast to the analogous strategy for 4-substituted prolines (see below, Scheme 34), catalytic hydrogenation of the coupling products leads to mixtures of diastereoisomers in various proportions.

Several groups have worked on the alkylation of 4-oxoproline to access to 3-substituted prolines. In the method developed by Holladay, the regioselective alkylation of C-3 is performed on an enamine.[54] A separable mixture of diastereoisomers is obtained, along with some dialkylation product (Scheme 21).

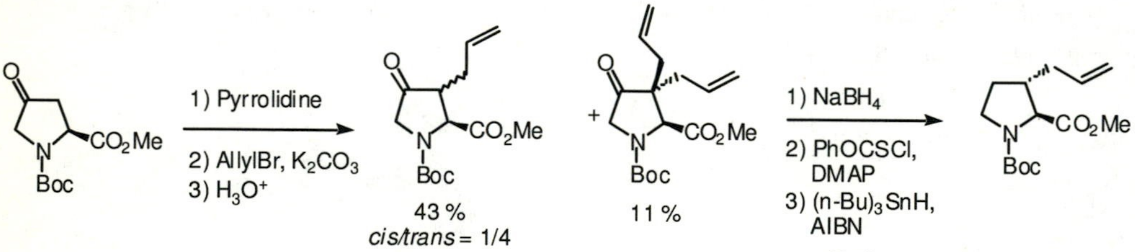

Scheme 21

Sharma[55] has demonstated that when the Boc group is replaced by a bulkier protecting group, 9-phenylfluoren-9-yl, the regioselective alkylation can be directly performed on the ketone derivative. This strategy led to the synthesis of a few gem-dialkylated derivatives.

Later, it has been shown that the monoaddition of the enolate on various aldehydes was highly diastereoselective at -78 °C.[56] Unfortunately, alkylations with alkyl electrophiles (even activated) require higher temperatures and there is a loss of selectivity due to a deprotonation-reprotonation process of the monoalkylated product.

A synthesis of optically pure *trans*-3-prolines substituted by aminoethyl and guanidinoethyl groups has been reported by Mamai *et al.*[57] The strategy is based on the diastereoselective conjugate addition of LiCH$_2$CN on an α,β-unsaturated lactam, obtained from (S)-pyroglutamic acid (Scheme 22).

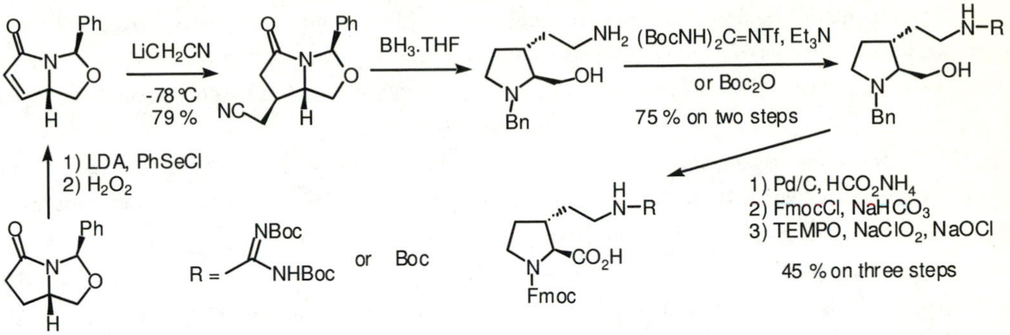

Scheme 22

Reduction of the lactam yields the *trans*-substituted pyrrolidine which gives access to the two amino acids orthogonally protected on the side chain and the secondary amine. This very efficient method has not yet been used to introduce other functionalities on position 3.

2.2.2. 3-Substituted prolines by C-C bond formation through anionic process

In 1988, Yoo and co-workers[58] described a racemic synthesis of 3-carboxymethyl-proline, as a mixture of two diastereoisomers, by *Michaël* addition of a sodium enolate to an α,β-unsaturated ester (Scheme 23).

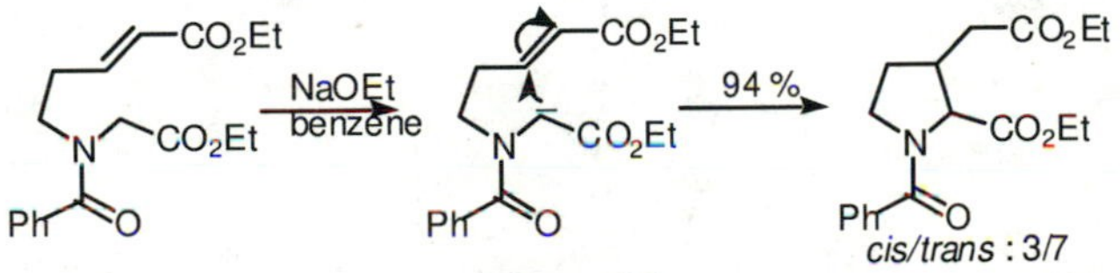

Scheme 23

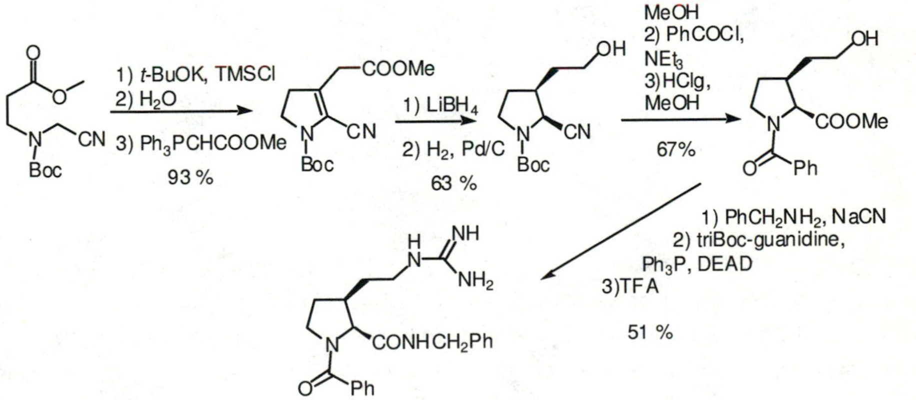

Scheme 24

227

More recently, a racemic synthesis of *cis*- and *trans*-3-substituted prolines with guanidinoethyl has been reported by Pellegrini and co-workers (Scheme 24).[59] After addition of the carbanion on the ester and trapping the enol by TMSCl, the 3-pyrrolidinone is obtained by hydrolysis.

Nitrile, as a highly electron-withdrawing group, allows the endocyclic migration of the doubled bond formed after the Wittig reaction. Thus, the *cis* derivative is obtained after catalytic hydrogenation of the double bond. After hydrolysis of the nitrile, the thermodynamically more stable *trans* derivative can be obtained by epimerisation of the α-center. This strategy has also been used to introduce aromatic rings on position 3.[60]

In 1997, Karoyan and co-workers[61] have reported the amino-zinc-ene-enolate cyclization[62] and applied the reaction to the synthesis of 3-substituted prolines bearing all types of natural amino acids side chains (Scheme 25).[63-69] Starting from, now commercially available, linear amine,[70] the deprotonation by LDA followed by transmetallation of the lithium enolate with zinc salts lead cleanly to the cyclic organozinc reagent. The reaction is highly stereospecific and stereoselective leading to the cyclic organozinc reagent with a *cis*-relative stereochemistry, the absolute configuration depending on the chiral auxiliary (*i.e. S* or *R* α-methylbenzylamine). *Trans* isomers were obtained by epimerisation of the α centre, the condition of epimerisation depending on the nature of the side chain.

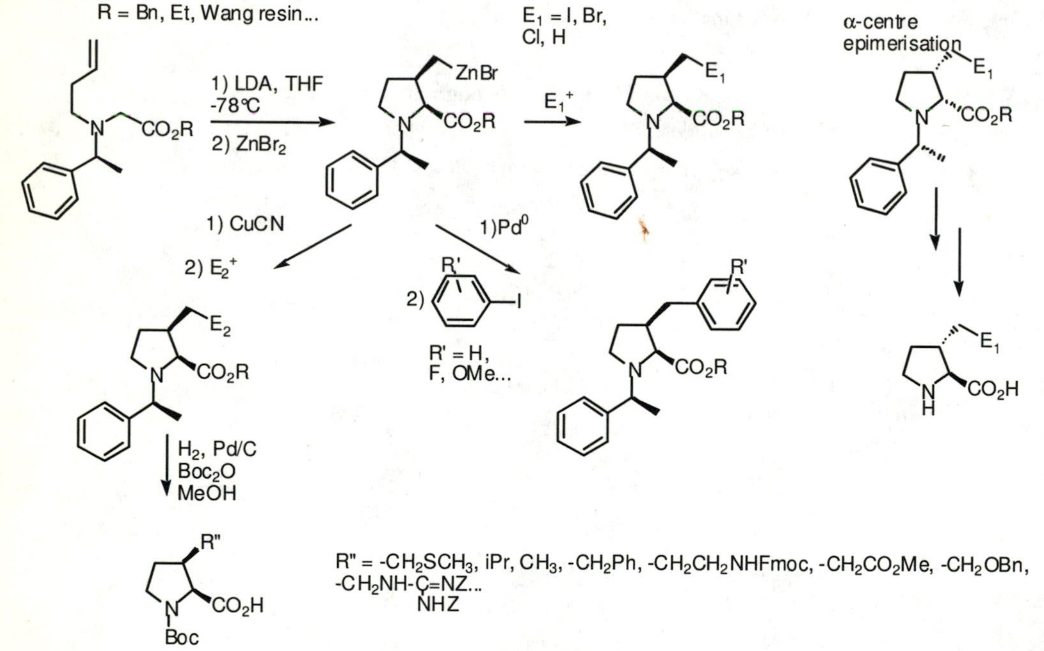

Scheme 25

The cyclic organozinc reagent was reacted with a few electrophiles (NIS, NBS, I_2, H_2O) or transmetallated into palladium- or copper-zinc species allowing the introduction of various functional groups. The methodology reported here allows, starting from the same precursors, the large scale asymmetric synthesis of the four stereoisomers of 3-substituted prolines bearing all types of substituents (alkyls, aromatics, basics, acids) in a few steps and suitably protected for peptide syntheses.

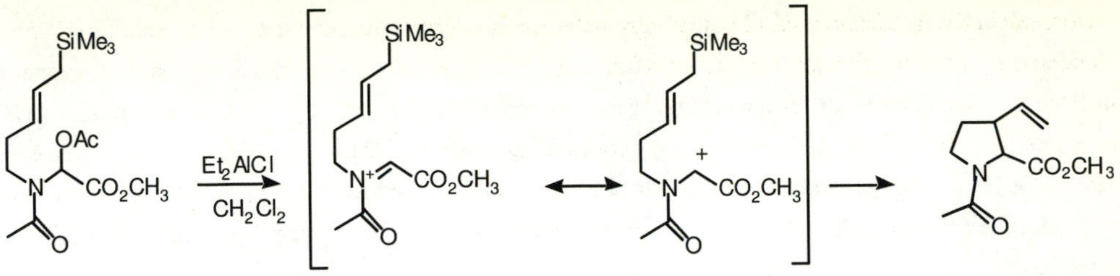
SiMe3
OAc
CO2CH3
Et2AlCl
CH2Cl2
CO2CH3

Scheme 26

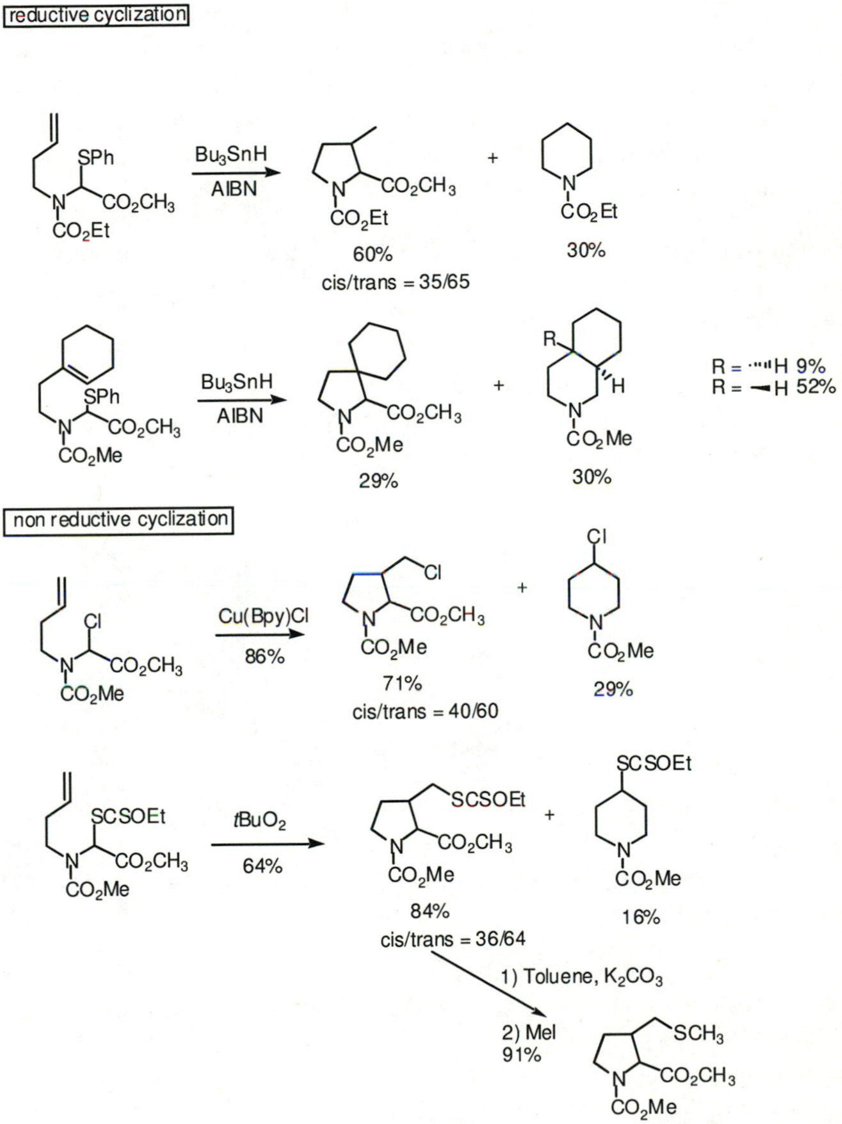
reductive cyclization
SPh
CO2CH3
CO2Et
Bu3SnH
AIBN
60%
cis/trans = 35/65
30%
R = ····H 9%
R = ◄H 52%
CO2Me
29%
30%
non reductive cyclization
Cl
Cu(Bpy)Cl
86%
71%
cis/trans = 40/60
29%
SCSOEt
tBuO2
64%
84%
cis/trans = 36/64
16%
1) Toluene, K2CO3
2) MeI
91%
SCH3

Scheme 27

2.2.3. 3-Substituted prolines by C-C bond formation through cationic process

There are several contributions from Hiemstra and Speckamp[71] on the synthesis of substituted pyrrolidines through cationic or radicalar processes (see below). The chemistry of the *N*-acyliminium cation is well documented and in 1987, Mooiver[72] reported the racemic synthesis of prolines derivatives by intramolecular cyclization reaction of propargyl- or allylsilane on *N*-acyliminium cation (Scheme 26).

The double bond activation through allylsilane avoids the 2-aza-Cope rearrangement usually observed in this approach.

2.2.4. 3-Substituted prolines by C-C bond forming through radicalar process

Intramolecular radicalar cyclization is suitable for carbocycle constructions and in the area of amino acids, glycine-derived radicals are well documented. Thus, Bachi and co-workers[73] have first reported their use in the synthesis of β-lactam and Hiemstra and co-workers[74-76] have reported the synthesis of 3-substituted prolines with reductive and non-reductive radicalar cyclization processes. In the former case, the atom transfer allows further functionalization. However, in all cases, products were obtained as racemate mixtures of five- and six-membered rings.

2.2.5. 3-Substituted prolines by intramolecular cyclization through C-N bond formation

In these cases, the stereochemistry is introduced before the cyclization step and we will primarily focus on the synthesis of the chiral linear precursors.

2.2.5.1. From aspartic or glutamic acid derivatives

Cotton reported the synthesis of *cis*- and *trans*-3-carboxy-prolines, suitably protected for peptide synthesis, from aspartic acid orthogonally protected on the two acid groups (Scheme 28).[77]

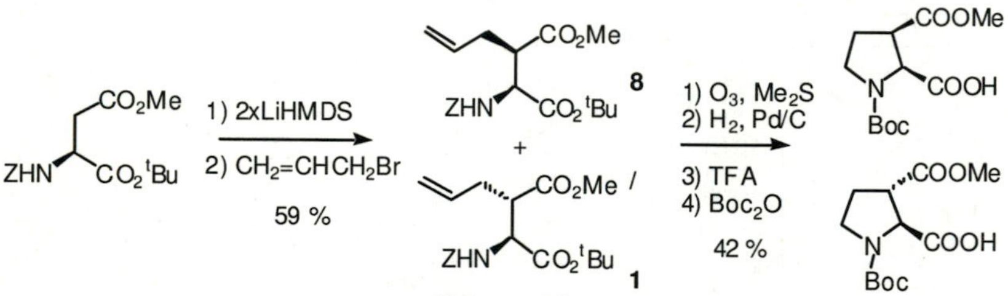

Scheme 28

Generation of the dianion of aspartic acid allows the regioselective alkylation of the β position. The use of a hindered base such as LiHMDS prevents deprotonation of the α-carbon.[78] The alkylation product is obtained as a mixture of two diastereoisomers, which can be separated by chromatography and lead to the *cis*- and *trans*-3-carboxyprolines.

2.2.5.2. Chiral substrates from Garner's aldehyde

A more recent example describes the diastereoselective synthesis of *cis*-3-methyl-, vinyl- and phenyl-prolines (Scheme 29).[79] The key step is the highly diastereoselective 1,4-addition of dialkylcuprates on a chiral oxazolidine α,β-unsaturated ester easily available from Garner's aldehyde. The resulting *cis* linear precursor leads to the corresponding *cis*-3-substituted prolines.

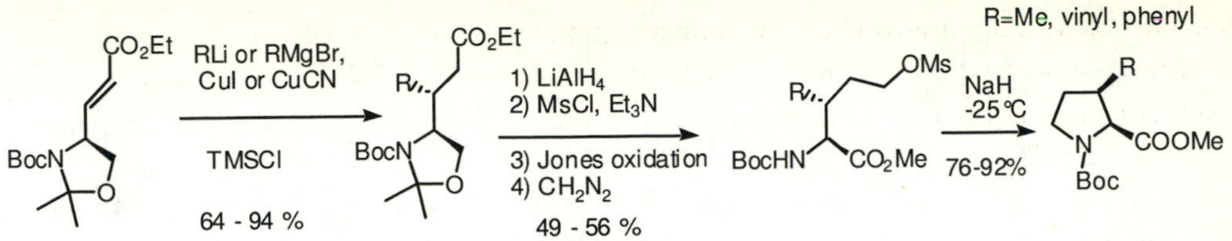

R=Me, vinyl, phenyl

Scheme 29

2.2.5.3. Michaël additions of nucleophilic glycine equivalent to chiral oxazolidinones

Michaël additions of an achiral Ni(II)-complex of glycine Schiff base to chiral 4-substituted oxazolidine-2-ones proceed in high yield and selectivity (Scheme 30).[80] The Ni(II)-complex assures the (Z)-configuration of the enolate derived from the Schiff base while the chiral moiety on the Michaël acceptor controls efficiently both face diastereoselectivity of the enolate and the C,C-double bond. Decomposition of the resulting complex affords the *trans*-3-substituted pyroglutamates in good yield and allows recovery of the chiral auxiliary. The compounds of the aryl series have been converted into the corresponding *trans*-3-arylprolines.

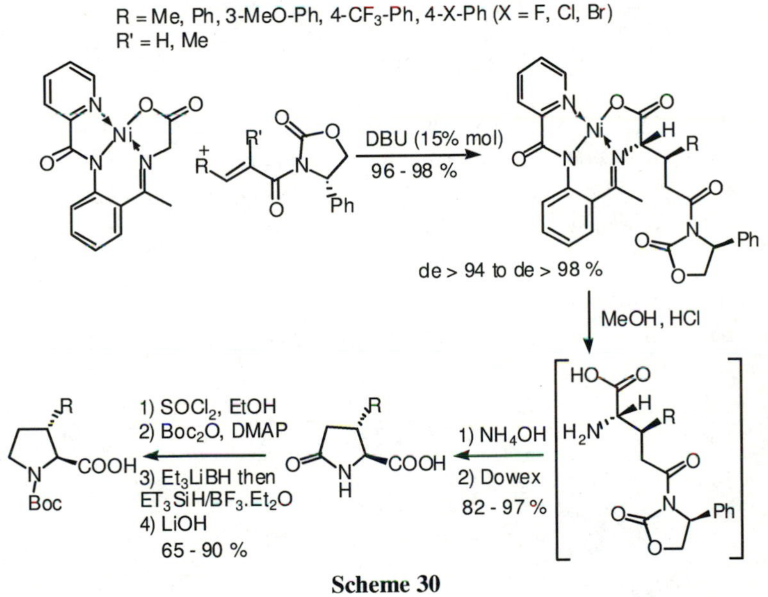

R = Me, Ph, 3-MeO-Ph, 4-CF₃-Ph, 4-X-Ph (X = F, Cl, Br)
R' = H, Me

Scheme 30

2.2.6. Synthesis of 3-substituted prolines by intermolecular cyclization

2.2.6.1. 1,3-Dipolar cyclization

Many studies describe 1,3-dipolar cycloadditions of azomethine ylides derived from aziridines to synthesize substituted pyrrolidines and prolines.[81-99] However, these methods severly lack regio- and endo/exo selectivity and are hardly efficient to synthesize substituted prolines. Among the asymmetric versions of these strategy,[92-97] a synthesis of carboxyproline derivatives has been described by Garner (Scheme 31).[94] Oppolzer's sultam, attached to the dipole, allows a good facial diastereoselectivity. However, as regio- and endo/exo selectivity are weak, a mixture of regio- and diastereoisomers are obtained.

231

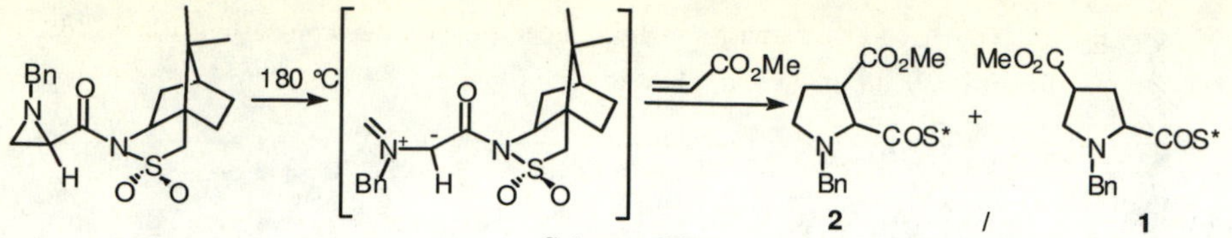

Scheme 31

2.2.6.2. Cyclization by Michaël addition/alkylation sequences

Several groups have worked on the condensation of diethylbenzyloxycarbonyl aminomalonate on α,β-unsaturated aldehydes (Scheme 32):[100-103]

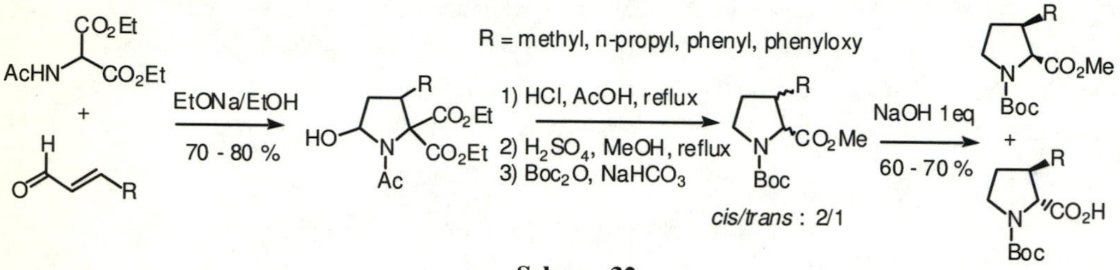

Scheme 32

After hydrogenolysis and decarboxylation, the compounds are obtained as a mixture of diastereoisomers. The steric hindrance of the ester in the *cis* derivative allows the selective saponification of the *trans* derivative and diastereoisomers can be separated. A resolution of the racemate by coupling to α-methylbenzylamine has also been described.[102]

2.2.7. Cyclization through dialkylation

In this method, the functionalization of the C-3 is achieved by an alkylation in the α position of a sulfone derived from serine (Scheme 33).[104]

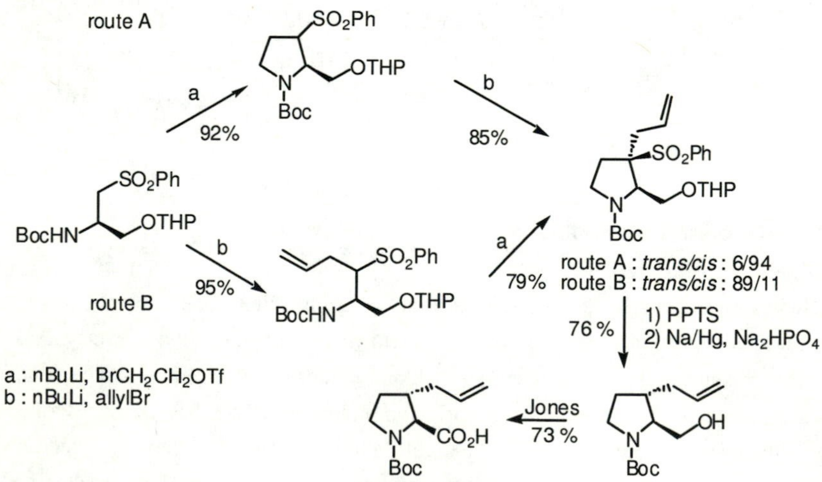

a : nBuLi, BrCH$_2$CH$_2$OTf
b : nBuLi, allylBr

Scheme 33

Then, the dianion generated with *n*-BuLi reacts with 2-bromoethyltriflate to allow the diastereoselective formation of the pyrrolidine cycle, giving preferentially the compound in which the allyl group is *trans* to the alcohol group. Moreover, partial epimerisation of C-3 during desulfonylation facilitates the obtention of the thermodynamically more stable *trans* derivative with good optical purity.[105]

If the allyl group is introduced after cyclization (route A), the diastereoisomeric ratio is reversed in favor of the *cis* isomer.[105,106] However, because of the partial epimerisation of C-3 during desulfonylation, the *cis*-3-substituted prolines are difficult to obtain.

2.3. 4-Substituted prolines syntheses
2.3.1. From 4-hydroxyproline
2.3.1.1. Direct functionalization

The commercially available *trans*-4-hydroxyproline has been widely used as chiral starting material to synthesize substituted prolines, mostly in position 4.

Cis-4-substituted prolines can be easily obtained by nucleophilic displacement of the tosylate or mesylate of *trans*-4-hydroxyproline[12,107-118] or by direct Mitsunobu reaction on *trans*-4-hydroxyproline (Scheme 34).[10,119,120]

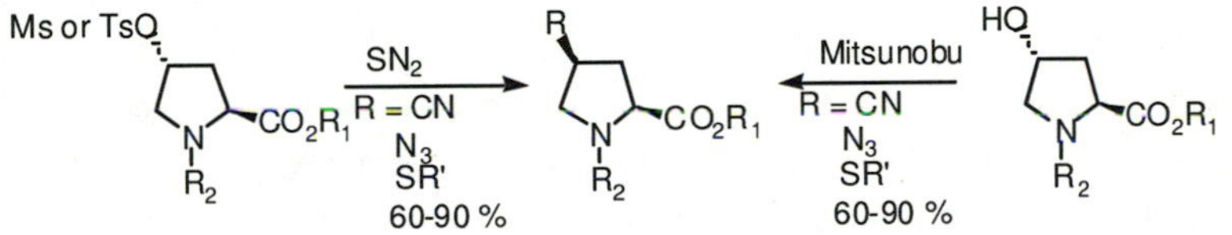

Scheme 34

SN$_2$		Mitsunobu
R=CN[12,112-114,118], N$_3$[109,110,112,114-117], SR'[107,108,111,112]		R=CN[10], N$_3$[120], SR'[119]

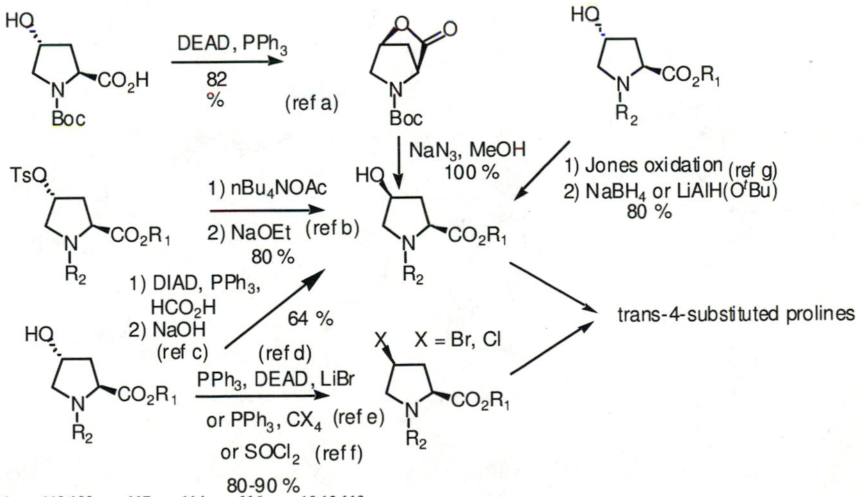

a:[120], b:[113], c:[119,122], d:[117], e:[114], f:[116], g:[10,12,113]

Scheme 35

233

Trans-4-substituted prolines can be obtained by the same strategies when the stereochemistry of C-4 is first inverted by SN$_2$[113,114,116] or Mitsunobu reaction[113,117,119,120] (Scheme 31). *Cis*-4-hydroxyproline can also be obtained by an oxidation/reduction sequence [10,12,113] as the reduction of the ketone is stereoselective, giving exclusively the *cis* isomer (Scheme 35).[121] All these methods seem similarly efficient.

Some of these studies led to important 4-substituted prolines directly usable for peptide synthesis such as analogues of arginine (*cis* and *trans*-4-guanidinoethylprolines,[10] *cis* and *trans*-4-guanidinomethylprolines,[114] *cis* and *trans*-4-guanidinoprolines[117]), ornithine (*cis*-4-aminoproline[115] and *trans*-4-aminomethylproline[118]), glutamate (*cis* and *trans*-4-carboxyprolines[12]) and methionine (*cis* and *trans*-4-methylthioproline[111,119]). Finally, direct fluorination of *cis*- and *trans*-4-hydroxyprolines allowed the synthesis of *trans*- and *cis*-4-fluoroprolines, respectively (Scheme 36).[122]

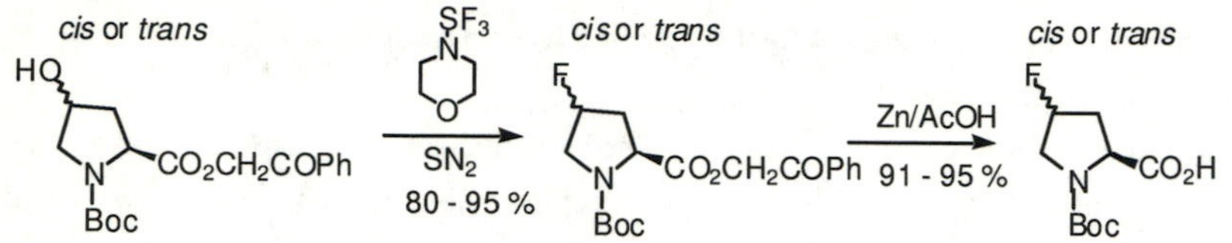

Scheme 36

2.3.2. From 4-oxoproline

4-Oxoproline has also been used to synthesize enantiopure 4-substituted prolines. Firstly, addition of phenylmagnesium bromide on 4-oxoproline gives the corresponding phenylcarbinol as a unique diastereoisomer (Scheme 37).[111,123] The yield is significantly improved when the bulky *tert*-butyl group is used for ester protection to avoid competitive reaction of the Grignard reagent on the ester.[123] Interestingly, another group describes a significant yield increase when CeCl$_3$ is used during the addition of 4-chlorophenylmagnesium on N-benzyl-4-oxoproline methyl ester.

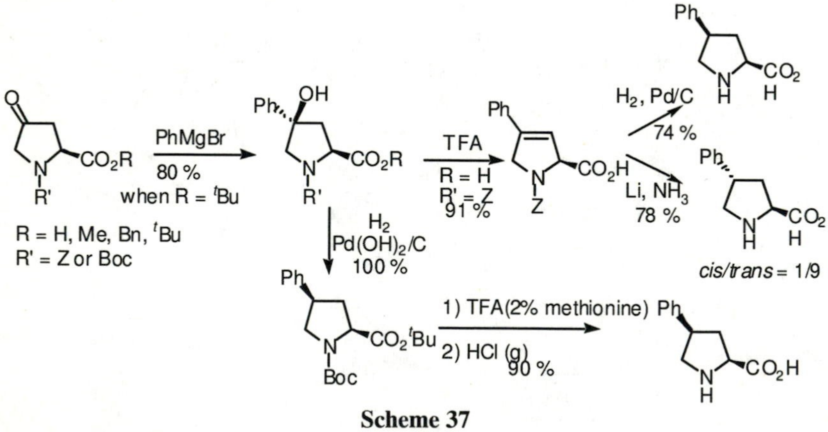

Scheme 37

After dehydration, catalytic hydrogenation gives exclusively *cis*-4-phenylproline. Alternatively, lithium/liquid ammonia reduction gives a 9:1 mixture in favour of the *trans* isomer.[111] Later, it has been shown that hydrogenation of the phenylcarbinol over Pd(OH)$_2$/C leads in a "one-pot" procedure and in high yield to *cis*-4-phenylproline (de > 95 %).[117]

Cis-4-arylprolines have also been obtained by palladium coupling to the enol triflate derivative from 4-oxoproline (Scheme 38).[124] Hydrogenation of the resulting alkenes gives selectively the *cis*-4-arylprolines.

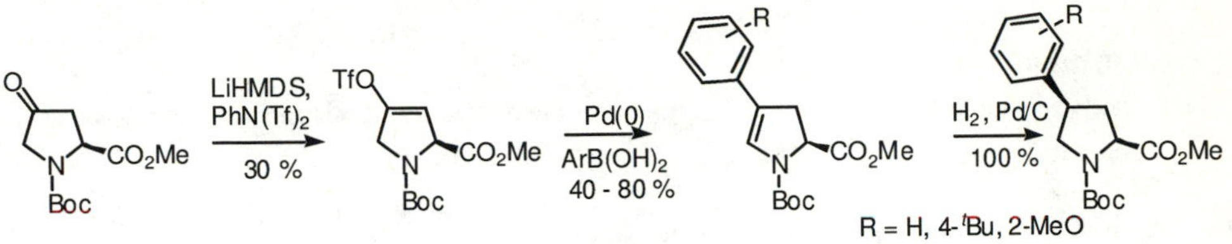

LiHMDS, PhN(Tf)$_2$ 30 %   Pd(0) ArB(OH)$_2$ 40 - 80 %   H$_2$, Pd/C 100 %   R = H, 4-^{t}Bu, 2-MeO

Scheme 38

Wittig reactions on 4-oxoproline followed by catalytic hydrogenation yield the 4-substituted prolines with a marked selectivity for the *cis* isomers, especially for bulky substituents (Scheme 39).[111,124-126]

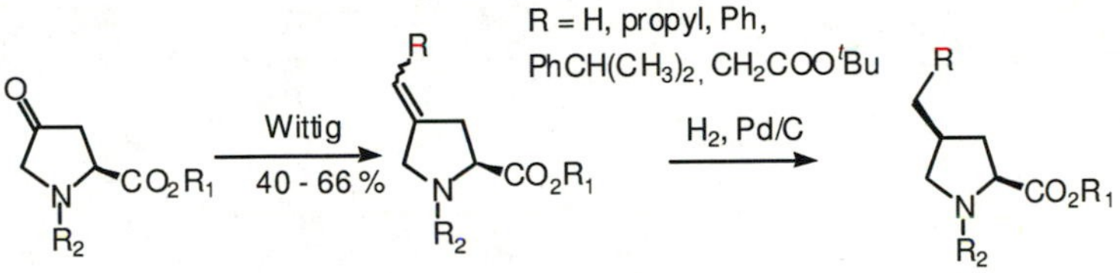

R = H, propyl, Ph, PhCH(CH$_3$)$_2$, CH$_2$COO^{t}Bu   Wittig 40 - 66 %   H$_2$, Pd/C

Scheme 39

Starting from the corresponding prolinol, Del Valle and Goodman use the different facial selectivity of hydrogenation of unprotected *vs* protected 4-alkyl/arylideneprolinols to obtain *cis*- or *trans*-4-substituted prolines (Scheme 40).[127]

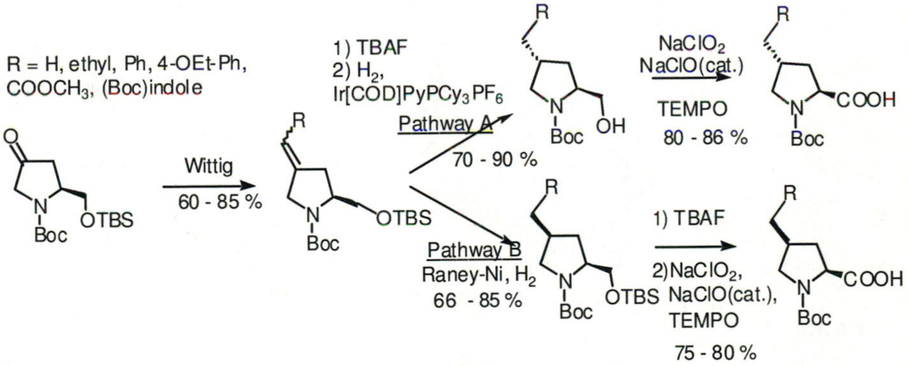

R = H, ethyl, Ph, 4-OEt-Ph, COOCH$_3$, (Boc)indole   Wittig 60 - 85 %   1) TBAF 2) H$_2$, Ir[COD]PyPCy$_3$PF$_6$ Pathway A 70 - 90 %   NaClO$_2$ NaClO(cat.) TEMPO 80 - 86 %   Pathway B Raney-Ni, H$_2$ 66 - 85 %   1) TBAF 2) NaClO$_2$, NaClO(cat.), TEMPO 75 - 80 %

Scheme 40

In pathway A, homogenous hydrogenation of the deprotected prolinol with the Crabtree catalyst (Ir[COD]PyPCy$_3$PF$_6$) affords the *trans* isomer with good to excellent selectivity (*trans/cis* > 40:1 except for R=COOCH$_3$:*trans:cis*=16:1). In pathway B, heterogeneous hydrogenation of the prolinol protected as a silyl ether occurs preferentially from the less hindered face as for corresponding proline derivatives (see above

235

Scheme 37), giving the *cis* isomer. Best selectivity was achieved using Raney-Ni as catalyst and this selectivity increases with the bulkiness of the substituents [*cis/trans* ratio from 3:1 for R=H to 22:1 for R=(Boc)indole]. For both isomers, oxidation to 4-substituted prolines in the presence of TEMPO, bleach and sodium chlorite occurs cleanly except for the indolyl prolinol, for which the corresponding prolines could not be obtained.

2.3.3. From pyroglutamic acid

The readily available L-pyroglutamic acid has also been used as chiral starting material to access to numerous substituted prolines by several strategies.

2.3.3.1. Alkylation of pyroglutamates derived enolates

Alkylation of lithium enolates of protected L-pyroglutamic acid generated with one equivalent of LiHMDS has been used to introduce substituents in position 4 without epimerisation of the α-centre (Scheme 41).[128-134]

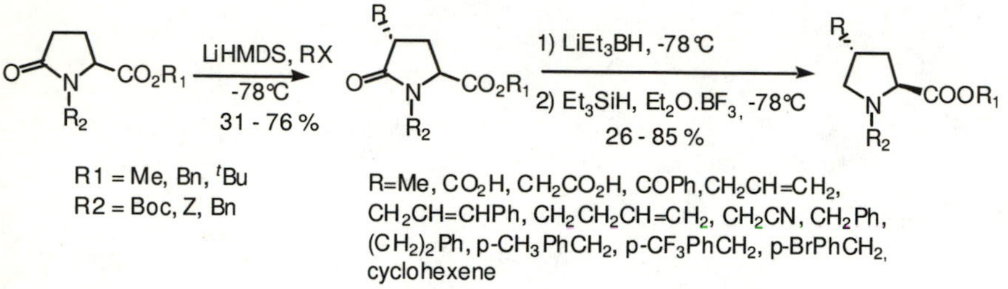

LiHMDS, RX
-78°C
31 - 76 %
1) LiEt3BH, -78°C
2) Et3SiH, Et2O.BF3, -78°C
26 - 85 %
R1 = Me, Bn, tBu
R2 = Boc, Z, Bn
R=Me, CO2H, CH2CO2H, COPh, CH2CH=CH2, CH2CH=CHPh, CH2CH2CH=CH2, CH2CN, CH2Ph, (CH2)2Ph, p-CH3PhCH2, p-CF3PhCH2, p-BrPhCH2, cyclohexene

Scheme 41

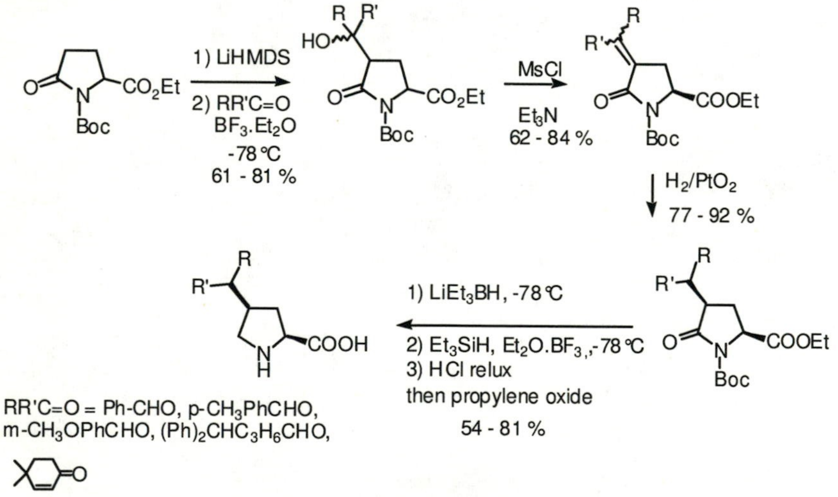

1) LiHMDS
2) RR'C=O
BF3.Et2O
-78 °C
61 - 81 %
MsCl
Et3N
62 - 84 %
H2/PtO2
77 - 92 %
1) LiEt3BH, -78 °C
2) Et3SiH, Et2O.BF3, -78 °C
3) HCl relux
then propylene oxide
54 - 81 %
RR'C=O = Ph-CHO, p-CH3PhCHO, m-CH3OPhCHO, (Ph)2CHC3H6CHO,

Scheme 42

When benzylic electrophiles are used, the *trans* isomer is selectively obtained.[128,129,132,134] With other electrophiles, the alkylation leads to a mixture of the two diastereoisomers and the selectivity in favour of the *trans* isomer increases with the bulkiness of the electrophiles but is independent of the ester protecting group.[128,129] Reduction of the lactam carbonyl group is best achieved with LiEt$_3$BH/Et$_3$SiH-Et$_2$O.BF$_3$ to give the corresponding 4-substituted prolines.[130,133,134] Condensation of the pyroglutamate lactam enolates with aldehydes and ketones also occurs cleanly in the presence of Lewis acid BF$_3$.Et$_2$O (Scheme 42).[135] Treatment of the diastereoisomeric mixture of the hydroxyl derivatives with mesyl chloride in the presence of triethylamine gives the corresponding 4-alkylidenepyroglutamates. Similarly to 4-alkylideneprolines, hydrogenation over Pd/C selectively affords the *cis* isomers, which lead to the corresponding *cis*-4-substituted prolines after reduction of the lactam carbonyl group.

2.3.3.2. Other strategies

Treatment of protected L-Pyroglutamic acid with Bredereck's reagent leads to an enaminone[136] which reacts with DIBAL or Grignard reagents to afford a variety of 4-alkylidenepyroglutamates (Scheme 43),[137-139] some of which have been converted into *cis*-4-substituted prolines *via* hydrogenation/carbonyl reduction pathways (see above).

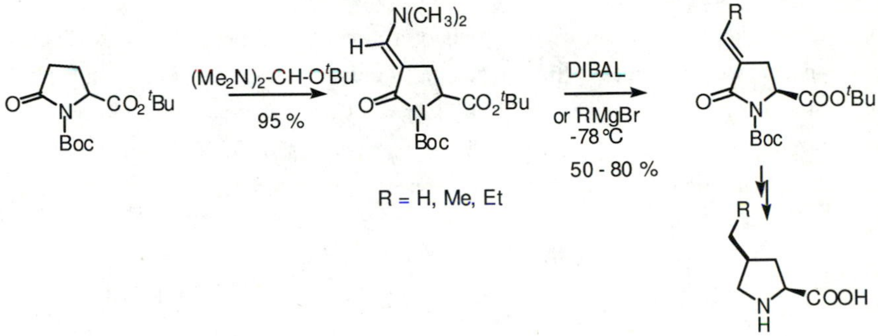

Scheme 43

The chiral *O,N*-acetal derived from L-Pyroglutamic acid can also be used in alkylation reactions (Scheme 44).[140-142] However, *endo:exo* selectivity highly depends on the electrophile.[142] Alkylation with small electrophiles (*e.g.* MeI) occurs preferentially from the more hindered *endo* face (*endo:exo* ratio of 5:1 for MeI) because the nitrogen lone pair directs the alkylation. For medium sized electrophiles (*e.g.* allylBr), steric effects compete with stereoelectronic effects affording 1:1 mixtures and large electrophiles alkylate preferentially the less hindered *exo* face.

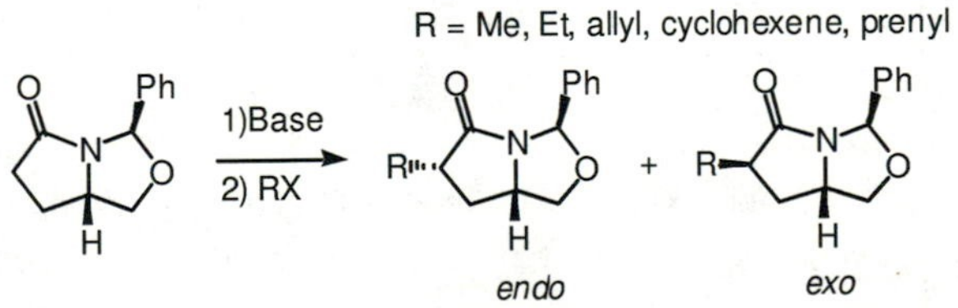

Scheme 44

A recent improvement of this strategy was to use *endo*-selective kinetic protonation.[142] In a one pot procedure, the chiral lactam is alkylated (with BnBr or allylBr) and treated with one more equivalent of

LDA at -78 °C to afford the *exo* isomer with high stereoselectivity (97:3). This method has been used to synthesize 4-*trans*-carboxymethylproline directly useable in peptide synthesis (Scheme 45).

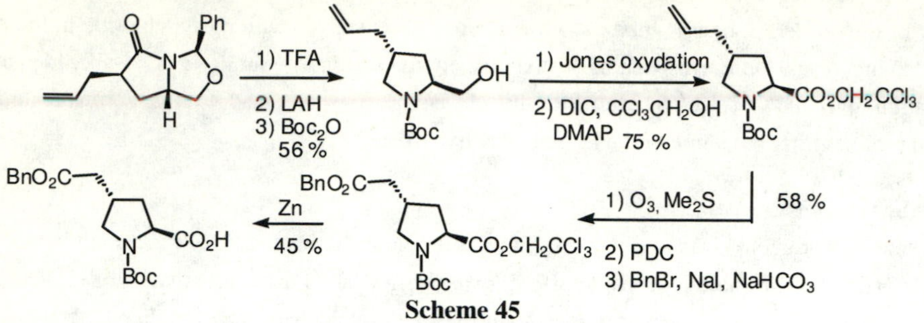

Scheme 45

2.3.4. Synthesis of substituted prolines by intramolecular cyclization with C-C bond formation

2.3.4.1. Cationic cyclization

Agami *et al.* described an asymmetric synthesis of a 4-substituted proline by aza-Cope/ene-iminium ion cyclization tandem reaction (Scheme 46)[143] When the chiral amino alcohol reacts with glyoxal in formic acid, the ene iminium system rearranges *via* a cationic aza-Cope process and a bicyclic hemiacetal is obtained after formic acid induced cyclisation. The hemiacetal is transformed into the *cis*-4-substituted proline (here with 2*R* configuration) by a two steps sequence.

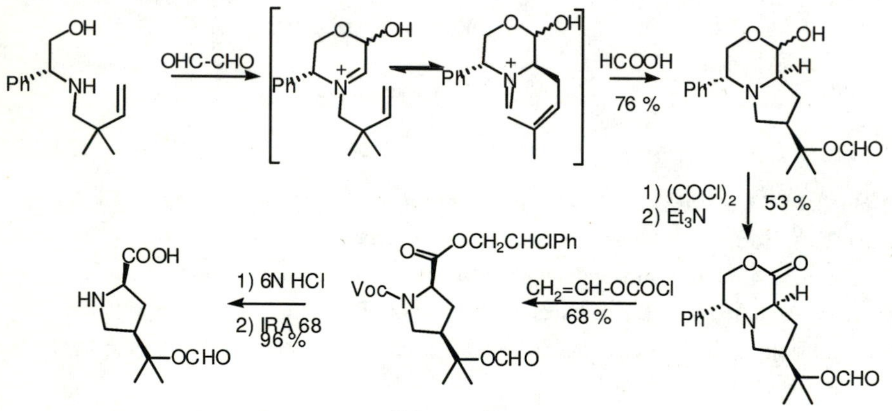

Scheme 46

2.3.4.2. Radical cyclization

Various methods have been described to synthesize substituted prolines by radical cyclization. Cobalt-mediated radical cyclization of the N-alken-2-yl-bromoester derived from L-serine leads to the corresponding 4-substituted proline (Scheme 47).[128,144] However, yield is quite moderate due to competing reduction of the bromide, which leads to an acyclic alanine derivative, and the reaction is not diastereoselective.

From similar starting materials, Basak *et al.* have shown that using the adequate sulphonamide as the amine protecting group, as well as a bulky ester protecting group, high selectivity could be obtained in the radical cyclization (Scheme 48).[145] Thus, *trans*-4-methylproline derivative can be obtained in high yield and selectivity (*cis/trans*: 1/33). Selectivity is significantly lower for the 4-benzylproline analogue (*cis/trans*: 1/3.5).

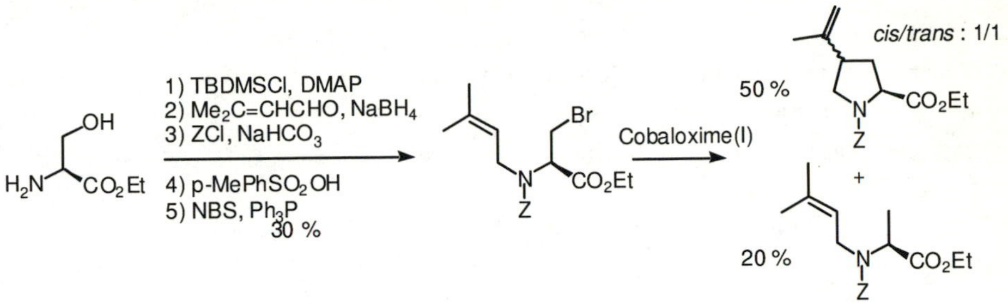

Scheme 47

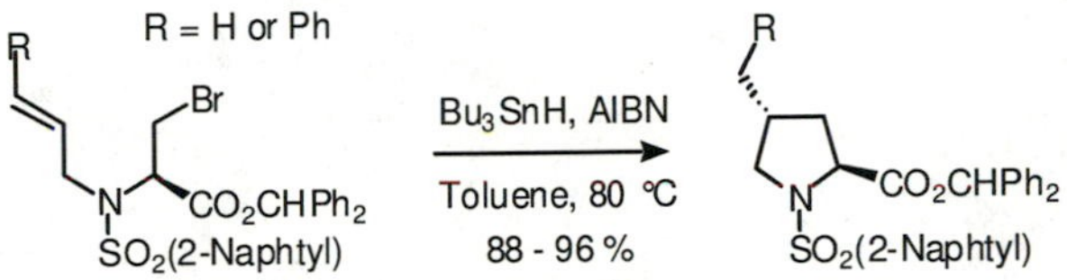

Scheme 48

N-alkylation of a β-lactone derived from Boc-L-serine followed by ring opening with sodium benzeneselenoate provides chiral substrates which undergo intramolecular radical cyclization to afford *cis/trans* mixtures (~ 60/40) of a few 4-alkylprolines (Scheme 49).[146]

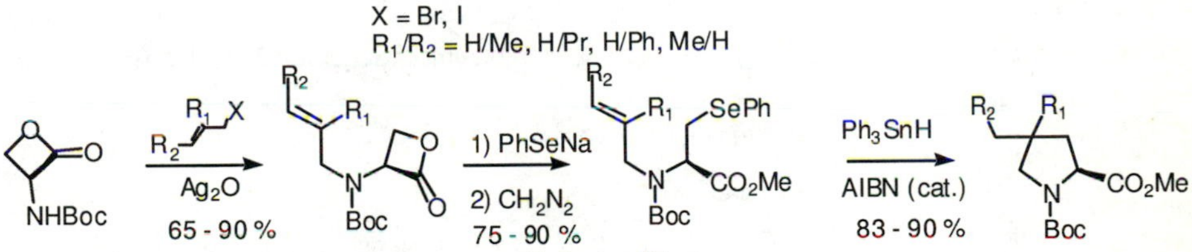

Scheme 49

Irradiation of α-silylmethylamino derivatives with UV/vis light in the presence of a sensitizer yields α-silylmethylamino radical cations, which readily undergo fragmentation by loss of the TMS group, leading to free α-aminomethyl radicals.[147] Thus, the α-silylmethylamino acid derived from racemic allyl glycine undergoes radical cyclization by photonelectron transfer (PET) to give *trans*-4-methylproline with high diastereoselectivity (Scheme 50).

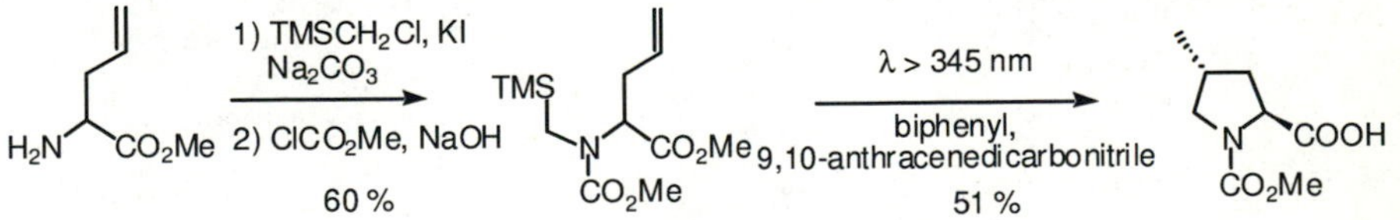

Scheme 50

In the same study, the authors have incorporated this α-silylmethylamino acid in various peptides. PET-catalysed cyclisation of these peptides leads to oligopeptides containing the 4-methylproline analogue but the diastereoselectivity is low in all cases.

Wang et al.[148,149] described a strategy for the asymmetric synthesis of 4-substituted prolines starting from (2R)-2,3-O-isopropylideneglyceraldehyde and sulfone[104] derived from serine (Scheme 51). After reductive amination and conversion to the corresponding epoxide, intramolecular alkylation leads to the pyrrolidine as a unique diastereoisomer. The stereochemistry at C-3 was not determined as it is lost later in the desulfonylation step. This pyrrolidine has been used to access to several 4-substituted prolines.[148,149]

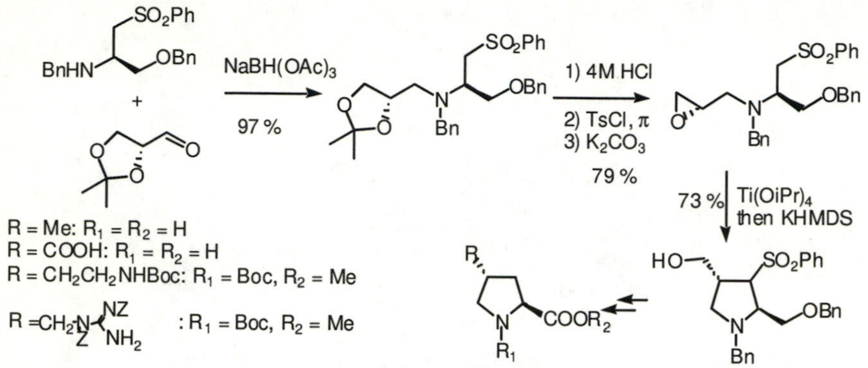

Scheme 51

2.3.5. From aspartic or glutamic acid derivatives

Alkylations at the γ position of glutamates have yielded chiral linear substrates to access to 4-susbtituted prolines.[150,151] Using *tert*-butyl as the α-ester protecting group, high selectivity is observed for the alkylations, yielding the *trans* product as a unique diastereoisomer (Scheme 52).[151] Corresponding *cis*-4-substituted prolines are then obtained in good yield.

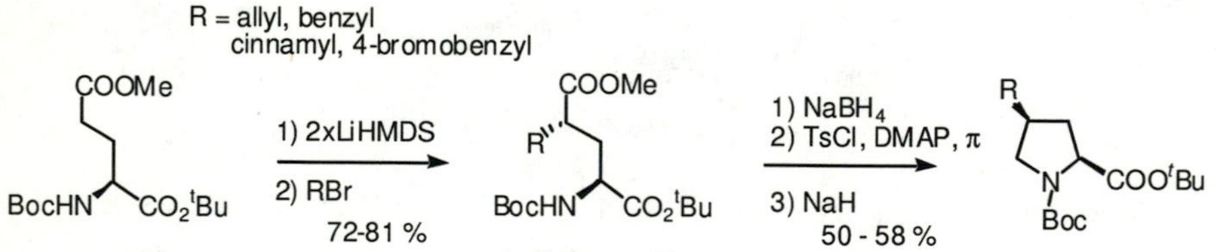

Scheme 52

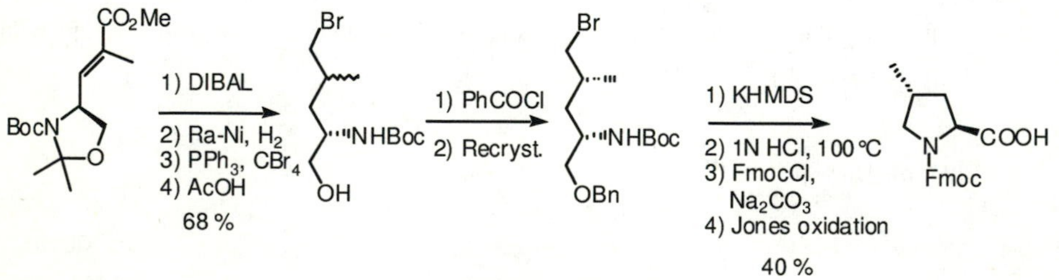

Scheme 53

240

Starting from a Garner's type aldehyde, an asymmetric synthesis of *trans*-4-methylproline was described (Scheme 53).[152] Hydrogenation of the double bond with Ra-Ni gives a *cis:trans* ratio of 14:86 and the *trans* isomer had to be separated by recristallization at a later stage, giving the *trans* pyrrolidine after cyclization.

Recently, 1,3-dipolar cycloadditions of chiral imidazolinium ylides with various alkenes have been described (Scheme 54).[153] The chiral auxiliary on the ylides is conformationally restrained by the heterocyclic ring, thus assuring excellent facial selectivity in the cycloaddition with methyl acrylate, yielding selectively the homochiral imidazole. No problem of regioselectivity or *endo/exo* selectivity was reported for this cycloaddition. Then, removal of the chiral template by a two steps reductive sequence gives the corresponding 4-substituted prolines.[153] However, due to partial epimerisation of C-4 during aminal reduction and the different stability of the *cis*- and *trans*-pyrrolidines under hydrogenation condition, a mixture of *cis*- and *trans*-4-carboxyproline is finally obtained.

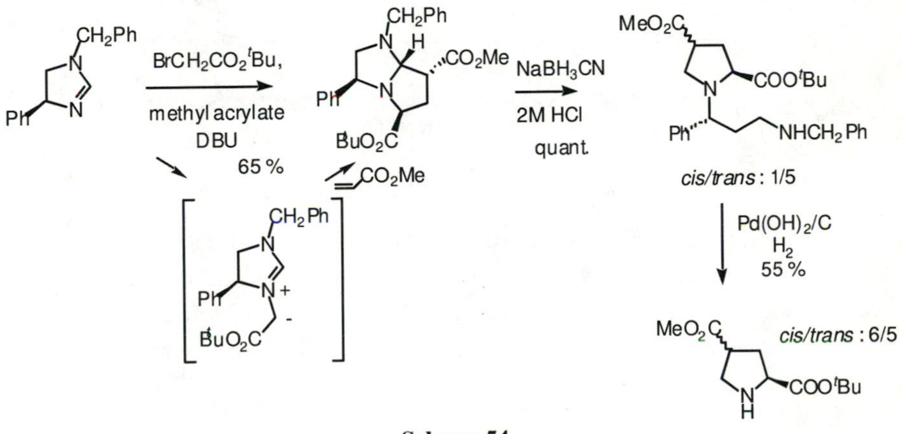

Scheme 54

Condensation of diethyl *N*-Cbz-aminomalonate with acrolein leads to *1*-Cbz-5-hydroxypyrrolidine-2,2-dicarboxylate which has been used to synthesize the four diastereoisomers of 4-carboxymethylproline which are then separated by coupling to (+)-menthol (Scheme 55).[154]

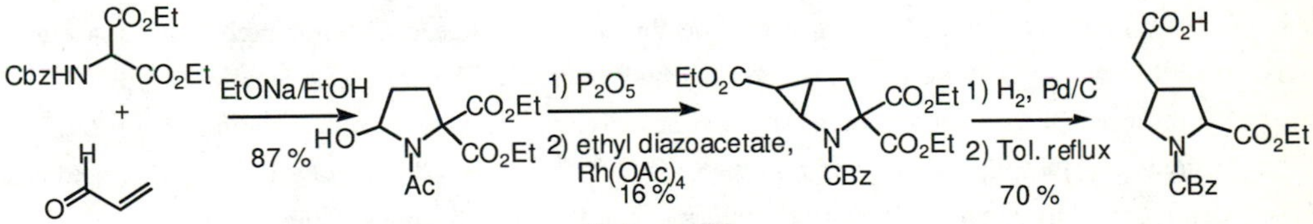

Scheme 55

2.4. 5-Substituted prolines syntheses

The synthesis of 5-substituted prolines has been accomplished by several pathways, involving mesylate displacement of a suitable substituted α-amino acid, nucleophilic attack on N-acyliminium ions or reduction of the corresponding imines. Few of them will be reviewed and the synthesis of C$_2$-symmetric 2,5-disubstituted pyrrolidines will not be discussed here.

241

2.4.1. 5-Substituted prolines syntheses starting from proline derivatives

In 1970, Gallina and co-workers[35] have reported the synthesis of *cis-N,N'*-dicarbobenzyloxy-5-amino-D,L-proline starting from *N*-carbobenzyloxy-D,L-pyrrolidine-2,5-dicarboxylic acid anhydride.

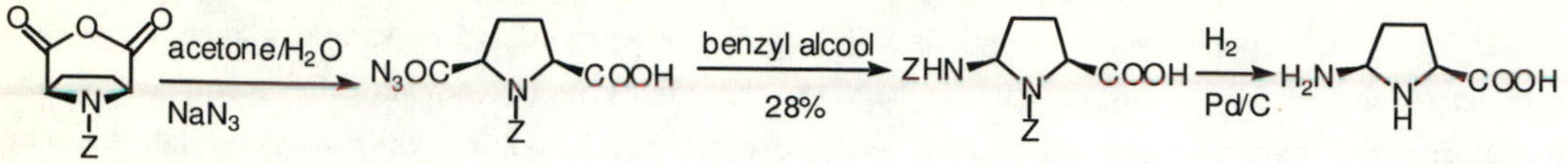

Scheme 56

The anhydride opening by sodium azide, followed by Curtius rearrangement reaction and catalytic hydrogenation over palladium charcoal led to the title compound in 28% yield.

Interested in lengthening the chain of dicarboxylic acid by one carbon atom, Della[155] has reported the synthesis of methyl 2-(1-benzyl-5-methoxycarbonylpyrrolidin-2-yl)-acetate through Arndt-Eistert homologation reaction (Scheme 57).

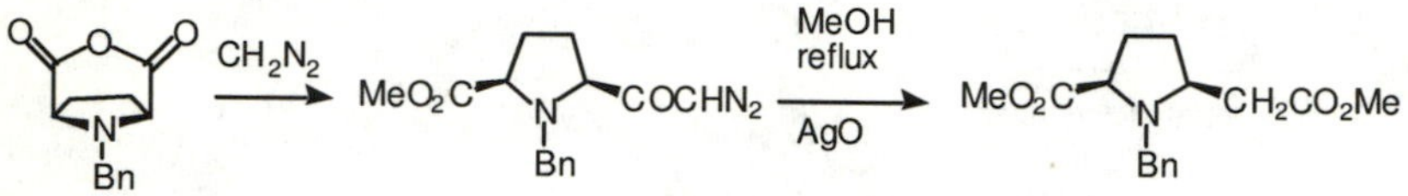

Scheme 57

Treatment of the cyclic anhydride with an excess of diazomethane yielded the diazoketone ester which was converted into the acetate derivative by refluxing methanol in presence of silver oxide. The major component (85%) was identified as the *cis*-isomer. Starting from pyroglutamate esters, Petersen and co-workers,[156] working on chirospecific synthesis of Anatoxin a, have reported the synthesis of the orthogonally protected diester (Scheme 58).

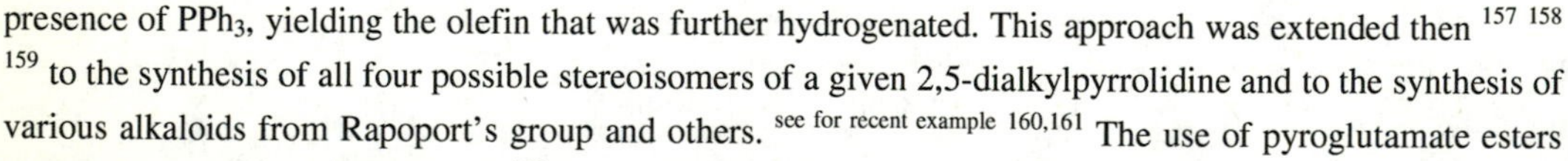

Scheme 58

The introduction of the C-5 carbon-carbon bond on the *tert*-butyl ester of *N*-benzyl pyroglutamic acid was realized by a sulfide contraction reaction. The thiolactam was reacted with methyl bromoacetate in presence of PPh$_3$, yielding the olefin that was further hydrogenated. This approach was extended then [157] [158] [159] to the synthesis of all four possible stereoisomers of a given 2,5-dialkylpyrrolidine and to the synthesis of various alkaloids from Rapoport's group and others. [see for recent example 160,161] The use of pyroglutamate esters has also been reported by Jain *et al.*[162] for the synthesis of chiral piperazinones (Scheme 59).

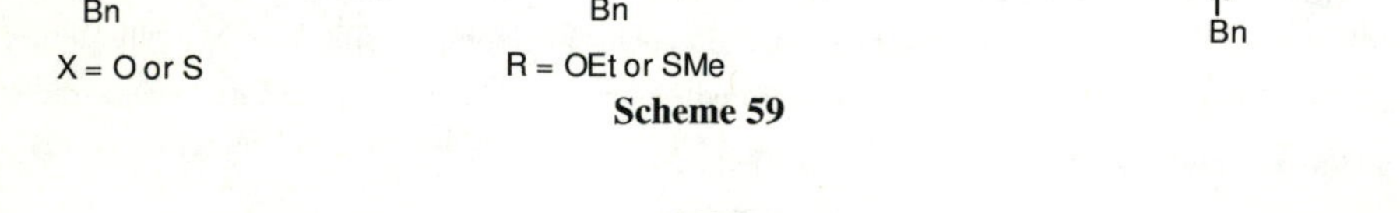

X = O or S   R = OEt or SMe

Scheme 59

242

Condensation of the iminium with nitromethane yielded the nitroenamine that could be further reduced in alkanamine. A somewhat modified approach, combining Rapoport's and Jain methodologies has been recently used by Honda and Kimura in the synthesis of (-)-Adalinine[163].

2.4.1.1. *5-Substituted prolines syntheses by nucleophilic attack on N-acyliminium ions*

In 1975, Shono and co-workers[164] reported the anodic oxidation as a convenient method to introduce a methoxy group to the position α to the nitrogen of cyclic and acyclic amides and carbamates. This methoxy function (hemiaminal) is a leaving group, allowing the introduction of various π-nucleophiles (Nu) *via* the corresponding *N*-acyliminium ion. The initiation step of the oxidation involves electron transfers from the lone pair electrons of the nitrogen to the anode (Scheme 60).

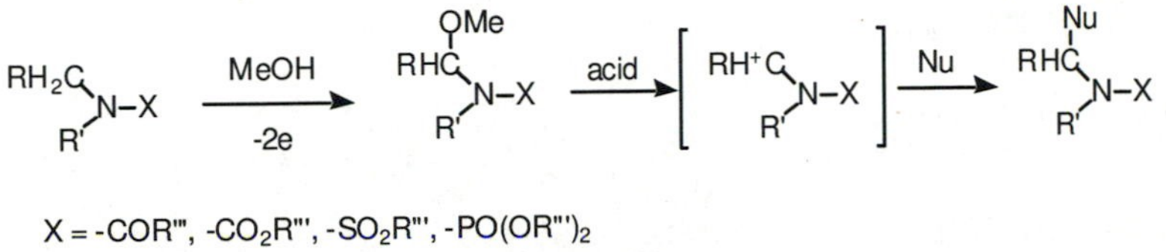

$$X = -COR''', -CO_2R''', -SO_2R''', -PO(OR''')_2$$

Scheme 60

The anodic oxidation of proline yielding α-methoxylated proline has also been reported by Shono[165,166] (Scheme 61).

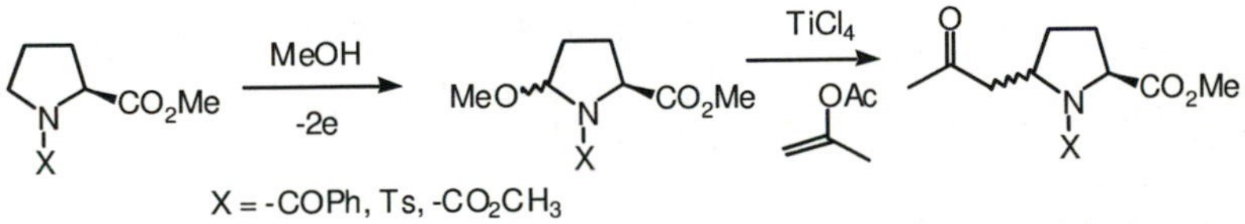

$$X = -COPh, Ts, -CO_2CH_3$$

Scheme 61

Treatment of the α-methoxylated compound with isopropenyl acetate in the presence of $TiCl_4$ gave a mixture of stereoisomers in 7:3 ratio in favor of the *cis* isomer, with 85% yield [167]. Similar results were reported by Malmberg and Nyberg for the corresponding N-formyl derivative[168]. Selectivity in favour of *trans* isomers were obtained in cobalt-catalysed amidocarbonylation.[169]

This reaction has been widely used by several authors to introduce functionalities in position five of the pyrrolidine ring. Thus, the introduction of an ethynyl group was reported by Manfré *et al.*, treating the methoxy derivative with bis(trimethylsilyl)-acetylene in the presence of $TiCl_4$, yielding *cis/trans* mixture in 4/3 ratio and 61% yield[170]. A phenyl group was introduced by the same author *via* a Friedel-Craft reaction catalysed by $AlCl_3$,[171] the reaction leading in that case to a 1:1 *cis:trans* mixture in 34% yield.

Skrinjar and Winstrand have reported the addition of RCu,BF_3, yielding a 96:4 mixture of diastereoisomers in favour of the *trans* isomer[172,173] (Scheme 62) [starting from the N-benzyloxycarbonyl protected methyl ester of proline, a similar selectivity was observed by Lhommet's group[174] with addition of organocopper $(CH_2=CHCH_2CH_2Cu)$].

Thus, addition of RCu gives a reversed and increased stereoselectivity compared to that observed for π-nucleophiles[175]. A mechanism involving nucleophilic attack on the less hindered face of a sterically biased *N*-acyliminium ion-copper complex was proposed to explain the stereochemical outcomes of the reaction.

Lower yields and selectivity were obtained by using other organocopper reagents (*i.e.*: R$_2$CuLi, RCuCNLi). In 1995, Kemp's group reported a minor modification of the Wistrand approach, using boron-trifluoride-mediated cuprate addition to the methoxyaminal intermediate.[176] For recent application, see for example 177,178,179

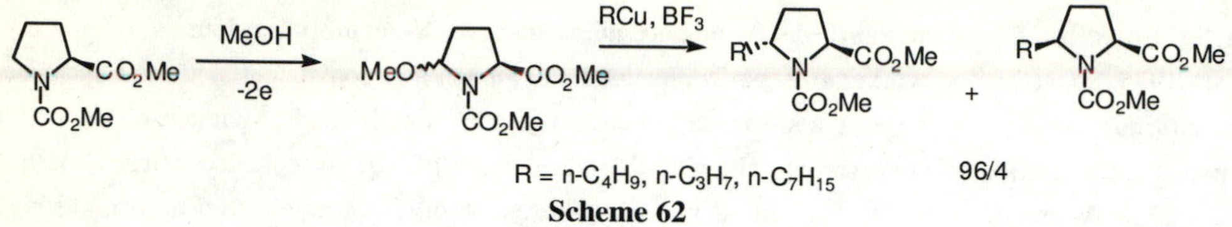

$$R = n-C_4H_9, n-C_3H_7, n-C_7H_{15} \qquad 96/4$$

Scheme 62

A radicalar approach has been reported by Barrett and Pilipauskas[180] for the functionalization of 5-methoxy-proline hemiaminal derivative (Scheme 63).

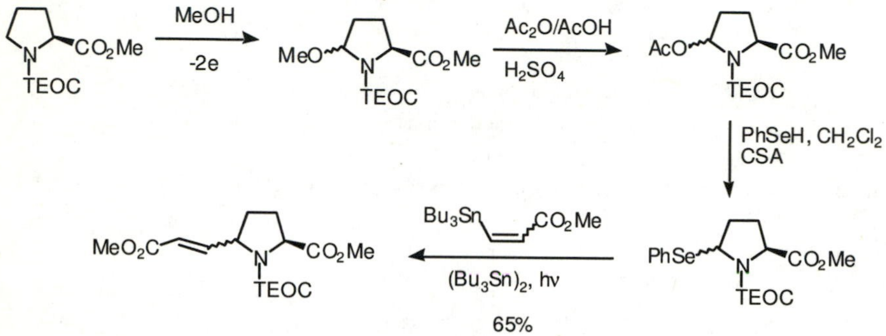

Scheme 63

The selenide derivatives, prepared from (S)-proline, gave the (methoxycarbonyl)ethenyl compound upon irradiation with hexabutyldistannane and methyl (Z)-2-tributylstannylacrylate. The title compound was obtained as a 1:3 mixture of *cis:trans* isomers.

An alternative to Shono's electrochemical oxidative procedure for amides have been reported by Magnus and Hulme[181] (Scheme 64), by Corey and co-workers[182] (Scheme 65) and by Kim.[183]

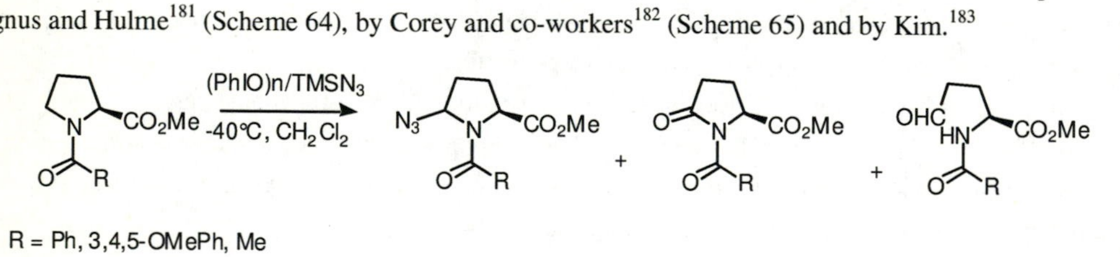

R = Ph, 3,4,5-OMePh, Me

Scheme 64

The oxidation is here realized with the iodosylbenzene/trimethylsilylazide reagent combination and led however to complexe mixture.

In 1990, Corey *et al.*[182] reported the synthesis of 5-methoxy-proline derivatives starting from *N*-protected pyroglutamic ester (Scheme 65).

Ethyl (S)-N-(ethoxycarbonyl)pyroglutamate was treated first with Dibal and then with PTSA in methanol to give the 5-methoxy-proline derivative as a mixture of stereoisomers. Reaction of this mixture with TMSCN and a catalytic amount of SnCl$_4$ yielded the nitrile derivative [for recent application, see 184]. Similar results have been reported by Ohta and co-workers,[185] starting from the methyl ester of the N-benzyloxycarbonyl pyroglutamate. The lactam reduction was in that case carried out with NaBH$_4$ in MeOH at −30 °C. 5-Substituted prolinates were obtained upon condensation of the 5-methoxypyrrolidine with benzylacetoacetate in presence of TiCl$_4$ (Scheme 65). Langlois *et al.*[186] have reported the synthesis of (2S,5S) pyrrolidine-2,5-dicarboxylic acid by a somewhat modified Corey's approach.[For recent application, see 187]

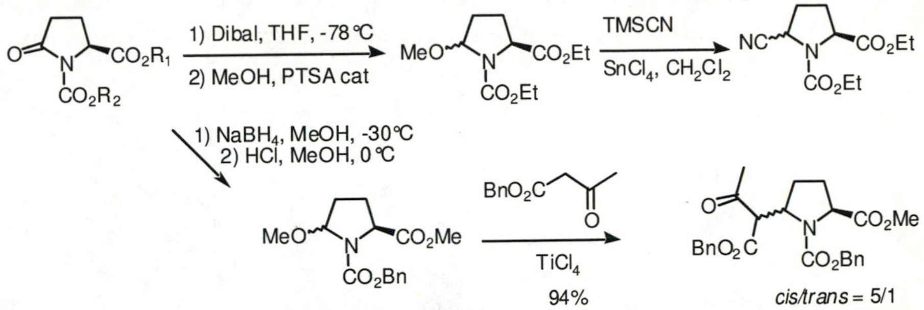

Scheme 65

The two-steps procedure to generate the methoxyaminal intermediate from pyroglutamate esters was also reported by Collado and co-workers in 1995[188] (Scheme 66).

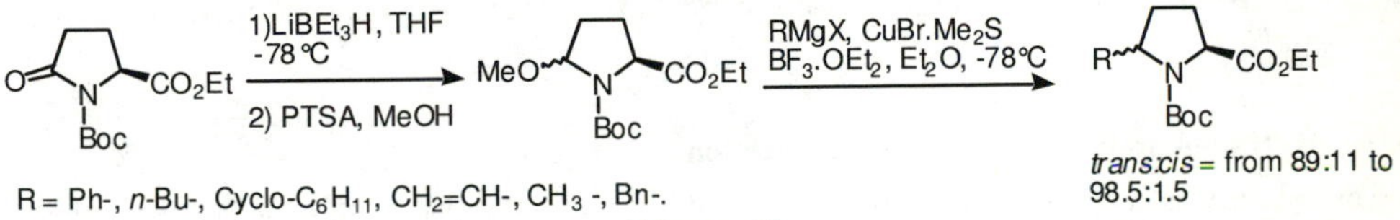

R = Ph-, *n*-Bu-, Cyclo-C$_6$H$_{11}$, CH$_2$=CH-, CH$_3$ -, Bn-.

Scheme 66

The reduction was here carried out using LiBEt$_3$H instead of Dibal and the organocopper reagents were generated *in situ* from stoichiometric amounts of a Grignard reagent, copper (I) bromide-dimethyl sulfide complex. The stereoselectivity observed here is in agreement with the mechanism previously proposed by Wistrand (see above). Oba used the same strategy to produce and deuterate in position five the methoxy aminal intermediate.[189 For recent application, see for example 190,191]

In 2001, Davies and Venkataramani[192,193] reported the synthesis of 2,5-disubstituted pyrrolidines by kinetic and double stereodifferentiation in C-H insertion of rhodium carbenoids (Scheme 67).

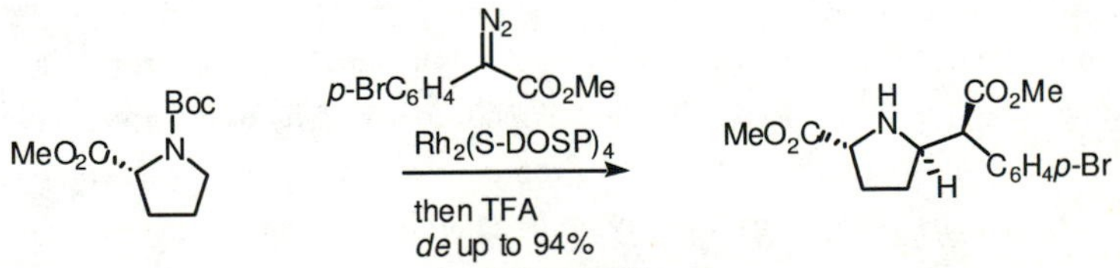

Scheme 67

This reaction can be considered as a surrogate for an asymmetric Mannich reaction. Starting from D-proline methyl ester, the reaction with *p*-bromophenyldiazoacetate (decomposed with Rh$_2$(S-DOSP)$_4$) led, after removal of the Boc group, to the C-H insertion product as a single diastereoisomer in 68% yield. In contrast, the same reaction performed with L-proline methyl ester led to a mixture of diastereoisomers of C-H insertion products. The reaction with D-proline methyl ester is the matched reaction, while the reaction with the L-derivative is the mismatched reaction. Kinetic resolution of the racemate led to a *de* of 86% and *ee* of 79% in 66% yield. The intramolecular version of the carbenoid chemistry has been reported by Davis *et al.* [194] (see below).

The synthesis of (2S,5R)-phenylproline methyl ester has been reported by Severino.[195] The title compound was obtained through Heck arylation of enecarbamate with diazonium salts (Scheme 68).

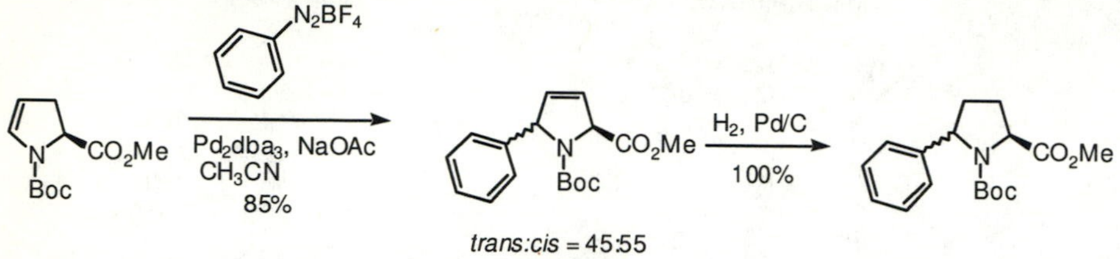

trans:cis = 45:55
Scheme 68

Heck arylation of the ene carbamate using phenyldiazonium tetrafluoroborate and Pd$_2$(dba)$_3$ led to the Heck adduct with good yield. The arylation proceeded with a low stereoselectivity providing an inseparable mixture of diatereoisomers. Hydrogenation over palladium charcoal yielded the 5-phenylproline as a 1:1 mixture of *cis:trans* isomers. Both diastereoisomers could be isolated in a pure form after removal of the Boc protecting group.

2.4.2. 5-Substituted prolines syntheses *via* cyclization

Several intramolecular cyclization reactions have been reported, involving nucleophilic displacement of a mesylate, Michaël addition, reduction of the corresponding cyclic imine or radicalar process.

Pyroglutamate esters were used as starting material by Ohta[196] (Schemes 69 and 70) and Collado[197] (Scheme 72). In 1987, Ohta and co-workers[196] reported the synthesis of 2,5-disubstituted pyrrolidines.

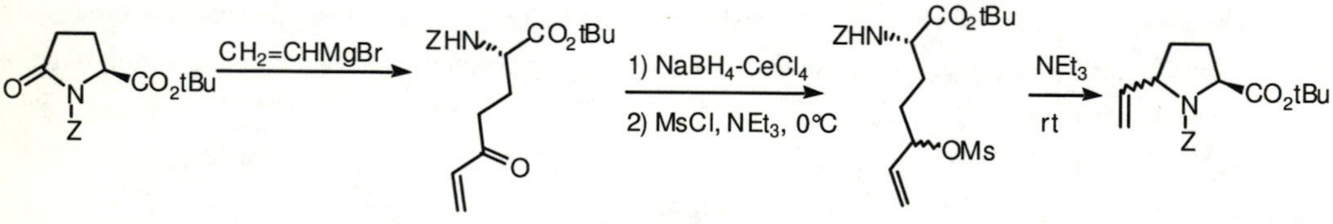

Scheme 69

N-Protected pyroglutamate ester was function-selectively reacted on treatment with Grignard reagent at low temperature yielding the enone. Reduction into alcohol followed by transformation into mesylate and cyclization furnished a *cis-trans* mixture of 5-vinyl prolinate (this method was recently used by Fournié-Zaluski *et al.*[198] for the synthesis of various 5-phenylproline derivatives, substituted or not on the phenyl ring and mesylate cyclization has been recently used by Gosselin and Lubell in the synthesis of pyrroloazapinone[199]).

The same function-selectivity was observed by addition of lithium enolate of esters to *N*-carbamoylpyroglutamates[200] (Scheme 70).

246

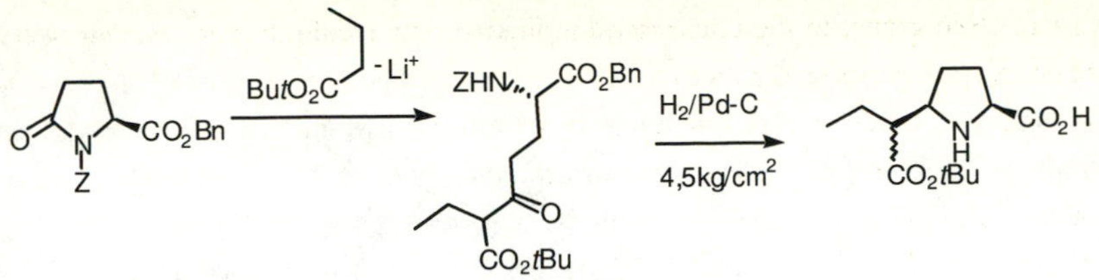

Scheme 70

Hydrogenolysis of the 5-oxoamino acid derivative with 10% Pd-C at medium pressure afforded 2,5-*cis* 5-substituted proline.For recent application, see for example 201

In contrast, Li *et al.*[202] have reported that trialkylaluminiums (AlR₃, R=Me, Et or *n*-Pr) add on benzyl *N*-methoxy- or allyloxycarbonyl-pyroglutamates esters to give the corresponding alkylated hemiaminal without ring opening (Scheme 71).

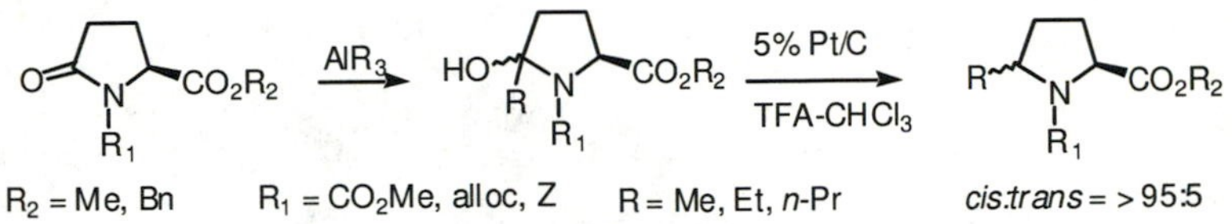

R_2 = Me, Bn R_1 = CO₂Me, alloc, Z   R = Me, Et, *n*-Pr   *cis:trans* = > 95:5
Scheme 71

The reduction of the alkylated hemiaminal was carried in various conditions and the best results were obtained with R_1=CO₂Me, R_2=Me by catalytic hydrogenation over Pt/C.

Collado[197] reported diastereoselective functionalization of 5-hydroxyprolinates by tandem Horner-Emmons-Michaël reaction (Scheme 72).

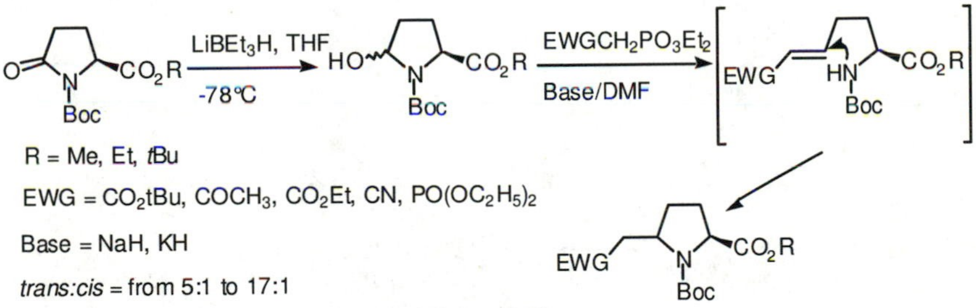

R = Me, Et, *t*Bu

EWG = CO₂tBu, COCH₃, CO₂Et, CN, PO(OC₂H₅)₂

Base = NaH, KH

trans:cis = from 5:1 to 17:1

Scheme 72

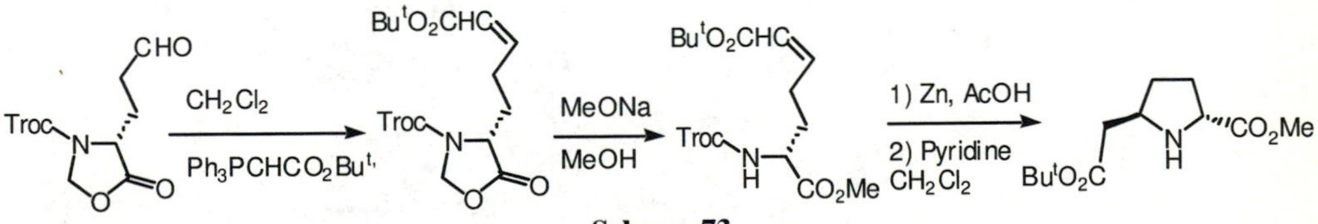

Scheme 73

Reduction of the amide carbonyl group was realized with LiBEt₃H in 99% yield. Several phosphonates stabilized with different electron withdrawing groups (EWG) were then reacted with the hemiaminal with very good yield (from 69 to 91%). Cyclization by Michaël addition spontaneously occurred and the best

247

results in term of stereoselectivity were obtained using hindered protecting groups on both phosphonate and carboxylic acid functions using KH as a base.[For recent example, see 203 204]

The combination of Wittig-Horner reaction and cyclization by Michaël addition has also been reported earlier by Bycroft.[205] The 5-substituted proline was obtained starting from (R)-glutamic acid γ-aldehyde derivative (Scheme 73).

Condensation of the aldehyde with the Wittig reagent derived from t-butyl bromoacetate afforded the *trans* alkene. Alcoholysis, followed by reductive deprotection of the Troc group released the α-amino function, which in the presence of pyridine underwent an intramolecular Michaël addition to yield the *trans*-proline derivative. The synthesis of enantio- and diatereoisomerically pure substituted prolines by Michaël addition of a Ni^{II} complex of the Schiff base derived from (S)-o-[(N-benzylprolyl)-amino]benzophenone and glycine to substituted acrylaldehydes followed by reductive cyclization of the corresponding iminium intermediate was described by Belokon and co-workers[206] (Scheme 74).

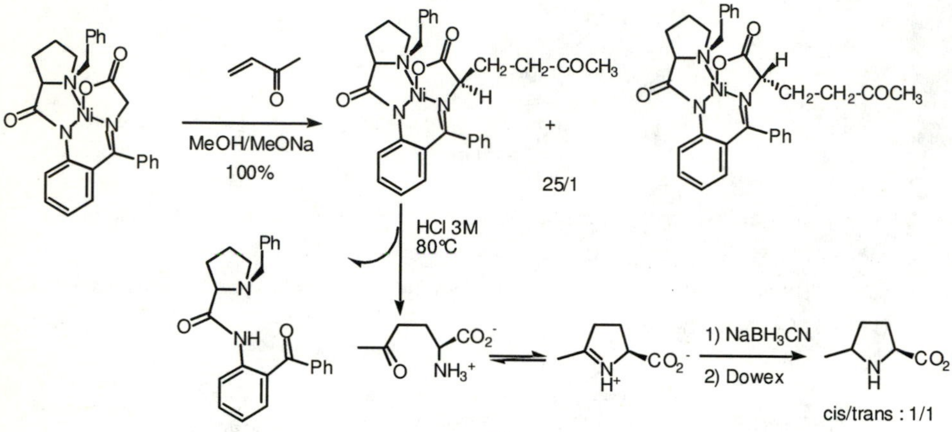

Scheme 74

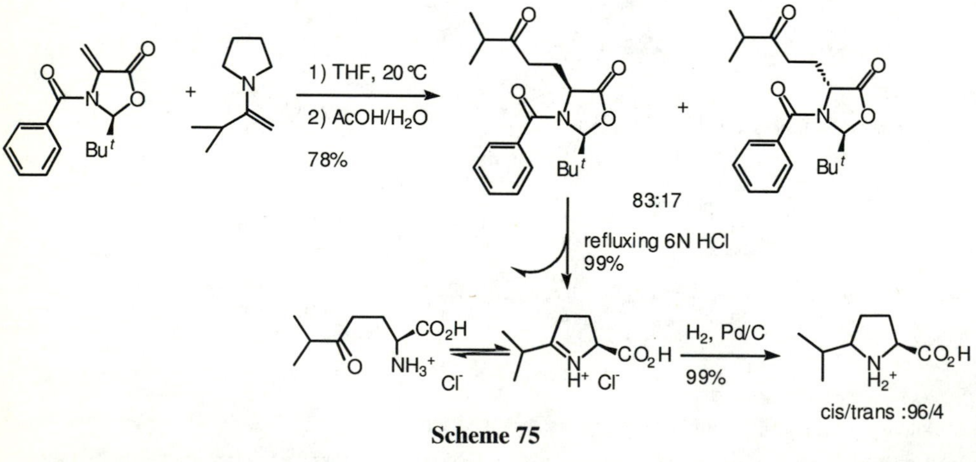

Scheme 75

A 1/1 *cis/trans* mixture of (2S)-5-methyl proline was obtained by addition of the Nickel complex to methyl vinyl ketone, followed by hydrolysis of the complex and reduction of the iminium with sodium borohydride. In the same manner, the synthesis of 3-or 4-substituted prolines were obtained by addition of the complex to variously substituted acrylaldehydes. The reductive cyclization of iminium has also been reported by Pyne and co-workers (Scheme 75).

The key intermediate was here obtained by reaction of the oxazolidinone with the enamine with moderate diastereoselectivity (83:17). Treatment of the diastereomerically pure oxazolidinone adduct in refluxing 6N HCl yielded the iminium salt in equilibrium with the corresponding ketone. The 5-substituted proline was obtained upon hydrogenation over palladium charcoal yielding the *cis*-5-isopropylproline as the major product (*cis:trans*=96:4).

There are several contributions from Lubell and co-workers in the field of unnatural amino acids syntheses. In 1993, Ibrahim and Lubell[207] reported the synthesis of enantiopure 5-substituted prolines *via* acylation of N-(phenylfluorenyl)glutamate enolate and reduction of the corresponding imine intermediate (Scheme 76).

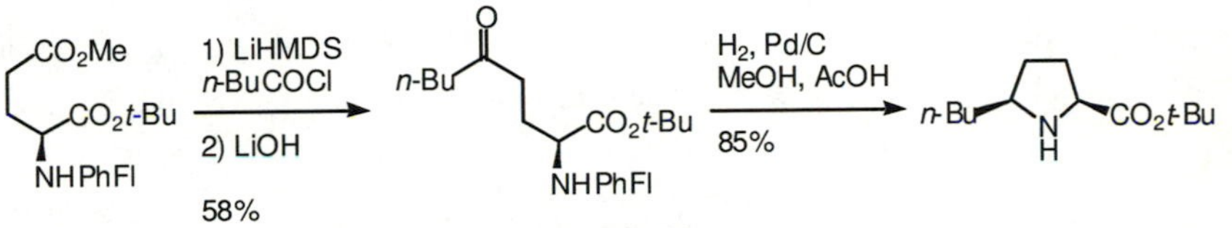

Scheme 76

The 5-*n*-butylproline *tert*-butyl ester was obtained by acylation with valeryl chloride of the glutamate derivative, subsequent decarboxylation with LiOH, followed by hydrogenation over palladium charcoal.

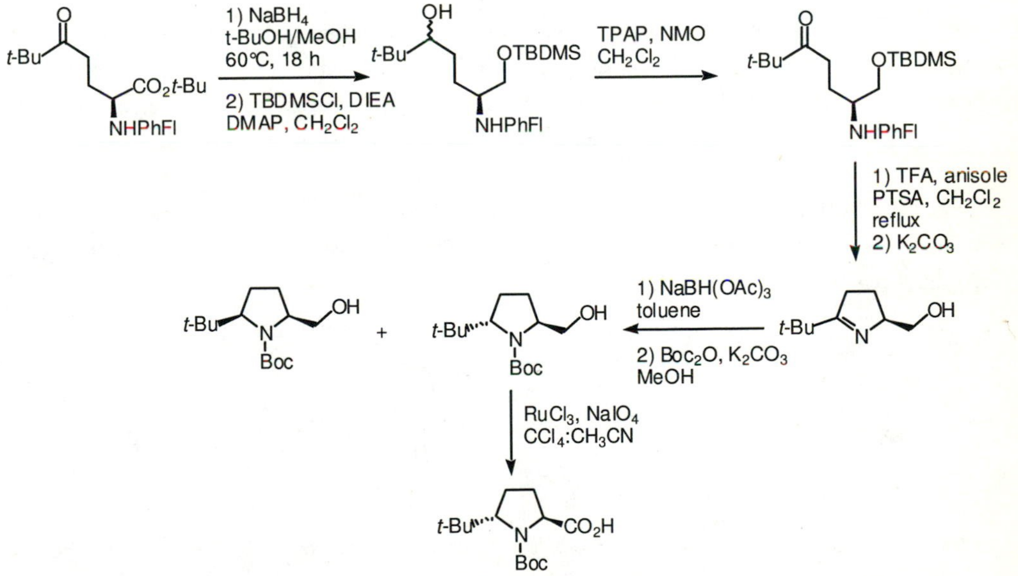

Scheme 77

This hydrogenation proceeded by cleavage of the phenylfluorenyl group, intramolecular imine formation, protonation and hydrogen addition to the less hindered face of the iminium ion to selectively furnish *cis*-(2*S*,5*S*)-5-*n*-butylproline *tert*-butyl ester in 85% yield. This methodology has been extended to the synthesis of all four stereoisomers of 5-*tert*-butylproline[208] as well as the synthesis of optically active *cis*-5-methyl proline [209]. The *cis*-isomers were obtained as described above, starting from L or D-glutamic acids, whereas the *trans* isomers were obtained by epimerisation of the *N*-(Boc)amino acid ester with KHMDS in 3:2 ratio in favour of the *cis*-isomer or by cleavage of the phenylfluorenyl group with TFA, followed by reduction with Me$_4$NBH(OAc)$_3$ of the intermediate iminium with 58% diastereoisomeric excess in favour of the *trans* isomer. Recently, an improved synthesis of *trans-tert*-butylproline was reported by Lubell's group[210] (Scheme 77).

In this approach, the selective reduction of the imine intermediate with triacetoxyborohydride followed by Boc protection yielded the alcohol as a >95:5 diastereoisomeric mixture in favor of the *trans* isomer. The final compound was obtained by oxidation using ruthenium trichloride and sodium periodate. The synthesis of this compound has been recently reported by Wallén[179] in high yield using the methodology developed by Wistrand.

Lathbuty and co-workers[211,212] have reported palladium (II)-mediated routes cyclization to functionalised heterocycles, starting from α-substituted amino acid derivatives (Scheme 78).

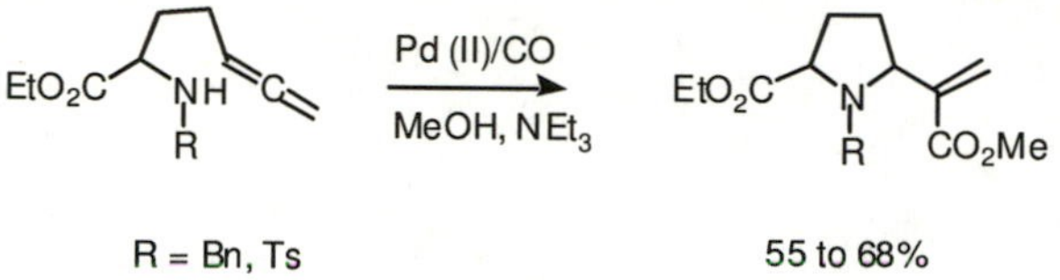

$$R = Bn, Ts \qquad\qquad\qquad 55 \text{ to } 68\%$$

Scheme 78

The allenic amine or sulfonamide was shown to undergo a palladium (II)-catalysed cyclisation, in the presence of carbon monoxide and methanol to give the heterocyclic acrylate as a racemate. The stereoselectivity in favor of *cis*-5-substituted proline in this cyclization process was described by Kinsman[213] using homogeneous silver (I)-catalysis. The mode of action of silver (I) in this catalytic reaction was not explained.

There are several contributions from Kemp and co-workers on the synthesis of unnatural amino acid able to induce strong conformational biases in peptide. Thus, the synthesis of *cis*-5-(hydroxymethyl)-proline[214] has been described (Scheme 79).

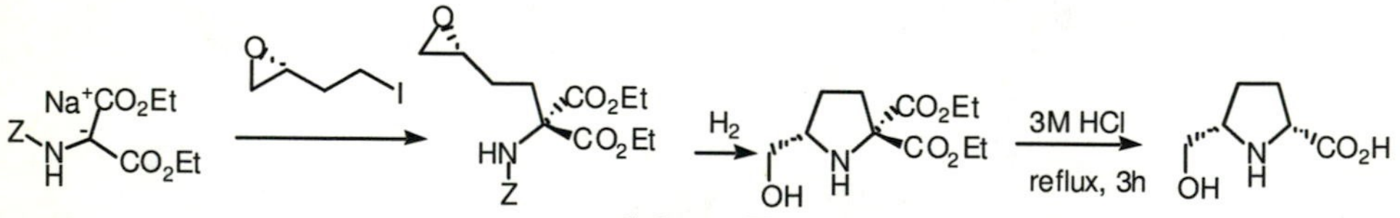

Scheme 79

Alkylation of the sodium salt of (*N*-carbobenzoxyamino)-malonate with (*R*)-4-iodo-1,2-epoxybutane, followed by hydrogenolysis of the Z group yielded the cyclic diester. Indeed, cyclization spontaneously occurred upon hydrogenolysis. The final compound was obtained after stereoselective decarboxylation in refluxing HCl. The enantioselective synthesis of both 2*R* and 2*S cis*-5-(hydroxymethyl)-proline has been reported by Björkling and co-workers[215] *via* enzyme catalysed hydrolysis of *cis*-N-benzyl-2,5-

250

bismethoxycarbonylpyrrolidine in H_2O/DMSO medium. The synthesis of the *trans* isomer was reported by Kemp and Curran[216,217] by *cis-trans* epimerisation and reduction of the *cis*-derivative of 1-benzyl-2,5-dicarbomethoxypyrrolidine. The synthesis of 5-substituted prolines by C-N bond formation through epoxide opening has also been reported by Manfré *et al.*[170] (Scheme 80).

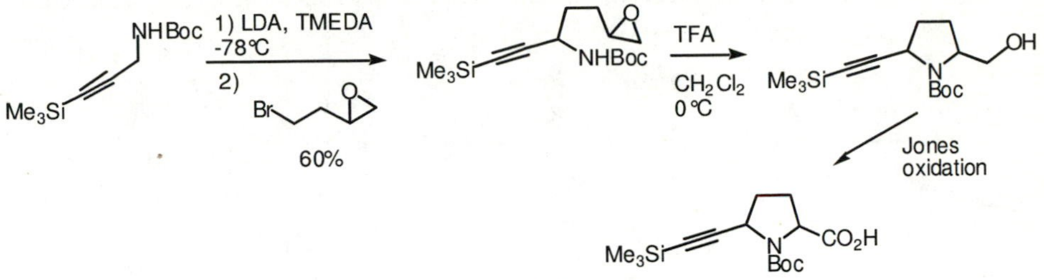

Scheme 80

A dianion was generated using LDA in the presence of TMEDA and alkylation with the bromoepoxide led to the epoxide carbamate in 60% yield as a mixture of unseparable stereoisomers. The C-N bond formation occurred upon cyclization of the epoxide in presence of TFA in CH_2Cl_2 at 0 °C. The removal of the nitrogen protecting group was not observed in this condition. Jones oxidation of the diastereoisomeric mixture of alcohols gave the corresponding acids. In 1989, Takano and co-workers[218] reported the synthesis of optically active *trans*-2,5-dioxymethylpyrrolidine (Scheme 81).

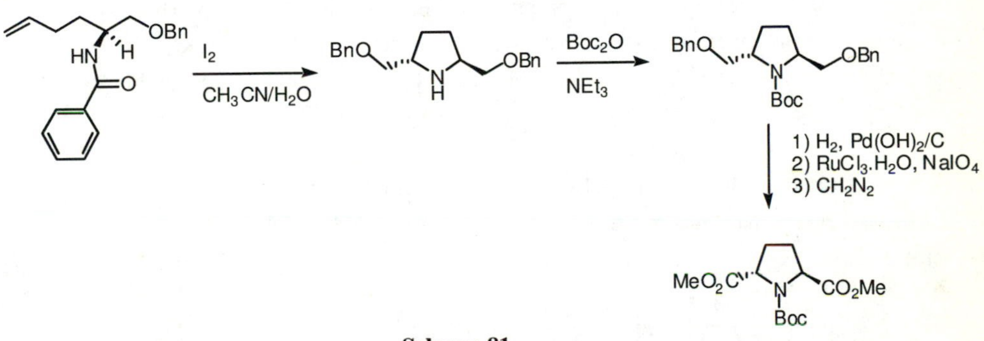

Scheme 81

The title compound was obtained by iodine-mediated cyclization of (*S*)-1-benzoylaminohex-5-ene, followed by Boc-protection, hydrogenolysis of the benzyl ether, oxidation of the alcohol and esterification with diazomethane.

The synthesis of substituted prolines through radical cyclization of aminyl radicals derived from amino acids have been reported by Bowma *et al.*[219] (Scheme 82).

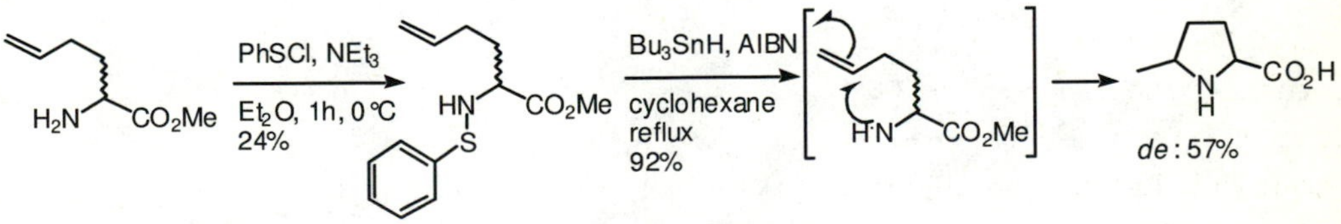

Scheme 82

251

The starting sulfenamide was obtained by reaction of the amino ester with benzenesulfenyl chloride with poor yield and the cyclization was carried using Bu$_3$SnH and AIBN. The proline derivative was obtained in good yield with a *de* of 57%.

Davis *et al.*[194] have reported the synthesis of 5-phenyl proline *via* cyclization mediated by intramolecular N-H insertion reaction of amino α-diazoketone catalysed by Rhodium metals (Scheme 83).

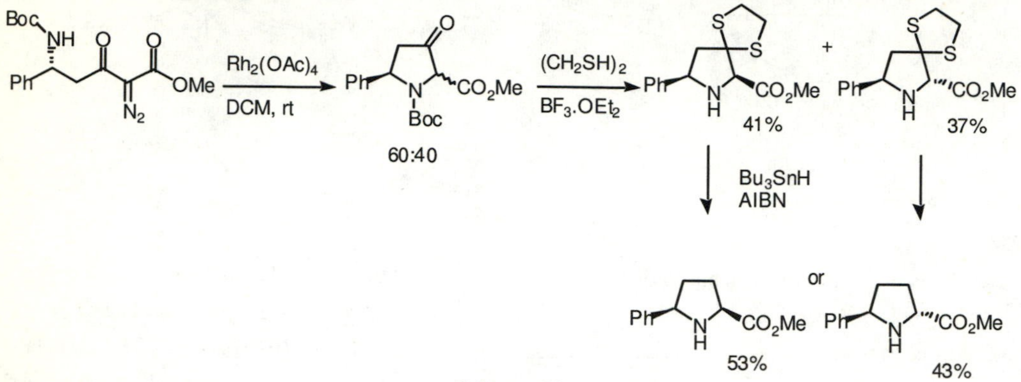

Scheme 83

The starting diazo compound was treated in DCM with Rh$_2$(OAc)$_4$ leading to the oxo-proline derivative as a single diatereoisomer, which proved to be unstable to work-up. Thus a 60:40 mixture of diastereoisomers was obtained on chromatographic purification. This mixture was treated with 1,2-ethanedithiol to give a separable mixture of thioketals. Desulfurization with tributyltin hydride gave the expected 5-phenyl proline derivatives.

2.4.3. 5-Substituted prolines syntheses by alternative methods

Magaard *et al.*[220] have reported the racemic synthesis of 5,5-dimethylproline through nitrone chemistry (Scheme 84).

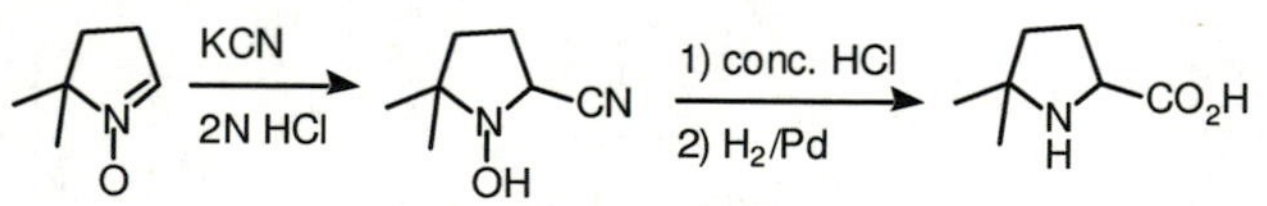

Scheme 84

The commercially available 5,5-dimethylpyrroline-1-oxide was treated with KCN in 2N HCl to give racemic N-hydroxy-2-cyano-5,5-dimethylpyrrolidine. Hydrolysis, followed by catalytic hydrogenation yielded the 5,5-dimethylproline.

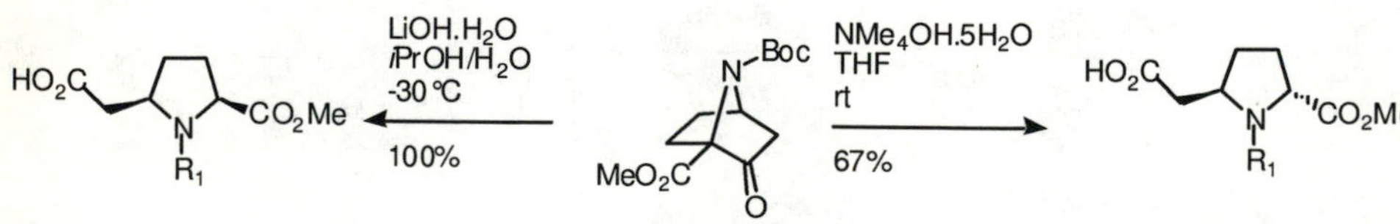

Scheme 85

252

In 2003, Avenoza et al.[221] proposed a retro-Dieckmann reaction for the enantiopure synthesis of all four stereoisomers of 5-(carboxymethyl)pyrrolidine-2-carboxylic acid methyl ester (Scheme 85).

Starting from previously reported keto-ester, the retro-Dieckmann reaction was promoted by LiOH.H$_2$O in isopropanol leading exclusively to orthogonally protected *cis* isomer of *N*-Boc-5-(carboxymethyl)-pyrrolidine-2-carboxylic acid methyl ester in 100% yield. The *trans* isomer was obtained in 67% yield, starting from the same keto-ester but using dry THF and NMe$_4$OH as the source of HO⁻. There are several contributions from Sasaki's group on the chemistry of substituted prolines and the synthesis of *N*-Boc-(2S,5R)-5-(1'-hydroxy-1'methylethyl)proline has been reported[222] (Scheme 86).

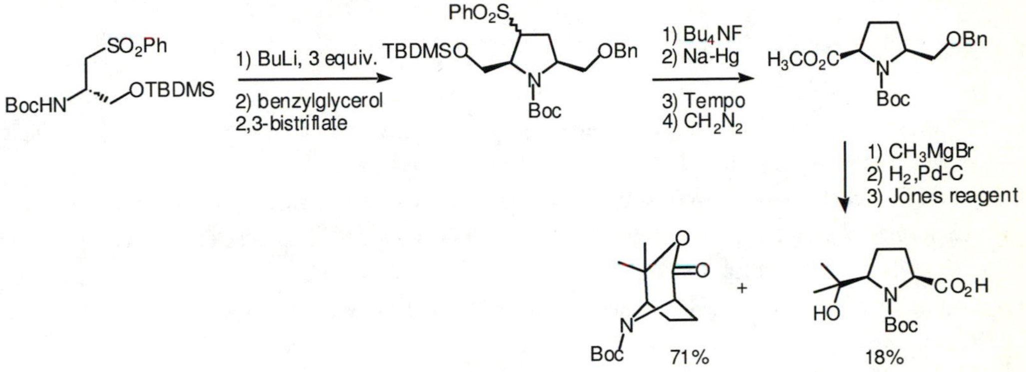

Scheme 86

The trianion, generated by treatment of the chiral sulfone (obtained in five steps from serine derivative) with 3 equiv. of BuLi, was reacted with (2)-1-benzylglycerol-2,3-bistriflate leading to the pyrrolidine. Removal of TBDMS group, followed by desulfonylation and oxidation gave the acid intermediate, which was immediately esterified with diazomethane. Nucleophilic addition of MeMgBr to the ester function gave the tertiary alcohol. Hydrogenolysis followed by Jones oxidation afforded the title compound in 18% yield, the major compound being the lactone (71%). A piperidine-pyrrolidine ring contraction leading to 5-substituted proline derivative has been reported recently by Tanaka et al.[223] (Scheme 87).

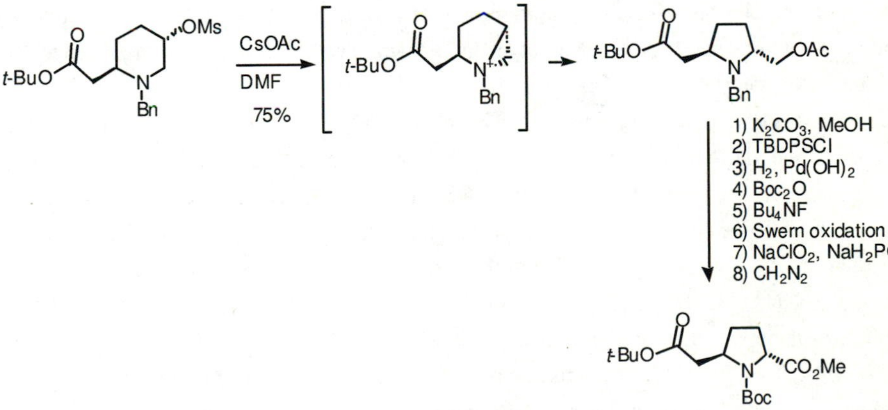

Scheme 87

253

The starting mesylate was obtained from 3-hydroxypyridine in eleven steps. The ring contraction occurred by treatment of the mesylate with cesium acetate in warm DMF to furnish the (2R,5R)-*trans*-disubstituted pyrrolidine acetate. The title compound was then obtained as described on Scheme 87 in eight steps. Collado has reported the synthesis of the enantiomer in two steps starting from pyroglutamate esters (see Scheme 72).

3. Effect on peptide conformation and structure

In this part, the nomenclature adopted to name the different positions in the pyrrolidine ring of modified prolines is as follows : α–, β–, γ–, δ– instead of 2-, 3-, 4- and 5-.

3.1. Conformational properties of proline

Proline exhibits unique conformational properties owing to its cyclic structure. The inclusion of backbone N and Cα atoms in the pyrrolidine ring places severe constraints on the φ torsion angle, corresponding to the rotation around the N-Cα bond. As a consequence, the φ angle restriction ($\approx$ -65°) strongly limits the conformational (φ,ψ) space of proline compared to other aminoacid residues. A survey of proline conformation in protein structures from the Protein Data Bank [224] indicates that the mean value of the φ angle is -63° ± 15° and that the ψ angle is clustered around two mean values of -35° and +150°, corresponding to the α_R (right-handed α helix) and the P_{II} (polyproline II) region of the Ramachandran diagram, respectively. Conformational studies on Ac-Pro-NHMe peptide show that proline prefers a P_{II} conformation [225] in aqueous solution. Proline can also adopt a stable γ-turn conformation (ψ $\approx$ +80°) in nonpolar organic solvents, characterized by a hydrogen bond between the N-terminal carbonyl and the C-terminal amide group (C_7, seven-membered hydrogen-bonded ring).[225]

Another unique feature of proline residues is that the preceding peptide bond has a less pronounced preference for the *trans* isomer. Indeed, about 6% of peptide bonds preceding proline residues in a set of protein structures from the PDB occur as the *cis* isomer, as compared to only 0.06% for other residues [224,226]. The proportion of *cis* isomer can be substantially larger ($\approx$ 25-30%) in proline model peptides [225] and in small peptides. This corresponds to an energy difference ΔG^0 between *cis* and *trans* isomers of the order of 1-2 kcal.mol^{-1}.[227] The activation energy barrier $\Delta G^{\ddagger}$ for *cis-trans* isomerisation is around 20 ± 3 kcal.mol^{-1} [227]. As a consequence, the *cis-trans* interconversion is slow on the NMR time scale and mixtures of *cis* and *trans* isomers can typically be observed in solution by NMR for peptides and proteins. *Cis* peptidyl-proline bonds often occur in turns and bends of the polypeptide chain. Type VI β-turns are characterized by a *cis*-proline residue in position i+2.[228]

Because of the covalent bond between the side chain $C^\delta H_2$ methylene group and amide nitrogen, the backbone φ angle and the four side chain χ1-χ2-χ3-χ4 dihedral angles are interdependent. Typically five-membered rings are puckered rather than planar and can adopt a variety of envelope or half-chair (twist) conformations, characterized by 1 or 2 atoms out of plane, respectively. In pyrrolidine rings, the lower barriers for rotation around C-N bonds relative to those around C-C bonds favor smaller values for dihedral angles around C^δ-N and N-C^α bonds.[229] As a consequence, C^γ and/or C^β atoms are usually puckered out of the mean plane of the pyrrolidine ring. Therefore, proline can rapidly interconvert in solution between two distinct puckerings. These two different forms are referred to as C^γ-exo/C^β-endo (also termed "up" or type A puckering) and C^γ-endo/C^β-exo ("down" or type B puckering), and correspond to χ1 dihedral angles about -

30° and +30°, respectively (Scheme 88). Quantum mechanical *ab initio* calculations[230] indicate that the C^γ-endo puckering is slightly favored in Ac-L-Pro-NHMe by ≈ 0.4 kcal.mol^{-1}. In proteins, these two conformers are almost equally probable for *trans*-proline residues[231]. However, the C^γ-endo puckering is preferred ($\approx$ 80%) for *cis*-proline residues.[232,233]

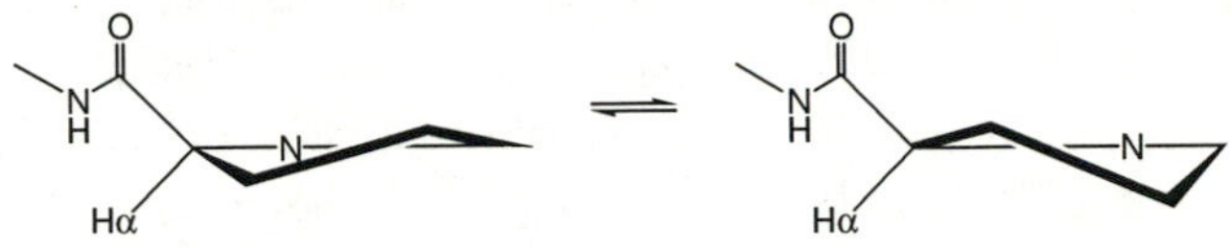

The two distinct puckerings of the pyrrolidine ring. C^γ-exo (up or type A) puckering corresponds to ($\chi1,\chi2,\chi3,\chi4$) angles around (-30,+30,-30,+10). C^γ-endo (down or type B) puckering corresponds to ($\chi1,\chi2,\chi3,\chi4$) angles around (+30,-30,+30,-10).

Scheme 88

Dependent on the ring puckering, the ϕ value ranges around -65° ± 25° in crystal structures of proline compounds[229]. There is a correlation between the ϕ angle and the puckering: ϕ values are clustered around -70° for C^γ-endo puckerings while those associated with C^γ-exo puckerings are around -60°. Largely negative values of $\phi \approx$ -80° are associated with $\chi1 \approx$ +30° puckering (C^β-exo envelope or C^β exo/C^γ endo twist) whereas for $\phi \approx$ -50°, $\chi1 \approx$ -30° puckering is favored (C^β endo envelope or C^β endo/C^γ exo twist).[229]

The puckering of the pyrrolidine cycle also influences the ψ dihedral angle. C^γ-endo puckering is associated with slightly larger values of ψ in minimum-energy conformers.[230] This effect can be accounted for by repulsion between the carbonyl and the $C^\gamma H_2$ group, which is relieved if ψ approaches 180°. In the case of C^γ-exo conformers, the repulsion between the $C^\beta H_2$ group and atom Ni+1 disfavors values of ψ too close to 180°. Computational studies[234] indicate that the ψ dihedral angle is less restrained for the C^γ-endo puckering. As a matter of fact, a survey of protein structures indicates that the ψ angle is clustered around 120-160° for the C^γ-exo puckering and 120-180° for the C^γ-endo puckering.[232]

Restriction of the conformational space not only concerns proline itself but also the preceding residue. Conformational energy map calculations[235,236] indicate that the (ϕ,ψ) restriction is caused by steric clashes between the NH and $C^\beta H_2$ atoms of the preceding residue and the $C^\delta H_2$ methylene group of the pyrrolidine ring. A conformational analysis of protein structures[224] reveals that residues preceding proline have a marked preference for the β region and avoid the α_R region (less than 10% residues). Even glycine turns out to be conformationally restricted in Gly-Pro sequences and adopts extended conformations.

3.2. Conformational effects of the introduction of substituents

3.2.1. Substituents in position 2

The conformational effects of α-methylation were investigated two decades ago by Delaney and Madison[225] on the model peptide Ac-L-2-MePro-NHMe, using NMR, CD, FT-IR spectroscopies and molecular mechanics (based on CFF forcefield). The 2-methyl group strongly stabilizes γ-turn conformers in non polar solvents (chloroform and CCl$_4$), as evidenced by intramolecular hydrogen-bonded NH groups in infrared spectra. In polar solvents, the proportion of γ-turn conformers decreases but the CD spectrum in aqueous solution may be interpreted as indicating that the γ-turn conformation remains significantly populated. Another noteworthy consequence of α-methylation is the strong destabilization of the *cis* isomer of the preceding peptide bond since virtually no *cis* isomer could be detected by NMR spectroscopy in any solvent. This effect can be ascribed to the strong steric repulsion between the C^{α} methyl group of 2-me-proline and the acetyl methyl group (equivalent to the C^{α} of the preceding residue).

The α-methylation is widely used to stabilize helical conformations of peptides. However, this modification has been rarely applied to proline, presumably due the low efficiency of coupling during peptide synthesis. A 2-methyl group was introduced on prolines to stabilize reverse turn conformations in several biologically active peptides, such as bradykinin (BK).[237] BK possesses three proline residues in its sequence in positions 2, 3 and 7 (Arg-Pro-Pro-Gly-Phe-Ser-Pro-Phe-Arg). The introduction of αMe-Pro at position 3 or position 7 stabilizes reverse-turn conformations in both analogues around residues Pro2-Phe5 and Ser6-Arg9, as evidenced by NMR. However the types of populated β-turns were not further investigated by the authors. The α-methylation of proline was shown to stabilize type I β-turns in the sequence Tyr-Pro-Tyr-Asp of a peptide antigene[238] and in tandem repeats of Asn-Pro-Asn-Ala motif.[239]

3.2.2. Substituents in position 3

A wide range of substituents have been inserted in position 3 of proline residues, from single methyl to bulky isopropyl and phenyl groups, in order to synthesize prolino–amino acid chimeras. The conformational effects strongly depend on the nature of the side chain and the configuration of the C3 carbon.

Madison and Delaney have analyzed the structural effects of 3-methylation in two model peptides, Ac-L-*trans*-3-MePro-NHMe and Ac-L-*cis*-3-MePro-NHMe.[240] The introduction of a single methyl group in the *trans* position has only minor effects on the conformational space of proline. Vicinal $^3J_{H\alpha-H\beta}$ coupling constant in chloroform (3.2 Hz) indicate that the C^{γ}-endo puckering is the more stable. The γ-turn conformation dominates in non polar solvents; there is a mixture of γ, α_R and P_{II} conformations in polar organic solvents such as acetonitrile whereas the P_{II} conformation is dominant in aqueous solution. In contrast, a methyl susbstituent in the *cis* position stabilizes the C^{γ}-endo puckering and strongly restricts the conformational space around the ψ angle, through steric interaction with the carboxamide group. In particular, the γ-turn conformation is strongly destabilized, even in weakly polar solvents (chloroform). Although this conformation is still populated in non polar solvents such as CCl_4, the low-frequency NH stretch band in infrared spectra indicates that the intramolecular hydrogen bond is weakened. The conformational effects of two methyl substituents were also investigated by FT-IR and molecular mechanics.[241] The γ-turn conformation was not observed for Ac-L-3,3-Me$_2$Pro-NHMe peptide in chloroform. Systematic conformational search using AMBER94 forcefield indicated that the ψ dihedral angle is restricted around 150°. Calculated low-energy minima corresponded to the C^{γ}-endo puckering.

Despite the large number of substituents inserted on the C3 position, there have been relatively few published studies aimed to analyze the local conformational effects on proline ring pucker and ϕ, ψ dihedrals. The conformations of *cis* and *trans* 3-isopropylproline (prolinoleucine) and 3-methylenethiomethyl (prolinomethionine) have been analyzed using molecular mechanics calculations on N-acetyl N'-methyl proline derivatives and NMR spectroscopy on substance P analogues.[13] Increasing the bulkiness of the *trans*-3-substituent gradually shifts the puckering equilibrium toward the C^{γ}-exo form. Indeed, the proportion of C^{γ}-exo conformers, estimated from $^3J_{H\alpha-H\beta}$ coupling constant, amounts to 50% for *trans*-3-methylenethiomethylproline and reaches 70% for *trans*-3-isopropylproline.[13] Similar conclusions can be drawn for the ϕ, ψ values associated with each ring pucker of *trans*-3-isopropylproline, as compared to proline: the C^γ-exo puckering corresponds to less negative values of ϕ and smaller values of ψ. In the case of *cis*-3-isopropylproline,[13] the C^γ-endo form is strongly stabilized. The isopropyl substituent constrains more the ψ torsion angle and an opposite relation is observed between ring pucker and ψ values in minimum-energy conformers: smaller values of ψ are associated with the C^γ-endo form. As observed for methyl substituent, the *cis* substitution destabilizes γ-turn conformations.

The effects of a 3-substituent on *cis-trans* isomerism of the preceding peptide bond have been investigated in a few studies.[240,241] The 3-methyl substituents marginally affect the proportion of peptide bond *cis* and *trans* isomers. Indeed, the proportion of *cis* isomers in Ac-L-3-MePro-NHMe and Ac-L-3,3-Me$_2$Pro-NHMe peptides ranged about 25-30% in water, as observed for other proline peptides.[225,240,241] However, the presence of two methyl substituents caused a 7-fold decrease in the rate of amide bond isomerization in water.[241] This effect may be due to an unfavorable geometry around the ψ angle ($\approx150°$) which disfavors prolyl amide pyramidalization by Coulomb interactions and therefore impedes amide bond isomerization[242]

3.2.2.1. Side chain conformation

Prolinoamino acid chimera (Scheme 86) bearing the side chain of a natural amino acid in position 3 of the pyrrolidine ring constitute valuable conformational tools to mimic bioactive conformations of unconstrained amino acids. The conformational space available to the side chain depends on the proline ring pucker and the *cis/trans* diastereoisomerism (Schemes 89 and 90).

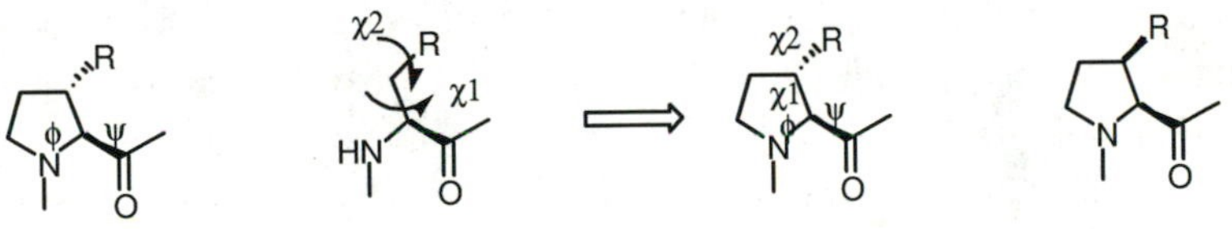

trans and cis 3-prolinoamino acides

Schematic representation of 3-prolinoamino acids.
Scheme 89

In *trans*-prolinoamino acids, the C^γ-endo puckering corresponds to a side chain $\chi1$ around -90°, value close to the gauche+ rotamer ($\approx$ -60°) of the unconstrained amino acid. The C^γ-exo puckering is associated with a side chain $\chi1$ around -150°, corresponding to the *trans* ($\approx$ 180°) rotamer. Oppositely, the C^γ-endo and C^γ-exo puckerings of *cis*-prolinoamino acids correspond to *trans* ($\chi1 \approx$ +150°) and gauche- ($\chi1 \approx$ +90°) rotamers, respectively. Therefore, only two of the three possible $\chi1$ rotamers are accessible to prolinoamino acids, the *trans* and a single gauche rotamer (g$^-$ or g$^+$ for *trans* or *cis*-prolinoamino acids, respectively). Although the geometrical constraint due to the cyclization induces a 30° deviation of $\chi1$ from ideal values, the conformers of prolinoamino acids fit well with the corresponding structures of unconstrained amino acids.[13]

3.2.3. Substituents in position 4

Ih the case of prolinoamino acid chimera bearing the side chain of a natural amino acid in position 4 of the pyrrolidine ring, the $\chi1$ is that of proline and is restrained to ± 30°, dependent on the ring pucker. Therefore, only the *gauche* $\chi1$ rotamers are accessible to this class of prolinoamino acids. The $\chi2$ dihedral angle is also restrained and depends on the proline ring pucker and the *cis/trans* diastereoisomerism, a similar situation as for the χ1 angle in 3-prolinoamino acids.

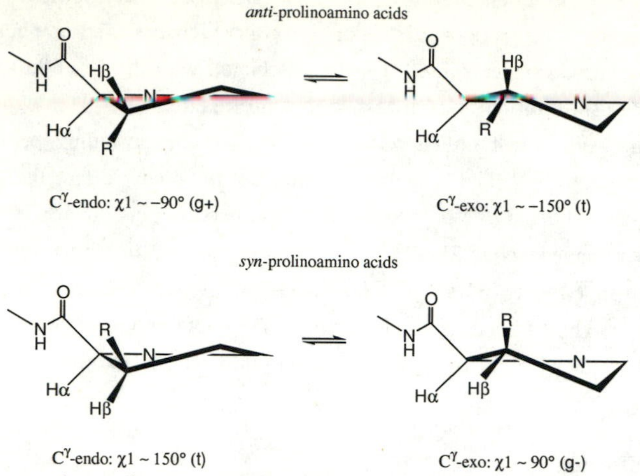

anti-prolinoamino acids

C$^\gamma$-endo: χ1 ~ −90° (g+)   C$^\gamma$-exo: χ1 ~ −150° (t)

syn-prolinoamino acids

C$^\gamma$-endo: χ1 ~ 150° (t)   C$^\gamma$-exo: χ1 ~ 90° (g-)

Relation between the proline ring pucker and the χ1 conformers of 3-prolinoamino acids.
Scheme 90

Among the different derivatives synthesized, the *cis*- and *trans*-4-mercaptoprolines have been incorporated in angiotensin II analogues[243] (see below): they can be considered as constrained analogues of homocysteine. The conformational space of 4-carboxylate[113] and 4-guanidinoethyl-prolines,[10] which are constrained analogs of glutamate and homoarginine, respectively, have also been compared to that of the unconstrained amino acid. 4-Hydroxy-prolines can be viewed as mimics of homoserine but are also proteinogenic amino acids. The conformational properties of 4-hydroxy-prolines and the 4-fluoro-proline analogues have been extensively studied in order to analyze their role in collagen stability.

4(R)-hydroxy-L-proline (Hyp) naturally occurs in collagen, the most abundant protein in vertebrates. The polypeptide chain of collagen is characterized by the succession of approximately 300 (Xaa-Yaa-Gly) repeats, where Xaa and Yaa are most often Pro and Hyp, respectively. The polypeptide backbone adopts a left-handed P$_{II}$ helix and three chains intertwine themselves to form a right-handed triple-helix coiled-coil structure. The crystal structure of (ProProGly)$_{10}$ triple helix[232] and molecular modeling studies of (Pro-Pro-Gly)n polypeptides[234] suggest that the alternation of C$^\gamma$-endo and C$^\gamma$-exo puckerings in Xaa and Yaa positions, respectively, is favored. The triple helix formation is inhibited by the presence of 4(S)-hydroxy-L-proline in either Xaa or Yaa positions. Collagen chains incorporating 4(R)-fluoro-L-proline in the Yaa position have a greatly increased stability.[244] Since fluorine is more electronegative but is a less effective hydrogen-bond acceptor than the hydroxyl group, this stresses the importance of stereoelectronic effects in the stability of collagen.

Stereoelectronic effects have been analyzed by means of quantum mechanical *ab initio* calculations, taking into account solvent effects.[230] The stereochemistry and the electronegativity of the 4-substituent influence the relative stability of C$^\gamma$-endo and C$^\gamma$-exo puckerings. The relative stability of the C$^\gamma$-exo conformer increases with the electronegativity of the 4(R)-substituent: it is favored by 0.5 kcal.mol^{-1} in Ac-4(R)-hydroxyproline-NHMe and 1.1 kcal.mol^{-1} in Ac-4(R)-fluoroproline-NHMe. These calculations are in agreement with the strong preference for the C^{γ}-exo puckering observed for Hyp and Flp in solution.[245] Therefore, the 4(R) electronegative substituents have the effect of reversing the preferred puckering of proline. The preference for the C^{γ}-exo puckering can be explained by interactions between the C^{γ}-X antibonding orbital and the vicinal C^{δ}-H_{ax} and C^{β}-H_{ax} bonding orbitals. The stabilization of the C^{γ}-exo conformer favors smaller values of the backbone dihedral angles ϕ ($\approx$ -60°) and ψ ($\approx$ 145°). In contrast, C^{γ}-endo puckering is stabilized by 4(S) electronegative substituents. The 4-substituents also affect *cis-trans* isomerism of the preceding peptide bond: *trans* isomers are stabilized by 4(R) electronegative substituents and *cis* isomers by 4(S) substituents.

The conformation of 4-substituted prolines has also been analyzed in dimeric and trimeric proline oligomers, in the case of isopentyl and carboxymethyl substituents.[141] These peptides preferentially adopt a P_{II} conformation in solution, as shown by NMR. This raises the possibility to synthesize functionalized P_{II} mimics based on 4-substituted prolines.

3.2.4. Substituents in position 5

The introduction of alkyl substituents in position 5 of the pyrrolidine cycle can destabilize the *trans* isomer of the preceding peptide bond because of unfavorable steric interactions. In Ac-5-MePro-NHMe peptides,[225,240] the 5-methyl in *trans* position caused a slight increase (5%) in the proportion of *cis*-isomer in aqueous solution. The 5-methyl in *cis* position had no effect on the *cis-trans* ratio but lowered the activation barrier for *cis-trans* isomerism in water. In a Phe-Pro dipeptide analogue incorporating two methyl substituents in position 5, the proportion of *cis* isomers was shown to be around 90%.[220] The insertion of 5,5-dimethylproline in tripeptide sequences derived from RNase A was also shown to stabilize type VI β-turns with a *cis* peptide bond.[246]

The conformational consequences of incorporating a bulky substituent such as a *tert*-butyl group were analyzed on Ac-5-*tert*-butylproline-NHMe model peptides by NMR, FT-IR and molecular mechanics.[247] The 5-*tert*-butyl substituent causes a strong increase in the proportion of *cis*-isomers in water, since it amounts to 48% and 66% for the *cis* and *trans* diastereoisomers respectively, as compared to 25% for Ac-Pro-NHMe. The *cis*, but not the *trans* diastereoisomer also lowers the activation energy barrier for amide isomerization by 3.9 kcal.mol^{-1}. The 5-*tert*-butyl substituent affects the conformational space around the ψ dihedral angle. (ω, ψ) conformational maps indicate that the ψ angle is more restricted in the case of *trans*-5-*tert*-butylproline. In particular the 5-*tert*-butyl substituent in *trans* position destabilizes the γ-turn conformation in chloroform, which is the preferred conformation for both Ac-Pro-NHMe and Ac-*cis*-5-*tert*-butylproline-NHMe peptides. Minima are found for $\psi \approx 0°$ in Ac-*cis*-5-*tert*-butylproline-NHMe and for $\psi \approx 125°$ in Ac-*trans*-5-*tert*-butylproline-NHMe. The preferred conformation of $\psi \approx 0°$ may explain the lower barrier for isomerization in the *cis* diastereoisomer, since it is the ideal conformation to stabilize the pyramidalized transition state of the reaction.[242]

The *cis* amide bond stabilization makes 5-*tert*-butylprolines attractive mimics to stabilize type VI β-turns in peptides. The crystal structure of Ac-Leu-*cis*-5-*tert*-butylPro-NHMe shows that this peptide adopts an ideal type VIa β-turn conformation with (ϕ,ψ) angles of (-61°, 139°) for Leu and (-95°, 19°) for Pro, and an ω angle of 17° for the Leu-Pro peptide bond.[248] In solution, the Ac-Ala-*cis*-5-*tert*-butylPro-NHMe and Ac-Leu-*cis*-5-*tert*-butylPro-NHMe peptides were shown to adopt type VI β-turn conformations, independent of solvent composition whereas the Ac-Ala-Pro-NHMe and Ac-Leu-Pro-NHMe peptides did not.[248]

259

4. Biological activity of peptides containing proline-amino acid chimeras

Not so many biologically active peptides have been modified with these substituted-prolines in structure-activity relationships studies. The most obvious reason for this underconsumption is the non-routine chemical syntheses of such amino acid chimeras. Among substituted prolines, (2S)-3- or (2S)-4-substituted-pyrrolidine-2-carboxylic acid (3- and 4-prolino-L-amino acid), that is amino acid chimeras combining the pyrrolidine ring constraint with the side-chain of a natural amino acid are thus not so examplified. However as described above, substitution in the correct position of the pyrrolidine ring can confer appropriate conformational restrictions in the χ space of a particular amino acid to explore the topographic requirements in biologically active peptides (Table 1). With (2S)-3-substituted-pyrrolidine-2-carboxylic acid, only the $\chi1$ angle is strictly restricted while both $\chi1$ and $\chi2$ are constrained in (2S)-4-substituted–pyrrolidine-2-carboxylic acid.

Using these conformational probes, biologically active peptides the most analyzed to date are ligands for G-protein coupled receptors: angiotensin, bradykinin, cholecystokinin, melanocortin, opioid and, tachykinin (Table 1). In all cases, the use of these amino acid chimeras led to structural information about the receptor-bound conformation(s) (at the backbone level) of the peptides. But in some cases, introduction in these peptides of cis- or trans-n-substituted-pyrrolidine-2-carboxylic acid gave also information about the $\chi1$ and sometimes $\chi2$ requirement for high-binding affinity and/or selectivity and/or activity of the bioactive peptide. In the following paragraphs we only report the studies with biological active peptide analogues (Table 1) but not pseudopeptides, although interesting, such as recently described for inhibitors of hepatitis C virus protease.[249]

Table 1.

Substitution position	R =	Nomenclature of the proline-amino acid chimeras	Peptide
χ2 — R, χ1 position (2S)-3-substituted	-CH₃	(2S)-3-methyl-pyrrolidine-2-carboxylic acid : 3-prolino-L-norvaline	PLG [264]
	-CH₂-CH₂-CH₃	(2S)-3-propyl-pyrrolidine-2- carboxylic acid : 3-prolino-L-norleucine	CCK₄ [54,119]; CCK₇ [251,252]
	-CH(CH₃)₂	(2S)-3-isopropyl-pyrrolidine-2-carboxylic acid : 3-prolino-L-leucine	SP [262], PLG [264]
	-S-CH₃	(2S)-3-methylsulfanyl-pyrrolidine-2-carboxylic acid	CCK₄ [119]
	-S-CH₂-CH₃	(2S)-3-ethylsulfanyl-pyrrolidine-2-carboxylic acid	CCK₄ [119]
	-S-CH₂-CH₂-CH₃	(2S)-3-propylsulfanyl-pyrrolidine-2-carboxylic acid	CCK₄ [119]
	-CH₂-S-CH₃	(2S)-3-methylsulfanylmethyl-pyrrolidine-2-carboxylic acid : 3-prolino-L-methionine	SP [258]
	(phenyl)	(2S)-3-phenyl-pyrrolidine-2-carboxylic acid : 3-prolino-L-phenylalanine	SP[260];MTII[11];PLG[264]
	-H₂C (phenyl)	(2S)-3-benzyl-pyrrolidine-2-carboxylic acid : 3-prolino-L-homophenylalanine	SP [260]
	(phenyl)-OH	(2S)-3-(4'-hydroxy)phenyl-pyrrolidine-2-carboxylic acid : 3-prolino-L-tyrosine	JOM-13 [101]
R, χ2 — χ1 position (2S)-4-substituted	-SH	(2S)-4-mercapto-pyrrolidine-2-carboxylic acid : 4-prolino-L-homocysteine	ATII [242], BK [250]
	-S-CH₃	(2S)-4-methylsulfonyl-pyrrolidine-2-carboxylic acid : 4-prolino-L-methionine	SP [257], CCK₄ [119]
	-COOH	(2S)-4-carboxy-pyrrolidine-2-carboxylic acid : 4-prolino-L-γ-glutamic acid	glutathion [12]
	-(CH₂)₂-NH-C(NH₂)=NH	(2S)-4-(2'-guanidinoethyl)-pyrrolidine-2-carboxylic acid : 4-prolino-L-arginine	ANP [10]
	-O-(phenyl)	(2S)-4-oxybenzyl-pyrrolidine-2-carboxylic acid	SP [257]
	-S-(phenyl)	(2S)-4-sulfonylbenzyl-pyrrolidine-2-carboxylic acid	SP [257]

4.1. Angiotensin, bradykinin

In the angiotensin II (ATII: Asp-Arg-Val-Tyr-Val-His-Pro-Phe) series, it has been suggested that the receptor-bound conformation possesses a turn around residues 3-5, either a γ- or a β-turn. Cyclization by side chain disulfide formation between Cys or Hcy residues in position 3 and 5 led to potent agonists and antagonists. To further constrain the peptide backbone and define the receptor-bound conformation(s), *cis*- or *trans*-(2S)-4-mercapto-pyrrolidine-2-carboxylic acid have been incorporated in position 3 or 5 of ATII (Table 1).[243] This amino acid combines the pyrrolidine constraint with the *cis* or *trans* homocysteine side chain (P$^C_4$Hcy and P$^t_4$Hcy). Biological activity of these analogues has shown interesting segregation towards the two ATII receptor types AT1 (rat liver) and AT2 (rabbit uterus). Indeed, ATII had identical nanomolar affinity for AT1 and AT2. The cyclic analogue c[Sar1, Hcy3, P$^t_4$Hcy5]ATII also recognized the two receptors with the same selectivity and affinity. But, all other analogues had better selectivity for the AT2 receptor (the affinity ratio IC$_{50}$(AT1)/IC$_{50}$(AT2) ranged from 12 to 390). The two ATII receptor types better recognized the *trans*-amino acid chimera when incorporated either in position 3 or 5 of the cyclic analogue. Molecular modeling of these analogues together with Ramachandran plot of model tripeptides (of type c[Hcy-Ala-P$^t_4$Hcy]-NHCH$_3$ and Ac-Ala-Ala-Pro-NHCH$_3$) led the authors to refute all the conformations previously described as those bound to the receptor. Alternatively, they proposed that Tyr^4 in ATII assumes torsional values associated with a left-handed helix, or adopts an «open turn» (Φ_4=-130°, Ψ_4=65°). Finally they concluded that common backbone conformations for ATII are involved in receptor interaction, the three-dimensional arrangement of critical side chains being the key to molecular recognition of these ATII analogues. Further NMR analysis of these analogues led to well-defined conformations confirming an «open turn» in the backbone of the Tyr4-Val5 residues, as well as the *gauche*$^+$ orientation for the Tyr4 residue [250]. The same type of analogues have been synthesized for bradykinin (Arg-Pro-Pro-Gly-Phe-Ser-Pro-Phe-Arg) establishing a β-turn centered on residues 4 and 5 as the receptor-bound conformation [251].

4.2. Cholecystokinin

CCK heptapeptide and tetrapeptide analogues have also been studied with these conformational probes.[54,119] In the CCK$_7$ series, by comparison with proline, the use of (2S)-3-propyl-pyrrolidine-2-carboxylic acid has evidenced the crucial role of norleucine31 side chain in the interaction of (des-NH$_2$)Tyr(SO$_3^-$)27-Nle28-Gly29-Trp30-Nle31-Phe32-NH$_2$ with the CCK-B receptor.[252,253] It has also led to analogues with increased selectivity towards CCK-B versus CCK-A receptors. However a more complete study has been done with CCK$_4$. The short analogue CCK$_4$ (Trp30-Met31-Asp32-Phe33-NH$_2$) has better affinity for cortical CCK-B (IC$_{50}$=25 nM) than for pancreatic CCK-A (IC$_{50}$=1800 nM). Replacing Met31 by Nle led to an analogue with only two- to three-fold less affinity for the two receptors, showing that the sulfur atom is not essential for receptor interaction.[54] However the [Pro31]CCK$_4$ analogue lost almost all affinity for the two receptors. Thus, the amino acid (2S)-3-propyl-pyrrolidine-2-carboxylic acid (3-prolino-L-norleucine) has been incorporated in position 2 of CCK$_4$ in order to re-introduce the norleucine side chain on the pyrrolidine ring.[54] The analogue containing the *trans*-(2S)-3-propyl-pyrrolidine-2-carboxylic acid had even better affinity (IC$_{50}$=1.9 nM) for the CCK-B receptor than CCK$_4$ and similar affinity (IC$_{50}$=2700 nM) for the CCK-A receptor. The *cis* diastereoisomer analogue showed 25-fold less affinity for the CCK-A receptor than the *trans* diastereoisomer.[54] Further analysis had then been performed with various *cis*- and *trans*-, 3- and 4-alkylthio-pyrrolidine-2-carboxylic acid incorporated in position 2 of CCK$_4$.[119] Like in angiotensin, only one diastereoisomer is still well recognized among analogues containing the 3- or 4-prolino amino acid chimera, that is the *cis* in the 4-substituted series and the *trans* in the 3-substituted one. These analogues also better recognized the CCK-B than the CCK-A receptor. In the conformational model

261

derived from all these analogues, it was suggested that the backbone peptide bonds lie within the same plane with an amphiphilic structure, accommodating a Z-like bend model as previously described.[119,254-256] The hydrophobic side chains of the Trp, Met and Phe residue would locate at one surface and the hydrophilic side chain of the Asp residue and the C-terminal carboxamide, at the other surface. In this model, the hydrophobic side chains can cluster together to fill in a suggested hydrophobic pocket within the CCK-B receptor, as seen when increasing the chain length of the alkyl substituent. Methyl to propyl substitution on 3-mercaptoproline led to increased conformational flexibility and increased potency, even for the D-aminoacids-containing peptide analogues.[119]

4.3. Melanocortin

The χ space of residue 3 in a melanocortin (MTII) analogue (Ac-Nle-c[Asp-His-D-Phe-Arg-Trp-Lys]-NH$_2$ has been examined for steric effects and selectivity for hMC3, hMC4 and hMC5 receptors, using (2S)-3-substituted-pyrrolidine-2-carboxylic acid.[11] The constrained *trans* (2S)-3-phenyl, -(*p*-chlorophenyl), -(*p*-bromophenyl), -(*p*-trifluoromethylphenyl), -(*p*-methoxyphenyl)-pyrrolidine-2-carboxylic acid have been incorporated in place of His3 and the biological activity at human melanocortin receptors MCR3, MCR4 and MCR5 examined. Although only the *trans* diastereoisomers have been studied, results have shown that there is a binding decrease at the hMC3 and hMC4 receptors in the order of increasing steric bulk of the *p*-substituted-3-phenyl-pyrrolidine-2-carboxylic acid.[11] This was not the case for hMC5 receptor. Thus these chimeras might be useful to design hMC5 selective ligands. Altogether the peptide backbone recognized by the three melanocortin receptors should essentially be the same, the χ space being of crucial importance for optimal ligand-receptor interaction.

4.4. Opioid

The δ-selective cyclic analogue Tyr-c[D-Cys-Phe-D-Pen]OH (JOM-13) has been further constrained with L or D, *cis*- or *trans*-(2S)-3-(4'-hydroxyphenyl)-pyrrolidine-2-carboxylic acid, that mimicks a chimera between proline and tyrosine, introduced in position 1 of the peptide.[101] Binding data on the μ and δ receptors have shown that the *trans* diastereoisomer is recognized in a manner identical to JOM-13 by the two receptor types. Although 3-fold less recognized, the *cis* diastereoisomer had higher selectivity towards the δ-receptor. Very complete computational studies led the authors to propose a model for δ-receptor-bound conformation of cyclic JOM-13.[101] The side chain conformation of residue 1 in high- affinity peptides is *trans* with $\chi 1=180°$ and $\chi 2=90°$. Further conformational analysis led the authors to propose a model for the δ-opioid receptor pharmacophore.[257] All three important pharmacophoric elements of JOM-13, *i.e.*, the Tyr1 NH$_3^+$ group and the aromatic rings of the Tyr1 and Phe3 residues, are situated on the same side of the molecule and form an almost continuous surface.[257] The opposite surface of the molecule consists of hydrophilic main chain carbonyl groups, which probably point toward the water solution from the binding site. Finally, the charged N- and C-terminal NH$_3^+$ and COO$^-$ groups of JOM-13 are oriented in opposite directions.[257]

4.5. Tachykinins

The use of proline chimeras in the SP (Arg-Pro-Lys-Pro-Gln-Gln-Phe-Phe-Gly9-Leu10-Met11-NH$_2$) series was first described for 4-substituted-pyrrolidine-2-carboxylic acid incorporated in position 9 or 11 of SP(6-11) [258]. With analogues containing *cis* or *trans* (2S)-4-oxybenzyl- or -sulfonylbenzyl-pyrrolidine-2-

carboxylic acid in position 9 of SP(6-11), only the one having the *trans*-(2S)-4-oxybenzyl-pyrrolidine-2-carboxylic acid conserved high potency in contractile activity on isolated guinea-pig ileum. By contrast, replacing the Met11 residue by (2S)-4-methylsulfonyl-pyrrolidine-2-carboxylic acid or 4-prolino-L-methionine, led to an analogue totally inactive. However further studies with 3-substituted-pyrrolidine-2-carboxylic acid in SP analogues allowed the proposal that in analogues containing the 4-prolino-L-methionine, χ2 is not in the correct orientation for high-affinity interaction with each of the two NK-1 binding sites.[259,260] Introduction of *trans*-(2S)-3-phenyl or –3-benzyl-pyrrolidine-2-carboxylic acid in the antagonist peptide analogue [D-Pro,[9] Pro,[10] Trp[11]]SP led to higher antagonist potency (in the nanomolar range) with increased length of the bulky side-chain.[261] These results confirmed the crucial role of residue 10 side chain in antagonist ligand-receptor interaction, similarly to the agonist series.[262]

Although the receptor-bound conformation of SP had been well-established as a highly flexible N-terminal domain and a helical structure from residues 4 to 8, there were divergent views on the more or less extended C-terminal domain, being a β-strand, a poly γ-turn or a PPII. In order to determine the C-terminal structure of SP, *cis* and *trans* (2S)-3-isopropyl- and (2S)-3-methylsulfonylmethyl-pyrrolidine-2-carboxylic acid have been introduced in place of Leu[10] and Met[11] residues.[259,263] In position 10, only the *trans* isomer was recognized by the NK-1 receptor with high-affinity whereas in position 11 the two diastereisomers were identically bound by the receptor with an affinity similar to SP. NMR analysis and molecular mechanics calculation led to determine the *gauche+* (χ1) and *trans* (χ2) orientation of the Leu10 side chain, and the *trans* orientation (χ1) for the Met11 side chain. Although quite precise, these results did not allow to discriminate between a β-strand or a PPII for the C-terminal tripeptide receptor-bound conformation of SP.[263] Two or three proline-amino acid chimeras were then incorporated in the SP sequence to observe the effects of additional constraints on the biological activity of the resulting analogues. Biological results have shown that, although describing similar conformational spaces, no more than two constraints were tolerated by the NK-1 receptor and only in position 10 and 11.[264] One plausible explanation for the limited additivity of these constraints is that the peptide backbone conformation is not the only key element for high-affinity binding of SP analogues. Therefore when introducing one constraint which would not be the perfect one in terms of receptor-bound conformation, there might be an adaptation of the conformation of the peptide backbone to position correctly the other crucial recognition side chains, giving an analogue with preserved high-affinity for the receptor by comparison with SP. Even though we observed that these analogues containing two or three proline-amino acid chimeras shared globally the same conformational space in solution, the conformational space of the doubly substituted analogue (residues 10 and 11) recognized by the receptor could not accommodate the conformational space of another additional constraint (residue 9). These data suggest a stepwise mutual adaptation of both the ligand and the receptor. However the use of these multi-constrained analogues are in favour of an extended (PPII or β-sheet) organization for the C-terminal part of SP. Beside these well-illustrated studies on GPCR ligands, a few papers have also reported the use of these proline chimeras in other biologically active peptides.

4.6. Melanostatin-PLG

The tripeptide Pro-Leu-Gly-NH$_2$ (PLG) is a positive allosteric modulator of dopamine receptors in the central nervous system. It has been shown that the peptide analogues Pro-Leu-Pro-NH$_2$ and even Pro-Pro-Pro-NH$_2$ possess good dopamine receptor modulating activity both *in vitro* and *in vivo*.[265] The conformation of the tripeptide in the receptor has been determined as a type II β-turn.[266] Therefore to conserve the β-turn mimic but to possibly access the hydrophobic pocket with which leucine side chain was thought to interact, *cis* and *trans* (2*S*)-3-methyl-pyrrolidine-2-carboxylic acid had been incorporated in position 2, to lead to Pro-P₃Leu-Pro-NH₂ analogues. However the two isomer analogues were less active than PLG itself. One possible explanation proposed by the authors is that the methyl group projects off the pyrrolidine ring in each case in such a manner that instead of accessing the hydrophobic pocket it actually protrudes into an area of the binding site that results in adverse steric interactions.[265]

4.7. Glutathione

The γ-glutamyl moiety in glutathione (GSH: γ-Glu-Cys-Gly) is the main binding determinant for GSH-glutathione S-transferase. This amino acid has been replaced with *cis* and *trans* (2*S*)-4-carboxy-pyrrolidine-2-carboxylic acid (a chimera between proline and γ-glutamic acid).[12] Biological results have shown that the two isomers are completely resistant to degradation by γ-glutamyltranspeptidase. But only the *trans* isomer is substrate of the glutathione S-transferase. The multifunctional roles of glutathione in biological processes render this tripeptide an appealing target for structure-activity relationships studies. These constrained analogues of GSH might be exploited as enzyme inhibitors such as glutathione S-transferase that are overexpressed in a number of solid tumors and thus dimishing the effectiveness of cytotoxic drugs.

4.8. Atrial natriuretic peptide

The optimized smallest analogue of atrial natriuretic peptide (miniANP: Met-c[Cys-His-Phe-Gly-Gly-Arg-Met-Asp-Arg-Ile-Ser-Cys]-Tyr-Arg-NH₂) had been studied by proline-scanning for binding to the natriuretic peptide transmembrane receptor A.[267] The receptor-bound conformation resulting from this scan has also been described as a turnlike conformation at residues 6-9. To further explore the importance of the arginine side chain, *cis* and *trans* (2*S*)-4-(2'-guanidinoethyl)-pyrrrolidine-2-carboxylic acid have been incorporated in position 7 of miniANP.[10] Biological analysis of these analogues have shown that the two isomers recovered the same potency as miniANP when compared with [Pro7]miniANP analogue.

5. Conclusions

In conclusion, such proline-amino acid chimeras proved to be very useful for confirming previously predicted backbone conformation of a biologically active peptide but also to analyse the χ space or orientation of amino acid side chains that are key elements in the interaction of the peptide with its protein partner. Although the reported results with singly substituted analogues have been quite clear to interprete, introduction of multiple constraints may lead to non-additivity, as observed in the case of SP analogues. As previously analyzed in thermodynamics terms,[268] one should not forget that the affinity constant of a peptide for a protein does not reflect independent binding of the peptide amino acids to protein subsites. Indeed with such oversimplified view, modification of one particular amino acid should affect the binding of the ligand with one subsite and only one, which is of course not the case. Cooperativity between the peptide amino acids but also between the binding subsites in the protein occurs. A chemical modification in the ligand will influence not only the interaction of the ligand with the protein but also the cooperativity within the ligand to bind to this latter. Therefore when measuring the macromolecular dissociation constant (Ki), what is really measured is a very complex sum of several phenomena such as i) the cooperative conformational changes within the protein upon binding of the ligand; ii) the deviations of the ligand from pure competitive behavior; iii) the cooperative intra-ligand interactions within the peptide in solution and in the peptide-protein complex etc. Therefore when introducing multiple constraints, interpretation could become difficult. Finally, even though one could thus say that «more is not always better», the results obtained with the introduction of multiple constraints in a peptide ligand are actually not predictable and deserve at least to be assayed.

Acknowledgments

Philippe Karoyan dedicates this paper to Gérard Chassaing on the occasion of his 60*th* birthday.

List of abbreviations
AcOH: acetic acid
AIBN: 2,2´-azo-bis-isobutyronitrile
Alloc: allyloxycarbonyl
ANP: atrial natriuretic peptide
ATII: angiotensin II
BK: bradykinin
Bn: benzyl
Boc: *tert*-butyloxycarbonyl
Bpy: 2,2'-bipyridine
BuLi: butyl lithium
CCK: cholecystokinin
CD: circular dichroism
COD: cyclooctadienyl
CSA: camphor sulfonic acid
Dba: dibenzylideneacetone
DBU: 1,8-diazabicyclo[5.4.0]undec-7-ene
DCM: dichloromethane
de: diastereomeric excess
DEAD: diethylazodicarboxylate
DIAD: diisopropyl azodicarboxylate
DIBAL: Diisobutylaluminium hydride
DIC: *N,N'*-diisopropylcarbodiimide
DMAP: N,N-dimethylaminopyridine
DMF: dimethylformamide
DMPU: dimethylpropylene urea
DMSO: dimethylsulfoxide
DOSP: *N*-dodecylbenzenesulfonyl)prolinate
ee: enantiomeric excess
EWG: electron withdrawing group
Flp: fluoroproline

265

Fmoc: 9-fluorenylmethoxycarbonyl

FT-IR: Fourrier transform-infrared

g: gas

GPCR: G protein-coupled receptor

GSH: glutathione

Hyp: hydroxyproline

IRA: ion-exchange resin amberlit

KHMDS: potassium hexamethyldisilazanide

LAH: lithium aluminium hydride

LiHMDS: lithium hexamethyldisilazanide

LDA: lithium diisopropylamide

Ms: mesylate

MSCl: methane sulfonyl chloride

MTII: melanocortin

NBS: N-bromo-succinimide

NIS: N-Iodo-succinimide

NMO: N-methylmorpholine N-oxide

NMR: nuclear magnetic resonance

NS: nucleophilic substitution

Nu: nucleophile

PDB: protein database

PDC: pyridinium dichromate

PET: photonelectron transfer

PLE: Pig Liver Esterase

p-Ts: para-toluene sulfonyl

PII or PPII : polyproline II

PPTS: pyridinium p-toluenesulfonate

PTSA: para-toluene sulfonic acid

rt: room temperature

SAR: structure-activity relationship

SP: substance P

TBAF: tetrabutyl ammonium fluoride

TBDMS: *tert*-butyldimethylsilyl

TBDPS: *tert*-butyldipropylsilyl

TEMPO: 2,2,6,6-tetramethylpiperidinyloxy

TEOC: trimethylsilylethyl chloroformate

Tf: triflate

TEA: triethylamine

TFA: trifluoroacetic acid

THF: tetrahydrofurane

TMEDA: N,N,N',N'-tetramethylethylenediamine

266

TMS: trimethylsilyl

TPAP-NMO: tetra-n-propylammonium permthenate

Ts: tosyl

Z: benzyloxycarbonyl

References

1. du Vigneaud, V.; Winestock, G.; Murti, V. V. S.; Hope, D. B.; Kimbrough, R. D. *J. Biol. Chem.* **1960**, *235*, 64.
2. Hope, D. B.; Murti, V. V. S.; du Vigneaud, V. *J. Biol. Chem.* **1962**, *237*, 1563.
3. Seebach, D.; Kalinowski, H.-O. *Nachr. Chem. Tech.* **1976**, *24*, 415.
4. Goodson, L. H.; Honigberg, I. L.; Lehman, J. J.; Burton, W. H. *J. Org. Chem.* **1960**, *25*, 1920.
5. Holladay, M. W.; Nadzan, A. M. *J. Org. Chem.* **1991**, *56*, 3900.
6. Giannis, A.; Kolter, T. *Angew. Chem., Int. Ed. Engl* **1993**, 1244.
7. Hruby, V. J.; Balse, P. M. *Curr. Med. Chem.* **2000**, *7*, 945.
8. Marshall, G. R. *Biopolymers* **2003**, *60*, 246.
9. Schafmeister, C. E.; Po, J.; Verdine, G. L. *J. Am. Chem. Soc.* **2000**, *122*, 5891.
10. Sugase, K.; Horikawa, M.; Sugiyama, M.; Ishiguro, M. *J. Med. Chem.* **2004**, *47*, 489.
11. Cai, M.; Cai, C.; Mayorov, A. V.; Xiong, C.; Cabello, C. M.; Soloshonok, V. A.; Swift, J. R.; Trivedi, D.; Hruby, V. J. *J. Pep. Res.* **2004**, *63*, 116.
12. Paradisi, M. P.; Mollica, A.; Cacciatore, I.; Di Stefano, A.; Pinnen, F.; Caccuri, A. M.; Ricci, G.; Dupre, S.; Spirito, A.; Lucente, G. *Bioorg. Med. Chem.* **2003**, *11*, 1677.
13. Quancard, J.; Karoyan, P.; Lequin, O.; Wenger, E.; Aubry, A.; Lavielle, S.; Chassaing, G. *Tetrahedron Lett.* **2004**, *45*, 623.
14. Chalmers, D. K.; Marshall, G. R. *J. Am. Chem. Soc.* **1995**, *117*, 5927.
15. Gibbs, A. C.; Bjorndahl, T. C.; Hodges, R. S.; Wishart, D. S. *J. Am. Chem. Soc.* **2002**, *124*, 1203.
16. Williamson, M. P. *Biochemistry* **1994**, *297*, 249.
17. Kay, B. K.; Williamson, M. P.; Sudol, M. *FASEB J.* **2000**, *14*, 231.
18. Chang, E.; Roberts, D.; Fillon, Y.; Chmielewski, J. *Biopolymers* **2003**, *71*, 417.
19. Mauger, A. B.; Wiktop, B. *Chem. Rev.* **1966**, *66*, 1966.
20. Seebach, D.; Boes, M.; Naef, R.; Schweizer, B. *J. Org. Chem.* **1983**, *105*, 5390.
21. Lewis, A.; Wilkie, J.; Rutherford, T. J.; Gani, D. *J. Chem. Soc., Perkin Trans. 1* **1998**, 3777.
22. Wang, H.; Germanas, J. P. *Synlett* **1999**, *1*, 33.
23. Matsumura, Y.; Kinoshita, T.; Yanagihara, Y.; Kanemoto, N.; Watanabe, M. *Tetrahedron Lett.* **1996**, *37*, 8395.
24. Arboré, A. P. A.; Cane-Honeysett, D. J.; Coldham, I.; Middleton, M. L. *Synlett* **2000**, *2*, 236.
25. Confalone, P. N.; Huie, E. M.; Ko, S. S.; Cole, G. M. *J. Org. Chem.* **1988**, *53*, 482.
26. Witter, D. J.; Famiglietti, S. J.; Cambier, J. C.; Castelhano, A. L. *Bioorg. Med. Chem. Lett.* **1998**, *8*, 3137.
27. Ward, P.; Ewan, G. B.; Jordan, C. C.; Ireland, S. J.; Hagan, R. M.; Brown, J. R. *J. Med. Chem.* **1990**, *33*, 1851.
28. Chan, C.-O.; Cooksey, C. J.; Crich, D. *J. Chem. Soc., Perkin Trans. 1* **1992**, 777.
29. Pfeifer, M.; Linden, A.; Robinson, A. A. *Helv. Chim. Acta* **1997**, *80*, 1513.
30. Cox, C.; Lectka, T. *J. Am. Chem. Soc.* **1998**, *120*, 10660.
31. Culbertson, T. P.; Sanchez, J. P.; Gambino, L.; Sesnie, J. A. *J. Med. Chem.* **1990**, *33*, 2270.
32. Sato, T.; Mori, T.; Sugiyama, M.; Ishibashi, H.; Ikeda, M. *Heterocycles* **1994**, *37*, 245.
33. Papillon, J. P. N.; Taylor, R. J. K. *Org. Lett.* **2000**, *2*, 1987.
34. Sato, T.; Kawasaki, S.; Oda, N.; Yagi, S.; El Bialy, S. A. A.; Uenishi, J.; Yamauchi, M.; Ikeda, M. *J. Chem. Soc., Perkin Trans. 1* **2001**, 2623.
35. Gallina, C.; Petrini, F.; Romeo, A. *J. Org. Chem.* **1970**, *35*, 2425.
36. Bajgrowicz, J.; El Achquar, A.; Roumestant, M.-L.; Pigière, C. *Heterocycles* **1986**, *24*, 2165.
37. Scott, W. L.; O'Donnell, M. J.; Delgado, F.; Alsina, J. *J. Org. Chem.* **2002**, *67*, 2960.
38. Schöllkopf, U.; Hinrichs, R.; Lonsky, R. *Angew. Chem., Int. Ed. Engl.* **1987**, *26*, 143.

39. Chinchilla, R.; Galindo, N.; Nàjera, C. *Tetrahedron Asymm.* **1998**, *9*, 2769.
40. Smissman, E. E.; Chien, P. L.; Robinson, R. A. *J. Org. Chem.* **1970**, *35*, 3818.
41. Ellington, J. J.; Honigberg, I. L. *J. Org. Chem.* **1974**, *39*, 104.
42. Shono, T.; Matsumura, Y.; Katoh, S.; Ohshita, J. *Chem. Lett.* **1988**, 1065.
43. Betsbrugge, J. V.; Tourwé, D.; Kaptein, B.; Kierkels, H.; Broxterman, R. *Tetrahedron* **1997**, *53*, 9233.
44. Kawabata, T.; Kawakami, S.; Majumdar, S. *J. Am. Chem. Soc.* **2003**, *125*, 13012.
45. Duhamel, P.; Kotera, M. *J. Org. Chem.* **1982**, *47*, 1688.
46. Donati, D.; Fusi, S.; Macripò, A.; Ponticelli, F. *J. Heterocyclic Chem.* **1987**, *24*, 481.
47. Trancard, D.; Tout, J.-B.; Giard, T.; Chichaoui, I.; Cahard, D.; Plaquevent, J.-C. *Tetrahedron Lett.* **2000**, 3843.
48. Shatzmiller, S.; Dolithzky, B.-Z.; Bahar, E. *Liebigs Ann. Chem* **1991**, 375.
49. Häusler, J.; Schmidt, U. *Liebigs Ann. Chem.* **1979**, 1881.
50. Häusler, J. *Liebigs Ann. Chem.* **1981**, 1073.
51. Moss, W. O.; Jones, A. C.; Wisedale, R.; Mahon, M. F.; Molloy, K. C.; Bradbury, R. H.; Hales, N. J.; Gallagher, T. *J. Chem. Soc., Perkin Trans. 1* **1992**, 2615.
52. Demange, L.; Cluseau, J.; Menez, A.; Dugave, C. *Tetrahedron Lett.* **2001**, *42*, 651.
53. Kamenecka, T. M.; Park, Y.-J.; Lin, L. S.; Lanza, T. J.; Hagmann, W. K. *Tetrahedron Lett.* **2001**, *42*, 8571.
54. Holladay, M. W.; Lin, C. W.; May, C. S.; Garvey, D. S.; Witte, D. G.; Miller, T. R.; Wolfram, C. A. W.; Nadzan, A. M. *J. Med. Chem.* **1991**, *34*, 455.
55. Sharma, R.; Lubell, W. D. *J. Org. Chem.* **1996**, *61*, 202.
56. Blanco, M.-J.; Paleo, M. R.; Penide, C.; Sardina, F. J. *J. Org. Chem.* **1999**, *64*, 8786.
57. Mamai, A.; Hughes, N. E.; Wurthmann, A.; Madalengoitia, J. S. *J. Org. Chem.* **2001**, *66*, 6483.
58. Yoo, S.-E.; Lee, S.-H.; Kim, N.-J. *Tetrahedron Lett.* **1988**, *29*, 2195.
59. Pellegrini, N.; Schmitt, M.; Guery, S.; Bourgignon, J.-J. *Tetrahedron Lett.* **2002**, *43*, 3243.
60. Pellegrini, N.; Schmitt, M.; Bourgignon, J.-J. *Tetrahedron Lett.* **2003**, *44*, 6779.
61. Karoyan, P.; Chassaing, G. *Tetrahedron Lett.* **1997**, *38*, 85.
62. Lorthiois, E.; Marek, I.; Normant, J.-F. *Tetrahedron Lett.* **1997**, *38*, 89.
63. Karoyan, P.; Chassaing, G. *Tetrahedron Asymm.* **1997**, *8*, 2025.
64. Karoyan, P.; Chassaing, G. In *peptides*; Martinez, J., Fehrentz, J.-A., Eds.; EDK: 2000, p 373.
65. Karoyan, P.; Chassaing, G. *Tetrahedron Lett.* **2002**, *43*, 253.
66. Karoyan, P.; Chassaing, G. *Tetrahedron Lett.* **2002**, *43*, 1221.
67. Karoyan, P.; Quancard, J.; Vaissermann, J.; Chassaing, G. *J. Org. Chem.* **2003**, *68*, 2256.
68. Karoyan, P.; Triolo, A.; Nannicini, R.; Giannotti, D.; Altamura, M.; Chassaing, G.; Perrotta, E. *Tetrahedron Lett.* **1999**, *40*, 71.
69. Quancard, J.; labonne, A.; Jacquot, Y.; Lavielle, S.; Chassaing, G.; Karoyan, P. *J. Org. Chem.* **2004**, *69*, 7940.
70. Genzyme *Eichenweg 1 Ch - 4410 Liestal Switzerland, Tel: +41-(0)61-906-5959 Fax: +41-(0)61-906-5958.*
71. Hiemstra, H.; Speckamp, W. N. *Comprehensive Organic Synthesis* **1991**, *2*.
72. Mooiver, H. H.; Hiemstra, H.; Fortgens, H. P.; Speckamp, W. N. *Tetrahedron Lett.* **1987**, *28*, 3285.
73. Bachi, M. D.; Frolow, F.; Hoornaert, C. *J. Org. Chem.* **1983**, *48*, 1841.
74. Esch, P. M.; Hiemstra, H.; Speckamp, W. N. *Tetrahedron Lett.* **1990**, *31*, 759.
75. Udding, J. H.; Tuijp, J. M.; Hiemstra, H.; Speckamp, W. N. *Bull. Soc. Chim. Belg.* **1994**, *103*, 329.
76. Udding, J. H.; Tuijp, J. M.; Hiemstra, H.; Speckamp, W. N. *Tetrahedron Lett.* **1994**, *35*, 1907.
77. Cotton, R.; Johnstone, A. N. C.; North, M. *Tetrahedron Lett.* **1995**, *51*, 8525.
78. Baldwin, J. E.; Moloney, M. G.; North, M. *J. Chem. Soc., Perkin Trans. 1* **1989**, 833.
79. Flamant-Robin, C.; Wang, Q.; Chiaroni, A.; Sasaki, A. *Tetrahedron* **2002**, *58*, 10475.
80. Soloshonok, V. A.; Ueki, H.; Tiwari, R.; Cai, C.; Hruby, V. J. *J. Org. Chem.* **2004**, *69*, 4984.
81. Huisgen, R. *Angew. Chem.* **1962**, *13*, 604.
82. Huisgen, R.; Scheer, W.; Huber, H. *J. Am. Chem. Soc.* **1967**, *89*, 1753.
83. Huisgen, R.; Scheer, W.; Mader, H. *Angew. Chem. Int. Ed.* **1969**, *8*, 602.
84. Hall, J. H.; Huisgen, R.; Roos, C. H.; Scheer, W. *Angew. Chem. Int. Ed.* **1969**, *8*, 1188.

85. Grigg, R.; Kemp, J. *Tetrahedron Lett.* **1978**, *38*, 2823.
86. Grigg, R.; Kemp, J.; Scheldrick, G.; Trotter, J. *J. Chem. Soc., Chem. Comm.* **1978**, 109.
87. Joucla, M.; Hamelin, J. *Tetrahedron Lett.* **1978**, *38*, 2885.
88. Benhaoua, H.; Texier, F.; Guerot, P.; Martelli, J.; Carrie, R. *Tetrahedron Lett.* **1978**, *34*, 1153.
89. Gelas-Mialhe, Y.; Hierle, R.; Vessiere, R. *Bull. Soc. Chim. Fr.* **1974**, 709.
90. Attia, M.; Gelas-Mialhe, Y.; Vessiere, R. *Chem. Lett.* **1979**, 1095.
91. Imai, N.; Terao, Y.; Achiwa, K.; Sekiya, M. *Tetrahedron Lett.* **1984**, *25*, 1579.
92. Padwa, A.; Chen, Y. Y.; Chiacchio, U.; Deny, W. *Tetrahedron* **1985**, *41*, 3529.
93. Garner, P.; Ho, W. B.; Shin, H. W. *J. Am. Chem. Soc.* **1992**, *114*, 2767.
94. Garner, P.; Ozdemir, D. *J. Org. Chem.* **1994**, *59*, 4.
95. Sisko, J.; Weinreb, S. M. *J. Org. Chem.* **1991**, *56*, 3210.
96. Tanako, S.; Iwabucchi, Y.; Ogasawara, K. *J. Am. Chem. Soc.* **1987**, *109*, 5523.
97. Henke, B. R.; Karklis, A.; Heathcock, C. H. *J. Org. Chem.* **1992**, *57*, 7056.
98. DeSong, P.; Kell, D. A.; Sidler, D. R. *J. Org. Chem.* **1985**, *50*, 2309.
99. Tsuge, O.; Kanemasa, S.; Ohe, M.; Yorozu, K.; Takenaka, S.; Ueno, K. *Bull. Chem. Soc. Jpn.* **1987**, *60*, 4067.
100. Cox, D. A.; Johnson, A. W.; Mauger, A. B. *J. Chem. Soc.* **1964**, 5024.
101. Mosberg, H. I.; Omnaas, J. R.; Lomize, A.; Heyl, D. L.; Nordan, I.; Mousigian, C.; Davis, P.; Porreca, F. *J. Med. Chem.* **1994**, *37*, 4371.
102. Chung, J. Y. L.; Wasicak, J. T.; Arnold, W. A.; May, C. S.; Nadzan, A. M.; Holladay, M. W. *J. Org. Chem.* **1990**, *55*, 270.
103. Damour, D.; Doerflinger, G.; Pantel, G.; Labaudinière, R.; Leconte, J. P.; Sablé, S.; Vuilhorgne, M.; Mignani, S. *Synlett* **1999**, 189.
104. Sasaki, A.; Hashimoto, C.; Chiaroni, A.; Riche, C.; Potier, P. *Tetrahedron Lett.* **1987**, *28*, 6069.
105. Sasaki, A.; Dockner, M.; Chiaroni, A.; Riche, C.; Potier, P. *J. Org. Chem.* **1997**, *62*, 765.
106. Sasaki, A.; Pauly, R.; Fontaine, C.; Chiaroni, A.; Riche, C.; Potier, P. *Tetrahedron Lett.* **1994**, *35*, 241.
107. Verbiscar, A. J.; Witkop, B. *J. Org. Chem.* **1970**, *35*, 1924.
108. Eswarakrishnan, V.; Field, L. *J. Org. Chem.* **1981**, *46*, 4182.
109. Abraham, D. J.; Mokotoff, M.; Sheh, L.; Simmons, J. E. *J. Med. Chem.* **1983**, *26*, 549.
110. Rosen, T.; Fesik, S. W.; Chu, D. T. W.; Pernet, A. G. *Synthesis* **1988**, 40.
111. Krapsho, J.; Turk, C.; Cushman, D. W.; Powell, J. R.; DeForrest, J. M.; Spitzmiller, E. R.; Karanewsky, D. S.; Duggan, M.; Rovnvak, G.; Schwartz, J.; Natarajan, S.; Godfrey, J. D.; Ryono, D. E.; Neubeck, R.; Atwal, K. S.; Petrillo Jr., E. W. *J. Med. Chem.* **1988**, *31*, 1148.
112. Smith, E. M.; Swiss, G. F.; Neustadt, B. R.; Gold, E. H.; Sommer, J. A.; Brown, A. D.; Chiu, P. J. S.; Moran, R.; Sybertz, E. J.; Baum, T. *J. Med. Chem.* **1988**, *31*, 875.
113. Bridges, R. J.; Stanley, M. S.; Anderson, M. W.; Cotman, C. W.; Chamberlin, A. R. *J. Med. Chem.* **1991**, *34*, 717.
114. Webb, T. R.; Eigenbrot, C. *J. Org. Chem.* **1991**, *56*, 3009.
115. Curran, T. P.; Chandler, N. M.; Kennedy, R. J.; Keaney, M. T. *Tetrahedron Lett.* **1996**, *37*, 1933.
116. Yanagi, T.; Kitajima, A.; Anzai, K.; Kodama, K.; Mizoguchi, J. I.; Fujiwara, H.; Sakiyama, H.; Kamoda, O.; Kamei, C. *Chem. Pharm. Bull.* **1999**, *47*, 1650.
117. Tamaki, M.; Han, G.; Hruby, V. J. *J. Org. Chem.* **2001**, *66*, 1038.
118. Renau, T. E.; Léger, R.; Filonova, L.; Flamme, E. M.; Wang, M.; Yen, R.; Madsen, D.; Griffith, D.; Chamberland, S.; Dudley, M. N.; Lee, V. J.; Lomovskaya, O.; Watkins, W. J.; Ohta, T.; Nakayama, K.; Ishda, Y. *Bioorg. Med. Chem. Lett.* **2003**, *13*, 2755.
119. Kolodziej, S. A.; Nikiforovich, G. V.; Skeean, R.; Lignon, M. F.; Martinez, J.; Marshall, G. R. *J. Med. Chem.* **1995**, *38*, 137.
120. Gomez-Vidal, J. A.; Silverman, R. B. *Org. Lett.* **2001**, *3*, 2481.
121. Patchett, A. A.; Wiktop, B. *J. Am. Chem. Soc.* **1957**, *79*, 185.
122. Doi, M.; Nishi, Y.; Kiritoshi, N.; Iwata, T.; Nago, M.; Nakano, H.; Uchiyama, S.; Nakazawa, T.; Wakamiya, T.; Kobayashi, Y. *Tetrahedron* **2002**, *58*, 8453.
123. Tamaki, M.; Han, G.; Hruby, V. J. *J. Org. Chem.* **2001**, *66*, 3593.

124. O'Connell, C. E.; Rowell, C. A.; Ackermann, K.; Garcia, A. M.; Lewis, M. D.; Kowalczyk, J. J. *Chem. Pharm. Bull.* **2000**, *48*, 740.
125. Margerlein, B. J.; Birkenmeyer, R. D.; Herr, R. R.; Kagan, F. *J. Am. Chem. Soc.* **1967**, *89*, 2459.
126. Kemp, D. S.; Carter, J. *J. Org. Chem.* **1989**, *54*, 109.
127. Del Valle, J. R.; Goodman, M. *Angew. Chem. Int. Ed.* **2002**, *41*, 1600.
128. Baldwin, J. E.; Miranda, T.; Moloney, M. G. *Tetrahedron* **1989**, *45*, 7459.
129. Ezquerra, J.; Pedregal, C.; Rubio, A.; Yruretagoyena, B.; Escribano, A.; Sanchez-Ferrando, F. *Tetrahedron* **1993**, *49*, 8665.
130. Pedregal, C.; Ezquerra, J.; Escribano, A.; Carreno, C.; Ruano, J. L. G. *Tetrahedron Lett.* **1994**, *35*, 2053.
131. Langlois, N.; Rojas, A. *Tetrahedron Lett.* **1993**, *34*, 2477.
132. Llinas-Brunet, M.; Bailey, M.; Fazal, G.; Ghiro, E.; Gorys, V.; Goulet, S.; Halmos, T.; Maurice, R.; Poirier, M.; Poupart, M. A.; Rancourt, J.; Thibeault, D.; Wernic, D.; Lamarre, D. *Bioorg. Med. Chem. Lett.* **2000**, *10*, 2267.
133. Chen, X.; Du, D. M.; Hua, W. T. *Tetrahedron Asymm.* **2002**, *13*, 43.
134. Ezquerra, J.; Escribano, A.; Rubio, A.; Remuinan, M. J.; Vaquero, J. J. *Tetrahedron Lett.* **1995**, *36*, 6149.
135. Ezquerra, J.; Pedregal, C.; Yruretagoyena, B.; Rubio, A. *J. Org. Chem.* **1995**, *60*, 2925.
136. Danishefsky, S.; Berman, E.; Clizbe, L. A.; Hirama, M. *J. Am. Chem. Soc.* **1979**, *101*, 4385.
137. Moody, C. M.; Young, D. W. *Tetrahedron Lett.* **1993**, *34*, 4667.
138. Moody, C. M.; Young, D. W. *Tetrahedron Lett.* **1994**, *35*, 7277.
139. Moody, C. M.; Young, D. W. *J. Chem. Soc., Perkin Trans. 1* **1997**, 3519.
140. Thottathil, J. K.; Moniot, J. L.; Mueller, R. H.; Wong, M. K. Y.; Kissick, T. P. *J. Org. Chem.* **1986**, *51*, 3140.
141. Zhang, R.; Brownewell, F.; Madalengoitia, J. S. *J. Am. Chem. Soc.* **1998**, *120*, 3894.
142. Zhang, R.; Brownewell, F.; Madalengoitia, J. S. *Tetrahedron Lett.* **1999**, *40*, 2707.
143. Agami, C.; Couty, F.; Lin, J.; Mikaeloff, A.; Poursoulis, M. *Tetrahedron* **1993**, *49*, 7239.
144. Baldwin, J. E.; Li, C. S. *J. Chem. Soc., Chem. Commun.* **1987**, 166.
145. Bazak, A.; Bag, S. S.; Rudra, K. R.; Barman, J.; Dutta, S. *Chem. Lett.* **2002**, 710.
146. Soucy, F.; Wernik, D.; Beaulieu, P. *J. Chem. Soc., Perkin Trans. 1* **1991**, 2885.
147. Jonas, M.; S. B.; Steckhan, E. *J. Org. Chem.* **2001**, *66*, 6896.
148. Wang, Q.; Sasaki, A.; Potier, P. *Tetrahedron Lett.* **1998**, *39*, 5755.
149. Wang, Q.; Sasaki, A.; Potier, P. *Tetrahedron* **1998**, *54*, 15759.
150. Koskinen, A. M. P.; Rapoport, H. *J. Org. Chem.* **1989**, *54*, 1859.
151. Zhang, J.; Wang, W.; Xiong, C.; Hruby, V. *J. Tetrahedron Lett.* **2003**, *44*, 1413.
152. Nevalainen, M.; Kauppinen, P. M.; Koskinen, A. M. P. *J. Org. Chem.* **2001**, *66*, 2061.
153. Jones, R. C. F.; Howard, K. J.; Snaith, J. S. *Tetrahedron Lett.* **1996**, *37*, 1707.
154. Pellicciari, R.; Arenare, L.; De Caprariis, P. D.; Natalini, B.; Marinozzi, M.; Galli, A. *J. Chem. Soc., Perkin Trans. 1* **1995**, 1251.
155. Della, E. W.; Kendall, M. *J. Chem. Soc., Perkin Trans. 1* **1973**, 2729.
156. Petersen, J. S.; Fels, G.; Rapoport, H. *J. Am. Chem. Soc.* **1984**, *106*, 4539.
157. Shiosaki, K.; Rapoport, H. *J. Org. Chem.* **1985**, *50*, 1229.
158. Koskinen, A. M. P.; Rapoport, H. *J. Med. Chem.* **1985**, *28*, 1301.
159. Koskinen, A. M. P.; Ghiaci, M. *Tetrahedron Lett.* **1990**, *31*, 3209.
160. Lin, R.; Castells, J.; Rapoport, H. *J. Org. Chem.* **1998**, *63*, 4069.
161. Li, R.; Stapon, A.; Blanchfield, J. T.; Townsend, C. A. *J. Am. Chem. Soc.* **2000**, *122*, 9296.
162. Jain, S.; Sujatha, K.; Rama Krishna, K. V.; Roy, R.; Singh, J.; Anand, N. *Tetrahedron* **1992**, *48*, 4985.
163. Honda, T.; Kimura, M. *Org. Lett.* **2000**, *2*, 3925.
164. Shono, T.; Hamaguchi, H.; Matsumura, Y. *J. Am. Chem. Soc.* **1975**, *97*, 4264.
165. Shono, T.; Matsumura, Y.; Tsubata, K.; Sugihara, Y.; Yamane, S.; Kanazawa, T.; Aoki, T. *J. Am. Chem. Soc.* **1982**, *104*, 6697.
166. Shono, T.; Matsumura, Y.; Tsubata, K.; Uchida, K.; Kanazawa, T.; Tsuda, K. *J. Org. Chem.* **1984**, *49*, 3711.

167. Shono, T.; Matsumura, Y.; Tsubata, K.; Uchida, K. *J. Org. Chem.* **1986**, *51*, 2590.
168. Malmberg, M.; Nyberg, K. *Acta Chem. Scand.* **1984**, *B38*, 85.
169. Asada, S.; Kato, M.; Asai, K.; Ineyama, T.; Nishi, S.; Izawa, K.; Shono, T. *J. Chem. Soc., Chem. Commun.* **1989**, 486.
170. Manfré, F.; Kern, J. M.; Biellmann, J. F. *J. Org. Chem.* **1992**, *57*, 2060.
171. Manfré, F.; Pulicani, J.-P. *Tetrahedron Asymm.* **1994**, *5*, 235.
172. Skrinjar, M.; Wistrand, L.-G. *Tetrahedron Lett.* **1990**, *31*, 1775.
173. Thaning, M.; Wistrand, L.-G. *Acta Chem. Scand.* **1992**, *46*, 194.
174. Célimène, C.; Dhimane, H.; Le Bail, M.; Lhommet, G. *Tetrahedron Lett.* **1994**, *35*, 6105.
175. Wistrand, L.-G.; Skrinjar, M. *Tetrahedron* **1991**, *47*, 573.
176. McClure, K. F.; Renold, P.; Kemp, D. S. *J. Org. Chem.* **1995**, *60*, 454.
177. Beal, L. M.; Moeller, K. D. *Tetrahedron Lett.* **1998**, *39*, 4639.
178. Ma, D.; Yang, J. *J. Am. Chem. Soc.* **2001**, *123*, 9706.
179. Wallén, E. A. A.; Christiaans, J. A. M.; Gynther, J.; Vepsäläinen *Tetrahedron Lett.* **2003**, *44*, 2081.
180. Barrett, A. G. M.; Pilipauskas, D. *J. Org. Chem.* **1991**, *56*, 2787.
181. Magnus, P.; Hulme, C. *Tetrahedron Lett.* **1994**, *35*, 8097.
182. Corey, E. J.; Yuen, P.-W.; Hannon, F.; Wierda, D. A. *J. Org. Chem.* **1990**, *55*, 784.
183. Kim, S.; Hyashi, K.; Kitano, Y.; Tada, M.; Chiba, K. *Org. Lett.* **2002**, *4*, 3735.
184. Gloannec, P.; Hervé, Y.; Brémand, N.; Lecouvé, J.-P.; Bréard, F.; De Nanteuil, G. *Tetrahedron Lett.* **2002**, *43*, 3499.
185. Ohta, T.; Shiokawa, S.; Iwashita, E.; Sato, N.; Sakurai, K.; Ineyama, T.; Izawa, H.; Izawa, K.; Nozoe, S. *Heterocycles* **1992**, *33*, 143.
186. Langlois, N.; Rojas, A. *Tetrahedron* **1993**, *49*, 77.
187. Clive, D. L.; Yeh, V. S. C. *Tetrahedron Lett.* **1998**, *39*, 4789.
188. Collado, I.; Ezquerra, J.; Pedregal, C. *J. Org. Chem.* **1995**, *60*, 5011.
189. Oba, M.; Terauchi, T.; Hashimoto, J.; Tanaka, T.; Nishiyama, K. *Tetrahedron Lett.* **1997**, *38*, 5515.
190. Tong, Y.; Fobian, Y. M.; Wu, M.; Boyd, N. D.; Moeller, K. D. *J. Org. Chem.* **2000**, *65*, 2484.
191. Hanessian, S.; Sailes, H.; Munro, A.; Therrien, E. *J. Org. Chem.* **2003**, *68*, 7219.
192. Davies, H. M. L.; Venkataramani, C. *Org. Lett.* **2001**, *3*, 1773.
193. Davies, H. M. L.; Venkataramani, C.; Hansen, T.; Hopper, D. W. *J. Am. Chem. Soc.* **2003**, *125*, 6462.
194. Davis, F. A.; Fang, T.; Goswami, R. *Org. Lett.* **2002**, *4*, 1599.
195. Severino, E. A.; Costerano, E. R.; Garcia, A. L. L.; Correia, C. R. D. *Org. Lett.* **2003**, *5*, 305.
196. Ohta, T.; Hosoi, A.; Kimura, T.; Nozoe, S. *Chemistry Letters* **1987**, 2091.
197. Collado, I.; Ezquerra, J.; Vaquero, J. J.; Pedregal, C. *Tetrahedron Lett.* **1994**, *35*, 8037.
198. Fournie-Zaluski, M. C.; Coric, P.; Thery, V.; Gonzales, W.; Meudal, H.; Turcaud, S.; Michel, J.-P.; Roques, B. P. *J. Med. Chem.* **1996**, *39*, 2594.
199. Gosselin, F.; Lubell, W. D. *J. Org. Chem.* **2000**, *65*, 2163.
200. Ohta, T.; Sato, T.; Kimura, T.; Nozoe, S. *Tetrahedron Lett.* **1988**, *29*, 4305.
201. Turner, S. C.; Zhai, H.; Rapoport, H. *J. Org. Chem.* **2000**, *65*, 861.
202. Li, H.; Sakamoto, T.; Kikugawa, T. *Tetrahedron Lett.* **1997**, *38*, 6677.
203. Mulzer, J.; Schülzchen, F.; Bats, J.-W. *Tetrahedron* **2000**, *56*, 4289.
204. Stapon, A.; Li, R.; Townsend, C. A. *J. Am. Chem. Soc.* **2003**, *125*, 8486.
205. Bycroft, B. W.; Chhabra, S. R. *J. Chem. Soc., Chem. Commun.* **1989**, 423.
206. Belokon, Y. N.; Bulychev, A. G.; Pavlov, V. A.; Fedorova, E. B.; Tsyryapkin, V. A.; Bakhmutov, V. A.; Belikov, V. M. *J. Chem. Soc., Perkin Trans. 1* **1988**, 2075.
207. Ibrahim, H. H.; Lubell, W. D. *J. Org. Chem.* **1993**, *58*, 6438.
208. Beausoleil, E.; L'Archevêque, B.; Bélec, L.; Atfani, M.; Lubell, W. D. *J. Org. Chem.* **1996**, *61*, 9447.
209. Wei, L.; Lubell, W. D. *Can. J. Chem.* **2001**, *79*, 94.
210. Halab, L.; Bélec, L.; Lubell, W. D. *Tetrahedron* **2001**, *57*, 6439.
211. Lathbury, D.; Vernon, P.; Gallagher, T. *Tetrahedron Lett.* **1986**, *27*, 6009.
212. Gallagher, T.; Davies, I. W.; Jones, S. W.; Lathbury, D.; Mahon, M. F.; Malloy, K. C.; Shaw, R. W.; Vernon, P. *J. Chem. Soc., Perkin Trans. 1* **1992**, 433.
213. Kinsman, R.; Lathbury, D.; Vernon, P.; Gallagher, T. *J. Chem. Soc., Chem. Commun.* **1987**, 243.

214. Kemp, D. S.; Curran, T. P. *J. Org. Chem.* **1986**, *51*, 2377.
215. Björkling, F.; Boutelje, J.; Hjalmarsson, M.; Hult, K.; Norin, T. *J. Chem. Soc., Chem. Commun.* **1987**, 1041.
216. Kemp, D. S.; Curran, T. P. *J. Org. Chem.* **1988**, *53*, 5729.
217. Kemp, D. S.; Curran, T. P. *Tetrahedron Lett.* **1988**, *29*, 4931.
218. Takano, S.; Moriya, M.; Iwabuchi, Y.; Ogasawara, K. *Tetrahedron Lett.* **1989**, *30*, 3805.
219. Bowman, W. R.; Broadhurst, M. J.; Coghlan, D. R.; Lewis, K. A. *Tetrahedron Lett.* **1997**, *38*, 6301.
220. Magaard, V. W.; Sanchez, R. M.; Bean, J. W.; Moore, M. L. *Tetrahedron Lett.* **1993**, *34*, 381.
221. Avenoza, A.; Barriobero, J.; Busto, J. H.; Peregrina, J. M. *J. Org. Chem.* **2003**, *68*, 2889.
222. Wang, Q.; Tran Huu Dau, M.-E.; Sasaki, A.; Potier, P. *Tetrahedron* **2001**, *57*, 6455.
223. Tanaka, H.; Sakagami, H.; Ogazawara, K. *Tetrahedron Lett.* **2002**, *43*, 93.
224. MacArthur, M. W.; Thornton, J. M. *J. Mol. Biol.* **1991**, *218*, 397.
225. Delaney, N. G.; Madison, V. *Int. J. Pept. Prot. Res.* **1982**, *19*, 543.
226. Morris, A. L.; MacArthur, M. W.; Hutchinson, E. G.; Thornton, J. M. *Proteins* **1992**, *12*, 345.
227. Dugave, C.; Demange, L. *Chem. Rev.* **2003**, *103*, 2475.
228. Richardson, J. S. *Adv. Protein. Chem.* **1981**, *34*, 167.
229. Madison, V. *Biopolymers* **1977**, *16*, 2671.
230. Improta, R.; Benzi, C.; Barone, V. *J. Am. Chem. Soc.* **2001**, *123*, 12568.
231. Milner-White, E. J.; Bell, L. H.; Maccallum, P. H. *J. Mol. Biol.* **1992**, *228*, 725.
232. Vitagliano, L.; Berisio, R.; Mazzarella, L.; Zagari, A. *Biopolymers* **2001**, *58*, 459.
233. Nemethy, G. G., K. D.; Palmer, K. A.; Yoon, C. N.; Paterlini, G.; Zagari, A.; Rumsey, S.; Scheraga, H. A. *J. Phys. Chem.* **1992**, *96*.
234. Improta, R.; Mele, F.; Crescenzi, O.; Benzi, C.; Barone, V. *J. Am. Chem. Soc.* **2002**, *124*, 7857.
235. Schimmel, P. R.; Flory, P. J. *J. Mol. Biol.* **1968**, *34*, 105.
236. Hurley, J. H.; Mason, D. A.; Matthews, B. W. *Biopolymers* **1992**, *32*, 1443.
237. Welsh, J. H.; Zerbe, O.; Philipsborn, W.; Robinson, J. A. *FEBS Lett.* **1992**, *297*, 216.
238. Hinds, M. G.; Welsh, J. H.; Brennand, D. M.; Fisher, J.; Glennie, M. J.; Richards, N. G. J.; Turner, D. L.; Robinson, J. A. *J. Med. Chem.* **1991**, *34*.
239. Bisang, C.; Weber, C.; Inglis, J.; Schiffer, C. A.; van Gunsteren, W. F.; Jelesarov, I.; Bosshard, H. R.; Robinson, J. A. *J. Am. Chem. Soc.* **1995**, *117*.
240. Delaney, N. G.; Madison, V. *J. Am. Chem. Soc.* **1982**, *104*, 6635.
241. Beausoleil, E.; Sharma, R.; Michnick, S. W.; Lubell, W. D. *J. Org. Chem.* **1998**, *63*, 6572.
242. Fischer, S.; Dunbrack, R. L.; Karplus, M. *J. Am. Chem. Soc.* **1994**, *116*, 11931.
243. Plucinska, K.; Kataoka, T.; Yodo, M.; Cody, W. L.; He, J. X.; Humblet, C.; Lu, G. H.; Lunney, E.; Major, T. C.; Panek, R. L.; *et al. J. Med. Chem.* **1993**, *36*, 1902.
244. Holmgren, S. K.; Bretscher, L. E.; Taylor, K. M.; Raines, R. T. *Chem. Biol.* **1999**, *6*, 63.
245. Gerig, J. T.; McLeod, R. S. *J. Am. Chem. Soc.* **1973**, *95*, 5725.
246. An, S. S. A.; Lester, C. C.; Peng, J. L.; Li, Y. J.; Rothwarf, D. M.; Welker, E.; Thannhauser, T. W.; Zhang, L. S.; Tam, J. P.; Scheraga, H. A. *J. Am. Chem. Soc.* **1999**, *121*, 11558.
247. Beausoleil, E.; Lubell, W. D. *J. Am. Chem. Soc.* **1996**, *118*, 12902.
248. Halab, L.; Lubell, W. D. *J. Org. Chem.* **1999**, *64*, 3312.
249. Perni, R. B.; Farmer, L. J.; Cottrell, K. M.; Court, J. J.; Courtney, L. F.; Deininger, D. D.; Gates, C. A.; Harbeson, S. L.; Kim, J. L.; Lin, C.; Lin, K.; Luong, Y. P.; Maxwell, J. P.; Murcko, M. A.; Pitlik, J.; Rao, B. G.; Schairer, W. C.; Tung, R. D.; Van Drie, J. H.; Wilson, K.; Thomson, J. A. *Bioorg. Med. Chem. Lett.* **2004**, *14*, 1939.
250. Nikiforovich, G. V.; Kao, J. L.; Plucinska, K.; Zhang, W. J.; Marshall, G. R. *Biochemistry* **1994**, *33*, 3591.
251. Kaczmarek, K.; K.-M., L.; Skeean, R.; Dooley, D.; Humblet, C.; Lunney, E.; Marshall, G. R. *Pept.: Chem., Struct. Biol., Proc. Am. Pept. Symp., 13th Meeting Date 1993* **1994**, 687.
252. Holladay, M. W.; Bennett, M. J.; Tufano, M. D.; Lin, C. W.; Asin, K. E.; Witte, D. G.; Miller, T. R.; Bianchi, B. R.; Nikkel, A. L.; Bednarz, L. *J. Med. Chem.* **1992**, *35*, 2919.

253. Holladay, M. W.; Bennett, M. J.; Tufano, M. D.; Lin, C. W.; Witte, D. G.; Miller, T. R.; Bianchi, B. R.; Nadzan, A. M. *Peptides: Chemistry and Biology. Proceedings of the 12th American Peptide Symposium; Smith, J.A., Rivier, J. Eds; Escom: Leiden* **1992**, 443.

254. Fournie-Zaluski, M. C.; Belleney, J.; Lux, B.; Durieux, C.; Gerard, D.; Gacel, G.; Maigret, B.; Roques, B. P. *Biochemistry* **1986**, *25*, 3778.

255. Nadzan, A. M.; Garvey, D. S.; Tufano, M. D.; Holladay, M. W.; Shiosaki, K.; Shue, Y. K.; Chung, J. Y. K.; May, P. D.; May, C. S.; Lin, C. W.; Miller, T. R.; Witte, D. G.; Bianchi, B. R.; Wolfram, C. A. W.; Burt, S.; Hutchins, C. W. *Peptides: Chemistry and Biology. Proceedings of the 12th American Peptide Symposium; Smith, J.A., Rivier, J. Eds; Escom: Leiden* **1992**, 100.

256. Goudreau, N.; Weng, J. H.; Roques, B. P. *Biopolymers* **1994**, *34*, 155.

257. Mosberg, H. I.; Omnaas, J. R.; Lomize, A.; Heyl, D. L.; Nordan, I.; Mousigian, C.; Davis, P.; Porreca, F. *J. Med. Chem.* **1994**, *37*, 4384.

258. Rubini, E.; Laufer, R.; Gilon, C.; Frey, J.; Selinger, Z.; Chorev, M. *peptide chemistry 1987: T. Shiba & S. Sakakibara (Ed.) Protein Research Foundation, Osaka* **1988**, 553.

259. Sagan, S.; Karoyan, P.; Chassaing, G.; Lavielle, S. *J. Biol. Chem.* **1999**, *274*, 23770.

260. Sagan, S.; Karoyan, P.; Lequin, O.; Chassaing, G.; Lavielle, S. *Curr. Med. Chem.* **2004**, *11*, 2799.

261. Lavielle, S.; Brunissen, A.; Carruette, A.; Garret, C.; Chassaing, G. *Eur. J. Pharmacol.* **1994**, *258*, 273.

262. Lavielle, S.; Chassaing, G.; Brunissen, A.; Rodriguez, M.; Martinez, J.; Convert, O.; Carruette, A.; Garret, C.; Petitet, F.; Saffroy, M.; Torrens, Y.; Beaujouan, J. C.; Glowinski, J. *Int. J. Pept. Prot. Res.* **1993**, *42*, 270.

263. Quancard, J.; Karoyan, P.; Sagan, S.; Convert, O.; Lavielle, S.; Chassaing, G.; Lequin, O. *Eur. J. Biochem.* **2003**, *270*, 2869.

264. Sagan, S.; Quancard, J.; Lequin, O.; Karoyan, P.; Chassaing, G.; Lavielle, S. *submitted* **2005**.

265. Baures, P. W.; Pradhan, A.; Ojala, W. H.; Gleason, W. B.; Mishra, R. K.; Johnson, R. L. *Bioorg. Med. Chem. Lett.* **1999**, *9*, 2349.

266. Yu, K. L.; Rajakumar, G.; Srivastava, L. K.; Mishra, R. K.; Johnson, R. L. *J. Med. Chem.* **1988**, *31*, 1430.

267. Sugase, K.; Oyama, Y.; Kitano, K.; Akutsu, H.; Ishiguro, M. *Bioorg. Med. Chem. Lett.* **2002**, *12*, 1245.

268. Epps, D. E.; Cheney, J.; Schostarez, H.; Sawyer, T. K.; Prairie, M.; Krueger, W. C.; Mandel, F. *J. Med. Chem.* **1990**, *33*, 2080.

FIVE-MEMBERED HETEROARYL AZIDES AND DERIVED IMINOPHOSPHORANES: USEFUL INTERMEDIATES FOR THE CONSTRUCTION OF b-FUSED PYRIDINE RING SYSTEMS

Maria Funicello[a] and Piero Spagnolo[b]

[a]*Dipartimento di Chimica, Università degli Studi della Basilicata, Via N. Sauro 85, I-85100 Potenza, Italy (e-mail: funicello@unibas.it)*
[b]*Dipartimento di Chimica Organica "A. Mangini", Università di Bologna, Viale Risorgimento 4, I-40136 Bologna, Italy (e-mail: spagnolo@ms.fci.unibo.it)*

Abstract. *This minireview provides a brief outline of the peculiar aspects of the reactivity and synthesis of azides derived from five-membered heteroarene systems containing one heteroatom: in particular, it deals with the conversion of these heteroaryl azides to iminophosphorane derivatives as well as with the utility of these latter intermediates in the construction of b-fused pyridine ring systems.*

Contents

1. Introduction
2. Five-membered heteroaryl azides
    2.1. General reactivity
    2.2. Synthesis
3. Iminophosphoranes
    3.1. Iminophosphoranes in the synthesis of (fused) pyridines
    3.2. N-(Five-membered heteroaryl)iminophosphoranes in the synthesis of b-fused pyridines
4. Conclusions
Acknowledgments
References

1. Introduction

Organic azides are important intermediates that have found extensive use in the production of a great variety of acyclic and, especially, heterocyclic nitrogen-containing compounds.

The utility of these versatile intermediates stands from their capability of reacting with both electrophilic and nucleophilic species, additionally acting as 1,3-dipoles in cycloaddition reactions as well as affording reactive nitrenes under thermal and photochemical conditions.[1]

Synthetic applications of azides under radical conditions are less documented, but recent studies have clearly shown that these azide reactions also provide useful synthetic routes to N-heterocycles.[2]

Since the pioneering work of Smith in 1964, especially in the last two decades, remarkable attention has been dedicated to the chemistry of five-membered heteroaryl azides, though these attractive intermediates remain to date much less investigated than their aryl counterparts.

In this paper we will briefly discuss peculiar aspects of the reactivity and synthesis of the azides derived from five-membered heteroarenes containing one heteroatom and, especially, will deal with our very recent studies of the conversion of these heteroaryl azides to iminophosphorane derivatives as well as of the utility of these latter compounds in the synthesis of b-fused pyridine ring systems.

2. Five-membered heteroaryl azides

2.1. General reactivity

Reported studies have uncovered the fact that the chemical reactivity of five-membered heteroaryl azides largely depends upon the position of the heterocyclic ring to which they are attached.[3]

In particular, α-azides (and -nitrenes) derived from five-membered rings containing one heteroatom show a pronounced tendency to suffer ring opening fragmentation under very mild conditions leading to the formation of a nitrilic hetero-1,3-diene product, which can be usefully trapped by alkene dienophiles (Scheme 1). Ring cleavage fragmentation reactions of five-membered heteroaryl α-azides have been extensively discussed in two reviews simultaneously appeared in 1993.[3a,b] Further documented instances have been reported since in the thiophene,[4a,g,i] selenophene,[4b] benzothiophene,[4c-e] indole[4f] and pyrrole[4h] series.

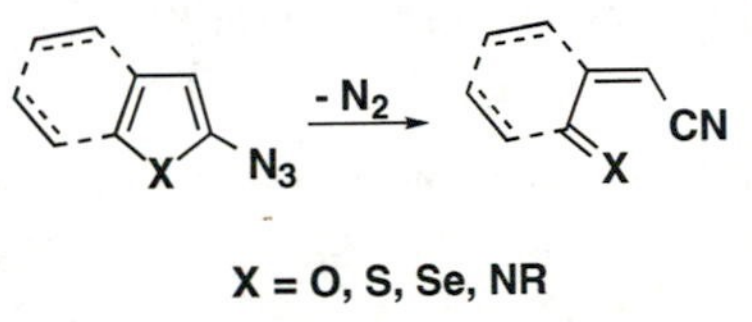

X = O, S, Se, NR

Scheme 1

On the other hand, β-azides (and -nitrenes) can normally display chemical behaviour not dissimilar from that of the aryl analogues. Especially in the thiophene and benzo[b]thiophene series there are several reports of β-azides bearing an adjacent aryl or α,β-unsaturated group which, on thermolysis or photolysis, lead smoothly to the formation of fused azoles by intramolecular 1,5-cyclization in a fashion strictly comparable to that of similarly substituted aryl azides (Scheme 2).[3,4d,e,5]

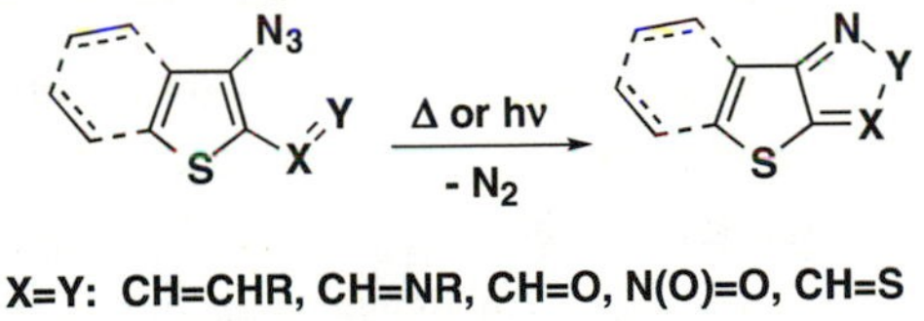

X=Y: CH=CHR, CH=NR, CH=O, N(O)=O, CH=S

Scheme 2

There is general evidence that the formal generation of a nitrene at the α-position of the thiophene[4a,g,i] and benzo[b]thiophene[3,4c,e] ring instead results in the preferential or exclusive formation of a ring-opened nitrilic enethione product. Indeed, there are only two definite examples of successful cyclization of 2-azidothiophene onto an adjacent substituent. These are represented by the very mild thermal cyclizations of methyl 2-azido-3-nitrothiophene-5-carboxylate to thienofurazan N-oxide[6] and 2-azidobenzo[b]thiophene-3-carbothialdehyde to benzothienoisothiazole.[4d]

The successful outcome of such thermal reactions is ascribable to a special ability of nitro and thioformyl group to intercept an *ortho*-azido function.

275

2.2. Synthesis

By far the most widely employed method for producing aryl azides is diazotization of the corresponding aniline followed by addition of sodium azide. This method, which has found application in the synthesis of azides derived from several types of azoles (particularly imidazoles, thiazoles, pyrazoles triazoles and isoxazoles)[3a] has a limited scope in five-membered heteroaromatic systems since especially amines derived from rings containing one heteroatom (*i.e.* furans, pyrroles, thiophenes, selenophenes, etc.) are unstable compounds, unless electron-withdrawing substituents are present, and commonly give difficulties in diazotization.

Nucleophilic displacement of a halo[3a] or nitro group[4c-e,7] by azide ion is an alternative method, but this is, however, only practicable provided that the heteroaromatic ring be sufficiently activated by the presence of suitable electron-withdrawing substituent(s).

This method is obviously unsuited to the preparation of the parent azides or those carrying electron-releasing groups.

Fortunately, electron-rich heteroaryl azides have become fairly accessible by the so-called "azide group transfer" method. This method, first reported by Spagnolo and Zanirato in 1978,[8] is essentially based on reaction of the lithiated heteroarene with tosyl azide, performed in dry diethyl ether at -78 °C, followed by slow fragmentation of the resulting triazene salt with aqueous sodium pyrophosphate (Scheme 3).

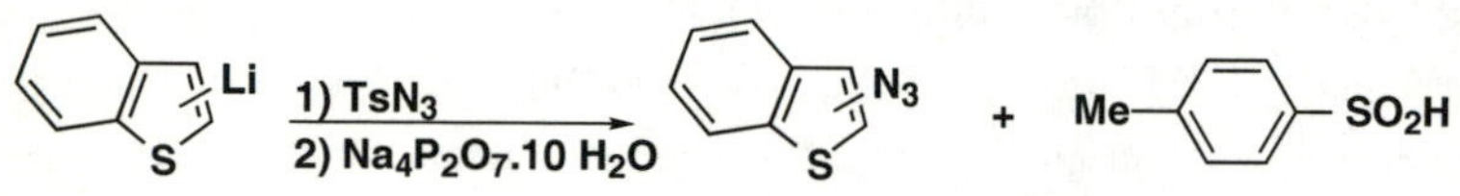

Scheme 3

In recent years, a large variety of α- and β-azides, including the parent ones, arising from thiophene,[4a,g,i,9] benzothiophene,[3a,4e] selenophene,[4b] benzofuran,[3a] and indole[3a,4f] have been successfully prepared by means of this azido group transfer reaction.

More recently, we have observed that a significant enhancement of isolated azide can be achieved by a slight modification of the original protocol, just consisting in final extraction of the azide product with ethyl acetate rather than with the original pentane.[10]

Aryl azides, bearing either electron-releasing or electron-withdrawing groups, are also conveniently prepared from aryl boronic acids through reaction with lead tetraacetate in the presence of catalytic amount of mercuric acetate, in $CHCl_3$ or CH_2Cl_2 solution at room temperature, followed by *in situ* treatment of the resultant aryllead triacetate with sodium azide in DMSO.[11]

Possible extension of such procedure to the preparation of the electron-rich heteroaromatic azides would be appealing especially since various heteroaryl boronic acids are now commercially available. Additionally, this procedure would be advantageous over azido group transfer since it would avoid use of a dangerous reagent such as explosive tosyl azide as well as of rather tedious experimental conditions.

We are currently investigating its possible application to the synthesis of α- and β-azides derived from thiophene, benzothiophene and indole.

Preliminary results have shown that β-azido-thiophene and –benzothiophene can actually be produced in useful yields, whereas for still unclear reasons the corresponding α-azides virtually fail to form owing to preferential formation of the parent compounds (Scheme 4).

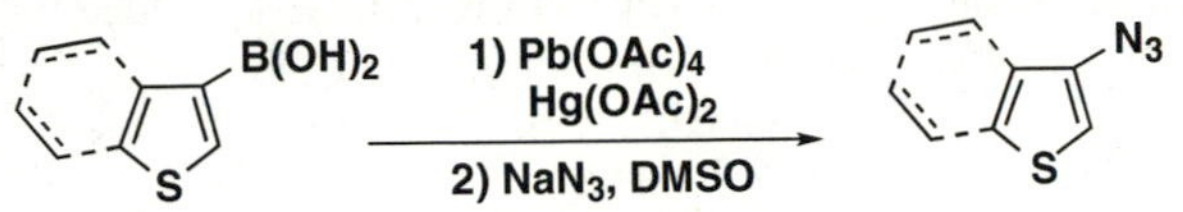

Scheme 4

. Iminophosphoranes

.1. Iminophosphoranes in the synthesis of (fused) pyridines

Iminophosphoranes were first prepared at the beginning of the last century by Staudinger by reacting organic azides with triphenylphosphine. Mechanistic studies involving kinetic and X-ray analyses and, very recently, also theoretical approach,[12] have revealed that the Staudinger reaction of phenyl azides with triphenylphosphine goes through two transition states which involve *trans-cis* isomerization of phosphazide and, finally, nitrogen extrusion via a four-center intermediate (Scheme 5).

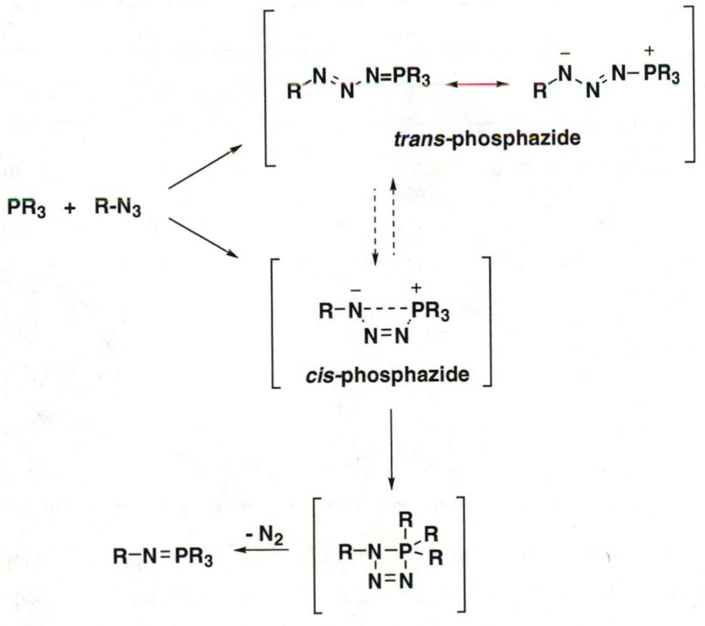

Scheme 5

Nowadays, iminophosphoranes have become a powerful tool in synthetic strategies directed toward the construction of nitrogen-containing heterocycles.[13]

Iminophosphoranes, often prepared from azides but also available from primary amines, undergo inter- and intra-molecular reactions with a wide variety of carbonyl compounds in a similar way to phosphonium ylids, leading to an excellent method for the construction of imine double bonds (aza Wittig reaction).

In particular, the aza Wittig reaction of iminophosphoranes derived from β-aryl (heteroaryl) vinyl azides with α,β-unsaturated aldehydes followed by thermally induced 6π-electrocyclization of the resultant 3-aza-1,3,5-hexatriene and eventual dehydrogenation has found useful application in the synthesis of simple pyridines.[13c,d]

Moreover, a modification of this strategy using saturated aldehydes or various heterocumulenes has been widely employed for the c-fusion of a pyridine ring onto both aromatic and heteroaromatic systems including inter alia furan, thiophene, indole and pyridine rings (Scheme 6).[13c,d,14]

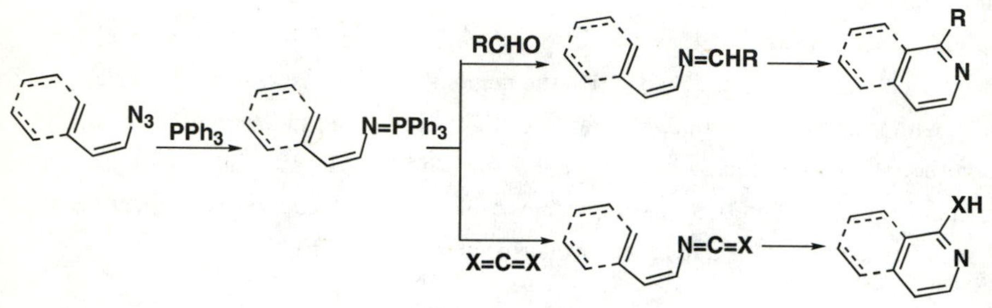

Scheme 6

Such a so-called tandem aza Wittig-electrocyclization strategy has initially found a limited application in the synthesis of b-fused pyridines. In fact, b-fused pyridines, including a number of quinoline[13c,15] and α-carboline[13c,16] derivatives, have first been prepared from heterocumulenes and those iminophosphorane produced from azides (or amines) bearing a vinylic ortho-substituent, which are not normally readily accessible (Scheme 7).

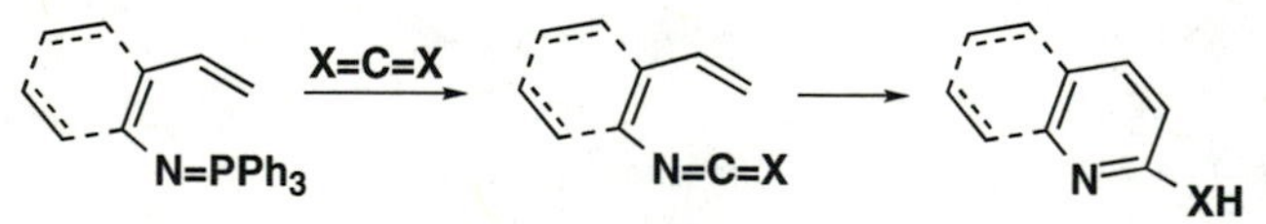

Scheme 7

3.2. N-(Five-membered heteroaryl)iminophosphoranes in the synthesis of b-fused pyridines

An important extension of the aza Wittig-electrocyclization strategy in the construction of b-fused pyridine ring systems can involve the use of α,β-unsaturated carbonyl compounds and iminophosphorane having ortho-unsubstituted five-membered heteroaryl N-substituents.

However, iminophosphoranes of that type still remain poorly explored, despite the fact that in principle their preparation has become feasible from readily available azido precursors rather than scarcely accessible and/or unstable amine precursors.

Our very recent (and current) studies have uncovered that various heteroaryl iminophosphoranes promptly prepared from 2-azido- and 3-azido-benzo[b]thiophene, 3-azidothiophene and 2-azido-1-methylindole actually are valuable precursors of b-fused pyridines.

Heteroaryl iminophosphoranes derived from five-membered rings containing one heteroatom were virtually unknown until 1996, when our long interest in the reactivity and synthetic utility of azidobenzothiophenes led us to prepare N-(3-benzo[b]thienyl)- and N-(2-benzo[b]thienyl)-iminotriphenyl-phosphorane, 1a and 2a, in order to examine their possible use in the synthesis of benzo[b]thieno[3,2-b]-pyridines and benzo[b]thieno[2,3-b]-pyridines, for which compounds the few reported synthetic methods were rather difficult and/or gave low yields.[17]

Benzothienopyridines are of pharmacological interest due to their isosterism with indolopyridines[18] and their reported activity as antibacterial,[19a] antiallergic[19b] and anxiolitic agents.[19c]

The iminotriphenylphosphoranes 1a, 2a were readily obtained in high yields by reacting 3-azido- and 2-azido-benzothiophene with triphenylphosphine in dry dichloromethane at 0 °C, according to the classical Staudinger method (Schemes 8 and 9). Both phosphoranes proved to be stable, solid compounds, in sharp contrast with the known instability of their azido precursors.[20]

Treatment of the compound 1a in toluene with acrylaldehyde, trans-crotonaldehyde, trans-cinnamaldehyde and methacrylaldehyde under mild thermal conditions (70 °C) directly furnished the aimed benzo[b]thieno[3,2-b]pyridines 3a-d in useful yields (40-70%).

Apparently, the phosphorane 1a smoothly reacted with the above aldehydes to give formal 3-aza-1,3,5-hexatriene intermediates. These intermediates then promptly underwent thermal electrocyclization finally leading to the isolated compounds 3a-d after further dehydrogenation of the cyclized dihydropyridine (Scheme 8).[20]

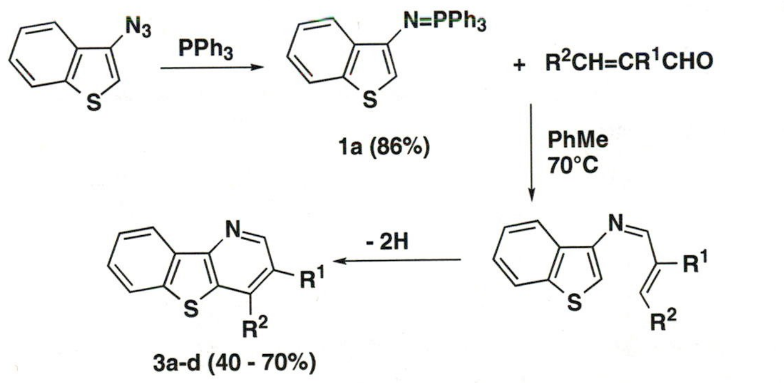

a: $R^1 = R^2 = H$; b: $R^1 = H$, $R^2 = Ph$; c: $R^1 = H$, $R^2 = Me$; d: $R^1 = Me$, $R^2 = H$

Scheme 8

Like the phosphorane **1a**, its positional isomer **2a** reacted with the same aldehyde compounds to eventually furnish the corresponding benzo[b]thieno[2,3-b]pyridines **4a-d** in comparable yields (Scheme 9).

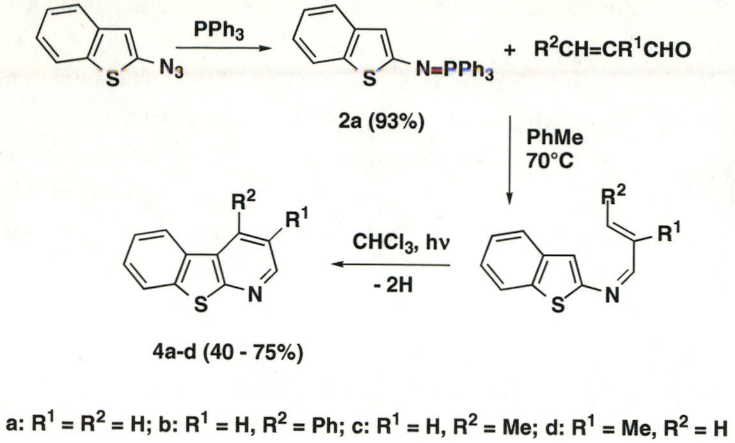

2a (93%)

PhMe
70°C

CHCl₃, hν
- 2H

4a-d (40 - 75%)

a: R¹ = R² = H; b: R¹ = H, R² = Ph; c: R¹ = H, R² = Me; d: R¹ = Me, R² = H

Scheme 9

However, in such cases the initial aza Wittig reaction generally proceeded somewhat more slowly and additionally, the ensuing azahexatrienes normally proved to be less prone to thermal ring closure. In fact total cyclization of these intermediates could be achieved upon further irradiation with a high pressure mercury vapour lamp.[20]

The believed involvement of the aza Wittig-electrocyclization mechanism was clearly supported by regiospecific occurrence of a single 4- or 3-substituted benzothienopyridine in the reactions with cinnamaldehyde, crotonaldehyde and methacrylaldehyde and, additionally, by successful detection of certain azahexatriene intermediates.[20]

Subsequent attempts to enlarge the scope of our novel annulation reactions by using but-2-en-3-one as the carbonyl substrate were frustrated by the fact that phosphorane **2a** and, especially, **1a** were found to be essentially inert toward that ketone.

Since the presence of electron-donating methyl group(s) on the phosphorus was known to normally enhance iminophosphorane reactivity[13d] we were led to examine the reactivity of phosphoranes **1** and **2** bearing methyl-P-substituent(s) towards both unsaturated ketones and aldehydes.

The mono-, di-, and trimethyl-P-substituted 3-benzothienyl phosphoranes **1b-d** were thus prepared in good yields by usual Staudinger reaction of 3-azidobenzothiophene with the appropriate phosphine (Figure 1).[21] The mono- and dimethyl-phosphoranes **1b,c** were isolated as fairly stable solid compounds, whereas the trimethyl analogue **1d** was obtained as a crude oil which showed a tendency to decompose and thence was directly used without purification.

Under usual conditions, these phosphoranes **1b-d** were found to react smoothly with both the ketone **5a-c** and the aldehydes **5d,e**, shown in Figure 1, but their general behaviour towards these substrates was nexpectedly found to be largely dissimilar from that of the triphenyl analogue **1a**. Indeed, their reactions with **5a-e** gave varying mixtures of the regioisomeric benzothienopyridines **6** and **7** as a result of preferential ttack of either imino nitrogen or α-thienyl carbon at the enone carbonyl group (Figure 1 and Table 1, entries -5, 7, 8, 10, 11, and 13-15).

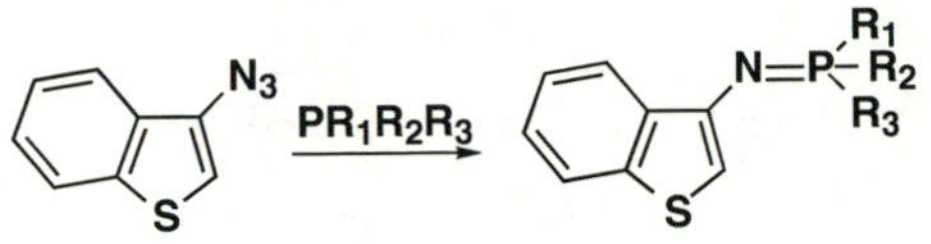

$R^1CH=CHCOR^2$

5a: R^1 = H, R^2 = Me
5b: R^1 = Ph, R^2 = Me
5c: R^1 = COOMe, R^2 = Me
5d: R^1 = Ph, R^2 = H
5e: R^1 = Me, R^2 = H

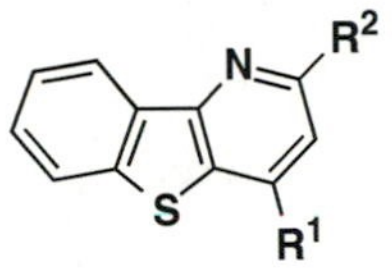

1b: R_1 = R_2 = Ph, R_3 = Me (80%)
1c: R_1 = Ph, R_2 = R_3 = Me (78%)
1d: R_1 = R_2 = R_3 = Me (75%)

6a: R^1 = H, R^2 = Me
6b: R^1 = Ph, R^2 = Me
6c: R^1 = COOMe, R^2 = Me
6d: R^1 = Ph, R^2 = H

7a: R^1 = Me, R^2 = H
7b: R^1 = Me, R^2 = Ph
7c: R^1 = Me, R^2 = COOMe
7d: R^1 = H, R^2 = Ph

Figure 1

Table 1. Reactions of the phosphoranes **1a-d** with α,β–unsaturated ketones **5a-c** and aldehydes **5d,e**.

ntry	Phosphorane	Enone	Benzothienopyridine(s) (% yield)
	1b	5a	**6a** (25) + **7a** (25)
	1c	5a	**6a** (15) + **7a** (60)
	1d	5a	**6a** (10) + **7a** (40)
	1b	5b	**6b** (26)
	1c	5b	**6b** (33) + **7b** (15)
	1a	5c	**6c** (30)
	1b	5c	**6c** (46) + **7c** (27)
	1c	5c	**6c** (47) + **7c** (42)
	1a	5d	**6d** (45)
0	1b	5d	**6d** (30) + **7d** (25)
1	1c	5d	**6d** (15) + **7d** (40)
2	1a	5e	**7a** (62)
3	1b	5e	**7a** (41) + **6a** (30)
4	1c	5e	**7a** (20) + **6a** (51)
5	1d	5e	**7a** (16) + **6a** (30)

The overall findings actually revealed that the progressive replacement of phenyl with methyl P substituent greatly enhanced the reactivity of phosphorane **1**, concomitantly enhancing the propensity of the phosphorane itself for addition to the enone by the α-thienyl carbon (Table 1, entries 1-15). In fact, with all the enones **5a-e** the methylated phosphoranes **1b** and, especially, **1c, d** showed a fair tendency to produce the respective benzothienopyridines **7a-d** (and **6a**) at the expense of the isomeric aza-Wittig electrocyclization ones **6a-d** (and **7a**). The compounds **1b-d** presumably led to the pyridines **7a-d** and **6a** through the reaction sequences outlined in Scheme 10. These involve primary addition of the α-thienyl carbon of **1b-d** to the carbonyl carbon of the enones **5a-e**.

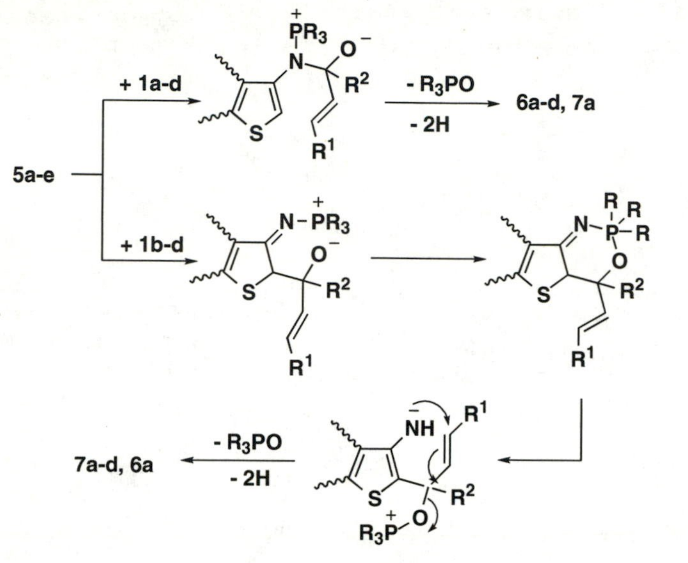

Scheme 10

Analogous type of reaction sequences have been invoked in the reaction of certain (vinylimino)phosphoranes with cinnamaldehydes similarly leading to cyclized pyridines.[22] Regardless from mechanistic implications, the above results were of special interest since indicated that the regiochemistry of the outcoming substituted benzothieno[3,2-b]pyridines can be largely governed by a proper choice of the phosphorane **1** reagent.[21]

Unlike the above phosphoranes **1b-d**, the next prepared mono- and di-methyl-P-substituted 2-benzothienyl phosphoranes **2b** and **2c** (Scheme 11) behaved towards enones in a fashion not dissimilar from that of their triphenyl congener **2a**.[10] In fact, under analogous conditions the compounds **2b,c** gave satisfactory yields of only isolable aza Wittig azahexatriene products in their successful reactions with both ketones and aldehydes. The produced azahexatriene intermediates could then quantitatively converted to the cyclized benzothienopyridines **8a-d** upon eventual photolysis in chloroform solution (Scheme 11).

In all cases examined the 2-benzothienyl phosphoranes **2b,c** provided no evidence at all for addition to enone carbonyl group by the β-thienyl carbon and thus, unlike the 3-benzothienyl counterparts **1b-d**, the generally failed to exhibit (vinylimino)phosphorane-like behaviour. The different reactivity mode displayed by the positionally isomeric compounds **1** and **2** probably results from different ability of their imino nitrogen to donate electron density to the attached thiophene ring.

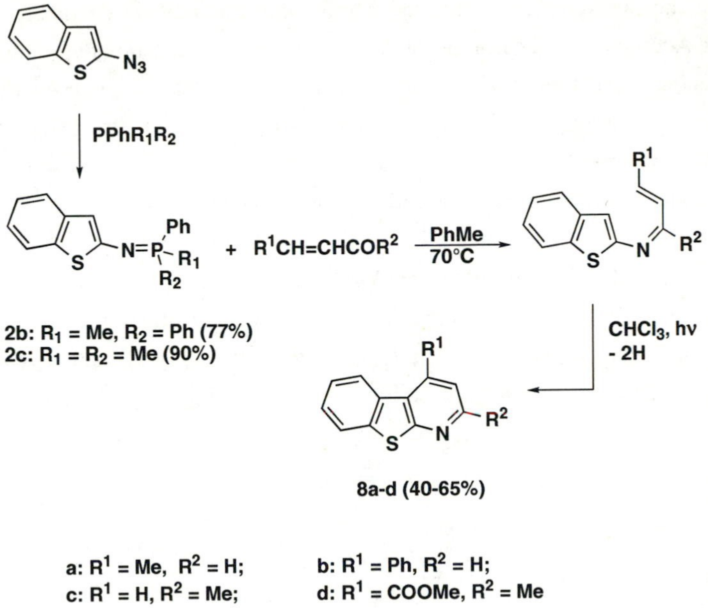

2b: R_1 = Me, R_2 = Ph (77%)
2c: R_1 = R_2 = Me (90%)

8a-d (40-65%)

a: R^1 = Me, R^2 = H;   b: R^1 = Ph, R^2 = H;
c: R^1 = H, R^2 = Me;   d: R^1 = COOMe, R^2 = Me

Scheme 11

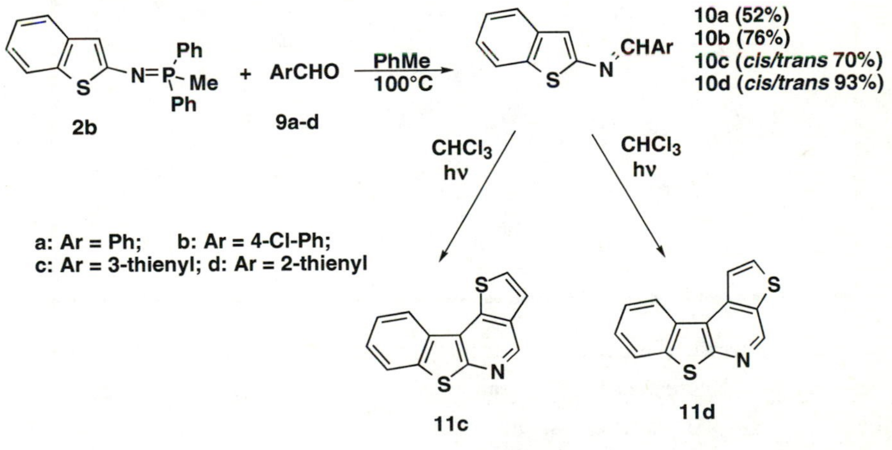

10a (52%)
10b (76%)
10c (*cis/trans* 70%)
10d (*cis/trans* 93%)

2b   9a-d

a: Ar = Ph;   b: Ar = 4-Cl-Ph;
c: Ar = 3-thienyl; d: Ar = 2-thienyl

11c   11d

Scheme 12

The dimethylphenylphosphorane **2b** was also found to react smoothly with benzaldehydes **9a,b** and thiophenecarboxaldehydes **9c,d**, though under more forcing conditions (100 °C), to give varying isolated yields of the corresponding aza Wittig imine products **10a-d** (Scheme 12).[10]

It is worth note that under our thermal conditions both thienylideneamines **10c** and **10d** occurred as mixtures of the *trans* and *cis* isomers, unlike subsequent observations that the aza Wittig reaction with other heteroaryl aldehydes normally results in exclusive production of the *trans* isomers over the 60-100 °C temperature range.[23] Attempted cyclization of the benzylideneamines **10a,b** under photochemical conditions only resulted in essential production of polymeric material, while under analogous conditions the *cis* isomers of the thienylideneamines **10c,d** were interestingly found to efficiently afford the respective tetracyclic pyridines **11c** and **11d** (Scheme 12).[10] Our methodology to furnish benzothienopyridines has been applied to the preparation of a novel 5-amino-4-methoxycarbonylbenzothieno[2,3-*b*]pyridine, which is of interest as a potential secondary peptide structure mimic.[24] Usual tandem aza Wittig-electrocyclization reaction of *N*-(4-methoxybenzothiophen-2-yl)iminomethyldiphenylphosphorane, easily obtained from 4-methoxybenzo-thiophene by a two-step sequence in an overall yield of 65%, with methyl *trans*-4-oxo-2-pentenoate, in chloroform solution at 50 °C, directly furnished the expected 5-methoxy-4-(methoxycarbonyl)benzothieno-pyridine in excellent yield. Demethylation with lithium bromide and subsequent Smiles rearrangement of the resultant phenolic compound eventually gave the aimed aminobenzothienopyridine (Scheme 13).[24]

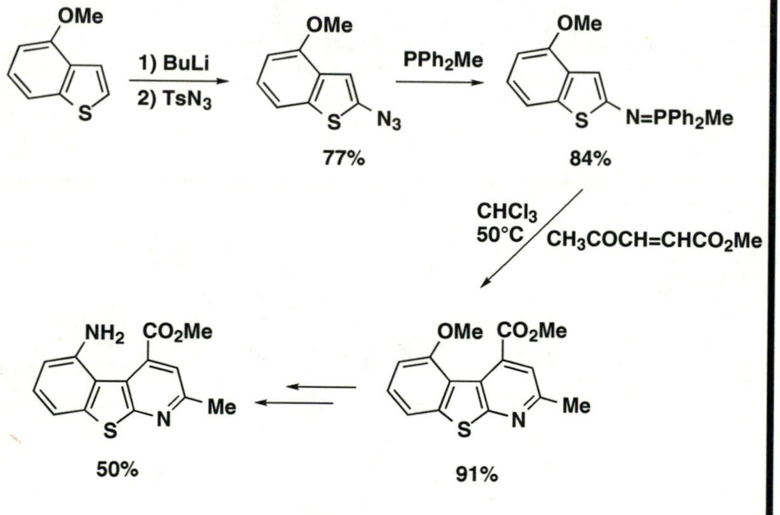

Scheme 13

We are currently interested in extending our methodology to using heteroaryl phosphoranes derived from other five-membered heteroarenes such as thiophene and indole. Preliminary results seem to indicate that 3-thienyl and 2-indolyl phosphoranes also are highly effective tools for achieving *b*-fusion of a pyridine ring onto thiophene and indole.

N-(3-Thienyl)iminotriphenylphosphorane **12**, newly prepared from 3-azidothiophene and triphenylphosphine (Scheme 14), has proved to undergo plain reaction with methyl *trans*-4-oxo-2 entenoate, under our usual conditions, affording the thienopyridine **13** as the main product along with small amounts of the isomeric aza Wittig-electrocyclization pyridine **14**.[25] The pyridine **13** is the expected product of initial addition of the adjacent α-thienyl carbon of **12** to the pentenoate ketone moiety (see Scheme 10).

Such chemical evidence first suggests that the 3-thienyl phosphoranes should have a much more pronounced propensity than the 3-benzothienyl counterparts **1** to adopt the β-(α-thienyl) carbon instead of the imino nitrogen. In fact, the 3-benzothienyl triphenylphosphorane **1a** furnished a modest yield of only aza Wittig-electrocyclization product in the corresponding reaction with the pentenoate (see Table 1, entry 6).

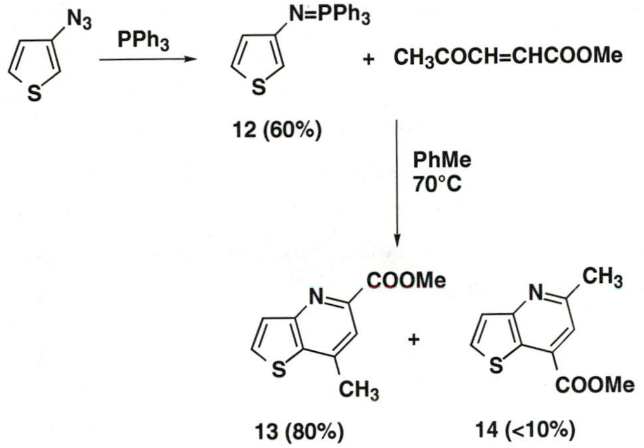

Scheme 14

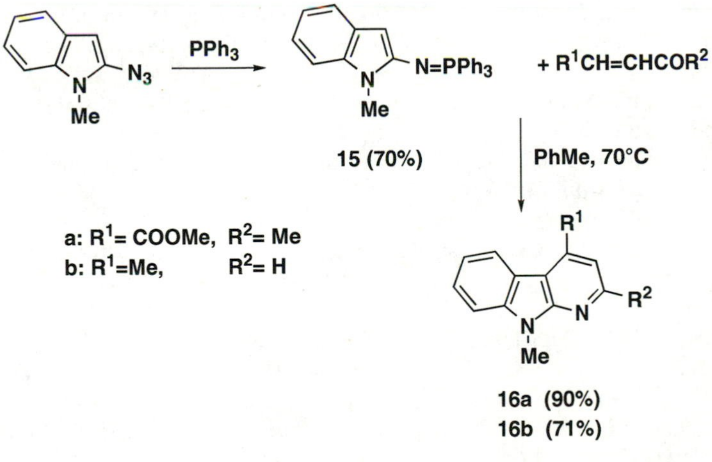

a: R^1 = COOMe, R^2 = Me
b: R^1 = Me, R^2 = H

Scheme 15

285

Moreover, *N*-(1-methylindol-2-yl)iminotriphenylphosphorane **15**, first produced in good yield from 2-azido-1-methylindole and triphenylphosphine (Scheme 15), has been shown to undergo highly efficient aza-Wittig-electrocyclization reaction with both methyl 4-oxo-2-pentenoate and crotonaldehyde leading to the respective α-carbolines **16a,b** in good to excellent yield (Scheme 15).[25] These data appear to indicate that 2-indolyl phosphoranes should react with enones in a similar manner to the 2-benzothienyl congeners **2**, but the outcoming pyridine products would occur in even higher yields.

4. Conclusion

Various *N*-(five-membered heteroaryl)iminophosporanes derived from benzo[*b*]thiophene, thiophene and indole are now promptly available through classical Staudinger reaction of corresponding azides with phosphines. These iminophosphorane derivatives have proved to be valuable tools in the tandem aza Wittig electrocyclization strategy for the construction of benzothieno-, thieno- and indole-[*b*]pyridines.

Such a strategy, in principle, should be of wide utility for achieving *b*-fusion of a pyridine ring onto five-membered heteroarenes by using easily accessible α- and β-azides as the nitrogen precursors.

Since our knowledge of the chemistry of the five-membered heteroaryl phosphoranes is to date highly limited, it is hoped that future studies will aim at uncovering the full synthetic potential of those attractive nitrogen intermediates in heterocyclic synthesis.

Acknowledgments

We gratefully acknowledge financial support from MIUR (PRIN 2003) and from the University of Bologna and Basilicata.

References

1. For general information concerning the chemistry of azides and nitrenes see: (a) Scriven, E. F V. Turnbull, K. *Chem. Rev.* **1988**, *88*, 297. (b) *Azides and Nitrenes: Reactivity and Utility*; Scriven, E. F V., Ed.; Academic Press: New York, 1984. (c) *The Chemistry of the azido group*; Patai, S., Ed.; Wiley Interscience: New York, 1971.

2. (a) Kim, S.; Joe, G. H.; Do, J. Y. *J. Am. Chem. Soc* **1993**, *115*, 3328. (b) Kim, S.; Joe, G. H.; Do, J. Y *J. Am. Chem. Soc* **1994**, *116*, 5521. (c) Benati, L.; Nanni, D.; Sangiorgi, C.; Spagnolo, P. *J. Org. Chem.* **1999**, *64*, 7836. (d) Moreno-Vargas, A. J.; Vogel, P. *Tetrahedron Lett.* **2003**, *44*, 5069. (e) Lizos, D. E.; Murphy, J. A. *Org. Biomol. Chem.* **2003**, *1*, 117. (f) Benati, L.; Bencivenni, G.; Leardini R.; Minozzi, M.; Nanni, D.; Scialpi, R.; Spagnolo, P.; Strazzari, S.; Zanardi, G.; Rizzoli, C. *Org. Lett* **2004**, *6*, 417 and refs therein.

3. For leading reviews on five-membered heteroaromatic azides see: (a) Funicello, M.; Spagnolo, P. Zanirato, P. *Acta Chem. Scand.* **1993**, *47*, 231. (b) Dehaen, W.; Becher, J. *Acta Chem. Scand.* **1993**, *47* 244.

4. (a) Spinelli, D.; Zanirato, P. *J. Chem Soc., Perkin Trans. 2* **1993**, 1129. (b) Gronowitz, S.; Zanirato, P *J. Chem Soc., Perkin Trans. 2* **1994**, 1815. (c) Capperucci, A.; Degl'Innocenti, A.; Scafato, P. Spagnolo, P. *Chem. Lett.* **1995**, 147. (d) Degl'Innocenti, A.; Funicello, M.; Scafato, P.; Spagnolo, P *Chem. Lett.* **1994**, 1873 (e) Degl'Innocenti, A.; Funicello, M.; Scafato, P.; Spagnolo, P.; Zanirato, P. *J Chem. Soc., Perkin Trans. 1* **1995**, 2141. (f) Foresti, E.; Di Gioia, M. T.; Nanni, D.; Zanirato, P. *Gazz Chim. Ital.* **1995**, *125*, 151. (g) Davies, D.; Spagnolo, P.; Zanirato, P. *J. Chem Soc., Perkin Trans.* **1995**, 613. (h) Nanni, D.; Zanirato, P. *J. Chem. Soc., Perkin 1* **1997**, 1003. (i) Valenti, F.; Zanirato, P *J. Chem. Soc., Perkin Trans. 2* **1999**, 623.

5. Dyall, L. K.; Sulfolk, P. M.; Dehaen, W.; L'Abbé, G. *J. Chem. Soc., Perkin Trans. 2* **1994**, 2115.

6. Noto, R.; Rainieri, R.; Arnone, C. *J. Chem. Soc., Perkin Trans. 2* **1989**, 127.

7. (a) Capperucci, A.; Degl'Innocenti, A.; Funicello, M.; Mauriello, G.; Scafato, P.; Spagnolo, P. *J. Org. Chem.* **1995**, *60*, 2254. (b) Capperucci, A.; Degl'Innocenti, A.; Funicello, M.; Scafato, P.; Spagnolo, P. *Synthesis* **1996**, 1185.

8. Spagnolo, P.; Zanirato, P. *J. Org. Chem.* **1978**, *43*, 3539.

9. Spagnolo, P.; Zanirato, P. *J. Chem. Soc., Perkin Trans. 1* **1996**, 963.

10. Bonini, C.; D'Auria, M.; Funicello, M.; Romaniello, G. *Tetrahedron* **2002**, *58*, 3507.

11. (a) Morgan, J.; Pinhey, J. T. *J. Chem. Soc., Perkin Trans. 1* **1990**, 715. (b) Huber, M. L.; Pinhey, J. T. *J. Chem. Soc., Perkin Trans. 1* **1990**, 721. (c) Morgan, J.; Pinhey, J. T. *J. Chem. Soc., Perkin Trans. 1* **1993**, 1673. (d) Morgan, J.; Buys, I; Hambley, T. W.; Pinhey, J. T. *J. Chem. Soc., Perkin Trans. 1* **1993**, 1677.

12. Tian, W. Q.; Wang, Y. A. *J. Org. Chem.* **2004**, *69*, 4299.

13. *Reviews*: (a) Barluenga, J.; Palacios, F. *Org. Prep. Proced. Int.* **1991**, 1. (b) Golologov, Y. G.; Kasukhin, L. *Tetrahedron* **1992**, *48*, 1353. (c) Molina, P.; Vilaplana, M. J. *Synthesis* **1994**, 1197. (d) Wamhoff, H.; Richardt, G.; Stolben, S. *Adv. Heterocyclic Chem.* **1995**, *64*, 125.

14. For very recent applications of the aza Wittig-electrocyclization strategy in the synthesis of natural isoquinoline and β-carboline alkaloids see: Fresneda, P. M.; Molina, P. *Synlett* **2004**, 1.

15. Molina, P.; Alajarin, M.; Vidal, A.; Sanchez Andrada, P. *J. Org. Chem.* **1992**, *57*, 929.

16. (a) Saito, T.; Ohmori, H.; Furuno, E.; Motoki, S. *J. Chem Soc., Chem Commun.* **1992**, 22. (b) Molina, P.; Fresneda, P. M. *Synthesis* **1989**, 878.

17. (a) Degl'Innocenti, A.; Funicello, M.; Scafato, P.; Spagnolo, P. *Tetrahedron Lett.* **1997**, *38*, 2171. (b) Levancher, V.; Boussad, N.; Dupas, G.; Bourguignon, J.; Queguiner, G.; *Tetrahedron* **1992**, *48*, 831. (c) Benoit, R.; Dupas, G.; Bourguignon, J.; Queguiner, G. *Synthesis* **1987**, 1124.

18. Roques, B. P.; Prange, T.; Oberlin, R. *Org. Magn. Reson.* **1977**, *9*, 185.

19. (a) Gorlitzer, K.; Kramer, C. *Pharmazie* **2000**, *55*, 645. (b) Fevrier, B.; Dupas, G.; Bourguignon, J.; Queguiner, G. *J. Heterocycl. Chem.* **1993**, *30*, 1085. (c) Benham, C. D.; Blackburn, T. P.; Johns, A.; Kotecha, N. R.; Martin, R. T.; Thomas, D. R.; Thompson, M.; Ward, R. W. *Bioorg. Med. Chem. Lett.* **1995**, *5*, 2455.

20. Degl'Innocenti, A.; Funicello, M.; Scafato, P.; Spagnolo, P.; Zanirato, P. *J. Chem. Soc., Perkin Trans. 1* **1996**, 2561.

21. Bonini, C.; Chiummiento, L.; Funicello, M.; Spagnolo, P. *Tetrahedron* **2000**, *56*, 1517.

22. Molina, P.; Pastor, A.; Vilaplana, M. J. *J. Org. Chem.* **1996**, *61*, 8094.

23. Bonini, C.; D'Auria, M.; Funicello, M.; Racioppi, R. manuscript in preparation

24. Bonini, C.; Funicello, M.; Scialpi, R.; Spagnolo, P. *Tetrahedron* **2003**, *59*, 7515.

25. Bonini, C.; Funicello, M.; Spagnolo, P. unpublished results.

1,3-THIAZOLIDINE-4-CARBOXYLIC ACIDS
AS BUILDING BLOCKS IN ORGANIC SYNTHESIS

Teresa M. V. D. Pinho e Melo

Departamento de Química, Faculdade de Ciências e Tecnologia
Universidade de Coimbra, P-3004-535 Coimbra, Portugal (e-mail: tmelo@ci.uc.pt)

Abstract. *The review aims to demonstrate the wide applicability of 1,3-thiazolidine-4-carboxylic acid derivatives as reagents for the synthesis of many well-known and novel heterocyclic systems.*

Contents

1. Introduction
2. Synthesis of 1,3-thiazolidine-annulated systems
3. Dipoles from 1,3-thiazolidine-4-carboxylic acids
4. 2-Aza-1,3-butadienes or their tautomers from 1,3-thiazolidine-4-carboxylic acids
5. Conclusion
References

1. Introduction

1,3-Thiazolidine-4-carboxylates are compounds of considerable importance, as some representatives show interesting biological activities.[1,2] They are well known as building blocks of natural penicillins and analogues,[2] as well as precursors of other pharmaceuticals. Some examples of dipeptides containing 1,3-thiazolidine-4-carboxylic acid have also been prepared.[3]

Thiazolidines are also a class of compounds with considerable interest as chiral heterocycles. Some derivatives have been used as chiral catalysts namely in the enantioselective reduction of aromatic ketones and in the enantioselective alkylation of aldehydes.[4] On the other hand, the existing chiral center(s) in 1,3-thiazolidine-4-caboxylic acids allows diastereoselective reactions to occur, and these have been widely used for various synthetic purposes.

The most frequently used method for the synthesis of 1,3-thiazolidine-4-carboxylates involves the cyclocondensation of cysteine with carbonyl compounds, aldehydes and ketones.

2-Substituted-1,3-thiazolidine-4-carboxylates are obtained from the reaction of aldehydes and *L*-cysteine esters in a process where a new chiral center at the C-2 position of the thiazolidine is created leading to a mixture of the (2*S*,4*R*)- and (2*R*,4*R*)-diastereoisomers.

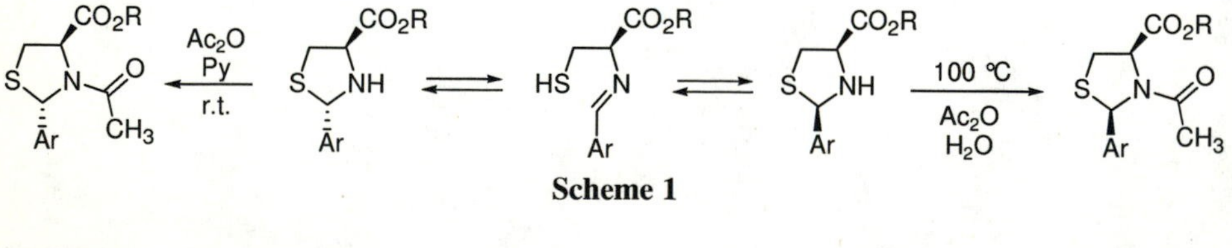

Scheme 1

The acylation of the diastereoisomeric mixture with acetic anhydride can lead to the selective synthesis of *N*-acyl-2-substituted-1,3-thiazolidine-4-carboxylates as pure stereoisomers with (2*R*,4*R*) or (2*S*,4*R*) stereochemistry depending on the reaction conditions. In fact 2-substituted-1,3-thiazolidine-4-carboxylates can undergo selective inversion at C-2 through a mechanism involving the opening of the ring but the protection with the acyl group prevents this epimerization and allows the isolation of pure diastereoisomers (Scheme 1). The same observation is valid for thiazolidines derived from *D*-cysteine.[5]

The selective synthesis of (2*R*,4*R*)-*N*-carbethoxy-2-phenyl-1,3-thiazolidine-4-carboxylic acid (**2**) and (2*S*,4*R*)-*N*-carbethoxy-2-phenyl-1,3-thiazolidine-4-carboxylic acid (**3**) has also been reported by Ferrario and co-workers (Scheme 2).[6] The *cis* derivative was prepared by treating the triethylamine salt of the thiazolidine **1** with ethyl chloroformate in tetrahydrofuran and the *trans* derivative was synthesised using ethyl chloroformate in dry pyridine.

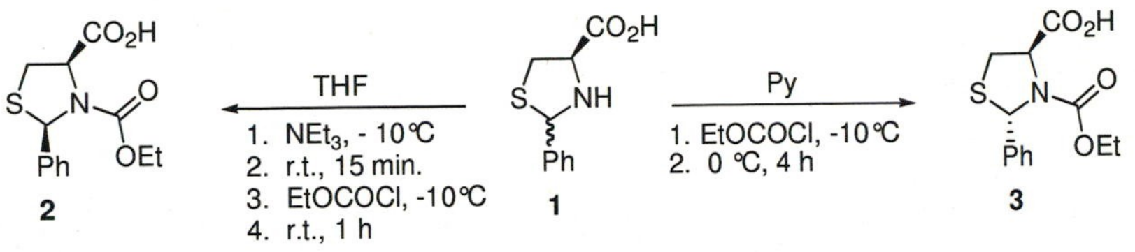

Scheme 2

This strategy of obtaining *N*-substituted-2-substituted-1,3-thiazolidine-4-carboxylates as single diastereoisomers from diastereoisomeric mixtures *via* the *ring opening-ring closure mechanism* has been exploited for various selective synthesis which are illustrated in several examples presented here.

The selective synthesis of 4-aminothiazolidines also takes advantage of the diastereoselective *in situ* *N*-acylation of the starting thiazolidine.[7] Using two equivalents of ethyl chloroformate in order to allow *N*-acylation and to activate the carboxy group for the reaction of **1** with sodium azide, the acyl azide intermediate **4** is obtained. This compound undergoes a Curtius rearrangement giving **5** without racemization (Scheme 3).

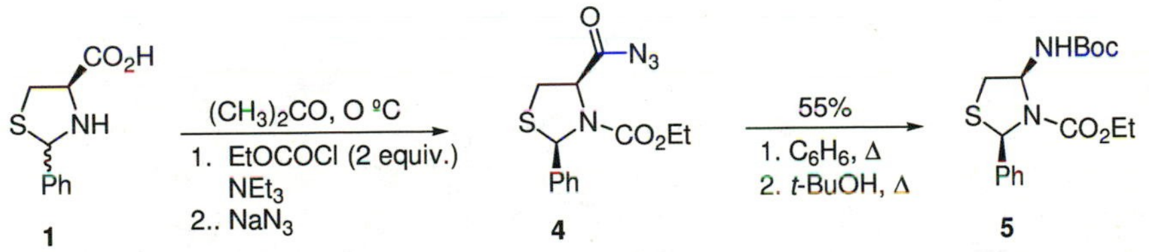

Scheme 3

Shiraiwa and co-workers reported the synthesis of the four stereoisomers of 1,3-thiazolidine-2,4-dicarboxylic acid.[8] The (2*R*,4*R*)-1,3-thiazolidine-2,4-dicarboxylic acid was diastereoselectively synthesized by condensation of *L*-cysteine with an equimolar amount of glyoxylic acid monohydrate in acetic acid under reflux. Under these conditions the reactions proceed under thermodynamic control to afford the thermodynamically stable (2*R*,4*R*) derivative. Carrying out the reaction with a slight excess of glyoxylic acid monohydrate the (2*S*,4*R*)-1,3-thiazolidine-2,4-dicarboxylic acid is obtained. These reactions starting from *D*-cysteine yield (2*R*,4*S*)- and (2*S*,4*S*)-1,3-thiazolidine-2,4-dicarboxylic acid. These thiazolidines undergo epimerization at the C-2 position *via* the *ring opening-ring closure mechanism* in aqueous solutions under acidic, basic and neutral conditions but not in dimethyl sulfoxide solution.

The condensation of two equivalents of cysteine with α,β-unsaturated aldehydes (*e.g.* acroleine, crotonaldehyde, 4-hydroxypentenal, polyconjugated aldehydes) proceeds *via* Michael addition of cysteine to the double bond followed by the addition of the second mole of cysteine to form the corresponding thiazolidine **8** (Scheme 4).[8] This type of thiazolidines exhibits increased antitumor activity and decreased toxicity in comparison with the free aldehydes.

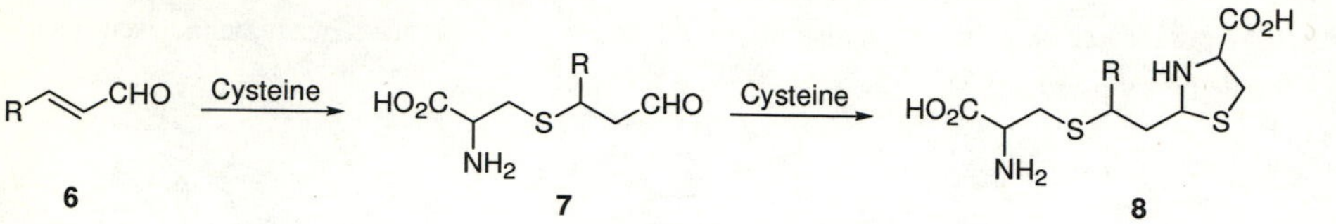

Scheme 4

The 1,3-dipolar cycloaddition of azomethine ylides with C=S dipolarophiles is a alternative approach to 1,3-thiazolidines. Gallagher and co-workers developed this strategy for the synthesis of penam skeleton.[9] β-Lactam based *N*-acyloxazolidinones **9** undergo a reversible two-step fragmentation: C5-O1 cleavage to give **10** followed by proton tautomerisation to give the azomethine ylide **11**. The thermolysis of **9** in the presence of thioketones gives penams **12** (Scheme 5).

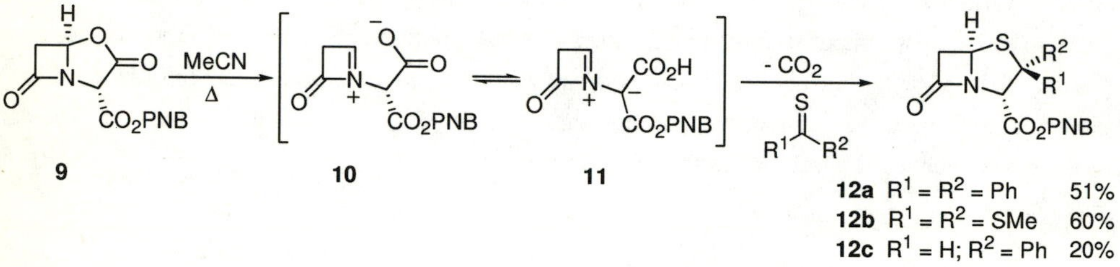

12a $R^1 = R^2 = Ph$ 51%
12b $R^1 = R^2 = SMe$ 60%
12c $R^1 = H; R^2 = Ph$ 20%

Scheme 5

Ar = 4-MeOC$_6$H$_4$

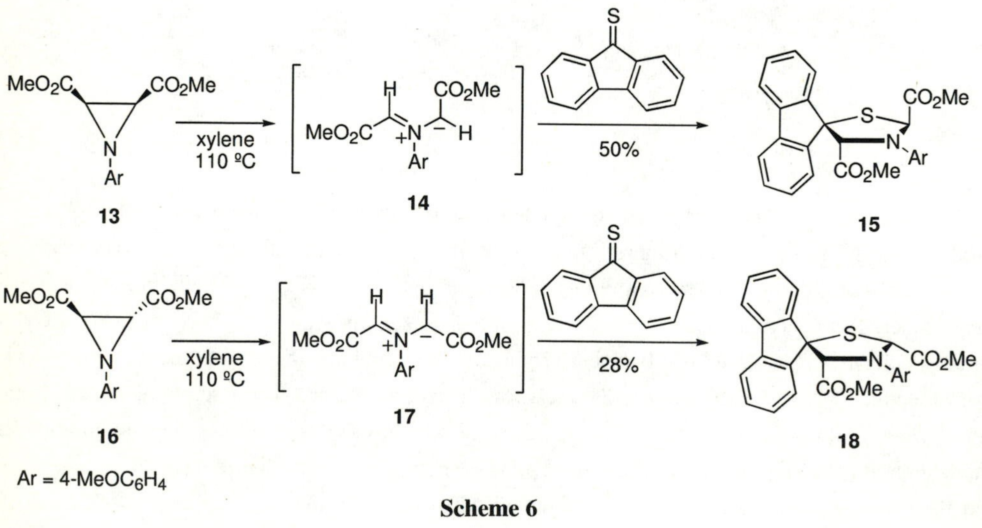

Scheme 6

(2*S*,4*R*)-1,3-Thiazolidine-2,4-dicarboxylic acid can also be prepared using the same concept as shown in Scheme 6.[10] The method to generate the azomethine ylides (**14** and **17**) was the stereoselective thermal ring opening of diethyl 1-(4-methoxyphenyl)aziridine-2,3-dicarboxylates (**13** and **16**) which, when reacted with 9*H*-fluorene-9-thione, give the thiazolidines (**15** and **18**).The (2*S*,4*R*)-1,3-thiazolidine-2,4-dicarboxylic acid can also be obtained from the reaction of *L*-cysteine with dichloroacetic acid under alkaline conditions.

The intermediate 2-amino-3-(carboxychloromethyl)-thiopropanoic acid **19** is formed and immediately undergoes intramolecular condensation giving the thiazolidine **20** as a diastereoisomeric mixture. The (2*S*,4*R*) derivative can be isolated by selective crystallisation (Scheme 7).[11]

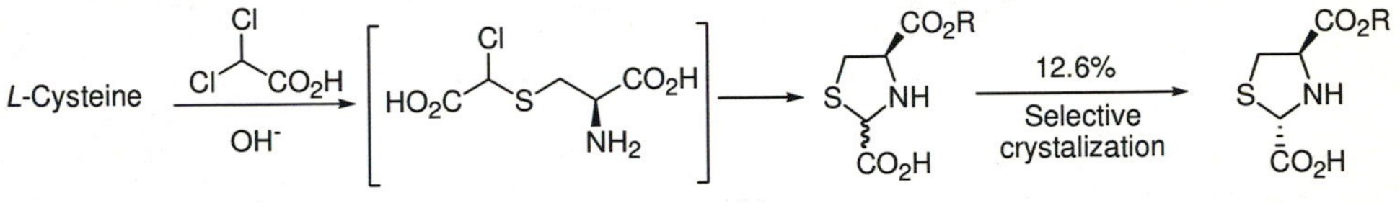

Scheme 7

This review will focus on the reactivity of 1,3-thiazolidine-4-carboxylic acids, in particular, their use as building blocks for the construction of 1,3-thiazolidine-annulated systems and acyclic species which in turn allow the synthesis of heterocycles.

2. Synthesis of 1,3-thiazolidine-annulated systems

The principle of *self-reproduction of chirality* can be illustrated with the α-alkylatin of thiazolidines.[12,13] In fact, cyclic amino acids such as thiazolidines can be alkylated without loss of optical activity. Seebach's approach involves bicyclic enolates (**24**) which can react with a wide variety of electrophiles at the *exo* face exclusively as illustrated in Scheme 8.

Pattenden and co-workers applied the same principle to prepare (*R*)-2-methylcysteine (**30**) and (*S*)-2-methylcysteine (**35**) hydrochloride (Scheme 9).[14] *N*-formylation of thiazolidine **26**, prepared from (*R*)-cysteine methyl ester hydrochloride, led to a single diastereoisomer **27**.

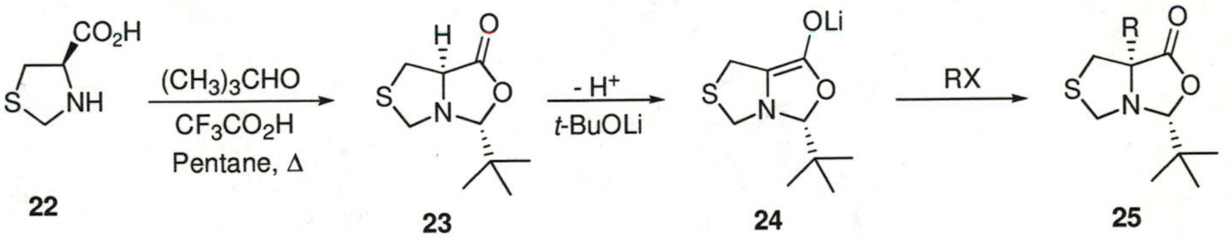

Scheme 8

Treatment of **27** with lithium diisopropylamide in the presence of DMPU at – 90 °C followed by reaction with iodomethane produced the corresponding 4-methylthiazolidine **29** containing the methyl and *tert*-butyl groups exclusively *anti* to each other. Hydrolysis of **29** with hydrochloric acid (5 M) afforded (*R*)-2-methylcysteine hydrochloride (**30**). Starting with (*S*)-cysteine methyl ester hydrochloride and using a similar synthetic strategy (*S*)-2-methylcysteine hydrochloride (**35**) was also prepared.

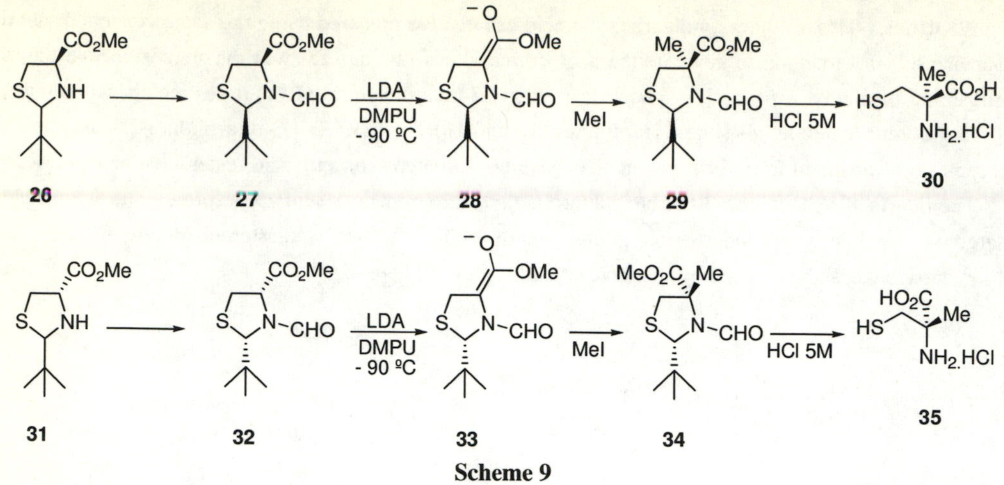
LDA
DMPU
- 90 ºC
MeI
HCl 5M
26
27
28
29
30
31
32
33
34
35

Scheme 9

(6*S*)-2-Oxo-8-thia-1,4,5-triazabicyclo[4.3.0]non-3-ene-6-carboxylates (**39**) can be obtained stereoselectively from 1,3-thiazolidine derivatives (Scheme 10).[15] The reaction of thiazolidine **36** with diketene and triethylamine or with methyl malonyl chloride and triethylamine (and a catalytic amount of DMAP) gives the acyl derivatives **37**. Under diazo-transfer conditions compounds **38** are obtained. The diazo esters **38** undergo a base induced cyclisation to give the bicyclic compounds **39**. The cyclisation is stereoselective, with retention of configuration of the chiral center, giving products with high enantiomeric purity.

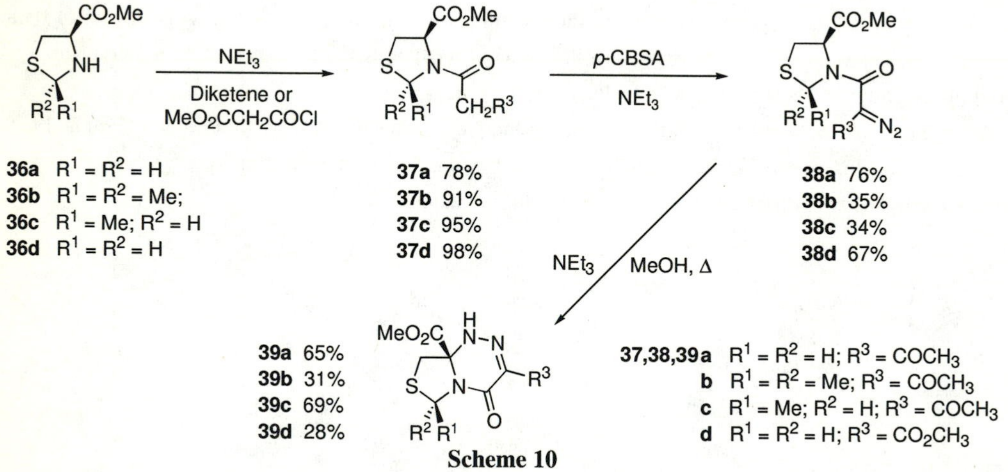
NEt3
Diketene or
MeO2CCH2COCl
p-CBSA
NEt3
NEt3
MeOH, Δ
36a R1 = R2 = H
36b R1 = R2 = Me;
36c R1 = Me; R2 = H
36d R1 = R2 = H
37a 78%
37b 91%
37c 95%
37d 98%
38a 76%
38b 35%
38c 34%
38d 67%
39a 65%
39b 31%
39c 69%
39d 28%
37,38,39a R1 = R2 = H; R3 = COCH3
b R1 = R2 = Me; R3 = COCH3
c R1 = Me; R2 = H; R3 = COCH3
d R1 = R2 = H; R3 = CO2CH3

Scheme 10

The authors[15] proposed that the reaction involves an ester enolate with axial chirality as intermediate. The stereoselectity can be rationalised considering a marked kinetic preference for the formation of one of the two possible enantiomeric ester enolates (Scheme 11). Considering the cyclisation of **38a** into **39a** the enolate **40** was considered to be involved and the kinetic preference for the diazo ester **38a** to undergo deprotonation to give **40** rather than its enantiomer **41** was attributed to the greater ease in attaining the required geometry for the deprotonation compared with **38a'** (having a destabilising interaction between the *N*-acyl substituent and the CO$_2$Me group). This cyclization in which the chirality of the starting material appears to be memorised in the enolate intermediate is referred as an example of *memory of chirality*.[16]

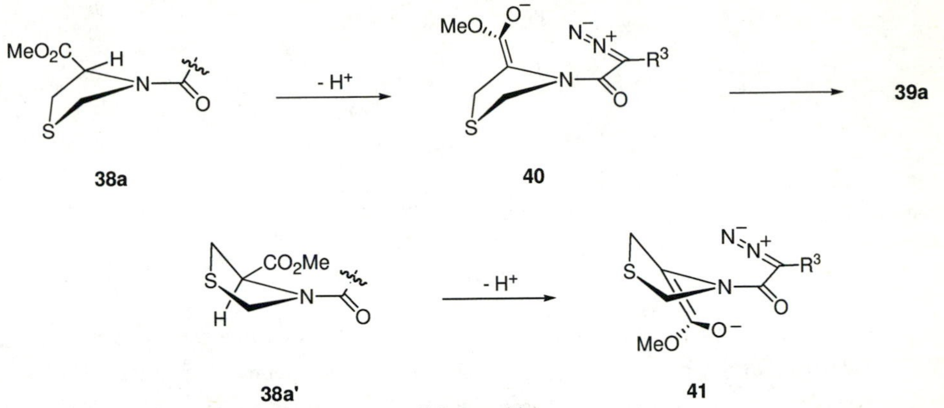

Scheme 11

The thiazolidine dioxide **42**, prepared from **38a** by oxidation with *m*-chloroperoxybenzoic acid, undergoes a similar stereoselective cyclisation, upon treatment with triethylamine in methanol under reflux, to give bicyclic sulfone **43** as single enantiomer in 86% yield. Methyl (1*R*,4*R*)-3-(2-diazo-3-oxobutanoyl)-1-oxo-1,3-thiazolidine-4-carboxylate **44** can also be obtained from **38a** in 25% yield using magnesium monoperoxyphthalate as oxidant. Compound **44** is converted, also with retention of configuration, into the bicycle sulfoxide **45** in 61% yield (Scheme 12).[17]

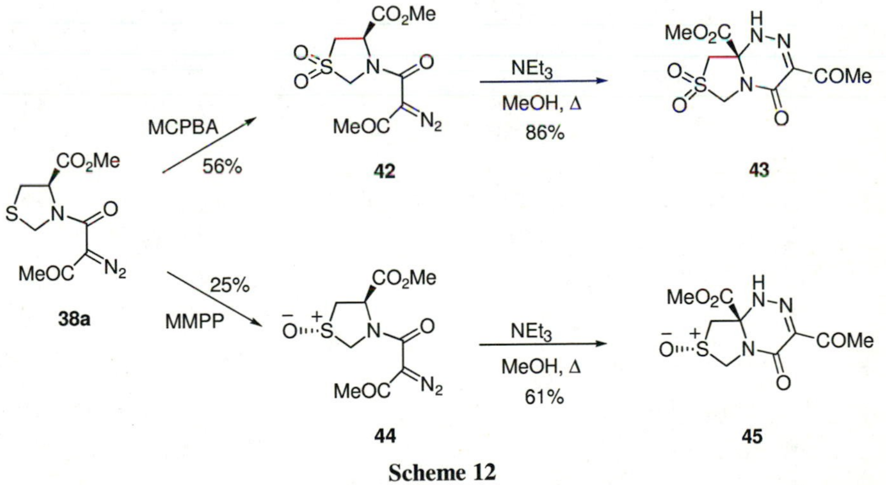

Scheme 12

Stoodley and co-workers have also reported that acetoacetylthiazolidine-4-carboxylates (**37a**, **46a** and **46b**) undergo an intramolecular aldol reaction in presence of KCN to give mixtures of the bicycles **47** and **48**. Based on HPLC analysis the authors showed that compounds **47c** and **48c** have ees of 99% suggesting

293

that an axially chiral enolate was involved as an intermediate. Sequential treatment of the mixtures **47a:48a** and **47c:48c** with Ac$_2$O-perchloric acid and DBU in CH$_2$Cl$_2$ gave thiazolidine-annulated hetorocycles **49** (Scheme 13).[18]

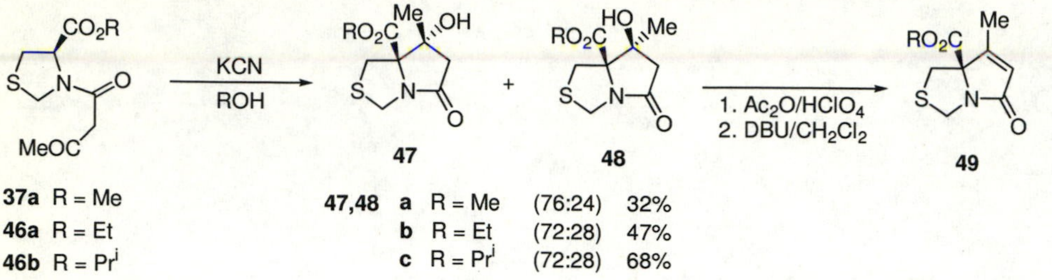

37a R = Me		**47,48 a** R = Me	(76:24)	32%
46a R = Et		**b** R = Et	(72:28)	47%
46b R = Pr^i		**c** R = Pr^i	(72:28)	68%

Scheme 13

Anthramycines (*e.g.* anthramycin methyl ether **50**) are naturally-occurring DNA-interactive antitumor antibiotics which structure incorporates the pyrrolo[1,2-*c*][1,4]benzodiazepine ring system. This prompted Erba and co-workers to study the synthesis of new compounds containing the pyrrolo[1,2-*c*][1,4]diazepine ring (Scheme 14).[19]

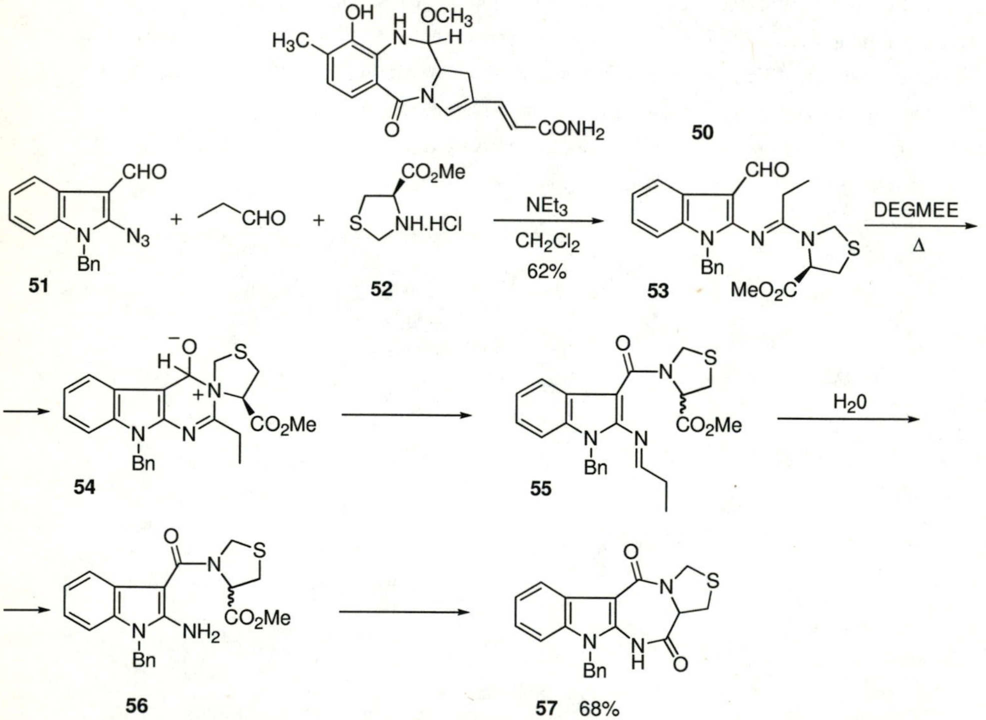

Scheme 14

The amidine **53** was obtained in good yield by reacting 1-benzyl-2-azido-indole-3-carbaldehyde (**51**) with propionaldehyde and methyl *L*-thiazolidine-4-carboxylate (**52**). Using diethyleneglycolmonoethyl ether

294

(DEGMEE) at reflux compound **53** afforded exclusively the new tetracyclic compound **57** in 68% yield. The formation of this pyrrolo[1,2-*c*][1,4]diazepine derivative can be explained considering the formation of a cyclic ammonium intermediate **54** which undergoes ring opening to give an imine **55**. The hydrolysis of **55** followed by cyclization gives the final product.

The synthesis of bisthiazolo[3,4-*a*:3',4'-*d*]pyrazine-5,10-dione derivatives from thiazolidine-4-carboxylic acids has been reported (Scheme 15).[20] The reaction of thiazolidines with phosgene affords the corresponding *N*-carboxy-α-amino acid anhydride (**59**) which leads to the formation of the first amide bond on reacting with ethyl (4*R*)-thiazolidine-4-carboxylate. The thermal treatment of the open dipeptide **60** allows the formation of the chiral tricyclic piperazine-2,5-diones (**61-63**).

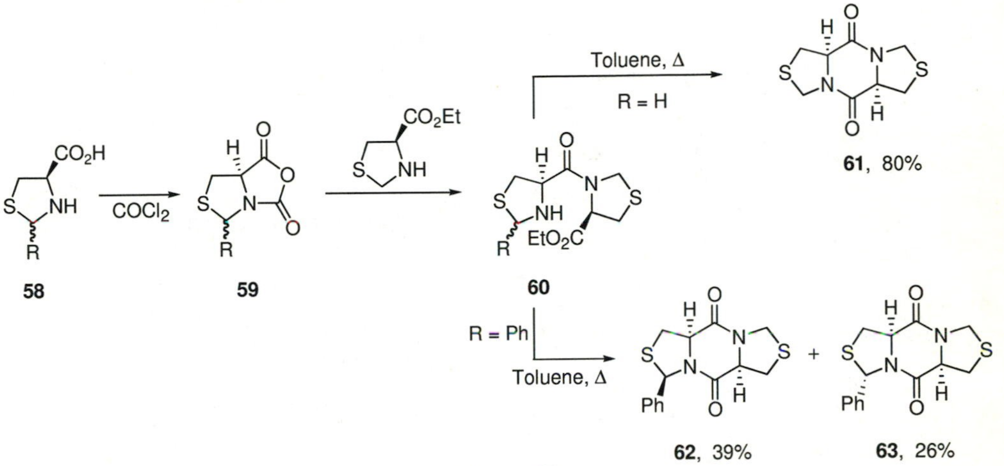

Scheme 15

A structural requirement for the antibacterial activity of β-lactam antibiotics is to have a suitable activated β-lactam ring. Considering that the β-acylaminopenems **64** have been considered too reactive for practical use as antibiotic, Baldwin and co-workers[21] decided to prepare a γ-lactam analogue **65** (Figure 1).

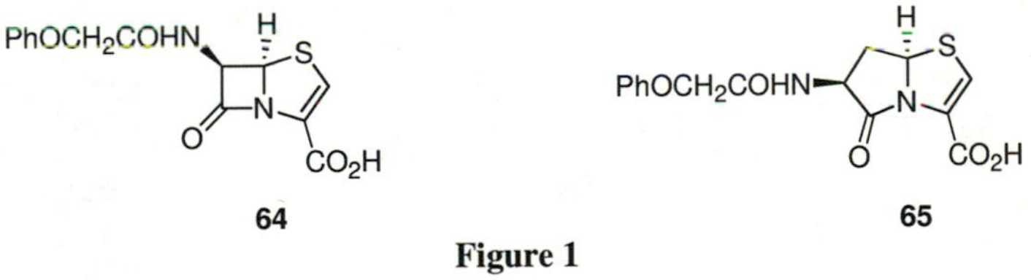

Figure 1

Condensation of aldehyde **66** with *D*-cysteine gave the crude bicyclic lactam **68** which was converted into the bicyclic ester **69** in 73% overall yield. Deprotection of the amino group followed by the reaction with phenoxyacetyl chloride in the presence of triethylamine gave **71** (92%).

Compound **71** was refluxed in benzene with benzoylperoxide (cat. cupric acetoacetate) leading to a single product, the *trans*-benzoate **72**, which undergoes a *syn*-elimination by thermolysis. The γ-lactam with the desired stereochemistry (**73**) was obtained in 47% yield although partial racemization was observed. Deprotection of the ester group gave the target molecule **65** (Scheme 16). However, compound **65** showed weak biological activity against *Staphylococcus aureus*.[21]

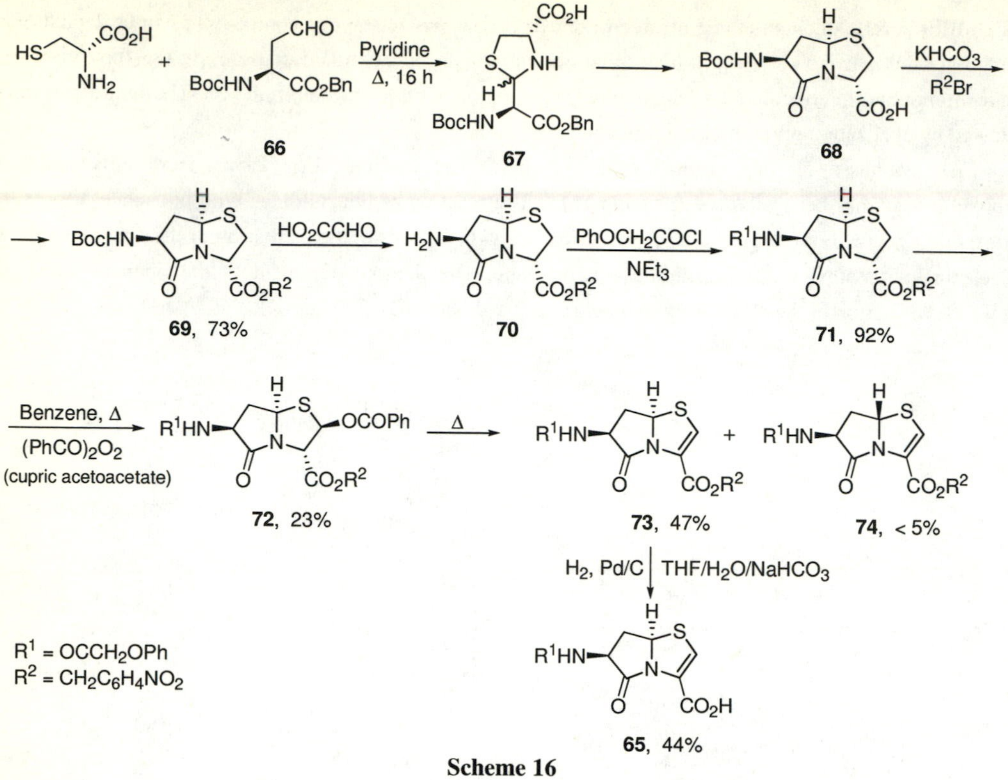

Scheme 16

HS—[CO$_2$H / NH$_2$] + BocHN—[CHO / CO$_2$Bn] →(Pyridine, Δ, 16 h) **66**

thiazolidine intermediate (CO$_2$H, S, NH, BocHN, CO$_2$Bn) **67**

BocHN— bicyclic (H, S, O, CO$_2$H) **68** →(KHCO$_3$, R^{2}Br)

→ BocHN— bicyclic (H, S, O, CO$_2$R^2) **69**, 73% →(HO$_2$CCHO)

H$_2$N— bicyclic (H, S, O, CO$_2$R^2) **70** →(PhOCH$_2$COCl, NEt$_3$)

R^{1}HN— bicyclic (H, S, O, CO$_2$R^2) **71**, 92% →

→(Benzene, Δ, (PhCO)$_2$O$_2$, (cupric acetoacetate)) R^{1}HN— bicyclic (H, S, OCOPh, O, CO$_2$R^2) **72**, 23% →(Δ)

R^{1}HN— bicyclic (H, S, O, CO$_2$R^2) **73**, 47% + R^{1}HN— bicyclic (H, S, O, CO$_2$R^2) **74**, < 5%

R^1 = OCCH$_2$OPh
R^2 = CH$_2$C$_6$H$_4$NO$_2$

73 →(H$_2$, Pd/C | THF/H$_2$O/NaHCO$_3$) R^{1}HN— bicyclic (H, S, O, CO$_2$H) **65**, 44%

Scheme 16

One approach to obtain information concerning the bioactive conformation of a peptide is to synthesise conformationally constrained analogues of the peptide, compounds to mimic certain secondary structural features of peptides which may play important roles in receptor recognition and biological activity. In this context Johnson and co-workers defined highly constrained spiro-bicyclic systems as their target to mimic the type II β-turn, a secondary structural feature found in many bioactive peptides.[22]

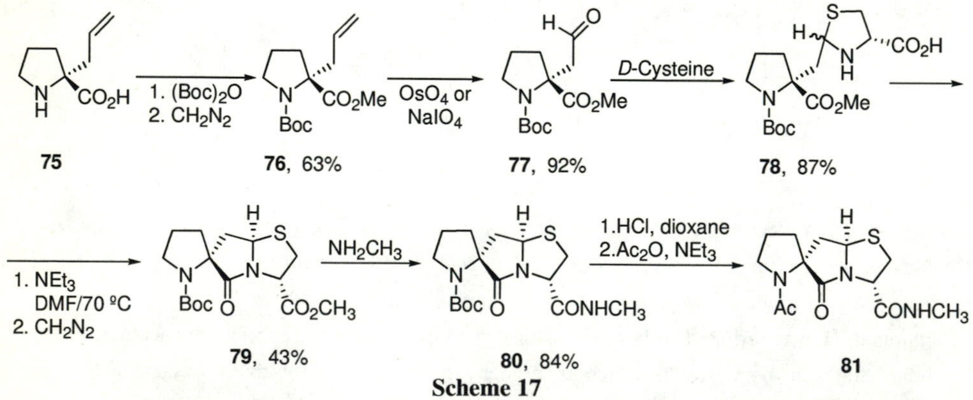

Scheme 17

allylproline (N-H, CO$_2$H) **75** →(1. (Boc)$_2$O, 2. CH$_2$N$_2$) allyl (N-Boc, CO$_2$Me) **76**, 63% →(OsO$_4$ or NaIO$_4$)

aldehyde (H, O, N-Boc, CO$_2$Me) **77**, 92% →(D-Cysteine) thiazolidine (S, N-H, CO$_2$H, N-Boc, CO$_2$Me) **78**, 87% →

→(1. NEt$_3$, DMF/70 ºC, 2. CH$_2$N$_2$) spiro-bicyclic (H, S, N-Boc, O, CO$_2$CH$_3$) **79**, 43% →(NH$_2$CH$_3$)

spiro-bicyclic (H, S, N-Boc, O, CONHCH$_3$) **80**, 84% →(1. HCl, dioxane, 2. Ac$_2$O, NEt$_3$) spiro-bicyclic (H, S, N-Ac, O, CONHCH$_3$) **81**

Scheme 17

The synthesis of spiro-bicyclic thiazolidine **81** is outlined in Scheme 17.[22a] The (R)-allylproline **75** was prepared from (S)-proline using Seebach's α-alkylation with self-reproduction of chirality.[23] Protection of

296

the nitrogen followed by the reaction with diazomethane gave **76**. Oxidative cleavage of the double bond afforded aldehyde **77** in good yield. This aldehyde was condensed with *D*-cysteine giving the corresponding thiazolidine (**78**) as a diastereoisomeric mixture. The authors used a modification of the procedure developed by Baldwin and co-workers[21] to cyclise the thiazolidine into the corresponding spiro-bicyclic obtained as a single stereoisomer. This compound was converted into ester **79** and the reaction with methylamine allowed the synthesis of **80**. Removal of the Boc protecting group followed by acylation with acetic anhydride gave spiro-bicyclic type II β-turn peptidomimetic **81**.The synthesis of spiro-bicyclic lactam *L*-proline-*L*-leucyl-glycinamide peptidomimetics **82** and **83** has also been reported.[22b] A synthetic strategy similar to the one described for **81** was used: in the synthesis of compound **82** homocysteine was used instead of *D*-cysteine and in the synthesis of **83** (*S*)-2-(3'-butenyl)proline was used instead of compound **75**. The construction of the bicyclic thiazolidines **84a-84b**[22c] was carried out using the general procedure described by Baldwin and co-workers.[21] Bicyclic thiazolidine lactam **84b** was designed as a compound that could not achieve a β-turn conformation. In order to enhance further the activity peptidomimetic of **84a** the synthesis of 6-substituted derivatives was also performed (**84c-84e**).[22d] The synthesis of peptidomimetic **87**[22c] (Scheme 18) was done using oxazolidinone aldehyde **85** as starting material which was prepared by the route described by Lee and Miller.[24]

The work developed by Johnson and co-workers[22] allowed to gather data in support of the hypothesis that the bioactive conformation of *L*-proline-*L*-leucyl-glycinamide is a type II β-turn.

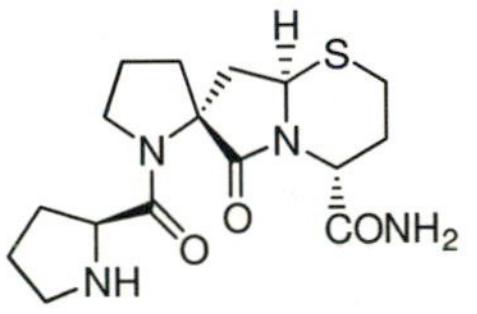

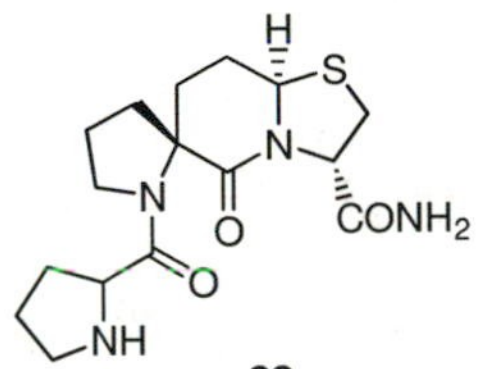

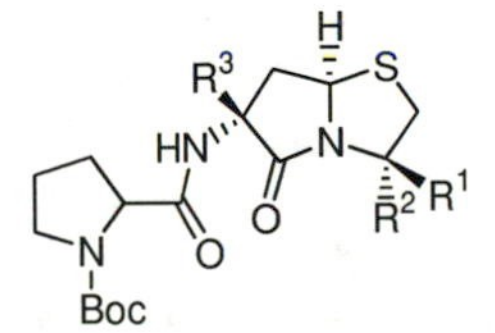

84a R^1 = H;R^2 = CONH$_2$; R^3 = H
84b R^1 = CONH$_2$;R^2 = H; R^3 = H
84c R^1 = H; R^2 = CONH$_2$; R^3 = CH$_2$CH(CH$_3$)$_2$
84d R^1 = H; R^2 = CONH$_2$; R^3 = (CH$_2$)$_3$CH$_3$
84e R^1 = H; R^2 = CONH$_2$; R^3 = CH$_2$Ph

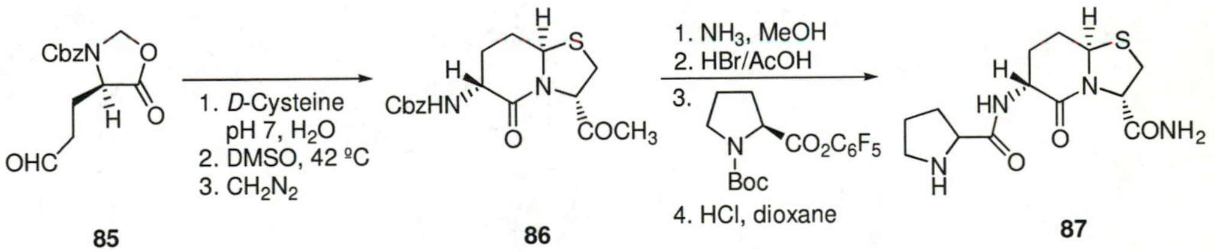

Scheme 18

Geyer and co-workers combined the rigidity of the bicyclic dipeptide isosters with the polyfunctional character of carbohydrate to give constrained polyol peptidomimetics (Schemes 19 and 20).[25] The authors found reaction conditions to prevent the synthesis of the thiazolidine **89** in favour of the bicyclic compound **92**. Condensation of *D*-glucurono-3,6-lactone and *L*-cysteine methyl ester carried out in a solvent mixture of water/pyridine (9:1) gave the 7,5-bicyclic thiazolidinelactam **92** with high diastereoselectivity. The thiazolidine **89** formed in this process is in equilibrium with the starting material whereas the annulation proceeds with irreversible amide bond formation followed by *N*-acyliminium ion (**91**) cyclization and addition of the thiol group. Thus, the bicyclic **92** can be obtained in high yield (Scheme 19).

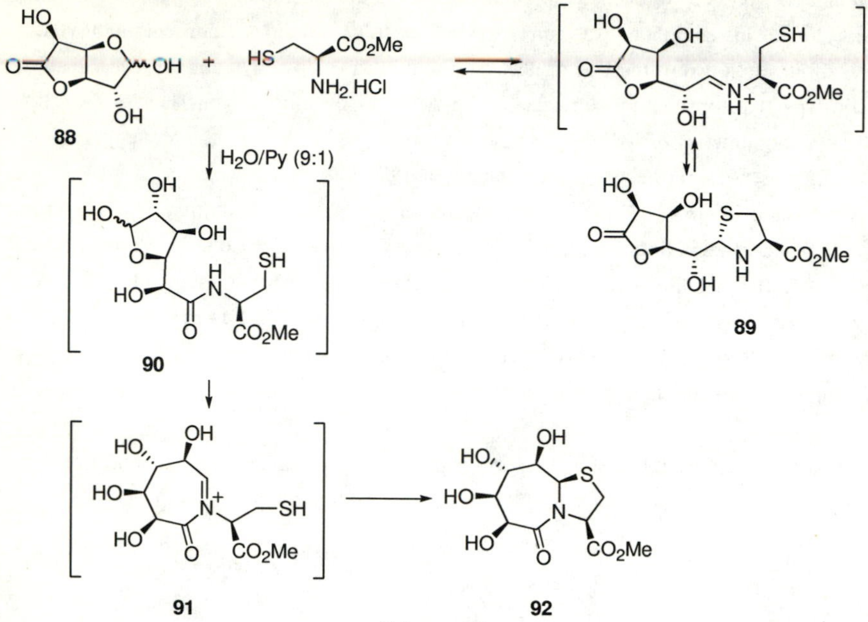

Scheme 19

The reaction of **92** with Boc-Ala-OH gives the depsipeptide **93** and elongation of the carboxy terminus yields **94** and **95** (Figure 2).

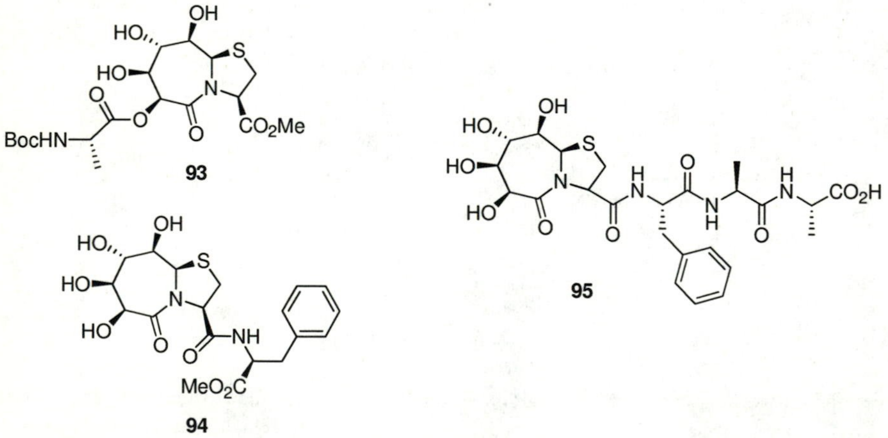

Figure 2

The α-hydroxy group of the *D*-glucurono-3,6-lactone (**88**) can be exchanged for an amino function and condensed with *L*-cysteine methyl ester to give the bicyclic peptidomimetic **98**. The amino group of this compound can be deprotected and the peptidic coupling allows the formation of **100** (Scheme 20).[25] These 7,5-bicyclic polyol thiazolidinelactams mimic hydroxy amino acids and can be further functionalized.

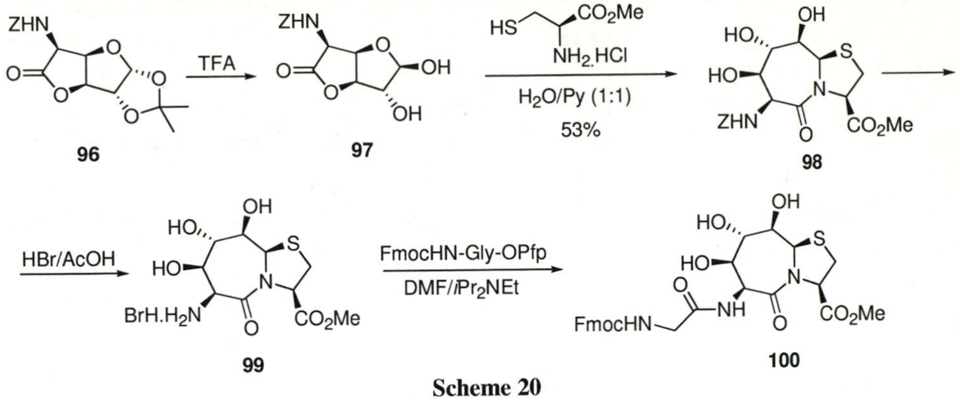

96   **97**   **98**

99   **100**

Scheme 20

Katritzky and co-workers[26] have described the synthesis of chiral hexahydropyrrolo [2,1-*b*][1,3]thiazoles (Scheme 21 and Table 1). The hexahydropyrrolo[2,1-*b*][1,3]thiazole ring system (**101**) was obtained *via* the Mannich condensation of *L*-cysteine ethyl ester hydrochloride with succindialdehyde and benzotriazole. The benzotriazolyl group of compound **101** undergoes nucleophilic substitution with a range of nucleophiles giving the hexahydropyrrolo[2,1-*b*][1,3]thiazoles **102**, in some cases with high diastereoselectivity.

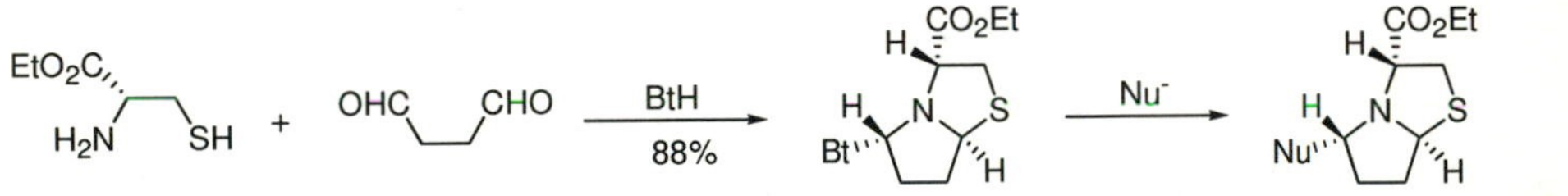

101, D.e. > 99%   **102** Major stereoisomer

Scheme 21

Table 1. Yields and D.e. values for compound **102**.

Reagents	Nu	**102**, Yield (%)	**102**, D.e. (%)
PhSH/NaH	PhS	62	81
NaCN	CN	61	> 99
PO(OEt)$_3$ (ZnBr$_2$)	PO(OEt)$_2$	56	> 99
CH$_2$=CHCH$_2$TMS*	CH$_2$=CHCH$_2$	72	54
CH$_2$=C(CH$_3$)CH$_2$TMS*	CH$_2$=C(CH$_3$)CH$_2$	77	80
CH$_2$C(Ph)OTMS*	PhCOCH$_2$	77	50
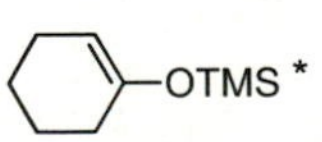	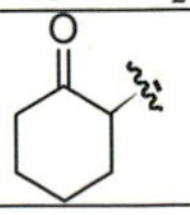	76	> 99
(CH$_3$)$_2$CC(OCH$_3$)OTMS*	C(CH$_3$)$_2$CO$_2$CH$_3$	82	> 99
PhZnCl	Ph	65	60

* in presence of BF$_3$.Et$_2$O.

299

The reaction of 2-hydroxymethylthiazolidines with phosgene was the strategy used by González and co-workers[27] to prepare dihydro-thiazolo[3,4-c]oxazol-3-ones (Scheme 22). The reaction of (R)-cysteine with paraformaldehyde gives dihydro-thiazolo[3,4-c]oxazol-1-one 103 in quantitative yield in one step. Treatment of 103 with methyllithium leads to the thiazolidine alcohol 104 (40%) and subsequent reaction with phosgene affords 105 in good yield. The synthesis of the unsubstituted parent system 106a was carried out in a similar way, by the reaction of 108a with phosgene. Dihydro-thiazolo[3,4-c]oxazol-3-one derivatives substituted at C-5 were also prepared. The thiazolidine alcohol 107b and 107c are obtained as diastereoisomeric mixtures but the reaction with phosgene proceeded with selectivity which can be explained by a ring opening-ring closure mechanism. In the case of the phenyl derivative (108b) the diastereoisomer with R configuration at C-5 was the major product (64%), the epimer being isolated in 14% yield. The tert-butyl analogue (108c) was obtained as single stereoisomer (73%) but its stereochemistry was not determined.

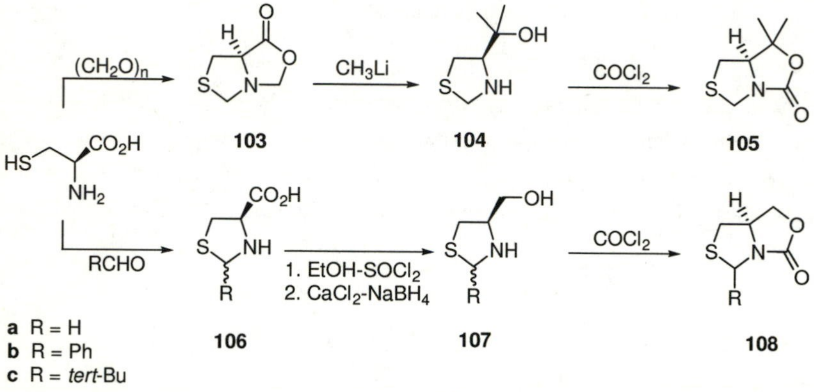

Scheme 22

The reaction of ethyl (S)-thiazolidine-4-carboxylate with isocyanates and isothiocyanates can be used to prepare bicyclic heterocycles containing an imidazolidine ring, a class of compounds known as biologically active namely compounds with fungicidal activity.[28] Carrying out the reaction of thiazolidine 109 with methyl isothiocyanate, phenyl isocyanate and phenyl isothiocyanate in acetonitrile at room temperature (1 hour) the corresponding 6-methyl-5-thioxo-tetrahydro-imidazo[1,5-c]thiazol-7-one (111b), 6-phenyl-dihydroimidazo-[1,5-c]thiazole-5,7-dione (111c), and 6-phenyl-5-thioxo-tetrahydroimidazo [1,5-c]thiazol-7-one (111d) were obtained directly. With the least reactive methyl isocyanate the thiazolidine derivative 110a was obtained but attempts to promote its cyclization were not successful (Scheme 23).

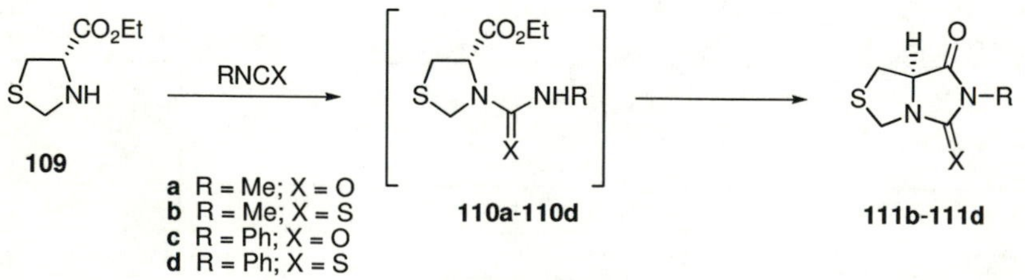

Scheme 23

300

The chemical industry also reported the synthesis of a range of 6-aryl-dihydro-imidazo[1,5-c]thiazole-5,7-dione derivatives, obtained from the reaction of thiazolidine-4-carboxylic acids or esters with aryl isocyanate, which are active against *Botrytis cinerea*.[29]

Györgydeák and co-workers[30] reported that the reaction of (2R,4S)- and (2S,4S)-2-(2-thienyl)-5,6-dimethylthiazolidine-4-carboxylic acid **112**, containing two chiral centers, with aryl isothiocyanates affords 5-aryl-3,3-dimethyl-1-(2-thienyl)-4-thioxo-tetrahydro-imidazo[1,5-c]thiazol-6-one **(116)** with (1R)-configuration. The thioureido acid intermediate **113** is formed as single diastereoisomer *via* a ring opening-ring closure mechanism. These acids were cyclized into **116** under thermal conditions. Compound **113** undergoes ring opening and subsequent epimerization-ring closure leading exclusively to the less strained fused ring system with (1R,4S) configuration (Scheme 24).

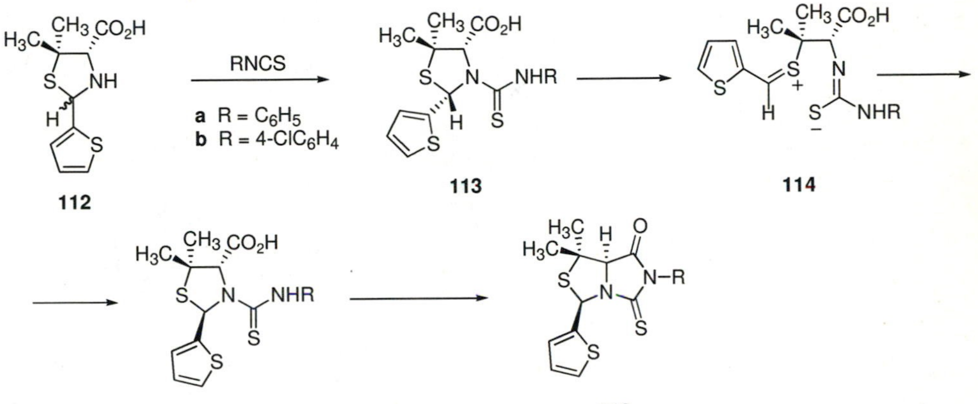

Scheme 24

Reactions of thiazolidine-2,4-dicarboxylic acids or their esters (**117**) with phenyl isothiocyanate lead to bicyclic compounds in a regio- and stereoselective manner involving the carboxylic group at C-4.[31,32] The 5,5-dimethylthiazolidine-2,4-dicarboxylate **122** shows the same regioselectivity in the reaction with phenyl isothiocyanate giving compound **123**. However, the bicyclic compound **121** obtained from 5,5-dimethylthiazolidine-2,4-dicarboxylic acid results from the cyclization involving the carboxylic group at C-2 (Scheme 25).[30,32] It is noteworthy that the configuration at C-1 of the bicyclic products is *R* and is independent of the configuration at C-4, the substituent at C-2, the configuration of the thiazolidine and the direction of the cyclization.

The methyl (4R)-thiazolidine-4-carboxylate can be converted into (8aS)-7-methoxycarbonylmethyl-5,8-dioxo-tetrahydrothiazolo[3,4-a]pyrazine **127** *via* the *N*-acyl thiazolidine **126** (R=H) (Scheme 26).[33] Couquelet and co-workers also reported that the *N*-acylation of thiazolidine-2,4-dicarboxylate with chloroacetyl chloride followed by the reaction with benzylamine gave two products the bicyclic compounds **128** (52%) and **129** (24%) although the stereochemistry of these compounds was not determined.[31]

The synthesis of tricyclic isoindole derivatives from *L*-cysteine has been reported. The reaction of *L*-cysteine methyl ester hydrochloride with 2-carboxybenzaldehyde in the presence of potassium hydrogen carbonate was carried out at room temperature (30 min.) and resulted in the direct diastereoselective synthesis of methyl (3*R*,9b*S*)-5-oxo-2,3,5,9b-tetrahydro-thiazolo[2,3-*a*]isoindole-3-carboxylate **130a** in 71% yield (Scheme 27).[34]

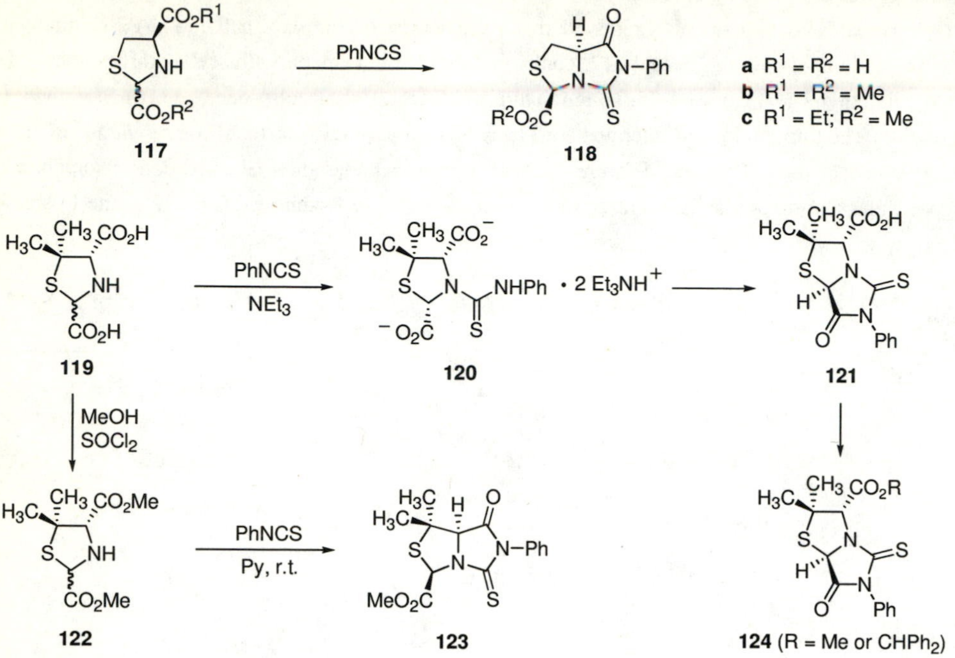

a R¹ = R² = H
b R¹ = R² = Me
c R¹ = Et; R² = Me

Scheme 25

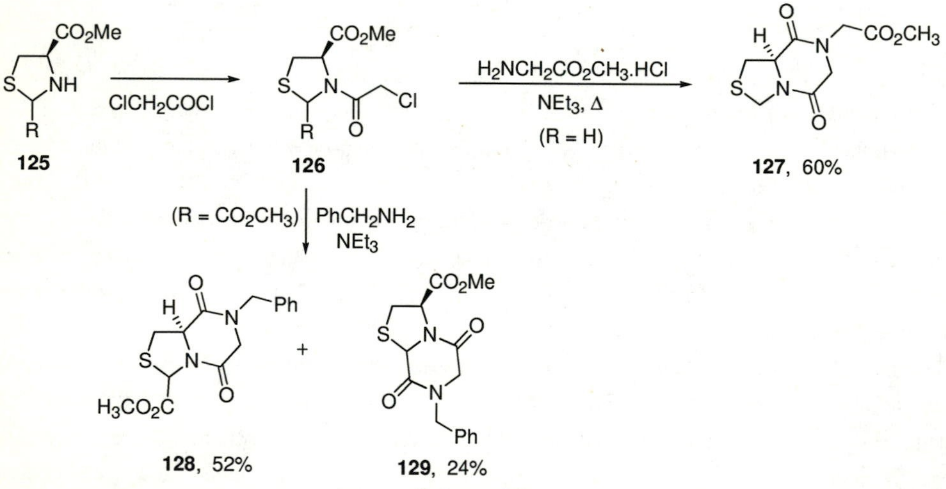

Scheme 26

The synthesis of compound **130a** has also been reported by Allin and co-workers using more drastic reaction conditions.[35] Compound **130a** was converted into the corresponding acid **131a** in 91% yield by the reaction with lithium iodide in ethyl acetate and treatment with aqueous HCl. Oliver and co-workers

302

reported the synthesis of 5-oxo-2,3,5,9b-tetrahydrothiazolo[2,3-*a*]isoindole-3-carboxylic acid **131a** directly from the reaction of 2-carboxybenzaldehyde with cysteine hydrochloride in the presence of pyridine.[36] The reaction of *L*-cysteine methyl ester with 2-acetylbenzoic acid was carried out in the presence of sodium acetate in refluxing toluene for 5 hours giving (3*R*,9b*S*)-9b-methyl-5-oxo-2,3,5,9b-tetrahydrothiazolo[2,3-*a*]isoindole-3-carboxylate **130b** in 75% yield. Compound **131b** was obtained from **130b** in 85% yield. The reaction of (3*R*,9b*S*)-5-oxo-2,3,5,9b-tetrahydrothiazolo[2,3-*a*]isoindole-3-carboxylic acid **131a** and (3*R*,9b*S*)-9b-methyl-5-oxo-2,3,5,9b-tetrahydrothiazolo[2,3-*a*]isoindole-3-carboxylic acid **131b** with acetic anhydride in a sealed tube was carrying out.[34] The solution was heated at 150 °C for four hours giving 3-methylene-2,5-dioxo-3*H*,9b*H*-oxazolo[2,3-*a*]isoindole **135a** (37%) and 9b-methyl-3-methylene-2,5-dioxo-3*H*,9b*H*-oxazolo[2,3-*a*]isoindole **135b** (64%). The mechanism proposed for the formation of compound **6** can be regarded as involving the formal elimination of the elements of SH from (3*R*,9b*S*)-5-oxo-2,3,5,9b-tetrahydrothiazolo[2,3-*a*]isoindole-3-carboxylic acid **131a** leading to *N*-acyliminium ion **134** followed by a 5-*endo*-trig cyclization. The isoindole derivatives (**135**) represent a new member to a class of compounds having a significant number of applications.[37]

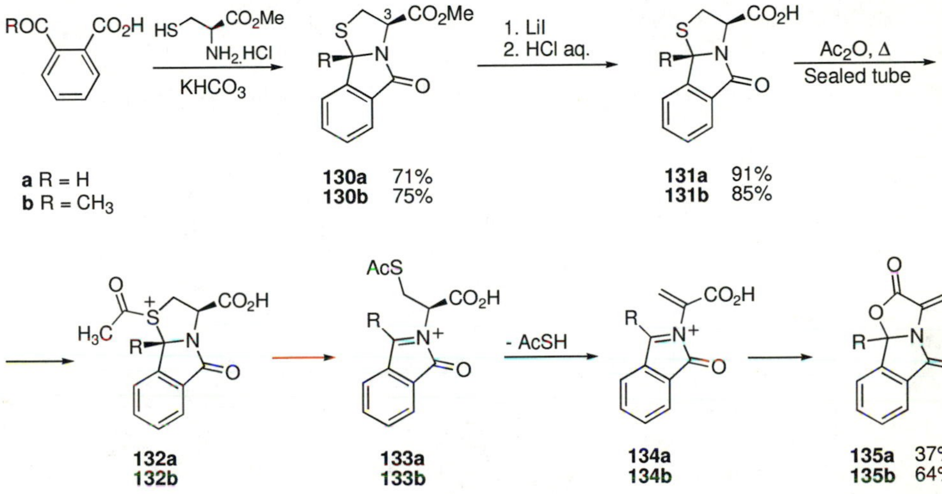

ROC
CO2H
HS
CO2Me
NH2.HCl
KHCO3
a R = H
b R = CH3
130a 71%
130b 75%
1. LiI
2. HCl aq.
131a 91%
131b 85%
Ac2O, Δ
Sealed tube
AcS
- AcSH
132a
132b
133a
133b
134a
134b
135a 37%
135b 64%

Scheme 27

On flash vacuum pyrolysis (600 °C/ $3x10^{-2}$-$4x10^{-2}$ mbar) (3*R*,9b*S*)-5-oxo-2,3,5,9b-tetrahydro-thiazolo[2,3-*a*]isoindole-3-carboxylic acids **131a** and **131b** undergo decarboxylation to the corresponding chiral (9b*S*)-5-oxo-2,3,5,9b-tetrahydrothiazolo[2,3-*a*]isoindoles (**136a** and **136b**) in moderate yields[34b] (Scheme 28). Compounds **136a** and **136b** have been prepared before from the reaction of carboxybenzaldehyde or 2-acetylbenzoic acid with 2-aminoethanethiol. However, they were obtained as racemic mixtures.[35,36,38] Some 5-oxo-2,3,5,9b-tetrahydrothiazolo[2,3-*a*]isoindoles substituted at C-9 with aryl and heteroaromatic groups have also been prepared as racemic mixtures although the separation of both enantiomers can be achieved by chromatography on cellulose triacetate.[37b]

The chemistry of [1,3]thiazolo[3,2-*c*][1,3]benzoxazines is an area of considerable interest since some derivatives show biological activity namely immunoactivating action which makes the synthesis of this ring system more appealing.[39] This tricyclic ring system can be prepared as outlined in Scheme 29.[34b] Thiazolidine 137 was prepared by condensing L-cysteine methyl ester hydrochloride with salicylaldehyde in presence of potassium hydrogen carbonate.

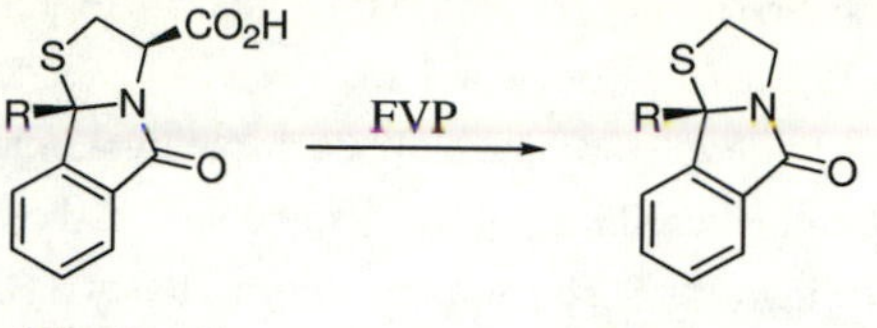

131a, R = H
131b, R = CH$_3$

136a, R = H   29%
136b, R = CH$_3$ 21%

Scheme 28

The approach to construct the six-membered ring was to react methyl 2-(2-hydroxyphenyl)thiazolidine-4-carboxylate 137 with phosgene. The reaction was carried out at room temperature and after six hours (2R,4R)-N-chlorocarbonyl-2-(2-hydroxyphenyl)thiazolidine-4-carboxylate 138 was obtained diastereoselectively in 67% yield. The methyl (3R,10bR)-5-oxo-2,3-dihydro-10bH-[1.3]thiazolo[3,2-c][1,3]benzoxazine-3-carboxylate 139 (87.5%) was obtained treating thiazolidine 138 with DBU. The subsequent reaction of compound 139 with lithium iodide in ethyl acetate followed by treatment with aqueous HCl allows the synthesis of (3R,10bR)-5-oxo-2,3-dihydro-10bH-[1.3]thiazolo[3,2-c][1,3]benzoxazine-3-carboxylic acid 140.

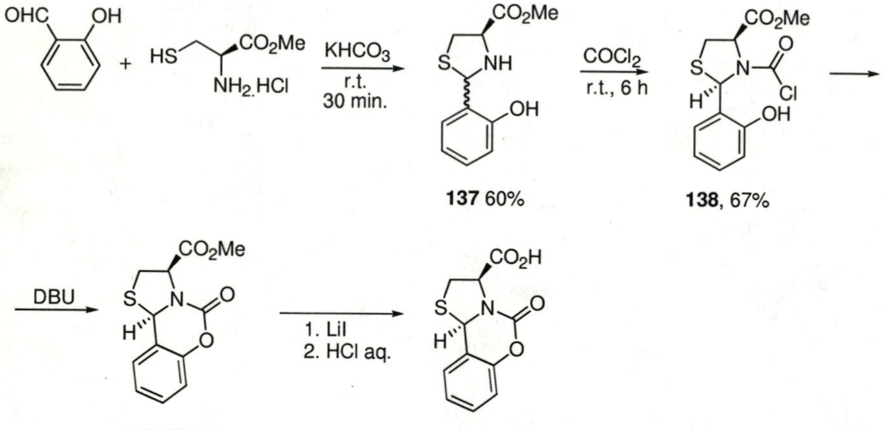

137 60%

138, 67%

139, 87.5%

140, 71%

Scheme 29

The thermolysis of 140 in acetic anhydride, carried out in a sealed tube, led to the synthesis of 5-acetyl-2-phenyl-2,3-dihydrothiazole 143 in low yield[34b] (Scheme 30). The formation of this product can be rationalised as involving a double decarboxylation giving 141 which is converted into 2,3-dihydrothiazole 142 through prototropy. Acylation of this intermediate gives compound 143.

(+)-Biotin is a water-soluble B vitamin that has attracted considerable attention as a synthetic target because of its biological properties for human nutrition and animal health. Several enantiospecific synthesis of (+)-biotin starting from L-cysteine have been reported and are covered in the De Clerq's review.[40]

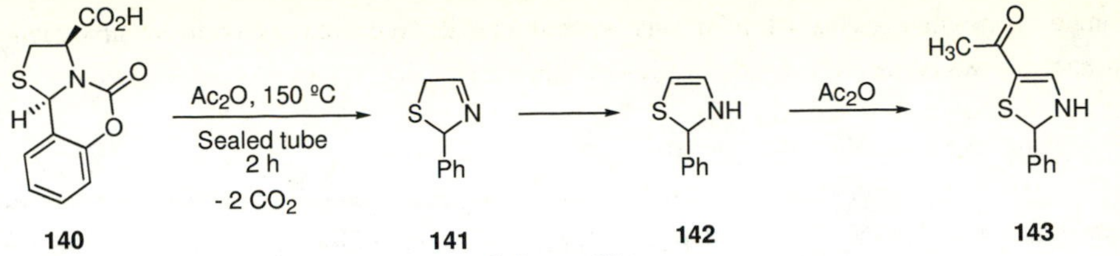

140   Ac₂O, 150 °C, Sealed tube 2 h, -2 CO₂   **141**   **142**   Ac₂O   **143**

Scheme 30

More recently Seki and co-workers synthesised (+)-biotin (**155**) from *L*-cysteine in 25% overall yield in a process involving eleven steps (Scheme 31).[41] (+)-Biotin possesses three contiguous stereocenters on the thiophane ring in the *all-cis* configuration. These asymmetric centers were formed through a stereoselective Lewis base-catalyzed cyanosilylation of aldehyde **147** giving **148**.

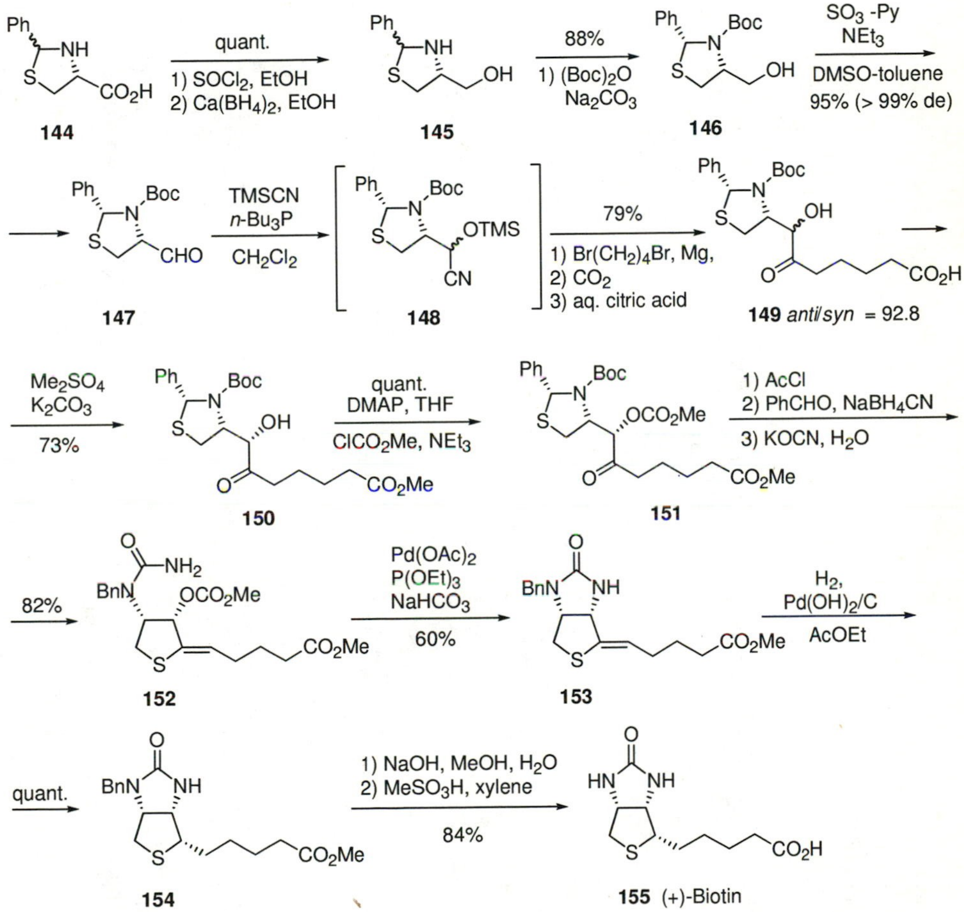

Scheme 31

The 4-carboxybutyl chain was formed by reacting **148** with a di-Grignard reagent derived from 1,4-dibromobutane and subsequent treatment with carbon dioxide. A palladium-catalyzed intramolecular allylic amination of *cis*–allylic carbonate **152** established the bicyclic ring system. Improvement of this general synthetic strategy was reported.[42]

3. Dipoles from 1,3-thiazolidine-4-carboxylic acids

1,3-Thiazolidine-4-carboxylic acids can be used for the generation of nonstabilized azomethine ylides by decarboxylative condensation with enolizable aldehydes (Scheme 32).[43] The reaction of thiazolidine-4-carboxylic acid **157** with methyl (*E*)-4-(2-formylphenoxy)-2-butenoate gives the *cis* fused cycloadducts **162** (68%) and **163** (22%). The major product is the result of the intramolecular cycloaddition of the *anti* ylide **160**. This ylide is generated from the iminium intermediate **158** which undergoes a rapid cyclization into 5-oxazolidinone intermediate **159** followed by the elimination of carbon dioxide. Ylide **160** equilibrates with *syn* ylide **161** and its intramolecular 1,3-dipolar cycloaddition leads to cycloadduct **163**.

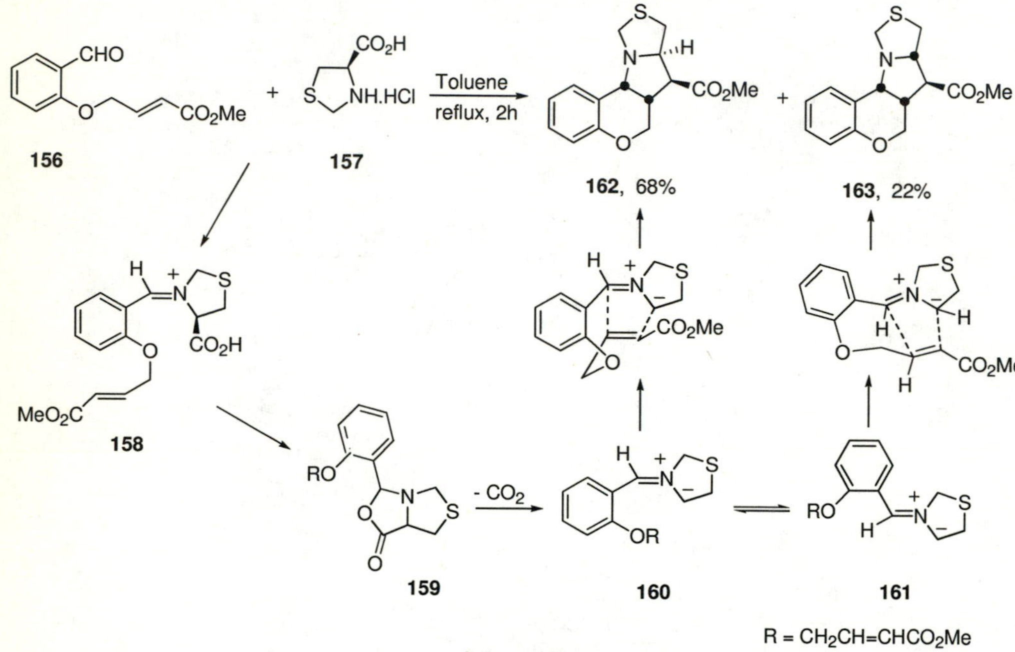

Scheme 32

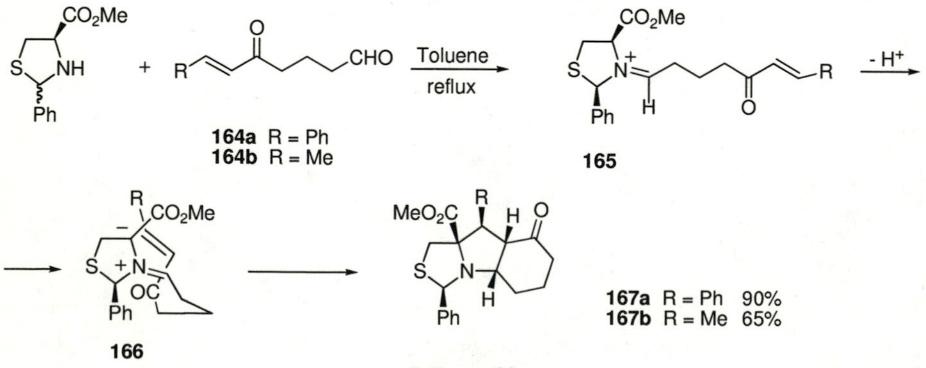

Scheme 33

306

Kanemasa and co-workers have also reported that cycloadducts **167a** and **167b** can be obtained as single stereoisomers in 90% and 65%, respectively, from the reaction of aldehydes **164a** and **164b** with methyl 2-phenylthiazolidine-4-carboxylate under reflux in toluene (Scheme 33).[44] The ester-stabilized azomethine ylide **166** is generated from deprotonation of the iminium intermediate (**165**) leading to the selective formation of products with *cis* ring junction.

In contrast with the preceding results the intramolecular 1,3-dipolar cycloaddition of non stabilized azomethine ylides generated from decarboxylative condensation of 2-phenylthiazolidine-4-carboxylic acid with **168a** and **168b** involves the carbonyl group as the dipolarophile giving, respectively, compounds **170a** and **170b** as single stereoisomers (Scheme 34).[44] The *trans* configuration of 4a-H and 8a-H is due to the selective generation of azomethine ylide **169** which leads to absolute diastereofacial selectivity. No addition to the carbon-carbon double bond is observed, which can be explained considering that a non statbilized azomethine ylide having a high-lying HOMO reacts preferentially with the carbonyl group characterised by a low-lying LUMO.

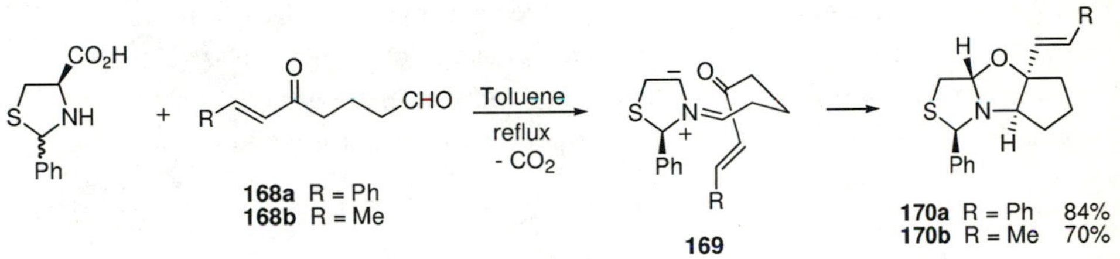

CO2H
S
NH
Ph
+
R
O
CHO
Toluene
reflux
- CO2
168a R = Ph
168b R = Me
S
N
+
Ph
O
R
169
H
O
R
S
N
H
Ph
170a R = Ph 84%
170b R = Me 70%

Scheme 34

Selecting an olefin aldehyde with one carbon shorter then **168** the intramolecular cycloaddition to carbonyl become extremely difficult. In fact, in the reaction of 2-phenylthiazolidine-4-carboxylic acid with **171** the carbon-carbon double bond acts as dipolarophile in the cycloaddition with the *in situ* generated azomethine ylide (Scheme 35).[44] Two regioisomers are obtained each as a single stereoisomer since the internal olefinic dipolarophile approaches the ylidic face from the side opposite to the R[1] substituent.

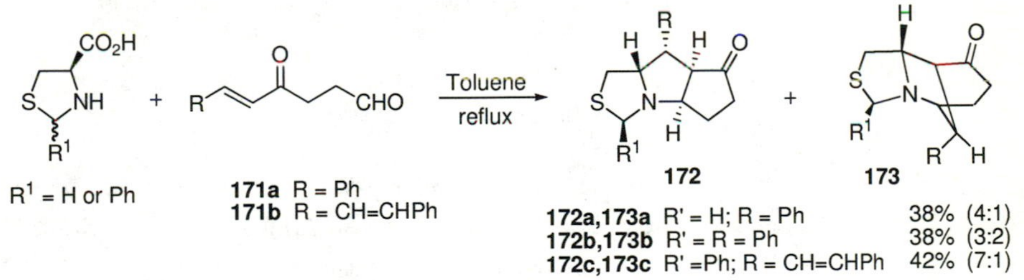

CO2H
S
NH
R1
+
R
O
CHO
Toluene
reflux
R1 = H or Ph
171a R = Ph
171b R = CH=CHPh
R
H
H
O
S
N
H
R1
172
+
H
O
S
N
R1
R
H
173
172a,173a R' = H; R = Ph 38% (4:1)
172b,173b R' = R = Ph 38% (3:2)
172c,173c R' =Ph; R = CH=CHPh 42% (7:1)

Scheme 35

A change in stereo- and regioselectivity is observed when olefin aldehyde bearing an ester group at the terminal olefinic carbon (**174**) was employed (Scheme 36).[44] Although no regioisomers were produced a 1:1 mixture of stereoisomeric cycloadducts (**175** and **176**) was obtained in 58% yield. Compound **176** results from the approach of the internal dipolarophile to the sterically hindered face of the azomethine ylide intermediate. In this case, the attractive secondary orbital interaction between the methoxycarbonyl group and the 2-Ph allows to overcome the repulsive interaction between them.

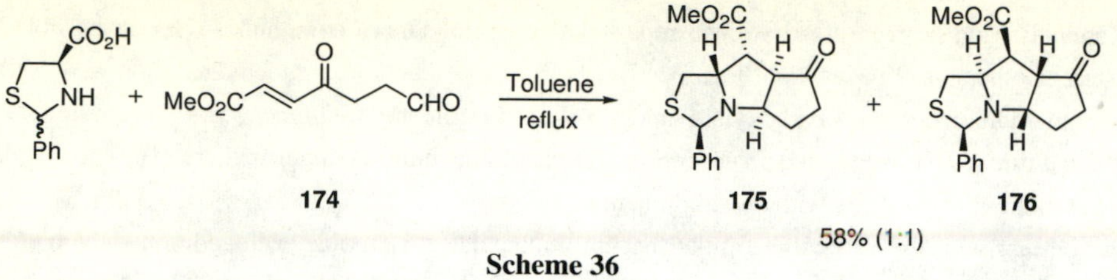

Scheme 36

The dipolar cycloaddition of münchnones (oxazolium-5-oxides) represents a particularly attractive approach to the synthesis of pyrroles. This mesoionic ring system acts as cyclic azomethine ylide in the reaction with dipolarophiles leading to the corresponding cycloadduct, followed by cycloreversion of carbon dioxide generating pyrrole derivatives. Bicyclic mesoionic ring system provide a route to heterocycles in which another ring system is annulated to pyrrole. This type of münchnones can be generated by cyclodehydration of cyclic N-acyl-α-amino acids with reagents such as acetic anhydride or carbodiimides.

Thus, N-acylthiazolidine-4-carboxylic acids have been used to generate bicyclic münchnones, *5H,7H*-thiazolo[3,4-*c*]oxazol-4-ium-1-olates which give some examples of intermolecular 1,3-dipolar addition, namely the addition to dimethyl acetylenedicarboxylate,[45] to α-chloroacrylonitrile[46] and to imines.[47] The products from dimethyl acetylenedicarboxylate are 1H,3H-pyrrolo[1,2-*c*]thiazole derivatives and this type of heterocyclic compounds can also be obtained from the reaction with α-chloroacrylonitrile and ethyl propiolate[46b] (generated *in situ* from ClCH$_2$(Cl)CHCO$_2$Et/NEt$_3$/TsCl).

The reaction of the bicyclic mesoionic compound **177** with N-phenyl methylenebenzenesulfonamide affords mixtures of diastereoisomeric spirocyclic β-lactams (**178**) and/or 1H-imidazo[1,5-*c*]thiazoles (**179**) (Scheme 37).[47]

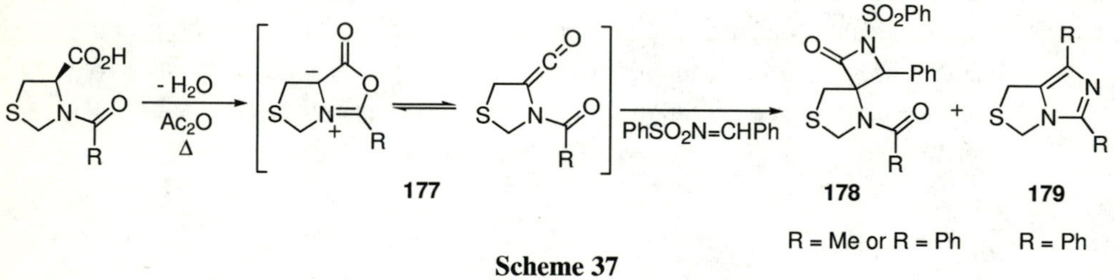

Scheme 37

Györgydeák and co-workers[45b] have shown that chiral 3-substituted pyrrolo[1.2-*c*]thiazole-6,7-dicarboxylates (**180**) can be obtained from 2-substituted-N-acyl-1,3-thiazolidine-4-carboxylic acids with high enantiomeric excess. The chirality at C-4 of the thiazolidine is lost and the chirality at C-2 (C-3 in the product) is retained (Scheme 38).

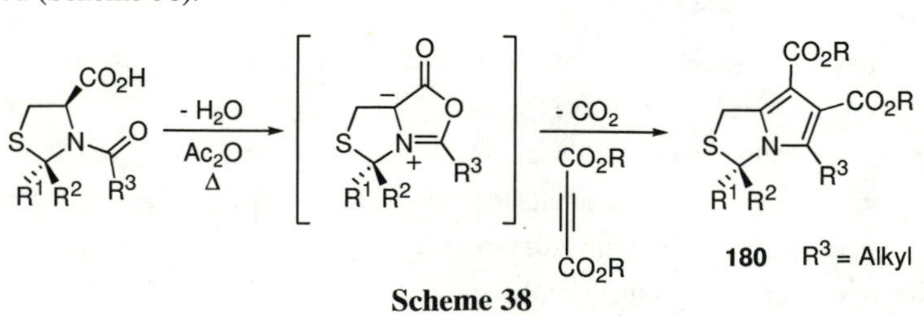

Scheme 38

308

Cycloaddition strategy to chiral pyrrolo[1,2-c]thiazoles, a class of compounds with potential biological activity,[45,46] has been further explored.

In our work we observed that starting from 2-phenylthiazolidine-4-carboxylic acid as a mixture of 2R,4R- and 2S,4R-diastereoisomers, chiral 1H,3H-pyrrolo[1,2-c]thiazole derivatives were obtained (Scheme 39).[48] The thiazolidine was heated in a solution of acetic anhydride in the presence of a dipolarophile. Under these reaction conditions the stereoselective N-acylation occurs in situ, followed by the intermolecular dipolar cycloaddition giving chiral products with R configuration (182 and 183). The reaction with methyl propiolate gave the regioisomer 183 exclusively. The münchnone involved in this process has a C-4 substituent tethered to the mesoionic nucleus and a C-2 substituent untethered and this feature can explain the observed regioselectivity. In fact, Coppola and co-workers[49] have reported that cycloadditions of methyl propiolate with bicyclic mesoionic compounds with single-tethered substituents is characterised by a regioselectivity where the β-carbon of the propiolate combines preferentially with the untethered center. This was attributed to the ability of the untethered center of the münchnone to became more pyramidalized in the transition state thereby allowing for a greater degree of bond formation with the β-carbon of the propiolate (Scheme 39).

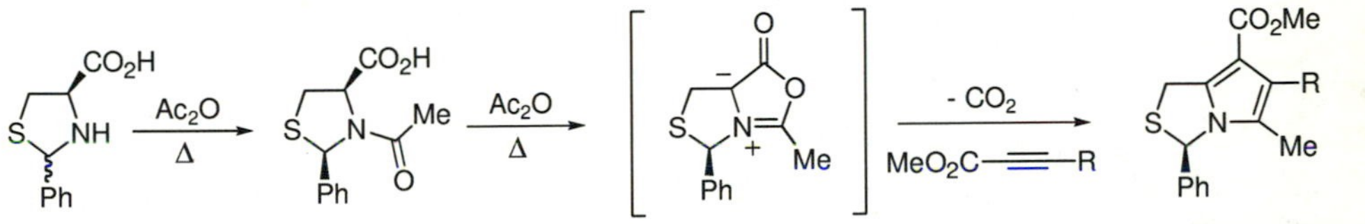

Scheme 39

Heating 2-phenylthiazolidine-4-carboxylic acid in acetic anhydride in the presence of methyl vinyl ketone two products are obtained, the expected methyl (3R)-7-acetyl-3-phenyl-5-methyl-1H,3H-pyrrolo[1,2-c]thiazole 184a in 25% yield, and the spiro compound 185a in 7% yield. The cycloaddition with ethyl vinyl ketone leads to synthesis of 1H,3H-pyrrolo[1,2-c]-thiazole derivative 184b in 10% yield and spiro compound (185b) in 8% yield (Scheme 40).[48]

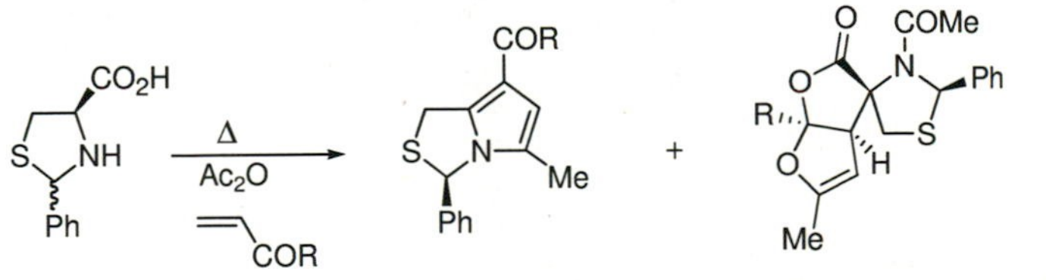

184a R = Me 25%   185a R = Me 7%
184b R = Et 10%   185b R = Et 8%

Scheme 40

We proposed that the mechanism involves the formation of the mesoionic species which reacts with alkyl vinyl ketone and acetic anhydride to give the intermediates 186 which lead to the spiro compounds 185 on eliminating acetic acid (Scheme 41).

309

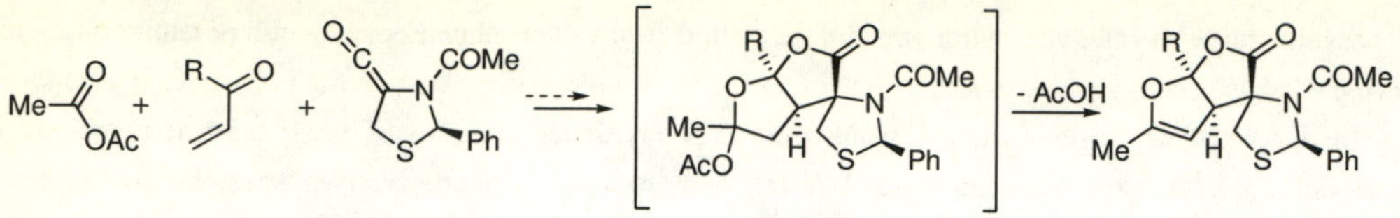

Scheme 41

We studied the generation of (5*R*)-3-aryl-5-phenyl-5*H*,7*H*-thiazolo[3,4-*c*]oxazol-4-ium-1-olates (**188a** and **188c**) and (5*S*)-3-aryl-5-phenyl-5*H*,7*H*-thiazolo-[3,4-*c*]oxazol-4-ium-1-olates (**188b** and **188d**) from the appropriated *N*-aroyl-2-phenylthiazolidine-4-carboxylic acid and their dipolar cycloaddition.[50] The cycloaddition with dimethyl acetylenedicarboxylate led to the synthesis of 1*H*,3*H*-pyrrolo[1,2-*c*]thiazoles (**189**) in very high yield. These reactions proved to be more efficient than the cycloaddition of (5*R*)-3-methyl-5-phenyl-5*H*,7*H*-thiazolo[3,4-*c*]oxazol-4-ium-1-olate with this dipolarophile (Scheme 42).

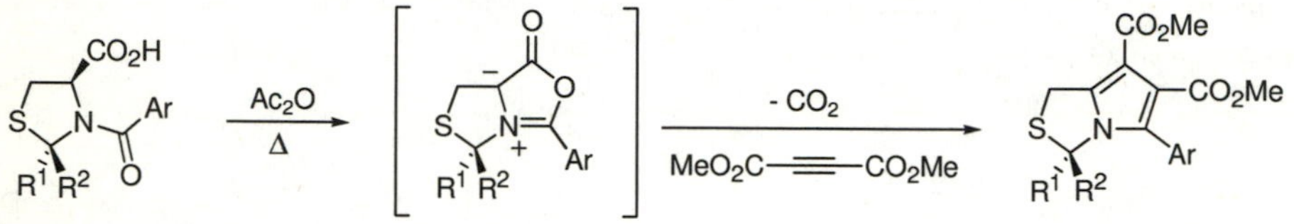

a R^1 = H; R^2 = Ph; Ar = Ph
b R^1 = Ph; R^2 = H; Ar = Ph
c R^1 = H; R^2 = Ph; Ar = C$_6$H$_4$OMe-*p*
d R^1 = Ph; R^2 = H; Ar = C$_6$H$_4$OMe-*p*

189 a 90%
    b 83%
    c 85%
    d 92%

Scheme 42

The dipolar cycloaddition of the bicyclic mesoionic ring systems **188a-188d** with the dipolarophile methyl propiolate led to 1*H*,3*H*-pyrrolo[1,2-*c*]thiazoles in similar yields to the ones obtained with (5*R*)-3-methyl-5-phenyl-5*H*,7*H*-thiazolo[3,4-*c*]oxazol-4-ium-1-olate although different regioselectivity was observed (Scheme 43).[48]

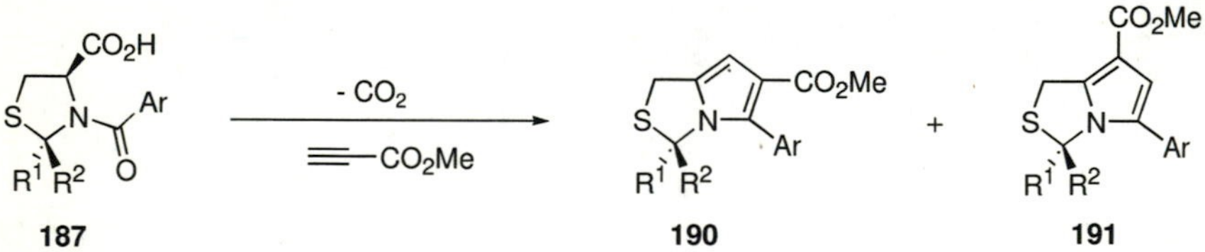

a R^1 = H; R^2 = Ph; Ar = Ph
b R^1 = Ph; R^2 = H; Ar = Ph
c R^1 = H; R^2 = Ph; Ar = C$_6$H$_4$OMe-*p*
d R^1 = Ph; R^2 = H; Ar = C$_6$H$_4$OMe-*p*

(**190:191**)
a 68% (62:38)
b 55% (64:36)
c 55% (70:30)
d 47% (79:21)

Scheme 43

In contrast with observations of Coppola and co-workers,[49] the dipolar cycloaddition of (5*R*)-3-aryl-5-phenyl-5*H*,7*H*-thiazolo[3,4-*c*]oxazol-4-ium-1-olates (**188a** and **188c**) and (5*S*)-3-aryl-5-phenyl-5*H*,7*H*-thiazolo-[3,4-*c*]oxazol-4-ium-1-olates (**188b** and **188d**) with methyl propiolate led to the formation of the two possible regioisomers, the major product being the result of a regioselectivity where β-carbon of the propiolate combines with the tethered center. In all four cases the major component (**190**) could be separated from the mixture (**190/191**) by selective crystallisation with ethyl ether-hexane. These bicyclic münchnones **188** are characterised by having an aryl group at the untethered center and this can prevent the pyramidalized of this center in the transition state thus explaining the observed regioselectivity.

In fact the regiochemistry involved in the cycloaddition of the mesoionic compounds **188a-188d** with methyl propiolate is similar to the one observed in the cycloaddition of this dipolarophile with monocyclic münchnones which also gives rise to a mixture of pyrrole regioisomers, the major product resulting from an interaction where the carbonyl-substituted terminus of the münchnone combines with the β-carbon of methyl propiolate.[51] The presence of the p-methoxy group in the mesoionic aryl substituent at C-3 leads to higher regioselectivity. The effect of this group on the selectivity can be attributed to the reduced pyramidalization of the C-3 dipole terminus in the transition state due to the contribution from the quinoid resonance form. The results obtained from the dipolar cycloaddition of münchnones **188a-188d** with methyl propiolate suggest that stereoelectronic transition state interactions between the substituents of the münchnone and the substituents of the dipolarophile play an important role in determining the regioselectivity. This observation is in agreement with the work of Gribble and co-workers on the 1,3-dipolar cycloaddition of unsymmetrical münchnones with 2- and 3-nitroindoles.[52]

The thermolysis of (2R,4R)-N-ethoxyoxalyl-2-phenylthiazolidine-4-carboxylic acid (**192**) in acetic anhydride leads to the synthesis of N-(1-ethoxycarbonyl-2-phenylvinyl)-2-phenyl-4-thioxo-1,3-thiazolidine (**193**) via the generation of a münchnone as intermediate (Scheme 44).[53]

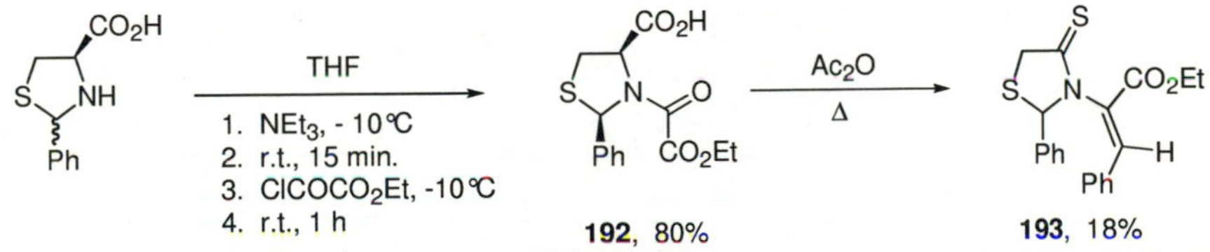

Scheme 44

It has been reported that thiobenzophenone and methyl thiobenzoate combine with 3-methyl-2,4-diphenyloxazolium-5-olate (**194**) to give N-thiobenzoylenamines (**197**) and CO_2. The process involves a 1,3-dipolar cycloaddition followed by cycloreversion of carbon dioxide giving **7**. This compound undergoes a rearrangement leading to the N-thiobenzoylenamines **197** (Scheme 45).[54]

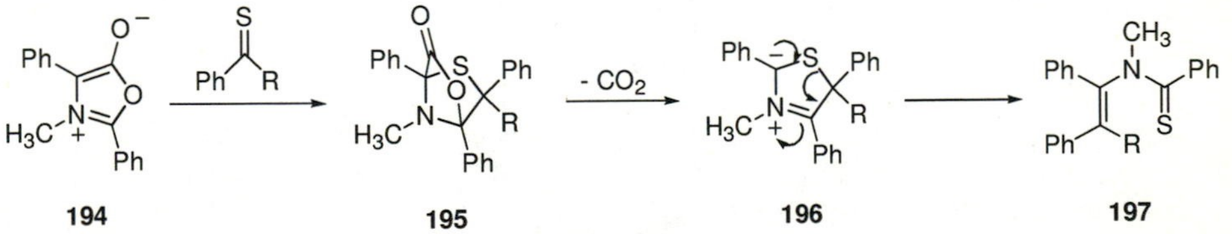

Scheme 45

The synthesis of compound **193** can be rationalized by a similar mechanistic sequence (Scheme 46). Thiobenzaldehyde is generated from thiazolidine **192** which could participate in the 1,3-dipolar cycloaddition with 3-carbethoxy-5-phenyl-5H,7H-thiazolo[3,4-c]oxazol-4-ium-1-olate **198** leading to **199**.

Cycloreversion of carbon dioxide followed by rearrangement allows the formation of *N*-(1-ethoxycarbonyl-2-phenylvinyl)-2-phenyl-4-thioxo-1,3-thiazolidine **193**.[53]

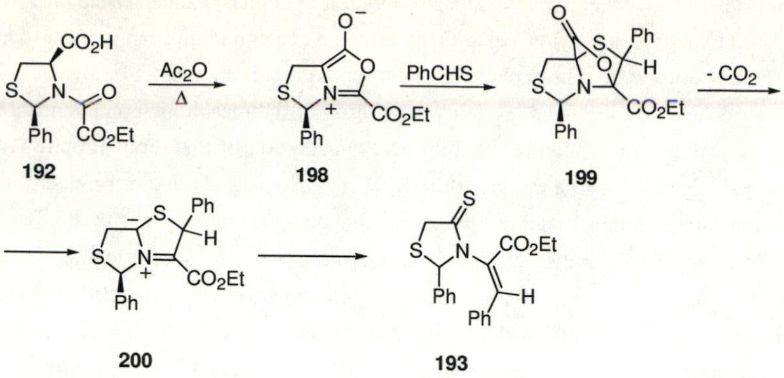

Scheme 46

The synthetic strategy for the generation of 5-substituted-*5H,7H*-thiazolo[3,4-*c*]oxazol-4-ium-1-olates containing internal dipolarophiles involves the *N*-acylation of 2-substituted-thiazolidine-4-carboxylates with the appropriate acid chloride.[55,56] The reaction conditions used allowed the selective synthesis of *N*-acylthiazolidines **201** with (*2R,4R*) stereochemistry. The reaction of compounds **201** with lithium iodide in ethyl acetate followed by treatment with aqueous HCl led to the formation of the corresponding thiazolidine-4-carboxylic acids **202**.

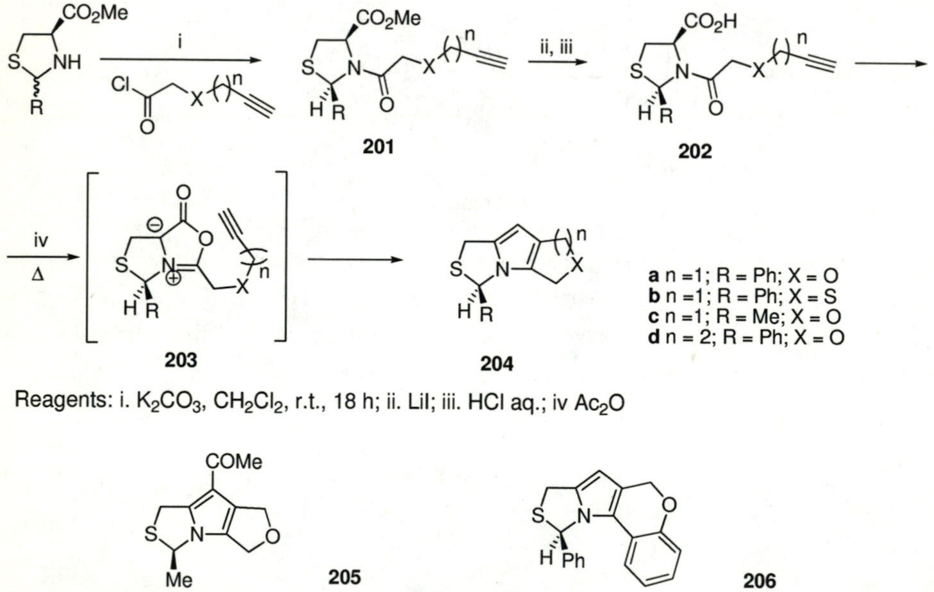

a n =1; R = Ph; X = O
b n =1; R = Ph; X = S
c n =1; R = Me; X = O
d n = 2; R = Ph; X = O

Reagents: i. K_2CO_3, CH_2Cl_2, r.t., 18 h; ii. LiI; iii. HCl aq.; iv Ac_2O

Scheme 47

These compounds were heated in a solution of acetic anhydride to generate the mesoionic species which participated in the intramolecular dipolar cycloaddition to give the chiral tricyclic pyrrolo[1,2-*c*]thiazole derivatives **204a-204d** with *R* configuration.

From the reaction of thiazolidine **202c** the *(7R)*-4-acetyl-7-methyl-1,3-dihydro-*5H,7H*-furo[3',4':2,3]pyrrolo[1,2-*c*]thiazole **205** was also obtained in 12% yield. Using a similar route (10*R*)-10-phenyl-*6H,8H*-chromeno[3',4':2,3]pyrrolo[1,2-*c*]thiazole **206** was also obtained (Scheme 47).The reaction of phenylglyoxal and *L*-cysteine methyl ester gives also a thiazolidine as a mixture of the (2*S*,4*R*)- and (2*R*,4*R*)-diastereoisomers (Scheme 48).[56]

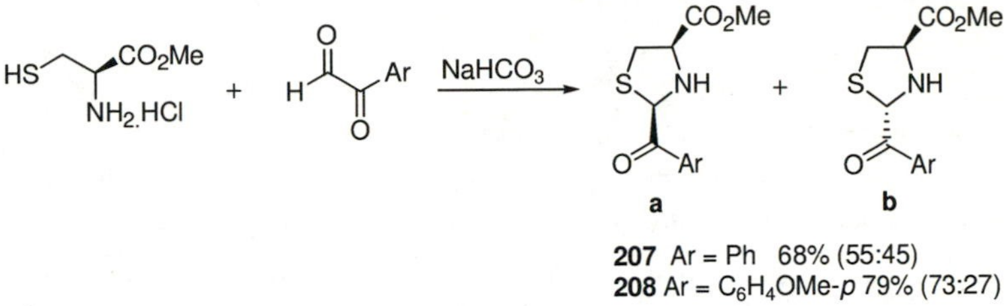

HS
CO2Me
NH2.HCl
H
Ar
NaHCO3
CO2Me
S
NH
Ar
a
b
207 Ar = Ph 68% (55:45)
208 Ar = C6H4OMe-p 79% (73:27)

Scheme 48

However, in this case the two diastereoisomers of methyl 2-benzoylthiazolidine-4-carboxylate (**207a** and **207b**) can be separated by chromatography proving that the interconversion of these isomers through a reversible ring opening mechanism is difficult. In the case of the reaction of *p*-methoxyphenylglyoxal, obtained by oxidation of 4-methoxyacetophenone with aqueous hydrobromic acid and dimethyl sulfoxide, with *L*-cysteine methyl ester a mixture of diastereoisomeric thiazolidines (**208a** and **208b**) is obtained which can also be separated but this time by selective crystallisation.

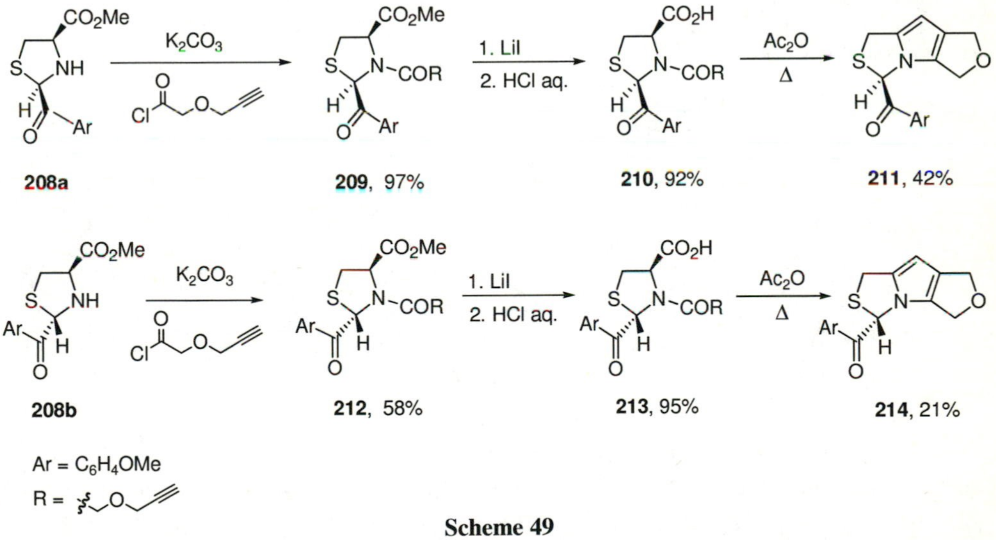

K2CO3
1. LiI
2. HCl aq.
Ac2O
Δ
208a
209, 97%
210, 92%
211, 42%
208b
212, 58%
213, 95%
214, 21%
Ar = C6H4OMe
R =

Scheme 49

The possibility of isolating pure diastereoisomers of 2-substituted thiazolidines characterised by an unfavorable interconversion between the (2*R*,4*R*)- and (2*S*,4*R*)-isomers led us to explore the use of **207a**, **208a**, **207b** and **208b** as precursors of *5H,7H*-thiazolo[3,4-*c*]oxazol-4-ium-1-olates with internal dipolarophiles and to promote the dipolar cycloaddition in the hope of retaining the configuration of the starting thiazolidine at C-2.[56] The reaction of thiazolidine **208a** with prop-2-ynyloxyacetic acid chloride in presence of potassium carbonate led to *N*-acylthiazolidine **209** meaning that retention of configuration at C-2 has occurred. Using our synthetic strategy chiral 5-benzoyl-*1H*-pyrrolo[1,2-*c*]thiazole derivative **211** was obtained with *R* configuration. The thiazolidine **212** was obtained from the acylation of thiazolidine **208b** with prop-2-ynyloxyacetic acid chloride. Thiazolidine **212** was converted into the corresponding carboxylic acid **213** which was heated in acetic anhydride giving the cycloadduct **214** the enantiomer of compound **211** (Scheme 49).

Our study showed that using methyl 2-(*p*-methoxybenzoyl)thiazolidine-4-carboxylate **208a** and **208b** as starting material 7-benzoyl-1,3-dihydro-*5H,7H*-furo[3',4':2,3]pyrrolo[1,2-*c*]thiazole derivatives can be obtained as single enantiomers with retention of configuration at C-2 of the starting thiazolidine.

In contrast with the results obtained with 2-(*p*-methoxybenzoyl)thiazolidines the acylation reaction of 2-benzoylthiazolidine **207a** and **207b** led to a product which showed identical ^{1}H NMR (Scheme 50). They both led to the same thiazolidine (**215**) with *R* configuration at C-2. This observation indicates that the acylation of thiazolidine **207a** occurs with retention of configuration whereas thiazolidine **207b** undergoes inversion of configuration at C-2. Compound **215** was converted into thiazolidine **216** which was heated in acetic anhydride. Two products were obtained, the chiral (*7R*)-7-benzoyl-1,3-dihydro-*5H,7H*-furo[3',4':2,3]pyrrolo[1,2-*c*]thiazole **217** in 17% yield, and pyrrolo[1,2-*c*]-[1,4]thiazine derivative **218** in 29% yield.[56]

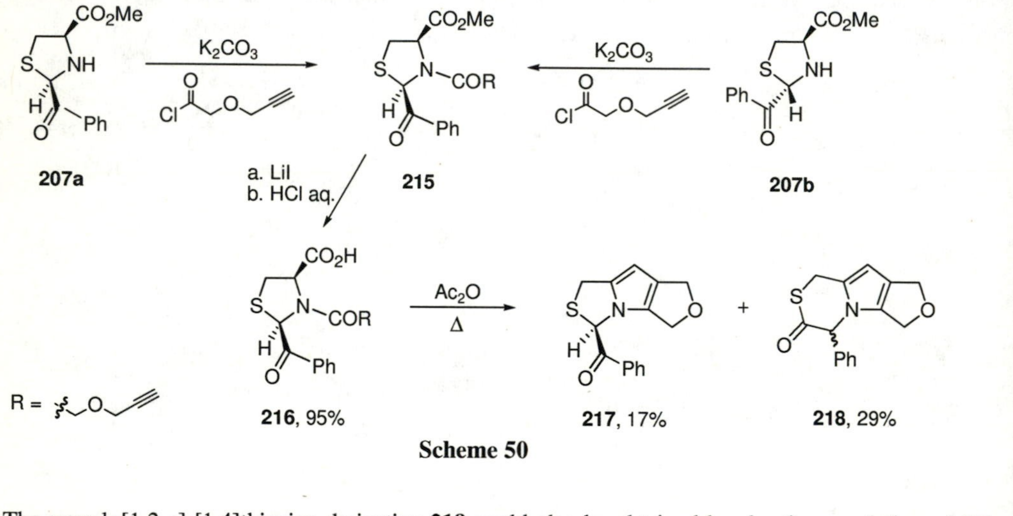

Scheme 50

The pyrrolo[1,2-*c*]-[1,4]thiazine derivative **218** could also be obtained by heating a solution of *1H*-pyrrolo[1,2-*c*]thiazole derivative **217** in acetic anhydride. This result clearly indicates that compound **218** was formed from pyrrolo[1,2-*c*]thiazole derivative **217** and not from the *N*-acyl-1,3-thiazolidine **216**.

The mechanism proposal for the synthesis of the pyrrolo[1,2-*c*]-[1,4]thiazine derivative **218** is outlined in Scheme 51. The first step is the acetylation of the sulfur atom with acetic anhydride giving **219**. The conversion of this compound into **222** can be explained by two possible ways. One route (Path A) involves the formation of the carbene **224** which undergoes a 1,2-nucleophilic migration giving a ketene intermediate (**225**). Cyclization of this species leads to compound **222**. An alternative route (Path B) involves the ring

314

opening through an imminium ion (**220**) and its cyclization giving **221**. This intermediate undergoes a nucleophilic rearrangement to compound **222** which gives the isolated product **218** on deacylation.

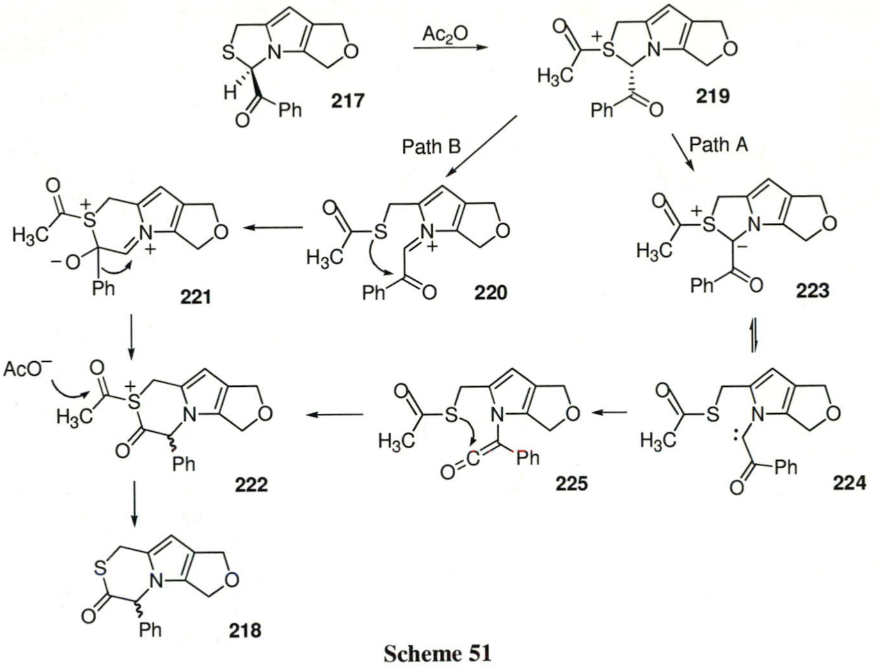

Scheme 51

It is known that 1,3-thiazolidines (**226**) can be converted into dihydro-1,4-thiazines (**230**) by action of chlorine (Scheme 52).[57] In this case it was postulated that a transient sulfeny chloride was formed followed by the ring opening with formation of an imminium ion.

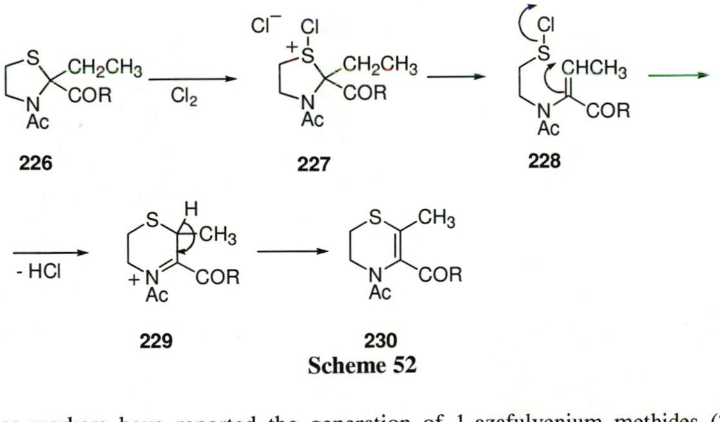

Scheme 52

Storr and co-workers have reported the generation of 1-azafulvenium methides (**231-234**) by the thermal extrusion of sulfur dioxide from pyrrolo[1,2-*c*]thiazole-2,2-dioxides.[58] They described the first evidence for trapping of transient 1-azafulvenium methide systems in pericyclic reactions. These extended

315

dipolar systems **231-233** undergo sigmatropic [1,8]H shifts giving vinylpyrroles (**235-237**) and the acyl derivatives **234** electrocyclise to give pyrrolo[1,2-*c*]-[1,3]oxazines (**238**) (Scheme 53).

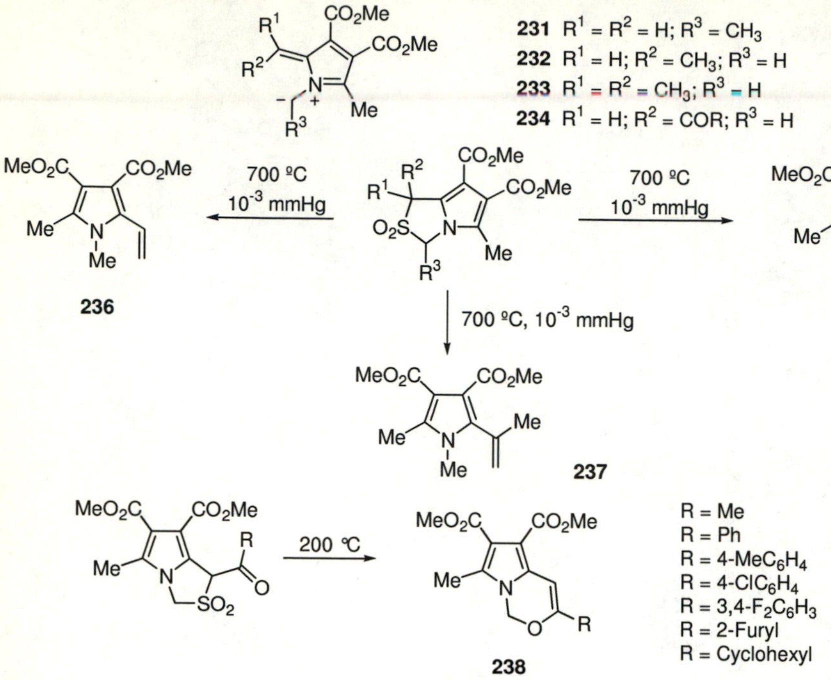

231 R^1 = R^2 = H; R^3 = CH$_3$
232 R^1 = H; R^2 = CH$_3$; R^3 = H
233 R^1 = R^2 = CH$_3$; R^3 = H
234 R^1 = H; R^2 = COR; R^3 = H

700 ºC
10^{-3} mmHg

700 ºC
10^{-3} mmHg

236

235

700 ºC, 10^{-3} mmHg

237

200 ºC

R = Me	52%
R = Ph	86%
R = 4-MeC$_6$H$_4$	85%
R = 4-ClC$_6$H$_4$	76%
R = 3,4-F$_2$C$_6$H$_3$	60%
R = 2-Furyl	72%
R = Cyclohexyl	68%

238

Scheme 53

NaNO$_2$
HCl

TFAA
Et$_2$O

DMAD

MCPBA

239

240

241

Δ

Me$_3$Si—≡—SiMe$_3$

242

243

244

prolonged heating

245

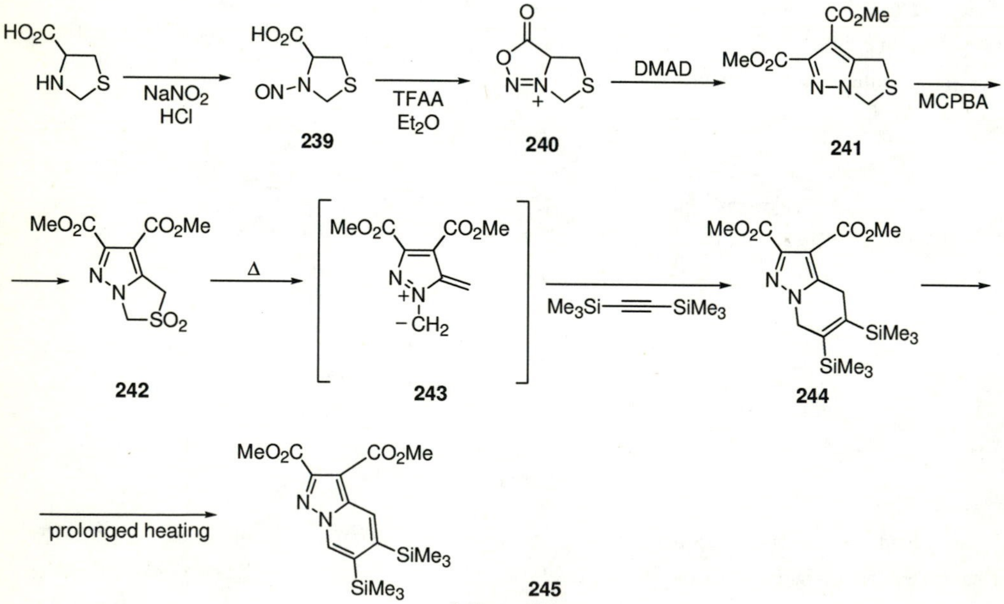

Scheme 54

The pyrazole derivative **242** can be obtained as shown and undergoes extrusion SO_2 in refluxing 1,2,4-trichlorobenzene (Scheme 54).[58] The 1,2-diazafulvenium methide **243** did not react with *N*-phenylmaleimide or dimethyl acetylenedicarboxylate but can be intercepted in $8\pi+2\pi$ cycloaddition with silylated acetylenes giving adducts resulting from the addition across the 1,7-positions. This behaviour is partially explained by Frontier MO theory.

The generation of azafulvenium methides from pyrrolo[1,2-*c*]thiazoles was further studied (Scheme 55).[59] The extrusion of SO_2 from pyrrolo[1,2-*c*]thiazole-2,2-dioxides **246** could be carried out in a sealed tube leading to styryl-1*H*-pyrrole **249** in good yield.

The formation of styryl-1*H*-pyrroles **249** can be explained considering the generation of azafulvenium methides **247** followed by an 1,7-electrocyclic reaction giving **248** which rearrange to the final products.

Attempts were made to trap **247a** by promoting the sealed tube thermolysis in the presence of DMAD and also in the presence of bis(trimethylsilyl)acetylene although no evidence was obtained for the formation of adducts and the only product was styryl-1*H*-pyrrole **249a**. Nevertheless, the synthesis of pyrroles **249** is a strong evidence of the generation of the new azafulvenium methides **247**.

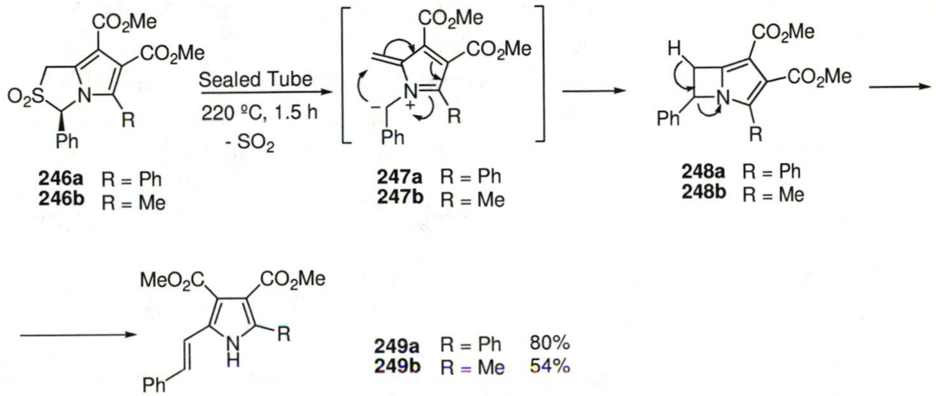

Scheme 55

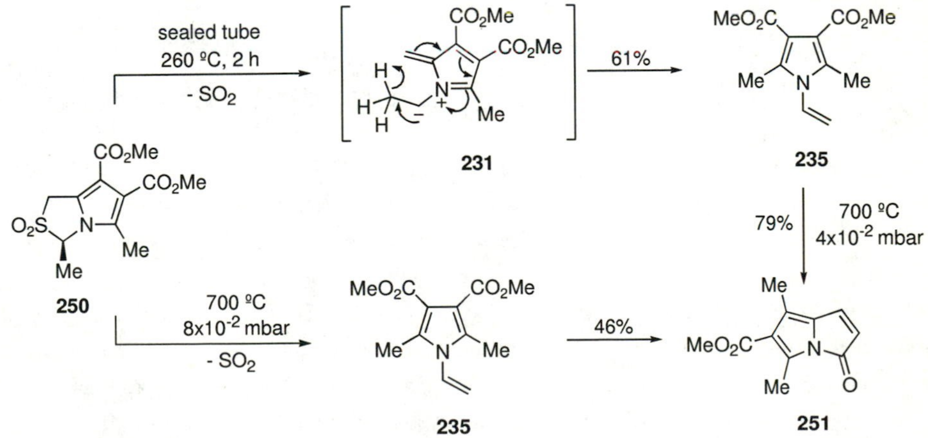

Scheme 56

317

The authors found that the same vinylpyrrole (**235**) could be obtained in 61% yield carrying out the reaction in a sealed tube allowing to conclude that sulfone **250** extrudes SO$_2$ without the need of FVP conditions (Scheme 56).[59]

The flash vacuum pyrolysis of sulfone **250** (700 °C/8x10^{-2} mbar) led to a different outcome than the previously reported result.[58] One product was obtained in 46% yield which was identified as being methyl 1,3-dimethyl-5-oxo-5*H*-pyrrolizine-2-carboxylate **251**. When the FVP of **250** was carried out at 700 °C/4x10^{-2} mbar a mixture of **251** (44%) and **235** (27%) was obtained. This suggested that the lower pressure reduces the period of time that the substance to be pyrolysed remains in the hot zone not allowing the complete conversion of **250** into compound **251**. The result of the sulfone **250** FVP (700 °C/1.3x10^{-3} mbar) described by Storr and co-workers[58] is also in agreement with this observation. Thus, vinylpyrrole **235** must in fact be an intermediate in the formation of methyl 1,3-dimethyl-5-oxo-5*H*-pyrrolizine-2-carboxylate **251** from sulfone **250**. In order to corroborate this mechanistic interpretation we performed the FVP of dimethyl 2,5-dimethyl-1-vinyl-1*H*-pyrrole-3,4-dicarboxylate **235**. In fact, the flash vacuum pyrolysis carried out at 700 °C/4x10^{-2} mbar led to the efficient synthesis of compound **251** (79%).

The formation of methyl 1,3-dimethyl-5-oxo-5*H*-pyrrolizine-2-carboxylate **251** from *N*-vinylpyrroles **235** can be rationalized as outlined in Scheme 57. It is known that 2-substituted 3-(pyrrol-2-yl)propionate methyl esters undergo concerted elimination of methanol on FVP to give pyrrol-2-ylideneketene intermediates which give pyrrolizinones by electrocyclisation.[60] Thus, 3-(4-methoxycarbonyl-3,5-dimethylpyrrol-2-yl)propionate methyl **255** must be an intermediate in the synthesis of **251**. In our case we envisage that pyrrole **255** is formed from **235** through a sequence of sigmatropic shifts.

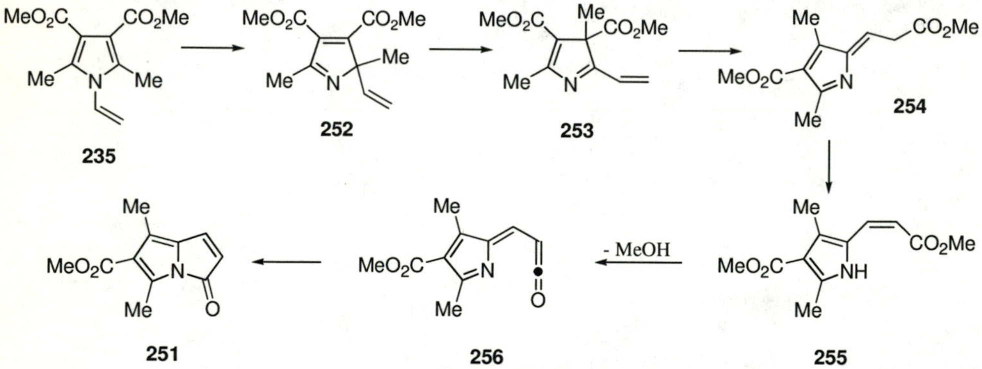

Scheme 57

5-Methyl-3-phenyl-1*H*-pyrrolo[1,2-*c*]thiazole-2,2-dioxide **257** is converted into methyl 2-methyl-4-oxo-1,4-dihydro-1-aza-benzo[*f*]azulene-3-carboxylate **258** on flash vacuum pyrolysis (Scheme 58).[59] Under these reaction conditions styryl-1*H*-pyrrole **241b** is formed and converted into a pyrrole fused to a benzocyclohepten-5-one ring system. This was confirmed by promoting the FVP of styryl-1*H*-pyrrole **241b** which also gave compound **258** (31%). The most likely mechanism for the formation of **258** is shown in Scheme 58. It has been reported that methyl pyrrole-2-carboxylate undergoes elimination of methanol to produce pyrrol-2-ylketene under FVP conditions.[60] In a similar manner styryl-1*H*-pyrrole **241b** generates pyrrol-3-ylketene **259** on eliminating methanol. Electrocyclisation of **259** followed by two sigmatropic H-shifts gives compound **258**.

The pyrrolo[1,2-c]thiazole generated by dehydration of diethyl 5-methyl-1H,3H-pyrrolo[1,2-c]thiazole-6,7-dicarboxylate 2-oxide acts as a thiocarbonyl ylide in cycloaddition with electron deficient alkenes but as an azomethine ylide with electron deficient alkynes.

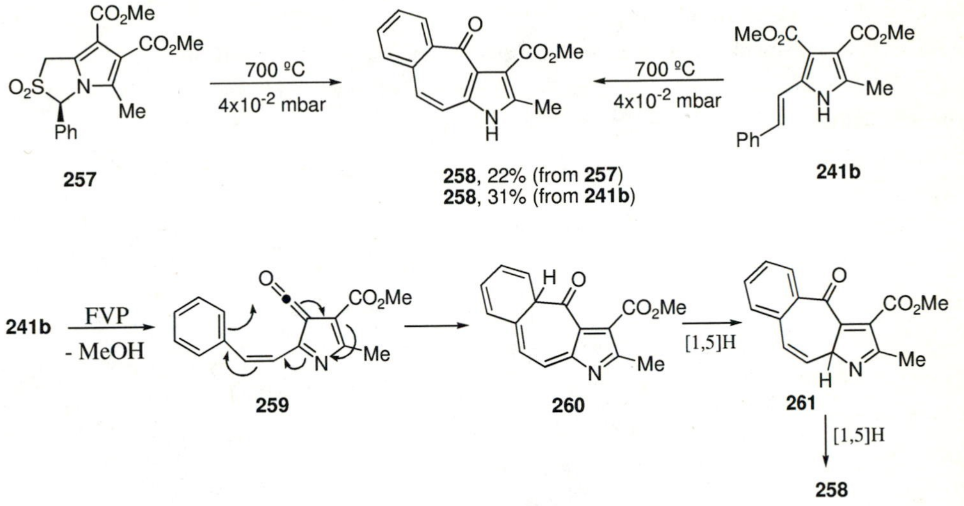

257

258, 22% (from **257**)
258, 31% (from **241b**)

241b

241b   FVP / - MeOH

259

260

261

258

Scheme 58

Kane showed that dehydration of **262** in boiling acetic anhydride in the presence of N-phenylmaleimide leads to a mixture of 1:1 cycloadducts, the *exo* and *endo* adducts (**264**), *via* cycloaddition of the transient 'non-classical' pyrrolo[1,2-c]thiazole **263**.[61]

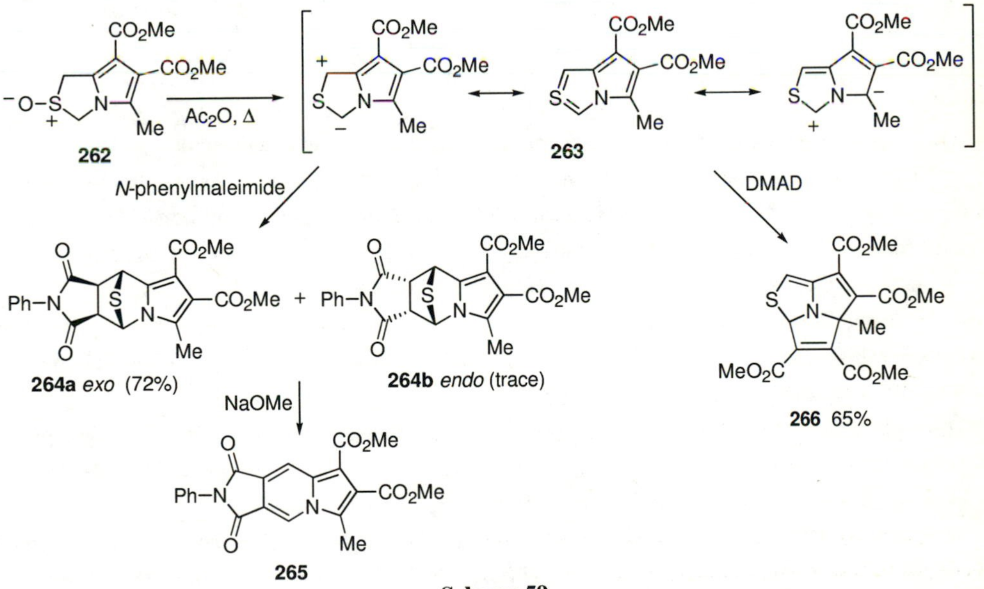

262

263

N-phenylmaleimide

DMAD

264a *exo* (72%)

264b *endo* (trace)

266 65%

NaOMe

265

Scheme 59

Gilchrist and co-workers confirmed this result and identified the major product as being the *exo* adduct **264a** (72%) and the *endo* adduct was the minor product (only trace amount by TLC).[62] Treatment of this mixture with sodium methoxide afforded indolizine **265**. Dehydration of sulfoxide in the presence of DMAD gives thiazolo[2,3,4-*cd*]pyrrolizine in 65% yield, which results from cycloaddition across the azomethine ylide portion of the 'non-classical' system **263**[62] (Scheme 59).

The analogous pyrazolo[1,5-*c*]thiazole **268** generated by dehydration of dimethyl 1*H*,3*H*-pyrazolo[1,5-*c*]thiazole-6,7-dicarboxylate 2-oxide (**267**) acts as thiocarbonyl ylide with both electron deficient alkenes and electron deficient alkynes (Scheme 60).[62]

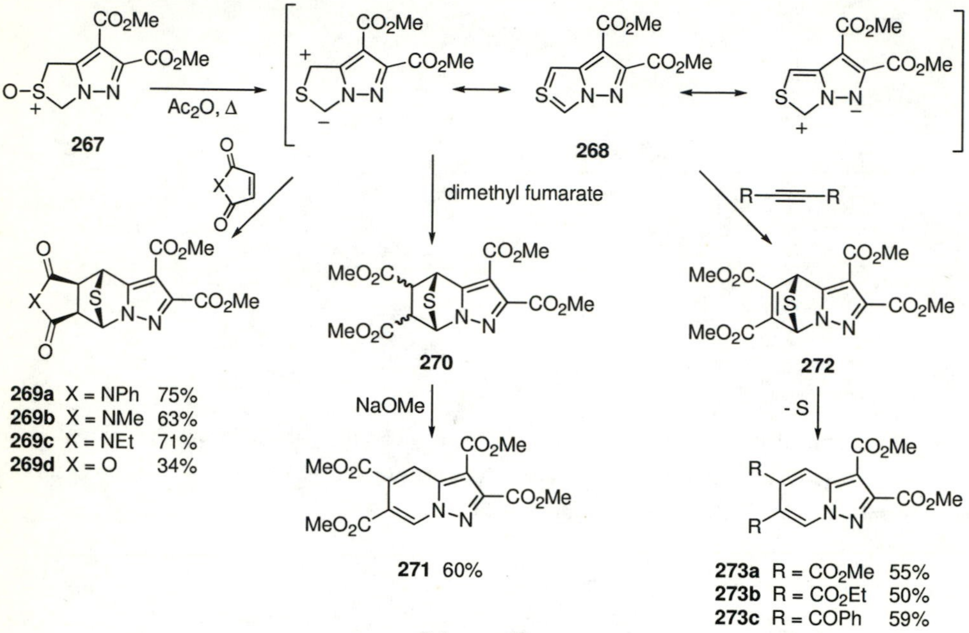

269a X = NPh 75%
269b X = NMe 63%
269c X = NEt 71%
269d X = O 34%

270

271 60%

272

273a R = CO$_2$Me 55%
273b R = CO$_2$Et 50%
273c R = COPh 59%

Scheme 60

4. 2-Aza-1,3-butadienes or their tautomers from 1,3-thiazolidine-4-carboxylic acids

The Diels-Alder reaction of acyclic 2-azadienes is a useful method for the preparation of pyridines, dihydropyridines and tetrahydropyridines. In 1986 Wulff and Böhnke reported the first example of [4+2] cycloaddition of *N*-arylidenedehydroamino esters.[63] These azadienes were generated by dehydration of Schiff bases of serine methyl ester with *N,N'*-carbonyldiimidazole and triethylamine and were found to dimerise by way of a [4+2] cycloaddition process giving tetrahydropyridines.[64] *N*-Arylidenedehydroalanine methyl esters had already been prepared in 1979 by Öhler and Schmidt, by the reaction of thiazolidine esters with silver carbonate and DBU.[65]

This synthetic strategy for the generation of the 2-azadienes **275a–275h** from the corresponding thiazolidines **2** was further explored (Scheme 61). These intermediates, except **275g** and **275h,** can participate as 2-azadienes in Diels-Alder reactions with a range of dienophiles.[66]

The azadiene **275a**, generated *in situ* from the thiazolidines **274a** in the presence of a large excess of but-3-en-2-one, led to the formation of three compounds in an overall yield of 76% (Scheme 62). Compounds **276** and **277** are the products which would result from prototropy (*i.e.*, imine to enamine tautomerization) of *endo* and *exo* cycloadducts. Compound **278** was derived from one or both compounds **277** and **278** by prototropy. In the presence of a base, the primary adduct **279** might exist in equilibrium with the dipolar species **280**; this might then isomerise to compound **278**.[66]

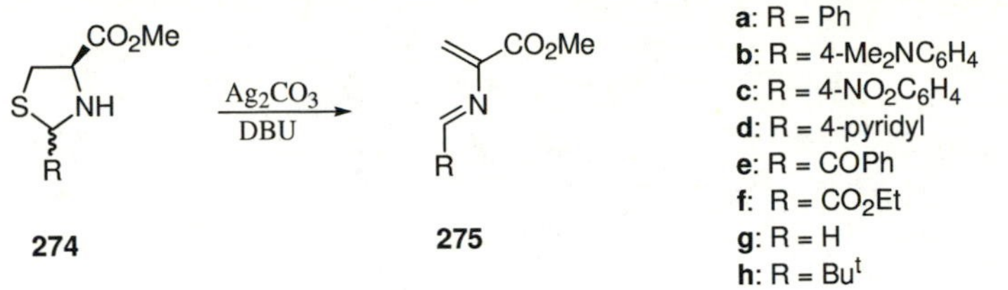

a: R = Ph
b: R = 4-Me$_2$NC$_6$H$_4$
c: R = 4-NO$_2$C$_6$H$_4$
d: R = 4-pyridyl
e: R = COPh
f: R = CO$_2$Et
g: R = H
h: R = Bu^t

274   Ag$_2$CO$_3$ / DBU   **275**

Scheme 61

275a   MeOC   **276** (20%)   +   **277** (5%)   +   **278** (51%)

279   **280**   **278**

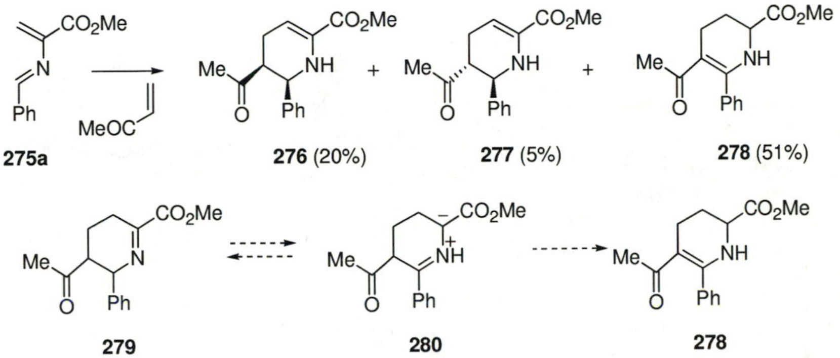

Scheme 62

Cycloaddition reactions of *N*-benzylidenedehydroalanine methyl ester **275a** (generated from the thiazolidine **274a**) were also attempted with other electron deficient dienophiles. Cycloadducts were isolated in moderate to low yield from several dienophiles including methyl acrylate, acrylonitrile, diethyl fumarate, ethyl propiolate and diethyl acetylenedicarboxylate. Thus, several examples of Diels-Alder reactions with electron deficient dienophiles have been discovered and adducts have been isolated in moderate to low yield. With one dienophile, but-3-en-2-one, adducts were obtained in good combined yield.[66]

The cycloaddition reactions of azadiene **275a** with enamines as electron rich dienophiles was also studied. The azadiene **275a** reacts with *N*-cyclohex-1-enylpyrrolidine with the formation of the cycloadducts **281** (37%) and **282** (20%). These compounds are the result of an *endo* and *exo* addition. The reaction of the *N*-phenylidenedehydroalanine methyl ester **275a** with *N*-cyclopenten-1-ylpyrrolidine gave a single adduct **283** in 35% yield (Scheme 63).[66] The above experiments demonstrated that the azadiene **275a** would undergo cycloaddition reactions with both electron deficient and electron rich alkenes. The more electron rich azadiene **275b** formed by reaction of the thiazolidine **274b** with silver carbonate and DBU underwent a Diels-Alder reaction with but-3-en-2-one.

When silver carbonate is used in excess (2.3 equivalents) a single product **284** was isolated in 39% yield (Scheme 64). On the other hand, when silver carbonate was used in equimolar amount a different product, the isomer **285**, was isolated in 37% yield; with 1.4 equivalents of silver carbonate both **284** (9%) and **284** (43%) were isolated.

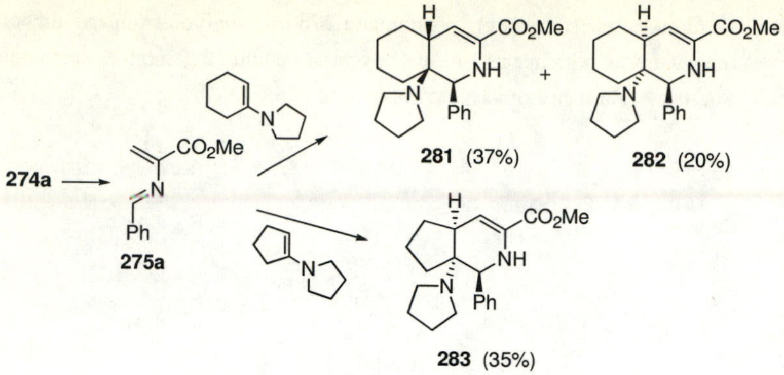

281 (37%) + **282** (20%)

283 (35%)

Scheme 63

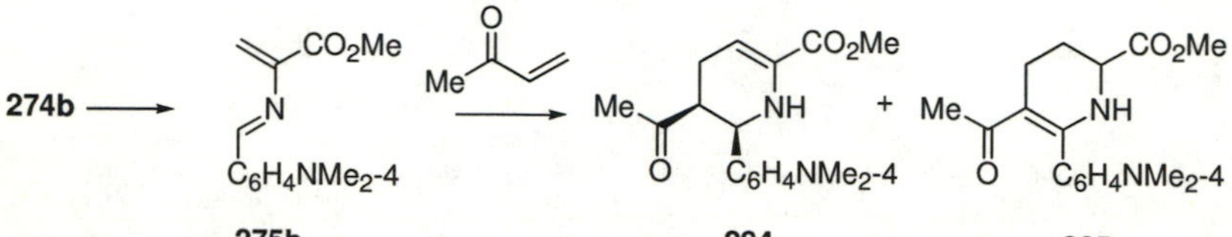

275b　　　　**284**　　　　**285**

Scheme 64

The azadienes **275c-275f**, more electron deficient then N-benzylidenedehydroalanine methyl ester **275a**, participate in the Diels-Alder reactions with N-cyclohexen-1-ylpyrrolidine and N-cyclopenten-1-ylpyrrolidine (Table 2).

Table 2. Cycloaddition reaction of azadienes **275c-275f** with N-cyclohexen-1-ylpyrrolidine and N-cyclopenten-1-ylpyrrolidine.

Azadiene	Dienophiles	Products (%)	
275c	N-cyclohexen-1-ylpyrrolidine	**287a** (20)	**286** (53)
275d	N-cyclohexen-1-ylpyrrolidine	**287b** (54)	
275e	N-cyclohexen-1-ylpyrrolidine	**287c** (14)	**288a** (35)
275f	N-cyclohexen-1-ylpyrrolidine	**287d** (26)	**288b** (26)
275c	N-cyclopenten-1-ylpyrrolidine	**287e** (53)	
275e	N-cyclopenten-1-ylpyrrolidine	**287f** (27)	
275f	N-cyclopenten-1-ylpyrrolidine	**287g** (51)	

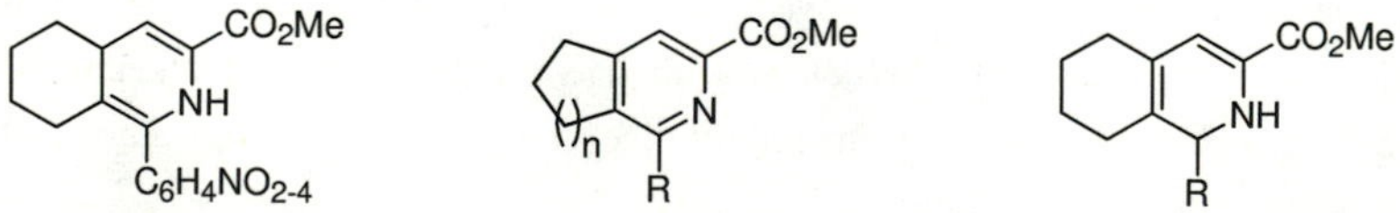

286

287a R = $C_6H_4NO_2$-4, n = 2
287b R = 4-pyridyl, n = 2
287c R = COPh, n = 2
287d R = CO_2Et, n = 2
287e R = $C_6H_4NO_2$-4, n = 1
287f R = COPh, n = 1
287g R = CO_2Et, n = 1

288a R = COPh
288b R = CO_2Et

The unusual reactivity of azadiene **275a**, characterized by the participation in both the *normal* and *inverse* Diels-Alder reaction, can be attributed to the fact that HOMO and LUMO energy levels are rather close as indicate by AM1 calculations. AM1 calculations of the energies of the frontier orbitals of azadienes **275b** - **275f** also show that among the studied molecules azadiene **275c** is the one having the lowest energy value for the LUMO orbital. This prediction agrees with the experimental findings where azadiene **275c** has been shown to be the most reactive towards electron rich dienophiles. The AM1 calculated polarization of the relevant frontier orbitals of azadiene **275a** shows a higher contour value at C-4 than at C-1 in both LUMO and HOMO orbitals. This polarization explains the observed regioselectivity in the *normal* and *inverse* Diels-Alder reactions of this compound. On the other hand, the calculations predict both orbitals are nearly symmetrical with respect to the C=N-C=C plane, a result which is in agreement with the fact that no *endo*/*exo* selectivity has been experimentally observed.[67]

The cycloadditions of 1-(benzylideneamino)acrylates **275a** with cyclopentadiene gave, in moderate yield, a product which contained two components: the norbornene ester **291a** and the tetrahydropyridine ester **292a**. When allowed to stand in solution at room temperature for 10 days the product was found to consist of only a single component, compound **292a**. Thus, the norbornene ester **291a** had isomerised to **292a**. In fact, compound **292a** was also prepared by heating a solution of the Schiff base **291a** in toluene. A similar result was obtained starting from thiazolidine **289** (Scheme 65).[68]

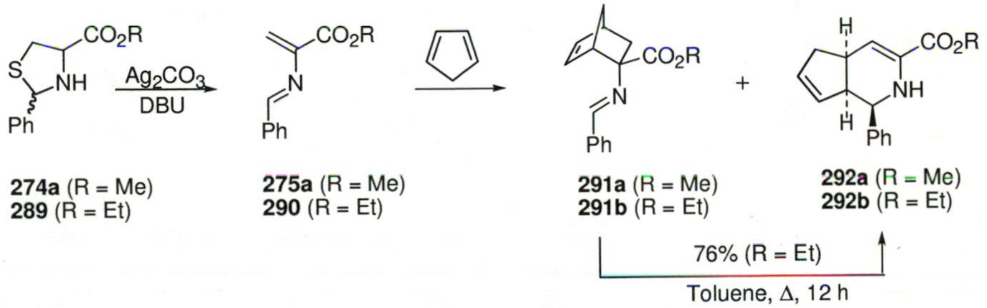

CO2R
S
NH
Ph
Ag2CO3
DBU
274a (R = Me)
289 (R = Et)
CO2R
N
Ph
275a (R = Me)
290 (R = Et)
CO2R
N
Ph
291a (R = Me)
291b (R = Et)
+
H
CO2R
NH
H
Ph
292a (R = Me)
292b (R = Et)
76% (R = Et)
Toluene, Δ, 12 h

Scheme 65

Thus, the azadienes (**275a** and **290**) act exclusively as dienophiles in the reaction with cyclopentadiene and the cycloadditon is selective, leading to the stereoisomer with the benzylideneamino group *endo* (**291**).

These compounds then undergo an aza Cope rearrangement (a 2-aza [3,3]-sigmatropic shift) at room temperature followed by prototropy to give **292** (Scheme 66).

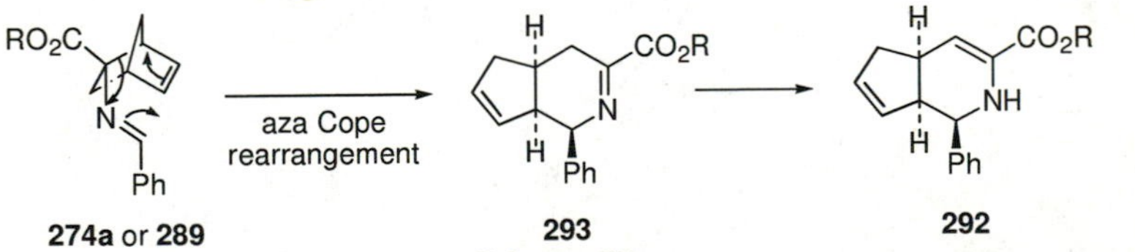

RO2C
N
Ph
274a or 289
aza Cope
rearrangement
H
CO2R
N
H
Ph
293
H
CO2R
NH
H
Ph
292

Scheme 66

Thiazolidines **294a–294d** react with silver carbonate and DBU to give isolable cross-conjugated bis(enamines) **296a–296d**, the more stable tautomers of the azadienes **295a–295d** (Scheme 67).[69] The diester **296a** has been also prepared from methyl β-halo-α-aminopropionate hydrohalides by reaction with bases.[70]

Other cross-conjugated bis(enamines) of this type have been produced by thermal rearrangement of vinylaziridines.[71] These compounds undergo an interesting photocyclization to 3,4-dihydropyrroles which can be intercepted, as 1,3-dipoles, in cycloaddition reactions with alkenes and alkynes.[70b,71,72] Compound **296a** is also reported to react as an electrophile with hydrazines[73] and with primary amines,[74] giving hydrazones and imines of methyl pyruvate as products. In an attempt to cyclize the diester **296a** directly to dimethyl pyrrole-2,5-dicarboxylate by using DDQ as an oxidant a single product was isolated in high yield. Instead of the pyrrole it proved to be a 1:1 adduct of the diester and DDQ (**297**) whose structure was established for the compound by X-ray crystallography (Scheme 67).[69]

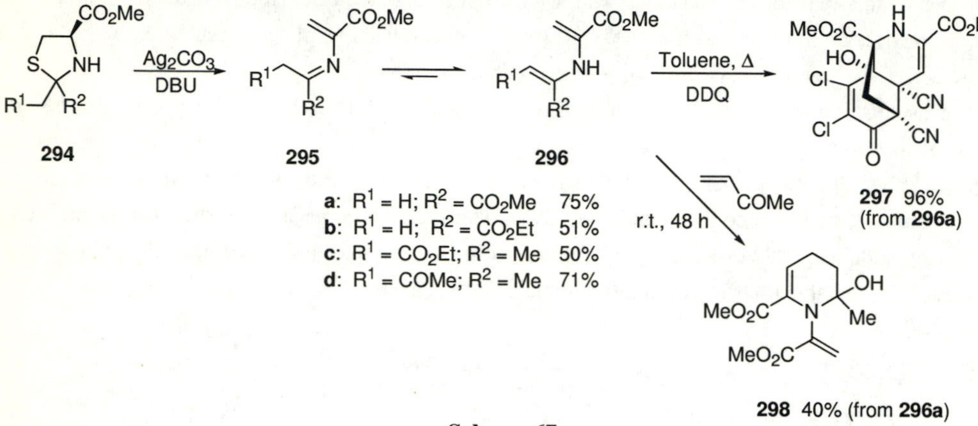

a: R^1 = H; R^2 = CO_2Me 75%
b: R^1 = H; R^2 = CO_2Et 51%
c: R^1 = CO_2Et; R^2 = Me   50%
d: R^1 = COMe; R^2 = Me   71%

Scheme 67

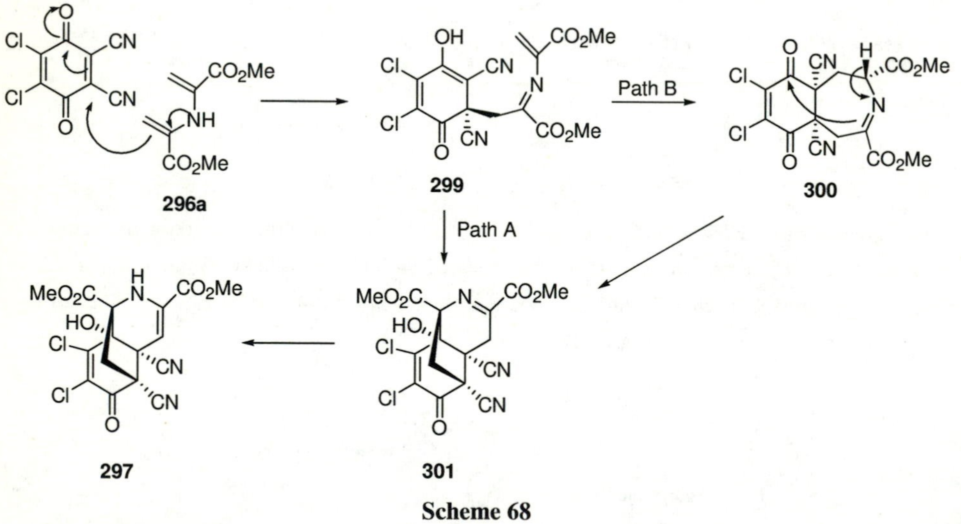

Scheme 68

In the Nenitzescu indole synthesis p-benzoquinones react with primary enamides to produce 5-hydroxyindoles. The first step is a conjugate addition of the enamide, through the β-carbon atom, to benzoquinone. In a similar way, **296a** first reacts with DDQ by conjugate addition to give the intermediate **299**. Two possible ways in which this intermediate could be converted into the final product **297**. The most direct route (path A) is an intramolecular Diels–Alder reaction in which the enol of the six-membered ring acts as the dienophile, giving compound **301** which would be expected to tautomerize to the final product **297**. An alternative (path B) is a second conjugate addition reaction to give the tetrahydroazepine **300** (Scheme 68).[69]

Reaction of the diester **296a** with methyl vinyl ketone gave the tetrahydropyridine **298** in moderate yield. The formation of this product can also be rationalized as a conjugate addition–cyclization sequence, somewhat analogous to the Hantzsch dihydropyridine synthesis (Scheme 67).

Diurno and co-workers reported the synthesis of two tricyclic systems having structures of benzo[*g*]isoquinoline-5,10-dione and dihydrothieno[2,3-b]naphtha-4,9-dione (Scheme 69).[75]

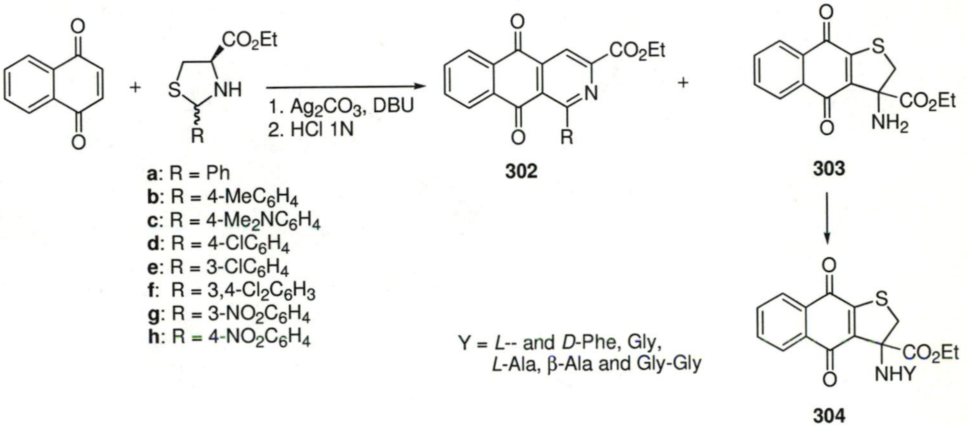
CO2Et
S
NH
R
1. Ag2CO3, DBU
2. HCl 1N
a: R = Ph
b: R = 4-MeC6H4
c: R = 4-Me2NC6H4
d: R = 4-ClC6H4
e: R = 3-ClC6H4
f: R = 3,4-Cl2C6H3
g: R = 3-NO2C6H4
h: R = 4-NO2C6H4
302
303
304
NH2
NHY
Y = L-- and D-Phe, Gly, L-Ala, β-Ala and Gly-Gly

Scheme 69

The 2-azadienes, generated *in situ* from the corresponding thiazolidine, reacted with naphthoquinone giving the fully aromatised Diels-Alder cycloadducts **302** (15-49%). From these reactions the racemic α-amino ester **303** was also obtained (0-30%). The coupling of **303** with the Boc protected amino acid derivatives followed by Boc-deprotection resulted in products as racemic or diastereoisomeric mixtures in high yield (70-75%).

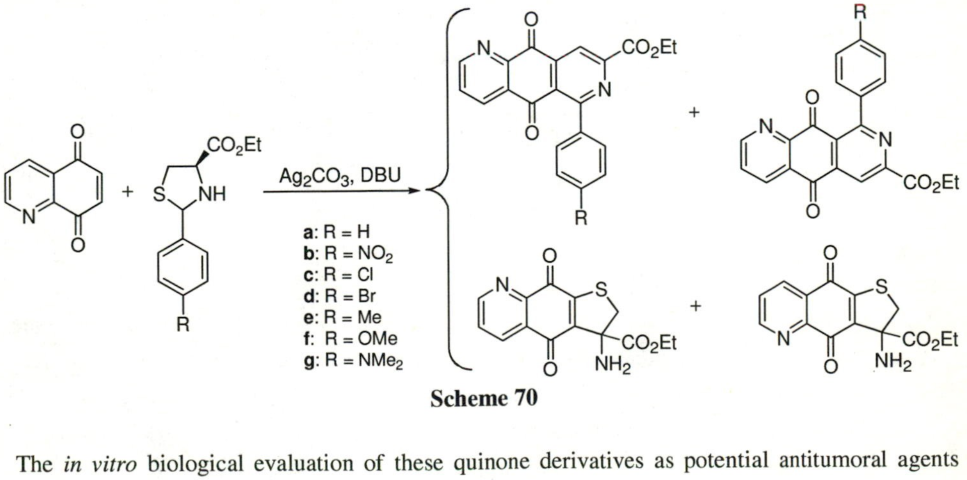
CO2Et
S
NH
R
Ag2CO3, DBU
a: R = H
b: R = NO2
c: R = Cl
d: R = Br
e: R = Me
f: R = OMe
g: R = NMe2
NH2

Scheme 70

The *in vitro* biological evaluation of these quinone derivatives as potential antitumoral agents was carried out. The dihydrothieno[2,3-b]naphtha-4,9-dione-Gly derivative showed *in vitro* antitumoral activity towards several human leukemia and solid tumour. Using a similar synthetic strategy a new series of 1-arylpyridoisoquinolindiones and dihydrothienoquinolindione derivatives have been reported, some of which have cytotoxic activity (Scheme 70).[76]

Nicolaou and co-workers described a chemical route for the construction of dihydropiperidine moiety **306** of the antibiotic thiostrepton, as shown in Scheme 71. The process involved the generation of 2-azadiene **305** from the corresponding thiazolidine which underwent aza-Diels-Alder dimerization in which one molecule acted as the 2-azadiene and the olefinic unit of another molecule acted as dienophile. The imine hydrolysis of compound **306** gave the final product.[77]

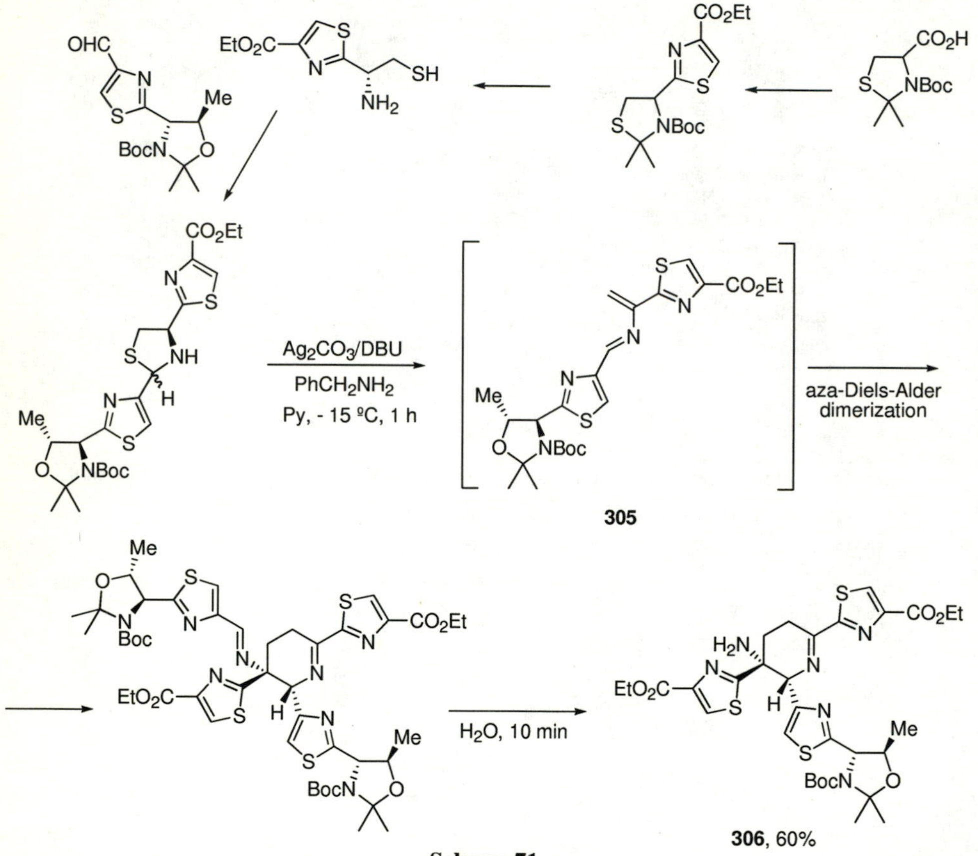

Scheme 71

5. Conclusion

1,3-Thiazolidine-4-carboxylic acids have proven to be useful building blocks in organic synthesis. The most interesting aspect of their reactivity is the possibility of carrying out diastereoselective reactions thus allowing the development of synthetic routes to chiral heterocycles. Of particular interest is also the use of 1,3-thiazolidine-4-carboxylic acids as precursors of reactive intermediates (*e.g.* azadienes or dipoles) which widen the type of structures that can be made available from these heterocycles.

References

1. Terzuoli, L.; Leoncini, R.; Pagani, R.; Guerranti, R.; Vannoni, D.; Ponticelli, F.; Marinello, E. *Life Sciences* **1998**, *63*, 1251.

2. (a) Cook, A. H.; Heilbron, I. M. In *The Chemistry of Penicillin*; Clarke, H. T.; Johson, J. R.; Robinson, R., Eds.; University Press: Princeton, 1949. (b) Dürckheimer, W.; Blumbach, J.; Lattrell, R.; Scheunemann, K. H. *Angew. Chem., Int. Ed. Engl.* **1985**, *24*, 180. (c) Gómez-Gallego, M.; Mancheño, M. J.; Sierra, M. A. *Tetrahedron* **2000**, *56*, 5743.

3. Fairweather, R.; Jones, J. H. *J. Chem. Soc., Perkin Trans. 1* **1972**, 2475.

4. (a) Xingshu, L.; Rugang, X. *Tetrahedron: Asymmetry* **1996**, *7*, 2779. (b) Calmes, M.; Escale, F.; Paolini, F. *Tetrahedron: Asymmetry* **1997**, *8*, 3691. (c) Meng, Q.; Li, Y.; He, Y.; Guan, Y. *Tetrahedron: Asymmetry* **2000**, *11*, 4255. (d) González, A.; Granell, J. R.; López, C. *J. Organomet. Chem.* **2001**, *637-639*, 116. (e) González, A.; Granell, J.; Pinielle, J. F.; Alvarez-Larena, A. *Tetrahedron* **1998**, *54*, 13313.

5. (a) Szilágyi, L.; Györgydeák, Z. *J. Am. Chem. Soc.* **1979**, *101*, 427. (b) Györgydeák, Z.; Kajtár-Peredy M.; Kajtár, J.; Kajtár, M. *Liebigs. Ann. Chem.* **1987**, 927.

6. Benedini, F.; Ferrario, F.; Sala, A.; Sala, L.; Soresinetti, P. A. *J. Heterocyclic Chem.* **1994**, *31*, 1343.

7. Braibante, M. E. F.; Braibante, H. S.; Costenaro, E. R. *Synthesis* **1999**, 943.

8. Shiraiwa, T.; Katayama, T.; Kaito, T.; Tanigawa, H.; Kurokawa, H. *Bull. Chem. Soc. Jpn.* **1998**, *71*, 1911.

9. (a) Planchenault, D.; Wisedale, R.; Gallagher, T.; Hales, N. *J. Org. Chem.* **1997**, *62*, 3438. (b) Brown, G. A.; Anderson, K. M.; Large, J. M.; Planchenault, D.; Urban, D.; Hales, N. J.; Gallagher, T. *J. Chem. Soc., Perkin Trans. 1* **2001**, 1897. (c) Brown, D.; Brown, G. A.; Martel, S. R.; Planchenault, D.; Turmes, E.; Walsh, K. E.; Wisedale, R.; Hales, N. J.; Fishwick, C. W. G.; Gallagher, T. *J. Chem. Soc., Perkin Trans. 1* **2001**, 1270.

10. (a) Mloston, G.; Urbaniak, K.; Heimgartner, H. *Helv. Chim. Acta* **2002**, *85*, 2056. (b) Mloston, G.; Urbaniak, K.; Linden, A.; Heimgartner, H. *Helv. Chim. Acta* **2002**, *85*, 2644.

11. (a) Esterbauer, H.; Ertl, A.; Scholz, N. *Tetrahedron* **1976**, *32*, 285. (b) Pini, E.; Nava, D.; Stradi, R. *Org. Prep. Proced. Int.* **2004**, *36*, 166.

12. Seebach, D.; Sting, A. R.; Hoffman, M. *Angew. Chem., Int. Ed. Engl.* **1996**, *35*, 2708.

13. (a) Seebach, D.; Weber T. *Tetrahedron Lett.* **1983**, *24*, 3315. (b) Seebach, D.; Weber T. *Helv. Chim. Acta* **1984**, *67*, 1650.

14. (a) Mulqueen, G. C.; Pattenden, G.; Whiting, D. A. *Tetrahedron* **1993**, *49*, 5359. (b) Pattenden, G.; Thom, S. M. *J. Chem. Soc., Perkin Trans. 1* **1993**, 1629. (c) Pattenden, G.; Thom, S. M.; Jones, M. F. *Tetrahedron* **1993**, *49*, 2131.

15. (a) Beagley, B.; Betts, M. J.; Pritchard R. G.; Schofield, A.; Stoodley, R. J.; Vohra, S. *J. Chem. Soc., Chem. Commun.* **1991**, 924. (b) Beagley, B.; Betts, M. J.; Pritchard R. G.; Schofield, A.; Stoodley, R. J.; Vohra, S. *J. Chem. Soc., Perkin Trans. 1* **1993**, 1761.

16. Kawabata, T.; Fuji, K. *Topics in Stereochemistry* **2003**, *23*, 175.

17. Betts, M. J.; Pritchard R. G.; Schofield, A.; Stoodley, R. J.; Vohra, S. *J. Chem. Soc., Perkin Trans. 1* **1999**, 1067.

18. Brewster, A. G.; Frampton, C. S.; Jayatissa, J.; Mitchell, M. B.; Stoodley, R. J.; Vohra, S. *J. Chem. Soc., Chem. Commun.* **1998**, 299.

19. Clerici, F.; Erba, E.; Pocar, D. *Tetrahedron* **2003**, *59*, 1667.

20. González, A.; Vorobeva, S. L.; Linares A. *Tetrahedron: Asymmetry* **1995**, *6*, 1357.

21. Baldwin, J. E.; Freeman, R. T.; Lowe, C.; Schofield, C. J.; Lee, E. *Tetrahedron* **1989**, *45*, 4537.

22. (a) Genin, M. J.; Johnson, R. L. *J. Am. Chem. Soc.* **1992**, *114*, 8778. (b) Khalil, E. M.; Ojala, W. H.; Pradhan, A.; Nair, V. D.; Gleason, W. B.; Mishara, R. K.; Johnson, R. L. *J. Med. Chem.* **1999**, *42*, 628. (c) Subasinghe, N. L.; Bontems, R. J.; McIntee, E.; Mishara, R. K.; Johnson, R. L. *J. Med. Chem.* **1993**, *36*, 2356. (d) Khalil, E. M.; Pradhan, A.; Ojala, W. H.; Gleason, W. B.; Mishara, R. K.; Johnson, R. L. *J. Med. Chem.* **1999**, *42*, 2977.

23. Seebach D.; Boes, M.; Naef, R.; Schweizer, W. B. *J. Am. Chem. Soc.* **1983**, *105*, 5390.

24. Lee, B. H.; Miller, M. J. *Tetrahedron Lett.* **1984**, *25*, 927.

25. Geyer, A.; Moser, F. *Eur. J. Org. Chem.* **2000**, 1113.

26. Katritzky, A. R.; Zhang, Y.; He, H.-Y. *ARKIVOC*, **2002**, *V*, 161.
27. González, A.; Lavilla, R.; Piniella, J. F.; Alvarez-Larena, A. *Tetrahedron* **1995**, *51*, 3015.
28. Eremeev, A. V.; Nurdinov, R.; Polyak, F. D.; Zolotoyabko, R. M.; Mishnev, A. F.; Bundule, M. F.; Bleidelis, Y. Y. *Institute of Organic Synthesis, Academy og Sciences of the Latvian SSR* **1985**, 1086. Translated from Khim. Geterotsikl. **1985**, *10*, 1327.
29. Shigematsu, T.; Yoshida, K.; Nakazawa, M.; Tsuda, M., Japanese Patent 79,157,590 (1979); *Chem. Abstr.* **1980**, *93*, 46682x.
30. Györgydeák, Z.; Kövér, K. E.; Miskolczi, I.; Zékány, A.; Rantal, F.; Luger, P.; Katona Strumpel, M. *J. Heterocyclic Chem.* **1996**, *33*, 1099.
31. Refouvelet, B.; Robert, J.-F.; Couquelet, J.; Tronche, P. *J. Heterocyclic Chem.* **1994**, *31*, 77.
32. Miskolczi, I.; Zékány, A.; Rantal, F.; Linden, A.; Kövér, K. E.; Györgydeák, Z. *Helv. Chim. Acta* **1998**, *81*, 744.
33. Malesic, M.; Krbavcic, A.; Stanovnik, B. *J. Heterocyclic Chem.* **1997**, *34*, 49.
34. (a) Pinho e Melo, T. M. V. D.; Santos, C. I. A.; Rocha Gonsalves, A. M. d'A.; Paixão, J. A.; Beja, A. M.; Silva, M. R. *Tetrahedron Lett.* **2003**, 44, 8285. (b) Pinho e Melo, T. M. V. D.; Santos, C. I. A.; Rocha Gonsalves, A. M. d'A.; Paixão, J. A.; Beja, A. M. *Tetrahedron* **2004**, *60*, 3949.
35. Allin, S. M.; Vaidya, D. G.; Page, M. I.; Slawin, A. M. Z. *ARKIVOC*, **2000**, *1*, 151.
36. Oliver, G. L.; Gates, Jr., J. W. *J. Am. Chem. Soc.* **1958**, 702.
37. (a) Takahashi, I.; Kawakami, T.; Hirano, E.; Yokota, H.; Kitajima, H. *Synlett* **1996**, 353. (b) Mertens, A.; Zilch, H.; König, B.; Schäfer, W.; Poll, T.; Kampe, W.; Seidel, H.; Leser, U.; Leinert, H. *J. Med. Chem.* **1993**, *36*, 2526. (c) Schäfer, W.; Friebe, W.-G.; Leinert, H.; Mertens, A.; Poll, T.; Saal, von der W.; Zilch, H.; Nuber, B.; Ziegler, M. L. *J. Med. Chem.* **1993**, *36*, 726. (d) Clercq, E. *J. Med. Chem.* **1995**, *38*, 2491. (e) Allin, S. M.; Vaidya, D. G.; Page, M. I.; Slawin, A. M. Z.; Smith, T. *Tetrahedron Lett.* **2000**, *41*, 2219.
38. Hiskey, R. G.; Dominianni, S. J. *J. Org. Chem.* **1965**, *30*, 1506.
39. Matsuda, A. H; Tatezaki, Y. R.; Mizuno, S. H.; Katori, T. T. US4, 564,615, SS, Pharmaceutical Co, Japan (1986).
40. De Clercq, P. *J. Chem. Rev.* **1997**, *97*, 1755.
41. Seki, M.; Mori, Y.; Hatsuda, M.; Yamada, S.-i. *J. Org. Chem.* **2002**, *67*, 5527.
42. (a) Seki, M.; Kimura, M.; Hatsuda, M.; Yoshida, S.-i.; Shimizu, T. *Tetrahedron Lett.* **2003**, 44, 8905. (b) Kimura, M.; Seki, M. *Tetrahedron Lett.* **2004**, 45, 1635.
43. Kanemasa S.; Saramoto, K.; Tsuge, O. *Bull. Chem. Soc. Jpn* **1989**, *62*, 1960.
44. Kanemasa S.; Doi, K.; Wada, E. *Bull. Chem. Soc. Jpn* **1990**, *63*, 2866.
45. (a) Kane, J. M. *J. Org. Chem.* **1980**, *45*, 5396. (b) Györgydeák, Z.; Szilágyi, L.; Kajtár, J.; Argay, G.; Kálmán, A. *Monatsh. Chem.* **1994**, *125*, 189. (c) Anderson, W. K.; Mach, R. H.; *J. Med. Chem.* **1987**, *30*, 2109. (d) Ladurée, D.; Lancelot, J.-C.; Robba, M.; Chenu, E.; Mathé, G. *J. Med. Chem.* **1989**, *32*, 456. (e) Padwa, A.; Fryrell, G. E.; Gasdaska, J. R.; Venkatramanan, M. K.; Wong, G. S. K. *J. Org. Chem.* **1989**, *54*, 644.
46. (a) Fabre, J. L.; Farge, D.; James, C.; Lavé, D. *Tetrahedron Lett.* **1985**, *26*, 5447. (b) Lavé, D.; James, C.; Rajoharison, H.; Bost, P. E.; Cavero, I. *Drugs of the Future* **1998**, *14*, 891.
47. (a) Dalla Croce, P.; Ferraccioli, R.; La Rosa, C. *Tetrahedron* **1995**, *51*, 9385. (b) Dalla Croce, P.; Ferraccioli, R.; La Rosa, C. *Tetrahedron* **1999**, *55*, 201.
48. Pinho e Melo, T. M. V. D.; Soares, M. I. L.; Barbosa, D. M.; Rocha Gonsalves, A. M. d'A.; Matos Beja, A.; Paixão, J. A.; Ramos Silva, M.; Alte da Veiga, L. *Tetrahedron* **2000**, *56*, 3419.
49. Coppola, B. P.; Noe, M. C.; Shih-Huang Hong, S. *Tetrahedron Lett.* **1997**, *38*, 7159.
50. Pinho e Melo, T. M. V. D.; Gomes, C. S. B.; Rocha Gonsalves, A. M. d'A.; Paixão, J. A.; Matos Beja, A.; Ramos Silva, M.; Alte da Veiga, L. *Tetrahedron* **2002**, *58*, 5093.
51. Coppola, B. P.; Noe, M. C.; Schwartz, D. J.; Abdon, R. L. II; Trost, B. M. *Tetrahedron* **1994**, *50*, 93.
52. Gribble, G. W.; Pelkey, E. T.; Simon, W. M.; Trujillo, H. A. *Tetrahedron* **2000**, *56*, 10133.
53. Pinho e Melo, T. M. V. D.; Gomes, C. S. B.; Soares, M. I. L.; Rocha Gonsalves, A. M. d'A.; Paixão, J. A.; Matos Beja, A.; Ramos Silva, M.; *J. Heterocycl. Chem.* **2004**, in press.
54. Funke, E.; Huisgen, R.; Schaefer, F. C. *Chem. Ber.* **1971**, *104*, 1550.

55. Pinho e Melo, T. M. V. D.; Barbosa, D. M.; Ramos, P. J. R. S.; Rocha Gonsalves, A. M. d'A.; Gilchrist, T. L.; Matos Beja, A.; Paixão, J. A.; Ramos Silva, M.; Alte da Veiga, L.; *J. Chem. Soc., Perkin Trans. 1* **1999**, 1219.

56. Pinho e Melo, T. M. V. D.; Soares, M. I. L.; Rocha Gonsalves, A. M. d'A.; Paixão, J. A.; Matos Beja, A.; Ramos Silva, M. *J. Org. Chem.* **2002**, *67*, 4045.

57. Lee, W. S.; Nam, K. D.; Hahn, H.-G.; Mah, H. D. *J. Heterocyclic Chem.* **1993**, *30*, 1105.

58. (a) Sutcliffe, O. B.; Storr, R. C.; Gilchrist, T. L.; Rafferty, P.; Crew, A. P. A. *Chem.Comm.* **2000**, 675. (b) Sutcliffe, O. B.; Storr, R. C.; Gilchrist, T. L.; Rafferty, P. *J. Chem. Soc., Perkin Trans. 1* **2001**, 1795.

59. Pinho e Melo, T. M. V. D.; Soares, M. I. L.; Rocha Gonsalves, A. M. d'A.; McNab, H. *Tetrahedron Lett.* **2004**, *45*, 3889.

60. (a) Campbell, S. E.; Comer, M. C.; Derbyshire, P. A.; Despinoy, X. L. M.; McNab, H.; Morrison, R.; Sommerville, C. C.; Thornley, C. *J. Chem. Soc., Perkin Trans. 1* **1997**, 2195. (b) McNab, H.; Parson, S.; Stevenson, E. *J. Chem. Soc., Perkin Trans. 1* **1999**, 2047.

61. Kane, J. M. *J. Org. Chem.* **1980**, *45*, 5396.

62. Sutcliffe, O. B.; Storr, R. C.; Gilchrist, T. L.; Rafferty, P. *Tetrahedron* **2000**, *56*, 10011.

63. (a) Wulff, G.; Böhnke, H. *Angew. Chem., Int. Ed. Engl.* **1986**, *25*, 90. (b) Wulff, G.; Lindner, H. G.; Böhnke, H.; Steigel, A.; Klinken, H. T. *Liebigs Ann. Chem.* **1989**, 527. (c) Wulff, G.; Klinken, H. T. *Tetrahedron* **1992**, *48*, 5985.

64. (a) Wulff, G.; Böhnke, H. *Angew. Chem., Int. Ed. Engl.* **1984**, *23*, 380. (b) Wulff, G.; Böhnke, H.; Klinken, H. T. *Liebigs Ann. Chem.* **1988**, 501.

65. Öhler, E.; Schmidt, U. *Chem. Ber.* **1979**, *112*, 107.

66. (a) Gilchrist, T. L.; Gonsalves, A. M. d'A. R.; Pinho e Melo, T. M. V. D. *Tetrahedron Lett.* **1993**, *34*, 4097. (b) Gilchrist, T. L.; Gonsalves, A. M. d'A. R.; Pinho e Melo, T. M. V. D. *Tetrahedron* **1994**, *50*, 13709. (c) Gilchrist, T. L.; Gonsalves, A. M. d'A. R.; Pinho e Melo, T. M. V. D. *Pure & Appl. Chem.* **1996**, *68*, 859.

67. Pinho e Melo, T. M. V. D.; Fausto, R.; Gonsalves, A. M. d'A. R.; Gilchrist, T. L. *J. Org. Chem.* **1998**, *63*, 5350.

68. Gilchrist, T. L.; Gonsalves, A. M. d'A. R.; Pinho e Melo, T. M. V. D. *Tetrahedron Lett.* **1993**, *34*, 6945.

69. Pinho e Melo, T. M. V. D.; Cabral, A. M. T. D. P. V.; Gonsalves, A. M. d'A. R.; Beja, A. M.; Paixão, J. A.; Silva, M. R.; Alte da Veiga, L. *J. Org. Chem.* **1999**, *64*, 7229.

70. (a) Mitsuhashi, K. *Asahi Garasu Kogyo Gijutsu Shoreikai Kenkyu Hokoku* **1973**, *23*, 355 (*Chem. Abstr.* **1975**, *82*, 86591. (b) Zaima, T.; Matsunaga, Y.; Mitsuhashi, K. *J. Heterocycl. Chem.* **1983**, *20*, 1.

71. Gelas–Mialhe, Y.; Mabiala, G.; Vessière, R. *J. Org. Chem.* **1987**, *52*, 5395.

72. (a) Zaima, T.; Matsuno, C.; Matsunaga, Y.; Mitsuhashi, K. *J. Heterocycl. Chem.* **1984**, *21*, 445. (b) Zaima, T.; Matsuno, C.; Matsunaga, Y.; Mitsuhashi, K. *Nippon Kagaku Kaishi* **1984**, 1293 (*Chem. Abstr.* **1984**, *101*, 230340).

73. Zaima, T.; Mitsuhashi, K. *Nippon Kagaku Kaishi* **1979**, 901 (*Chem. Abstr.* **1979**, *91*, 140303).

74. Zaima, T.; Matsuno, C.; Mitsuhashi, K. *Nippon Kagaku Kaishi* **1983**, 152 (*Chem. Abstr.* **1983**, *98*, 179141).

75. (a) Gomez-Monterrey, I. M.; Campiglia, P.; Mazzoni, O.; Novellino, E; Douino, M. V. *Tetrahedron Lett.* **2001**, *42*, 5755. (b) Gomez-Monterrey, I.; Campiglia, P.; Grieco, P.; Douino, M. V.; Bolognese, A.; Colla, P. L.; Novellino, E. *Bioorg. Med. Chem.* **2003**, *11*, 3769.

76. Bolognese, A.; Correale, G.; Manfra, M.; Lavecchia, A.; Mazzoni, O.; Novellino, E.; Colla, P. L.; Sanna, G.; Loddo, R. *J. Med. Chem.* **2004**, *47*, 849.

77. Nicolaou, K. C.; Nevalainen, M.; Safina, B. S.; Zak, M.; Bulat, S. S. *Angew. Chem., Int. Ed.* **2002**, *41*, 1941.

ADVANCES IN ASYMMETRIC SYNTHESIS OF BIOLOGICALLY ACTIVE MOLECULES USING HETEROCYCLES AS BUILDING BLOCKS

Pietro Spanu* and Fausta Ulgheri

Istituto di Chimica Biomolecolare del Consiglio Nazionale delle Ricerche, Sezione di Sassari

Traversa La Crucca 3, Baldinca, I-07040 Li Punti- Sassari, Italy (e-mail: p.spanu@icb.cnr.it)

Abstract. *The stereoselective and versatile approach to enantiomerically pure bioactive substances where heterocyclic building blocks are involved in the stereoselective key carbon-carbon bond formation and then transformed in order to introduce suitable functional groups in the target molecules, has widely shown its utility. In this review we report recent works focused on new asymmetric procedures for the synthesis of enantiopure biofunctional molecules based on this methodology.*

Contents

1. Introduction

Many efforts of the organic chemists directed to the synthesis of substances that play an important role in biological systems, involve heterocyclic compounds because of their high biological relevance. Nevertheless, in this article addressed to the use of heterocyclic compounds for the synthesis of biofunctional molecules, we decided to review another important aspect that involves heterocycles in the synthesis of natural and "unnatural" compounds with biological interest. In fact stereoselective methodologies for preparing bioactive substances where heterocyclic compounds are used as building blocks, and are involved in the stereoselective carbon-carbon bond formation key step of the total synthesis of biologically active molecules and where they are subject to a transformation in order to achieve the final target molecules, have widely shown their utility. In these cases heterocycles were used to successful control the stereochemistry of the products in the key step and to introduce proper functional groups into the target molecules through their transformation. The concept of heterocycle-functional group equivalence as a powerful tool for new synthetic approach was first introduced by Meyers in 1974[1] and a large number of application to the stereoselective synthesis of bioactive molecules and analogues by using achiral heterocyclic compounds or heterocycles embodied with stereogenic centres such as 2-substituted-furans, -pyrroles, -thiophenes,[2] thiazoles,[3] oxazolines,[4] oxazolidinones,[5] imidazoles,[6] piperazines,[7] pyrimidines,[8] 1,3-dithianes,[9] etc have been reported in previous years.

The present article mainly highlights the most recent stereoselective approaches to biofunctional molecules, including our achievements, which utilise heterocycles as building blocks and meet two requirements:

a) The heterocycle is directly involved in the stereoselctive key carbon-carbon bond formation;

b) Precursor-to-target interconversion involves heterocycles transformation (ring opening, ring enlargement, etc.).

Methodologies that use chiral precursors, auxiliaries or catalysts to introduce chirality are included, while procedures based on enzyme catalysis to introduce optical activity are excluded.

The review is organised into twelve sections depending from the heterocyclic compounds exploited in the synthesis. The literature cited covers the period from January 2003 to September 2004.

2. Furans

2.1. 2-Trialkylsilyloxyfurans

The methodology to assemble chiral bioactive molecules by homologative technique that exploits, as the key operation, the vinylogous Mukaiyama-aldol addition between furan- pyrrole- and thiophene-based siloxy dienes and carbonyl precursors obtainable from the chiral pool, has been extensively investigated by Casiraghi, since 1989.[2] Recently, Casiraghi and Rassu have introduced a variable strategy toward enantiopure carbasugars and relatives using furan-, pyrrole- and thiophene-based dienoxysilane in the sequential execution of two strategic carbon-carbon bond-forming reactions, a vinylogous crossed aldol addition and a silylative cycloaldolization as depicted in Scheme 1.[10]

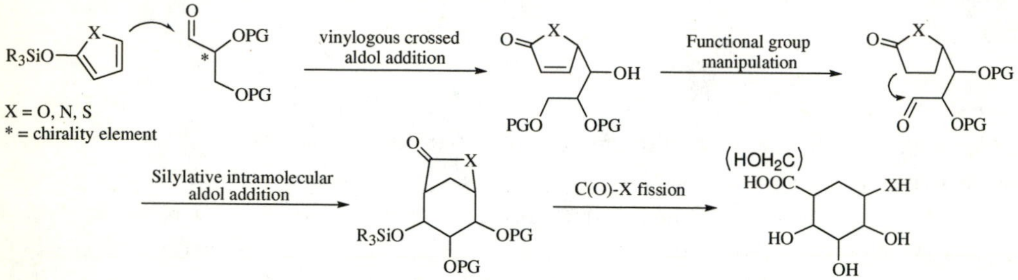

Scheme 1

Applying this synthetic approach to carbocyclic structures the high diastereoselective synthesis of 2-deoxy-2-amino-5a-carba-β-L-mannopyranuonic acid **8** and 2-deoxy-2-amino-5a-carba-β-L-mannopiranose **9**, a positional stereoisomer of validamine, was reported in 12 and 13 steps and in 28% and 24% overall yield respectively (Scheme 2).[11a]

The synthesis involves the use of a Lewis acid promoted vinylogous crossed Mukaiyama-aldol reaction between 2-[(*tert*-butyldimethylsilyl)oxy]furan **1** and (2S)-2,3-*O*-isopropilideneglyceraldehyde *N*-benzyl imine **2** to give the adduct **3** as a single isomer. Compound **3** was converted into the piperidinone **4** by a ring enlargement protocol with DBU and further transformed into the aldehyde **5** *via* direct Swern oxidation of triethylsilyl-protected terminal diol. The silylative intramolecular cycloaldolization reaction gave the *trans*-configured trihydroxy-azabicyclooctanone **6** as a single isomer. To complete the synthesis of both carbasugars **8** and **9**, the benzyl substituted bicycle **6** was transformed into the *N*-Boc protected derivative that after hydrolysis of the heterocycle ring furnished the protected amino acid **7** in good yield.

Alternative elaboration of this adduct furnished the desired carbasugars **8** and **9** in enantiopure form.

A short entry to novel C(2)-methyl branched 4a-carbafuranoses and carbaseptanose derivatives was recently described by the same authors using the same synthetic strategy.[11b,c]

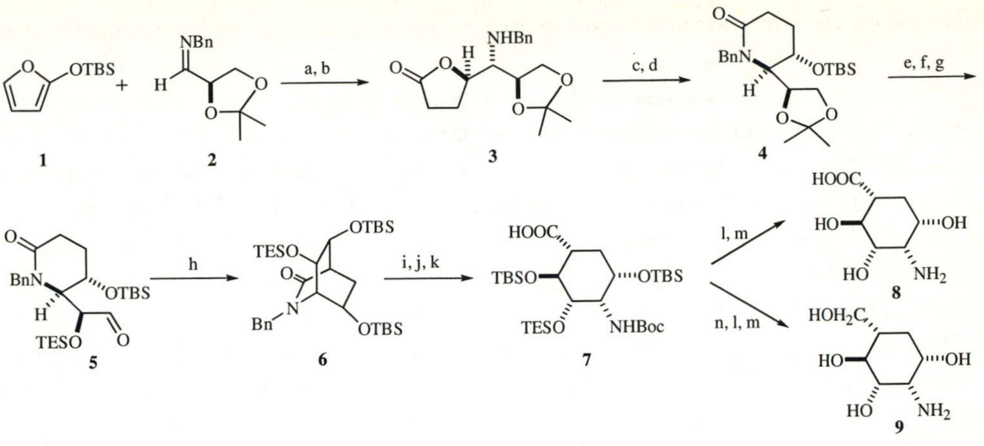

^{a} (a) TBSOTf, CH$_2$Cl$_2$; (b) NaBH$_4$, NiCl 68% for 2 steps; (c) DBU, 140 °C, 96%; (d) TBSOTf, 2,6-lutidine, 92%; (e) aq AcOH, 80 °C, 79%; (f) TESOTf, pyridine, DMAP, 96%; (g) Swern, 91%; (h) TBSOTf, DIPEA, 89%; (i) Na, NH$_3$ liq., 97%; (j) Boc$_2$O, DMAP, 87%; (k) LiOH aq, THF, 93%; (l) HCl, THF, MeOH; (m) DOWEX, 98%; (n) BH$_3$·Me$_2$S, THF, 86%.

Scheme 2

The synthesis of fluorinated piperidine derivatives **14** *via* an highly diastereoselective vinylogous Mannich reaction of trimethylsilyloxyfuran with fluorinated aldimines was recently reported by Crousse, Bonnet-Delpon (Scheme 3).[12] The reaction between CF$_3$-aldimine **10** and trimethyl silyloxyfuran **11** was studied in the presence of three different Lewis acids in dichloromethane at -78 °C. Starting from aldimine **10** with R= Bn and allyl, and when BF$_3$·Et$_2$O or TMSOTf (0.5 equiv) were used as catalyst, only the *anti*-configured butenolide **12** was obtained.

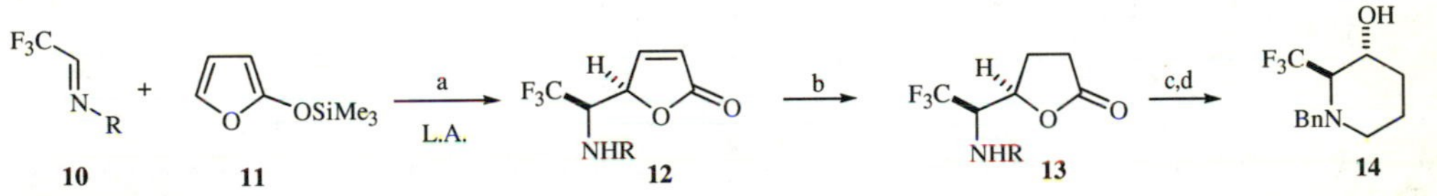

R=Bn, Allyl, PMP   L.A.=Yb(OTf)$_3$, BF$_3$·Et$_2$O, TBSOTf

^{a} (a) CH$_2$Cl$_2$, L.A., 0.5 equiv, -78 °C; (b) Pd/C/ H$_2$, THF, 85%; (c) H$_2$SO$_4$, MeOH, 80%; (d) LiAlH$_4$/AlCl$_3$, THF, 90%

Scheme 3

The *anti* isomers are also the most frequently observed in nonfluorinated series, although reaction time are longer and the reaction is slightly less stereoselective (80%). In order to obtain the trifluoromethyl-substituted piperidine derivative, butenolide **12** was hydrogenated to give the butyrolactone **13** without removal of the *N*-benzyl protective group. Compound **13** was then dissolved in methanol and heated at reflux overnight in the presence of a catalytic amount of sulfuric acid to give an hydroxy-lactam. Complete reduction of the carbonyl function with LiAlH$_4$/AlCl$_3$ gave the target CF$_3$-substituted piperidine **14** in very good yield.

2.2. 2-Imido-furans

A new strategy for *epi*-zephyranthine, an alkaloid of the *Amaryllidaceae* family, was reported by Padwa. The synthesis is based on an extraordinarily simple intramolecular Diels-Alder reaction of a 2-imido

333

substituted furan and a Rh(I)-catalyzed ring opening of the oxabicyclic adduct for the construction of hexahydroindolinone core (Scheme 4).[13] The synthesis started from the preparation of 2-imido-furan **17**, obtained by addition of lithium carbamate **16** to **15**, which rapidly reacted at r.t. to give the Diels-Alder cycloadduct **18** in good yield. The oxabicyclic intermediates have been used as valuable intermediates for the synthesis of a variety of natural products because of the large number of selective transformations possible for the oxabicyclic system. Using Lauten's conditions ([Rh(COD)Cl]$_2$, DPPF), the regioselective ring opening was performed employing phenol and *N*-methyl-aniline as nucleophilic reagents to give the *cis* alcohol **19** with high diastereoselectivity (from 5:1 to 20:1). When Rh(I)-catalyzed reaction was carried out with phenyl boronic acid and without added base, the *cis*-boronate **21** was formed. Then, boronate **21** was cleaved to the corresponding diol which was transformed into dioxolane **22** that was easily converted into the corresponding benzamide **24** by removing the *t*-Boc group followed by benzoylation with the acid chloride **23**.

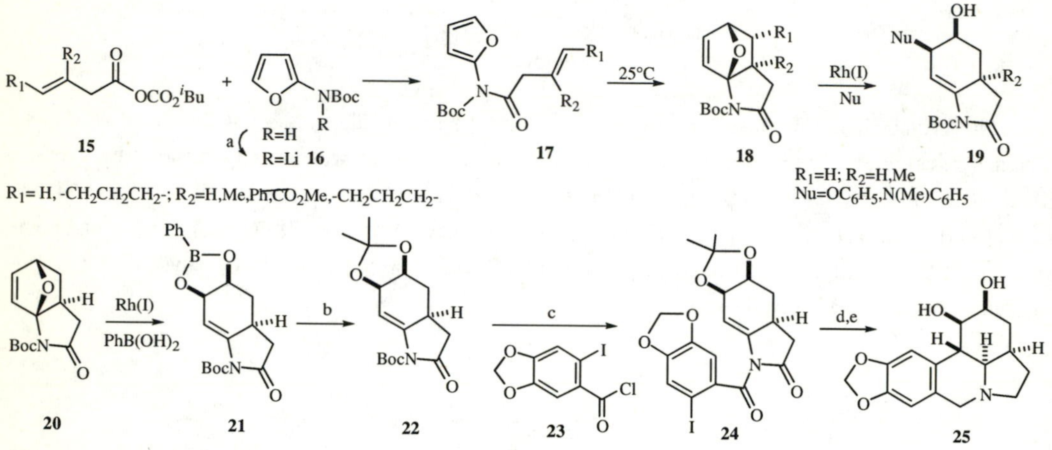

a (a) *n*-BuLi, THF, 0 °C; (b) SnCl$_2$, (CH$_3$)$_2$CO; (c) Mg(ClO$_4$)$_2$, CH$_3$CN; (d) AIBN, Bu$_3$SnH, C$_6$H$_6$, reflux; (e) BH$_3$·THF then HCl, MeOH

Scheme 4

The tetracyclic product was then obtained *via* exposure of **24** to *n*-Bu$_3$SnH in benzene at reflux in the presence of AIBN. Reduction of the two carbonyl groups followed by hydrolysis of the 1,3-dioxolane afforded the *epi*-zephyranthine target **25** in 14.5% overall yield for 7 steps.

2.3. 2-Carboxy-furans

A facile asymmetric synthesis of the bi- and tricyclic lactones core nuclei of biologically active xanthanolides, guaianolides and eundesmanolides was described by Reiser.[14] The key step of the synthesis is the preparation of disubstituted γ-butyrolactones by using the asymmetric copper(I)-bis(oxazolines)-catalyzed cyclopropanation of furan-2-carboxylic esters **26** (Scheme 5). By this way, either enantiomers of **27** can be readily prepared on a multigram scale in pure form. Ozonolysis of **27** followed by reductive workup leads to the aldehyde **28**, which undergoes highly diasteroselective addition with nucleophiles to **29** followed by retroaldol/lactonization cascade to **30**. An example of the use of this methodology is the synthesis of tricyclic framework found in guaianolides with all-*trans* stereochemistry at the ring junctions

(Scheme 5). By using allylsilane **31** as nucleophile in the addition to **28**, the aldehyde **32** was obtained in excellent *anti*-selectivity (95:5) of the substituents placed on the lactone ring, while the *anti*-relationship was observed between the two five-membered rings exclusively. Introduction of a second allyl group by BF$_3$-mediated reaction with allylsilane **33** take place with moderate diastereoselectivity (4:1) to yield the 1,8-diene **34** as the major diastereomer. The ring-closing metathesis step was achieved after conversion of **34** to the TMS-ether affording **35** in 34% yield.

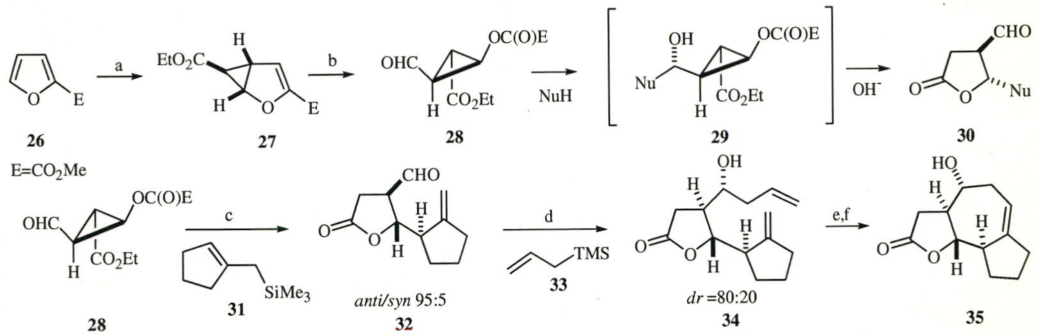

26   **27**   **28**   **29**   **30**

E=CO$_2$Me

28   **31**   **32** *anti/syn* 95:5   **33**   **34** *dr* =80:20   **35**

^{a} (a) 1. ethyl diazoacetate, Cu(OTf)$_2$, (*S,S*)-^{t}Bu-box, PhNHNH$_2$, CH$_2$Cl$_2$, 91% *ee*, 2. recrystallization (pentane), >99% *ee*, 53%; (b) 1. O$_3$, CH$_2$Cl$_2$, -78 °C, 2. dimethyl sulfide, 94%; (c) 1. BF$_3$·Et$_2$O, CH$_2$Cl$_2$, -78 °C, 2. Ba(OH)$_2$, MeOH, 0 °C, 83% ; (d) allyl-TMS, CH$_2$Cl$_2$, then BF$_3$·Et$_2$O, 66%; (e) TMSCl, Et$_3$N,74%; (f) (4,5-dihydroIMES)(PCy$_3$)Cl$_2$Ru=CHPh, CH$_2$Cl$_2$, reflux, 34%, then HCl.

Scheme 5

2.4. 3-Sulfinylfuran-2(5*H*)-ones

The asymmetric 1,3-dipolar cycloaddition reaction of 3-sulfinylfuran-2(5*H*)-ones with 11*H*-dibenzo[*b,e*]azepine 5-oxide **36** has been recently used by Garcia Ruano and Martin for the synthesis of biologically interesting pyrroloazepines and isoxazoloazepines.[15] The authors studied the behaviour of the chiral dipolarophiles **37** and **38** in reaction with nitrone **36** to obtain the optically pure furoisoxazoloazepines equipped with the proper substituents to be transformed into highly functionalized isoxazoloazepines and pyrroloazepines (Scheme 6). The addition of nitrone **36** to enantiomerically pure furanone **37** required less than 5 min at room temperature to afford *anti*-**39**-*endo* as the only product isolated in 85% yield after crystallization from acetone. The reaction of the same nitrone **36** with sulfoxide **38** afforded **40** as a mixture of stereoisomers in different ratio depending from temperature and reaction time. The amount of the adduct *anti*-**40**-*endo* grows larger as the time increases and mainly when the temperature raises. The highest stereoselectivity was observed using toluene at 100 °C for 5 min, yielding a 72:19:9 mixture of the diastereomers **40**, **41**, and **42**. Formation of the *anti*-**40**-*endo*-adduct in a 77% isolated yield after crystallization from this reaction mixture suggested an easy cycloreversion during the purification process. Comparison of the results obtained for the reaction of nitrone **36** with a furanone bearing no sulfinyl group, indicates that the sulfinyl group significantly increases the reactivity, it is the main controller of the *endo/exo* selectivity, and it modulates the π-facial selectivity mainly governed by the configuration at C-5. The synthesis of pyrroloazepines from the corresponding isoxazolines was obtained by conversion of the adduct **40** into the corresponding diol **44** and mesylation of the hydroxyl groups. Then, the aminoalcohol formed in the opening of isoxazolidine ring by hydrogenolysis of N-O bond evolved spontaneously into **46** which subsequently reacted with 4-benzylpiperidine to afford the corresponding enantiopure aminoalcohol target **47** in high overall yield.

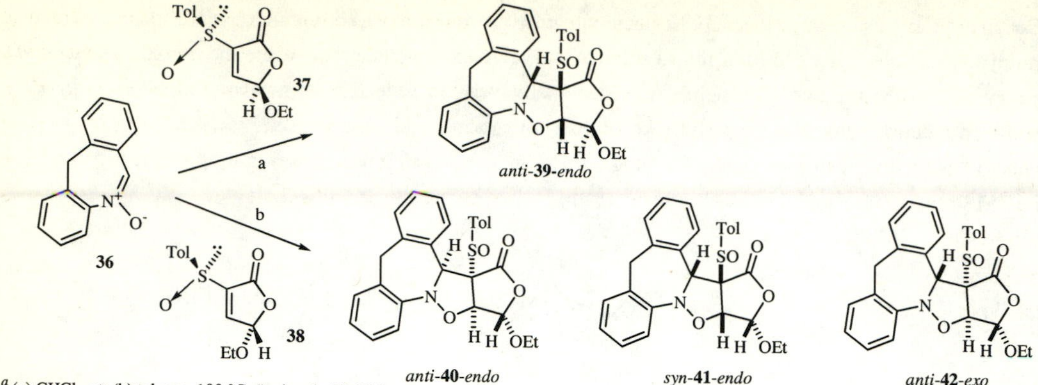

^{a} (a) CHCl$_3$, rt; (b) toluene, 100 °C, 5 min, dr 72:19:9

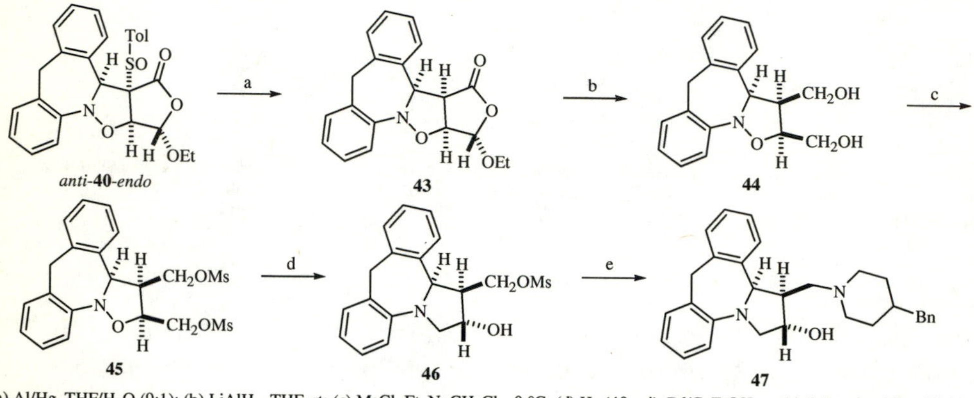

^{a}(a) Al/Hg, THF/H$_2$O (9:1); (b) LiAlH$_4$, THF, rt; (c) MsCl, Et$_3$N; CH$_2$Cl$_2$, 0 °C; (d) H$_2$ (40 psi), Pd/C, EtOH, rt; (e) 4-Bn-piperidine, 70 °C.

Scheme 6

2.5. 2(5H)-Furanones

An highly efficient and diastereoselective synthesis of (+)-Lineatin pheromone from the 2(5H)-furanone **48** was reported by Alibès and Font.[16]

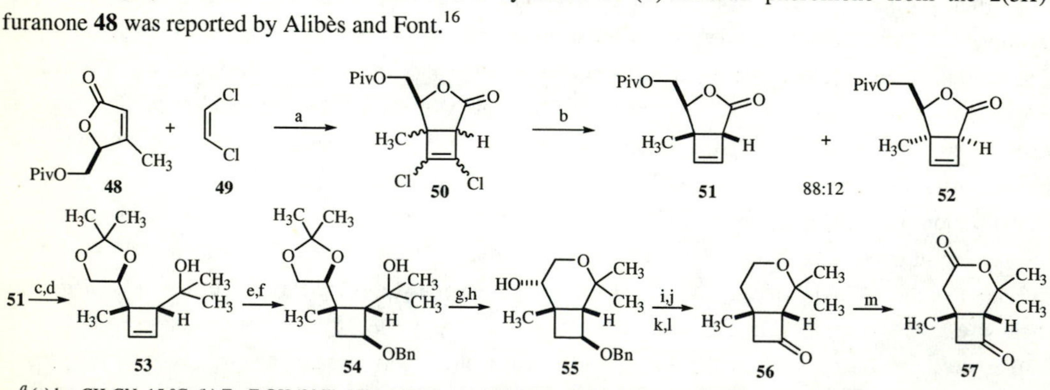

^{a} (a) hv, CH$_3$CN,-15 °C, (b) Zn, EtOH (80%), MW, 105 °C; (c) MeLi, THF, -78 °C, 86%; (d) p-TsOH, acetone, 91%; (e) 1. Hg(OAc)$_2$, THF/H$_2$O, 2. NaBH$_4$, NaOH, 92%; (f) NaH, BnBr, THF, 79%;(g) TFA, MeOH/H$_2$O, 94%; (h) TsCl, DMAP(cat.), py,reflux, 84%; ((i) TCDI, THF, 70 °C 90%; (j) Bu$_3$SnH, AIBN, toluene, 93%; (k) H$_2$ Pd/C, HOAc, EtOAc, 98%; (l) Dess-Martin periodinane, CH$_2$Cl$_2$; (m) RuCl$_3$, NaIO$_4$, CCl$_4$/H$_2$O, 81% 2 steps.

Scheme 7

336

The key steps of this synthetic approach feature the diastereoselective construction of a cyclobutane through a photochemical [2+2] cycloaddition and a regiocontrolled oxymercuration reaction (Scheme 7). Irradiation of lactone **48** with (Z)-1,2-dichloroethylene **49** in acetonitrile furnished a mixture of seven isomeric cycloadducts and the reductive dehalogenation of this mixture under microwave irradiation gave cyclobutanes **51** and **52** in 88:12 dr in a good 61% yield from **48**. Treatment of **51** with an excess of MeLi in THF resulted in the addition of two methyl groups to the lactone carbonyl function with simultaneous removal of the pivaloyl group. Regioselective oxymercuration-demercuration process afforded, after protection of the newly formed hydroxyl group, **54** as a single isomer. After removing of acetonide protecting group an oxycyclization step afforded the bicyclic derivatives **55**.

The secondary hydroxyl group was then removed by Barton-McCombie procedure and the product obtained was oxidate to the keto lactone **57** that has previously been transformed into (+)-Lineatin pheromone by carbonyl reduction and acetalization.

2.6. Furan-2,5-diones

Taking advantage from the known photocycloaddition of maleic anhydride to *trans*-1,4-dichloro-2-butene Birman has described a concise and versatile synthesis of the bioactive marine alkaloids (±)-sceptrin and (±)-dibromosceptrin (Scheme 8).[17]

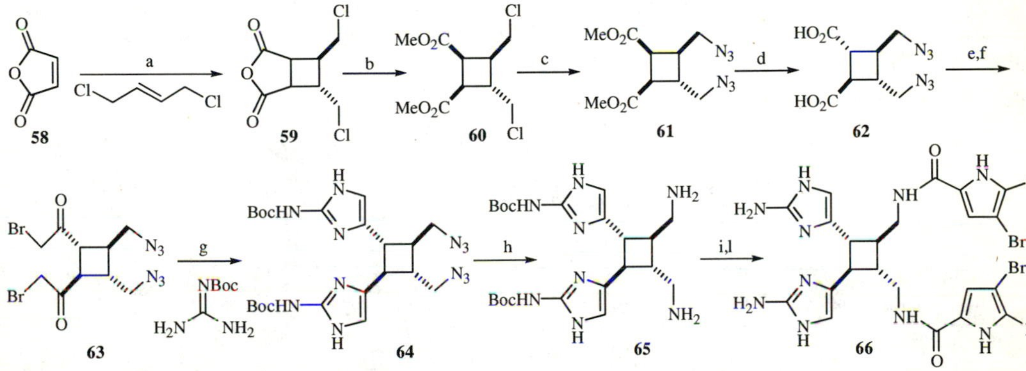

[a] (a) Ph$_2$CO, hv, rt; (b) MeOH, H$_2$SO$_4$, reflux, 78%; (c) NaN$_3$, KI, DMF, 80 °C, 82%; (d) KOH, MeOH, rt, quant.; (e) SOCl$_2$, reflux ; (f) 1. CH$_2$N$_2$, 0 °C, 2. HBr, -10 °C, 72% 3 steps; (g) DMF, rt, 30%; (h) PPh$_3$, THF, reflux, NH$_4$OH, 93%; (i) 2-trichloro-acetyl-4,5-dibromo- pyrrole, DMF, rt, X=H, 82%, X=Br 49%; (l) TFA, CH$_2$Cl$_2$ quant.

Scheme 8

The benzophenone-sensitised photo-addition of maleic anhydride to *trans*-1,4-dichloro-2-butene afforded **59** in 76% yield. Compound **59** was converted into dimethyl diester **60** and treated with sodium azide affording the diazide **61** without significant epimerization. The required all-*trans* stereochemistry was achieved in the next step, during the base hydrolysis to the diacid **62**. The diester obtained from **62** was transformed into the bis-bromomethyl ketone and then converted into the bis-aminoimidazole by reaction with *N*-Boc guanidine. The completion of the total synthesis of sceptrin and dibromosceptrin required reduction of the azide groups to diamine **65** and acylation with 2-trichloroacetyl-4-bromopyrrole or 2-trichloroacetyl-4,5-dibromopyrrole respectively in DMF followed by TFA deprotection to give **66** in 10.5 and 6.3 % overall yield.

3. Pyrroles

3.1. *N*-Boc-pyrroles

Based on the "naked sugar" methodology which transforms enantiomerically pure furan derived intermediates into a large variety of sugars and rare analogues,[18] Vogel has reported the synthesis of 7-azanorbornanones as key intermediates for a new "naked aza-sugar" chemistry. As first application of this methodology, the synthesis of 3,4-dihydroxy-5-hydroxymethylproline and conformationally constrained bicyclic diamine glycosidase inhibitors was reported (Scheme 9).[19]

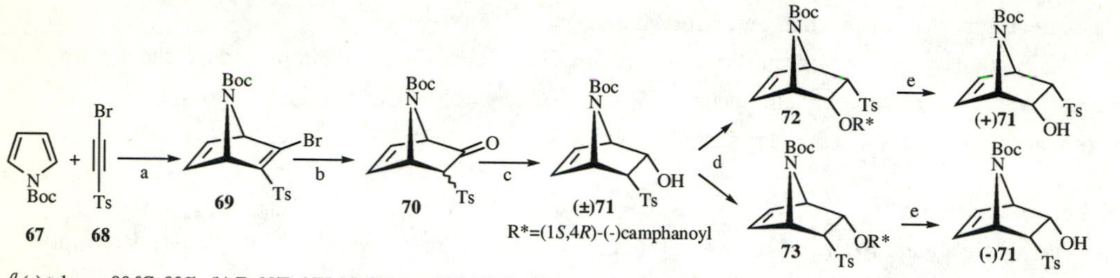

[a] (a) toluene, 90 °C, 80%; (b) Et$_3$N/Et$_2$NH, MeCN then HCl (10%), 74%; (c) LiBH$_4$, THF, -78 °C, 85%; (d) (1*S*,4*R*)-(-)Camphanic acid chloride, Et$_3$N, DMAP, CH$_2$Cl$_2$; (e) NaOMe, MeOH-THF 88-90%

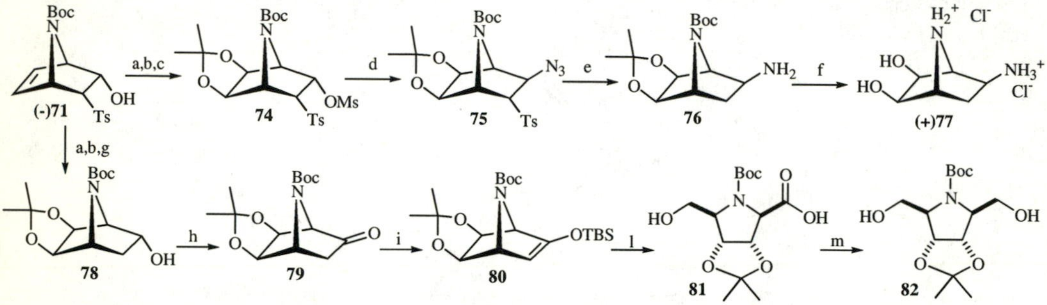

[a] (a) OsO$_4$ (cat), Morpholine *N*-oxide, acetone-H$_2$O, 83%; (b) (MeO)$_2$CMe$_2$, p-toluensulphonic acid (cat), acetone, 91%; (c) MsCl. DMAP, Pyridine, 82%; (d) TMSN$_3$, TBAF, THF, 63%; (e) 1. H$_2$, Pd/C (10%), MeOH, 2. Na-Hg (5%), THF-MeOH 60%; (f) HCl (1M)/THF (1:1) 80 °C,quant.; (g) Na-Hg (5%), THF-MeOH, 75%; (h) Swern oxidation, 91%; (i) CF$_3$CON(Me)TBS, Et$_3$N, DMF, 60 °C, 86%; (l) 1. O$_3$, CH$_2$Cl$_2$-MeOH, 2. NaBH$_4$, 88%; (m) BH$_3$-Me$_2$S, THF, reflux

Scheme 9

The synthesis started from the known Diels-Alder reaction between *N*-Boc pyrrole and 2-bromoethynyl p-tolyl sulfone to give the diene **69** that was transformed into the ketone **70** in good yield as a mixture of isomers *endo/exo*=1.2. Reduction of compound **70** with LiBH$_4$ at -78 °C gave exclusively the *endo*-alcohol **71**. The stereospecificity of this reduction was explained by involving first *endo*-epimerization of the tosyl group under basic conditions that facilitates the *exo*-attack of hydride onto the ketone moiety. The key intermediate β-hydroxysulfone **71** was resolved in enatiomerically pure +**71** and -**71** employing (-)-(1*S*,4*R*)-camphanic acid chloride. Starting from (-) **71** the synthesis of the target diamino azabicyclo compound was undertaken. Stereoselective dihydroxylation of **71** followed by protection of the *exo*-diol group afforded, after S$_N$2 displacement of the activate OH group by azide anion, the *exo*-5-azido product **75**. Formation of minor amounts of the 5-*endo*-azido-6-*exo*-tosylsulfonyl isomer was observed, probably owed to a competitive β-elimination of methanesulfonic acid induced by the basic conditions, to produce an alkene intermediate. Catalytic hydrogenation followed by desulfonylation and hydrolysis of protecting group gave the target diamine (+)-**77**, as dihydrochloride. The synthesis of proline derivative **82** was instead performed transforming compound **71** into the 7-azanorbornanone key intermediates **79** *via* oxidation of the desulfonilated product **78**. Compound **79** was then treated with CF$_3$CON(Me)TBS to give silyl enol ether **80**. Ozonolysis of **80** at -78 °C followed by reduction of the ozonide and the aldehyde intermediate, gave compound **81** in good overall yield. Reduction of **81** gave the known pyrrolidine **82**. Compound **81** was also incorporated into a tripeptide following the Fmoc strategy for peptide synthesis.

3.2. 2-Trialkylsilyloxypyrroles

Casiraghi and Rassu continuing their efforts to exploit furan-, pyrrole-based dienoxysilane for the synthesis of enantiopure carbasugars and relatives have also investigated the use of silyloxypyrrole derivatives for the synthesis of carbaseptanose, carbaoctanose and cyclopentanedicarboxylic acid.[11c,20] As previously described, the key steps of the synthetic methodology are a vinylogous crossed aldol addition and the silylative cycloaldolization. As a remarkable extension of this versatile strategy by using pyrrole-based building blocks, we will describe here the synthesis of aminocycloheptane derivatives (Scheme 10).

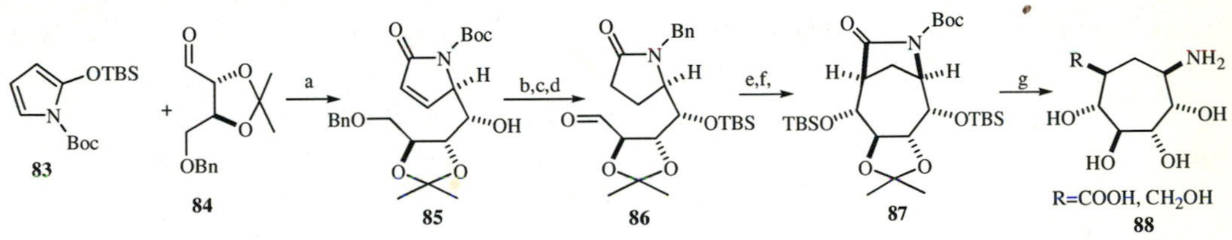

^{a} a) SnCl$_4$, Et$_2$O, -80 °C, 80%; (b) NaBH$_4$, NiCl$_2$ then TBSOTf, 2,6-Lutidine, 83%; (c) CAN, MeCN, reflux then BnCl, KH, 60 °C, 69%; (d) H$_2$, Pd(OH)$_2$ then Swern, 78%; (e) TBSOTf, DIPEA, rt, 85%; (f) Na/NH$_3$, THF, -70 °C then Boc$_2$O, DMAP, CH$_3$CN, 93%; (g) for R=COOH: 6N HCl reflux, 85%; for R= CH$_2$OH: NaBH$_4$, THF, H$_2$O then 6N HCl, THF, MeOH, 84%.

Scheme 10

The synthesis of carbasugar **88** started from the SnCl$_4$ assisted vinylogous aldol reaction between L-threose **84** and 2-[(*tert*-butyldimethylsilyl)oxy]pyrrole **83** to provide the 4,5-*threo*-5,6-*erythro* configured unsaturated lactam **85** in high yield as only diastereomer. A Felkin-Ahn-type induction was called upon to justify the facial (5,6-*erythro*) diastereoselectivity, whereas a sterically and an electronically favorable Diels-Alder-like transition state was contemplated to account for the simple (4,5-*threo*) diastereoselectivity observed. Chemoselective saturation of the lactam double bond in **85** followed by silylation delivered a protected lactam that was transformed in the aldehyde **86** *via* a three step procedure. At first the N-Boc protecting group was converted into N-Bn group, the primary hydroxyl function was selectively deprotected and then subjected to a Swern oxidation. Silylative intramolecular aldol reaction was then performed to give as the single diastereomer, the tricyclic adduct **87**. Because of the reluctance of N-Bn lactams to undergo both hydrolytic and reductive amide fission, compound **87** was transformed into N-Boc derivative and hydrolysed to give the target carbasugar **88** (R=COOH) or alternatively reduced and hydrolysed affording compound **88** (R=CH$_2$OH).

3.3. 1,5-Dihydro-2H-pyrrol-2-ones

Inspired on the Schlessinger's chemistry of tetronic acid acid, Huang has described the asymmetric approach to 5-alkyl tertamic acid derivatives based on the use of (S)-**89** as the first synthetic equivalent to the chiral non racemic tetramic acid 5-carbanionic synthon and their use in the synthesis of (3S,4S)-4-amino-3-hydroxy-5-phenylpentanoic acid (AHPPA) and (+)-preussin (Scheme 11).[21]

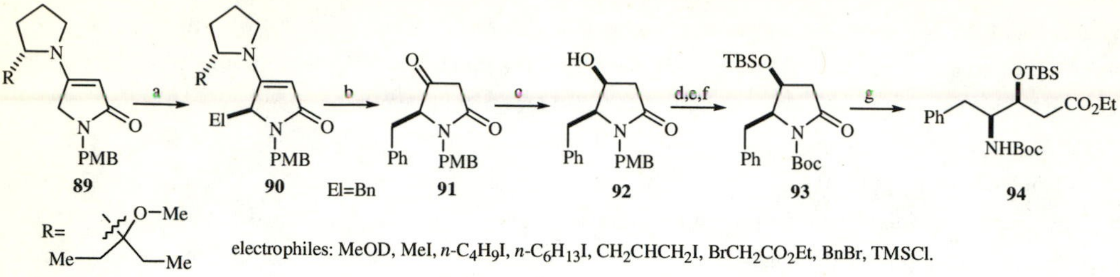

R=

electrophiles: MeOD, MeI, n-C$_4$H$_9$I, n-C$_6$H$_{13}$I, CH$_2$CHCH$_2$I, BrCH$_2$CO$_2$Et, BnBr, TMSCl.

[a] (a) t-BuLi, THF, HMPA, -78 °C, electrophiles; (b) 10% HCl, THF; (c) NaBH$_4$, MeOH, 82% for 2 steps; (d) TBSCl, DMF, DMAP, imidazole, 90%; (e) CAN, MeCN/H$_2$O (3:1), 75%; (f) Boc$_2$O, Et$_3$N, CH$_2$Cl$_2$, 97%; (g) KCN, EtOH, THF, 91%

Scheme 11

After preparation of **89** from N-protected tetramic acid and (S)-proline derivative, a series of alkylation reactions were performed. The reaction of the lithiated **89** with electrophiles proceeds with excellent C-5 regioselectivities and high *anti*-diastereoselectivities (>97%) to give the 5-alkyl tetramic acid derivatives in high yields. The synthesis of AHPPA started from **91** derived from **90** (El=Bn) by removal of the chiral auxiliar. The tetramic acid **91** was reduced to give *cis*-**92** with *cis/trans* ratio 20:1 that was transformed through a conventional chemistry into the lactam **93**. Potassium cyanide catalyzed ring opening gave the protected AHPPA **94** in high yield.

3.4. Pyrrolidine-2,5-diones

The same author has also introduced a flexible non-amino acid-based asymmetric synthesis of the precursor **99** of the naturally occurring antimicrobial dienol **100** (Scheme 12).[22] The synthesis is based on an high regioselective Grignard addition to (S)-malimide followed by a *trans*-diastereoselective reductive deoxygenation.

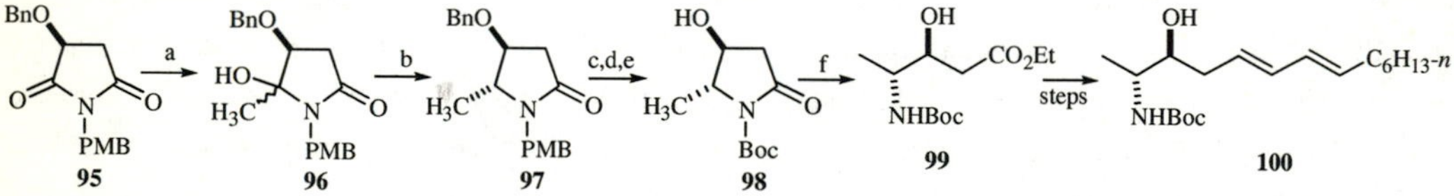

[a] (a) MeMgI, THF, -78 °C, 82%; (b) Et$_3$SiH, BF$_3$·Et$_2$O, CH$_2$Cl$_2$, -78 °C-rt, 89%; (c) CAN, CH$_3$CN-H$_2$O (3:1), 70%; (d) (Boc)$_2$O, DMAP, Et$_3$N, CH$_2$Cl$_2$, 81%; (e) H$_2$, 1 atm, 10% Pd/C, EtOH, 89%; (f) KCN, EtOH, THF, 97%.

Scheme 12

Reaction of an excess of methyl magnesium iodide to malimide **95** occurred regioselectively at the C-2 position, leading to **96** as a diastereomeric mixture in a ratio of 18:82. This mixture was then subjected to Lewis acid mediated ionic hydrogenation affording the *trans*-(+)-**97** as the only isolable diastereomer in 89% yield. The fact that starting from an 18:82 diastereomeric mixture of **96**, was obtained only one diastereomer **97**, was explained by a transformation of **96** into **97** *via* intermediacy of an N-acyliminium. Oxidative N-deprotection followed by N-protection as N-Boc gave the activated lactam. The alcohol **98** derived from catalytic hydrogenolysis was finally submitted to potassium cyanide promoted lactam ring opening to give N-Boc-(3S,4R)-4-amino-3-hydroxypentanoic acid ethyl ester **99**, intermediate for the synthesis of **100**. Furthermore, a study of the protecting group effect on regio and diastereoselective reductive alkylation of malimides was performed.

4. Pyrones

4.1. 2H-Pyran-2-ones

A concise synthesis of carbocyclic α-amino acid derivatives *via* Diels-Alder cycloaddition of 3-phenylamino-5-bromo-2-pyrone was proposed by Cho (Scheme 13).[23]

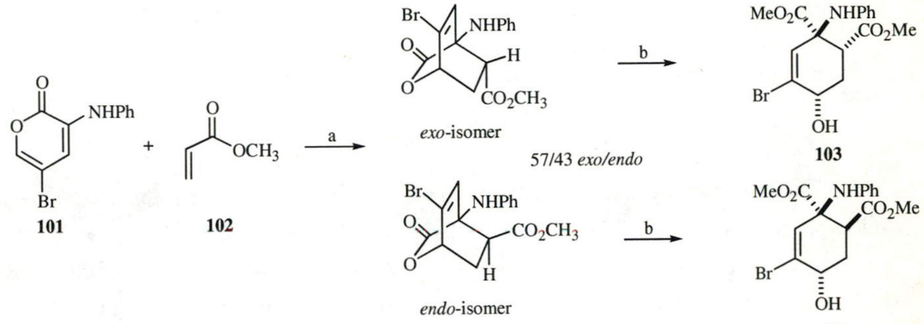

a (a) CH₂Cl₂, 100 °C, sealed tube, 30 min; (b) MeOH, NaOMe, 0 °C.

Scheme 13

The amino-substituted 2-pyrones are of particular interest as they can be used as synthetic building blocks for various constrained carbocyclic α-amino acids through cycloaddition and subsequent lactone ring opening. Due to the presence of the amino group on the pyrone unit, 3-phenylamino-5-bromo-2-pyrone underwent only normal electron demand Diels-Alder cycloaddition with electron deficient dienophiles (Scheme 13). Attempted Diels-Alder cycloaddition with electron rich dienophiles, for example, benzyl vinyl ether, gave only a trace of cycloadducts. A series of reactions were performed with electron deficient dienophiles (CH₂=CH-CO₂Me, CH₂=CH-COMe, CH₂=CHCN, CH₂=C(CH₃)CO₂Me, CH₂=CH-Ph-*p*-Br, MeO₂C-CH=CH-COMe...). Unlike to the parent 3,5-dibromo-2-pyrone or other 2-pyrone derivatives, Diels-Alder cycloadditions of **101** exhibited moderate to high *exo*-selectivity. The presence of aryl group may impose an additional steric repulsion with the incoming dienophile to destabilize the otherwise favorable *endo*-transition state. Lactone ring opening of the cycloadducts with NaOMe provided various substituted carbocyclic amino acid methyl esters.

5. Pyridines

5.1. 1,4-Dihydropyridines

Reiser recently developed a conveniently synthesis of biologically interesting β-aminocyclopropane-carboxylic acids from N-Boc pyrrole *via* cyclopropanation with methyl diazoacetate.[24a] In extension of this strategy, novel conformationally restricted β- and γ-amino acids containing cyclopropane ring were stereoselectively synthesised from readily available dihydropyridine or indoles by copper(I)-catalyzed cyclopropanation with methyl diazoacetate followed by oxidative cleavage of the resulting adducts (Scheme

341

14).[24b] In the synthesis reported by Reiser, the copper(I) triflate was generated *in situ* from copper(II)triflate and phenylhydrazine.

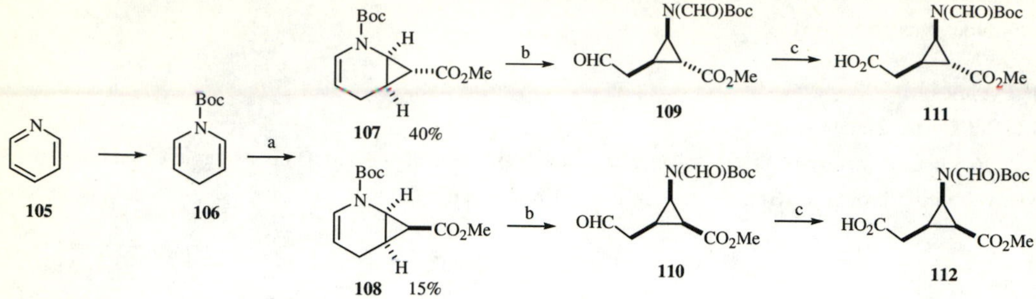

a (a) N$_2$CHCO$_2$Me, Cu(OTf)$_2$, PhNHNH$_2$, CH$_2$Cl$_2$, rt, 55%; (b) O$_3$, DMS, CH$_2$Cl$_2$, -78 °C to rt, 91%; (c) 1. NaClO$_2$, H$_2$O$_2$, CH$_3$CN, 0 °C to rt; 2. 2-diethylaminoethylamine, CH$_3$CN, rt, yield for two steps (111) 84%, (112) 72%.

Scheme 14

Cyclopropanation of dihydropyridine **106** with methyl diazoacetate gave both, *exo* and *endo* (ratio 8:3) monocyclopropanated adducts **107** and **108** respectively which were separated by chromatography on silica gel. Ozonolysis of the individual isomers followed by reductive workup cleanly produced the aldehydes **109** and **110**. Oxidation using sodium chlorite/hydrogen peroxide proceeded in almost quantitative yield to the new -γ-amino acids **111** and **112**.

6. Dioxolanes

6.1. 1,3-Dioxolanes

Alonso performed the asymmetric synthesis of α-amino acid derivatives by the diastereoselective intermolecular addition of 1,3-dioxolanyl radical to enantiopure N-acyl aldohydrazones (Scheme 15).[25] The author studied the reaction using different alkyl and aryl substituents on the aldohydrazone, and different reaction conditions. With an electron-withdrawing activating substituent (R=COOEt) the best results were obtained at -78 °C, the product **115** was synthesised in quantitative yield in 15 min, and dr 80:20 without the use of chelating InCl$_3$. With a not activating alkyl substituent (R=Et) the best results were obtained at -78 °C, 87% yield after 75 min, and dr 91:9, with 2 eq. of InCl$_3$.

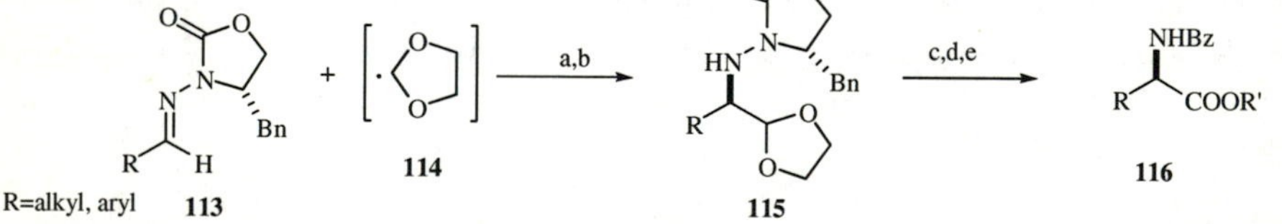

a(a) 1,3 dioxolane, *p*-TsOH (cat), rt, 1h; (b) Ph$_2$CO, hv, InCl$_3$, -78 °C; (c) *n*-BuLi, THF, -78 °C, BzCl, 83%; (d) SmI$_2$, THF, HMPA, 20 min, 90%; (e) O$_3$, AcOEt, -78 °C, 94%.

Scheme 15

Using a one pot protocol with different aldehydes they found the best results for the propionaldehyde, where they obtained the product **115** in 93-99% yield, and dr 99:1, at -78 °C with InCl$_3$. The reaction was carried out mixing one pot the aldehyde, (S)-3-amino-4-benzyl-1,3-oxazolan-2-one with 1,3-dioxolane in the presence of a catalytic amount of p-TsOH for 1h, followed by UV irradiation (1-1.5h) in the presence of Ph$_2$CO and in some cases InCl$_3$ at -78 °C. The adducts were then transformed in the desired amino acids after deblocking of the auxiliar by conventional chemistry.

6.2. 1,3-Dioxolan-4-ones

In extension of the use of mandelic acid as synthetic equivalent of benzoyl carbanion, Pedro has recently reported the enantioselective synthesis of α-hydroxy-α,β-diaryl-γ-lactams.[26] The strategy employed to exert stereochemical control in the newly created stereogenic centres, involved the use of (S)-mandelic acid as a source of chiral information through its previous conversion into (S,S)-cis-1,3-dioxolan-4-one **118**, according to the Seebach's principle of self-regeneration of stereocentres (Scheme 16).[27] Diastereoselective Michael addition of the lithium enolate of **118** to several aromatic nitroalkenes in the presence of HMPA proceeds readily to give only two adducts **119** and **120** in good yields and diastereoselectivities (from 73/27 to 93/7). Reduction of the nitro group with Zn/HCl/EtOH/H$_2$O with concomitant intramolecular aminolysis of the acetal moiety leads to enantiomerically pure biologically interesting α-hydroxy-α,β-diaryl-γ-lactam **121**.

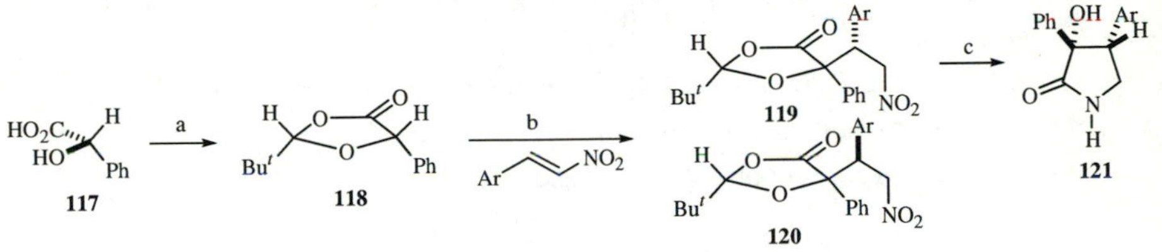

Ar= Ph, 4-MeO-Ph, 3,4-(MeO)$_2$-Ph, 2,4-(MeO)$_2$Ph, 4-OH, 3-MeO-Ph, 4-Br-Ph, 4-CF$_3$O-Ph, 4-CF$_3$-Ph
^{a} (a) Bu^{t}CHO, H$^+$; (b) LDA, TFA-HMPA, -78 °C; (c) Zn/HCl, EtOH/H$_2$O

Scheme 16

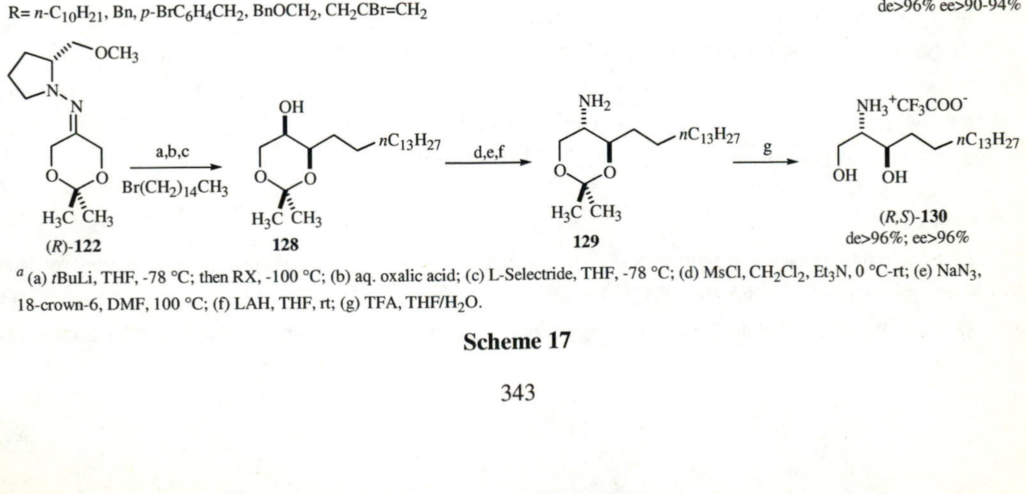

R= n-C$_{10}$H$_{21}$, Bn, p-BrC$_6$H$_4$CH$_2$, BnOCH$_2$, CH$_2$CBr=CH$_2$

^{a} (a) tBuLi, THF, -78 °C; then RX, -100 °C; (b) aq. oxalic acid; (c) L-Selectride, THF, -78 °C; (d) MsCl, CH$_2$Cl$_2$, Et$_3$N, 0 °C-rt; (e) NaN$_3$, 18-crown-6, DMF, 100 °C; (f) LAH, THF, rt; (g) TFA, THF/H$_2$O.

Scheme 17

343

7. Dioxanes

7.1. 1,3-Dioxan-5-ones

A practical method for the highly enantio and diastereoselective synthesis of 2-amino-1,3-diols and D-erythro-sphinganine, starting from easily available dihydroxy acetone equivalent 2,2,-dimethyl-1,3-dioxan-5-one was reported by Enders (Scheme 17).[28]

The stereogenic centres are generated by an highly stereoselective α-alkylation key step using SAMP/RAMP hydrazone methodology and diastereoselective reduction of the ketone (S)-124 with L-selectride. The resulting alcohols (S,S)-125 are converted into amines (S,R)-127 by nucleophilic substitution with sodium azide followed by reduction with lithium aluminium hydride. The products were obtained in high diastereomeric and enantiomeric excess (de>96%, ee=90-94%). By employing this methodology, the D-erythro-sphinganine (R,S)-130 was synthesised via alkylation of the RAMP hydrazone (R)-122 with pentadecyl bromide in 47% overall yield and with excellent diasteromeric and enantiomeric excess (de and ee >96%). This methodology provides a flexible route to 2-amino-1,3-diols and allows a broad range of modification in the side chain of sphinganines. In addition, both anti-configured enantiomers are accessible.

8. Oxazoles

8.1. 1,3-Oxazolidines

In the course of the study towards new synthetic routes for sphingosine and structurally related Myriocin and Mycestericin A, Koskinen reported a highly diastereoselective aldol addition to oxazolidine 131 to obtain α-substituted-β-hydroxy amino acid (Scheme 18).[29]

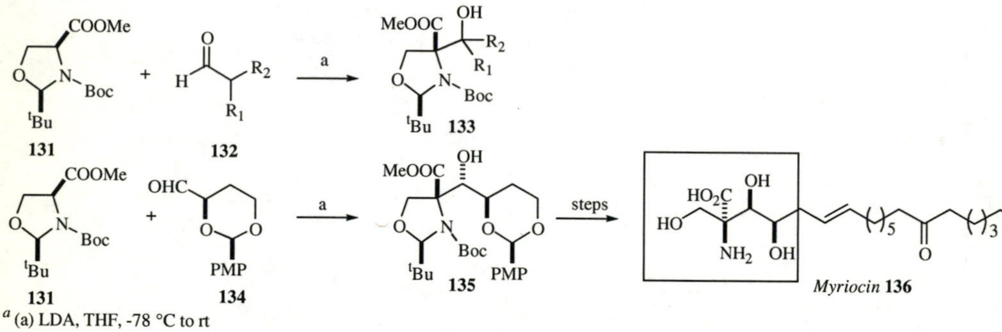

^{a} (a) LDA, THF, -78 °C to rt

Scheme 18

Although aldol addition of Li-enolate of N-formyl and N-benzyl protected five ring N,O-acetals was previously reported the authors expected to obtain even better diastereoselectivities with highly substituted N-Boc protected N,O-acetals. While the aldol addition of achiral aliphatic and aromatic carbonyl compounds to lithium enolate of 131 leads to a mixture of diastereomers, the double sterodifferentiation of chiral aldehyde 134 gave a single syn-Felkin diastereomer 135 but in only 25% yield.

8.2. Oxazolines

An highly enantioselective synthesis of biologically important α-alkyl serines was proposed by Jew and Park via phase-transfer-catalytic (PTC) alkylation of 2-phenyl-2-oxazoline-4-carboxylic acid tert-butyl ester 137 (Scheme 19).[30] The authors chosen the oxazoline moiety as the best substrate for PTC alkylation because this functional group not only acts as an excellent protecting group for both amino and hydroxy groups of the serine ester, but also because enhances the acidity of the α-proton of the ester. The reaction was studied using RX=PhCH$_2$Br with different catalyst and reaction conditions in order to increase yield and diastereoselectivity. Using (S)-binaphtol derived catalyst at 0 °C the (S)-**138** was obtained with very high enantioselectivity (>99% *ee*) and yield (99%). Hydrolysis of the oxazoline ring with HCl 6N led after purification through an ion-exchange resin to the optically pure (S)-(+)-α-benzylserine in 98% yield.

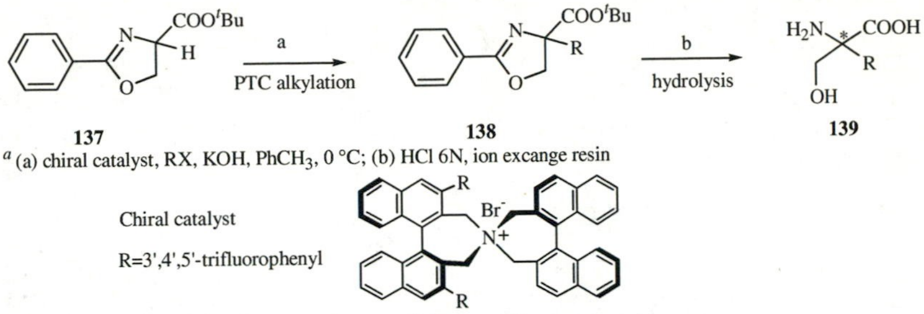

^{a} (a) chiral catalyst, RX, KOH, PhCH$_3$, 0 °C; (b) HCl 6N, ion exchange resin

Chiral catalyst

R=3',4',5'-trifluorophenyl

Scheme 19

Further enantioselective phase-transfer alkylation with various alkyl halides were investigated under the optimised reaction conditions and the same catalyst. The very high *ee* values obtained show that this reaction is a very efficient method for the preparation of α-alkyl serines.

8.3. 1,3-Oxazoles

Griesbeck carried out the diastereoselective synthesis of *erythro*-α-amino β-hydroxy carboxylic acid esters by a photocycloaddition of aromatic and aliphatic aldehydes to 5-methoxyoxazoles (Scheme 20).[31] At first the reaction of 5-methoxyoxazoles (R=H) with a group of aldehydes R^1=Ph, 2-Nph, BnCH$_2$, Et, *i*-Pr, *i*-Bu was investigated. Photolyses were performed at λ=300+/-10nm; high conversions were achieved (>90%) and in all cases only the regioisomer bicyclic oxetane **142** with *exo* configuration was detected with a very high diastereoisomeric ratio (dr 98:2). The *exo/endo* selectivity was excellent both for aromatic and aliphatic aldehydes, and except for the propionaldehyde product, the diastereoisomeric ratio of the ring opened product matched the dr of oxetane precursor, with the major diastereomer **143** showing *erythro* configuration. The authors investigated also a second group of 5-methoxyoxazoles with R=Me, Et, Pr, *i*-Pr, *i*-Bu with the same aldehydes, and in the same reaction conditions.

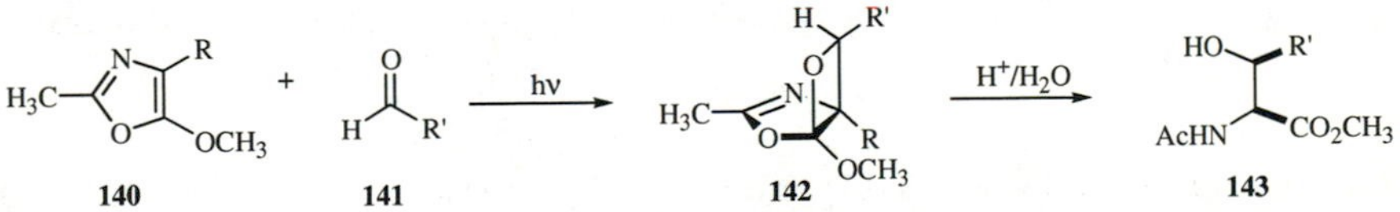

Scheme 20

The primary photoadducts were obtained with high to excellent diastereoselectivities except for the benzaldehyde addition to oxazoles with bulky substituents, also in these cases an *exo*-selectivity was observed. Then they performed the reaction using enantiomerically pure isoleucine-derived oxazole in order to study the facial selectivity. The photocycloaddition of all aldehydes investigated proceeded with an high *exo*-selectivity but no facial selectivity was observed.

8.4. 1,3-Oxazolidin-5-ones

The first enantioselective synthesis of a phenylalanine-based tetrahydroxypyridazinone **150** and its conversion to the dipeptido-mimetic scaffold **151**, an important component of extended β-strand mimetics, was reported by Abell (Scheme 21).[32] The key of this synthesis is the use of chiral oxazolidinone chemistry pioneered by Seebach,[5] to generate the α,α-dialkylated amino acid **147**.

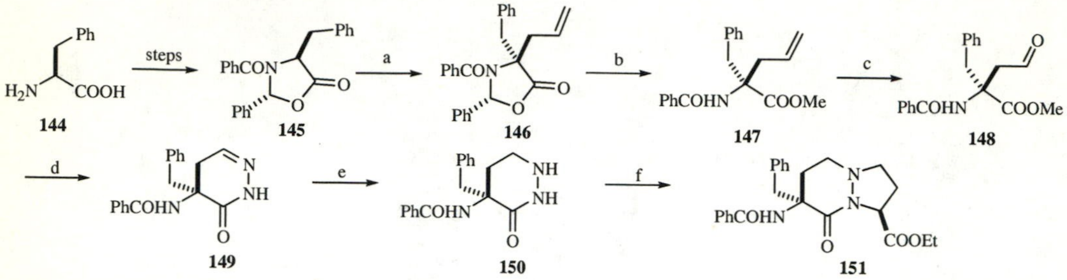

[a] (a) LiHMDS, -78 °C, THF, allyl Bromide, 93%; (b) NaOH, MeOH reflux followed by CH$_2$N$_2$, 99%; (c) O$_3$, CH$_2$Cl$_2$/MeOH (3:1), -78 °C, 96%; (d) hydrazine reflux, 85%; (e) NaBH$_3$CN, MeOH/HCl, 0 °C-rt, 73%; (f) CH$_2$O, reflux, ethyl acrylate, 21%.

Scheme 21

The starting phenyloxazolidinone **145** was prepared in three steps from (S)-phenylalanine. Deprotonation of **145** at C-4 with LiHMDS and alkylation of the resulting anion with allyl bromide gave the alkylated oxazolidinone **146** in high yield as a single isomer. Hydrolysis of oxazolidinone ring followed by addition of diazomethane gave the optically active amino acid **147** in high yield. Ozonolysis of the olefin gave the corresponding aldehyde **148** which upon reflux with hydrazine gave the cyclic hydrazone **149** in good yield. Proceeding in the construction of the heterocyclic core of the peptidomimetic **151**, reduction of **149** was performed to give tetrahydropyridazinone **150**. Finally treatment of **150** with formaldehyde and subsequent heating with an excess of ethyl acrylate gave rise to a 1,3-dipolar cycloaddition. The desired bicyclic template **151** was obtained in modest yield.

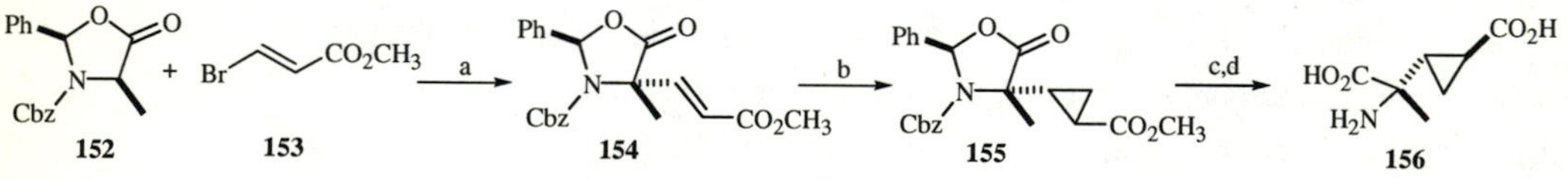

[a] (a) KHMDS, Toluene, -78 °C, 62%; (b) Pd(OAc)$_2$, Et$_2$O, CH$_2$N$_2$, 67%; (c) LiOH, THF/H$_2$O, 4 °C, quant.; (d) HBr, AcOH, 85%

Scheme 22

Also the conformationally restricted matabotropic glutamate receptor antagonist 2-methyl-2-(carboxycyclopropyl)glycine has been synthesised by Pajouhesh in a stereoselective manner (>99% ee) by using a "self-reproduction of chirality" process (Scheme 22).[33] Starting from the known (2R,4R) oxazolidinone **152**, alkylation with methyl-(E)-3-bromo propanoate took place with high streoselectivity and moderate yield to give the dialkylated oxazolidinone **154**. This compound was cyclopropanated with diazomethane in the presence of palladium acetate, to give **155** as only isomer. Subsequent deprotection and hydrolysis yielded the target **156** as a single stereoisomer.

8.5. 1,3-Oxazol-4(5*H*)-ones

Pursuing previous work on the use of azalactones for the synthesis of quaternary amino acids,[34a] Trost performed a very interesting enantio- diastereo- and regioselective synthesis of α-hydroxyacids starting from 5*H*-oxazol-4-ones *"oxalactims"* as new building blocks by a Molybdenum-catalyzed asymmetric allylic alkylation (Scheme 23).[34b] The authors examined the Mo-AAA reaction using **157** with R=CH$_3$ and **158** with Ar=Ph by using HMDS as the base. Lithium counterion gave the higher yield, regio-, diastereo- and enantioselectivity over sodium and potassium, while X=carbonate was better leaving group then phosphate. In this conditions the adduct **159** was obtained in 91% yield and very high stereoselectivities branched/linear ratio 99:1, dr 11.5:5 and ee >99%. They also studied the stereoselectivity of the reaction with respect to changing R and Ar substituents obtaining in almost all cases excellent results about stereoselectivity and yield. Products obtained from the Mo-AAA reaction are easily opened to α-hydroxy amides by treatment with NaOH in EtOH and then transformed in the corresponding α-hydroxy acids by diazotation with isoamyl nitrite in anhydrous 1,4-dioxane and 2M HCl in Et$_2$O. Further combining the AAA with a ring closing metathesis provides access to chiral cyclic products.

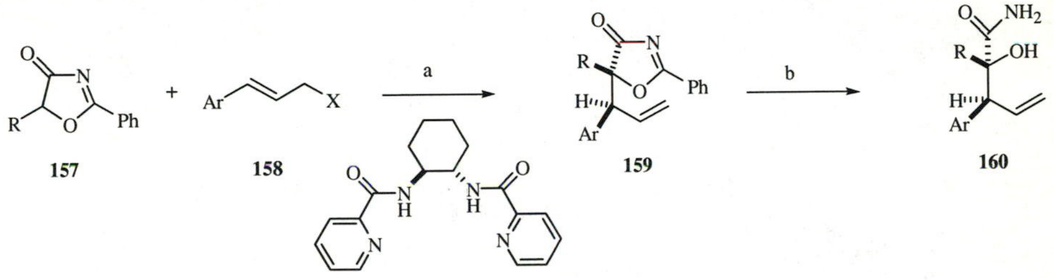

^{a} (a) 10% Mo(CO)$_3$C$_7$H$_8$, base, THF, 65 °C; (b) 1-2.5N, NaOH, EtOH-H$_2$O, 60 °C, 2-3h.

Scheme 23

8.6. 1,3-Oxazol-5(4*H*)-ones

A new stereocontrolled approach to 3-formylcyclopent-3-enyl and 3-carboxycyclopentylglycine derivatives *via* retro-aldol or retro-Claisen reaction was reported by Gelmi (Scheme 24)[35]. The synthesis of these biologically interesting molecules started from norbornane or norbornene derivatives with an appropriate electrophilic functionalization at the C-2 and C-3 that can be transformed into 1,3-disubstituted cyclopentene or cyclopentane derivatives by cleavage of the C2-C3 bond. Diastereomeric *exo*-**163** and *endo*-**164** were then the starting materials for the preparation of the cyclopentylglycines and were easily available *via* a Diels-Alder reaction of cyclopentadiene **162** and the ethyl 2-phenyl-5-oxo-oxazol-4-methylene-carbonate (Z)-**161**. Using Mg(ClO$_4$)$_2$ as catalyst for the cycloaddition reaction, cycloadducts were obtained in a good yields 87% and *exo*-**163**/*endo*-**164** 70:30. To realise the synthesis of cyclopentylglycine derivatives *exo*-**167**, *endo*-**168** and *exo*-**171**, *endo*-**172** were used as intermediates.

Heterocycle ring opening using bis-(dibutylchlorotin)oxide as the catalyst gave the esters *exo*-**165** and *endo*-**166** in good yield from **163** and **164**, while the synthesis of the keto derivatives *exo*-**171** and *endo*-**172** was performed *via* PCC oxidation of the saturated hydroxy esters *exo*-**169** and *endo*-**170**. For the preparation of cyclopentylglycine derivatives functionalized with a formyl group at C-3, retro-aldol reaction was performed refluxing the mixture *exo*-**165** and *endo*-**166** or the pure *exo*-**165** in EtOH with sodium carbonate to give **173** and **174** as, in all cases, a 1:1 mixture of the diastereomeric aldehydes. The aldehydes **173** and

347

174 were also directly obtained allowing *exo*-**165** to react with methanol in presence of sodium carbonate. Although obtaining the compound **173** and **174** was an interesting result, the loss of chirality at C-3 by migration of the double bond was unfortunate. Retro-Claisen reaction was then investigated starting from the saturated keto derivatives *exo*-**171** or a mixture of *exo*-**171** and *endo*-**172**. Using pyridine as base and solvent and adding water to the reaction mixture cyclopentanecarboxylic acids **175** and **176** were obtained in a 1:1 ratio. Partial control of the stereochemistry at C-α was realised by using NaHCO₃ in a mixture of acetone and DMF. In this case the ketone *exo*-**171** afforded compounds **175/176** in a 1:2 ratio probably because of the formation of the intermediate E, which can evolve to enolate D or proceed directly to the carboxy compound **176** *via* a six-membered intermediate that permits retention of stereochemistry at C-α.

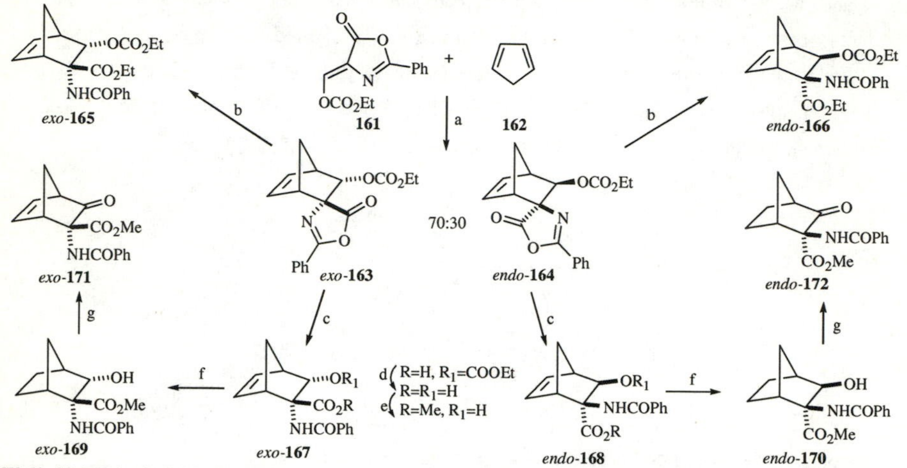

^{a} (a) CH₂Cl₂, Mg(ClO₄)₂; (b) EtOH, (Bu₂ClSn)₂O, reflux; (c)THF, HCl; (d) Me₂NH, EtOH, reflux; CH₂Cl₂, CH₂N₂, -20 °C; (f) H₂, Pd/C, EtOH; (g) CH₂Cl₂, PCC

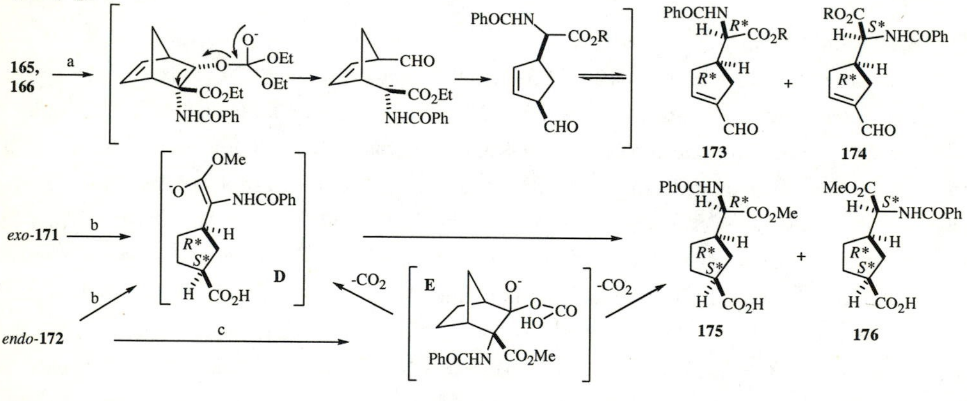

^{a} (a) EtOH, Na₂CO₃, reflux; (b) pyridine/H₂O, reflux, then H₃O⁺; (c) NaHCO₃, acetone/DMF, reflux, then H₃O⁺

Scheme 24

9. Imidazoles

9.1. Imidazolidin-4-ones

Highly stereoselective cycloaddition reaction between a new homochiral imidazolone-derived nitrone as a chiral glycine equivalent and alkenes was utilised by Baldwin for the synthesis of γ-lactones and γ-hydroxy-α-amino acids (Scheme 25).[36] The nitrone **178** was conveniently prepared from Seebach's *tert*-butyl-substituted imidazolidinone **177** by direct oxidation and undergoes cycloaddition reaction with a wide range of alkenes (CH$_2$=CHPh, CH$_2$=CH-4-'BuPh, CH$_2$=CHCy, CH$_2$=CMeCOOMe, MeCH=CHCOOMe) in excellent yields. In all cases studied, stereochemical outcome of the cycloaddition reaction is the result of a reaction on the less hindered α-face of nitrone **178** and places the isoxazolidine 5-alkyl/aryl substituent of the major stereisomer in the *exo* position as indicated for the reaction of **178** with styrene. Stereochemical ratios (*exo/endo*) ranged from 5:1 to 20:1.

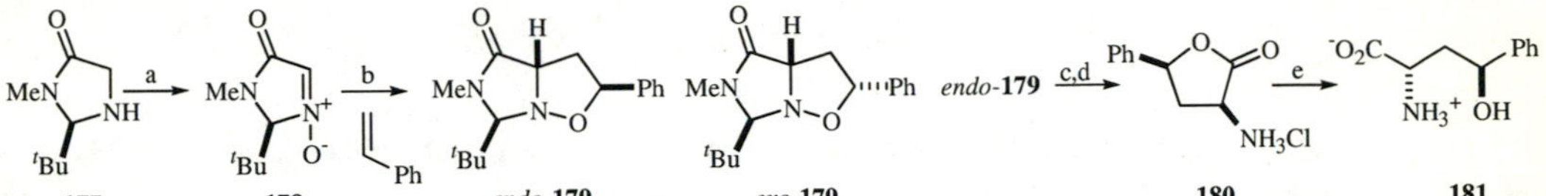

a (a) MTO (5.1%)/UHP (5eq), CH$_2$Cl$_2$, 80%; (b) DCE, reflux, 89%; (c) 1. Zn/HOAc/Ac$_2$O; 2. K$_2$CO$_3$/MeOH reflux, 68%; (d) 6N HCl , THF, reflux, 99%; (e) NaOH/MeOH pH 10, Dowex 50W-X-8, 75%.

Scheme 25

In all cases the major isomer was isolated in a pure form after crystallization or flash chromatography. The conversion of the γ-lactones and γ-hydroxy-α-amino acids involves initial N-O bond hydrogenolysis followed by imidazolidinone hydrolysis using acidic conditions. Lactones were then conveniently converted to the ring opened α-amino acids by careful base hydrolysis followed by ion exchange chromatography on Dowex 50W-X8 and lyophilization.

9.2. Imidazolidine-2,4-diones

A recent work of our laboratory has demonstrate the potentiality of 1,3-dibenzyl-imidazolidin-2,4-dione as homologating reagent for the diastereoselctive chain extension of enantiopure aldehydo sugars. Enantiomerically pure 5-(alditol-1-*C*-yl)-hydantoin derivatives with variable polyol chain extension were synthesised and used as precursors of polyhydroxylated-α-amino acids (Scheme 26).[37a]

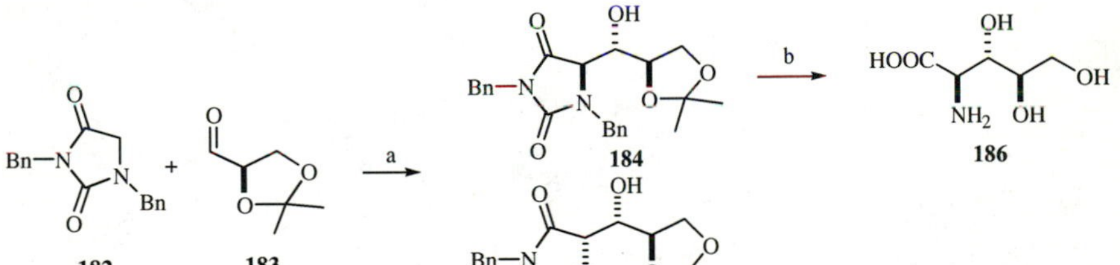

a (a) LiHMDS, THF, -80 °C; (b) 57% HI, 110 °C

Scheme 26

The aldol-type addition of the lithium enolate of the protected hydantoin **182** to 2,3-*O*-isopropylidene-D-glyceraldehyde was first investigated under kinetic control conditions furnishing D-*ribo*-configured 5-(alditol-1-*C*-yl)-hydantoin **184** and its C-5 epimer **185** in a 80/20 isomer ratio. Formation of 5,1'-*anti*-1',2'-*anti* aldol **184** and 5,1'-*syn*-1',2'-*anti* aldol **185** results from *unlike* (*Re* enolate, *Si* aldehyde) and *like* (*Si* enolate, *Si* aldehyde) approach respectively, of the reaction partners in the addition step. In both cases the *Si*-

face of the aldehyde reacts selectively. Starting from D-ribo configured derivatives **184** the synthesis of the biologically important 2-amino-2-deoxy-D-ribonic acid **186** was carried out.

When compound **184** was treated with 57%HI, benzyl protecting groups were removed with concomitant hydrolytic cleavage of the heterocycle ring to give the polyhydroxylated amino acid **186**. To demonstrate the synthetic versatility of this procedure for the synthesis of 5-(alditol-1-*C*-yl)-hydantoin derivatives, the aldol-type additon of **182** lithium enolate to protected D-xylose was conducted under the same reaction conditions affording D-glycero-L-talo configured 5-(alditol-1-*C*-yl)-hydantoin with good diastereoselectivity.

10. Thiazoles

A recent interesting extension of the well known Dondoni's thiazole chemistry for the synthesis of biologically active molecules,[3] is the new synthetic approach to α-(1→5)-D-arabinofuranosyl *C*-oligosaccharides reported by Dondoni and Marra.[38]

These carbon-linked arabinofuranosyl oligosaccharides are potential therapeutic agents against mycobacterial infections such as tuberculosis and leprosy. Three designed arabinofuranose building blocks allowed the diasteroselective synthesis of the oligosaccharides by Wittig olefination. While the two monofunctionalized sugar moieties **188** and **190** were suitable precursor to the head and tail of the oligosaccharidic chain, the difunctionalized compound **189** was envisaged as repeating unit. Interestingly, all *C*-monosaccharides **188-190** appeared to be accessible from the same perbenzylated thiazolyl *C*-arabinoside **187** exploiting the well established thiazole-to-formyl conversion (Scheme 27).

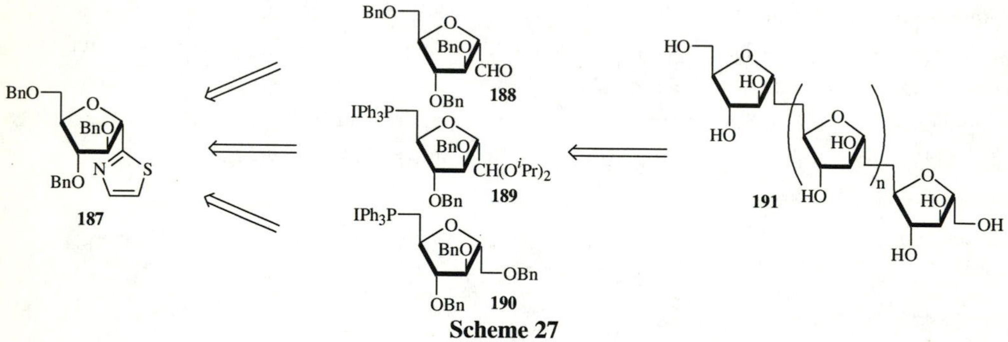

Scheme 27

The synthesis started from the addition at low temperature of 2-lithiothiazole to the lactone **192**, followed by acetylation of the crude thiazolylketose, to give **193** as a mixture 4:1 of anomers (Scheme 28).

After deoxygenation of **193** with Samarium(II) iodide in the presence of ethylene glycol, the desired α-D-*C*-glycoside **187** (60%) was easily separated from the β-anomer (16%). The key intermediate **187** was then submitted to the usual unmasking sequence involving *N*-methylation, hydride reduction, and silver assisted hydrolysis of the thiazolidine intermediate to give the α-linked formyl *C*-arabinoside **188** in high yield. The second monofunctionalized building block **190** was easily obtained from the aldehyde **188** by reduction to alcohol, iodination and treatment with neat triphenylphosphine. The preparation of the third building block from **187** requires selective removal of the 5-*O*-benzyl group, silylation of the resulting alcohol, and conversion into the formyl *C*-arabinofuranoside **189**. After protection of the crude aldehyde, desililation gave the alcohol **195** that was transformed into the target phosphonium salt **189** *via* the iodide derivative. To complete the synthesis the three building blocks were assembled. The Wittig coupling of the aldehyde **188** with the phosphorane derived from **189**, gave a mixture of the (*Z,E*)-*C*-disaccharide. The masked aldehyde was then regenerated upon treatment with trifluoroacetic acid and then coupled with tribenzylated phosphorane prepared from **190**. The *C*-trisaccharides **197** was hydrogenated to afford the α-D-arabinotriose methylene isostere target **198**.

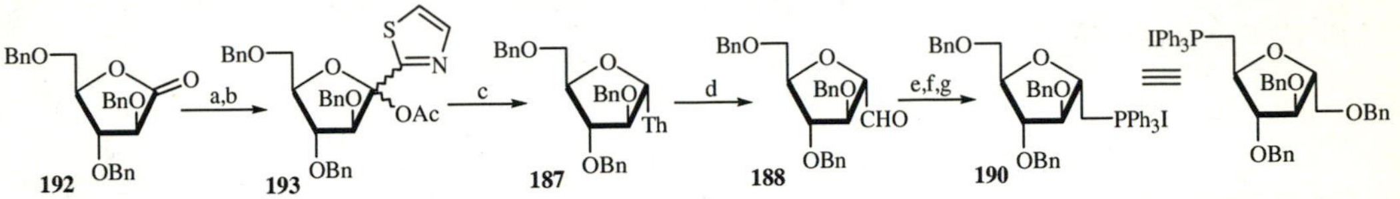

^{a} (a) 2-lithiothiazole, dry Et$_2$O, -75 °C; (b) Ac$_2$O, Et$_3$N, CH$_2$Cl$_2$, 84% 2 steps; (c) SmI$_2$, (CH$_2$OH)$_2$, dry THF, 60%; (d) MeOTf, CH$_3$CN then NaBH$_4$, MeOH then AgNO$_3$, CH$_3$CN, H$_2$O, 85%; (e) NaBH$_4$, Et$_2$O, MeOH; (f) I$_2$, PPh$_3$, imidazole, toluene, 80 °C, 80% 2 steps; (g) neat PPh3 120 °C, 86%.

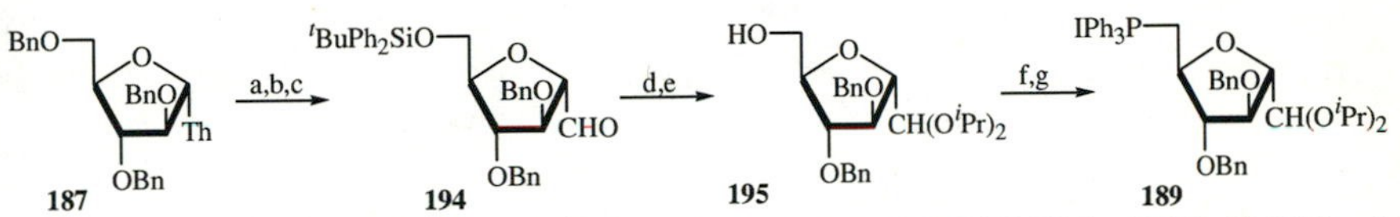

^{a}(a) Ac$_2$O, AcOH, H$_2$SO$_4$ then Et$_3$N, MeOH, H$_2$O, 82%; (b) ^{t}BuPh$_2$SiCl, Py, DMAP; (c) MeOTf, CH$_3$CN then NaBH$_4$, MeOH then AgNO$_3$ CH$_3$CN, H$_2$O; (d) CH(Oi-Pr)$_3$, BF$_3$ Et$_2$O, dry CH$_2$Cl$_2$; (e) Bu$_4$NF, THF, 43% 4 steps; (f) I$_2$, PPh$_3$, imidazole, 80°C, 84%; (g) neat PPh$_3$, 120 °C, 91%.

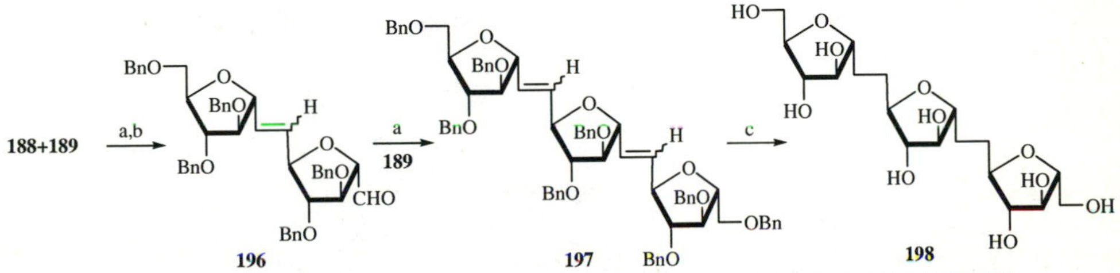

^{a} (a) BuLi, 3:1 dry THF-HMPA, 4 A MS, -20 °C, 49%; (b) TFA, H$_2$O, THF, rt, 100%; (c) H$_2$, Pd(OH)$_2$/C, AcOEt-MeOH, 100%

Scheme 28

11. Oxazines

11.1. 1,4-Oxazin-2-ones

A new chiral heterocycle glycine equivalent **203** was recently proposed by Ley as a new building block for the stereoselective synthesis of enantiopure *N*-protected α-amino acids (Scheme 29).[39] Enantiomerically and diastereomerically pure glycine anion equivalent **203** was synthesised starting from commercially available (*S*)-glycidol **199** using a chiral memory protocol that employs the chirality of the amino alcohol **200** to establish the chirality of the 2,3-mixed acetal aminal functionality.

Different alkyl halides were added to the lithium enolate of **203** under kinetic control conditions. The alkylated products **204** were synthesised with good to excellent yield and high diastereoselectivity (dr from 2:1 to 22:1). Noe and X-ray diffraction studies on the major diastereomers confirmed that the alkyl group was located in the equatorial position. This suggested an attack of the alkyl halide on the enolate carbon atom was occurring from the side opposing the 1,3-related axial methoxy group. Finally an acidic heterocycle ring opening afforded the desired monosubstituted enantiopure D-α-amino acids in good yields.

351

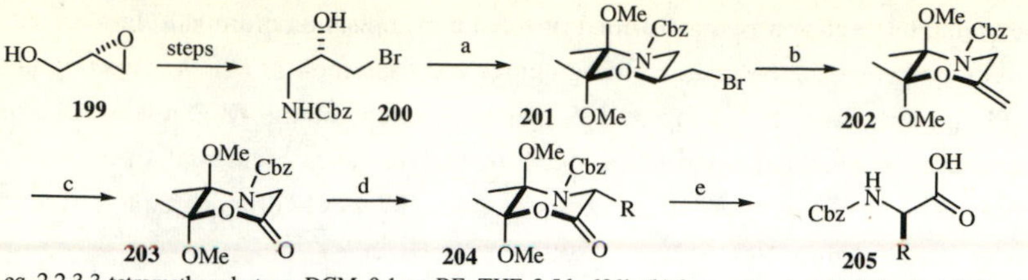

^{a} (a) 5 eq. 2,2,3,3-tetramethoxybutane, DCM, 0.1 eq. BF$_3$·THF, 2.5 h, 69%; (b) 2 eq. potassium bis(trimethylsilyl)amide, THF, -78 °C to rt, 64%; (c) RuCl$_3$, NaIO$_4$, MeCN, H$_2$O, CCl$_4$, 85%; (d) 1.1 eq. LDA, THF, 1.1 eq. HMPA, -78 °C then 3 eq. RX, -55 °C then 1.1 eq. AcOH followed by Et$_2$O; (e) 9:1 TFA-H$_2$O.

Scheme 29

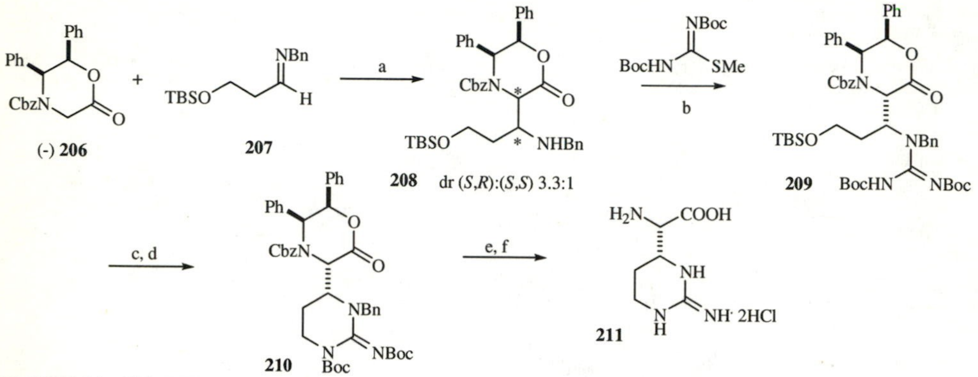

^{a} (a) LHMDS, Me$_2$AlCl, THF, -78 °C, 50-60%; (b) Et$_3$N, AgOTf, DMF, 74%; (c) 1.7% HF in MeCN, 70-91%; (d) DIAD, PPh$_3$, THF, 0 °C, 87%; (e) H$_2$, PdCl$_2$, 115 psi, 4d; (f) 0.5 M HCl, reflux, 95%.

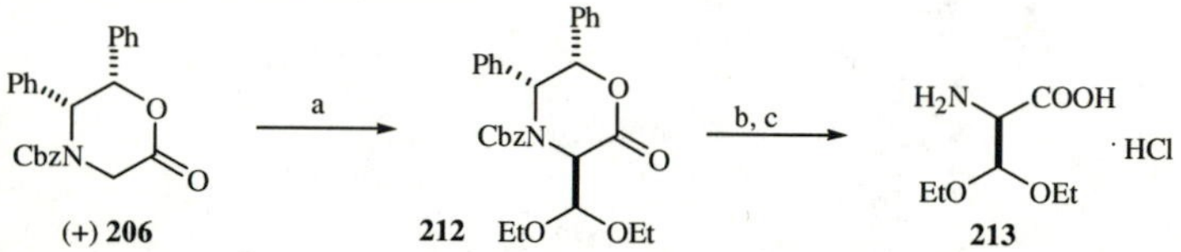

^{a} (a) TiCl$_4$/Et$_3$N, (EtO)$_3$CH, CH$_2$Cl$_2$, -78°C-0°C, 85%; (b) H$_2$ (90 psi), Pd(OH)$_2$, 3:1 THF:MeOH, 2d, 99%; (c) 3N EtOH·HCl, 96%.

Scheme 30

Continuing to explore the potentialities of oxazinone templates for the synthesis of bioactive molecules,[40] Williams performed the asymmetric synthesis of (2S,3R)-capreomycidine, a non-proteinogenic amino acid, in order to be used for the total synthesis of capreomycin IB, a tuberculostatic cyclic peptide antibiotic (Scheme 30).[41a] The key step of the synthesis is a novel enolate-aldimine reaction between chiral glycinate **206** and benzyl imine **207**. The addition of imine resulted in a 50-60% yield of Mannich product as an inseparable 3.3:1 mixture of diastereomers epimeric at the β-carbon with (S,R) and (S,S) configuration respectively. The mixture of **208** was used for the next step to give the product **209** as only diastereomer, because only the major diastereomer of the mixture was able to react, as resulted from NMR analysis. This diastereomer was then transformed in the desired (2S,3R)-capreomycidine **211** via cyclization and deprotection steps. For the same purpose the synthesis of enantiomerically pure (R)-α-formylglycine **213** was also accomplished in two steps starting from the the chiral glycinate **206** by adapting the titanium enolate chemistry. Asymmetric synthesis of highly functionalized α-amino acid derivatives by using the oxazinone as glycine anion equivalent were also reported by Roos, Singh, and Williams.[41b,42]

As a further remarkable extension of this chemistry for the synthesis of biofunctional compounds, a concise asymmetric synthesis of the marine hepatotoxin 7-Epicylindrospermopsin **219** was also proposed by Williams (Scheme 31).[41c] The synthesis is based on a 1,3-dipolar cycloaddition of the nitrone **215** to yield the tricyclic compound **216** endowed with three contiguous new stereogenic centres. This product was conveniently prepared starting from the crotylglicine derivative **214** prepared in high yield and 99% ee from oxazinone template (+) **206**. Acidic removal of the Boc group with concomitant methyl ester formation followed by reduction with lithium aluminium hydride gave the optically pure alcohol that was transformed into the free morpholinone and then oxidated to give the oxazinone *N*-oxide **215**. Exposure of **215** to elevated temperatures gave the triciclyc isoxazolidine **216** as a 10:1 mixture with the minor product assumed to arise from *endo* approach of alkene to the nitrone.

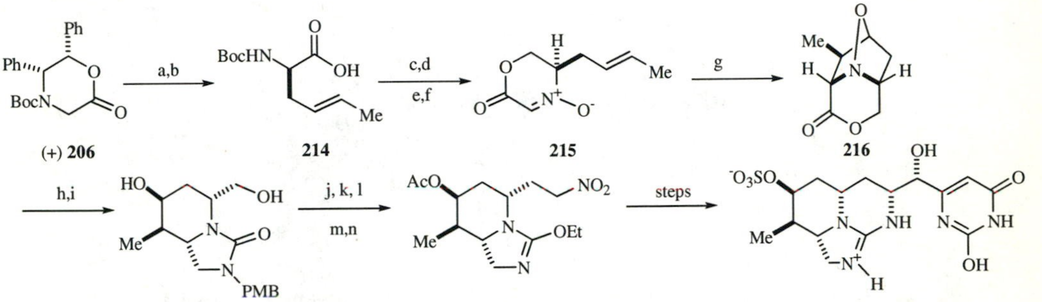

^{a} (a) KHMDS, (*E*)-crotyl iodide, THF,-78 °C, 82%,ee >99%; (b) Li, NH$_3$, THF, EtOH, 68-87%; (c) AcCl, MeOH, 0 °C to rt; (d) LiAlH$_4$, THF, 65% (2 steps); (e) BrCH$_2$COOPh, ^{i}Pr$_2$NEt, MeCN, 63-80%; (f) MCPBA, Na$_2$HPO$_4$, CH$_2$Cl$_2$, -78 °C, 84%; (g) PhMe, 200 °C (sealed tube), 78%; (h) DIBAL-H, CH$_2$Cl$_2$, -78 °C, 87%; (i) PMBNH$_2$, H$_2$/Pd/C, EtOAc, then (*p*-NO$_2$PhO)$_2$CO, MeCN, 81%; (j) TEMPO, PhI(OAc)$_2$, MsOH, CDCl$_3$, 75%; (k) MeNO$_2$, *n*BuLi, THF, 84%; (l) Ac$_2$O, DMAP, CH$_2$Cl$_2$, then NaBH$_4$, EtOH, 67%; (m) TFA, reflux, 80%; (n) Et$_3$OBF$_4$, Cs$_2$CO$_3$, CH$_2$Cl$_2$, 78%,

Scheme 31

Tandem reductive amination/N-O bond cleavage gave a free diamine which was more readily handled as the urea **217**. Selective oxidation of the primary alcohol with TEMPO gave aldehyde derivative that was nitromethylated upon treatment with the lithium salt of nitromethane in THF. The resulting nitroalcohol was then transformed in the nitroalkane *via* nitroalkene reduction. *O*-ethylation with triethyloxonium tetrafluoborate of the ureido functionality gave the stable isourea **218**. A four steps procedure; nitroaldol reaction with 2,6-dimethoxypyrimidine-4-carboxyaldehyde with TBAF, reductive guanidilation of the nitro group (1:0.8 mixture in favour of the diastereomer required for the synthesis), acidic hydrolysis of the pyrimidine and sulfonation of the C12 hydroxy group transforms then the isourea **218** in the desired 7-Epicylindrospermopsin **219**.

Williams also reported the asymmetric total synthesis of Spirotryprostatin A, an antimitotic arrest agent, starting from 6-methoxy-isatin by an azomethine ylide dipolar cycloaddition reaction as a key step.[41d]

Starting from the oxazinone (+) **206** Saba reported a remarkable, rapid and convenient asymmetric synthesis of hydroxylated pyrrolidine glycosidase inhibitor, through a high asymmetric cascade process with a radical key step (Scheme 32).[43a] The synthesis is based on a [1,2]-sigmatropic rearrangement of a spirocyclic ammonium ylide with a ring expansion sequence. The synthesis started from an α-diazo carbonyl compound tethered to a chiral morpholinone **221** prepared by conjugate addition of commercially available enantiopure morpholinone (+) **206** to ethyl 2-diazo-3-ketopent-4-enoate **220**. The Copper(II)-catalysed decomposition of **221**, through the cascade evolution of the spirocyclic ammonium ylide formed, provided oxazepinone **222** as a single diastereomer. The intermediate spiranic ylide **222** results from a carbenoid attack to nitrogen on the same face of both phenyl substituents. Transesterification of lactone **222** in methanol gave directly the 3-prolynone bicyclic hemiacetal **223** while the LiAlH$_4$ gave the substituted enantiomerically pure pyrrolidine **224**.

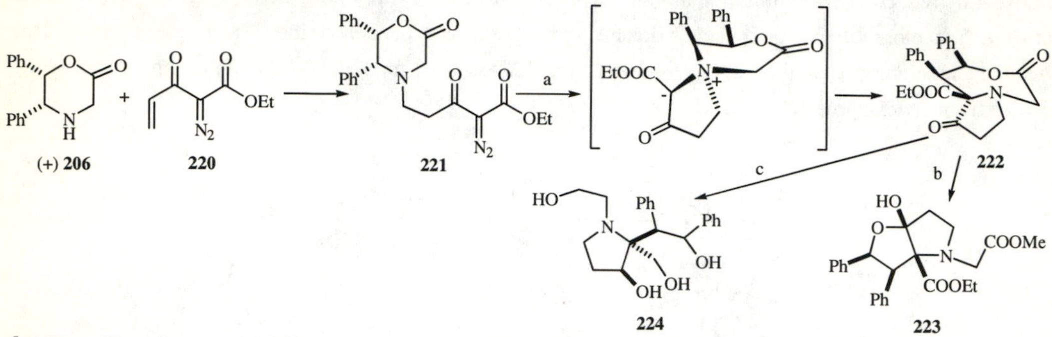

^{a} (a) Copper(II)acetylacetonate 1%, boiling Toluene; (b) Ts(OH), MeOH rt; (c) LiAlH$_4$, THF.

Scheme 32

Saba recently reported also the synthesis of indolizidine alkaloids by using the same tandem metallo-carbenoid/ylide/[1,2]-Stevens rearrangement sequence starting from L-proline.[43b]

Naito introduced an highly diastereoselective intermolecular carbon radical addition to a nitrone **225** for the asymmetric synthesis of α-amino acids.[44] They first investigated ethyl radical addition to chiral glyoxylic nitrone **225**. A high degree of stereocontrol was observed using Et$_3$B as an ethyl radical source (>98% de). The reductive cleavage of the *N-O* bond of **226** with Mo(CO)$_6$, followed by hydrogenolysis in the presence of Pd(OH)$_2$, and treatment with CbzCl afforded the enantiomerically pure (R)-*N*-Cbz amino acid **227**. Furthermore the tin-free reaction of nitrone **225** with alkyl radicals under iodine atom-transfer reaction conditions using Et$_3$B as radical initiator was investigated. Good chemical yields and high diastereoselectivity of the desired alkylated products were observed.

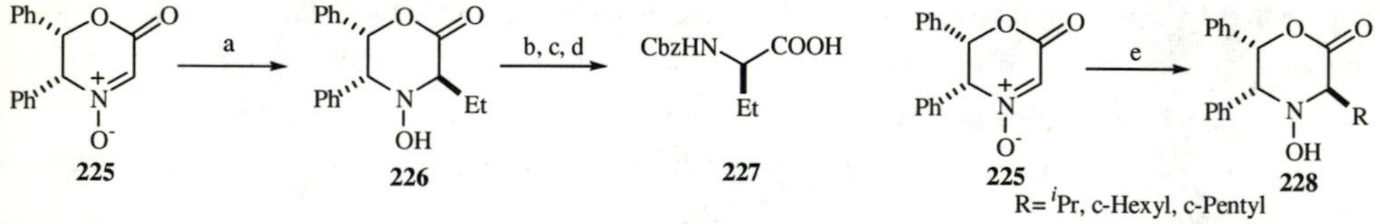

^{a} (a) Et$_3$B 5 eq., Bu$_3$SnH 1.2 eq., CH$_2$Cl$_2$, -78°C, 64%, de >98% ; (b) Mo(CO)$_6$, H$_2$O-MeCN, 84%; (c) H$_2$, Pd(OH)$_2$/C, MeOH; CbzCl, Na$_2$CO$_3$, acetone-H$_2$O, 38% 2 steps; (e) RI, benzene 3 : 1 , Et$_3$B 5 eq., reflux, de >98%.

Scheme 33

Wanner has recently reported the asymmetric synthesis of α-monosubstituted and α,α-disubstituted α-amino acid using 6-*tert*-butyl-5-methoxy-6-methyl-3,6-dihydro-2*H*-1,4-oxazine-2-one **230** as a new chiral glycine equivalent (Scheme 34).[45] (S)-**230** and (R)-**230** were synthesised starting from (S)-**229** and (R)-**229** carboxylic acid. The monoalkylation of **230** was performed under standard conditions (THF, -78 °C, sBuLi)

354

with several alkyl halides to give the corresponding alkylation product with good to excellent yields and *trans*-diastereoselectivities (from 96.8:3.2 to 99.7:0.3 dr) and very small amounts of dialkylated product. A second alkylation reaction was performed starting from the mixtures of diasteromers **231** and **232** to give the dialkylated products **233** in good yield and high *trans*-diastereoselectivity. Both enantiomers of the final product (**234**) were easily accessible simply by changing the sequence of the introduction of the C-3 substituent. A two-steps sequence was employed for the oxazinone ring hydrolysis of the monosubstituted (**231**, **232**) and disubstituted (**233**) derivatives. These compounds were treated with TFA at 60 °C for 20h resulting in complete cleavage of imidate function and then with NaOH to hydrolyse the remaining ester function. For monoalkylated products some significant racemization was detected by chiral HPLC analysis.

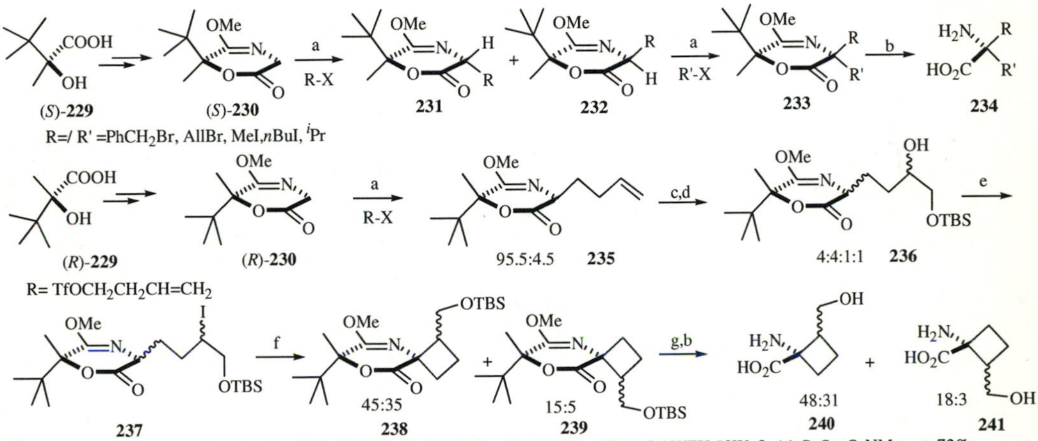

^{a} (a) *s*BuLi, THF, -80 °C (95.5:4.5 *dr*), 69%; (b) TFA, 60 °C 2d then NaOH 24h, 60 °C; DOWEX 50Wx8; (c) OsO$_4$, O-NMe$_3$, rt, 72%; (d) TBDMSCl, imidazole, DMAP, DMF, 88%; (e) PPh$_3$, I$_2$, imidazole 0 °C, 90%; (f) ^{t}Bu-P$_4$,-80 °C, 72%; (g) TBAF, rt, 81%.

Scheme 34

Starting from the glycine equivalent (*R*)-**230** the same author reported the synthesis of four stereoisomers of cyclobutane amino acid. Deprotonation of **230** with *s*BuLi and but-3-enyl triflate generated *in situ* were the best alkylation conditions. Also in this case the *trans*-diastereoselectivity was high (95.5:4.5). Next, the terminal double bond of the butenyl side chain was transformed into diol surprisingly in a ratio of ca. 4:4:1:1 indicating that epimerization at the 3-position of the oxazine ring must have occurred. The mixture of diastereomers was used unchanged for the following reaction. After protection of the primary hydroxy group as TBS-ether the secondary hydroxy group was transformed in iodide. The desired silylated iodohydrins were obtained in high yield with an unchanged ratio of stereisomers. Finally the cyclization step was performed *via* deprotonation of **237** with phosphazenic base to give a mixture of the four diastereomeric spirocyclic compounds in 45:35:15:5 ratio, with a prevalence of *trans* isomers, *i.e.* **238**/**239** ~8/2. Removal of the silyl protecting group and oxazine ring opening *via* the known two step procedures gave the amino acid targets **240** and **241**.

12. Pyrazines

12.1. Piperazine-2,5-diones

Porzi and Sandri have recently reported a study directed to the stereoselective synthesis of novel analogues of biologically active 2,6-diaminopimelic acid based on the well studied chemistry of piperazine derivatives.[46a] They reported an efficient synthesis of the γ-methylene and alkyl γ-methylene analogues of 2,6-DAP. The synthesis started from glycine derived chiral synthon **242** that was alkylated with 2-chloromethyl-3-chloropropene employing two equivalents of base to give **243** and **244** in very good yield and diastereomeric ratio 30:70 respectively. The bicyclic diastereomers were submitted to the Birch reduction to remove the chiral inductor. The intermediate enantiomers **246** and **249** were then transformed in γ-methylene analogues of 2,6-DAP via hydrolysis in HCl at 60 °C.

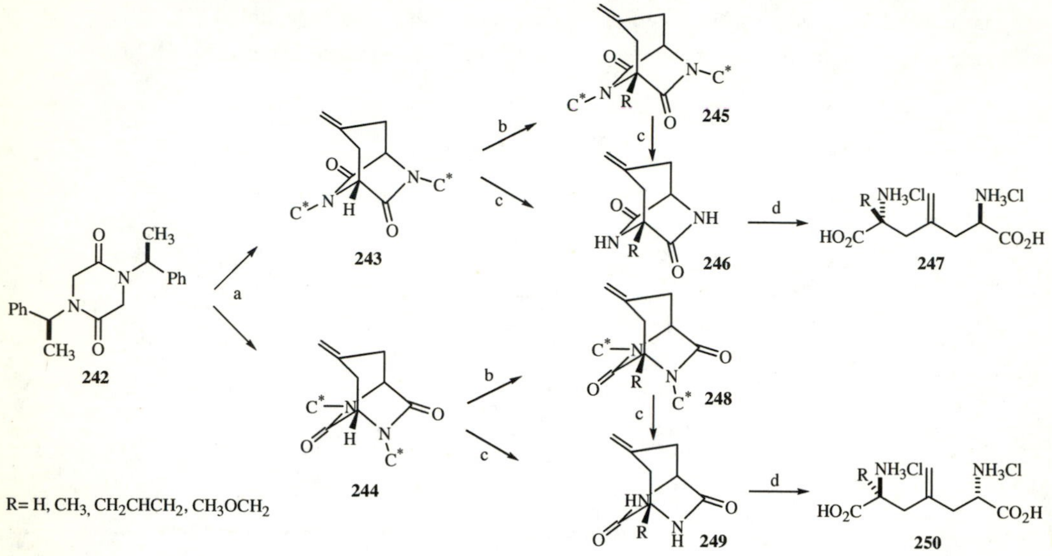

a (a) 2-chloromethyl-3-chloropropene, THF, LHMDS, 2 eq., -78 °C; (b) CH₃Li or n-C₄H₉Li , RCl, THF, -78 °C; (c) Li/NH₃ THF/*t*-butanol; (d) 2N HCl 60 °C 40h.

Scheme 35

Alternatively, alkylation of **243** and **244** by using CH₃Li occurred exclusively at the bridgehead position giving the bicyclic diastereomers **245** and **248**. While the formation of **245** appears to be accomplished irrespective of the bulkiness of the base used, CH₃Li or n-C₄H₉Li, **248** was obtained in lower yield when a bulkier base was used. The authors also proposed an alternative procedure for the synthesis of pseudo-tripeptides incorporating a γ-methylene derivative of 2,6-DAP by a high diastereoselective (de>94%) alkylation of glycine derived chiral lactim **251** with 2-iodomethyl-3-iodopropene and subsequent hydrolysis of the lactim rings.[46b]

12.2. 3,6-Dihydropyrazin-2(1*H*)ones

The same authors reported the stereocontrolled synthesis of the peptide **255** and analogues, structural variants of 2,6-diaminopimelic acid containing a proline residue fused to a diketopiperazine ring (Scheme 36).[46c]

The synthesis started from the mono-lactim ether **251** easily synthesised from L-valine. Deprotonation of **251** at-78°C followed by alkylation with 0.5 equivalent of 1,3-diiodopropane, afforded the diastereomer **252** in 87% yield. The largely prevalent diastereomer **252** was converted into the corresponding enolate and then alkylated with 1,3-dichloro propane in good yield and with a total 1,4-*trans* induction with respect to the isopropyl group. By submitting the intermediate **253** to the Finkelstein reaction (NaI, acetone reflux), the byciclic derivative **254** was isolated in good yield. The peptide **255** was then obtained as hydrochloride after removal of the benzyl group followed by hydrolysis of the lactim ring under mild acid conditions.

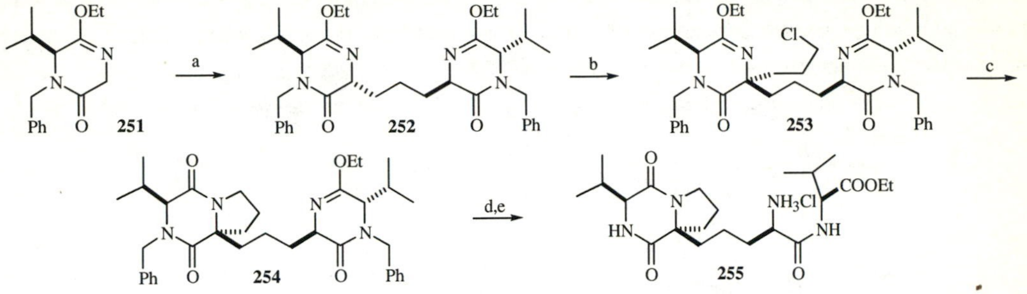

^{a} (a) LiHMDS, THF, -78 °C, I-(CH$_2$)$_3$-I;(b) LiHMDS, THF, -78 °C, Cl-(CH$_2$)$_3$-Cl; (c) NaI refluxing acetone; (d) Li/NH$_3$; (e) HCl 0.5 M, EtOH

Scheme 36

12.3. 2,5-Dihydropyrazines

Starting from Schöllkopf's bis-lactim ether,[7] Undheim have extensively investigated the synthesis of rigidified α-amino acid derivatives some of which depicted in the Scheme 37.[47] The key step of the synthesis was the bisalkylation of the chiron **256**, that provided substrates with two different enantiopure *gem*-substituents. After the first alkylation step, monoalkylated intermediates **257**, **258**, and **259**, were all lithiated at -78 C and treated with acetaldehyde, racemic propylene oxide, or 4-bromo-1-butene respectively. Once lithiated, the stereochemistry at the 2-carbanionic site is lost. The new stereogenic centre in the heterocyclic 2-position was formed in all cases with high *trans*-selectivity while low diastereoselctivity was observed at the carbonyl carbon in the aldol formation. In the preparation of functionalised cycloalkene α-quaternary α-amino acid derivatives like **272**, the construction of the carbocyclic ring was effected *via* spiroannulation reaction of the oxidated products **263** and **264**. The internal aldol condensation was performed under mild alkaline conditions in acetonitrile under reflux. Finally heterocycle ring opening in mild acidic conditions gave the desired carbocyclic amino acid like **272**. The Rh(II)-catalysed carbenoid cyclization reaction of diazoketones **265** and **266** is the key step in the construction of the rigidified five or six heterocycle ring of the target structure **273**, **274** and **275**. Intramolecular rhodium(II)-catalysed reaction in geminally disubstituted derivatives chiron, occurred with complete chemoselectivity at the adjacent annular nitrogen in preference to carbon-carbon double bond addition or C-H insertion. The products were four, five and six-membered annulated rings. Controlled acid hydrolysis provided the target rigidified cyclic α-amino acid derivatives **273**, **274** and **275**.

The same author has recently reported another synthesis of carbacyclic α-amino acids *via* 5-exo-trig-spiroannulation Pd-catalysed to chiral pyrazine derivatives obtained from the dialkylation of the Schöllkopf chiron.[47d]

12.4. Pyrazin-2(1*H*)-ones

Stolz performed the total synthesis of (⁻)-Lemonomycin, a member of the tetrahydroisoquinoline family of antitumor antibiotics starting from 6-methyl-4-*N*-benzyl-1*H*-pyrazin-2-one bromide **276** (Scheme 38).[48]

This high convergent synthesis features an asymmetric Joule dipolar cycloaddition between the compounds **276** and **277** to give diazabicycle **278** where the absolute stereochemistry derived from an Oppolzer sultam acrylamide as auxiliary controller. The elaboration of the enamide **278** followed by a Suzuki coupling to connect the diazabicycle to the aryl subunit, afforded enamide **281** that was then transformed into the suitable substrate **284** for the key stereoselective Picted-Spengler reaction, that directly incorporates the aminoglycoside subunit. Cyclization of the threonine derived aminoglycosyloxy aldehyde and the TFA salt of aminotriol **284** gave the target product **285** in 15 steps starting from compound **276**.

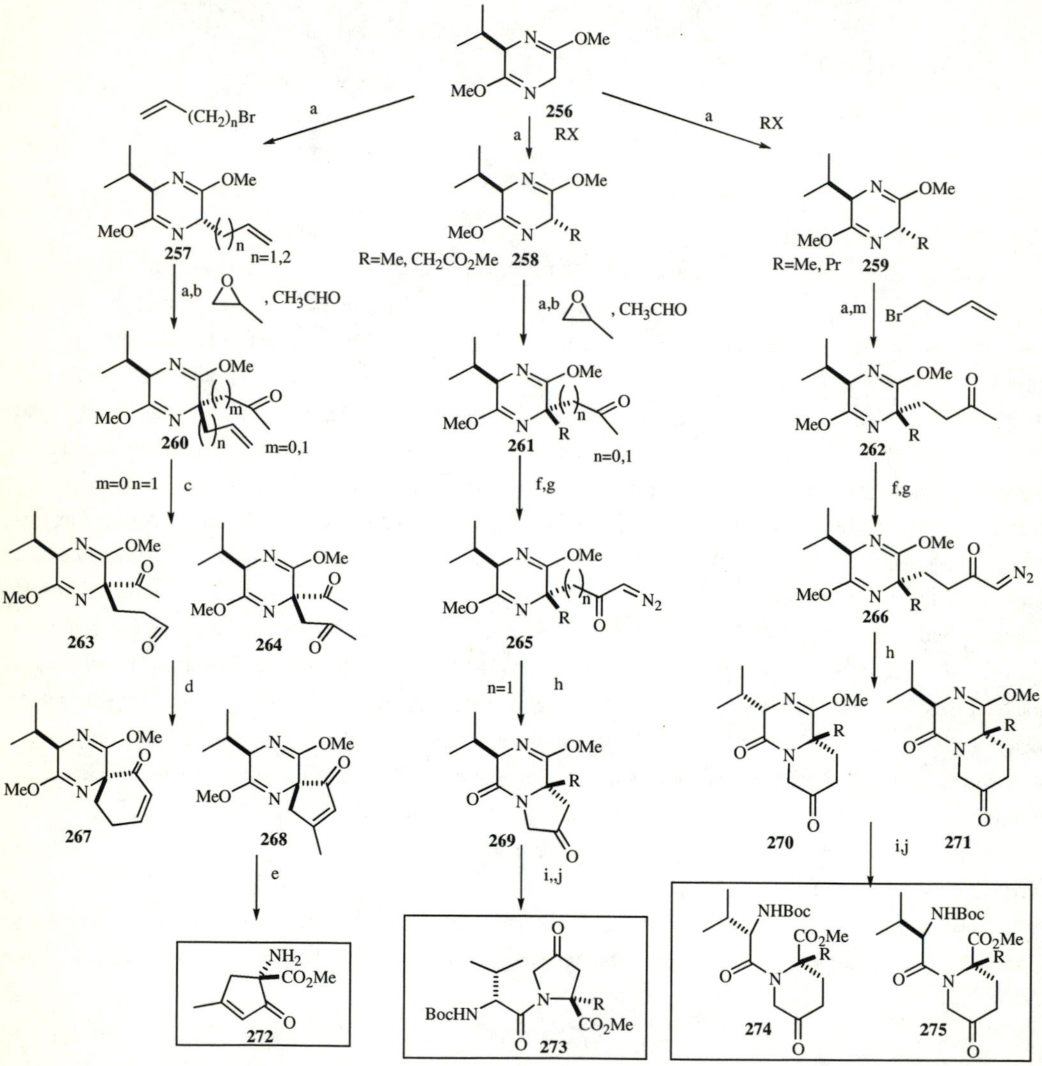

^{a} (a) nBuLi, THF, -78 °C; (b) Dess-Martin oxidation; (c) PdCl$_2$, CuCl, H$_2$O 7:1; (d) CsCO$_3$, MeCN, reflux; (e) 0.1 M TFA, H$_2$O:MeCN 1:1, 3d; (f) CF$_3$CO$_2$CH$_2$CF$_3$, LiHMDS, THF, -78 °C; (g) TsN$_3$, Et$_3$N, MeCN, rt; (h) Rh$_2$(OAc)$_4$, CH$_2$Cl$_2$, rt; (i) 3N HCl, THF; (l) Boc$_2$O, Et$_3$N, CH$_2$Cl$_2$;

Scheme 37

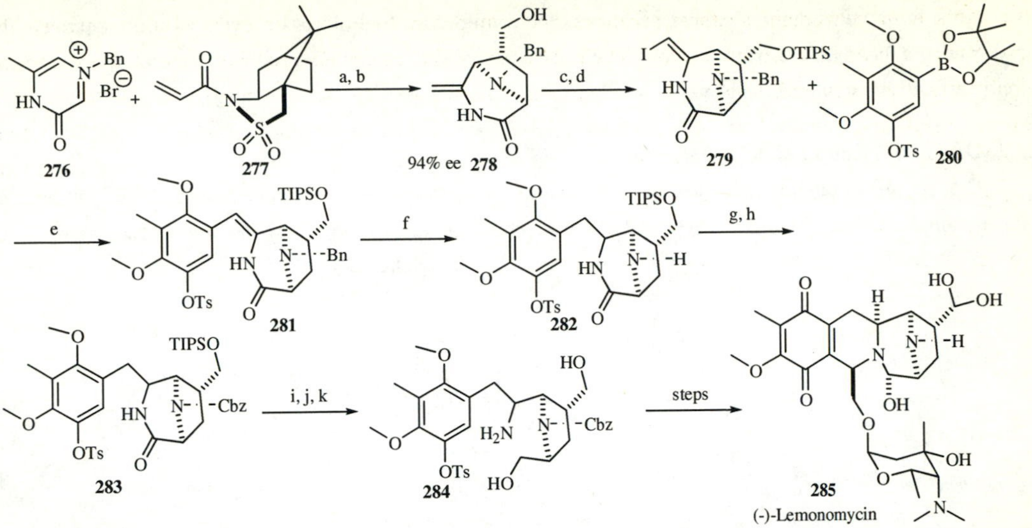

^{a} (a) NMM, CH$_3$CN; (b) NaBH$_4$, EtOH, 72% yield 2 steps; (c) TIPSOTf, CH$_2$Cl$_2$, 2,6-lutidine, 82%; (d) ICl, CH$_2$CL$_2$, 0 °C, 81%; (e) Pd(PPh$_3$)$_4$, K$_2$CO$_3$, PhH, MeOH, 70 °C, 69%; (f) Pd/C, H$_2$ 1000 psi, TFA, EtOH, 72%; (g) CbzCl, DMAP, CH$_3$CN; (h) KOTMS, CH$_3$CN, 83% 2 steps; (i) (Boc)$_2$O, DMAP, CH$_3$CN; (j) NaBH$_4$, EtOH; (k) HCl, MeOH, 81% 3 steps.

Scheme 38

13. Pyrimidines

13.1. Pyrimidin-4(1H)-ones

A simple synthesis of (+)-(1S,2R) and (−)-(1R,2S)-2-aminocyclobutane-1-carboxylic acid based on a [2+2] photocycloaddition reaction of a chiral N-1-substituted uracil mimic with ethylene was recently reported by Aitken (Scheme 39).[49]

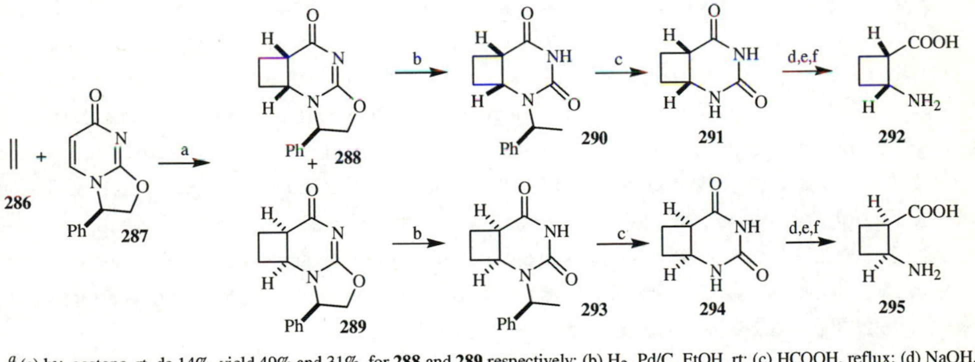

^{a} (a) hv, acetone, rt, de 14%, yield 49% and 31% for **288** and **289** respectively; (b) H$_2$, Pd/C, EtOH, rt; (c) HCOOH, reflux; (d) NaOH, rt; (e) NaNO$_2$, HCl; (f) ion exchange resin, 20% and 33% overall yield for **292** and **295** respectively.

Scheme 39

The chiral auxiliary equipped uracil (R)-**287** was submitted to [2+2] photocycloaddition reaction with ethylene affording a mixture of photoadducts diastereomers **288** and **289** with *cis* configuration in only 14% de. After chromatographic separation, the selective opening of the five heterocyclic rings of **288** and **289** was achieved by catalytic hydrogenation. To complete the synthesis a two steps transformation of the uracil

359

ring into β-amino acid derivative was performed *via* a mild base hydrolysis followed by a diazotation. The final products **292** and **295** were obtained in 33 and 20 % yield respectively, without trace of epimerization, after purification on ion exchange resin.

13.2. Dihydropyrimidin-2,4(1*H*,3*H*)-diones

Proceeding in our programme aimed to develop new stereoselective procedures for the synthesis of biofunctional molecules we have investigated the potentialities of dihydrouracil as homologating reagent for the chain extension of 2,3-*O*-isopropylidene-D-glyceraldehyde (Scheme 40).[37b]

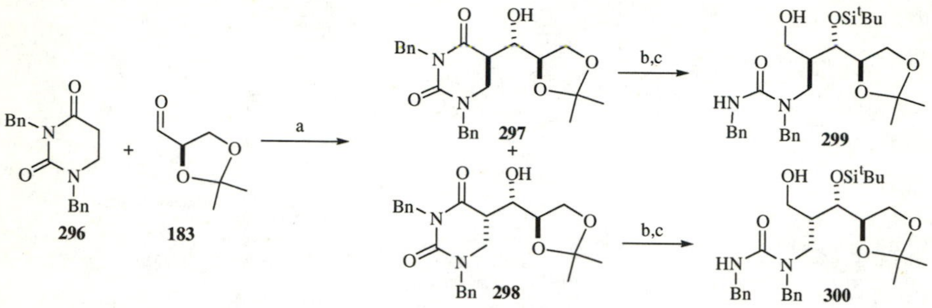

a (a) LDA, THF, -80°C, 88%; (b) TBSTf, CH$_2$Cl$_2$, 2,6-lutidine; (c) NaBH$_4$, EtOH/H$_2$O (3:1), 68% and 61% for 2 steps from **297** and **296**

Scheme 40

The synthesis of enantiomerically pure 5-trihydroxypropyl-dihydrouracil derivatives as key intermediates for the synthesis of nitrogen-containing molecules of biological interest was realised in high yield and good diastereoselectivities. In order to explore this potentiality, the synthesis of enantiopure β-ureido polyols was then performed. The aldol-type addition of the lithium enolate of **296** to 2,3-*O*-isopropylidene-D-glyceraldehyde **183** in THF under kinetic control conditions afforded **298** and its C-5 epimer **297** in 81/16 dr with a 3% of an other diastereomers. Formation of isomers (5,1'-*syn*-1',2'-*anti*) **298** and (5,1'-*anti*-1',2'-*anti*) **297** results by an *unlike* (*Re* enolate, *Si* aldehyde) and *like* (*Si* enolate, *Si* aldehyde) approach of the reaction partners in the addition step. It should be noted that in both cases the *Si*-face of aldehyde **183** reacts selectively according to the Felkin-Anh model. Reductive ring cleavage of the DHU heterocycle ring of compounds **298** and **297** after protection of the secondary hydroxyl group, gave the β-ureido polyols **300** and **299**. The effect of the solvent and of a binary reagent system composed of a lithium enolate and a Lewis-acid was investigated, the results showed the preferential formation of diastereomer **297** with respect to **298**, dr=69/26, in Et$_2$O and in the presence of SnCl$_4$.

14. Conclusions

This review highlighted the research area directed at finding new stereoselective procedures for the synthesis of biofunctional molecules by using heterocyclic compound as building blocks that are also latent functional group equivalents for the construction of the target architectures. The large number of heterocycles exploited in this chemistry, the number or the complexity of the chiral structures of biological interest that were object of the syntheses as well as the chemistry involved have demonstrated the usefulness of this approach. In many cases "old" well studied heterocycles were used in the asymmetric synthesis of new target bioactive molecules but were also exploited in new streoselective key procedures, in other cases overlooked and underutilized "new" heterocyclic compounds have been object of interest and have demonstrated their potentiality as building blocks. A large number of articles on the stereoselective preparation of bioactive molecules by using this approach were found by us on searching the literature, and we tried to cover the majority of these in this review reporting the very last advances from January 2003 to September 2004. Nevertheless we apologise for inevitable omissions.

References

1. Meyers, A. I. In *Heterocycles in Organic Synthesis*; Taylor, E. C., Weissberger, A., Eds: Wiley: New York, 1974.

2. For selected articles, see: (a) Casiraghi, G.; Colombo, L.; Rassu, G.; Spanu, P. *Tetrahedron Lett.* **1989**, *30*, 5325. (b) Casiraghi, G.; Colombo, L.; Rassu, G.; Spanu, P.; Gasparri Fava, G.; Ferrari Belicchi, M. *Tetrahedron* **1990**, *46*, 5807. (c) Casiraghi, G.; Rassu, G.; Spanu, P.; Pinna, L. *J. Org. Chem.* **1992**, *57*, 3760. (d) Spanu, P.; Rassu, G.; Pinna, L.; Battistini, L.; Casiraghi, G. *Tetrahedron: Asymmetry* **1997**, *8*, 3237. (e) Spanu, P.; Rassu, G.; Ulgheri, F.; Zanardi, F.; Battistini, L.; Casiraghi, G. *Tetrahedron* **1996**, *52*, 4829. (f) Rassu, G.; Pinna, L.; Spanu, P.; Zanardi, F.; Battistini, L.; Casiraghi, G. *J. Org. Chem.* **1997**, *62*, 4513. For reviews see: (g) Casiraghi, G.; Zanardi, F.; Rassu, G.; Spanu, P. *Chem. Rev.* **1995**, 1677. (h) Casiraghi, G.; Rassu, G. *Synthesis* **1995**, 607. (i) Casiraghi, G.; Zanardi, F.; Appendino, G.; Rassu, G. *Chem. Rev.* **2000**, 1929. (l) Battistini, L.; Casiraghi, G.; Rassu, G.; Zanardi, F. *Chem. Soc. Rev.* **2000**, *29*, 109.

3. (a) Dondoni, A.; Marra, A. *Chem. Rev.* **2004**, *104*, 2557. (b) Dondoni, A. *Synthesis* **1998**, 1681. (c) Application in carbohydrate chemistry: Dondoni, A.; Marra, A. In *Preparative Carbohydrate Chemistry*; Hanessian, S. Ed.; Marcel Dekker: New York; **1997**, 173. (c) Dondoni, A.; Marra, A. *Chem. Rev.* **2000**, *100*, 4395.

4. (a) Meyers, A. I.; Mihelic, E. D. *Angew. Chem., Int. Ed. Engl.* **1976**, *15*, 270. (b) Meyers, A. I.; Mihelic, E. D. *New Synthetic Methods*; Verlag Chemie: Weinhein, 1979; Vol. 5, p. 105. (c) Meyers, A. I. *J. Heterocycl. Chem.* **1988**, *35*, 991. (d) Cardillo, G.; Gentilucci, L.; Tolomelli, A. *Aldrichim. Acta* **2003**, *36*, 39.

5. (a) Seebach, D.; Imwinelried, R.; Weber, T. *Modern Synthetic Methods*, R. Scheffold, Ed, Vol. 4, p. 125, Springer, Berlin 1986. (b) Seebach, D.; Naef, R.; Calderari, G. *Tetrahedron* **1984**, *40*, 1313.

6. (a) Seebach, D.; Roggo, S.; Zimmerman, J. *Stereochemistry of Organic and Bioorganic Transformations*, Bartmann, W.; Sharpless, K. B., Eds.; Verlag Chemie: Berlin, 1986, p. 85. (b) Seebach, D.; Naef, R. *Helv. Chim. Acta* **1981**, *64*, 2704. (c) Seebach, D.; Aebi, J. A.; Naef, R.; Weber, T. *Helv. Chim. Acta* **1985**, *68*, 144. (d) Seebach, D.; Juaristi, E.; Miller, D. D.; Schickli, C.; Weber, T. *Helv. Chim. Acta* **1987**, *70*, 237.

7. (a) Schollkopf, U. *Tetrahedron* **1983**, *39*, 2085. (b) Grauert, M.; Schollkopf, U. *Liebigs Ann. Chem.* **1985**, 1817. (c) Kobayashi, S.; Furuta, T.; Hayashi, T.; Nishijima, M.; Hanada, K. *J. Am. Chem. Soc.* **1998**, *120*, 908. (d) Kobayashi, S.; Furuta, T. *Tetrahedron* **1998**, *54*, 10275. (e) Ruiz, M.; Ojea, V.; Ruanova, T. M.; Quintela, J. M. *Tetrahedron: Asymmetry* **2002**, *13*, 795. (f) Ruiz, M.; Ojea, V.; Quintela, J. M. *Synlett* **1999**, 204. (g) Ruiz, M.; Ruanova, T. M.; Ojea, V.; Quintela, J. M. *Tetrahedron Lett.* **1999**, *40*, 2021.

8. (a) Juaristi, E.; Quintana, D.; Lamatsch, B.; Seebach, D. *J. Org. Chem.* **1991**, *56*, 2553. For reviews see: (b) Juaristi, E.; Quintana, D.; Escalante, J. *Aldrichim. Acta* **1994**, *27*, 3. (c) Cole, D. C. *Tetrahedron* **1994**, *50*, 9517. (d) Liu, M.; Sibi, M. P. *Tetrahedron* **2002**, *58*, 7991.

9. (a) Corey, E. J.; Seebach, D. *Angew. Chem., Int. Ed. Engl.* **1965**, *4*, 1075 and 1077. (b) Seebach, D. *Synthesis* **1969**, 17. (c) Yus, M.; Najera, C.; Foubelo, F. *Tetrahedron* **2003**, *59*, 6147.

10. (a) Rassu, G.; Auzzas, L.; Pinna, L.; Battistini, L.; Zanardi, F.; Marzocchi, L.; Acquotti, D.; Casiraghi, G. *J. Org. Chem.* **2000**, *65*, 6307. (b) Rassu, G.; Auzzas, L.; Pinna, L.; Zambrano, V.; Zanardi, F.; Battistini, L.; Marzocchi, L.; Acquotti, D.; Casiraghi, G. *J. Org. Chem.* **2002**, *67*, 5338 and references quoted therein.

11. (a) Rassu, G.; Auzzas, L.; Zambrano, V.; Burreddu, P.; Pinna, L.; Battistini, L.; Zanardi, F.; Casiraghi, G. *J. Org. Chem.* **2004**, *69*, 1625. (b) Rassu, G.; Auzzas, L.; Zambrano, V.; Burreddu, P.; Battistini, L.; Curti, C. *Tetrahedron Asymmetry* **2003**, *14*, 1665. (c) Rassu, G.; Auzzas, L.; Pinna, L.; Zambrano, V.; Zanardi, F.; Battistini, L.; Gaetani, E.; Curti, C.; Casiraghi, G. *J. Org. Chem.* **2003**, *68*, 5881.

12. Spanedda, M. V.; Ourèvitch, M.; Crousse, B.; Bèguè, J.-P.; Bonnet-Delpon, D. *Tetrahedron Lett.* **2004**, *45*, 5023.

13. (a) Wang, Q.; Padwa, A. *Org. Lett.* **2004**, *6*, 2189. Earlier report on IMDAF see: (b) Padwa, A.; Brodney, M. A.; Dimitroff, M. *J. Org. Chem.* **1998**, *63*, 5304. (c) Bur, S. K.; Lynch, S. M.; Padwa, A. *Org. Lett.* **2002**, *4*, 473. (d) Ginn, J. D.; Padwa, A.*Org. Lett.* **2002**, *4*, 1515.

14. Nosse, B.; Chhor, R. B.; Jeong, W. B.; Böhm, C.; Reiser, O. *Org. Lett.* **2003**, *5*, 941.

15. Garcìa Ruano, J. L.; Gil, J. I. A.; Fraile, A.; Martin Castro, A. M.; Martin, M. R. *Tetrahedron Lett.* **2004**, *45*, 4653.

16. Alibès, R.; de March, P.; Figueredo, M.; Font, J.; Racamonde, M.; Parella, T. *Org. Lett.* **2004**, *6*, 1449.

17. Birman, V. B.; Jiang, X.-T. *Org. Lett.* **2004**, *6*, 2369.

18. For reviews on the "Naked sugar" methodology and application see: (a) Vogel, P. *Curr. Org. Chem.* **2000**, *4*, 455. (b) Vogel, P. In *Glycoscience, Chemistry and Biology*; Fraiser-Reid, B., Tatsuta, K., Thiem, J., Eds.; Springer-Verlag: Berlin, 2001; Vol. II, Chapter 4.4, p. 1023.

19. (a) Moreno-Vargas, A. J.; Robina, I.; Petricci, E.; Vogel, P. *J. Org. Chem.* **2004**, *69*, 4487. (b) Moreno-Vargas, A. J.; Schütz, C.; Scopelliti, R.; Vogel, P. *J. Org. Chem.* **2003**, *68*, 5632.

20. Battistini, L.; Curti, C.; Zanardi, F.; Rassu, G.; Auzzas, L.; Casiraghi, G. *J. Org. Chem.* **2004**, *69*, 2611.

21. Huang, P.-Q.; Wu, T.-J.; Ruan, Y.-P. *Org. Lett.* **2003**, *5*, 4341.

22. He, B. Y.; Wu, T.-J.; Yu, X.-Y.; Huang, P.-Q. *Tetrahedron Asymmetry* **2003**, *14*, 2101.

23. Kim, W.-S.; Lee, J.-H.; Kang, J.; Cho, C.-G. *Tetrahedron Lett.*, **2004**, *45*, 1683.

24. (a) Beumer, R.; Bubert, C.; Cabrele, C.; Vielhauer, O.; Pietzsch, M.; Reiser, O. *J. Org. Chem.* **2000**, *65*, 8960. (b) Gnad, F.; Poleschak, M.; Reiser, O. *Tetrahedron Lett.* **2004**, *45*, 4277.

25. Fernandez, M.; Alonso, R. *Org. Lett.* **2003**, *5*, 2461.

26. (a) Blay, G.; Fernàndez, I.; Monje, B.; Pedro, J. R. *Tetrahedron* **2004**, *60*, 165. (b) Blay, G.; Cardona, L.; Fernàndez, I.; Michelena, R.; Pedro, J. R.; Ramirez, T.; Ruiz-Garcìa, R. *Synlett* **2003**, 2325.

27. Seebach, D.; Sting, A. R.; Hoffmann, M. *Angew. Chem., Int. Ed. Engl.* **1996**, *35*, 2708.

28. Enders, D.; Müller-Hüwen, A. *Eur. J. Org. Chem* **2004**, 1732.

29. Brunner, M.; Koskinen, A. M. P. *Tetrahedron Lett.* **2004**, *45*, 3063.

30. Jew, S.-s.; Lee, Y. J.; Lee, J.; Kang, M. J.; Jeong, B.-S.; Lee, J.-H.; Yoo, M.-S.; Kim, M.-J.; Choi, S.-h.; Ku, J.-M.; Park, H.-g. *Angew. Chem. Int. Ed.* **2004**, *43*, 2382.

31. Griesbeck, A. G.; Bondock, S.; Lex, J. *J. Org. Chem.* **2003**, *68*, 9899.

32. Gardiner, J.; Abell, A. D. *Tetrahedron Lett.* **2003**, *44*, 4227.

33. Pajouhesh, H.; Curry, K.; Pajouhesh, H.; Meresht, M. H.; Patrick, B. *Tetrahedron Asymmetry* **2003**, *14*, 593.

34. (a) Trost, B. M.; Dogra, K. *J. Am. Chem. Soc.* **2002**, *124*, 7256. (b) Trost, B. M.; Dogra, K.; Franzini, M. *J. Am. Chem. Soc.* **2004**, *126*, 1944.

35. Clerici, F.; Gelmi, M. L.; Pellegrino, S.; Pilati, T. *J. Org. Chem.* **2003**, *68*, 5286.

36. Baldwin, S. W.; Long, A. *Org. Lett.* **2004**, *6*, 1653.

37. (a) Ulgheri, F.; Orrù, G.; Crisma, M.; Spanu, P. *Tetrahedron Lett.* **2004**, *45*, 1047. (b) Ulgheri, F.; Bacsa, J.; Nassimbeni, L.; Spanu, P. *Tetrahedron Lett.* **2003**, *44*, 671.

38. (a) Dondoni, A.; Marra, A. *Tetrahedron Lett.* **2003**, *44*, 4067. (b) Another recent application of the thiazole chemistry was the synthesis of tetrabenzylated formyl β-D-*C*-glucopyranosides: Dondoni, A.; Marra, A *Tetrahedron Lett.* **2003**, *44*, 13.

39. (a) Harding, C. I.; Dixon, D. J.; Ley, S. V. *Tetrahedron* **2004**, *60*, 53. (b) Dixon, D. J.; Harding, C. I.; Ley, S. V.; Tilbrook, D. M. G. *Chem. Commun.* **2003**, 468.

40. (a) Williams R. M., *Synthesis of Optically Active a-Amino Acids*, Pergamon, Oxford 1989. (b) Williams R. M., *Advances in Asymmetric Synthesis*, A. Hassner, Ed., JAI Press 1995. (c) Williams R. M., *Peptidomimetics Protocols*, Kazmierski, W., Ed., *Methods in Molecular Medicine* **1999**, Vol. 23, Chapter 19, p. 339, Humana Press.

41. (a) DeMong, D. E.; Williams, R. M. *J. Am. Chem. Soc.* **2003**, *125*, 8561. (b) Jin, W.; Williams, R. M. *Tetrahedron Lett.* **2003**, *44*, 4635. (c) Looper, R. E.; William, R. M. *Angew. Chem. Int. Ed.* **2004**, *43*, 2930. (d) Onishi, T.; Sebahr, P. R.; Williams, R. M. *Org. Lett.* **2003**, *5*, 3135.

42. (a) Roos, G. H. P.; Dastlik, K. A. *Synth. Comm.* **2003**, *33* 2197. (b) Singh, S.; Pennington, M. W. *Tetrahedron Lett.* **2003**, *44*, 2638.

43. (a) Saba, A. *Tetrahedron Lett.* **2003**, *44*, 2895. (b) Muroni, D.; Saba, A.; Culeddu, N. *Tetrahedron Asymmetry* **2004**, *15*, 2609.

44. Ueda, M.; Miyabe, H.; Teramachi, M.; Miyata, O.; Naito, T. *Chem. Comm.* **2003**, 426.

45. (a) Koch, C.-J.; Höfner, G.; Polborn, K.; Wanner, K. T. *Eur. J. Org. Chem.* **2003**, 2233. (b) Koch, C.-J.; Šimonyiovà, S.; Pabel, J.; Kärten, A.; Polborn, K.; Wanner, K. T. *Eur. J. Org. Chem.* **2003**, 1244.

46. (a) Piccinelli, F.; Porzi, G.; Sandri, M.; Sandri, S. *Tetrahedron Asymmetry* **2003**, *14*, 393. (b) Balducci, D.; Porzi, G.; Sandri, S. *Tetrahedron Asymmetry* **2004**, *15*, 1085. (c) Galeazzi, R.; Garavelli, M.; Grandi, A.; Monari, M.; Porzi, G.; Sandri, S. *Tetrahedron Asymmetry* **2003**, *14*, 2639.

47. (a) Andrei, M.; Römming, C.; Undheim, K. *Tetrahedron Asymmetry* **2004**, *15*, 1359. (b) Andrei, M.; Esfkind, J.; Viljugrein, T.; Römming, C.; Undheim, K. *Tetrahedron Asymmetry* **2004**, *15*, 1301. (c) Andrei, M.; Undheim, K. *Tetrahedron Asymmetry* **2004**, *15*, 53. (d) Moller, B.; Undheim, K. *Eur. J. Org. Chem.* **2003**, 332.

48. Ashley, E. R.; Cruz, E. G.; Stoltz, B. M. *J. Am. Chem. Soc.* **2003**, *125*, 15000.

49. Gauzy, C.; Pereira, E.; Faure, S.; Aitken, D. J. *Tetrahedron Lett.* **2004**, *45*, 7095.

SYNTHESES OF POLYHYDROXYLATED AZEPANES

Soledad Pino-González,* Carmen Assiego and Noé Oñas
Departamento de Bioquímica, Biología Molecular y Química Orgánica, Facultad de Ciencias,
Universidad de Málaga, E-29071 Málaga, Spain (e-mail: pino@uma.es)

Dedicated to the memory of Dr. Fidel Jorge López Herrera

Abstract. *This review describes the methods of preparation and some properties of trihydroxy, tetrahydroxy and pentahydroxyazepanes. A bibliographic revision of the reported compounds with this structure type is well referenced through the text.*

Contents

1. Introduction

Over the last 30 years, iminocyclitols or iminosugars,[1] with a basic nitrogen instead of an oxygen in the ring, have attracted considerable attention from synthetic and medicinal chemists, biologists, and clinical researchers as a result of their potent inhibition of glycoprotein and glycolipid processing enzymes such as glycosidases and glycosyltransferases. These iminosugars are arousing great interest as potential therapeutic agents against HIV infection,[2,3] cancer,[4,5] diabetes[5] and other genetic or metabolic disorders.[6,7] Some of them have already found clinical application.[8,9] In recent years, the seven-membered ring iminosugars or polyhydroxyazepanes also called polyhydroxyperhydroazepines or seven-membered iminocyclitols, have also proved their utility as inhibitors of these enzymes.

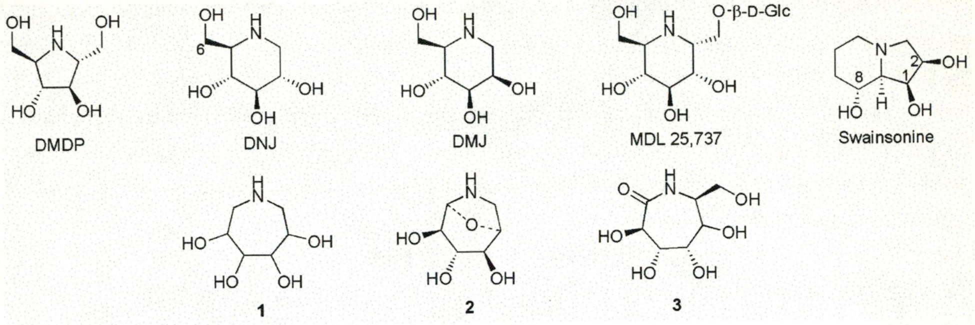

Figure 1

Polyhydroxylated azepanes type **1** were prepared for the first time by Paulsen and Todt in 1967[10a] from 6-amino-6-deoxy-aldoses. The previously reported 1,6-anhydrosugar with a nitrogen-containing ring **2**[10b] can be considered as a precursor of this type of iminosugar. In 1966 was obtained the destomic lactam **3** by hydrolysis of the antibiotic destomycin A[11a] (Figure 2). Compounds **2** and **3** would lead to tetrahydroxy (see Scheme 20) and pentahydroxyazepane, respectively, by reduction reactions. To our knowledge no monocyclic polyhydroxyazepane has been directly isolated from natural sources, but there are related azepane structures in natural products, two samples are:[11b] balanol, a popular synthetic target with remarkable inhibitory activity against protein kinase C, and ophiocordin, an antibiotic exhibiting antifungal activity; which could be considered as derived from *trans*-3-amino-4-hydroxy-azepane. (Figure 2).

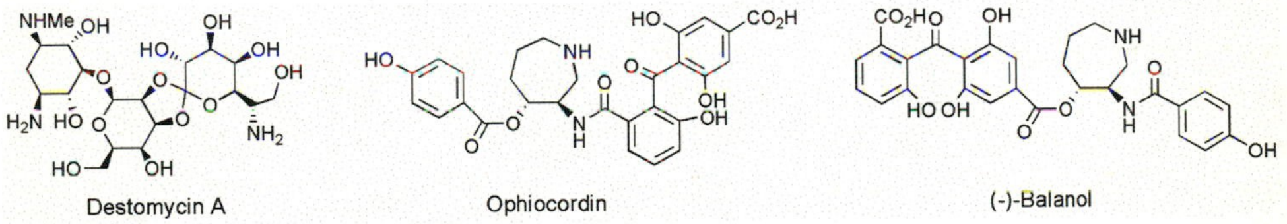

Figure 2

In spite of being recognised structures in the 60´s, in the following years less attention was given to the azepane analogues than to other iminosugars. Most work on the design and synthesis of glycosidase inhibitors has focused on five- and six-membered iminocyclitols, which are considered to mimic the substrate transition states with oxacarbenium ion character and a distorted six-membered ring. Polyhydroxyazepanes received little attention before Wong and co-workers[12] revealed that tetrahydroxy azepanes inhibited a broad range of glycosidases. Recently, polyhydroxylated azepanes were reported as new motifs for DNA minor groove binding agents (MGBLs) that were shown to have modest growth inhibitory activity in cancer cell lines and are capable of binding to DNA even at low pH.[13]

Polyhydroxyazepanes have several properties that make them potentially useful as drug candidates: **Flexibility**.- A seven-membered ring is more flexible than five- or six-membered rings and azepanes may adopt a quasi-flattened conformation. Additionally, in the polyhydroxylated rings, this flexibility allows the hydroxyl groups to adopt a variety of positions increasing the probability of forming hydrogen bonds with structural motifs found within DNA framework, or improving binding to the active site of the enzyme.

High water solubility.- The polyhydroxylation in azepanes is a great advantage to circumvent the problem of poor bioavailability which exists in other less hydroxylated iminoderivatives, with different ring sizes.

Chirality and high level of functionalisation.- The distinct routes to polyhydroxyazepanes allow different diastereomers syntheses, with a plethora of functionalised derivatives. During the last years, there have been increased structural activity/relationship (SAR) studies to improve selectivity against enzymes and bioavailability. Moreover, the affinity of an iminosugar can be significantly increased by inclusion of an appropriate aglycon moiety.

The present review summarises the progress which has been made and the current state of this field from the first reported synthesis[10a] to October 2004. In this work we will focus on synthetic methods but also describe biological properties when appropriate. Reports on the synthesis of polyhydroxyazepanes are still scarce. In some of those reported polyhydroxyazepanes have not been obtained as single products and usually are accompanied by their corresponding six-membered ring derivatives, requiring separation.

We describe here diverse syntheses of tetra- tri- penta- and hexahydroxyazepanes and derivatives.

2. Synthesis of tetrahydroxyazepanes

2.1. From bis-epoxides

The formation of tetrahydroxyazepanes from bis-epoxides by a nucleophilic opening-aminocyclisation has been described by several research groups.[12,14-17] Formally this approach involves a first completely regioselective opening of one of the epoxide functions by ammonia or a primary amine followed by a spontaneous ring closure to furnish seven-membered and/or six-membered cyclic azasugars (Scheme 1). Thus, the reaction of 1,2:5,6-dianhydro-3,4-O-dibenzyl-D-mannitol (**4**) with benzylamine[14a,b] gave a mixture of azepane **6** and piperidine **5** in different ratios depending of the solvent employed.

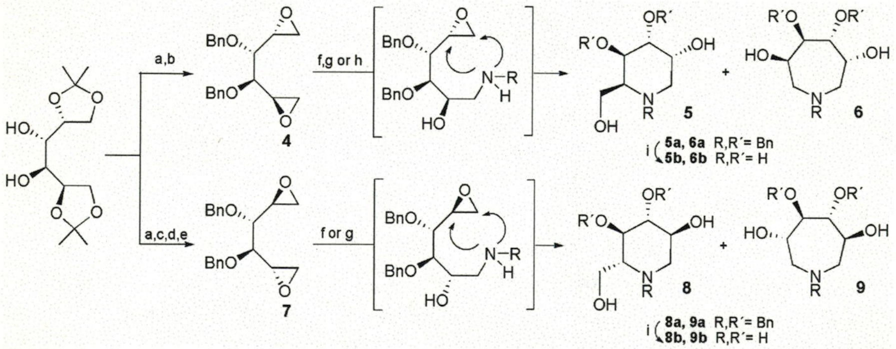

(a) NaH, BnBr, nBu₄NI, THF then aq AcOH, 87%; (b) Ph₃P, DIAD, 130 °C, 86%; (c) TBDMSCl, imidazole, DMF, 0 °C, 80%; (d) MsCl, NEt₃,CH₂Cl₂, 98%; (e) HCl, MeOH, then aq NaOH, 75%; (f) RNH₂ (10 eq), HClO₄ (5 eq), H₂O, rt, 4 h; (g) RNH₂ (5 eq), CHCl₃, Δ, 48 h; (h) RNH₂ (4 eq), Et₃N (5 eq), H₂O, 80 °C, 15 h; (i) H₂, Pd black, AcOH, 100%

Scheme 1

Le Merrer *et al.* reported[14a] that this reaction gave mainly the 6-*exo* adduct in aprotic medium, whereas in protic medium or aprotic medium with Lewis acid the 7-*endo* adduct was mainly obtained. Later,[14b] the aminocyclisation was performed with other primary amines giving the corresponding *N*-protected piperidines and azepanes (Table 1).

Table 1. Cyclisation with several amines.

Amine RNH₂	Conditions (Scheme 1)	Total yield from 4	Piperidine/Azepane 5:6	Total yield from 7	Piperidine/Azepane 8:9
PhCH₂NH₂	f	95%	3:7	75%	1:7
PhCH₂NH₂	g	95%	10:9	78%	4:3
PhCH₂CH₂NH₂	f	97%	1:2		
AcHN(CH₂)₄NH₂	g	80%	3:2		
BnONH₂.HCl	h	73%	3:7		
(Bn)₂NNH₂	g	53%	9.7		
t-BuO₂CCH₂NH₂.HCl	g+Et₃N(4eq)	73%	6:5		
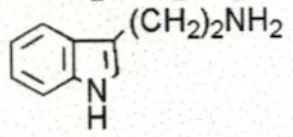 [indole]-(CH₂)₂NH₂	f	90%	2:5		
	g	95%	10:9		

Lohray and co-workers[15a] found that the reaction of 1,2:5,6-dianhydro-3,4-di-*O*-isopropylidene-D-sorbitol (**10a**)[15b] with benzylamine gave only the seven-membered azasugar. This fact was attributed to the lack of free rotation around the C3-C4 bond due to the presence of a *trans*-acetonide group in **10a**. Additionally, the isopropylidene derivatives of D-mannitol **10b**[15b] and L-iditol **10c**[14c] gave only the azepane derivatives, (Scheme 2) showing that the *trans*-acetonide group exerted a dramatic influence on the selectivity of the reaction leading exclusively to a 7-*endo-tet*-type cyclisation process. Attempts to isolate the intermediate were not successful. Transfer hydrogenation of **11a-c** proceeded smoothly and removal of the acetonide protecting group with methanolic HCl afforded tetrahydroxyazepanes **12**, **6b** and **9b** as hydrochloride salts, in high yields. The same group published the synthesis of new *N*-derivatives[15c] by reaction of bis-epoxides with various primary amines. The obtained azepanes (Figure 3A) inhibited β-glucosidases in micromolar range and also exhibited anticancer activity in various cancer cell lines with GI₅₀ values in the range of 2 to 9x10⁻⁵ M. These azepanes are poor inhibitors of HIV protease and their IC₅₀ values are in the 2x10⁻⁴ M range. Later, other amino or azido derivatives[18] were patented as α-glucosidase inhibitors. Protected hydroxylated azepanes, linked to an indole ring, have also been described (Figure 3B).[16]

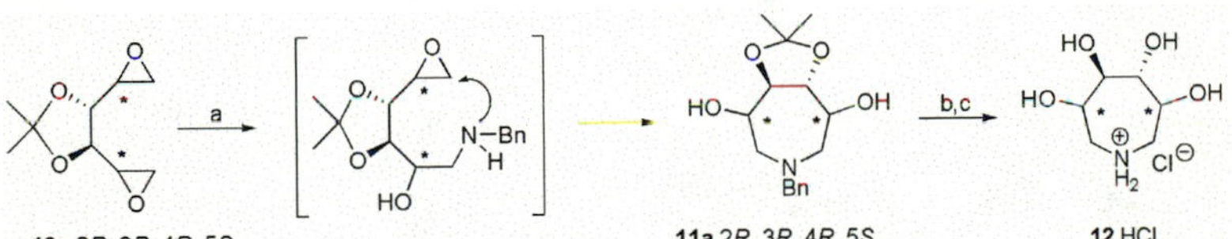

10a 2R, 3R, 4R, 5S
10b 2R, 3R, 4R, 5R
10c 2S, 3R, 4R, 5S

11a 2R, 3R, 4R, 5S
11b 2R, 3R, 4R, 5R
11c 2S, 3R, 4R, 5S

12.HCl
6b.HCl
9b.HCl

(a) BnNH₂, THF or dioxane, Δ, 8-50 h, 64-88%; (b) 10% Pd/C, HCO₂NH₄, MeOH, 2 h, Δ, 92-100%; (c) HCl(g)-MeOH, 0.5h, 25 °C, 90-98%.

Scheme 2

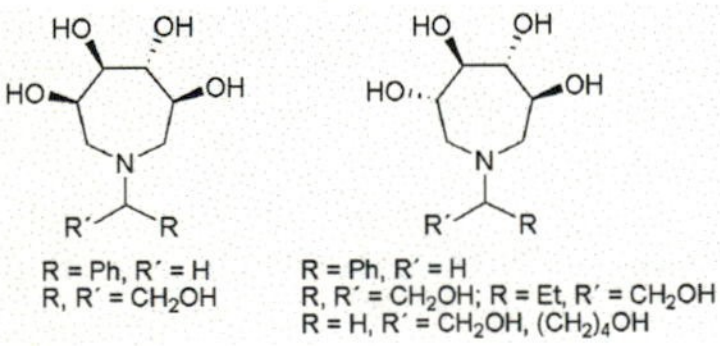

R = Ph, R´ = H
R, R´ = CH₂OH

R = Ph, R´ = H
R, R´ = CH₂OH; R = Et, R´ = CH₂OH
R = H, R´ = CH₂OH, (CH₂)₄OH

Figure 3A

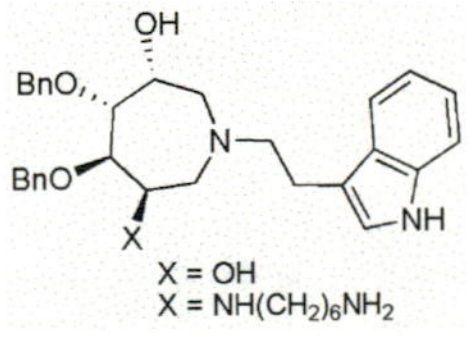

X = OH
X = NH(CH₂)₆NH₂

Figure 3B

2.1.1. Isomerisations (from or to) azepanes

Skeletal rearrangement of β-chloro-piperidines or-pyrrolidines, *via* an aziridinium salt, is a known reaction, but this method has found little application in polysubstituted heterocycles, and only a few cases were described in synthesis before 1996.[19,20] Hydroxyl group activation has also been performed either by mesylation, or by transformation into an alkoxyphosphonium salt under Mitsunobu conditions. Le Merrer and colleagues[21,14b] described a novel isomerisation of polyhydroxylated piperidines with the unmasked 2,6-hydroxyl groups. The L-*gulo*-piperidine **5a** reacts instantaneously with MsCl to afford the dimesylate **5c** (Scheme 3). Nucleophilic displacement by AcOCs/DMF, implying the formation of intermediate aziridinium salts, followed by methanolysis gave three compounds: the L-*gulo*-piperidine **5a** (18%), the D-*manno*-azepane **6a** (8%) and the L-*ido*-pyrrolidine **A** (45%), which were easily separated by flash chromatography. In the same conditions, the D-*gluco* piperidine, where all substituents are in equatorial direction, afforded a *manno*-pyrrolidine resulting from ring contraction. The reaction of the enantiopure C2-symmetric polyhydroxylated azepanes **6a** or **9a**[21,14b] with mesyl chloride gave chloromethyl-piperidines **5d** or **5e** respectively (Scheme 4). Reaction of **6a** under Mitsunobu conditions using benzoic acid, yielded a mixture of two monobenzoyl derivatives **6d** and **5f**.

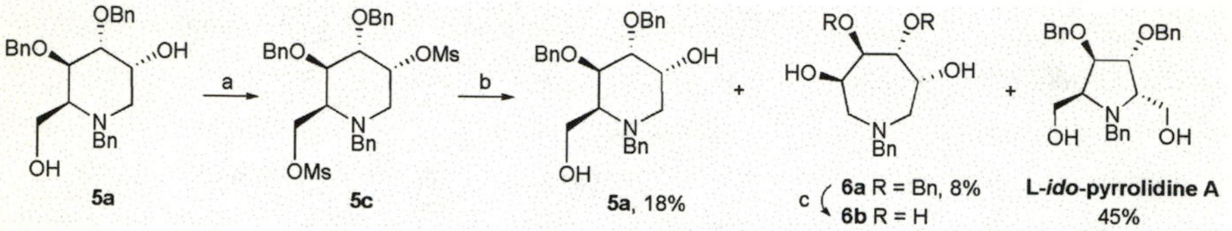

(a) MsCl (2.3 eq), Et₃N, CH₂Cl₂, 0 °C, 100%; (b) AcOCs, DMF, 40-50 °C then MeOH, K₂CO₃; (c) H₂, Pd-black, AcOH, 100%.

Scheme 3

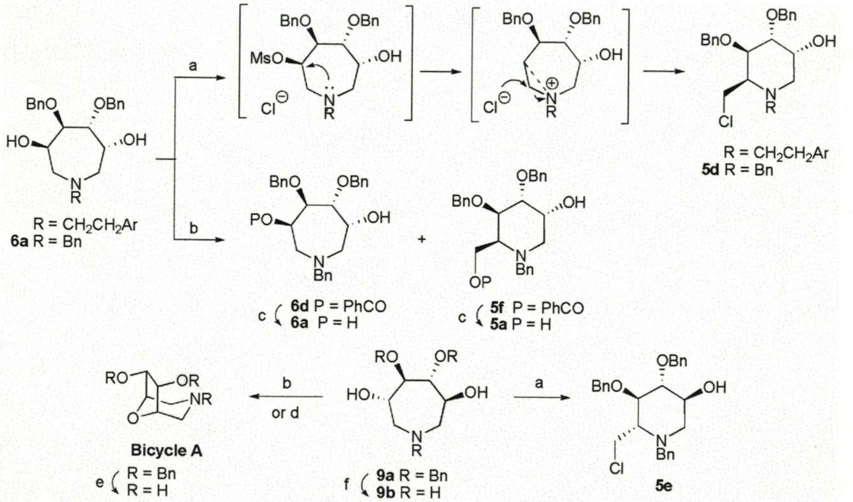

(a) MsCl (1.4 eq), Et₃N, CH₂Cl₂, 0 °C, 60%; (b) 1.2 eq [Ph₃P-DEAD-PhCO₂H], THF, 0 °C, overall yield from **6a**: 64%; from **9a**: 80%; (c) MeOH, K₂CO₃; (d) 1.2 eq [Ph₃P-DEAD], THF, 0 °C, 94%; (e) H₂, Pd-black, then DOWEX®, 100%; (f) H₂, Pd-black, AcOH, 100%.

Scheme 4

It is possible that in the latter case the formation of the two products results from a kinetic control whereas in the former case, two products (chloroazepane and chloromethyl-piperidine) could be initially formed, being the 3-chloroazepane further converted, *via* the aziridinium ion, to the more stable rearranged product **5d**. The previous Mitsunobu conditions applied to the azepane **9a** gave only bicycle **A**.

Other similar isomerisations were reported (Scheme 5).[22] Thus, isopropylidene azepane **11c**, prepared by the reaction of 1,2:5,6-dianhydro-3,4-O-isopropylidene-L-iditol **10c** (see Scheme 6) and benzylamine, undergoes a skeletal rearrangement to furnish the piperidine derivative **5g** under Mitsunobu conditions. Conversely, *N*-benzyl-3,6-diprotected-4,5-dihydroxyazepane **9c** under identical conditions furnishes a pyrrolidine derivative **B**. On the other hand, azepane **9d** undergoes an intramolecular nucleophilic substitution in the absence of an external nucleophile (benzoic acid) under Mitsunobu conditions to give the bicycle **B**.

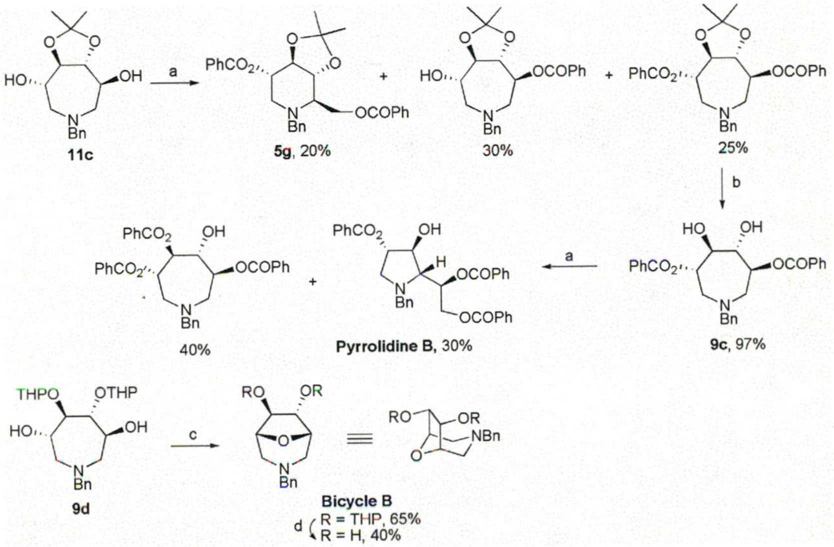

(a) PPh$_3$, DEAD, PhCO$_2$H, THF; (b) TFA, H$_2$O; (c) PPh$_3$, DEAD, THF; (d) *p*-TsOH, H$_2$O, MeOH.

Scheme 5

2.1.2. Solid phase synthesis

A solid phase synthesis of azepanes was performed with the aim of using them as scaffolds in the synthesis of new peptidomimetics libraries.[23] The strategy was based on the opening of L-iditol bis-epoxide **10c** by an amine-type resin,[24] to provide resin-bound azepane **13** (Scheme 6). A key feature was the use of a readily cleavable linker, as the Rink linker **16**, but a crowded amine might hinder the opening of bis-epoxide **10c**. In order to evaluate this fact, these authors carried out the reaction in liquid-phase on **10c** with amines **17**, **18** and **19** (Figure 4). The corresponding *N*-substituted azepane was obtained by reaction in DMF at 80 °C, or most easily in methanol at rt in about 80% yield after chromatographic purification, even with the more crowded amine **19**, which mimics the Rink resin. The azepane was formed on the Rink resin (Table 2) in DMF at 80 °C or in MeOH/CH$_2$Cl$_2$ at room temperature. Although the best loading in azepane was obtained by reaction in DMF after heating during 5 days (entry 2), a degradation of the resin beads occurred, lowering the yield of the free azepane obtained after protection of the secondary alcohols of **13** and

369

subsequent cleavage from the resin. Thus, although the overall yields were similar, conditions described in entry 3, (R: PhCO), appeared to be the best ones. The secondary alcohols were converted into esters or carbonates (Table 3), revealing the utility of resin-bound **13** in combinatorial chemistry.

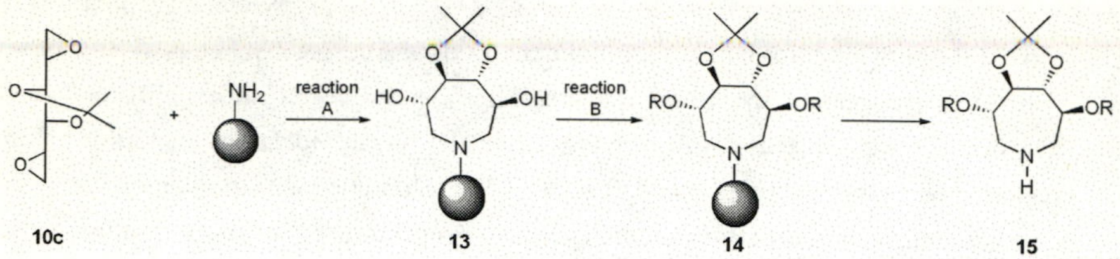

Scheme 6

Table 2. Solid-phase synthesis of azepane.

Entry	Reaction conditions A	Loading in azepane 13[a]
1	**10c** (2eq.),[b] DMF, 80 °C, 2 days	0.20 mmol/g
2	**10c** (2eq.),[b] DMF, 80 °C, 5 days	0.45 mmol/g
3	**10c** (5eq.),[b] MeOH/CH₂Cl₂ 1/1, rt, 2 days	0.23 mmol/g
4	**10c** (5eq.),[b] MeOH /CH₂Cl₂ 1/1, rt, 5 days	0.21 mmol/g
5	**10c** (5eq.),[b] MeOH/CH₂Cl₂ 1/4, rt, 3 days	0.12 mmol/g

[a] Initial loading of Rink resin **16**: 0.50 mmol/g.
[b] Equivalent of bis-epoxide **10c** for one Rink linker on solid support.

Table 3. Reaction B and cleavage.

Entry	R	Reaction conditions B[a]	Yield after cleavage[b]
1	FmocNHCH₂CO	Fmoc-Gly-OH, DIC,[c] DMAP, CH₂Cl₂	90%
2	Ac	Ac₂O, pyridine, CH₂Cl₂	64%
3	PhCO	PhCOCl, pyridine, CH₂Cl₂	88%
4	BnNHCO	CDI,[d] THF, then BnNH₂	80%
5	[-(CH₂)₅-]NCO	CDI,[d] THF, then piperidine	53%
6	PhNHCO	PhNCO, Et₃N, CH₂Cl₂	93%

[a] 10 eq. of reactant for one azepane on solid support. [b] Yields based on support-bound azepane **13**, the final products are obtained without purification with a purity ≥ 90% (based on ¹H-NMR). [c] D1C: N,N'-diisopropylcarbodiimide. [d] CDI: 1,1'-carbonyldiimidazole.

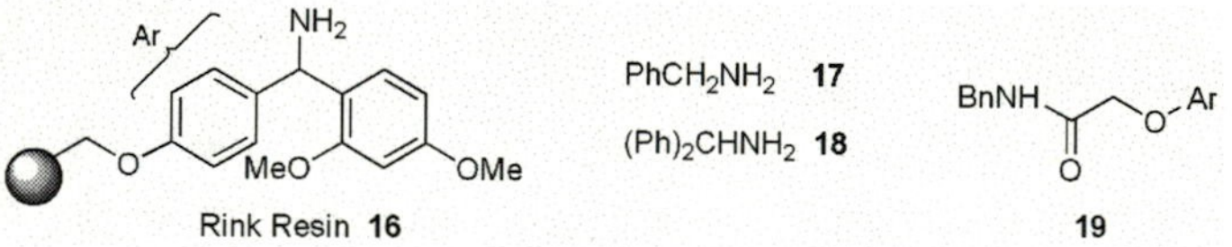

Rink Resin **16**   PhCH₂NH₂ **17**   BnNH CH₂—O—Ar

(Ph)₂CHNH₂ **18**   **19**

Figure 4

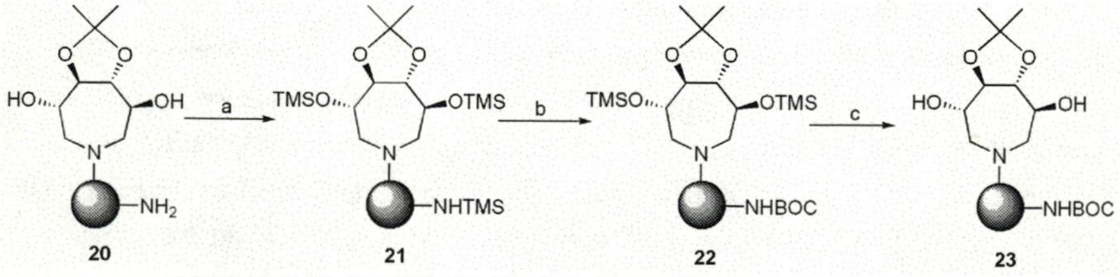

(a) TMSCl, Et₃N, THF; (b) (Boc)₂O, Et₃N, CH₂Cl₂; (c) TBAF, THF.

Scheme 7

370

A modification of this procedure was developed to protect the amine-linkers which were not involved in azepane formation, in order to prevent further secondary reactions (Scheme 7). Silylation of secondary hydroxyl groups and free amine-linkers followed by treatment with (Boc)$_2$O gave the fully protected compound **22**, which was desilylated with tetrabutylammonium fluoride.

2.1.3. Synthesis of pseudo, aza di-, tri- or oligo- saccharides

A chemo-enzymatic synthesis of new *N*-containing malto-oligosaccharides was reported by a japanese group.[25] Among these compounds, maltooligosaccharides having (3*R*,4*R*,5*R*,6*S*)-hexahydro-3,4,5,6-tetrahydroxy-1*H*-azepine residue at the reducing end showed strong inhibitory activities for human pancreatic and salivary amylases (HPA and HSA). The administration of (3*R*,4*R*,5*R*,6*S*)-hexahydro-3,5,6-trihydroxy-1*H*-azepine-4-yl-*O*-α-D-glucopyranosyl-(1→4)-α-D-glucopyranoside to ICR mice suppressed postprandial hyperglycemia.

In order to obtain new and stable inhibitors with increased specificity for selected glycosidases, Le Merrer *et al.*[26] delineated a synthesis of *N*-substituted iminosugars linked to a sugar analogue aglycon moiety by a nonhydrolyzable bond, synthesizing pseudo-di-(or tri-)saccharides (Scheme 8). Azadisaccharides **26c** and **27c** were obtained from the flexible bis-epoxide **4** (Scheme 8). Thus, reaction of **4** with sodium azide and silica gel under heating enabled the one step preparation of the azidomethyl-D-*gluco*-C-furanoside **24** in 95% yield. Silylation of **24** and reduction of the azide **25a** gave the amine **25b** which was reacted with bis-epoxide **4** to obtain a mixture of *N*-methylfurano azepane **26a** and *N*-methylfurano piperidine **27a**, which could be easily separated by flash chromatography. Removal of the silyl group and total hydrogenolysis of benzyl protecting groups gave the desired azadisaccharides **26c** and **27c**, which were isolated as their acetate salts after purification by flash chromatography.

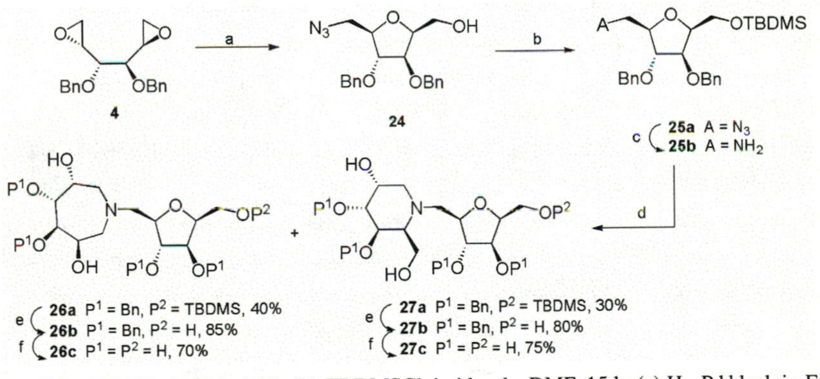

(a) NaN$_3$, SiO$_2$, CH$_3$CN, Δ, 48 h, 95%; (b) TBDMSCl, imidazole, DMF, 15 h; (c) H$_2$, Pd black in EtOAc; (d) **4**, MeOH, 8 days; (e) Bu$_4$NF, THF; (f) H$_2$, Pd black, AcOH.

Scheme 8

The protected azatrisaccharides **28a** and **29a** were prepared from the more rigid 3,4-*O*-acetonide D-*manno* **10b** by bis-nucleophile opening with azepane **6b** or piperidine **5b**, respectively (Scheme 9). Acid hydrolysis gave the pseudo-trisaccharides **28b** and **29b** in 50% and 75% overall yields.

The new aza-di-(or tri-)saccharides were screened against four common glycosidases (α-glucosidase, β-D-glucosidase, D-mannosidase and α-L-fucosidase). The pseudo-disaccharides **26c** and **27c** were totally inactive on all the glycosidases studied (K_i = 10^{-3} M) and consequently less potent inhibitors than the parent iminosugars **6b** and **5b**. Whereas pseudo-trisaccharide **29b** showed no inhibition against all the glycosidases, **28b** displayed competitive and selective inhibition towards α-L-fucosidase with a K_i = 15 μM, in the same order of magnitude of the D-*manno* azepane **6b**; no inhibition was found at 1 mM on the other enzymes assayed.

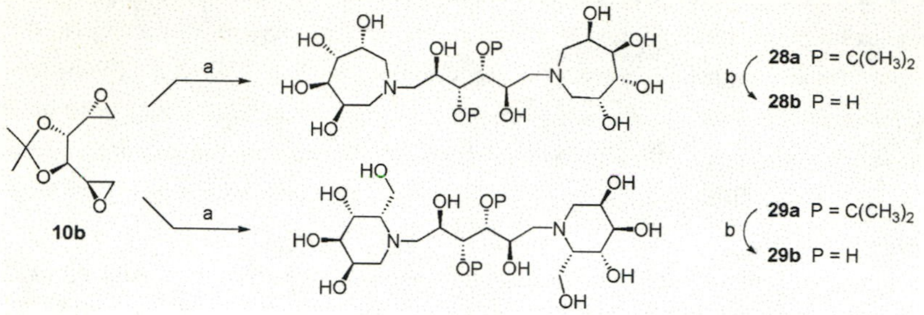

(a) **6b** or **5b**, 2 eq in MeOH, 12 h, 20 °C; (b) TFA/H_2O, 4/1, 20 °C

Scheme 9

2.1.4. Tetrahydroxyazepane derivatives as potential MGBL´s

Polyhydroxyazepanes have several properties that make them potentially useful as new motifs for DNA minor groove binding agents (MGBL's).[13] Their flexibility would allow the hydroxyl groups to adopt a variety of positions increasing the probability of them forming hydrogen bonds with the N-3 of the purine, the urea carbonyl of the pyrimidine bases (H-bond acceptors), or the 2-amino of guanine (H-bond donor) which point into the minor groove. The aromatic diamidines 1,5-bis-[4-amidinophenoxy)pentane (pentamidine, **30**) and 1,3-bis-[4-amidinophenoxy)propane (propamidine, **31**) have been shown to bind in the minor groove of DNA. However, their clinical use has been restricted by their significant toxicity. Johnson and Thomas[13] prepared bis-azepanes that share the dicationic nature and core structure of propamidine, allowing a direct comparison of their DNA binding abilities and biological activity. To investigate the effects of different stereochemistry on DNA binding two different C2-symmetric bis-epoxides were selected (**10b** and **10c**). The bis-epoxides **10b** and **10c** were reacted with either 1,3-bis-(4-aminophenoxy)propane or 1,3-bis-(4-aminomethylphenoxy)propane. However, in neither case was the required bis-azepane isolated. Rather than undergoing intramolecular opening, the second epoxide appeared to have reacted with another 1,3-bis-(4-aminoaryloxy)propane molecule leading to polymer formation. Changes in the reaction conditions and the ratio of the bis-epoxide to bis-amine did not give any of the required products. In an alternative strategy (Scheme 10) the 3,4,5,6-tetrahydroxyazepane **32** and its D-*ido* diastereomer were synthesised from 4-acetoxyphenylamine and bis-epoxides **10b** (D-*manno*) or **10c** (L-*ido*), respectively, by heating in water. Deacetylation of **32** and treatment of the phenol **33**, or its isomer, with potassium carbonate in dry ethanol, followed by addition of 1,3-dibromopropane, afforded the required di-isopropylidene bis-azepane **34** in 40% yield, (27% for L-*ido*). Deprotection of **34** with hydrochloric acid gave 1,3-bis-[(3R,4R,5R,6R)-3,4,5,6-tetrahydroxyazepane-N-p-phenoxy]propane dihydrochloride **35** or the (3S,4R,5R,6S)-diastereoisomer (from protected L-*ido*). The benzyl bis-azepane homologues, 1,3-bis-[(3R,4R,5R,6R)-3,4,5,6-tetrahydroxyazepane-N-p-benzyloxy]-propane dihydrochloride **37a** and its (3S,4R,5R,6S)-diastereoisomer were produced by a similar route (Scheme 11). The propamidine **31** together with the bis-azepanes **34, 35, 37a, 37b** and their diastereoisomers were tested against a number of cell lines *in vitro* biological tests. The results demonstrated that compounds **34, 37a, 37b** were more effective than chlorambucil, as active as propamidine and much less effective than doxorubicin in inhibiting the growth of colon cancer cell lines. One of them, bis-azepane **37a** is a better growth inhibitor than any of the drugs in a doxorubicin resistant breast cancer cell line. The bis-azepane **37b**, with the best DNA binding affinity showed the poorest inhibitory activity of those tested. Therefore, unless the acetonide groups on **37a** facilitate transport across the cell membrane and are then hydrolyzed within the cell to give the active compound **37b**, it is unlikely that the biological activity in this serie is linked to DNA binding, which may have very poor cellular uptake due to its increased polarity. The thermal denaturation studies indicated that **37b** had a stabilizing affect on the DNA, very weak compared with that of propamidine **31**.

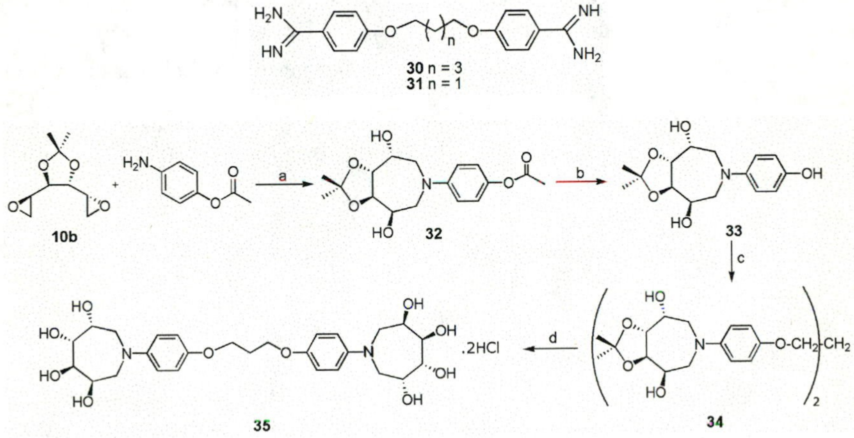

(a) H_2O, 95 °C, 40%; (b) 2M NaOH, MeOH, 75%; (c) $Br(CH_2)_3Br$, K_2CO_3, EtOH, reflux, 4 days, 40%; (d) 0.5 M HCl, 95%.

Scheme 10

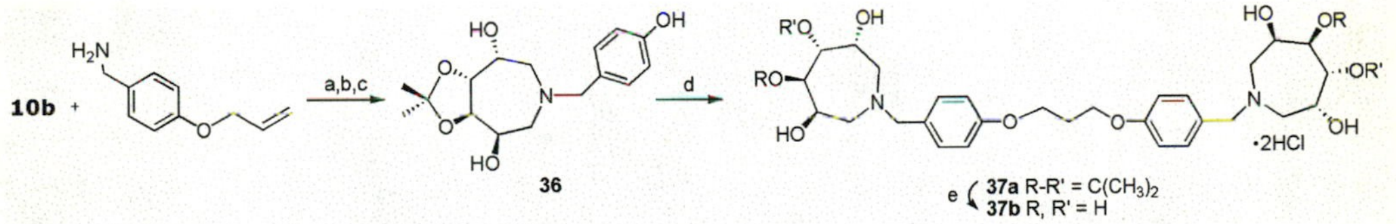

(a) H_2O, 60 °C, 69%; (b) *t*-BuOK, DMSO, 100 °C, 81%; (c) HgO, $HgCl_2$ in 10:1 acetone/water, 87%; (d) K_2CO_3, $Br(CH_2)_3Br$, EtOH, reflux for 4 days, 31%; (e) 0.5 M HCl, 100%.

Scheme 11

2.2. From azido or aminosugars

The first synthesis of tetrahydroxyazepanes from 6-amino-6-deoxy-D-glucose and D-galactose was published in 1967,[10a] as cited in the introduction, and in the last 15 years, several syntheses have been reported using azido or aminosugars as azepane precursors, and the reductive amination reaction as key step.[12,27,28]

Thus, in 1990, the azepane 1,6-dideoxy-1,6-imino-D-glucitol (**12**) and 1,5-dideoxy-1,5-imino-D-glucitol (1-deoxynojirimycin or DNJ, see Scheme 1) were synthesised by simple sequences in six and seven

373

steps, respectively, from D-glucofuranurono-6,3-lactone, *via* the azide derivatives.[27] Several years later, Wong *et al.*[12] obtained several polyhydroxyazepanes either by chemo-enzymatic or chemical synthesis and revealed that tetrahydroxyazepanes exhibited promising glycosidase inhibitory profiles against a broad range of glycosidases. The **enzymatic syntheses** of iminocyclitols involved the combined use of aldolases and isomerases.[12a] Condensation of (±)-3-azido-2-hydroxypropanal (**38**) and dihydroxyacetone phosphate (DHAP), in the presence of a DHAP-dependent aldolase, followed by treatment with acid phosphatase and an isomerase, gave a 6-azido-6-deoxy-aldopyranose, which upon reductive amination afforded the corresponding 3,4,5,6-tetrahydroxyazepane (Scheme 12). The ketose-aldose isomerisation step proceeded well, using glucose and fucose isomerases (GlcI and FucI), leading to azidosugars **40** and **41**, but rhamnose isomerase (RhaI) was unable to isomerise azidoketose **39**. Azido aldoses **40** and **41** were subjected to reductive amination to give (3*R*,4*R*,5*R*,6*S*)-tetrahydroxyperhydroazepine (**12**)[27,12a] and the *meso*-iminocyclitol (3*S*,4*R*,5*S*,6*R*)-tetrahydroxyperhydroazepine **42**.[10a] An attempted enzymatic synthesis of the methoxy derivatives of these azasugars was unsuccessful, leading, however, to both enantiomers of 1-deoxy-2-*O*-methylmannojirimycin.

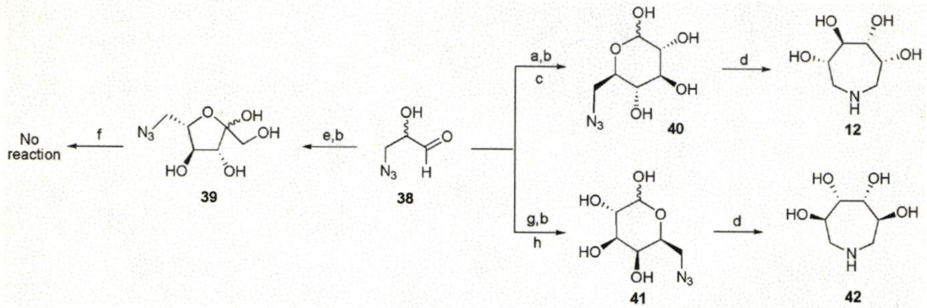

(a) pH 6.7, DHAP, FDPA; (b) pH 4.5, Pase; (c) pH 7.2, TAKASWEET (Glcl), 26% for **40**; (d) H$_2$ (50 psi), Pd/C, H$_2$O, 2 days, 91-94%; (e) pH 6.7, DHPA, RhaA; 30% for **39**; (f) pH 7.2, RhaI; (g) pH 6.7, DHAP, FucA; (h) pH 7.2, Fucl 21% for **41**.

Scheme 12

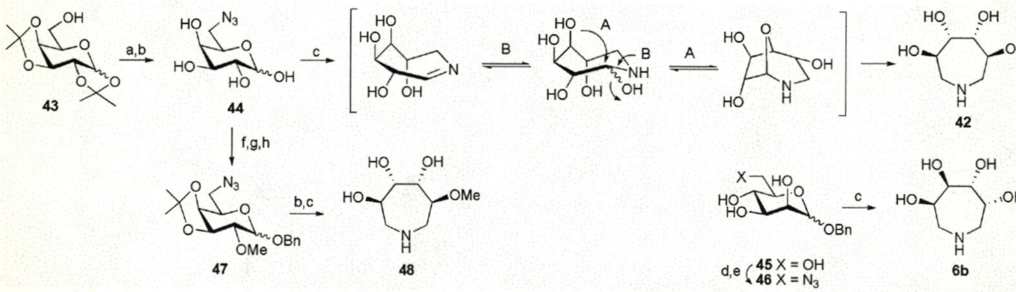

(a) i. Ph$_3$P, DEAD, THF, 0 °C; ii. (EtO)$_2$P(O)N$_3$, 80%; (b) 80% AcOH, 70-80 °C, 95%; (c) H$_2$ Pd/C, H$_2$O, (50 psi for **42** and **48**, 2 days, 90%; 1 atm for **6b**, 12 h, 85%); (d) TsCl, Py, 0 °C, 12 h, 61-68%; (e) NaN$_3$ (5 eq), NH$_4$Cl, (5 eq), EtOH/H$_2$O, 9:1, reflux, 12 h, 70%; (f) BnOH, 80 °C, BF$_3$.Et$_2$O, 75%; (g) 2,2-dimethoxypropane, *p*-TSA, DMF, rt, 95%; (h) MeI, NaH, THF, rt, 93%.

Scheme 13

Some of these azepanes display significant activity as glycosidase inhibitors, with Ki values in the micromolar range. Although all these iminocyclitols do not inhibit the HIV protease, a tetrafluoro-3,6-

dibenzyl derivative[12a] showed moderate inhibition. The X-ray structure indicates a pseudochair conformation for these products.

The **chemical synthesis** starts with 6-azido-6-deoxysugars which can be obtained by manipulations of aldopyranoses, protected as benzyl glycosides or diisopropylidene ethers. The key step is the reductive amination of 6-azido-6-deoxyaldohexopyranoses, (Scheme 13). Thus, the D-galactose derivative **43** led to the *meso* azepane **42** and benzyl D-mannopyranoside **45** led to azepane **6b** with a C2 symmetry axis. Desymmetrisation of the *meso* azasugar **42** was carried out from **44** by selective protection sequence, to yield a 6-methoxy-3,4,5-trihydroxyazepane **48**.

Several years later, Wang and co-workers[29] described a new chemo-enzymatic synthesis using benzyl glicosides as starting material and obtained a variety of polyhydroxyazepanes (see Section 4, Scheme 33 and Table 6).

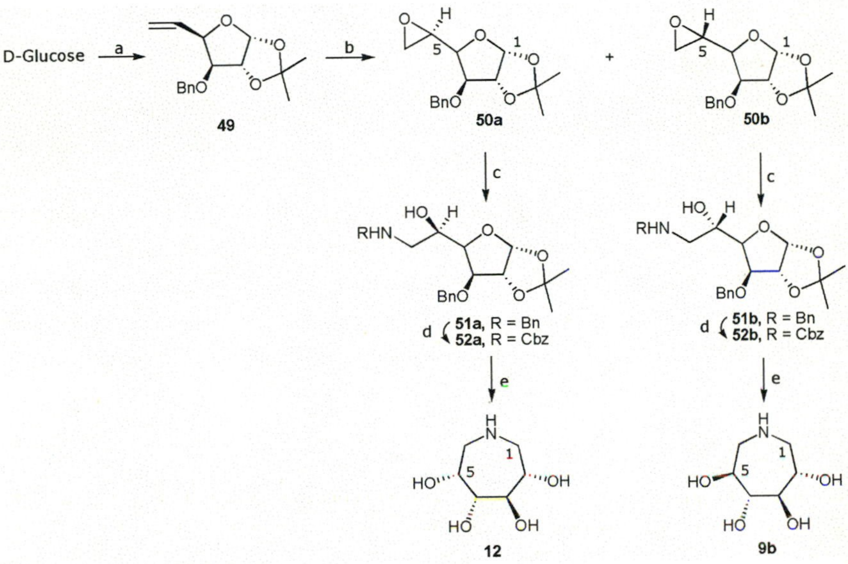

(a) Ref. 29; (b) mCPBA, CH$_2$Cl$_2$, 25 °C, 36 h; (c) for **50a**, BnNH$_2$, neat, 25 °C, 12 h; for **50b**, BnNH$_2$, 80 °C, 12 h; (d) i) HCOONH$_4$, 10% Pd–C, MeOH, reflux, 40 min; ii) ClCOOBn, MeOH–H$_2$O, 0–25 °C, 2 h; (e) i) TFA–H$_2$O, 25 °C, 2 h; ii) 10% Pd–C, MeOH, H$_2$, 80 psi, 25 °C, 24 h.

Scheme 14

Dhavale and co-workers[28] reported a divergent synthetic route to (2S,3R,4R,5R)-1,6-dideoxy-1,6-imino-D-glucitol **12**[15a,27] and (2S,3R,4R,5S)-1,6-dideoxy-1,6-imino-L-iditol **9b**,[10a,14a,b] using the chiron approach (Scheme 14). D-Glucose was converted to 5,6-dideoxy-1,2-O-isopropylidene-3-O-benzyl-α-D-xylo-hex-5-enofuranose (**49**) by a known method.[30] Epoxidation of **49** with mCPBA in dichloromethane at room temperature afforded a diastereomeric mixture of 5,6-epoxides **50a** and **50b** in the ratio 55:45, respectively. The regioselective nucleophilic attack of *N*-benzylamine, at the terminal carbon of the epoxy ring, afforded **51a**. Hydrogenolytic removal of the *N* and *O*-benzyl protecting groups in **51a**, with ammonium formate and 10% Pd–C in methanol under reflux, gave an aminoalcohol, that was directly subjected to selective *N*-Cbz protection with benzylchloroformate, in the presence of sodium bicarbonate in methanol–water, to give **52a**. Removal of the 1,2-acetonide group in **52a** with trifluoroacetic acid–water

375

(3:2) followed by catalytic hydrogenation using 10% Pd–C in methanol afforded **12**. A similar synthetic sequence afforded **51b** from 5,6-anhydro-β-L-*ido*-furanose **50b**. Removal of the protecting groups gave **9b**.

2.3. Synthesis from inositols

Painter and Falshaw[31] reported an unusual approach to tetrahydroxyazepanes starting from readily available D- and L-*chiro*-inositols, *via* the dialdehyde and using the double reductive amination reaction. In a previous study[31a] azepanes (-)-**6b** and (+)-**6b** were synthesised from D- and L-*chiro*-inositol respectively, by cis-α-diol protection (**53**), *trans*-diol oxidative cleavage (**54**), reductive amination of the resulting dialdehyde to give the azepane **55**, and deprotection to afford azepane **6b**, (Scheme 15).

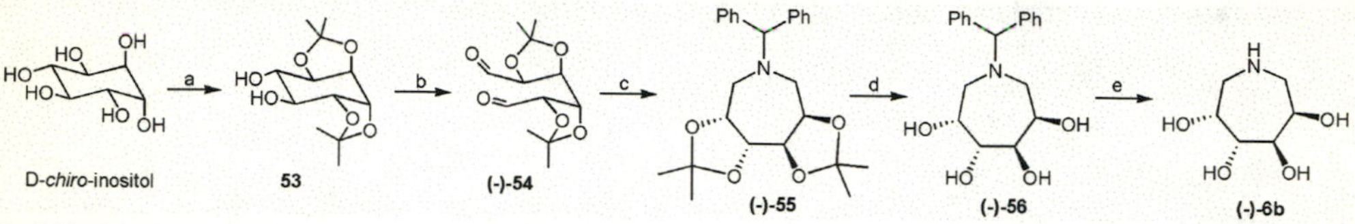

(a) Ref 33a; (b) NaIO4-activated silica, CH2Cl2, 86%; (c) NaBH3CN, (Ph)2CHNH2, AcOH (2 eq), 3 A molecular sieves, MeOH, -78 °C to rt, 71%; (d) HCl, MeOH-CH2Cl2, reflux, 89%; (e) Pd(OH)2/C, H2, MeOH, 94%.

Scheme 15

In further studies[31b] similar methodology was employed to synthesise the *ido*-configured azepanes (+)-**9b** and (-)-**9b** from D- and L-*chiro*-inositol, (Schemes 16 and 17). Protection of the four equatorial hydroxyl groups of D-*chiro*-inositol was prepared by a procedure known to favour the protection of 1,2-diequatorial *trans*-diol on six-membered rings.[32] Attempted oxidative cleavage of the diol moiety of the diacetal (-)-**57** with sodium periodate or lead tetraacetate under various reaction conditions failed, presumably due to a *trans*-diaxial relationship of the vicinal diol.

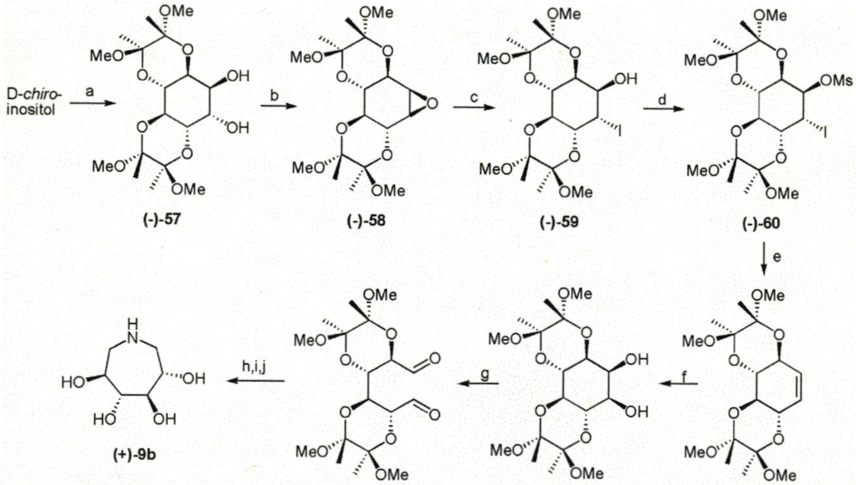

(a) butanedione, CH(OMe)3, H+, MeOH, reflux (80%); (b) Ph3P, DEAD, THF, 0 °C (68%); (c) AlCl3, MeCN, NaI, 0 °C (95%); (d) MsCl, Et3N, CH2Cl2 (57%); (e) Zn dust, AcOH, DMF, 80 °C (92%); (f) NaIO4, RuCl3, EtOAc, MeCN, H2O (98%); (g) NaIO4, activated silica, CH2Cl2 (90–95%); (h) NaCNBH3, MeOH, (Ph)2CHNH2, 3A° sieves, -78 °C–rt; (i) TFA, H2O; (j) H2, Pd(OH)2/C, MeOH (h–j, 42%).

Scheme 16

376

A new route through Mitsunobu epoxide formation (**58**), followed by regioselective epoxide ring opening with iodide ion, mesylation and reductive elimination, led to alkene (-)-**61** which was regioselectively *cis*-dihydroxylated to obtain (-)-**62**. Oxidative cleavage of (-)-**62** proceeded readily to afford dialdehyde **L-63** in nearly quantitative yield. Reductive amination of this product, followed by deprotection, gaved the previously reported tetrahydroxyazepane (+)-**9b**.[12,14b,15a] The new D-*ido*-azepane (-)-**9b** was prepared in a similar way starting from L-*chiro*-inositol (Scheme 17), with a significant improvement in the formation of **61**.

An alternative synthesis of (+)-**9b** was developed from the known tetra-O-benzyl-myo-inositol (-)-**64**.[33b] Oxidation with periodate afforded dialdehyde (+)-**65** that was subjected directly to reductive amination [NaBH(OAc)$_3$ gaved better results than NaCNBH$_3$ in the reduction step]. The pentabenzyl azepane (+)-**66** was isolated in good yield and deprotected to give (+)-**9b**, (Scheme 17). The azepane (-)-**9b** was synthesised by the same method from (+)-**64**.

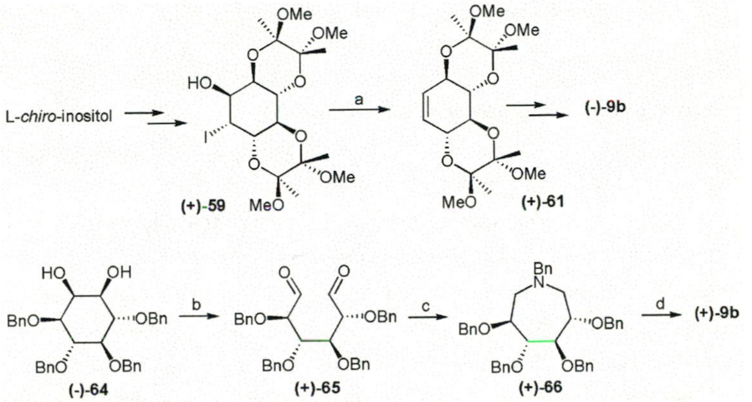

(a) i. Tf$_2$O, Py, DMF, ii. Zn, AcOH (i-ii, 95%); (b) NaIO$_4$/SiO$_2$, CH$_2$Cl$_2$; (c) NaBH(OAc)$_3$, BnNH$_2$, CH$_2$Cl$_2$, (b-c, 86%); (d) H$_2$, Pd(OH)$_2$/C, MeOH/AcOH (trace) (92%).

Scheme 17

The inhibitory properties of the four tetrahydroxyazepanes (-)-, (+)-**6b**, and (-)-, (+)-**9b** against eight enzymes were determined. The most active, but least selective isomer, the L-*ido*-compound (+)-**9b**, gave mainly analogous results to those already reported[12] for this inhibitor, showing activity with α- and β-glucosidase, mannosidase, α- and β-galactosidases and α-fucosidase, but showed no inhibition of β-D-N-acetylglucosaminidase. However, Wong *et al.*[12b] found that it also inhibited this enzyme (K_i = 22.7 μM). The enantiomer (-)-**9b** was much less potent. The L-*manno*-compound (+)-**6b** is a novel selective inhibitor of α-D-galactosidase, while the (-)-**6b** enantiomer inhibits α-L-fucosidase selectively. In some cases there is a strong correlation between the configurations of the inhibitors and those of the substrates of the inhibited enzymes, but this is not always true. Some of the apparent anomalies may be explained by the increased flexibility of tetrahydroxyazepanes as compared with pyranoses. Wong and colleages had also indicated[12a] that the introduction of hydrophobic substituents on the hydroxyl groups at the 3,6-positions of these compounds can alter their bioactivities making them protease rather than glycosidase inhibitors. This finding stimulated Painter´s group[31c] to prepare the 3,6-di-O-benzyl compounds **72** (Scheme 18) and **76**. (Scheme 19). The diallyl ether **67** was benzylated at the equatorial C-2, C-5 hydroxyl groups *via* a bis-stannylene intermediate, to give compound **68** which, as expected, was shown to adopt the $^4C_1$ chair conformation with the 1,6-diol oriented diaxially. Deallylation gave the dibenzyl ether **69** (32% from the diisopropylidene diol) that, alternatively, could be obtained in low yield, by reductive cleavage of the acetal groups of 1,2:5,6-di-*O*-benzylidene-D-*chiro*-inositol with DIBAL in dichloromethane. Tetraol **69** was selectively oxidised with sodium periodate at the diequatorial diol site to give dialdehyde **70** in high yield. Reductive amination followed by selective *N*-debenzylation afforded the desired dibenzyl ether **72**.

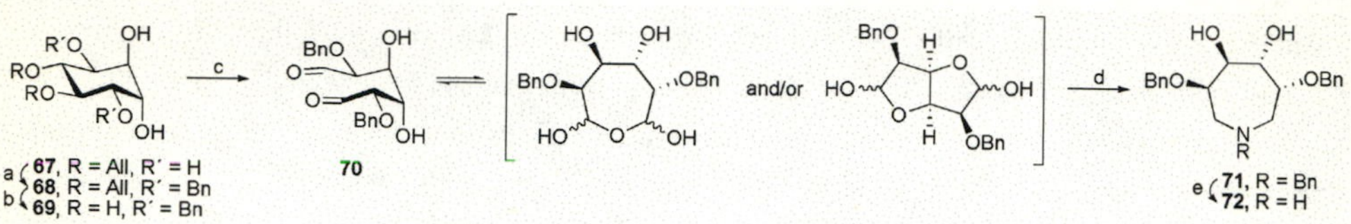

(a) (Bu₃Sn)₂O, PhMe, 110 °C, 1.5 h, then BnBr, Bu₄NBr, PhMe, 110 °C, 2 h, 53%; (b) Pd/C, TsOH, MeOH, 64 °C, 12 h, 60%; (c) NaIO₄, H₂O, MeCN, rt, 15 h, 86%; d) BnNH₂, NaCNBH₃, AcOH, sieves, MeOH, -78 °C-rt, 10 h, 60%, (e) Pd(OH)₂, aq. NH₃, MeOH, H₂, rt, 12 h, 57%.

Scheme 18

In order to obtain the tetrahydroxyazepane derivative **76**, the diallyldibenzyl compound **68**, which has a 1,6-diaxial diol in the preferred chair conformation, was tested for its susceptibility to periodate oxidation. Upon treatment with excess NaIO₄ in refluxing acetonitrile, the diol was very slowly cleaved affording 3,4-di-*O*-allyl-2,5-di-*O*-benzyl-L-*ido*-hexodialdose (**73**) but in only 35% yield (Scheme 19, route A). Reductive amination with benzylamine and sodium cyanoborohydride afforded the *N*-benzyl compound **74**. Deallylation and selective *N*-debenzylation gave the target L-*ido* azepane **76**. Given the propensity for bulky silyloxy groups to cause conformational change, the allyl protecting groups of compound **68** were exchanged for *t*-butyldimethylsilyl groups with the intention of increasing the susceptibility of the 1,6-diol to oxidative fission. To this end, diol **68** was converted to the dibenzoate **77**, which was deallylated with iridium (I) to give diol **78** (Scheme 19 route B).

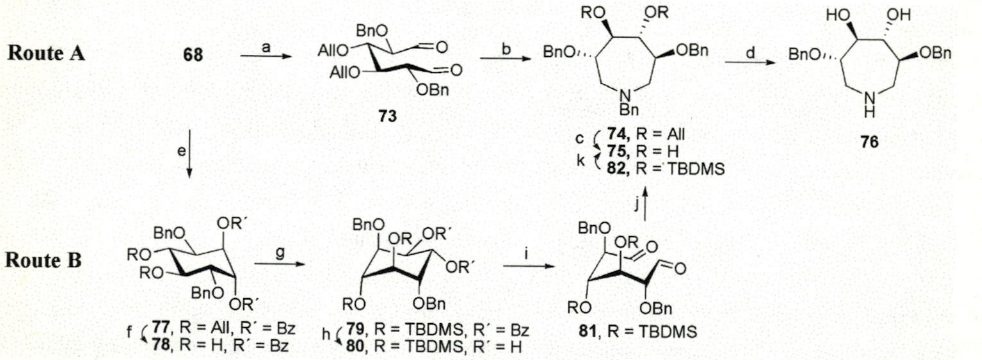

(a) NaIO₄, H₂O, MeCN, 90 °C, 20 h, 35%; (b) BnNH₂, NaCNBH₃, AcOH, sieves, MeOH, -78 °C-rt, 14 h, 58%; (c) Pd/C, TsOH, MeOH, 64 °C, 5 h, 50%; (d) Pd(OH)₂, NH₄OH, MeOH, H₂, rt, 2 h (74%); (e) BzCl, DMAP, py, 0 °C-rt, 0.5 h, 53%; (f) Ir(I), THF, rt, 1.5 h, then AcCl, MeOH, then Et₃N, 60%; (g) TBDMSCl, imidazole, DMF, 85 °C, 18 h, then TBDMSOTf, lutidine, 65 °C, 18 h, 60%; (h) MeONa, MeOH, rt, 18 h, 83%; (i) NaIO₄, H₂O, MeCN, 65 °C, 6 h, 82%; (j) BnNH₂, NaCNBH₃, AcOH, sieves, MeOH, -78 °C-rt, 18 h, 59%; (k) MeOH, conc HCl, rt, 2 h, 93%.

Scheme 19

378

The introduction of TBDMS group was difficult but, under forcing conditions, compound **79** was obtained. Finally, the benzoate groups were removed to afford diol **80**, which was oxidized efficiently albeit at elevated temperature (65 °C) with sodium periodate to dialdehyde **81** (82% yield). From this dialdehyde the L-*ido*-tetrahydroxyazepanes **82** and hence **75** were made by normal reductive amination followed by desilylation. The route B to compound **76** is longer but more efficient than route A.

2.4. Synthesis from glycosyl enamines

Fuentes *et al.*[34] have prepared polyhydroxyperhydroazepines starting from *N*-diethoxycarbonylvinyl glycosylamines. Glycosylenamines **83–96** are suitable for the formation of the 1,6-anhydro azasugars **97–106**. This method to form the C6-N bond is a shorter valuable alternative to the SN2 displacement of a sulphonyloxy or cyclic sulphate group with sodium azide. The β-D-aldopyranosylenamines **83–85** with D-*gluco*, D-*galacto*, and D-*manno* configurations[35] and the partially *O*-acylated β-D-glucosylenamines **89** and **90**, prepared[34c] from **83**, were chosen as starting materials. The 6-*O*-mesyl derivatives **86–88, 91, 92, 94** and **96** were prepared from the corresponding *N*-diethoxycarbonylvinyl-β-D-glycopyranosyl amines by per-*O*-mesylation with mesyl chloride (**86–88** from **83–85**), or 6-*O*-mesylation (**91, 92** from **89, 90**; and **94, 96** from **93, 95**, respectively).

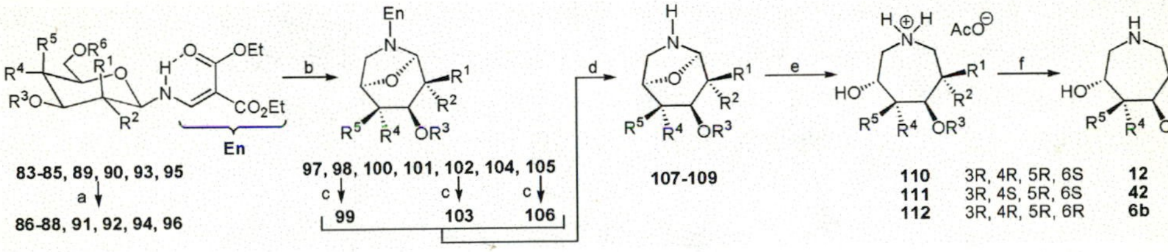

Scheme 20 (chemical reaction scheme with structures labeled 83-85, 89, 90, 93, 95; 97, 98, 100, 101, 102, 104, 105; 99, 103, 106; 107-109; 110, 111, 112 and conditions a–f)

(a) MsCl, py; (b) MeONa/DMF, rt, 20 mm Hg, 15 min; (c) MeONa/MeOH, rt, 4 h; (d) Cl₂/CH₂Cl₂, rt, 15 min; (e) NaCNBH₃/AcOH, rt, 24 h; (f) Dowex-50W. NH₄⁺ form. 1x20 cm.

Scheme 20

Table 4.

Compounds	R¹	R²	R³	R⁴	R⁵	R⁶ (for 83-96)	83-109
83ᵃ, 99, 107, 110, 12	H	OH	H	OH	H	H	D-*gluco*
84ᵃ, 103, 108, 111, 42	H	OH	H	H	OH	H	D-*galacto*
85ᵃ, 106, 109, 112, 6b	OH	H	H	OH	H	H	D-*manno*
86, 97	H	OMs	Ms	OMs	H	Ms	D-*gluco*
87, 101	H	OMs	Ms	H	OMs	Ms	D-*galacto*
88, 104	OMs	H	Ms	OMs	H	Ms	D-*manno*
89ᵇ, 98	H	OBz	Bz	OBz	H	H	D-*gluco*
90ᵇ,	H	OAc	Ac	OAc	H	H	D-*gluco*
91	H	OBz	Bz	OBz	H	Ms	D-*gluco*
92, 100	H	OAc	Ac	OAc	H	Ms	D-*gluco*
93, 102	H	OBz	Bz	H	OBz	H	D-*galacto*
94	H	OBz	Bz	H	OBz	Ms	D-*galacto*
95, 105	OBz	H	Bz	OBz	H	H	D-*manno*
96	OBz	H	Bz	OBz	H	Ms	D-*manno*

ᵃ See ref 35. ᵇ See ref 34c.

Intramolecular nucleophilic displacement of the 6-OMs group in basic media gave the 1,6-iminosugar derivatives (**97-105**). Debenzoylation of **98, 102,** and **105** afforded **99, 103,** and **106**, respectively, which

379

were *N*-deprotected with chlorine in chloroform to give the corresponding derivatives **107–109** in virtually quantitative yield. Compounds **107–109** were characterized by their FABMS spectra and used directly in the reduction step to obtaine the iminocyclitol ammonium acetates **110–112**. A possible mechanism for the formation of **110–112** was proposed. Treatment of **110–112** with Dowex–50W resin produced the target polyhydroxyazepanes **12**, **42** and **6b**. Attempts to carry out similar sequences from **97**, **100**, **101**, and **104** were either low-yielding or unsuccessful. A detailed NMR study for all of the synthesizing products was made. The inhibitory potential of iminocyclitols (**97**, **99**, **103**, and **106**), was determined on various glycosidases (α-glucosidase, β-galactosidase and β-glucosidase). All of the compounds assayed were either inactive or only weak inhibitors of glycosidases, even in the millimolar range. For β-galactosidase, the most active inhibitors were **103**, **99**, and **106**, whereas for β-glucosidase, the most active inhibitors were **97** and **106**.

2.5. Synthesis of lactams as azepane precursors

Sugar lactams having 7-membered rings[36] were prepared by reductive cyclisation of appropriate azido lactones or carboxylic acids and are potential precursors of polyhydroxyazepanes. Although some of these lactams have not been reduced to the corresponding azepanes, we consider them of interest as key compounds in the synthesis of seven-membered iminocyclitols.

Fleet *et al.*[37] prepared the seven-membered polyhydroxylactam **116** from product **113a**, obtained from D-galactono-1,4-lactone in 78% yield. Mesylation of **113a**, then azido substitution, hydrolysis of acetonide groups and catalytic hydrogenation afford to **116** (Scheme 21).

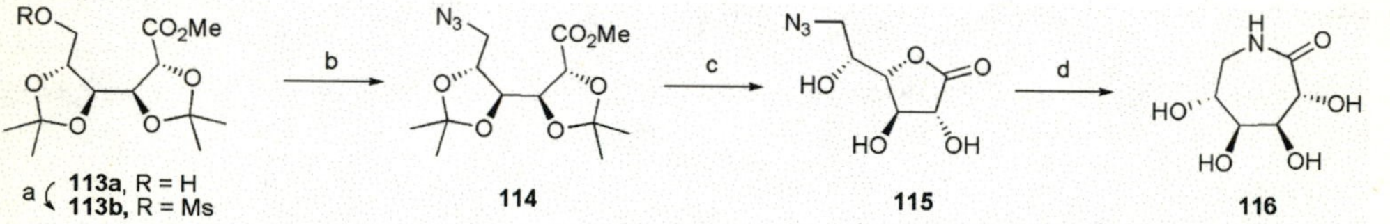

(a) MsCl, py, DMAP; (b) NaN$_3$, DMF; (c) aq. TFA; (d) H$_2$, Pd-black

Scheme 21

Other lactams were also synthesised from lactones by Chittenden´s group[38] and were reduced to the corresponding azepanes (Scheme 22).

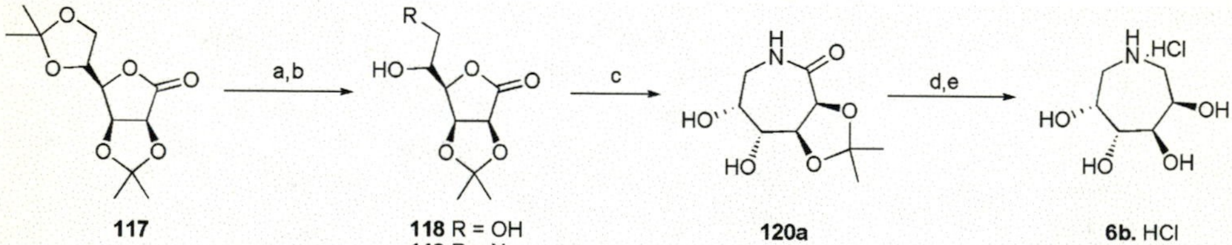

(a) AcOH, 81.5%; (b) Ph$_3$P/CBr$_4$, LiN$_3$, DMF, 72%; (c) H$_2$, Pd/C, EtOH, 69%; (d) BH$_3$-THF, THF; (e) HCl, MeOH, (d-e, 80%).

Scheme 22

The acetonide of D-mannono 1,4-lactone **117** was the key starting material for the synthesis of D-mannoazepane **6b**. Partial hydrolysis of **117** and one pot introduction of an azide function at C-6, gave **119**.

Catalytic hydrogenation proceeded with concomitant ring closure to produce the crystalline compound **120a**. Borane reduction of lactam **120a** followed by treatment with conc. HCl, gave **6b** as the crystalline hydrochloride salt.

The epimeric azepane **12** was prepared from D-glucono-1,5-lactone **121a** *via* the acetylated polyhydroxy ester **122** (Scheme 23). Treatment of the ester **122** with TMSCl/NaI, yielded the primary iodide **123**, which was converted to the crystaline azide **124**. Deacetylation and reductive cyclisation of the resulting crude product gave the expected pure lactam **126a**. In another experiment, reduction of **124** at lower pressure afforded the known uncyclised 6-amino-6-deoxy-D-gluconic acid **125**. The lactam **126a** was converted to the soluble, base stable derivative **126b** which was reduced with BH$_3$-Me$_2$S in boiling THF and treated with conc. HCl/MeOH to give pure **12** as a non crystalline hydrochloride salt. This azepane **12** (*S,R,R,R*) is a better inhibitor of β-*N*-acetylglucosaminidase than DNJ, which is often taken as a standard.

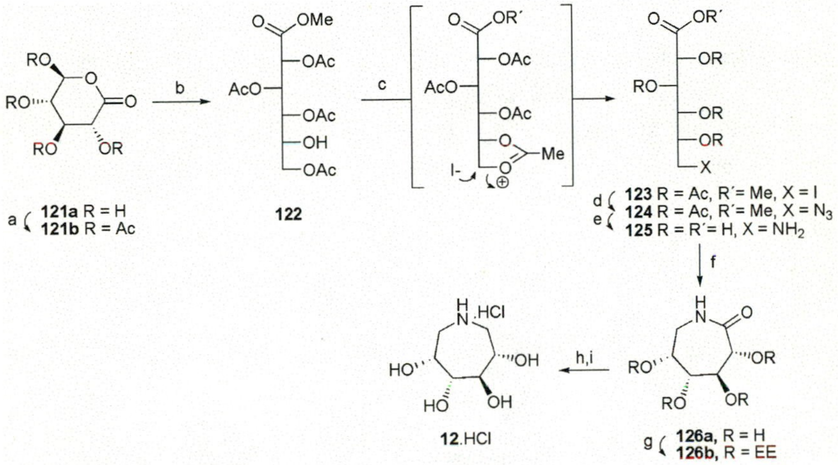

(a) Ac$_2$O/TFA; (b) MeOH, TsOH, 97%; (c) Me$_3$SiCl/NaI, CH$_2$Cl$_2$, 87%; (d) LiN$_3$/DMF; (e) i. KCN/MeOH, ii. H$_2$, 1 atm, Pd/C, EtOH; (f) H$_2$, 3 atm, 10% Pd/C, EtOH, (g) ethyl vinyl ether/DMF, TsOH; (h) 1M BH$_3$-Me$_2$S, THF; (i) conc. HCl/MeOH, (h-i, 67%).

Scheme 23

Recently, a direct and improved synthetic route to 6-amino-6-deoxy-D-galactono- and D-mannono-1,6-lactams (**116**) and (**120b**) has been reported[39] from unprotected D-galactono- and D-mannono-1,4-lactones, in three steps (Schemes 24 and 25). The selective bromination of the primary hydroxyl group in **127** using PPh$_3$/CBr$_4$ in pyridine gave the 6-bromo lactone **128** in 82% yield. Introduction of azide group, followed by catalytic hydrogenation of **129**, produced quantitatively the desired lactam **116**.

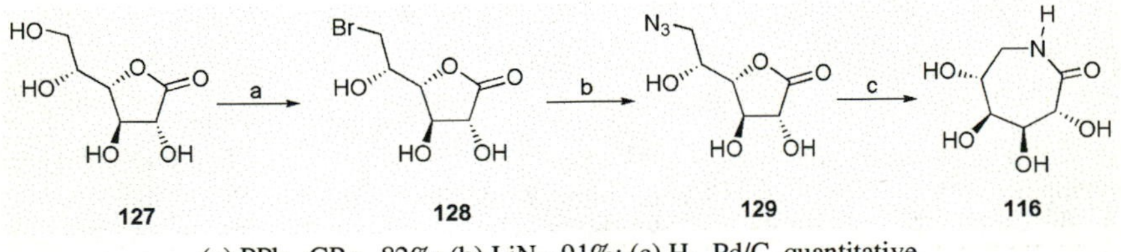

(a) PPh$_3$, CBr$_4$, 82%; (b) LiN$_3$, 91%; (c) H$_2$, Pd/C, quantitative.

Scheme 24

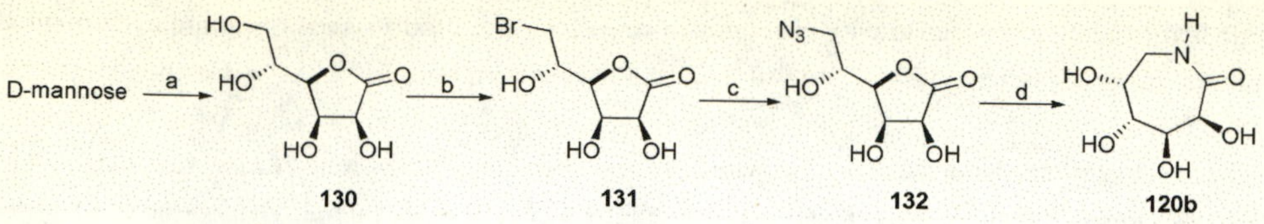

(a) Br_2, H_2O, $NaHCO_3$, quantitative; (b) PPh_3, CBr_4, 69%; (c) LiN_3, 98%; (d) H_2, Pd/C, quantitative.

Scheme 25

In order to obtain the diastereoisomeric lactam **120b**, a similar procedure was developed starting from readily available D-mannono 1,4-lactone (**130**). This lactone was isolated in quantitative yield by D-mannose oxidation, using bromine and sodium hydrogen carbonate. Bromination of **130**, azido introduction and catalytic hydrogenation produced the lactam **120b** in good yield.

2.6. Synthesis *via* ring closing methatesis

Diverse stereoisomers of seven-membered iminocyclitols were prepared in good yield using ring-closing olefin metathesis (RCM) as a key step followed by dihydroxylation of the ring double bond.[40] The azepane precursors were obtained from L-serine through the sequence depicted in Scheme 26.

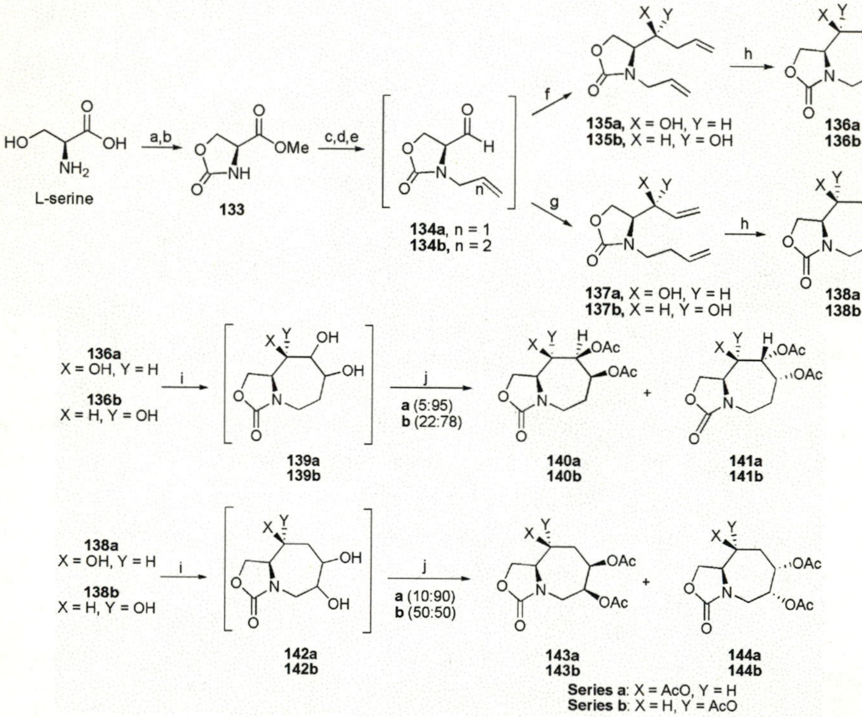

(a) $SOCl_2$, MeOH, reflux, overnight; (b) K_2CO_3, triphosgene, H_2O/toluene, 87%; (c) 60% NaH, allyl bromide or homoallyl bromide, DMF; (d) $NaBH_4$, MeOH; (e) DMSO, $(COCl)_2$, CH_2Cl_2, Et_3N; (f) allyl magnesium bromide; (h) Grubb's catalyst (10 mol%) CH_2Cl_2, reflux, 12 h; (i) OsO_4, NMO, acetone/H_2O, 8:1; (j) Ac_2O, py, DMAP, CH_2Cl_2, rt; (i-j): 80-86%).

Scheme 26

382

Aldehydes **134a** and **134b** were synthesised in several steps (esterification, oxazolidine formation, *N*-alkenylation, reduction and oxidation). Addition of alkenyl Grignard reagents to the crude aldehydes **134a** or **134b**, led to the corresponding alcohols **135a** and **135b** or **137a** and **137b**, which were cyclised using Grubb's catalyst, Cl$_2$(PCy$_3$)$_2$Ru=CHPh, to furnish the required intermediates, **136a** and **136b** or **138a** and **138b**. Dihydroxylation and acetylation afforded triacetates **140-144**. The configurations of compounds **140** and **141** were determined by ^{1}H-NMR, while those of **143b** and **144b** were assigned by comparing their ^{1}H-NMR with those of their enantiomers, obtained from D-serine. Moreover, X-ray crystallographic analysis was used to determine the stereochemistry of **143** and **144**. Dihydroxylation of allylic alcohols of cyclic compounds using OsO$_4$/NMO generally results in dihydroxylation with *anti*-diastereofacial selectivity relative to the preexisting hydroxyl group (or heteroatom) on the adjacent carbon atom. This appears to be the case for **136a** and **138a** whereas poor diastereofacial selectivities were observed for dihydroxylation of **136b** and **138b**, where the two heteroatoms, O and N have an *anti*-relationship.

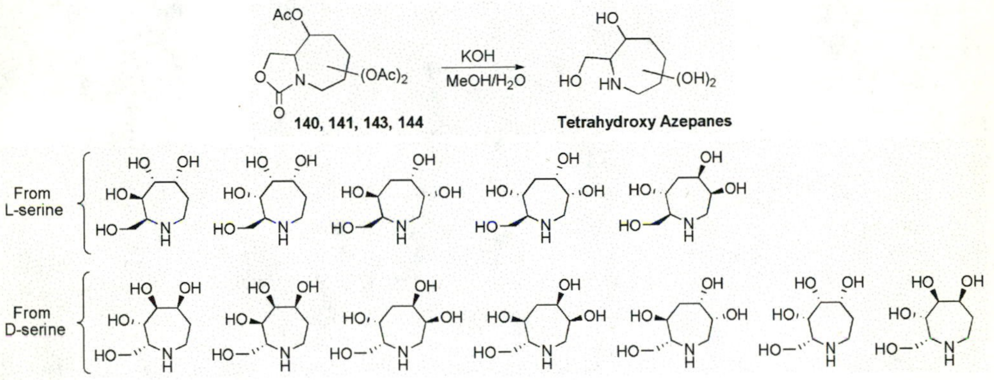

Scheme 27

Although these hydroxymethyl derivatives have been included in this part of the review because of their tetrahydroxy functionalisation, it is to be noted that azepanes with hydroxy methyl group have been previously reported by Martin[41] (pentahydroxyazepanes, see Section **6**) and by Mehta[42] (trihydroxy azepanes, Section **4**).

3. Synthesis of trihydroxyamino azepanes

Among the earliest publications about polyhydroxyazepanes, the work of Farr *et al.*[43] is to be mentioned. These authors synthesised a trihydroxyaminoazepane from Fleet's azido alcohol **145**.[44] Hydrogenation of **145** with Pd/C gave the amine **146a** which was protected as its Boc derivative **146b** (Scheme 28). Mesylation of the primary alcohol and displacement of the mesylate with sodium azide in DMF gave the azido ether **147b**. Oxidation of the benzyl ether followed by saponification of the resulting benzoate, gave lactol **148b**. Surprisingly, catalytic hydrogenation of **148b** with Pd black gave the stable bicyclic hemiaminal **149**. This same hemiaminal was prepared directly from azido ether **147b** by catalytic hydrogenation in HOAc using Pd/C, but the yield of this conversion was low. Reductive ring opening of the hemiaminal **149** with NaBH$_3$CN in HOAc gave the protected hexahydro-1-*H*-azepine **150** in 93% yield. Deprotection with methanolic HCl gave **151** as the dihydrochloride salt. The ^{1}H-NMR of **151**.2HCl showed the seven-membered ring adopted a half chair conformation with the amino group *equatorial*, The conformation in solution showed no significant change when **151**.2HCl was converted to the free base. Subsequent calculations using the SPARTAN electronic structure program confirmed that either as the monohydrochloride or as the free base, the conformation with the amino group *equatorial* was 1.4-1.6 kcal/mol more stable than the conformation with the amino group *axial*. Unfortunately, hexahydro-1*H*-azepine **151** (either as the hydrochloride or the free base) was inactive against jack-bean α-mannosidase.

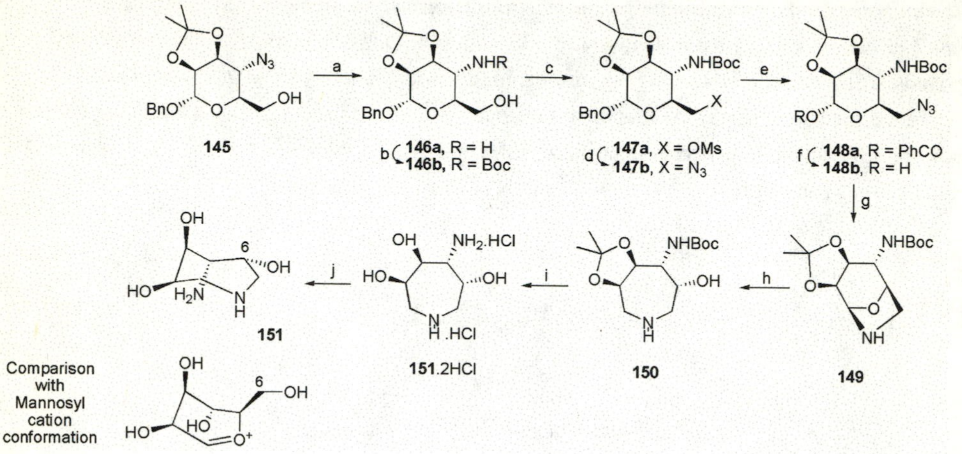

(a) H$_2$, Pd/C, ref. 44; (b) (Boc)$_2$O, THF; (c) MsCl, DMAP, py, CH$_2$Cl$_2$; (d) NaN$_3$, DMF; (e) NaIO$_4$/RuO$_2$xH$_2$O; (f) MeONa/MeOH; (g) H$_2$, Pd/C, EtOH; (h) NaBH$_3$CN, AcOH; (i) HCl/MeOH; (j) 30% NaOH.

Scheme 28

The lack of activity of **151** may be a result of its solution conformation which is not recognized by the enzyme, even though the ring could flip into the modeled conformation in the enzyme active site. Alternatively, the lack of perfect correlation of the 6-OH of the mannosyl cation and the corresponding 6-OH of hexahydro-1*H*-azepine **151** may be crucial. Winkler and Holan conclude that the "equivalent of the 6-OH appears to assist in binding of inhibitors into the active site, but it is not essential for activity".[45] Farr assumed that the position of the equivalent of the 6-OH group of the mannosyl cation in a potential inhibitor is a critical determinant of activity and that a better predictor of α-mannosidase activity is the close correlation of this 6-OH group with the 8-OH group of the very potent α-mannosidase inhibitor Swainsonine (Figure 1). A 6-acetamidoiminocyclitol, α-fucosidase inhibitor, was obtained[12a] from *N*-acetylglucosamine. A trihydroxyaminoazepane derivative (see Figure 3B[16]) was prepared as potential non-peptide mimic of somatostatin/Sandostatin® and other trihydroxyaminoazepanes were chemo-enzymatically synthesised (Section 4, Table 6[29]). In the last work, the biological activity was not determined.

4. Synthesis of trihydroxyazepanes

Several conceptually different syntheses of trihydroxyazepanes have been reported: intramolecular reductive amination from azido or amino compounds,[10a,46,47] from oximes,[29] lactam reduction,[46] nitrone addition to unsaturated sugar derivatives[48] and synthesis from norbornyl compounds.[42]

Lundt and colleagues[46] designed a route to prepare two diastereomeric trideoxy-1,6-iminohexitols **154** and **158** (Scheme 29), which are structurally related to established inhibitors of D-glycosidases such as

1-deoxymannojirimycin (**DMJ**) and swainsonine (Figure 1). Furthermore, two additional epimers (compounds **159** and **163**) and a regioisomer (compound **165**) were synthesised in order to estimate the importance for biological activity of the presence and orientation of different hydroxyl groups on the ring. The starting products in this methodology, 6-azido-3,6-dideoxysugars, can be easily prepared from partially protected 3-deoxysugars. For example, the required 3-deoxy-D-*ribo*-hexose (**152**)[49] is readily obtained from 1,2:5,6-di-*O*-isopropylidene-α-D-glucofuranose. Other 1,2:5,6-protected 3-deoxyhexoses are more conveniently prepared by conventional reduction of 3-deoxy-hexono-1,4-lactones, followed by protection of the resulting free 3-deoxyaldoses. Regioselective deprotection of the exocyclic diol in the 1,2:5,6-di-*O*-isopropylidene-protected 3-deoxyhexoses **152**, **156R**, **156S** and **161**, followed by 6-*O*-sulfonylation, gave the primary tosylates **153b**, **157b(R,S)** and **162b**, respectively. Subsequent displacement with azide ion led to the corresponding 6-azidodeoxy derivatives **153c**, **157c(R,S)** and **162c**, which furnished the corresponding free 6-azido-3,6-dideoxyhexoses with D-*ribo*, D-*arabino*, D-*xylo*, and L-*xylo* configurations, respectively, in good yields. Catalytic reduction using hydrogen in the presence of palladium-on-charcoal (5%) and concomitant intramolecular reductive amination of the intermediary 6-aminodeoxy sugars led to the desired 2,4,5-trihydroxyazepanes **154**, **158**, **159** and **163** (Scheme 29).

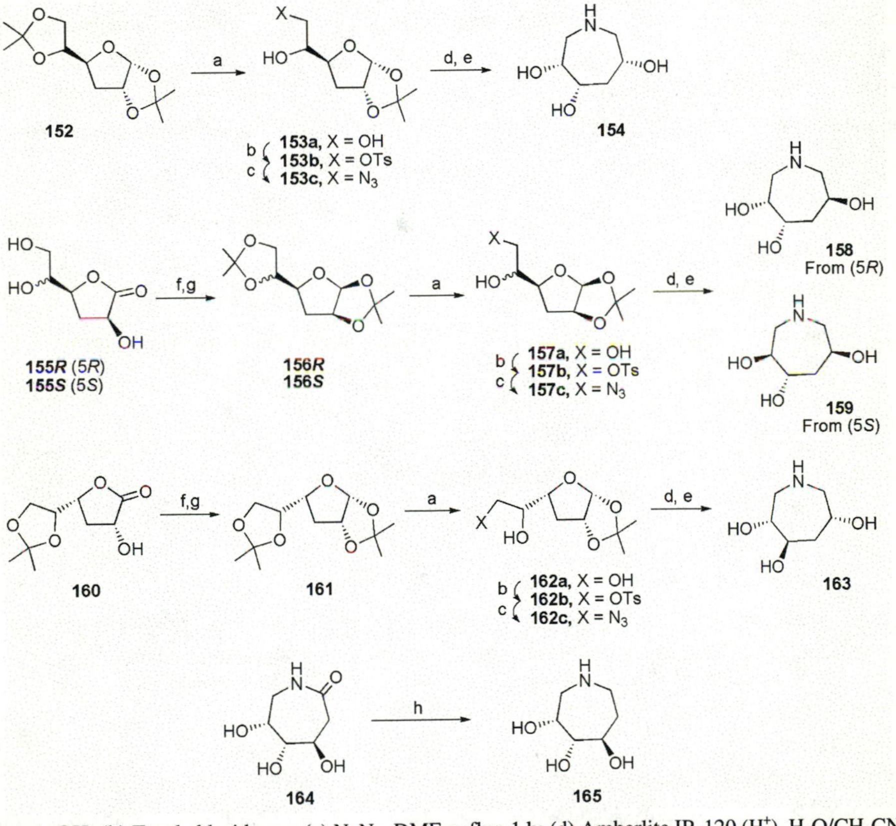

(a) 90% aq AcOH; (b) Tosyl chloride, py; (c) NaN$_3$, DMF, reflux 1 h; (d) Amberlite IR-120 (H$^+$), H$_2$O/CH$_3$CN, 40 °C; (e) 5% Pd/C, H$_2$, MeOH; (f) Disiamylborane, THF; (g) Acetone, camphor sulfonic acid (cat); (h) BH$_3$-Me$_2$S.

Scheme 29

The 3,4,5-trihydroxyazepane **165**, a regioisomer of **158** with D-*arabino* configuration, was prepared by reduction of the lactam **164**, readily obtainable from 2,6-dibromo-2,6-dideoxy-D-mannolactone.[50]

Inhibition of glycosidases.- Trihydroxyazepanes **154, 158, 159** and **163** did not exhibit any appreciable inhibitory activity against a variety of glycosidases. Apart from the lack of symmetry as compared with active azepanes such as **12c**, the preferred conformations neither match the motifs of substrates, nor those of proven inhibitors or proposed transition states such as the putative 'flap-up' mannopyranosyl oxocarbonium ion. Azepane **165**, which is a good inhibitor, can adopt conformations that can be superimposed to varying degrees with established glycosidase inhibitors. Superposition of azepane **165** with the glucosidase inhibitor 1-deoxynojirimycin (**DNJ**, Figure 1) shows good alignment of the secondary hydroxyl groups of both molecules. These authors concluded that the selectivity of **165** for α-glucosidases could be related to the lack of a hydroxyl group matching OH-6 of **DNJ**, and in these azepanes, three vicinal hydroxyl groups appeared to be essential for basic glycosidase inhibitory activity.

Gallos *et al.*[48] reported the 1,3-dipolar cycloaddition of the nitrone $CH_2=N(\rightarrow O)Bn$ to pent-4-enofuranoside **166**, readily available from D-ribose in three steps, giving cycloadduct **167** as a single diastereomer (Scheme 30). Raney nickel hydrogenation of **167** afforded **168**, which was treated with $NaCNBH_3$ to afford **169** in good yield. In order to prepare a bicyclic imino cyclitol, cyclic nitrones were used. Good diastereoselectivity was observed in the addition of the chiral nitrone **170** to **166**, giving **171** in good yield, together with the C-3a epimer, (8.5:1). Further hydrogenation gave directly the desired protected octahydropyrrolo[1,2-a]azepinepentol **173** (56%), together with its ketone precursor **172** (43%). The latter was quantitatively reduced to **173** with $NaBH_4$ and isolated as an inseparable mixture of epimers in ca. 1:1 ratio, due to the non-selective reduction of the carbonyl group of **172** by Raney Ni or $NaBH_4$.

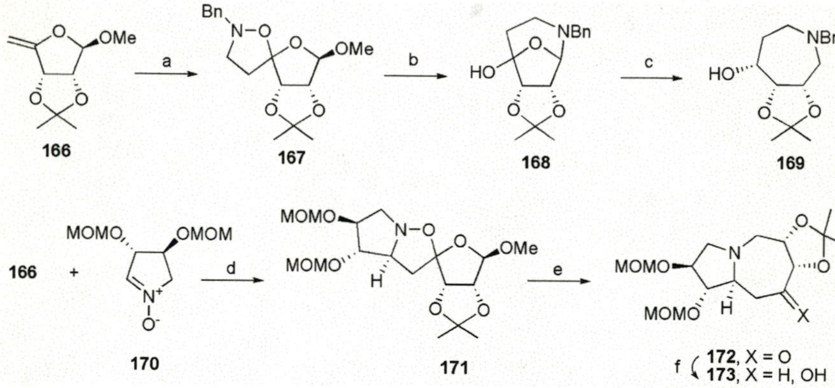

(a) $CH_2=N(\rightarrow O)Bn$, toluene, reflux, 72 h, 59% (30% of **166** recovered); (b) Raney Ni, H_2, MeOH, H_3BO_3 (20 eq), $MgSO_4$, 20 °C, 2 h, 56%; (c) $NaBH_3CN$, AcOH (gl), 20 °C, 24 h, 49%; (d) toluene, reflux, 48 h, 96%; (e) *idem* as (b) 24 h, 43% of **172** and 56% of **173**; (f) $NaBH_4$, EtOH, 20 °C, 4 h, 98% (1:1).

Scheme 30

A variant of the seven-membered iminocyclitols featuring an hydroxymethyl arm as additional binding site was prepared from the norbornyl framework.[42] The cyclohexenoid **175** was obtained by fragmentation of **174a**.[51] Elaboration of **175** to azepanes required oxidative cleavage of the cyclohexanoid ring to set-up either inter- and intramolecular *N*-alkylations or a double reductive amination (Scheme 31). Compound **175** was

386

transformed to the diol **176** *via* LiAlH$_4$ reduction, hydroxyl group protection and stereoselective dihydroxylation. Oxidative cleavage to the dialdehyde, reduction, mesylation and treatment with p-toluenesulfonamide under phase transfer conditions furnished the azepane **177a**. *N*-tosyl group reduction in **177a** and further acetylation permitted the characterisation of **177b** as a mixture of two rotamers. Deprotection of **177b** led to a homoisofagomine derivative **178a**. An isomeric derivative was also prepared from **175** through a tactically altered route. Reduction of the ester function in compound **175**, dihydroxylation and acetylation gave **179**, which was subjected to acetonide deprotection and then to periodate cleavage to deliver a dialdehyde, which was transformed by a double-reductive amination to the azepane **180** as the major product. Acetate hydrolysis in **180** gave the *N*-benzyl azepane and reductive removal of the benzyl group and deprotection gave the trihydroxyazepane **181b**. Another related route from the acetate **174b** started with a regioselective Baeyer-Villiger oxidation (87:13 mixture of regioisomers).[52] Reduction, acetylation and acetonide deprotection in **182** led to diol **183**. Oxidation of diol **183** to the dialdehyde, followed by double reductive amination with benzyl amine gave the triacetate **184** as the major product. Acetate hydrolysis of **184** gave the *N*-benzyl trihydroxy derivative **185a**, while reductive removal of benzyl group and acetate deprotection furnished the trihydroxyazepane **185b**.

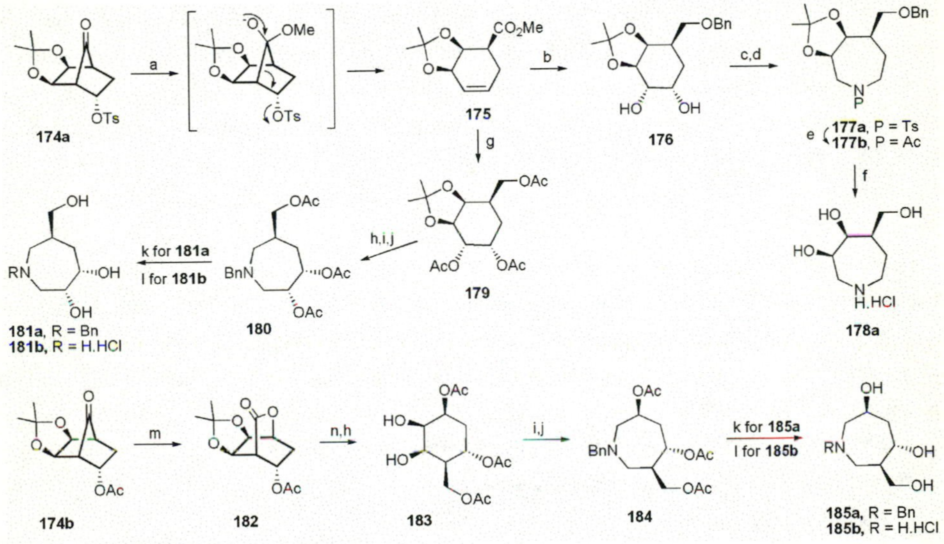

(a) Ref 51; (b) i. LAH, THF, 0-5 °C, 1 h, 90%, ii. BnBr, NaH, THF, rt, 6 h, 95%, iii. OsO$_4$, NMMO, Me$_2$CO-H$_2$O (4:1), rt, overnight, 83%; (c) i. NaIO$_4$-silicagel, DCM, 0 °C, 2 h, ii. NaBH$_4$, MeOH, 0 °C, 1 h, iii. MsCl, Et$_3$N, DCM, 0 °C, 1.5 h, 55% (three steps); (d) p-TsNH$_2$, KOH, TBAI, C$_6$H$_6$-H$_2$O (9:1), reflux, 24 h, 60%; (e) i. Na-naphthaledine, DME, -60 °C, 1 h, ii. Ac$_2$O, py, rt, overnight, 67% (two steps); (f) i. H$_2$, Pd/C, EtOH, 12 h, ii. 1N HCl, 90%, 85% (two steps); (g) See (b) i. and iii., then Ac$_2$O, DMAP, DCM, rt, 10 h, 78%; (h) Amberlyst-15, THF-H$_2$O (2:3), rt, 12 h, 85%; (i) NaIO$_4$-silicagel, DCM, 0 °C, 2 h; (j) BnNH$_2$, AcOH, NaCNBH$_3$, MeOH, 12-15 h, 30-35% (two steps); (k) K$_2$CO$_3$, MeOH, rt, 6 h, 95%; (l) i. H$_2$, Pd/C, EtOH, rt, 5h, ii. 1N HCl, 90 °C, 24 h, 90%; (m) Ref 52; (n) See (b) i. 3 h, 70%, then Ac$_2$O, DMAP, DCM, rt, 12 h, 89%.

Scheme 31

The new homoisofagomine analogues **178a**, **181a**, **181b**, **185a** and **185b** were assayed for glycosidase inhibition against a set of six commonly used enzymes (α- and β-glucosidases, galactosidases and mannosidases) following standard protocols. All the substrates examined exhibited weak but selective inhibition with **181a** and **181b** showing relatively better activity against β-glucosidase (K_i = 470 μM) and α-galactosidase (K_i = 600 μM), respectively. Similar selectivity was observed in the case of **185a** and **185b**, with the former exhibiting weak inhibition of β-glucosidase and the latter inhibiting α-glucosidase. These results reveal the importance of *N*-substitution in modulating selectivity and inhibition efficacy.

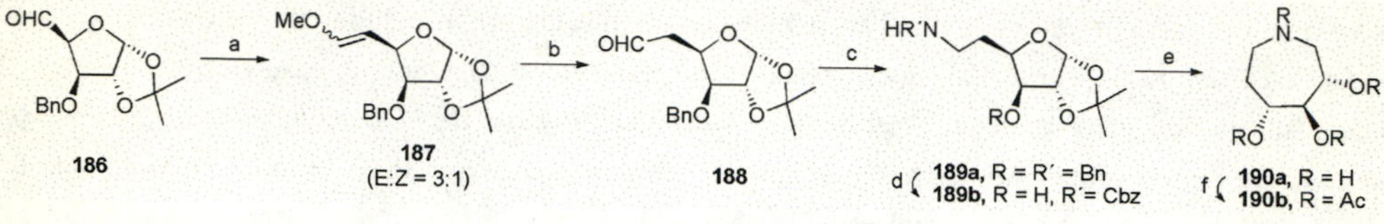

(a) *t*-BuOK (1.5 eq), MeOCH$_2$PPh$_3$Cl (1.2 equiv.), *t*-BuOH/THF, –40 °C, 2 h, 78%; (b) 3N HCl/THF (1:10), 25 °C, 3 h, 88%; (c) BnNH$_2$ (1.1 equiv.), NaCNBH$_3$ (1.5 eq), cat. CH$_3$CO$_2$H, MeOH, –78 °C, 2 h then 25 °C, 24 h, 86%; (d) i. HCO$_2$NH$_4$ (7 eq), 10% Pd/C, MeOH, reflux, 40 min, ii. ClCO$_2$Bn (1.5 eq), NaHCO$_3$ (2.8 eq), MeOH/H$_2$O, 0–25 °C, 2 h, 76%; (e) i. TFA/H$_2$O (3:2), 25 °C, 2 h, ii. 10% Pd–C, MeOH, H$_2$, 80 psi, 25 °C, 24 h, 85%; (f) Ac$_2$O, py, 25 °C, 72 h, 45%.

Scheme 32

Another efficient method[47] for the synthesis of a trihydroxyazepane **190** was based on the reductive amination of a suitably protected 6-amino 1,4-furanose, which can be obtained by a one carbon homologation of 3-*O*-benzyl-1,2-*O*-isopropylidene-α-D-*xylo*-pentodialdose **186** using the Wittig olefination and further transformations. Thus, reaction of compound **186** with the Wittig reagent, prepared from methoxymethyltriphenylphosphonium chloride and potassium-*t*-butoxide in *t*-butanol–THF, afforded a geometrical mixture of (*E*)-**187** and (*Z*)-**187** (*E:Z* = 3:1) which on mild acid hydrolysis gave aldehyde **188** (Scheme 32). Reductive amination of **188** using *N*-benzylamine and sodium cyanoborohydride afforded amino derivative **189a**. Hydrogenolytic removal of the *N*- and *O*-benzyl protecting groups in **189a** with ammonium formate and 10% Pd–C in methanol gave an amino alcohol, which was directly subjected to selective *N*-Cbz protection with benzyl chloroformate to give **189b**. Treatment of **189b** with TFA–water afforded an anomeric mixture of hemiacetals, which on hydrogenation using 10% Pd–C in methanol afforded trihydroxyazepane **190a** as a thick oil. The ^{1}H and ^{13}C NMR spectra and analytical data were in agreement with the proposed structure. The compound **190a**[10a] was treated with acetic anhydride in pyridine for three days to obtain the tetra-acetyl derivative **190b** as a solid.

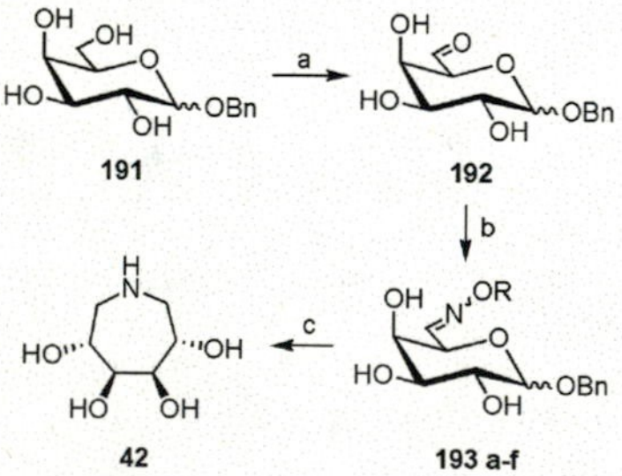

Table 5 Hydrogenolysis of **193 a-f**

Entry	Compound	R	Yield[c]
1	193a[b]	H	98%
2	193b	Me	98%
3	193c	Et	98%
4	193d	*t*-Bu	98%
5	193e	Bn	98%
6[a]	193f	Allyl	90%

[a] Hydrogenolysis halted at 24 h.
[b] Completed on a 5 g scale.
[c] Yield without further purification

(a) Galactose Oxidase Catalase, CuSO$_4$ 5H$_2$O, 50 mM Phosphate buffer pH 7.0; (b) RCHO:NH$_2$OR 1:1 w/w, MeOH/pyridine; (c) H$_2$, (5% w/w) Pd(OH)$_2$/C Degussa type, MeOH : H$_2$O : THF, 4 : 4 : 1, 60 psi/2 days.

Scheme 33

388

A variety of polyhydroxylated azepanes were synthesised by a chemo-enzymatic method.[29] Several benzyl pyranosides were chosen as starting products and easily prepared from aldopyranoses according to two different procedures in good to excellent yields.[53] In Scheme 33 is depicted the synthesis of azepane **42** from benzyl galactopyranoside **191**. The enzymatic oxidation reaction with Galactose oxidase (GAO) that selectively oxidizes exposed primary hydroxyl groups in nonreducing galactoses, permitted the formation of C-6 aldehydes. This reaction was monitored by NMR. To test the feasibility of generating seven membered rings, a series of benzyl-D-galactose C-6 oximes were synthesised (Table 5, entries 1–6). Table 6 shows the conversion of several GAO substrates to different polyhydroxyazepanes obtained by these authors. The claimed advantages for this method were: inexpensive starting materials, few synthetic steps, aqueous media and the possibility of proceeding on gram quantities.

Table 6. Substrate conversions to polyhydroxyazepanes.

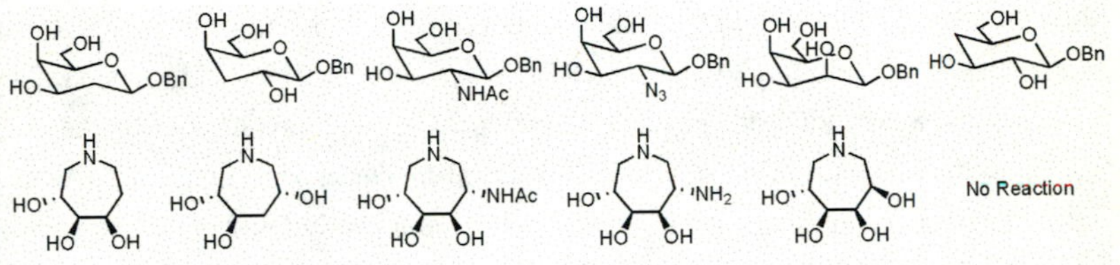

Substrate						
Product						No Reaction
$[\alpha]_D^{22}$	+40	-16	-9	-14	+30	
Yield without purif.	94(%)	92(%)	80(%)	75(%)	97(%)	

6. Synthesis of pentahydroxyazepanes

The first synthesis of pentahydroxyazepane derivatives was described by Martin[41] and more recently a limited number of procedures have been described.[54,55,56a]

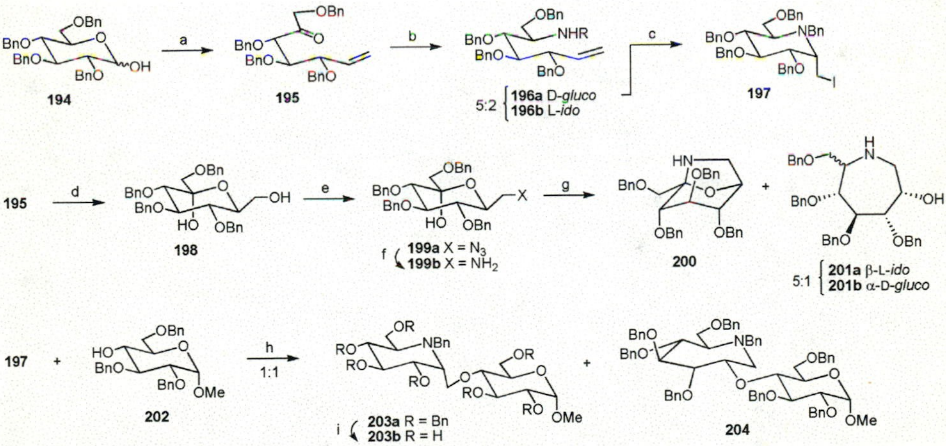

(a) i. Ph$_3$P=CH$_2$; ii. Swern oxid, 87%; (b) BnNH$_2$-AcOH, NaBH$_3$CN, MeOH, 80%; (c) NIS, CH$_2$Cl$_2$, 83%; (d) OsO$_4$, NMO; (e) i. TsCl; ii. LiN$_3$; (f) Ph$_3$P, H$_2$O; (g) NaBH$_3$CN, MeOH, 52%; (h) NaH, DMF, 60%; (i) H$_2$, Pd/C.

Scheme 34

These 1,6-dideoxy-1,6-iminoheptitols could be potential glycosidase inhibitors, because of the combination of azepane ring flexibility and –CH$_2$OH functionality. The key compound in the synthesis of

389

the azepanes **201**, was a 7-amino-7-deoxy-2-octulose derivative **199b** (Scheme 34).[41] Compound **199b** was obtained by way of the highly stereoselective hydroxylation of the alkenyl function of the unsaturated ketone **195**, followed by a functional group exchange. The reaction of **199b** with NaBH$_3$CN in MeOH gave a mixture of the internal O,N-acetal **200** and the desired azepane derivatives, the L-epimer **201a** (L-*glycero*-D-*gulo*, equivalent to "β-L-*ido*") being predominant. Compounds **201a** and **201b**, with a free OH group at C3, are useful precursors for the synthesis of a wide variety of azaglycoside mimics; and in fact, an azepane derivative was obtained in the synthesis of the homoazadisaccharide **203** by a ring expansion process, *via* an aziridinium cation. Our group, in a previous communication, reported the synthesis of an azepane derivative from an α,β-epoxyamide obtained from D-ribose in three steps.[54a] In order to prepare epoxide **210**, the key compound in the synthesis of **211** and **212**, we selectively protected the hydroxyl group at C3 in **206** as a silyl ether. When a *t*-butyldimethylsilyl ether was formed, we observed silyl group migration in basic medium. This migration was avoided when the diol **206** was regioselectively protected as triisopropylsilyl ether (TIPS), (Scheme 35). Mesylation of **207a** followed by deprotection of the trityl group afforded **209**, which was treated with sodium methoxide in CHCl$_3$ to give epoxide **210**. Azide group reduction with Ph$_3$P in CHCl$_3$, followed by water addition, gave the 6-*exo* product **212** favoured over the 7-*endo* product **211** (*exo/endo* 4:1). Reduction with Ph$_3$P in THF led to a mixture enriched in the desired azepane **211** (*exo/endo* 2:3).[54c] Compound **211** is a precursor of 1,6-dideoxy-1,6-iminoheptitol and of α-aminoacids.

In order to obtain the azepane as sole product, an alternative synthesis from **205** was performed (Scheme 36).[54b] The hydroxyl groups of **205** were protected as benzyl ethers as usual, obtaining, after 6 days at rt, a mixture of **213** and monobenzylated products that could be isolated and further re-benzylated. Hydrolysis of the trityl group, mesylation, and reduction of the azido group with Ph$_3$P followed by water addition, afforded the intermediate acyclic amine, which cyclised to the azepane **215a**, isolated in 75% yield. Catalytic hydrogenation of **215a** gave the debenzylated **215b.** The formation of other azepane derivatives starting from different epoxyamides is actually been performed in our laboratory.

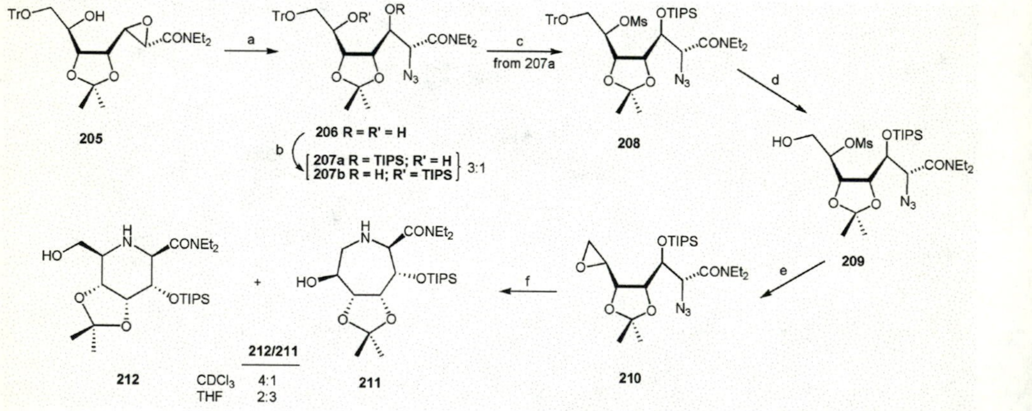

(a) NaN$_3$, AcOH, DMF, 80%; (b) TIPSOTf, 2,6-lutidine, CH$_2$Cl$_2$, 0 °C; (c) MsCl, py, 0 °C, 90%, (d) 2% TFA in CH$_2$Cl$_2$; (e) 1M solid MeONa in CHCl$_3$; (f) i. Ph$_3$P, THF or CDCl$_3$; ii. H$_2$O.

Scheme 35

The first examples of unprotected pentahydroxylated azepanes were reported by Blériot[55] and almost simultaneously, Dhavale published the synthesis of two new compounds.[56a]

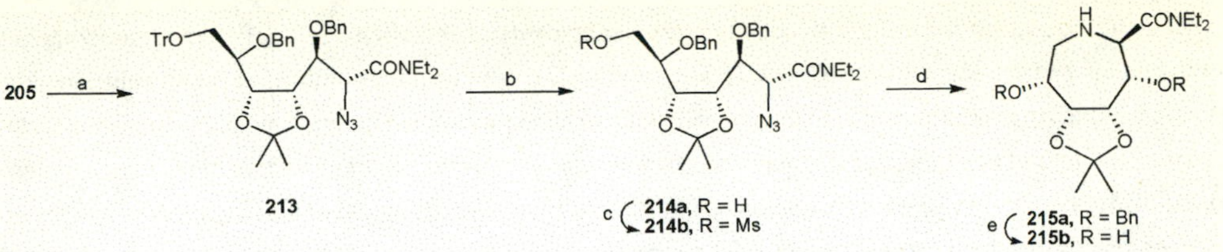

(a) BnBr, NaH, TBAI, THF, several days, 40%; (b) 5% TFA in CH_2Cl_2, 64%; (c) MsCl, py, 0 °C, 15 h, 81%; (d) Ph_3P, CH_2Cl_2, 17 h; ii. H_2O, K_2CO_3, 48 h, 71% (e) H_2, Pd/C, 75%.

Scheme 36

The synthetic strategy of the french group[55] was based on the ring-closing alkene metathesis (RCM) methodology. Reductive amination of the known ketone **216**,[57] easily available from D-arabinose, with allylamine and acetic acid in the presence of $NaBH_3CN$ gave the D-*arabino* and L-*xylo* N-allylaminohexenitols **217** and **218** in 58% yield (ratio 3 : 2) (Scheme 37). The aminodiene **217** was protected with a benzyloxycarbonyl group to afford carbamate **219** in 90% yield. Subsequent olefin metathesis of the diene **219** using Grubbs' catalyst proceeded very well to afford didehydroazepane **220** in 91% yield.

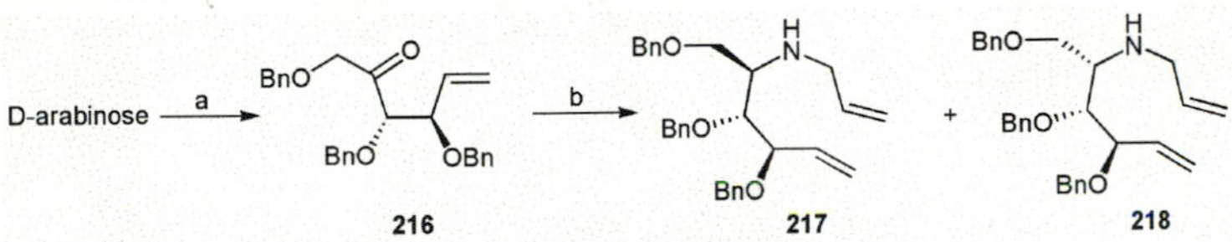

(a) Ref 57; (b) allylamine, AcOH, $NaBH_3CN$, CH_2Cl_2, 30 °C, 58%.

Scheme 37

Dihydroxylation of **220** (OsO_4, NMO) proceeded smoothly with modest facial diastereoselectivity to give *cis* diols **221** and **222** in 96% yield (3 : 7 ratio). Hydrogenolysis of the benzyl ethers **220-222** afforded the target 1,6-dideoxy-1,6-iminoheptitols **223–225** in quantitative yield (Scheme 38). The same sequence was applied to the L-*xylo* N-allyl-aminohexenitol **218** to afford the 1,6-dideoxy-1,6-iminoheptitols **230–232** (Scheme 38). The pseudo β-D-*manno*, α-D-*gluco*, α-L-*gulo* and β-L-*ido* configurations of compounds **224**, **225**, **231** and **232**, respectively, were established by the value of the vicinal $^3J_{H,H}$ coupling. The configurational assignments were confirmed when **232** was obtained by a different route and its structure unambiguously established by X-ray crystallography. Comparison of the NMR data of stereoisomers **224**, **225**, **231** with those of **232** confirmed the previous structural assignments

The 1,6-dideoxy-1,6-iminoheptitols **223**, **224**, **225**, **231**, and **232**, were assayed for their inhibitory activity toward 24 commercially available glycosidases. Three (**224**, **225** and **232**) of the six iminoalditols synthesised showed potent glycosidase inhibition in the low micromolar range, displaying a new inhibition profile compared to the previously reported polyhydroxylated azepanes. The best results were obtained with compound **225**, which is a selective and potent green coffee bean α-galactosidase inhibitor. The inhibition profile observed for each seven-membered ring iminoalditol, which cannot be correlated with its relative configuration, could be explained by a) the relative flexibility of these iminoalditols adopting a specific conformation in the binding pocket that favorably orientates the hydroxyl groups in the enzyme active site and/or b) by the different positioning of these compounds in the enzyme active site compared to the corresponding parent sugar.

In a subsequent publication,[58] the conformational analysis of this series of configurational isomers was carried out, in order to try to clarify SARs with respect to glycosidase inhibition. These authors used a combination of experimental NMR spectroscopic data, assisted by modeling methods. The seven-membered rings are flexible and may assume a variety of conformations that can interconvert with relative low energy barriers. NMR spectra obtained under acidic conditions were of higher quality than the ones recorded under neutral conditions and allowed a more precise measure of the key J values and NOE parameters. The polyhydroxyazepane glycomimetics **224**, **225**, **231** and **232**, may adopt two conformations in solution which display some selectivity towards different glycosidases. On this basis, they have been docked in the binding sites of three selected enzymes whose X-ray coordinates are available. In all cases, a extremely good correlation was observed between the docked energies of the models and the percentages of inhibition of the tested enzymes. This method may be used to understand the inhibition ability of glycomimetics, predicting the conformations in the binding state.

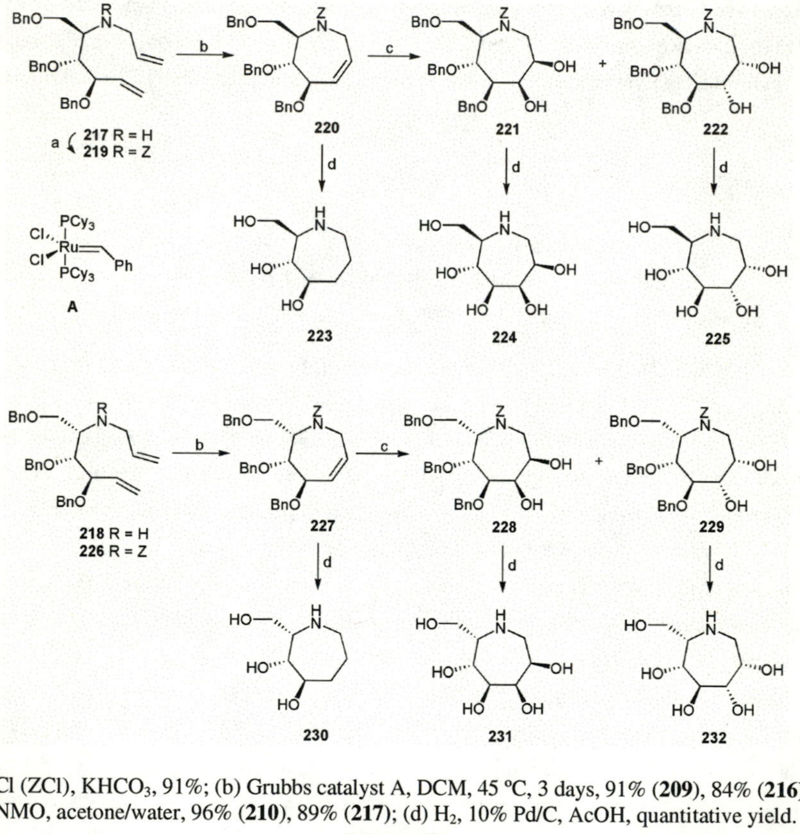

(a) BnOCOCl (ZCl), KHCO$_3$, 91%; (b) Grubbs catalyst A, DCM, 45 °C, 3 days, 91% (**209**), 84% (**216**); (c) OsO$_4$, NMO, acetone/water, 96% (**210**), 89% (**217**); (d) H$_2$, 10% Pd/C, AcOH, quantitative yield.

Scheme 38

A new synthetic approach[56a] chooses as the key intermediate, the 6-aminoheptoses **237**, wich could be obtained by applying the asymmetric dihydroxylation, cyclic sulfate formation, and nucleophilic azide ring opening protocol to the easily available D-glucose-derived α,β-unsaturated ester **233**,[56b] (Scheme 39). Dihydroxylation of **233** afforded a diastereomeric mixture of vicinal diols **234a** and **234b** in the ratio 2:1.

The formation of **234a** as a major product is in accordance with Kishi's empirical rule.[59] The diastereoselectivity in the formation of **234a** and **234b** was improved by using cinchona alkaloids as chiral ligands. Thus, the use of [(DHQ)$_2$PHAL] in osmylation afforded **234a** with high diastereoselectivity (d.e. 94%), while the use of [(DHQD)2PHAL] gave **234a:234b** in the ratio 32:68 as determined by the ^{1}H-NMR of the crude mixture (Table 7). The utility of **234a** and **234b** is shown in Scheme 39. This sequence of reactions gave 1,6-dideoxy-1,6-imino-(2S,3R,4R,5R,6R)-L-*glycero*-D-*gluco*-heptitol **238a** or 1,6-dideoxy-1,6-imino-(2S,3R,4R,5S,6S)-D-*glycero*-L-*ido*-heptitol **238b** as hydrochloride salts. Although the physical and spectral data of **238a** and **238b** were found to be consistent with the structures, the configurational assignment at each carbon atom in **238a** was based on the X-ray analysis of **234a**, while in the case of **238b** was tentatively made on the expected *syn*-dihydroxylation from the other face of **233** leading to **234b**. An additional structural confirmation was made by converting compounds **234a** and **234b** to the corresponding six-membered piperidine analogues: 1-deoxyaltronojirimycin and 1-deoxy-L-nojirimycin, wherein the configurational assignment at each carbon atom were clearly evident from NMR data, with either $^4C_1$ or $^1C_4$ conformations, and the same configurational assignments were applicable to azepane analogues.

Table 7. Asymmetric Dihydroxylation of **233**.

Entry	Ligand	Ratio of **234a/234b**	Yield
1	no ligand	67:33	91%
2	(DHQ)$_2$PHAL	97:03	88%
3	(DHQD)$_2$PHAL	32:68	82%

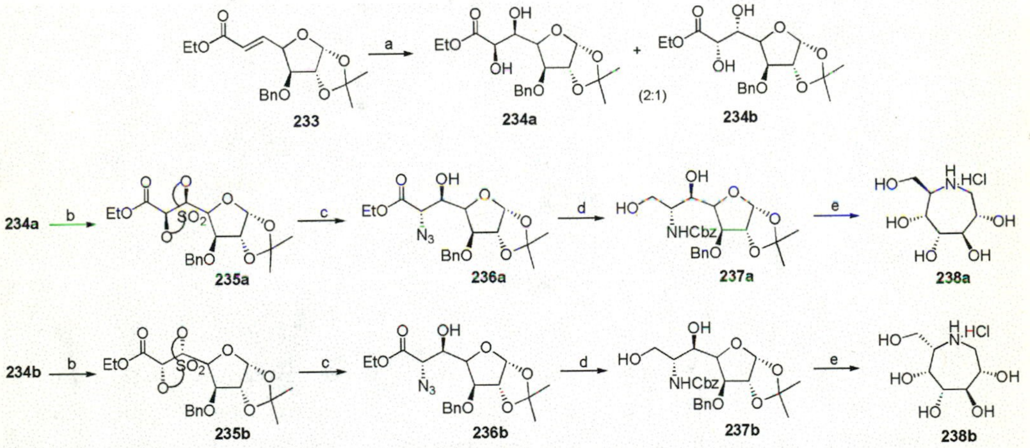

(a) OsO$_4$, K$_4$Fe(CN)$_6$, K$_2$CO$_3$, MsNH$_2$; (b) i. SOCl$_2$, Py, CH$_2$Cl$_2$, 0 °C, 30 min; ii. NaIO$_4$, RuCl$_3$.3H$_2$O, CH$_3$CN:H$_2$O (3:1), 0 °C, 10 min, 86%; (c) i. NaN$_3$, acetone:water (4:1), 0 to 25 °C, 2 h; ii. 20% H$_2$SO$_4$, ether:water (6:1), 25 °C, 6 h, 82-89%; (d) i. LAH, THF, 0 to 25 °C, 3 h; ii. HCOONH$_4$, Pd/C, MeOH, 80 °C, 1 h; iii. CbzCl, MeOH:H$_2$O (9:1), 0 to 25 °C, 3.5 h, 52-57%; (e) i. TFA:H$_2$O (2:1), 0 to 25 °C, 2.5 h; ii. H$_2$, Pd/C, MeOH:HCl (9:1), 80 psi, 24 h, 75-80%.
Scheme 39

6. Synthesis of hexahydroxyazepanes

A synthesis of *meso*-persubstituted azepanes has been reported[60a] from cyclooctatetraene (COT). The key product was the diepoxy-cyclooctene-diol **240**, which was reacted with MeNH$_2$ to give the bicyclic compound **241**. Treatment with O$_3$, reductive workup and acetylation afforded *meso*-azepane **242b**.

393

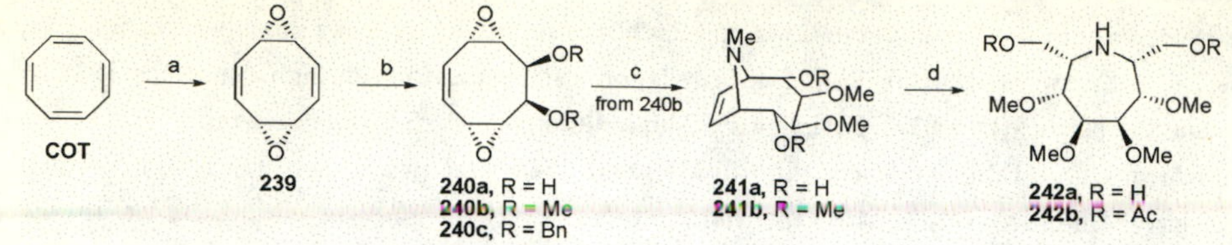

(a) CF$_3$CO$_3$H, ref 60b; (b) OsO$_4$, Me$_2$CO/H$_2$O, NMO, rt, 24 h; (c) MeNH$_2$; (d) i. O$_3$; ii. Reduction; iii. Ac$_2$O, py.

Scheme 40

7. Related products

In last years, compounds related to polyhydroxyazepanes have been synthesised (Figure 5). The first example of an eight-membered iminoalditol, (2R,3R,4R,5S)-2-hydroxymethyl-azocane-3,4,5-triol (**243**) has been recently[61] prepared from benzylated glucopyranose by way of a ring-closing metathesis and is a weak inhibitor of glycosidases. Compound **244**, a polyhydroxylated tetrahydro-4H-1,2,3-triazolo[1,5-a]azepin derivative[62] was found to be a very weak inhibitor of *E. coli* α-galactosidase and of isomaltase (α - glucosidase) from baker's yeast. Imidazolo azepanes have also been described.[63] C$_2$-Symmetric guanidino sugars **245** and **246**, analogues of tetrahydroxyazepanes and of 1,4-iminoalditols, have been synthesised from D-mannitol.[64] One of them (**246**) (4S,5R,6R,7S) is a selective inhibitor of α-L-fucosidase from bovine kidney. A trihydroxy diazepane[65] was synthesised from D-xylose. Azepane pseudo-disaccharides **247**, less hydroxylated than those described in other sections of this review, have been reported.[66] These glycosides inhibited growth of *Staphylococcus aureus*, including aminoglycoside-resistant strains and showed target binding and translation inhibition in the low micromolar range.

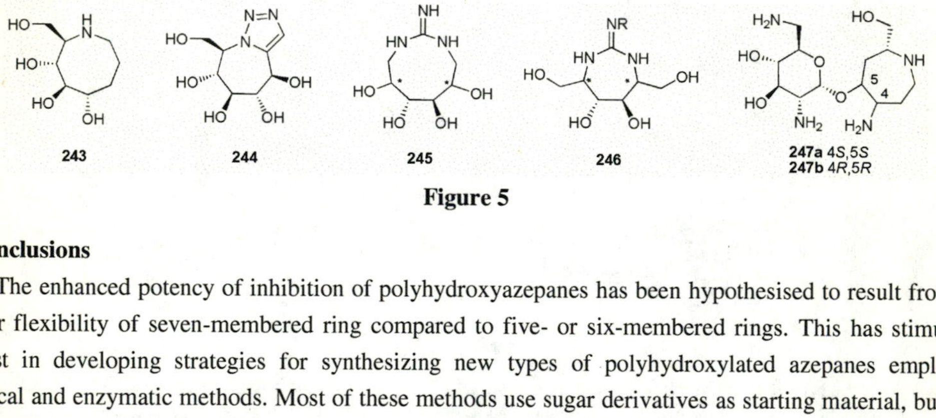

Figure 5

8. Conclusions

The enhanced potency of inhibition of polyhydroxyazepanes has been hypothesised to result from the greater flexibility of seven-membered ring compared to five- or six-membered rings. This has stimulated interest in developing strategies for synthesizing new types of polyhydroxylated azepanes employing chemical and enzymatic methods. Most of these methods use sugar derivatives as starting material, but new synthetic methods have been reported from non sugar compounds.

As regards to the inhibitory activity, tetrahydroxyazepane derivatives have been the most tested against different glycosidases and HIV proteases. Farr[43] assumed that the position of the equivalent of the 6-OH group of the mannosyl cation is a critical determinant of activity. Lundt[46] proposed that three vicinal hydroxyl groups could be essential for basic glycosidase inhibitory activity. N-substitution was considered of great importance to modulate selectivity and inhibition efficacy,[42] but it is not a general rule. Introduction of hydrophobic substituents, at the 3,6-positions can alter the bioactivities of these products.[12a] In some cases, the inhibition profile observed for polyhydroxyazepanes, can not be correlated to its relative configuration. In the case of pentahydroxy azepanes, the inhibitory results could be explained by the relative flexibility of these iminoalditols and/or by the different positioning of these compounds in the enzyme active site compared to the corresponding parent sugar[55] A combination of NMR data and modeling methods was proposed to understand the inhibition ability of these glycomimetics, predicting the conformations in the binding state.[58] The field of the pentahydroxyazepanes offers good perspectives.

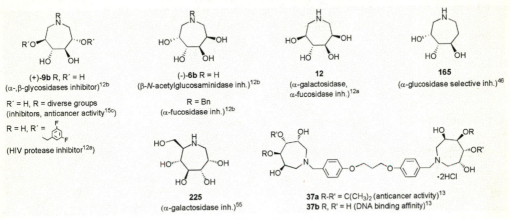

Figure 6

The preparation and study of new derivatives may lead to the finding of novel potent and highly selective glycosidase or protease inhibitors with applications in the control of several diseases (diabetes, HIV, cancer...).

Acknowledgments

We thank Dirección General de Investigación Científica y Técnica (Ref. BQU2001-1576) and Dirección General de Universidades e Investigación, Consejería de Educación y Ciencia, Junta de Andalucía (FQM 0158), for its financial support. We wish also to thank other colleagues in the field, whose names are found in the references of this review.

References

1. (a) Asano, N.; Nash, R. J.; Molyneux, R. J., Fleet, G. W. J. *Tetrahedron: Asymmetry* **2000**, *11*, 1645. (b) *Iminosugars as Glycosidase Inhibitors: Nojirimycin and Beyond*; Stütz, A. E., Ed.; Wiley-VCH: Weinheim, 1999. (c) Asano, N. *Glycobiology* **2003**, *13*, 93R.
2. Gruters, R. A.; Neefjes, J. J.; Teresmette, M.; de Goede R. E. Y.; Tulp, A.; Huisman, H. G.; Miedema, F.; Ploegh, H. L. *Nature* **1987**, *330*, 74.
3. Tyms, A. S.; Taylor, D. L.; Sincere, P. S.; Kang, M. S. *Desing of Anti-AIDS Drugs*. Clerq, E., Ed.; Elsevier: Amsterdam, 1990; p. 257.
4. Nishimura, Y. *Curr. Top. Med. Chem.* **2003**, *3*, 575.
5. Asano, N. *J. Enzyme Inhib.* **2000**, *15*, 215.
6. Rhinehart, B. L.; Robinson, K. M.; King, C. H. R.; Liu, P. S. *Biochem Pharmacol.* **1990**, *39*, 1537.
7. Compain, P., Martin, O. R. *Curr. Top. Med. Chem.* **2003**, *3*, 541.
8. Mitrakou, A.; Tountas, N.; Raptis, A. E.; Bauer, R. J.; Schulz, H.; Raptis, S. A. *Diab. Med.* **1998**, *15*, 657.

9. Cox, T.; Lachmann, R.; Hollak, C.; Aerts, J.; van Weely, S.; Hrebicek, M.; Platt, F. M.; Butters, T. D.; Dwek, R.; Moyses, C.; Gow, I.; Elstein, D.; Zimran, A. *Lancet* **2000**, *355*, 1481.

10. (a) Paulsen H; Todt K. *Chemische Berichte* **1967**, *100*, 512. (b) Paulsen, H.; Todt, K. *Angew. Chem.* **1965**, *77*, 589.

11. (a) Kondo, S.; Akita, E.; Sezaki, M. *J. Antibiot., Ser. A* **1966**, *19*, 137. (b) Yadav, J. S.; Srinivas, Ch. *Tetrahedron* **2003**, *59*, 10325.

12. (a) Moris-Varas, F.; Qian, X.-H.; Wong, C.-H. *J. Am. Chem. Soc.* **1996**, *118*, 7647. (b) Qian, X.-H; Moris-Varas, F.; Wong, C.-H. *Bioorg. Med. Chem. Lett.* **1996**, *6*, 1117. (c) Qian, X.-H.; Moris-Varas, F.; Fitzgerald, M. C.; Wong C.-H. *Bioorg. Med. Chem.* **1996**, *4*, 2055.

13. Johnson, H. A.; Thomas, N. R. *Bioorg. Med. Chem. Lett.* **2002**, *12*, 237.

14. (a) Poitout, L.; Le Merrer, Y.; Depezay, J-C. *Tetrahedron Lett.* **1994**, *35*, 3293. (b) Merrer, Y. L.; Poitout, L.; Depezay, J.; Dosbaa, I.; Geoffroy, S.; Foglietti., M. *Bioorg. Med. Chem.* **1997**, *5*, 519. (c) Le Merrer, Y.; Dureault, A.; Greck, C.; Micas-Languin, D.; Gravier, C.; Depezay, J.-C., *Heterocycles* **1987**, *25*, 541.

15. (a) Lohray, B. B.; Jayamma, Y.; Chatterjee, M. *J. Org. Chem.* **1995**, *60*, 5958. (b) Lohray, B. B.; Jayamma, Y.; Chaterjee, M. *Synth. Commun.* **1997**, *27*, 1711. (c) Lohray, B. B.; Prasuna, G.; Jayamma, Y.; Raheem, M. A. *Indian J. Chem., Sect. B*, **1999**, *38*, 1311.

16. Damour, D.; Barreau, M.; Blanchard, J.-C.; Burgevin, M.-C.; Doble, A.; Herman, F.; Pantel, G.; James-Surcouf, E.; Vuilhorgne, M.; Mignani, S. *Bioorg. Med. Chem. Lett.* **1996**, *6*, 1667.

17. Bernotas, R. C.; Ganem, B. *Tetrahedron Lett.* **1984**, *25*, 165.

18. (a) Kasai, K.; Okada, K.; Saito, S.; Tokutake, M.; Tobe, K. Pat JP 99-170371, 19990617. (b) Kasai, K.; Okada, K.; Saito, S.; Tokutake, M.; Tobe, K. Pat. JP 99-155051, 19990602.

19. (a) Cossy, J.; Dumas, C.; Michel, P.; Gomez-Pardo, D. *Tetrahedron Lett.* **1995**, *36*, 549.

20. (b) Furneaux, R. H.; Mason, J. M.; Tyler, P. C. *Tetrahedron Lett.* **1995**, *36*, 3055.

21. (a) Poitout, L.; Le Merrer, Y.; Depezay, J.-C. *Tetrahedron Lett.* **1996**, *37*, 1609. (b) Poitout, L.; Le Merrer, Y.; Depezay, J.-C. *Tetrahedron Lett.* **1996**, *37*, 1613.

22. Lohray, B. B.; Prasuna, G.; Jayamma, Y.; Raheem, M. A. *Indian J. Chem., Sect. B* **1997**, *36B*, 220.

23. Gauzy, L.; Le Merrer, Y.; Depezay, J.; Clerc, F.; Mignani, S. *Tetrahedron Lett.* **1999**, *40*, 6005.

24. (a) Katritzky, A. R.; Xie, L.; Zhang, G.; Griffith, M.; Watson, K.; Kiely, J. S. *Tetrahedron Lett.* **1997**, *38*, 7011. (b) Dankwardt, S. M.; Newman, S. R.; Krstenansky, J. L. *Tetrahedron Lett.* **1995**, *36*, 4923.

25. Uchida, R.; Nasu, A.; Tokutake, S.; Kasai, K.; Tobe, K.; Yamaji, N. *Chem. Pharm. Bull.* **1999**, *47*, 187.

26. (a) Le Merrer, Y.; Sanière, M.; McCort, I.; Dupuy, C.; Depezay, J.-C. *Tetrahedron Lett.* **2001**, *42*, 2661. (b) McCort, I.; Sanière, M.; Le Merrer, Y. *Tetrahedron* **2003**, *59*, 2693.

27. Dax, R.; Gaigg, B.; Grassberger, B.; Koelblinger, B.; Stütz, A. E. *J. Carbohydr. Chem.* **1990**, *9*, 479.

28. Tilekar, J. N.; Patil, N. T.; Harishchandra S. J.; Dhavale, D. D. *Tetrahedron.* **2003**, *59*, 1873.

29. Andreana, P. R.; Sanders, A.; Janczuk, J. I. W.; Wang, P. G. *Tetrahedron Lett.* **2002**, *43*, 6525.

30. Josan, J. S.; Eastwood, F. W. *Carbohydr. Res.* **1968**, *7*, 161.

31. (a) Painter, G. F.; Falshaw, A. *J. Chem. Soc., Perkin Trans. 1* **2000**, 1157. (b) Painter, G. F.; Eldridge, P. G.; Falshaw, A. *Bioorg. Med. Chem.* **2004**, *12*, 225. (c) Painter, G. F.; Falshaw, A.; Wong, H. *Org. Biomol. Chem.* **2004**, *2*, 1007.

32. Grice, P.; Ley, S. V.; Pietruszka, J.; Priepke, H. W. M.; Warriner, S. L. *J. Chem. Soc., Perkin Trans. 1* **1997**, 351.

33. (a) Paulsen, H.; Von Deyn, W.; Roeben, W. *Liebigs Ann. Chem.* **1984**, 433. (b) Shvets, V. I.; Klyashchitskii, V. A.; Stepanov, A. E.; Evstigneeva, R. P. *Tetrahedron* **1973**, *29*, 331.

34. (a) Fuentes, J.; Olano, D.; Pradera, M. A. *Tetrahedron Lett.* **1999**, *40*, 4063. (b) Fuentes, J.; Gasch, C.; Olano, D.; Pradera, M. A.; Repetto, G.; Sayago, F. J. *Tetrahedron: Asymmetry*, **2002**, *13*, 1743. (c) Fuentes, J.; Fernandez-Bolanos, J.; Garcia Fernandez, J. M.; Moreda, W.; Ortiz, C.; Pradera M. A.; Robina, I.; Welsh, C. *Carbohydr. Res.* **1992**, *232*, 47.

35. Gómez Sánchez, A.; Gómez Guillén, M; Cert Ventulá, A.; Scheidegger, U. *An. Quim.* **1968**, *64B*, 579.

36. (a) Havlicek, J.; Kefurt, K.; Husak, M.; Novotny, J.; Kratochvil, B. *Collect. Czech. Chem. Commun.* **1993**, *58*, 1600 and ref. cited. (b) Takeda, Y.; Akimoto, T.; Kyogoku, Y. *Carbohydr. Res.* **1982**, *106*, 175. (c) Hanessian, S. *J. Org. Chem.* **1969**, *34*, 675.

37. Long, D. D.; Stetz, R. J. E.; Nash, R. J.; Marquess, D. G.; Lloyd, J. D.; Winters, A. L.; Asano, N.; Fleet, G. W. J. *J. Chem. Soc., Perkin Trans. 1* **1999**, 901.
38. Joseph, C. C.; Regeling, H.; Zwanenburg, B.; Chittenden, G. J. F. *Tetrahedron* **2002**, *58*, 6907.
39. Chaveriat, L.; Stasik, I.; Demailly, G.; Beaupere, D. *Tetrahedron* **2004**, *60*, 2079.
40. Lin, C.-C.; Pan, Y.; Patkar, L. N.; Lin, H.-M.; Tzou, D.-L. M.; Subramanian, T.; Lin, C.-C. *Bioorg. Med. Chem.* **2004**, *12*, 3259.
41. Martin O. R. *Carbohydrate Mimics: concepts and methods*; Chapleur, Y., Ed.; Wiley-VCH, 1998.
42. Mehta, G.; Lakshminath, S. *Tetrahedron Lett.* **2002**, *43*, 331.
43. Farr, R. A.; Holland, A. K.; Huber, E. W.; Peet, N. P.; Weintraub, P. M. *Tetrahedron* **1994**, *50*, 1033.
44. Bashyal, B. P.; Fleet, G. W. J.; Gough, M. J.; Smith, P. W. *Tetrahedron* **1987**, *43*, 3083.
45. Winkler, D. A.; Holan, G. *J. Med. Chem.* **1989**, *32*, 2084.
46. Andersen, S. M.; Ekhart, C.; Lundt, I.; Stütz, A. E. *Carbohydr. Res.* **2000**, *326*, 22.
47. Dhavale, D. D.; Chaudhari, V. D.; Tilekar, J. N. *Tetrahedron Lett.* **2003**, *44*, 7321.
48. Gallos, J. K.; Demeroudi, S. C.; Stathopoulou, C. C.; Dellios, C. C. *Tetrahedron Lett.* **2001**, *42*, 7497.
49. Iacono, S.; Rasmussen, J. R. *Org. Synth.* **1986**, *64*, 57.
50. Bock, K.; Lundt, I.; Pedersen, C. *Acta Chem. Scand., Ser. B* **1987**, *B41*, 435.
51. Mehta, G.; Mohal, N; Lakshminath, S. *Tetrahedron Lett.* **2000**, *41*, 3505.
52. Mehta, G.; Talukdar, P.; Mohal, N. *Tetrahedron Lett.* **2001**, *42*, 7663.
53. Ballou, C. E. *J. Am. Chem. Soc.* **1957**, *79*, 165. (b) Gomez, A. M.; Danelon, G. O.; Valverde, S.; Lopez, J. C. *Carbohydr. Res.* **1999**, *320*, 138.
54. (a) Pino-González, M. S.; Assiego, C.; López-Herrera, F. J. *Tetrahedron Lett.* **2003**, *44*, 8353. (b) Pino-González, M. S.; Assiego, C.; López-Herrera, F. J. *Tetrahedron Lett.* **2004**, *45*, 2611. (c) Pino-González, M. S.; López-Herrera, F. J.; Assiego, C.; Oñas, N. *Abstracts XXIX Bienal de la R. S. E. Q.* Madrid, **2003**, p. 33.
55. Li, H.; Blériot, Y.; Chantereau, C.; Mallet, J.-M; Sollogoub, M.; Zhang, Y.; Rodriguez-Garcia, E.; Vogel, P.; Jimenez-Barbero, J.; Sinay, P. *Org. Biomol. Chem.* **2004**, *2*, 1492.
56. (a) Dhavale, D. D.; Markad, S. D.; Karanjule, N. S.; Prakasha Reddy, J. *J. Org. Chem.* **2004**, *69*, 4760. (b) Patil, N. T.; Tilekar, J. N.; Dhavale, D. D. *J. Org. Chem.* **2001**, *66*, 1065.
57. (a) Sellier, O.; Van de Weghe, P.; Le Nouen, D.; Strehler C.; Eustache, J. *Tetrahedron Lett.* **1999**, *40*, 853. (b) Sellier, O.; Van de Weghe, P.; Eustache, J. *Tetrahedron Lett.* **1999**, *40*, 5859. (c) Blériot, Y.; Giroult, A.; Mallet, J.-M.; Rodriguez, E.; Vogel, P.; Sinaÿ, P. *Tetrahedron: Asymmetry* **2002**, *13*, 2553.
58. Martínez-Mayorga, K.; Medina-Franco, J. L.; Mari, S.; Cañada, F. J.; Rodriguez-García, E.; Vogel, P.; Hongqing, L.; Blériot, Y.; Sinaÿ, P.; Jimenez-Barbero, J. *Eur. J. Org. Chem.* **2004**, 4119.
59. (a) Cha, J. K.; Christ, W. J.; Kishi, Y. *Tetrahedron* **1984**, *40*, 2247. (b) Brimacombe, J. S.; Kabir, A. K. M. S. *Carbohydr. Res.* **1986**, *150*, 35. (c) Jarosz, S. *Carbohydr. Res.* **1988**, *183*, 209.
60. (a) Armbruster, J.; Stelzer, F.; Landenberger, F. P.; Wieber, C.; Hunkler, D.; Keller, M.; Prinzbach, H. *Tetrahedron Lett.* **2000**, *41*, 5483. (b) Anastassiou, A. G.; Reichmanis E. *J. Org. Chem.* **1973**, *38*, 2421.
61. Godin, G.; Garnier, E.; Compain, P.; Martin, O. R.; Ikeda, K.; Asano, N. *Tetrahedron Lett.* **2004**, *45*, 579.
62. Tezuka, K.; Compain, P.; Martin, O. R. *Synlett* **2000**, 1837.
63. Weinberg, K., Jankowski, S.; Le Nouen, D.; Frankowski, A. *Tetrahedron Lett.* **2002**, *43*, 1089.
64. Le Merrer, Y.; Gauzy, L.; Gravier-Pelletier, C.; Depezay, J.-C. *Bioorg. Med. Chem.* **2000**, *8*, 307.
65. Ernholt, B. V.; Thomsen, I. B.; Jensen, K. B.; Bols, M. *Synlett* **1999**, 701.
66. Barluenga, S.; Simonsen, K. B.; Littlefield, E. S.; Ayida, B. K.; Vourloumis, D.; Winters, G. C.; Takahashi, M.; Shandrick, S.; Zhao, Q.; Han, Q.; Hermann, T. *Bioorg. Med. Chem. Lett.* **2004**, *14*, 713.

MASS SPECTROMETRY OF SIMPLE INDOLES.
PART 2: FOURIER TRANSFORM MASS SPECTROMETRY, FAST ATOM BOMBARDMENT, LASER DESORPTION AND ELECTROSPRAY IONISATION

David Bongiorno, Lorenzo Camarda, Leopoldo Ceraulo* and Mirella Ferrugia

Dipartimento di Chimica e Tecnologie Farmaceutiche e Rete di Spettrometria di Massa del C.N.R.,
Università di Palermo, Via Archirafi 32, I-90123 Palermo, Italy
(e-mail: lceraulo@unipa.it)

Abstract. *The review is devoted to the gas-phase ion chemistry studies of simple indoles by using several mass spectrometric methods. A screening of the papers concerning mass spectrometry studies of indoles to select those related to their behaviour under different ionisation (fast atom bombardment, laser desorption and electrospray) conditions as well as that performed by Fourier transform mass spectrometry has been done and the results have been critically discussed and reported.*

Contents

1. Introduction
2. Fourier transform mass spectrometry (FTMS)
    2.1. Gas phase acidity of indole carboxylic acids
    2.2. Gas-phase basicities of indole and tryptophan
3. Fast atom bombardment (FAB)
    3.1. Indoleamines and tryptamines
    3.2. Tryptophan and tryptophan derivatives
    3.3. Indole-3-acetic acid and indole-3-acetic acid conjugates
4. Laser Desorption (LD)
    4.1. Arylindoles
5. Electrospray ionisation (ESI)
    5.1. 1-Substituted indoles
    5.2. 3-Substituted indoles
    5.3. Other indole derivatives
6. Conclusions
Acknowledgments
References

1. Introduction

In the first part of this review the fundamental aspects of the gas-phase ion chemistry of simple indoles under electron ionisation (EI), photo-ionisation (PI) and electron capture ionisation (ECI) conditions have been summarised and critically discussed.[1] It has been reported that the high stability of the indole moieties also as radical cations obtained by EI is evidenced by the scarcity of indole ring fragmentation reactions from the molecular ions that are essentially observed for the simple indole itself and for arylindoles.

Generally, breakdown of the nucleus occurs following the more favoured fragmentations involving the substituents that allow an easy identification of their nature. Distinctive fragmentation reactions for hydroxyindole and the tautomeric oxindole form have been individuated. PI/MS studies have evidenced that the adiabatic ionisation energy is strictly related to the position of the methyl group in methylindoles. Finally, it has been determined that the radical anions formed by ECI do not produce ring cleavage reactions and it has been noted that the negative charge is mainly retained in the indole moiety.[1]

The aim of the present second part of the review is to report the gas-phase ion chemistry of indoles under ionisation techniques involving matter transfer (protonation, cationisation or proton loss) as FAB, LD and ESI as well as some intrinsic properties of indole derivatives determined by FTMS. Also in this case the review cannot be considered comprehensive, *i.e.* indolenine and indoline derivatives, fused indoles, indole alkaloids and the papers strictly concerning analytical applications have not been considered.

2. Fourier Transform Mass Spectrometry

FTMS is a greatly developing technique firstly described by Comisarow and Marshall[2,3] and successively revised by Amster[4] and Marshall *et al.*[5] The technique consists of simultaneously exciting all the ions present in a cyclotron cell by a fast scan of a large frequency range within a very short time (1 μs). A FT needs to transform the complex wave detected as a time-depending function of the ion image currents into a frequency-depending function, in its turn transformed in mass spectrum. Although its generally high cost of purchase and maintaining are little discouraging, the results obtained with this technique are almost unique. The possibility to trap the ions for times extremely long, allows to establish a reaction equilibrium even in the gas phase, thus binding constants or gas phase basicities can be measured. In addiction extremely slow fragmentation reactions, unobservable in traditional MS experiments can be seen and seek during the time.[6] These features are accompanied by a vertiginous resolution power ($> 10^6$),[7] respect to all the other traditional MS experiments, which also allows unit mass spacing for $^{12}C/^{13}C$ isotopic peaks of the multiply charged ions produced by ESI, so that the number of them in a unit of mass to charge ratio represents the number of charges.[8,9] Finally, using an ESI source coupled with an ion cyclotron resonance (ICR) FTMS an ion with a mass as large as $5 \cdot 10^6$ Da (with 2610 charge units) was observed.[10]

2.1. Gas-phase acidity of indole carboxylic acids

In aqueous solution indole **1** behaves as a weak acid (pKa = 16.97).[11] In water solution indole 2-carboxylic acid **2** (pKa = 3.87^{12} or 3.82^{13}) and indole-3-carboxylic acid **3** (pKa = $5.29)^{13}$ have the same behaviour of the carboxylic acids. However, it is not clear *a priori* if in the gas phase these compounds could differently act, varying the ionisation site with the substitution of carboxylic group at the indole ring. Under the appropriate experimental conditions[14,15] it was possible to determine gas phase acidity of indole derivatives bearing a carboxylic group on the side chain that constituted an intriguing task, being the indole NH itself an acidic residue.

It has been reported that the gas phase acidity (expressed as Gibbs energy change for reaction 1) of indole-2-carboxylic acid **2** and indole-3-carboxylic acid **3** is 328.4 kcal mol^{-1} and 325.9 kcal mol^{-1}, respectively[14] (Table 1).

$$AH\,(g) \rightleftharpoons A^-\,(g) + H^+\,(g) \qquad\qquad \Delta G^{\circ}_{acid}(g) \qquad\qquad (1)$$

Table 1. Gas phase acidity of compounds **2-5** taken from ref. 14.	
Compound	Δ G°/ kcal mol^{-1}
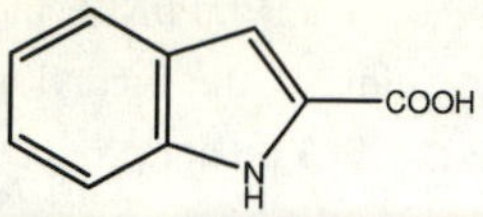 **2**	328.4
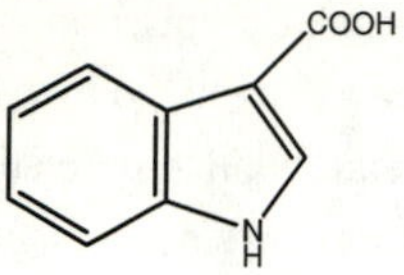 **3**	325.9
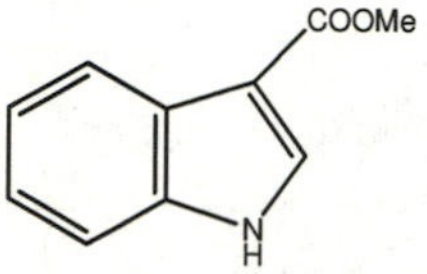 **4**	330.6
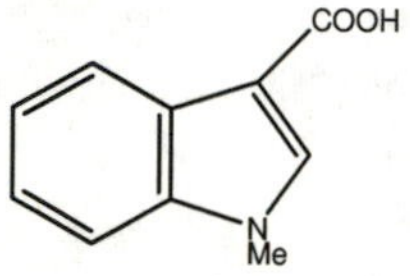 **5**	336.1

This result has been achieved by the determination of the equilibrium constant for reaction 2 through FTMS experiments, using a suitable reference gas with known $\Delta G°_{acid\,(ref)}$ (g).

$$AH\,(g) + A^{-}_{ref}\,(g) \rightleftharpoons A^{-}\,(g) + AH_{ref}(g) \qquad Kp,\ \delta\Delta G°_{acid}(g) \qquad (2)$$

Ab-initio calculations show that the NH acidity of compound **3** is very close to the experimental value, while the situation of the isomer **2** is not so clear. The analysis of methyl indole-3-carboxylate **4** and 1-methylindole-3-carboxylic acid **5**, both bearing only one possible deprotonation site, shows that the methylation of the carboxylic group of **3** decreases the acidity by only 2.2 kcal mol^{-1}, while the N-methylation decreases the acidity by 7.7 kcal mol^{-1}. This finding supports that the preferential deprotonation site for **3** is the NH group.[14]

2.2. Gas-phase basicities of indole and tryptophan

Parallel studies has been developed, by both traditional MS and FTMS experiments, to determine gas phase basicities (GB) of various aminoacids including tryptophan **6**. The gas phase basicity GB of a substance B, here represented in the relation (3), is correlated to its proton affinity (PA) by the relations (4) and (5).[16]

$$B + H^{+} \rightarrow BH^{+} \qquad \Delta G° = GB_{B}; \qquad -\Delta H° = PA_{B} \qquad (3)$$

substituting PA_B and GB_B in the Gibbs equation is obtained

$$PA_B = GB_B - T\Delta S° \qquad (4)$$

where

$$\Delta S° = \Delta S°_{transl} + \Delta S°_{rot} + \Delta S°_{vib} \qquad (5)$$

Some differences had been reported for the gas phase basicities determined by equilibrium[17] and reaction bracketing[18] techniques developed by FTMS and kinetic method[19] measurements by classical ion beam MS techniques.[16] The equilibrium and reaction bracketing approaches are more strictly concerning free energy variation and then the measured value is gas phase basicity.

FTMS equilibrium experiments require to measure accurately the ion signals in equilibrium conditions and the concentration of the neutral substances involved in the reaction indicated in equation (6).

$$B_1H^+ + B_2 \rightarrow B_1 + B_2H^+ \qquad (6)$$

where B_1 and B_2 are the bases involved in the gas phase equilibrium

It is necessary to establish ladders of relative GB which should include one or more reference compounds for which the gas phase basicity is known and, where possible, determined by other methods. Reaction bracketing techniques allow the reaction of the specie B_1H^+ with a variety of bases of known PA, assuming for rapid proton transfer a negative $\Delta H°$ (exothermic) and for slow proton transfer a positive $\Delta H°$ (endothermic) it is possible to bracket PA of B_1 within 3 kcal mol^{-1}. Recent studies however assumed that, under the FTMS condition, the reaction bracketing itself yields gas phase basicity (GB) instead of proton affinity (PA).[20]

The kinetic method[19] involves studies of weakly bound cluster ions, evaluating the competitive fragmentation reactions (7) and (8).

$$[B_1-H-B_2]^+ \rightarrow B_1H^+ + B_2 \qquad (7)$$
$$[B_1-H-B_2]^+ \rightarrow B_2H^+ + B_1 \qquad (8)$$

where B_1 and B_2 are the substances for one of which the PA value is known.

Through the application of absolute rate theory the equation (9) is obtained:

$$\ln([B_1H]^+/[B_2H]^+) = [PA (B_1) - PA (B_2)]/RT_{eff} \qquad (9)$$

where R is the Boltzmann constant and T_{eff} is the effective temperature.

However, various approximations must be done and imply the close similarity of the B_1 and B_2 species as the most important factor to avoid excessive difference of entropy between the two reactions. Further, a negligible value of reverse activation energies for equations (7) and (8) is required. The values of GB and PA for indole 1, are 213.8 kcal mol^{-1} [20] and 223.1 kcal mol^{-1},[21] respectively. These values are comparable with the PA of 212.4 kcal mol^{-1} calculated for the protonation at 3-position of 1 (Table 2).[22]

GB and PA for tryptophan 6 obtained by both FTMS[17,18] and kinetic method[23-25] together with the PA estimated by *ab initio* calculations[26] are also reported in Table 2. PA calculated for tryptophan 6 should take into consideration the entropy effects involved in the formation of intramolecular hydrogen bonds. However, in earlier reviews, PA of tryptophan **6** has been calculated from both arbitrary assumptions[16] of: $\Delta S°_{rot}= 0$ ($PA_{tryptophan}= 223.9$) and $\Delta S°_{rot}= -10$ cal mol^{-1} K^{-1} ($PA_{tryptophan}= 226.9$). Some approaches[25, 27] derived by variation of the most used kinetic method equation used to determinate PAs, lead to a minimize the effect of variation of the partition function, and then of entropy on the PA measurements. The values reported however differ not too much and the differences are, in some cases, within the experimental errors. PA values estimated by means of *ab initio* calculations are also in good agreement with experimental results.[26]

Table 2. GB and PA of indole **1** and tryptophan **6**.

Compound	GB/kcal mol^{-1}	PA/kcal mol^{-1}
1	213.8 (Ref. 20)	223.1 (Ref. 21)
		212.4 (Ref. 22)
6	217.6 (Ref. 17)	223.5 (Ref. 23)
	216 (Ref. 18)	225.4 (Ref. 21)
	212.9 (Ref. 24)	220.8 (Ref. 24)
	215.7 (Ref.16)	221.6 (Ref.25)
		220.7 (Ref. 26)

3. Fast Atom Bombardment

The introduction of FAB[28] and liquid secondary mass spectrometry (LSIMS)[29] techniques, that consist of focusing on the sample (dissolved in a non volatile liquid matrix) a high primary current beam of atoms or ions respectively, allowed the direct analysis of highly polar indole derivatives.

3.1. Indoleamines and tryptamines

In order to investigate the behaviour of the protonated molecular species of the biogenic indoleamines tryptamine **7**, serotonine **8**, 5,6-dihydroxytryptamine **9** and 5,7-dihydroxytryptamine **10**, Traldi and co-workers studied their FAB mass spectra (Table 3) and the CID MIKE spectra of the related MH$^+$ species.[30]

As expected, the behaviour of **7-10** under FAB conditions was very different from that observed in EI. Interestingly, in spite of the lower internal energy involved in FAB ionisation, the same fragmentation processes of the side-chain were observed, together with new fragmentation processes, consisting of NH$_2$ and NH$_3$ losses. The NH$_3$ loss was responsible for the most intense peak of the collision induced decomposition (CID) mass analysed ion kinetic energy (MIKE) spectra for **7-10**. This should indicate that the protonation site for these compounds is on the amine nitrogen atom. It is also worth noting that FAB/MS allows to an easy characterisation of the isomers **9** and **10**.

Table 3. FAB mass spectra of compounds **7-10** from ref. 30.

Compound	Th (I%)
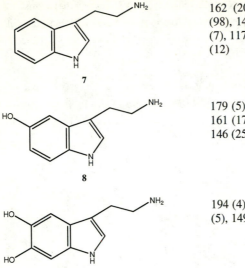 **7**	162 (20), 161 (100), 160 (15), 145 (18), 144 (98), 143 (20), 132 (18), 131 (30), 130 (33), 118 (7), 117 (15), 116 (5), 115 (10), 77 (8), 76 (4), 75 (12)
8	179 (5), 178 (30), 177 (100), 176 (25), 175 (10), 161 (17), 160 (75), 159 (14), 148 (11), 147 (10), 146 (25), 145 (17), 143 (6), 133 (7), 130 (5)
9	194 (4), 193 (80), 192 (5), 177 (15), 163 (5), 162 (5), 149 (5), 115 (12), 114 (100), 113 (8)
10	194 (24), 193 (80), 192 (28), 191 (12), 177 (15), 176 (60), 175 (20), 150 (15), 149 (100), 147 (14), 132 (20), 131 (30), 130 (12), 129 (21)

Peaks due to matrix ions have been omitted.

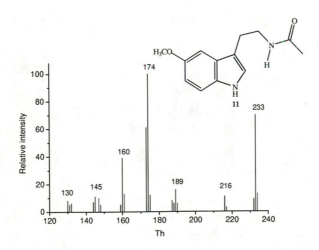

Figure 1. FAB mass spectrum of melatonin **11**.

A similar behaviour is observed for melatonin **11**. The positive ion FAB/MS spectrum of melatonin **11** (Figure 1) has been discussed in works concerning the inclusion complexes with cyclodextrin.[31,32] It shows abundant MH[+] ions (233 Th) and the main fragment ions at 174 Th due to the formal elimination of acetamide from the protonated molecule. A quite abundant ion at 160 Th attributable to the side chain CH_2-CH_2 bond cleavage is also observed.

3.2. Tryptophan and tryptophan derivatives

Van de Weert *et al.*,[33] in a paper concerning the mass spectrometric analysis of oxidised tryptophan and tryptophan-containing peptides, reported the CID-MS/MS spectra of the FAB generated MH[+] ions of tryptophan **6** (205 Th), of 2-hydroxytryptophan **12** (221 Th) and/or its tautomeric form oxindolylalanine **12'** (Scheme 1).

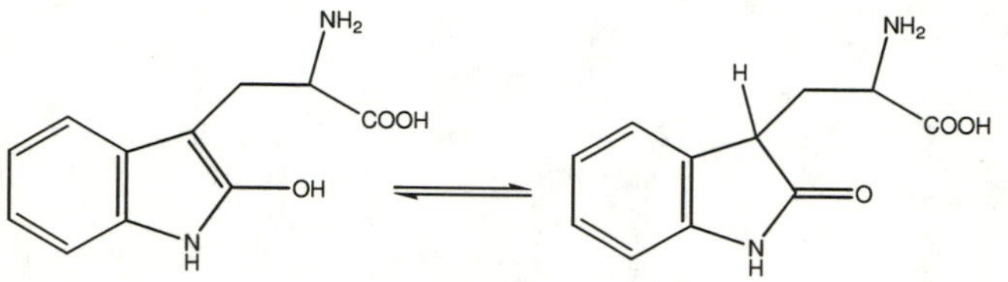

Scheme 1

The CID-MS/MS spectra of **6** and **12** are quite different (Table 4). In the spectrum of compound **6**, the most intense peaks (excepted for the quasi-molecular ion peak) are due to the ions at 130 Th (due to C_α-C_β bond cleavage) and at 188 Th (probably formed by NH_3 loss from the MH[+] ions). The abundance ratio of the ions at 130 Th and 117 Th for **6** and that of the corresponding ions at 146 Th and 133 Th for **12** is different. This ratio is about 5 for compound **6**, indicating a higher stability of the C_β-C_γ bond with respect to C_α-C_β bond. Instead, the ratio of ions at 146 Th and 133, in 12 is about 1, that means that in this case no significant differences occur between C_β-C_γ and C_α-C_β bond strength. This behaviour agrees with the presence in the gas-phase of the protonated oxindolylalanine molecule **12'** instead of that 2-hydroxytryptophan **12**. This finding evidences that, likely to EI/MS,[1] also CID MS/MS of FAB generated MH[+] ions allow to distinguish 2-hydroxyindole and oxindole tautomeric forms.

Table 4. Main peaks of CID MS/MS spectra of the FAB generated MH[+] ions of compounds
6 and 12 (and/or 12') from ref. 33.

Compound	Th (I%)
6	205 (100), 204 (6), 188 (93), 170 (2), 159 (4), 146 (4), 132 (4), 131 (5), 130 (29), 117 (6), 115 (3), 103 (2), 77 (2)
12 (and/or 12')	221 (100), 220 (35), 204 (13), 203 (38), 175 (96), 174 (7), 158 (23), 157 (7), 146 (33), 133 (30), 132 (35), 130 (15), 129 (34), 128 (8), 104 (7), 103 (6)

Surprisingly, the origin of the abundant ions at 175 Th as well as of that of the ions at 203 Th in the CID MS/MS spectrum of **12'** was not discussed. These ions are, in our opinion, of interest as they could be due to a different protonation site or hydrogen migration from amine nitrogen to hydroxyl oxygen of carboxylic group, which could produce the loss of H_2O (affording to the ions at 203 Th) followed by the CO loss (giving the ions at 175 Th) (Scheme 2).

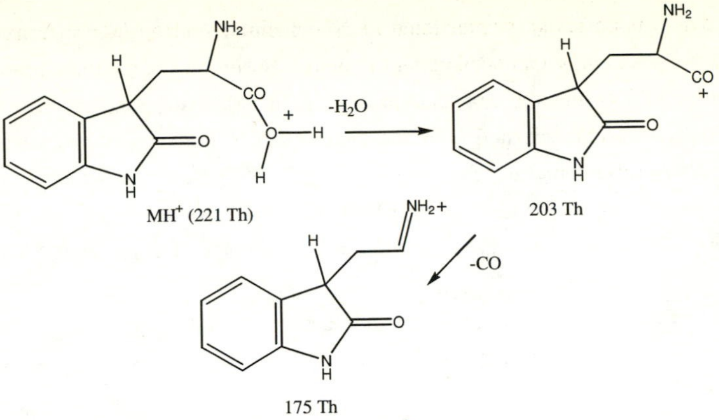

MH$^+$ (221 Th)   -H$_2$O   203 Th

175 Th

Scheme 2

3.3. Indole-3-acetic acid and indole-3-acetic acid conjugates

Indole-3-acetic acid **13** is an endogenous plant hormone regulating various aspect of plant growth and development, which in vegetative plant tissues is metabolised by conjugation with sugars and amino acids, and by hydroxylation and/or oxidation at various position in the indole ring. The FAB mass spectra of indole-3-acetic acid **13** and some of its hydroxylated analogues and sugar and amino acid conjugates were reported by Ostin *et al.*[34] The main peaks of FAB mass spectra of indole-3-acetic acid **13** and of its amino acid conjugates **14-20** are reported in Table 5.

Table 5. FAB mass spectra of indole-3-acetic acid **13** and its aminoacid conjugates **14-20** from ref. 34.

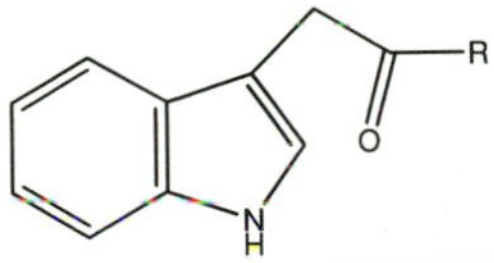

Compound	R	MH$^+$ Th (I%)	Q$^+$ Th (I%)	Other ions Th (I% < 10)
13	OH	176 (100)	130 (80)	117, 146
14	Gly	233 (100)	130 (65)	117, 144, 157, 175
15	Asp	291 (100)	130 (70)	117, 144, 157, 175
16	Glu	305 (100)	130 (80)	117, 144, 157, 175
17	Ala	247 (100)	130 (65)	117, 144, 157, 175
18	Val	275 (100)	130 (75)	117, 144, 157, 175
19	Ileu	289 (100)	130 (80)	117, 144, 157, 175
20	Phe	323 (100)	130 (93)	117, 144, 157, 175

The peaks related to the matrix adduct ions are not reported

As usual with FAB ionisation, the spectra of the indoles **13-20** contained abundant MH$^+$ ions with prominent adducts from glycerol (G) matrix represented by the addiction of one and two glycerol to MH$^+$, *i.e.* MHG$^+$ and MHG$_2^+$ ions (not reported). Dimerisation affording to M$_2$H$^+$ ions is also observed for **13-20** (not reported).

405

Compound **13** and its amino acid conjugates **14-20** generate abundant quinolinium ions Q$^+$ (130 Th) by β-cleavage of the side chain probably followed by ring expansion of the pyrrole moiety. The ions Q$^+$ at 130 Th are also responsible of the most abundant peak in the daughter ion spectra of the MH$^+$ ions by B/E linked scanning. By scanning the magnetic (B) and electrostatic (E) field simultaneously at constant B/E ratio, the daughter ions formed from a selected parent ion in the first field-free region can be detected.[35]

Another fragmentation observed for MH$^+$ of **14-20** in FAB spectra involves amine bond cleavage with hydrogen migration yielding the ions at 175 Th, which could correspond to protonated indole-3-acetic acid amide.

The FAB mass spectra of three isomers **21-23** (not distinguished) corresponding to **13** conjugated with different hydroxyl groups of inositol and that of indole-3-acetyl-β-D-glucose **24,** were also reported and discussed [33] (Table 6).

Table 6. FAB mass spectra of compounds **21-28** from ref. 33.

Compound	MH$^+$ Th (%)	Q$^+$ Th (%)	Other ions Th (%)
21 (isomer 1)	338 (92)	130 (100)	176 (15)
22 (isomer 2)	338 (82)	130 (100)	176 (16)
23 (isomer 3)	338 (100)	130 (85)	176 (27)
24	338 (7)	130 (75)	337 (11) 268 (4) 176 (100)

The peaks related to the matrix adduct ions are not reported

Compounds **21-24** behave under FAB ionisation likely to the amino acid conjugates **14-20**. In fact, in addition to the MH$^+$ ions they show abundant Q$^+$ ions and cleavage of the ester bond (leaving the oxygen linked to indole moiety) with hydrogen migration. This fragment is of diagnostic value to distinguish between amino acid conjugates (yielding the ions at 175 Th) and sugar conjugates (yielding the ions at 176 Th).

The main peaks of FAB mass spectra of oxygenated derivatives 5-hydroxy-indole-3-acetic acid **25,** oxindole-3-acetic acid **26,** 7-hydroxy-oxindole-3-acetic acid **27** and 3-hydroxy-oxindole-3 acetic acid **28,** are shown in Table 7.

The Q$^+$ ions provide useful information on the number of hydroxyl or carbonyl groups on the indole ring. However, FAB spectra do not allow to distinguish between isomers **25** and **26**.

Table 7. FAB mass spectra of compounds **25-28** from ref. 33.

Compound	MH$^+$ Th (I%)	Q$^+$ Th (I%)	Other ions Th (I%)
25	192 (100)	146 (56)	
26	192 (100)	146 (22)	
27	208 (100)	162 (77)	190 (12)
28	323 (100)	146 (90)	307 (52) 306 (48) 134 (12)

The peaks related to the matrix adduct ions are not reported

Distinct daughter ion spectra were instead obtained from the MH$^+$ ions of the isomers **25** and **26** by scanning magnetic field (B) and electrostatic field (E) at constant B/E. In fact, the daughter ions spectrum of **26** is dominated by the peak at 174 Th (loss of H$_2$O) and shows a very small peak at 146 Th, while a reversed situation is observed for **25**.

4. Laser Desorption Ionisation

The development of the matrix-assisted laser (generally UV) desorption ionisation (MALDI) technique, mainly by Karas and co-workers[36,37] and Tanaka (Nobel prize in Chemistry 2003) and co-workers,[38] made the time-of-flight mass spectrometry (TOF-MS) an important tool for the determination of the molecular masses of synthetic polymers[39] and biopolymers.[40]

4.1. Arylindoles

The presence of abundant matrix-related ions in the low-mass range makes very difficult the analysis of small-size molecules, even if a method for the analysis of low-mass analytes by MALDI-TOF-MS has been recently reported.[41] However the chemical and physical properties of the matrix as well as the wavelength of the irradiating laser source are the most critical parameters in order to achieve good analytical results for the analyte of interest. While the dependence on wavelength has been frequently taken into account, attention to the chemical structure related to the capability to be a good matrix compounds is less frequent. Hence, Nonami *et al.*[42] examined several heterocyclic compounds including 2-phenylindole **29**, 2-(2-pyridyl)-indole **30**, 1-methyl-2-(2-pyridyl)-indole **31**, 2-(3-pyridyl)-indole **32**, 2-(4-pyridyl)-indole **33** and 3-(4-pyridyl)-indole **34**, as potential MALDI matrices. Morphology, electronic spectra, diffuse fluorescence spectra, ground-state geometry and heats of formation calculated by semi-empirical parameterised PM3 method and the nitrogen LD-TOF-MS in positive ion and in negative ion modes (Table 8) have been reported and discussed.

Table 8. LD/TOF-MS of compounds **29-34** in positive and negative mode from ref. 42.

Compound	MW	Th (I%) Positive mode	Th (I%) Negative mode
2-phenylindole **29**	193	387(13), 194 (100), 69 (4)	192 (100), 115 (19), 64 (12)
2-(2-pyridyl)-indole **30**	194	195 (100)	193 (100), 112 (22)
1-methyl-2-(2-pyridyl)-indole **31**	208	209 (100)	261 (62), 190 (12), 149 (18), 110 (100)
2-(3-pyridyl)-indole **32**	194	195 (100)	193 (33), 113 (38), 97 (100), 81 (18), 69 (27)
2-(4-pyridyl)-indole **33**	194	195 (100)	171 (6), 110 (100), 68 (86)
3-(4-pyridyl)-indole **34**	194	195 (100)	193 (100)

The mass spectra in positive mode show only the MH$^+$ ions excepted for **29** which shows also the corresponding protonated dimeric form M$_2$H$^+$ (387 Th) with an intensity of 13%.

The mass spectra in negative mode are different for the considered compounds: in some cases the [M - H]$^-$ ions are present (**29, 30, 32, 34**), but with different intensity, and for the 3-(4-pyridyl)indole **34** is the only ion observed. The negative ions at 261 Th for the 21-methyl-2-(2-pyridyl)- indole **31** are probably due to fragmentation of the dimeric form.

It has been determined that all compounds that have both an indole NH and a pyridine N (**29, 30, 32-34**) acted as matrices for proteins, being very good for proteins of low molecular mass.

Good results have been obtained using **29, 30, 32, 34** (mainly for compound **30**) as matrix for synthetic polymers (polyethyleneglycole, polyamide) as well as for positive and negative ions MALDI-MS of carbohydrates (cyclodextrins).

5. Electrospray Ionisation

ESI is a technique which generates in the gas phase multiply charged ions as well as single charged ions (in the positive ion mode, a good correlation is observed between the maximum charge state and the basic sites present in the structure)[35] with the capability to maintain quite intact their structure. This allows the ESI/MS analysis of either large or small molecules. Early works on the development of electrospray were performed by Dole and co-workers[43] and Fenn (Nobel prize in Chemistry 2003) and co-workers[44-46] developed electrospray as a sample introduction and an ionisation technique for mass spectrometry.

Likely to other so called soft ionisation techniques, in ESI is very difficult to establish the mass spectrometry behaviour of a molecule or of a class of compounds. In fact, the spectrum pattern is strictly related to the internal energy of the molecule and its structure, and in ESI it widely depends from solvent, pH, temperature, geometry of the ion source and from a series of instrumental parameters that are still difficult to control or standardize properly. However, even with serious limitations implied with the actual implementation of ESI/MS, which mainly reflects on the reproducibility of the spectrum pattern from one instrument to another, we make an attempt to rationalise the fragmentation pattern of a limited series of 3 substituted indoles comprehending several biologically active molecules. This approach could make order in a series of mass spectrometric analysis that look ESI/MS mainly as detection technique.

5.1. 1-Substituted indoles

Negative-ion ESI/MS and ESI/MS/MS spectra have been used for the characterisation of indole carbamate **35** (lithium salt).[47] The ESI/MS shows, beside the peak at 160 Th due to the intact anion A^- and that at 116 Th due to $[A - CO_2]^-$ ions, the peaks at 321 Th (base peak) and at 327 Th attributed to the $[2A + H]^-$ and $[2A + Li]^-$ ions, respectively.

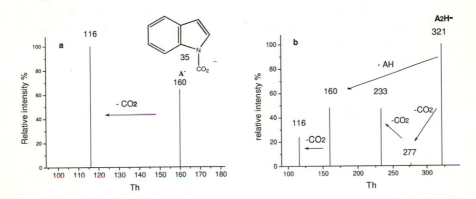

Figure 2. CID-MS/MS spectra of indole carbamate **35** at 5 eV collision energy. Selected precursor ions (**a**) A^- at 160 Th and (**b**) A_2H^- at 321 Th, adapted from ref. 47.

The CID MS/MS (5 eV collision energy) spectrum of the ions at 160 Th evidences the exclusive loss of CO_2 affording to the 116 Th ions (Figure 2a), while that of the ions at 321 Th (Figure 2b) shows two competitive fragmentation pathways, i) consecutive CO_2 losses affording to the 277 Th and 233 Th ions, ii) elimination of indole carbamic acid (HA) followed by CO_2 loss, affording to the ions at 160 Th and 116 Th, respectively.

5.2. 3-Substituted indoles

Looking at the simple ESI mass spectra of the 3-substituted indoles, like tryptophan **6**, tryptamine **7**, serotonine **8**, melatonin **11**, indole-3-acetic acid **13** and 5-hydroxytryptophol (5-hydroxy-3-(2-ethanol)-indole) **36** reported by Numan et al.,[48] it is worth to note that the MH^+ ions of these molecules are poor or absent in the spectra, despite of the softness of the technique (Table 9).

409

Table 9. ESI mass spectra (positive mode) of compounds **6-8**, **11**, **13** and **36**, from ref. 48.

Compound	MW	Th (I%)
6	204	205 (10, MH$^+$), 188 (100, [MH – NH$_3$]$^+$),170 (15), 146 (90),145 (42), 118 (40), 91 (5)
7	160	144 (100, MH$^+$), 117 (5)
8	176	160 (100, [MH – NH$_3$]$^+$), 132 (8), 114 (18)
11	232	233 (12, MH$^+$),174 (100, [MH – CH$_3$CONH$_2$]$^+$), 160 (15, Q$^+$), 143 (5)
13	175	176 (5, MH$^+$), 130 (100, Q$^+$), 103 (9)
36	177	178 (8, MH$^+$), 160 (100, [MH - H$_2$O]$^+$), 132 (10), 115 (5)

This behaviour could be determined by a relatively high cone voltage in the source, that makes the fragmentation processes favoured. Another aspect that can be inferred is that, as expected, fragmentations mainly involve the side chain substituents, while indole moiety remains quite intact. In fact the base peaks of ESI/MS of both the related amines as tryptamine **7** and serotonine **8** are due to the [MH - NH$_3$]$^+$ ions at 144 Th and 160 Th, respectively, that of melatonin **11** at 174 Th to the loss of acetamide elements, that of the hydroxytryptophol **36** at 160 Th to the [MH - H$_2$O]$^+$ ions, that of the indole-3-acetic acid **13** at 130 Th to [MH - CH$_2$O$_2$]$^+$ quinolinium ions. The mass spectrum of aminoacid tryptophan **6** is dominated by the presence of two major ions of quite equal amounts, at 188 Th and at 146 Th. Numan *et al.* suggested that the ion [MH-NH$_3$]$^+$ is responsible of the peak at 188 Th.[48] They attributed the ion at 146 Th to a further loss of ketene giving the protonated indole-3-aldehyde **37** (Scheme 3), on the ground of a similar behaviour previously reported to rationalise the presence of the base peak at 160 Th in the CID-MS/MS spectrum of the ESI generated MH$^+$ ions of the tryptophan methyl ester **38**.[49]

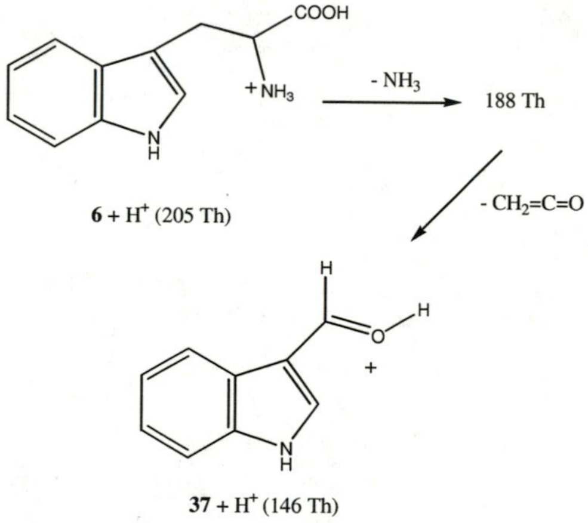

Scheme 3

The loss of ammonia from protonated amino acids has been well established by CI/MS,[50] desorption MS,[51,52] FAB/MS/MS[53] and ESI/MS/MS[54] investigations. However, in the case of tryptophan **6**, some doubts on the interpretation of the formation of the ions at 146 Th, by ketene loss from the [MH - NH$_3$]$^+$ ion at 188 Th, should occur.

Really, Prinsen *et al.*[49] reported, in a work focused on a rapid and reliable method to analyse indole-3-acetic acid metabolism in bacteria, the ESI/MS and the CID-MS/MS spectra of the MH$^+$ ions formed by ESI

410

of the indole derivatives **37-43**. The ESI/MS are mainly characterised by the presence of abundant protonated MH$^+$ and cationated [M + NH$_4$]$^+$ and [M + Na]$^+$ ions. The CID-MS/MS spectra are reported in Figure 3a-g.

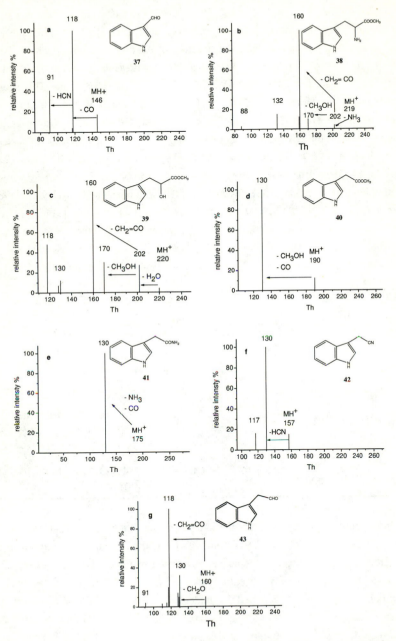

Figure 3. CID MS/MS spectra (collision energy 20 eV) of the ESI generated MH$^+$ ions of compounds **37-43**, adapted from ref. 49.

411

The MH$^+$ ions of tryptophan methyl ester **38** lose NH$_3$ giving the ions at 202 Th, while the loss of CH$_3$OH is not observed. The CH$_3$OH elimination follows the NH$_3$ one giving the ion at 170 Th. Similarly, the consecutive losses of H$_2$O and CH$_3$OH are observed for **39**, affording to ions at 202 Th and 170 Th, respectively. This should confirm that OH$^.$loss from MH$^+$ of tryptophan **6** does not contributes to the 188 Th ions. Concerning the formation of the ions at 160 Th present in the CID-MS/MS spectra of both **38** and **39**, this has been attributed[49] to the ketene loss from the ions at 202 Th ([MH - NH$_3$]$^+$ and [MH - H$_2$O]$^+$, respectively). The resulting ions should have the structure of the protonated molecule **43**. However, while the CID-MS/MS spectrum of the MH$^+$ of **39** below 160 Th presents a pattern similar to that of **43**, completely different ions are present in CID-MS/MS spectrum of the MH$^+$ of **38**. This should rule out a structural identity between the ions at 160 Th arising from the MH$^+$ ion of **38** and those formed by the MH$^+$ ions of both **43** and **39**. The formation of the ions at 160 Th in the CID-MS/MS spectrum of **38**[49] as well as that of the ions at 146 Th in the ESI-MS spectrum of **6**[48] should involve other routes not yet clear. In fact, the presence of the ions at 162 Th in the ESI-CID-MS/MS spectrum of the MH$^+$ ions (221 Th) of 5-hydroxytryptophan **44** (Table 10)[55] could correspond (with the appropriate mass shift) to the ions at 146 Th of **6**. Its formation involves NH$_3$ and COOH losses from MH$^+$ accompanied by two hydrogen migration. The other abundant ions in the ESI-CID-MS/MS of **44** at 204 Th and 175 Th are due to the loss of NH$_3$ and HCOOH respectively from MH$^+$, while the ions at 132 Th have been attributed to 5-hydroxyindolyl cation by loss of the whole side chain. The CID-MS/MS spectrum of the ion at 162 Th has been also recorded, but the reported data and their interpretation by Williamson *et al.*[55] are not enough for its structural assessment.

Table 10. CID-MS/MS spectra of the [MH]$^+$ at 221 Th and 162 Th ions of 5-hydroxytryptophan **44** from ref. 55.

Compound	Selected precursor ion Th	Daughter ions Th (I%)
44	221 MH$^+$	204 (100, [MH − NH$_3$]$^+$), 186 (6), 175 (10, [MH − HCOOH]$^+$), 162 (50, [MH − CO$_2$NH]$^+$), 149 (5), 132 (10)
	162 [MH -CO$_2$NH]$^+$	134 (80), 116 (40), 106 (100), 78 (20)

Finally, in our opinion, the formation of the ions at 160 Th in the CID-MS spectrum of **39** should be better rationalised by combined loss of CH$_3$OH and CO (formal elimination of formic acid methyl ester) from the MH$^+$ rather than by the proposed[49] ketene elimination from the [MH - H$_2$O]$^+$ ions at 202 Th.

On the other hand, in a recent work entirely dedicated to the study of the gas-phase reactions of protonated tryptophan **6**, Lioe *et al.*[56] reported that under low energy CID-MS/MS conditions (ion trap), the MH$^+$ ions (generated by ESI) of tryptophan **6** (205 Th), tryptophan O-methyl ester **38** (219 Th), tryptophanyl-glycine **45** (262 Th) and tryptophanyl-glycine O-methyl ester **47** (276 Th) fragment exclusively by loss of ammonia. In the CID-MS/MS spectrum of the protonated tryptophan **6** the only other peak observed was that at 159 Th (due to the combined loss of water and CO) with a very low relative abundance (Figure 4a). These Authors focused their work on the mechanism for the ammonia loss. In fact, although as reported above the loss of ammonia from protonated amino acids was a known process,[50-54] in some cases the mechanisms were obscure or in contention. In particular, for tryptophane **6**, two different mechanisms were previously proposed. The first was suggested by Prokai *et al.*[57] by using the fixed charge tryptophane derivative **48** under FAB conditions and AM1 semiempirical calculations. They suggested a cyclisation by nucleophilic attack from the C-2 position of the indole ring, followed by ring expansion (Scheme 4). The stability of the final ion was suggested as the driving force of the process.

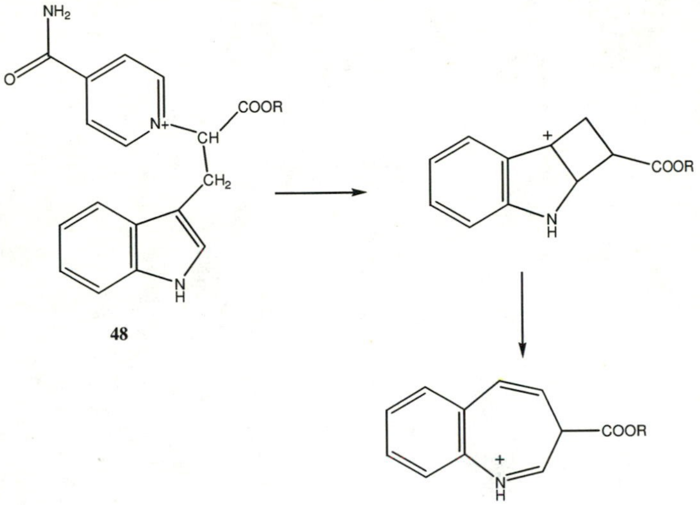

Scheme 4

The second was proposed by Rogalewicz *et al.*[54] that investigated the gas-phase reactions of tryptophan **6** by CID-MS/MS using a triple quadrupole mass spectrometer and deuterium labelling experiments. They found that **6** exclusively loses ammonia and that when all the NH and OH hydrogens are replaced by deuterium the losses of ND_3, ND_2H and NHD_2 occur, evidencing a hydrogen-deuterium scrambling in the gas-phase. This scrambling was rationalised by an initial protonation of the indole ring and subsequent isomerisation reactions through 1,2-hydride shifts around the indole ring followed by ammonia loss *via* nucleophilic attack from the C-4 (Scheme 5, Path A). However, several alternative mechanisms may be possible for the loss of ammonia from protonated tryptophan. In fact, a nucleophilic attack from the C-3 position of the indole ring affording to a spirocyclopropane derivative (Scheme 5, Path B), or the attack of C-8 or C-9 positions of the indole affording to tricyclic ions (Scheme 5, Path C and Path D, respectively) could be also considered.

In particular, the condensed phase isomerisation of a tryptophan analogue has been proposed to occur through the formation of a spirocyclopropane intermediate as in path B (Scheme 5)[58,59] and a related spirocyclopropane product ion formed by the loss of ammonia from protonated tyrosine has been proposed[53] and supported by theoretical calculations.[58] Furthermore, other alternative mechanisms could account the scrambling observed for the loss of ammonia from deuterium-labelled tryptophan, *i.e.* intramolecular proton transfer from the protonated amino group to either the C-2 or C-4 position of the indole ring. In particular, a mechanism involving scrambling of the hydrogen atom linked at C-4 position has been evidenced under photolysis in condensed phase.[60-63]

Hence, further CID MS/MS experiments on the MH$^+$ ions of tryptohan **6** and its deuterium labelled derivatives **6d$_1$**, **6d$_2$**, and the MD$^+$ ions of **6d$_4$** and **6d$_7$** (Figure 4a-e), were performed.[56] The MD$^+$ ions were obtained by using CH$_3$OD/D$_2$O instead of unlabelled solvent. NMR experiments demonstrated that the aromatic protons do not exchange in this condition.[56]

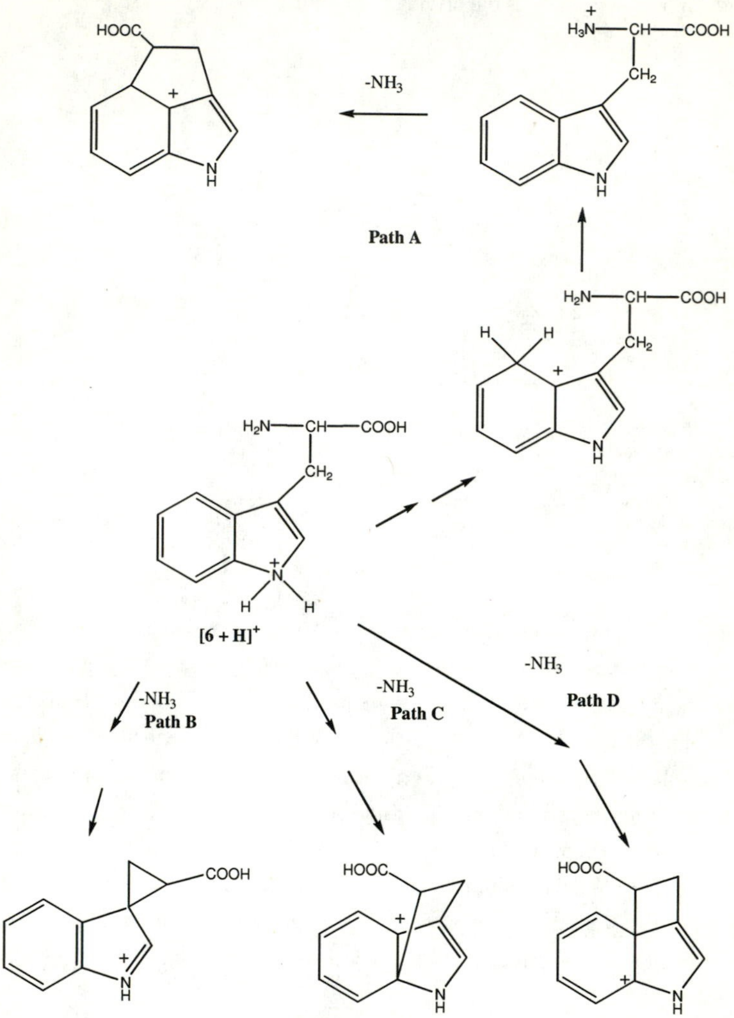

Scheme 5

The losses of ND$_3$, ND$_2$H and NH$_2$D in the ratio 2:4:1 from the MD$^+$ ions of **6d$_4$** are in agreement with the results of Rogalewicz.[54] This evidences that H/D scrambling, occurring in the gas phase *via* intramolecular proton transfer, is energetically more favoured than NH$_3$ loss. Further, the abundant loss of ND$_2$H from MH$^+$ ions of both **6d$_1$** and **6d$_2$**, with a slight preference for **6d$_1$**, evidences that proton transfer

414

occurs from C-2 and C-4 indole carbons. Lioe and co-workers[56] suggested the scrambling between the amine hydrogens (or deuterium atoms) and the C-2 and C-4 hydrogens (Scheme 6).

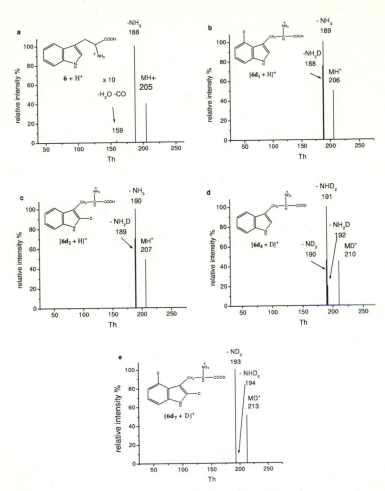

Figure 4. CID MS/MS spectra of MH$^+$ ions of **(a)** tryptophan **6**, **(b)** monodeuterotryptophan **6d$_1$**, **(c)** dideuterotryptophan **6d$_2$** and of MD$^+$ ions of **(d)** tetradeuterotryptophan **6d$_4$** and **(e)** heptadeuterotryptophan **6d$_7$**, adapted from ref. 56.

In fact, considering that previous computational studies predicted as preferred site of protonation of tryptophan **6** the amino nitrogen[64] rather than the indole nitrogen, they believed to be unlikely the H/D scrambling *via* a series of 1,2-hydride shifts from N-indole protonated tryptophan proposed by Rogalewicz *et al.* (Scheme 5).[54]

This was also supported by molecular orbital calculations B3LYP/6-31G* for optimised structures of the lowest energy conformers of tryptophan protonated at the amino nitrogen and at the indole nitrogen. The latter resulted the much higher in energy (+ 22.8 kcal/mol).

The transition state energy, calculated at B3LYP/6-31G* level of theory revealed that the average value of transition state barriers for 1,2-hydride shifts is 32.65 kcal/mol (Scheme 5). This value is considerably higher than those for direct proton transfer between the amino nitrogen and the C-2 and C-4 indole position (Scheme 6) that are 11.5 and 17 kcal/mol, respectively. These findings also reflect the very small relative abundance for the loss of ND$_2$H compared to ND$_3$ observed for the MD$^+$ ions of **6d$_7$**.

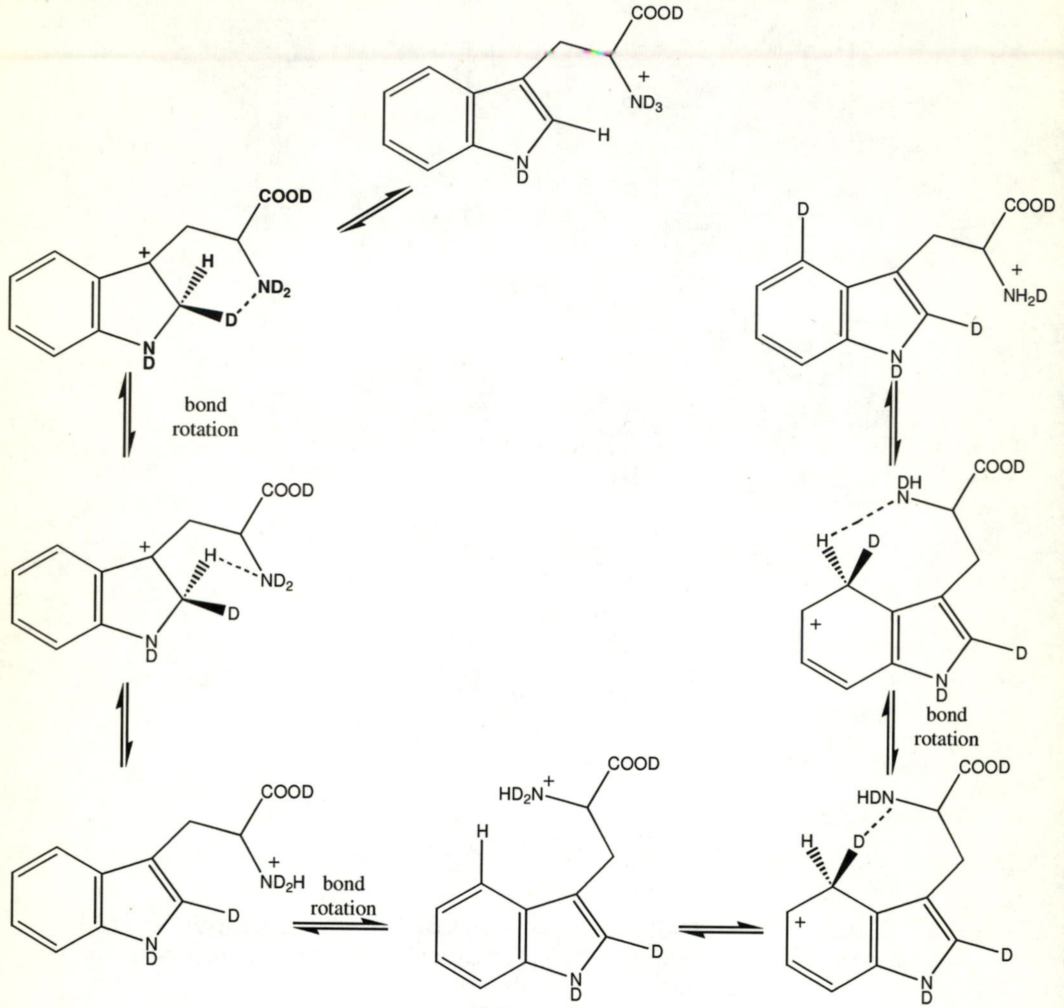

Scheme 6

They also calculated the energies of the isomeric ion structures formed upon loss of ammonia from protonated tryptophan [M + H −NH$_3$]$^+$ and compared these against those proposed by other researchers,[50-54,57] formed by nucleophilic attack from position 2, 3, 4, 8 and 9 of the indole side chain (Table 11).

The calculations indicate that the structure **D**, formed by nucleophilic attack from the C-3 of indole ring has the greater thermodynamic stability. Further, the relative energy of the transition state relative to the most stable protonated tryptophan conformer was found + 23.3 kcal/mol, *i.e.* 6.3 kcal/mol higher than the highest transition state barrier for the intramolecular proton transfer (Scheme 6), and this accounts for the occurrence of H/D scrambling, prior to ammonia loss, observed for deuterium labelled compounds. Finally, the competing reaction affording to loss of H_2O and CO requires proton transfer from the amino nitrogen (the thermodynamically favoured protonation site) to hydroxyl group of the carboxylic group, and the very high transition state barrier calculated (+ 50.0 kcal/mol) agrees with the small relative abundance of the resulting ions.

Table 11. B3LYP/6-31G* predicted total energies and zero point vibrational energies (ZPVE) for various $[6 + H - NH_3]^+$ ions from reference 56.

Structure	Total energies (Hartrees) B3LYP/6-31G*	Total energies (Hartrees) ZPVE	Relative energies (kcal/mol)
A	-630.115341	0.187946	19.3
B	-630.130581	0.188585	10.1
C	-630.115783	0.188732	19.5
D	-630.146511	0.188382	0.0
E	-629.993912	0.185989	94.3
F	-630.080520	0.187308	40.7

Also the gas-phase reaction of protonated melatonin **11** has been investigated in some extent. Its ESI/MS is fundamentally characterised by the formal loss of acetamide from the MH$^+$ ions at 233 Th affording to the ions at 174 Th.[48]

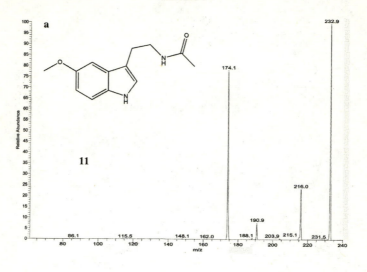

233 Th
[11 + H]⁺

174 Th

Scheme 7

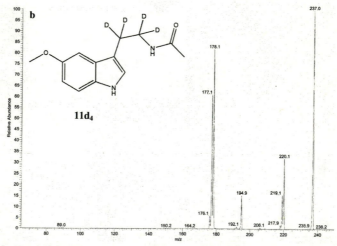

Figure 5. CID-MS/MS spectra of the ESI generated MH⁺ ions of (**a**) melatonin **11** and (**b**) α,α,β,β tetradeuteromelatonin **11d₄**.

We also observed the same fragmentation, even if in our hands the MH$^+$ ions at 233 Th largely predominated. This fragmentation is also occurring in chemical ionisation (CI, methane) and FAB conditions,[31,32] as well as by CID on the protonated molecule generated either in ESI[65,66] or CI experiments.[66] Xie *et al.* [65] suggested that the ions at 174 Th (63 %) in ESI-MS/MS experiments is formed by loss of HN=C(OH)-CH$_3$ from the protonated molecule through a four-membered transition state (Scheme 7). Even if not discussed, the spectrum also shows peaks at 216 Th (20%) and 191 Th (3%), due to the losses of OH radical and ketene, respectively.

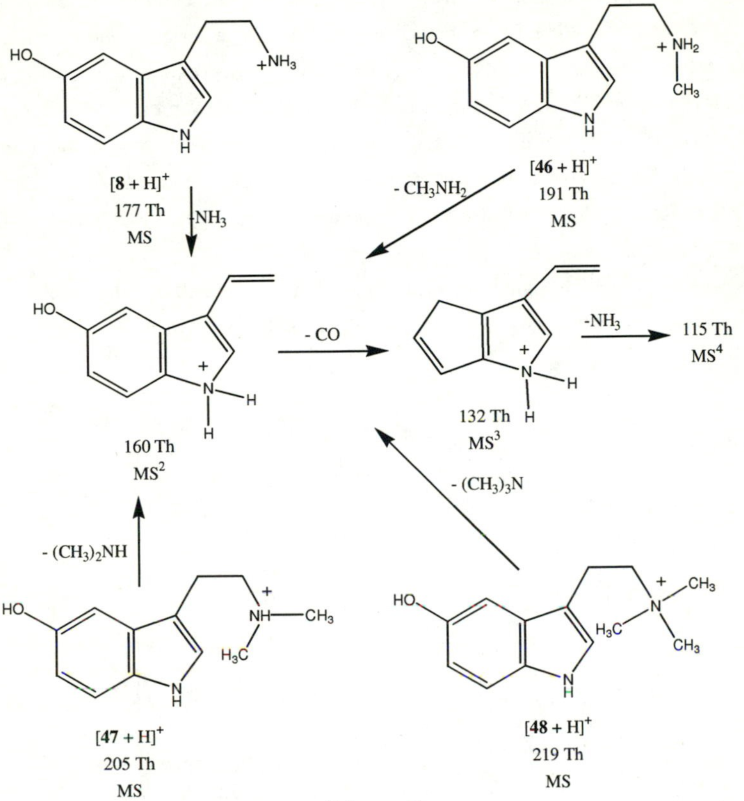

Scheme 8

However, the elimination of acetamide elements from the MH$^+$ ions of melatonin **11** cannot be rationalised by the simple mechanism reported in Scheme 7. In fact, the ESI CID-MS/MS and the CI/MS (methane) spectra of α,α–β,β-tetradeuteromelatonin **11d$_4$** show, besides the expected peaks at 237 Th and at 178 Th due to the protonated molecule and to the formal elimination of (unlabelled) acetamide, an intense peak at 177 Th attributable to the formal elimination of monodeuteroacetamide.[66] The CID-MS/MS spectra (ion trap) of the ESI-generated MH$^+$ ions of **11** and **11d$_4$** are reported in Figure 5. They evidence the occurrence of competitive reactions and/or H/D scrambling involving the side chain deuterium atoms for the acetamide elimination. In fact, the loss of unlabelled acetamide (178 Th), monodeuteroacetamide (177 Th) and, in a lesser extent, dideuteroacetamide (176 Th) from the MH$^+$ ions of **11d$_4$** at 237 Th, is observed. Of course, the same mechanisms should occur for the acetamide elimination from the MH$^+$ ions at 233 Th of the

419

native melatonin **11**. Further piece of information can be obtained by the ESI CID-MS/MS spectra of the MD$^+$ ions at 237 Th and 240 of the isotopomers of **11** and **11d$_4$** (fully deuterium exchanged at NH and OH) obtained by using CH$_3$CN/D$_2$O as ESI solvent. Both show the elimination of the elements corresponding to unlabelled acetamide (CH$_3$CONH$_2$), monodeuteroacetamide (CH$_3$CONHD), and dideuteroacetamide (CH$_3$COND$_2$). This suggests the occurrence of quite complicate mechanisms involving also H/D scrambling reactions, that are still under investigation.[67]

The fragmentation profiles for tryptamine **7**, 5-hydroxytryptamine (serotonine) **8**, *N'*-methyl 5-hydroxytryptamine **46**, *N',N'*-dimethyl 5-hydroxytryptamine (bufotenine) **47**, *N',N',N'*-trimethyl 5-hydroxytryptamine **48** and *N',N*-dimethyl 5-methoxytryptamine **49** (a series of indole amines isolated from frog skin) were carefully determined by McClean *et al.*[68] by ESI-MS^n experiments, using an ion trap mass spectrometer. Their results on MS^1 to MS^4 experiments for **8, 46-48** are summarised in Scheme 8.

This study also confirms the tendency of serotonine **8** to lose ammonia to give the fragment ions at 160 Th. This pattern and the ion at 160 Th are also common for the other compounds **46-49**. The subsequent step in the fragmentation for 5-hydroxy substituted compounds (**8, 44-46**) is the loss of CO to generate a fragment ion at 132 Th.

Analogously, *N',N*-dimethyl 5-methoxytryptamine **49** loses dimethylamine, and the resulting ion at 174 Th competitively loses formaldehyde and a methyl radical, which in turn undergoes CO elimination (Scheme 9).

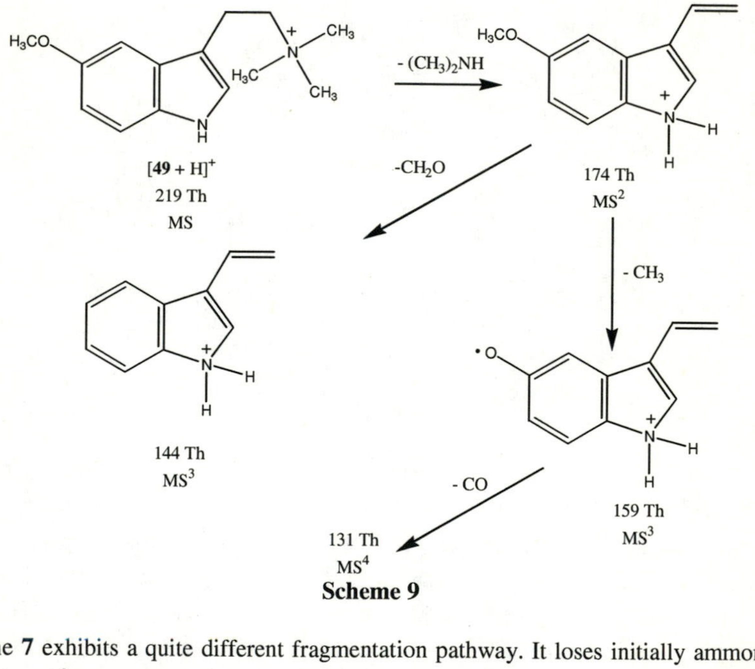

Scheme 9

Tryptamine **7** exhibits a quite different fragmentation pathway. It loses initially ammonia to give the ions at 144, but in MS^3 experiments it eliminates HCN to generate an ion at 117 Th, thus parallels the simple ESI spectra behaviour reported elsewhere.[48]

Even more complex 3-substituted indoles have been investigated. In fact, Watt *et al.*[69] studied: 3-(3-{4-[2-(3-fluoro-phenyl)-ethyl]-piperazin-1-yl}-propyl)-5-[1,2,4]triazol-4-yl-1*H*-indole **50**. This shows a product ion spectrum of the MH⁺ ions at 433 Th which is relatively rich in fragments, mainly derived from side chain losses, whose structure has been assigned by high resolution MS. The main fragmentation routes observed in the CID MS/MS spectrum of the MH⁺ ions of **50** involve cleavage reactions of the 1,3,5-triazole ring (Scheme 10).

The accurate masses of 26 fragment ions and their corresponding elemental compositions were determined, using the high resolution capabilities of the orthogonal quadrupole time-of-flight (Q-TOF) mass spectrometry. This, in addition to the analysis of second-generation products by ion-source fragmentation at high cone voltage, allowed determination of a consistent number of sequential fragmentation pathways, through software interpretation and chemical intuition. However, fragmentation of the indole ring never occurs.

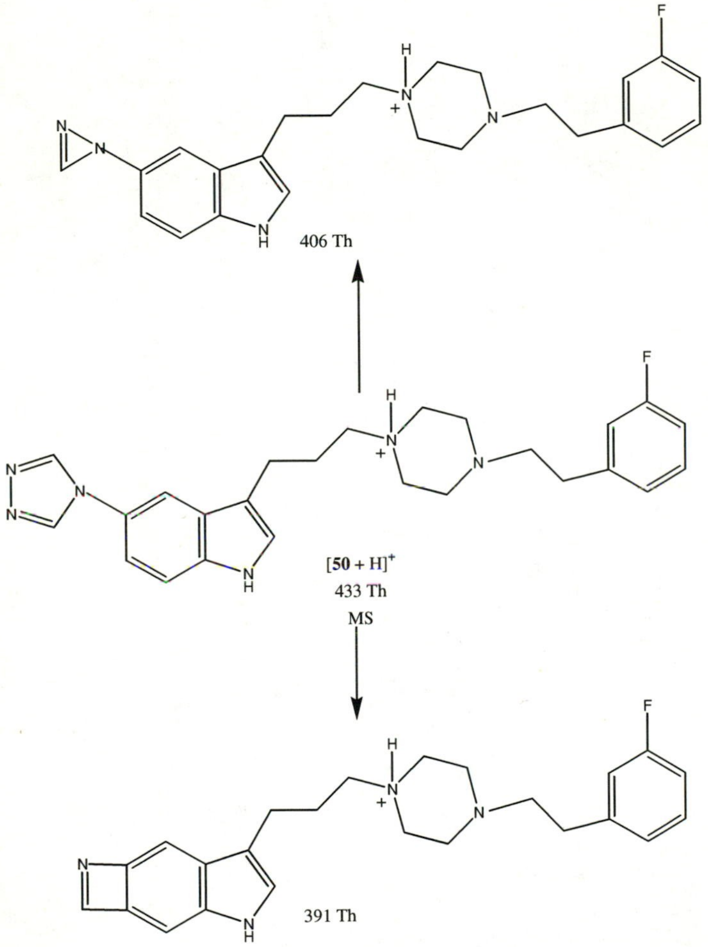

Scheme 10

5.3. Other indole derivatives

A differently substituted indole, (racemic) pindolol **51**[70] (a β-adrenergic antagonist), in simple ESI/MS experiments exhibits only the protonated molecule peak at 249 Th. The CID-MS/MS spectrum shows a base

421

peak at 172 Th corresponding presumably to the loss of water and isopropyl amine residue. The formation of the ions at 146 Th is easily rationalised, while the ions at 116 Th could arise by cleavage of one of the two ether C-O bond, which affords to isobaric ions (Scheme 11).

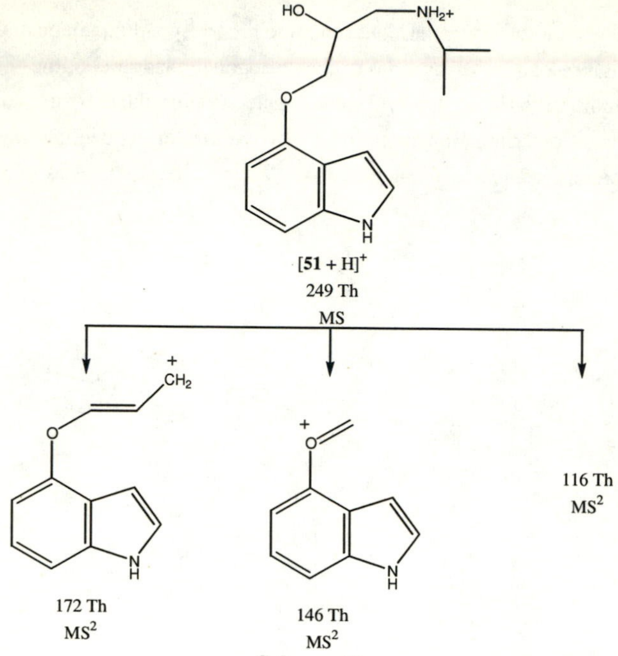

Scheme 11

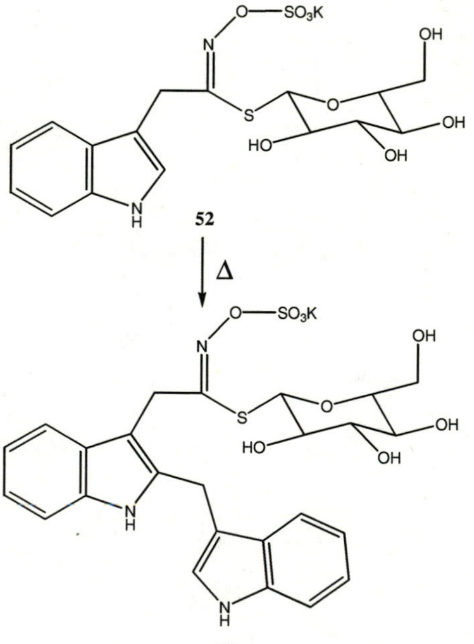

Scheme 12

ESI/MS[71] and ESI/MS[n] (using an ion trap mass spectrometer) of positive and negative ions[72] have been used for structure elucidation of 2-(3'-indolylmethyl)glucobrassicin 53, a thermal breakdown product of glucobrassicin 52 which represents the most widespread indole glucosinolate present particularly in cruciferous vegetables of the *Brassica* genus[73,74] (Scheme 12).

Glucobrassicin 52 and other glucosinolates and their breakdown products are known for their goitrogenic, antinutritional and cancer-preventing activities.[75] The CID-MS/MS spectrum of the ESI generated parent ions of 53 at 578 Th shows the ion at 498 Th as the only daughter ion, due to the loss of the sulphate moiety. By sequential MS[3] experiment, this ion gives the 381 Th, 369 Th, 336 Th (base peak) and 219 Th ions, that could be rationalised as shown in Scheme 13.

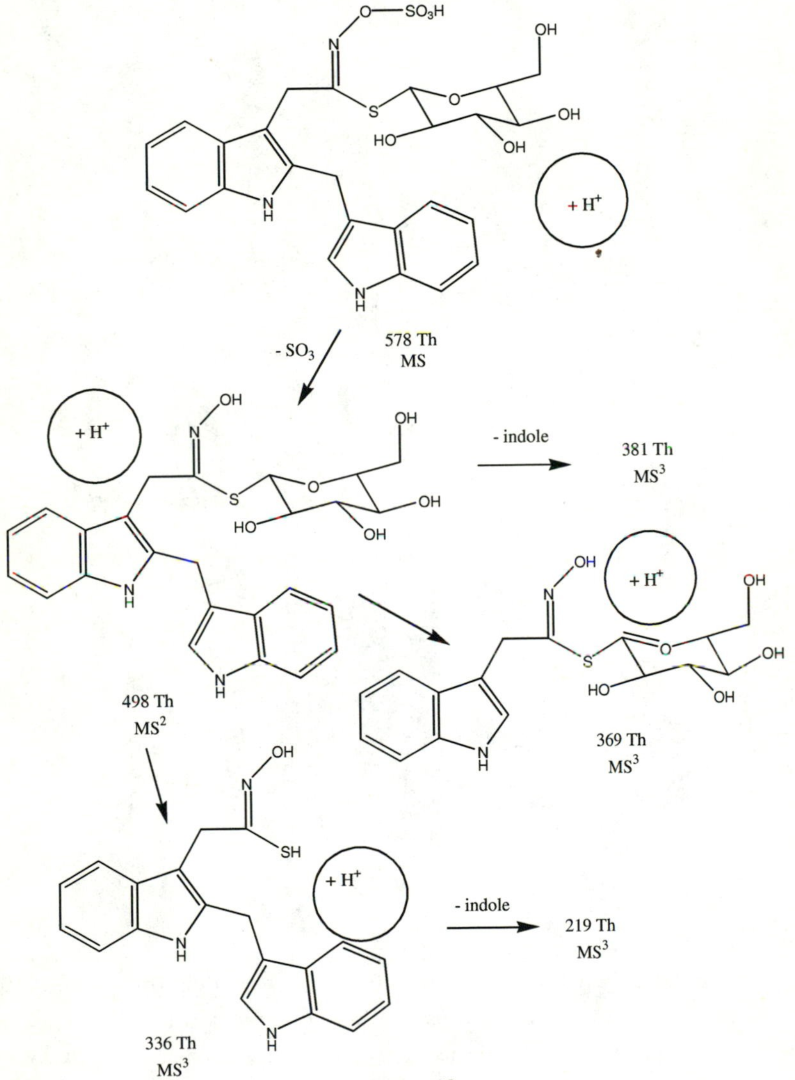

Scheme 13

423

The CID-MS/MS spectrum of the parent ion at 576 Th of **52** generated by ESI in negative mode shows fragmentation involving exclusively the side chain at 3 position (Scheme 14).

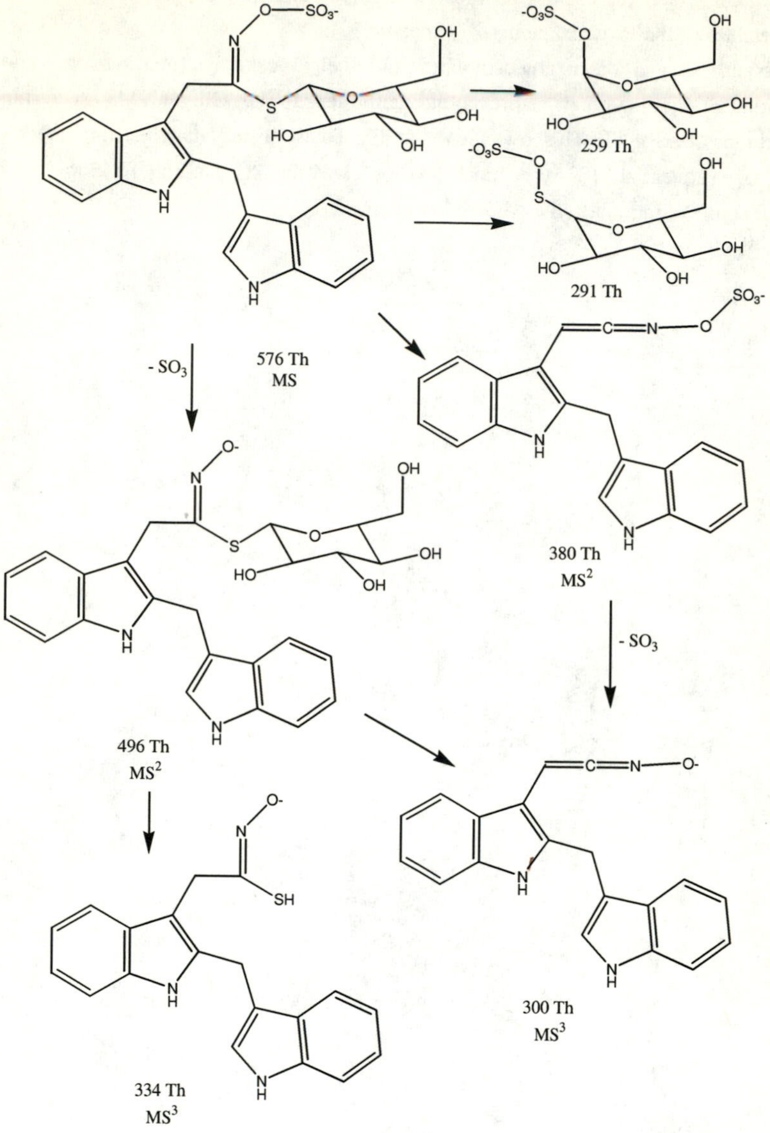

Scheme 14

6. Conclusions

Soft ionisation techniques have been widely used for indoles, even if the most of the studies described some mechanistic aspects in works that are focused mainly on analytical aspect of mass spectrometry. However, very recent papers have been oriented to accurate investigation (with the aid of labelling experiments and high resolution measurements) of the gas-phase ion chemistry of positive and negative ions

424

generated by soft ionisation methods. Interestingly, some apparently simple elimination reactions (ammonia from protonated tryptophan **6** or acetamide from protonated melatonin **11**) occur through quite complex mechanisms which also involve H scrambling reactions.

In general, owing their higher internal energy, FAB generated ions fragment more extensively than ESI generated ions. In both techniques the most of the fragmentation reactions involve the side chain substituents. In particular, ESI/MS and ESI/MS/MS experiments on substituted indoles, either in positive or in negative mode, leave quite intact the aromatic indole portion of molecule, so reflecting the high stability of the indole system. Another evidence is that during fragmentation reactions the charge (positive or negative) is quite always retained by the moiety containing the indole ring, and this is probably due to the indole ring intervention on charge stabilisation.

Finally, FT-MS studies leaded to the determination of gas phase acidity and gas phase basicity of some simple indole derivatives.

Acknowledgments

Financial support for our research program in the area of ion chemistry through mass spectrometric methods by the University of Palermo (Fondi R.S. ex 60%) is gratefully acknowledged.

References

1. Part 1: Bongiorno, D.; Camarda, L.; Ceraulo, L.; Ferrugia, M. In *Targets in Heterocyclic Systems*; Attanasi, O. A.; Spinelli, D., Eds; Italian Chemical Society: Rome, 2003; Vol. 7, p. 174.
2. Comisarow, M. B.; Marshall, A. G. *Chem. Phys. Letter* **1974**, *25*, 282.
3. Comisarow, M. B.; Marshall, A. G. *Chem. Phys. Letter* **1974**, *26*, 489.
4. Amster, I. J. *J. Mass Spectrom.* **1996**, *31*, 1325.
5. Marshall, A. G; Hendrickson, C. L.; Jackson, G. S. *Mass Spectrom.Rev.* **1998**, *17*, 1.
6. Bowers, M. T. *Gas Phase Ion Chemistry*; Academic Press, New York, 1979; Vol. 2.
7. Guan, S.; Marshall, A. G. *Rapid Commun. Mass Spectrom.* **1993**, *7*, 857.
8. Beu, S. C.; Senko, M. V; Quiinn, J. P.; McLafferty, F. W. *J. Am. Soc. Mass Spectrom.* **1993**, *4*, 190.
9. Kelleher, N. L.; Senko, M. W.; Little, D.; O'Connor, P. B.; McLafferty, F.W. *J. Am. Soc. Mass Spectrom.* **1995**, *6*, 220.
10. Smith, R. D.; Cheng, X.; Bruce, J. E.; Hofstadler, S. A; Anderson, G. A. *Nature* **1994**, *369*, 137.
11. Yagil, G. *J. Phys. Chem.* **1967**, *71*, 1034.
12. Lumme, P.; Seppalainen, N. *Suomen Kemistilehti B* **1962**, *35*, 123; *Chem. Abstr.* **1963**, *58*, 1056h.
13. Cativiela, C.; Dejardin, J. L.; Elguero, J.; Garcia, J. L.; Gonzalez, E.; Mayoral, J. A. *Collect. Czech. Chem. Commun.* **1990**, *55*, 72.
14. Notario, R.; Abboud, J. L. M.; Cativiela, C.; Garcìa, J.; Herreros, M.; Homan, H.; Mayoral, J. A.; Salvatela, L. *J. Am. Chem. Soc.* **1998**, *120*, 13224.
15. Abboud, J. L. M.; Herreros, M.; Notario, R.; Esseffar, M.; Mó, O.; Yáñez, M. *J. Am. Chem. Soc.* **1996**, *118*, 1126.
16. Harrison, A. G. *Mass. Spectrom. Rev.* **1997**, *16*, 201.
17. Locke, M. J.; McIver, Jr, R. T. *J. Am. Chem. Soc.* **1983**, *105*, 4226.
18. Gorman, G. S.; Speir, J. P.; Amster, I. J. *J. Am. Chem. Soc.* **1992**, *114*, 3986.
19. Cooks, R. G.; Patrick, J. S.; Kotiaho, T.; McLuckey, S. A. *Mass Spectrom. Rev.* **1994**, *13*, 287.
20. Lias, S. G.; Liebman, J. F.; Levin, R. D. *J. Phys. Chem. Ref. Data* **1984**, *13*, 695.
21. Hunter, E. P.; Lias, S. G. *J. Phys. Chem. Ref. Data* **1998**, *27*, 413.
22. Somers, K. R. F.; Kryachko, E. S.; Ceulemns, A. *Chem. Phys.* **2004**, *301*, 61.
23. Bojesen, G.; Brendahl, P. *J. Chem. Soc., Perkin Trans. 2* **1994**, 1029.
24. Li, X.; Harrison, A. G. *Org. Mass Spectrom.* **1993**, *28*, 366.
25. Shama, P.; Mirza, S. P.; Prabhacar, S.; Vairamani, M. *Rapid. Commun. Mass Spectrom.* **2001**, *15*, 957.

26. Maksić, Z. B.; Kovačević, B. *Chem. Phys. Lett.* **1999**, *307*, 497.
27. Cheng, X.; Wu, Z.; Fenselau, C. *J. Am. Chem. Soc.***1993**, *115*, 4844.
28. Barber, M.; Bordoli, R. S.; Sedgwick, R.D.; Tyler, A. N. *J. Chem. Soc., Chem. Commun.* **1981**, 325.
29. Alberth. W.; Straub, K. M.; Burlingame, A. M.; *Anal. Chem.***1982**, *54*, 2029.
30. Bertazzo, A.; Catinella, S.; Traldi, P. *J. Mass Spectrom.* **1996**, *31*, 735.
31. Ceraulo, L.; Mele, A.; Panzeri, W.; Selva, A. *Adv. Mass Spectrom.* **2001**, *15*, 791.
32. Buongiorno, D.; Ceraulo, L.; Mele, A.; Panzeri, W.; Selva, A.; Turco Liveri, V. *J. Mass Spectrom.* **2001**, *36*, 1189.
33. Van de Weert, M.; Lagerwerf, F. M.; Haverkamp, J.; Heerma, W. *J. Mass Spectrom.* **1998**, *33*, 884.
34. Ostin, A.; Moritz, T.; Sandberg, G. *Biol Mass Spectrom.* **1992**, *21*, 292.
35. Hoffmann, E.; Stroobant, V. *Mass spectrometry principles and application*; John Wiley & Sons, Eds.; Chichester, 2002.
36. Karas, M.; Bachmann, D.; Bahar, U.; Hillenkamp, F. *Int. J. Mass Spectrom. Ion Process* **1987**, *78*, 53.
37. Karas, M.; Hillenkamp, F. *Anal. Chem.* **1988**, *60*, 2299.
38. Tanaka, K.; Waki, H.; Ido, Y.; Akita, S.; Yoshida, Y.; Yoshida, T. *Rapid Commun. Mass Spectrom.* **1988**, *8*, 151.
39. Montaudo, G.; Lattimer, R. P. *Mass Spectrometry of Polymers*, CRC Press: Boca Raton, London, New York, Washington, D.C., 2002.
40. Vestal, M. L. In *Selected Topics in Mass Spectrometry in the Biomolecular Sciences. NATO ASI Series*; Caprioli, R.; Malorni, A.; Sindona, G., Eds.; Kluwer Acad. Pub.: London, 1979; Vol. 504, p. 243.
41. Guo, Z.; Zhang, Q.; Zou, H.; Guo, B.; Ni, J. *Anal. Chem.* **2002**, *74*, 1637.
42. Nonami, H.; Wu, F.; Thummel, R. P.; Fukuyama, Y.; Yamaoka, H.; Erra-Balsells, R. *Rapid Commun. Mass Spectrom.* **2001**, *15*, 2354.
43. Gieniec, J.; Mack, L. L.; Nakamae, K.; Gupta, C.; Kumar, V.; Dole, M. *Biomed. Mass Spectrom.* **1984**, *11*, 259.
44. Fenn, J. B.; Mann, M.; Meng, C. K.; Wong, S. F.; Whitehouse, C. M. *Science* **1989**, *246*, 64.
45. Whitehouse, C. M.; Dreyer, R. N.; Yamashita, M.; Fenn, J. B. *Anal. Chem.* **1985**, *57*, 675.
46. Fenn, J. B. *J. Am. Chem. Soc.* **1993**, *4*, 524.
47. Gross, H. J.; Eckert, A.; Siebert, W. *J. Mass Spectrom.* **2002**, *37*, 541.
48. Numan, A.; Danielson, N. D. *Anal. Chim. Acta* **2002**, *460*, 29.
49. Prinsen, E.; Donger, W. V.; Esmans, E. L.; Onckelen, H. A. V. *J. Mass Spectrom.* **1997**, *32*, 12.
50. Milne, G. W.; Axenrod, T.; Falex, H. M. *J. Am. Chem. Soc.* **1970**, *92*, 5170.
51. Parker, C. D.; Hercules, D. M. *Anal. Chem.* **1985**, *57*, 698.
52. Bouchonnet, S.; Denhez, J. P.; Hoppilliard, Y.; Mauriac, C. *Anal. Chem.* **1992**, *64*, 743.
53. Dookeran, N. N.; Yalcin, T.; Harrison, A. G. *J. Mass Spectrom.* **1996**, *31*, 500.
54. Rogalewicz, F.; Hoppilliard, Y.; Ohanessian, G. *Int. J. Mass Spectrom.* **2000**, *195/196*, 565.
55. Williamson, B. L.; Johnson, K. L; Tomlinson, A. J.; Gleich, G. J.; Naylor, S. *Toxicol. Lett.* **1998**, *99*, 139.
56. Lioe, H.; O'Hair, A. J.; Reid, G. E. *J. Am. Soc., Mass Spectrom.* **2004**, *15*, 65.
57. Prokai, L.; Prokai-Tatrai, K.; Pop, E.; Bodor, N.; Lango, J.; Roboz, J. *Org. Mass Spectrom.* **1993**, *28*, 707.
58. Shoeib, T.; Cunje, A.; Hopkinson, A. C.; Siu, K. V. M. *J. Am. Soc., Mass Spectrom.* **2002**, *13*, 408.
59. Johansen, J. E.; Christie, B. D.; Rapoport, H. *J. Org. Chem.* **1981**, *46*, 4914.
60. Saito, I.; Sugiyama, H.; Yamamoto, A.; Muramatsu, S.; Matsuura, T. *J. Am. Chem. Soc.* **1984**, *106*, 4286.
61. Saito, I.; Muramatsu, S.; Sugiyama, H.; Yamamoto, A.; Matsuura, T. *Tetrahedron Lett.* **1985**, *26*, 5891.
62. Shizuca, H.; Serizawa, M.; Shimo, T.; Saito, I.; Matsuura, T. *J. Am. Chem. Soc.* **1988**, *110*, 1930.
63. Cozens, F.; McClelland, R. A.; Steenken, S. *Tetrahedron Lett.* **1992**, *33*, 173.
64. Maksik, Z. B.; Kovacevic, B. *Chem. Phys. Lett.* **1999**, *307*, 497.
65. Xie, F.; Wong, P.; Yoshioka, K.; Cooks, R. G.; Kissinger, P. T. *J. Liq. Chromatogr. Rel. Technol.* **1998**, *21*, 123.

66. Bongiorno, D.; Ceraulo, L.; Ciofalo, M.; Mele, A.; Selva, A.; Turco Liveri, V. *Massa 2002*, Cetraro, 27th June – 1st July 2002, Proceedings P. 113.

67. Note in preparation.

68. McClean, S.; Robinson, R. C.; Shaw, C.; Smith, W. F. *Rapid Commun. Mass Spectrom.* **2002**, *16*, 346.

69. Watt, A. P.; Pike, A.; Morrison, D. *J. Am. Soc., Mass. Spectrom.* **2001**, *12*, 1145.

70. Motoyama, A.; Suzuki, A.; Shirota, O.; Namba, R. *J. Pharmac. Biomed. Anal.* **2002**, *28*, 97.

71. Chevolleau, S.; Gasc, N.; Rollin, P.; Tulliez, J. *J. Agric. Food Chem.* **1997**, *45*, 4290.

72. Chevolleau, S.; Debrauwer, L.; Boyer, G.; Tulliez, J. *J. Agric. Food Chem.* **2002**, *50*, 5185.

73. Fenwick, G. R.; Heaney, R. K.; Mullin, N. *CRC Crit. Rev. Food Sci. Nutr.* **1983**, *18*, 123.

74. McDanell, R.; McLean, A. E. M.; Hanley, A. B.; Heaney, R. K.; Fenwick, G. R. *Food Chem. Toxicol.* **1988**, *26*, 59.

75. Fahey, J. W.; Zalcmann, A. T.; Talalay, P. *Phytochemistry* **2001**, *56*, 5; Corrigendum **2002**, *59*, 237.

SYNTHESIS OF BENZO[c]QUINOLIZIN-3-ONES: DISCOVERY AND DEVELOPMENT OF NOVEL INHIBITORS OF HUMAN STEROID 5α-REDUCTASE

Antonio Guarna, Fabrizio Machetti and Ernesto G. Occhiato

Dipartimento di Chimica Organica "U. Schiff" Università di Firenze and Istituto di Chimica dei Composti Organometallici – CNR

Via della Lastruccia 13, I-50019 Sesto Fiorentino (Firenze), Italy (e-mail: antonio.guarna@unifi.it)

Abstract. *Dihydrotestosterone (DHT) is produced by the NADPH-dependent reduction of testosterone (T) under catalysis of the enzyme steroid 5α-reductase (5αR) (EC 1.3.99.5). The DHT production is in many cases related to the maintenance of some pathological human diseases and endocrine disorders, so that the use of 5αR inhibitors for the possible control or suppression of DHT formation is a therapeutic target. Two different DNA-encoded isoenzymes of 5α-reductase, named type 1 and type 2 (5αR-1 and 5αR-2) are known, which transform T into DHT with different efficacy. This isozyme, involved in the transformation of testosterone to dihydrotestosterone mainly in human skin and scalp tissues, is possibly causing the development and maintenance of disorders such as acne and androgenic alopecia in men, and hirsutism in women. The reported inhibitors are benzo[c]quinolizin-3-one derivatives bearing at position 1, 4, 5, and/or 6 a methyl group and at position 8, i.e. on the aromatic ring, a full range of diverse substituents. All these compounds were tested toward 5αR-1 and 5αR-2 expressed in CHO cells resulting, in most cases, selective inhibitors of the type 1 isoenzyme, with inhibitory potencies (IC_{50}) ranging from 7.6 to 9100 nM. A 3D QSAR model has been developed, which could be very helpful in interpreting the biological data as well as in providing new insight for the design of new inhibitors. These selective 5αR-1 inhibitors are candidates for the development of a drug for the treatment of acne and androgenic alopecia in men, hirsutism and polycystic ovarian syndrome in woman. Moreover, by the appropriate substitution at the position 8, also some potent dual inhibitors of 5α-reductase, with IC_{50} ranging between 93 and 166 nM for both isozymes, were found. These dual inhibitors could be developed as drugs in the treatment of 5α-reductase related diseases where total suppression of the circulating DHT is necessary, as for example in the treatment of BPH.*

Contents

1. Introduction

1.1. Steroid 5α-reductase

The control of the biological action of steroids through the inhibition of specific enzymes involved in their metabolism, without significant changes in the overall profile of the other hormones has represented an attractive pharmaceutical target during the last twenty years. For example, several androgen-dependent disorders and diseases such as prostate cancer, benign prostatic hyperplasia (BPH), acne and (androgenic) alopecia in men, and hirsutism (associated with polycystic ovarian syndrome) in women appear to be related to the 5α-dihydrotestosterone (DHT) production.[1,2,3,4,5,6] For this reason, high interest has been paid to the synthesis of inhibitors of steroid 5α-reductase (5αR, EC 1.3.99.5),[7,8] a membrane bound, NADPH dependent enzyme which catalyses the selective, irreversible reduction of 4-ene-3-oxosteroids to the corresponding 5α-H 3-oxosteroids (Scheme 1).[9,10] Two isozymes of 5α-reductase in particular have been cloned, expressed and characterised (5αR-1 and 5αR-2),[9-10] whose homology is poor (50% ca.) and they have different chromosomal localisation, enzyme kinetic parameters and tissue expression patterns. For example the most important substrate, testosterone, has high affinity for the type 2 isozyme (Km=4-50 nM) while the affinity for the 5αR type 1 is considerably lower (Km=1-5 μM).

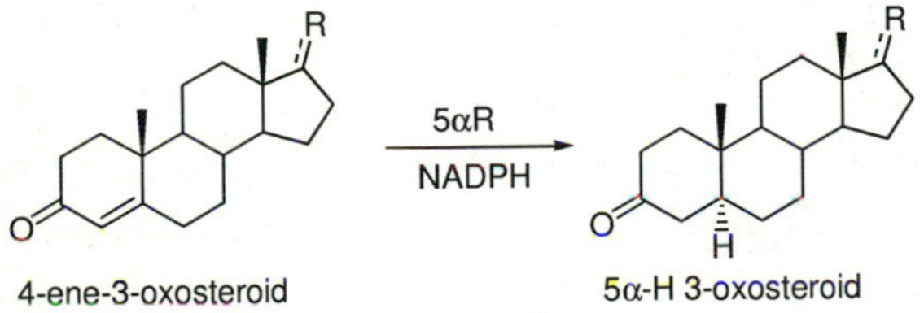

4-ene-3-oxosteroid   5α-H 3-oxosteroid

Scheme 1

1.2. Inhibitors of 5α-reductase

As a consequence of the discovery that two different DNA-encoded isoenzymes of 5α-reductase (5αR-1 and 5αR-2), transform T into DHT with different efficacy, and that these two isozymes are not equally distributed in the human tissues, 5αR-1 being present mainly in scalp, skin and liver, and 5αR-2 in the prostate,[11,12,13,14,15,16] different therapeutic approaches were developed. The synthesis and use of selective 5αR-2 inhibitors was initially envisioned for the specific treatment of a prostate disease such as BPH, although later the use of (dual) 5αR-1 and 5αR-2 inhibitors became a more efficacious therapy to completely reducing the circulating DHT.[17] The diffuse presence of 5αR-1 (although its role is not completely clear) in the scalp and skin of men suffering from pattern baldness, alopecia, or acne, and skin of women suffering from hirsutism and polycistic ovarian syndrome (PCOS), suggested a strong implication of 5αR type 1 in these disorders.[18] Therefore, the possibility of reducing the DHT level in those tissues by using selective 5αR-1 inhibitors (not affecting 5αR-2, mainly located in prostate tissue), is currently a new therapeutic approach for the treatment of the above mentioned skin disorders. In particular, the use of a "pure" 5αR-1 inhibitor is essential for the treatment of hirsutism and PCOS because of the well known risks of pseudohermafroditism for the male fetus associated to the 5αR-2 blockade in pregnant women.

Because the only information available on the structure of the 5α-reductase isozymes is the primary sequence estimated from their c-DNAs, the first inhibitors that have been synthesised were designed by modifying the structure of natural substrates, including the substitution of one carbon atom of the A or B ring of the steroids by an heteroatom. This has led to the discovery of potent inhibitors of human 5α-reductase such as 4-azasteroids (among which finasteride, and dutasteride, both marketed for the treatment of BPH),[19,20,21] 6-azasteroids,[22,23] 10-azasteroids[24,25,26,27,28] and steroidal carboxylic acid inhibitors (Scheme 2).[29,30,31,32]

Due to the potential undesired hormonal action exhibited by steroidal compounds, the research toward the discovery of nonsteroidal inhibitors has gained great importance in the last years, and several pharmaceutical and academic groups have pursued the discovery of nonsteroidal compounds that inhibit human 5α-reductases. Nonsteroidal inhibitors reported so far can be classified according to their structure. They have in fact emerged from the design of compounds mimic of (aza)steroidal inhibitors, generally by the formal removing of one or more rings from the (aza)steroidal structure. These compounds are generally thought to act all as competitive inhibitors vs. testosterone with exception of epristeride analogues which are uncompetitive inhibitors. They include benzo[f]quinolinones,[33,34,35,36] pyridones and quinolinones mimics of 4-azasteroid inhibitors,[37] benzo[c]quinolinones mimics of 6-azasteroids,[38] and benzo[c]quinolizinones mimics of 10-azasteroids.[39,40,41] The most potent and selective inhibitors of human 5αR-1 are found among these classes of compounds. Almost all the other nonsteroidal inhibitors can be grouped as carboxylic acid (generally butanoic acid) derivatives which are thought to act as noncompetitive inhibitors vs. testosterone.[42]

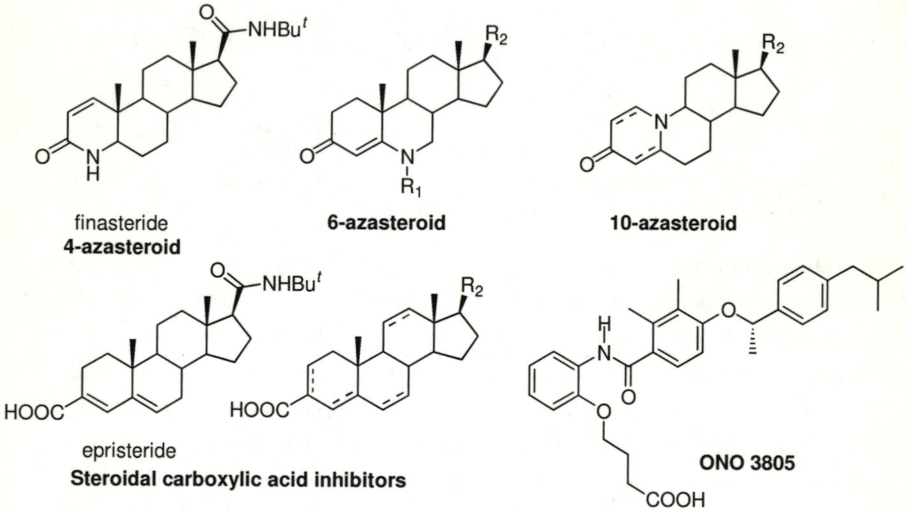

finasteride
4-azasteroid

6-azasteroid

10-azasteroid

epristeride
Steroidal carboxylic acid inhibitors

ONO 3805

Scheme 2

2. Synthesis of benzo[c]quinolizinones

2.1. Synthetic strategies. From steroidal to non steroidal inhibitors of 5αR

The preparation of benzo[c]quinolizin-3-one inhibitors (Figure 1) has required the study and development of synthetic strategies different from that employed in the synthesis of 19-nor-10-azasteroids inhibitors.[24,25,26,27,28] The latter have been prepared through the tandem thermal rearrangement-annulation of suitably functionalised isoxazoline-5-spirocyclopropanes[24,43,44,45] a methodology developed in our laboratory which allows for the sequential construction of the A and B rings of the azasteroids (Scheme 3).[46] However it is a matter of course that this procedure is inapplicable to the synthesis of benzo[c]quinolizin-3-ones due to the presence of an aromatic C ring in the structure.

As an alternative to the isoxazoline methodology, we therefore developed a different synthesis of 19-nor-10-azasteroids that was based on the TMSOTf-promoted tandem Mannich-Michael reaction of 2-silyloxy-1,3-dienes with N-(acyloxy)iminium ions generated in situ from the corresponding N-Boc-α-ethoxy derivatives (Scheme 3).[26,27,28,40]

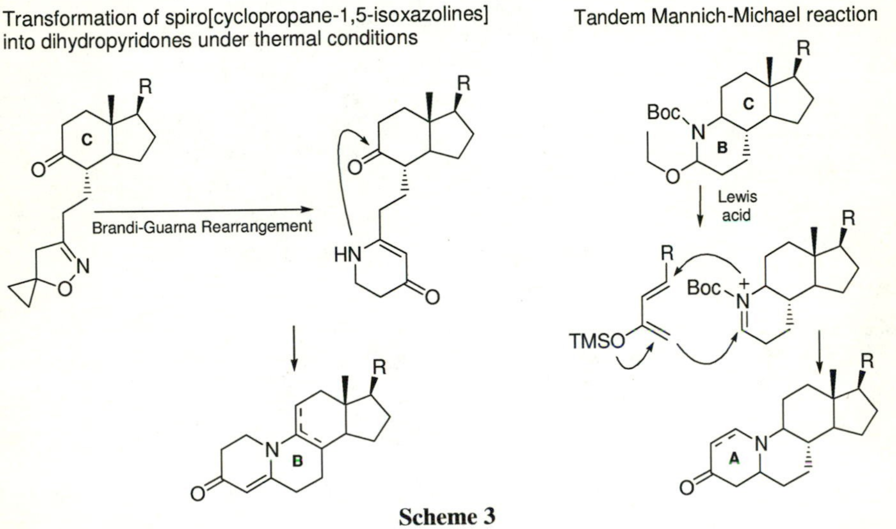

Scheme 3

2.2. First synthesis of benzo[c]quinolizinones as non-steroidal inhibitors of 5αR-1

Since the appearance in the literature of non-steroidal inhibitors of 5αR-1 we targeted two types of novel inhibitors of 5αR-1 having a benzo[c]quinolizinone skeleton (Figure 1). Surprisingly, these two novel compounds in spite of their simple chemical structure have never been previously synthesised. We based their synthesis on the tandem Mannich-Michael cyclisation between 2-silyloxy-1,3-butadienes and N-Boc iminium ion, as depicted in Scheme 4.

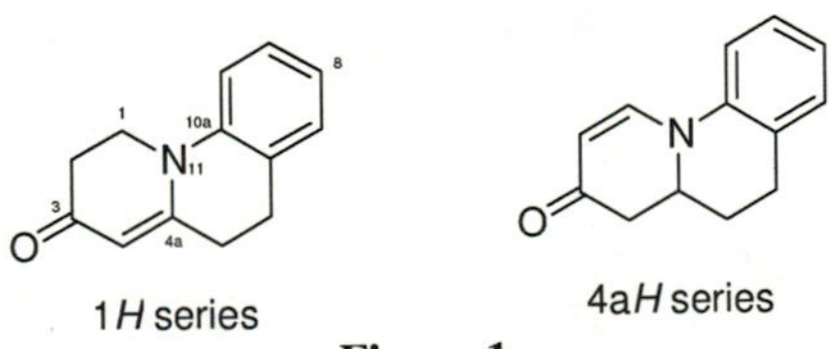

1H series   4aH series

Figure 1

The commercially available dihydroquinoline-2-one **3** was protected as N-Boc, then reduced to the ethoxy derivative **4** by treatment with NaBH$_4$ in EtOH at -25 °C followed by the slow addition of HCl 2N up to pH 3-4, according to a reported method. The key step of the synthesis was the Lewis acid catalysed tandem Mannich-Michael cyclisation of *N*-Boc iminium ion **5** with a silyloxidiene. The generation *in situ* of the *N-Boc* iminium ions from *N*-Boc-α-ethoxy derivatives such as **4**, can be promoted by different Lewis acids but in our hand the best choice was the use of TMSOTf according to the methodology reported by Pilli *et al.*[47]

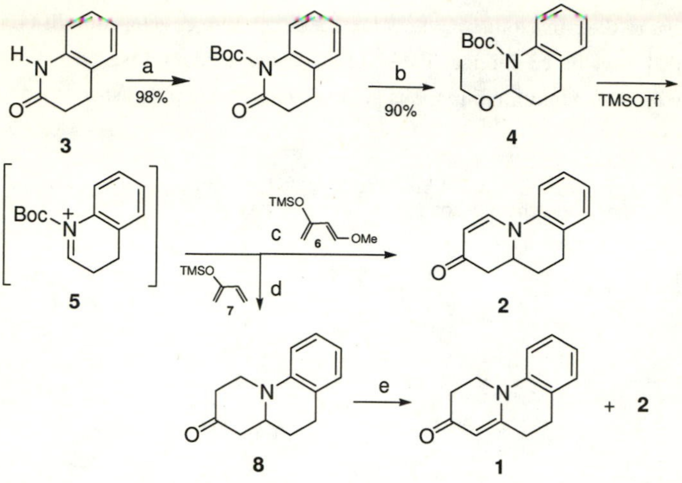

Reagents and conditions: a) Boc$_2$O, Et$_3$N, DMAP (cat.) CH$_2$Cl$_2$, 16 h, 25 °C; b) NaBH$_4$ in EtOH, -25 °C, 4 h; then HCl 2N in EtOH, pH 3-4, 0 °C, 1.5 h; c) Danishefsky's diene **6**, Et$_3$N, TMSOTf in CH$_2$Cl$_2$, 0 → 25 °C, 30 min; then NaHCO$_3$ (sat), 36 h; d) 2-silyloxy-1,3-butadiene **7**, Et$_3$N, TMSOTf in CH$_2$Cl$_2$, 0 → 25 °C, 30 min; then NaHCO$_3$ (sat), 36 h; e) Hg(OAc)$_2$, EDTA tetrasodium salt, 5% CH$_3$COOH (aq), 90 °C, 2 h.

Scheme 4

The use of Danishefsky's diene **6** led directly to 4a*H*-benzo[c]quinolizin-3-one **2** in fair yield (30%). Saturated compounds **8** was instead obtained by reaction of **5** with 2-silyloxy-1,3-butadiene **7** (prepared as reported or generated *in situ* from methyl vinyl ketone, TMSOTf and Et$_3$N). The introduction of the double bond at the 4-4a position was achieved by Hg(OAc)$_2$ oxidation, leading to target 1*H* compound **1** in 20% yield, besides a minor amount (ca. 10%) of the corresponding 4*H* derivative **2**.

2.3. Synthesis of 8-chloro-benzo[c]quinolizin-3-ones: Mannich-Michael tandem reaction of N-Boc iminium ions with 2-silyloxybutadienes

We decided to investigated the effect of the 8-chloro substituent on the inhibitory potency toward 5αR-1 of our benzo[c]quinolizinones **1** and **2** inspired by the results in term of inhibitor potencies obtained by Ely Lilly researchers.[33]

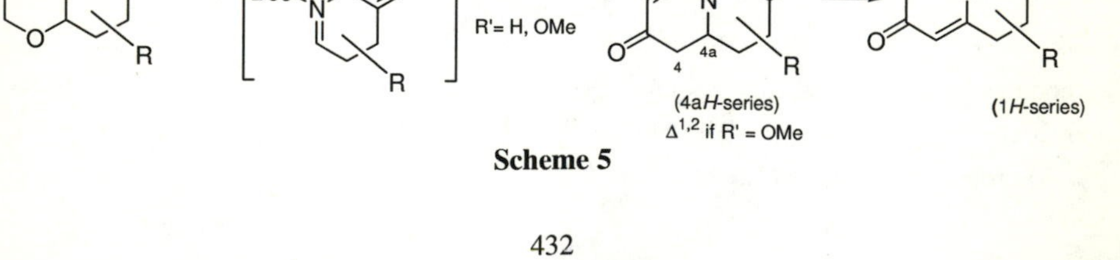

Scheme 5

432

As described in section 2.2 we based the synthesis of these compounds on the TMSOTf-promoted tandem Mannich-Michael reaction[47] of the appropriated *N-Boc* iminium ion (Scheme 5) with two different 2-silyloxy-1,3-butadienes. Unfortunately, the yield of the cyclisation was very low (< 18%) due to the rapid degradation of the intermediates to the corresponding quinolines.

We employed other Lewis acids [TiCl₄, SnCl₄, AlCl₃, TiCl₂(*O-i*-Pr)₂] to promote the cyclisation reaction and, studying in particular the reaction between **10c**, prepared as shown in Scheme 6, and **7**, we found that only TiCl₄ promoted the iminium ion formation and the cyclisation step better than TMSOTf, providing the corresponding 8-chlorobenzo[c]quinolizin-3-one **9g** in 35% yield.

The TiCl₄ procedure was then applied to the synthesis of compounds **9c-j** (Scheme 6) from 3-methyl- and 4-methyl substituted 2-ethoxy carbamates **10b** and **10c**, using diene **7** or Danishefsky's diene **6** (R'=OMe) as partners in the cyclisation reactions.

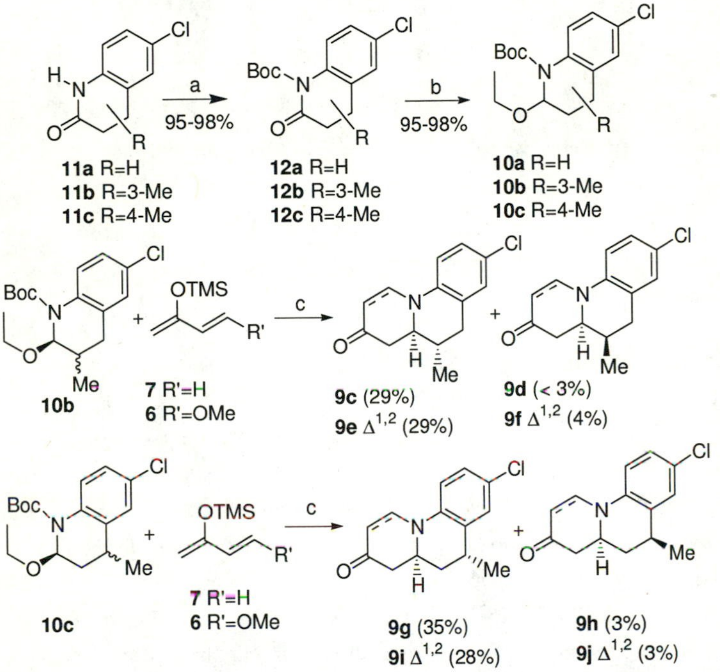

Reagents and conditions: a) Boc₂O, Et₃N, DMAP, CH₂Cl₂, 25 °C, 18 h; (b) 6 equiv NaBH₄, EtOH, -25 °C, 6 h, then 2 N HCl in EtOH, -25 °C → 25 °C, 2-4 h; (c) 1 M TiCl₄ in CH₂Cl₂, -30 °C, 10 min, then **7** or **6**, -30 → 25 °C, 30 min; then NaHCO₃ (satd), 40 min.

Scheme 6

Lactams **11a-c** were known or prepared according to procedures already described for similar compounds.[48,49] After protection of the lactams as *N-Boc* derivatives **12a-c**, these were transformed into the iminium ion precursors **10a-c** in 90-96% overall yield.

The reaction of iminium ions from **10b-c** with diene **7** was stereoselective, providing as the major products the isomers **9c** and **9g** having the methyl group and the bridgehead 4a proton in cis relative position. Thus, compounds **9c** and **9g** were isolated in 29 and 35% yield, respectively, after chromatographic purification (CH₂Cl₂, 0.5% Et₃N), while their isomers **9d** and **9h** were only detected (< 3%) by ¹H NMR

433

analysis of the crude reaction mixtures. The reactions of **10b,c** with Danishefsky's diene **6**, which afforded $\Delta^{1,2}$ unsaturated compounds **9e** and **9i**, occurred with similar stereoselectivity. Since strong Lewis acids are able to remove the *N-Boc* protection,[50] the stereochemical outcome of these reactions could be explained by the formation, after the addition of TiCl$_4$ (2 equiv) to the α-ethoxy carbamate, of a planar imine in which the N atom coordinates a titanium complex.[27] The methyl group at position 3 or 4 then leads to a preferred less hindered *anti* attack by the dienes. Oxidation of compounds **9b**, **9c**, and **9g** (Scheme 7), to prepare the 4-4a unsaturated compounds, was performed by Hg(OAc)$_2$ in 5% aqueous acetic acid, obtaining **13b-c**, **13d** in 36-54% yield, together with the corresponding compounds of the 4a*H*-series **14o**, **9e**, and **9i** in 27-51%.

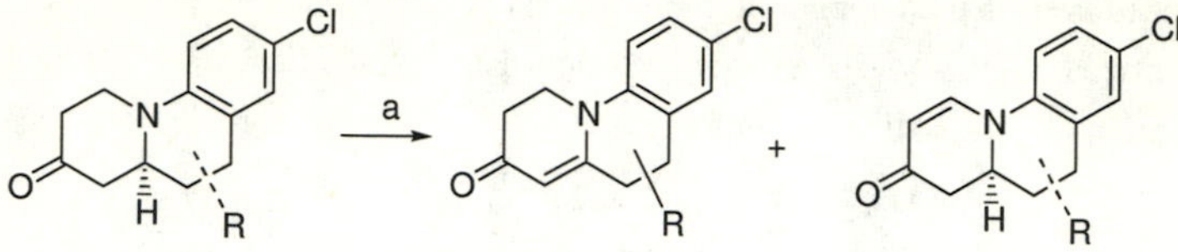

9b R= H	13b R= H (54%)	14o R= H (27%)
9c R= 5-Me	13c R= 5-Me (43%)	9e R= 5-Me (29%)
9g R= 6-Me	13g R= 6-Me (36%)	9i R= 6-Me (51%)

Reagents and conditions: Hg(OAc)$_2$, EDTA tetrasodium salt, 5% AcOH (aq), 90 °C, 2 h.

Scheme 7

With these results we have demonstrated that TiCl$_4$ can be used to promote the tandem Mannich-Michael reaction of silyloxydienes with iminium ion from aromatic precursors with more efficiency than TMSOTf and considerably shorter reaction times.

2.4. Synthesis of more substituted benzo[c]quinolizinones

If applied to the synthesis of benzo[c]quinolizinones **8, 9, 15-28** (Scheme 8), the methodology described in 2.2.1 would require the preparation, from lactams **3, 11b** and **49-52**, of bicyclic *N*-Boc-α-ethoxy quinolines **4, 10a, 10b, 10c, 53** to be reacted in turn with the suitable substituted silyloxydienes in the presence of a Lewis acid. This would generate the corresponding iminium ion which undergoes tandem Mannich-Michael attack by the silyloxydiene. In the case Danishefsky's diene **6** (R^1=OMe, R^2=H) the reaction would lead directly to 1,2-unsaturated compounds of 4a*H* series (wherein R^1=H, R^2=H) otherwise oxidation of intermediates **8, 9b, 9c, 15-28** is necessary to introduce a double bond that conjugates the bridgehead *N* and the 3-oxo group. As we have already found in 19-nor-10-azasteroids, the lack of conjugation between N and C=O causes a loss of inhibitory potency and this, if the inhibition mechanism is similar for the two classes of compounds, is likely to occur also in benzo[c]quinolizinones **8, 9b, 9c, 15-28**.

The synthesis of *N*-Boc-α-ethoxyquinolines is reported in Scheme 9. Lactams **11b** and **49-52** have been prepared according to known procedures,[48,49] lactam **3** (R^3, R^4, R^5=H) is commercially available. Aldehyde **56** (Scheme 9), treated with 2-(triphenylphosphanylidene)-propionic acid methyl ester[51] gave cynnamate **57** as a mixture of diastereoisomers in 90% yield, which was converted into 3-methyl substituted lactam **11b** by hydrogenation over PtO$_2$ in AcOH. 4-Substituted anilines **58-59** were reacted with 3-chloropropanoyl or 3-chlorobutanoyl chlorides to give the corresponding β-chloroamides **60-63** in quantitative yield. Lactams were obtained after the intramolecular Friedel-Craft alkylation that amides **60-63** underwent in the presence of AlCl$_3$ at 120-130 °C.

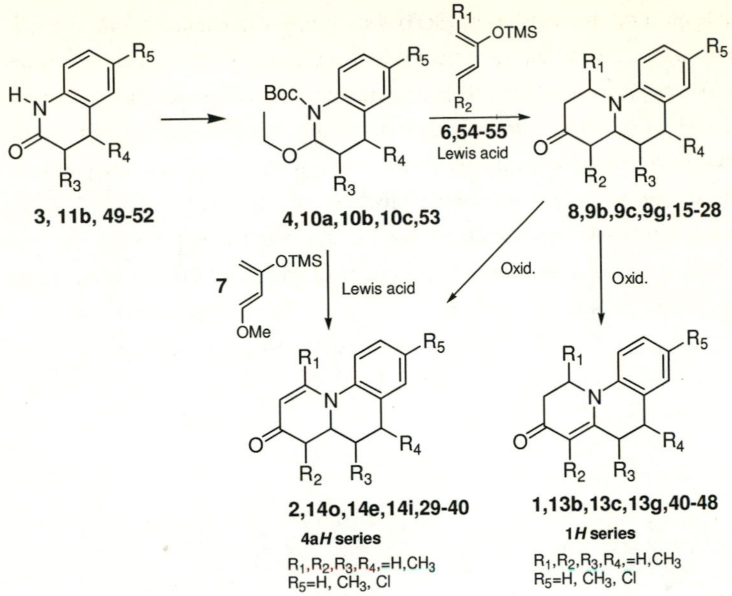

3, 11b, 49-52 → **4,10a,10b,10c,53** → **8,9b,9c,9g,15-28**

6,54-55
Lewis acid

7

Lewis acid

Oxid.

Oxid.

2,14o,14e,14i,29-40
4aH series
$R_1,R_2,R_3,R_4,=H,CH_3$
$R_5=H, CH_3, Cl$

1,13b,13c,13g,40-48
1H series
$R_1,R_2,R_3,R_4,=H,CH_3$
$R_5=H, CH_3, Cl$

Scheme 8

In the case of 6-methyl substituted amides **62** and **63** a strict control of the temperature was necessary to avoid migration of the methyl group on the aromatic ring and thus the formation of isomers. After protection of N atom as *N*-Boc to give compounds **12a, 12c, 66, 67**, reduction of the 2-oxo group by NaBH$_4$ in ethanol at –25 °C, followed by acidic quench with 2 N HCl in ethanol, afforded *N*-Boc-2-ethoxy derivatives in 94-100% overall yield. In the case of 3- and 4-methyl substituted derivatives **10b** and **10c**, these compounds were obtained as 1:1 mixtures of diastereoisomers.

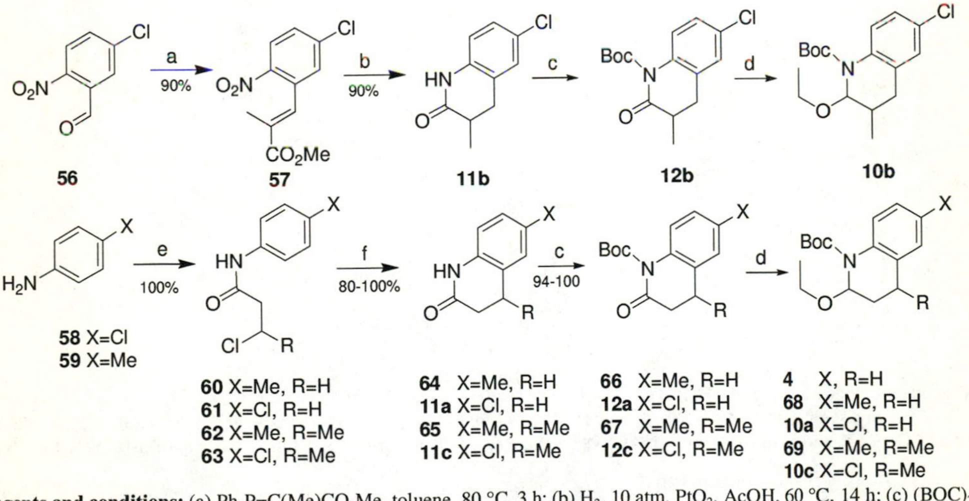

56 →(a, 90%) **57** →(b, 90%) **11b** →(c) **12b** →(d) **10b**

58 X=Cl
59 X=Me
→(e, 100%) →(f, 80-100%) →(c, 94-100) →(d)

60 X=Me, R=H
61 X=Cl, R=H
62 X=Me, R=Me
63 X=Cl, R=Me

64 X=Me, R=H
11a X=Cl, R=H
65 X=Me, R=Me
11c X=Cl, R=Me

66 X=Me, R=H
12a X=Cl, R=H
67 X=Me, R=Me
12c X=Cl, R=Me

4 X, R=H
68 X=Me, R=H
10a X=Cl, R=H
69 X=Me, R=Me
10c X=Cl, R=Me

Reagents and conditions: (a) Ph$_3$P=C(Me)CO$_2$Me, toluene, 80 °C, 3 h; (b) H$_2$, 10 atm, PtO$_2$, AcOH, 60 °C, 14 h; (c) (BOC)$_2$O, Et$_3$N, DMAP, CH$_2$Cl$_2$, 18 h, 25 °C;(d) NaBH$_4$, EtOH, -25 °C, 2-5 h, then 2 N HCl in EtOH, pH 3-4, -25 f 0 °C, 1.5-7 h; (e) RCH(Cl)CH$_2$COCl, acetone, reflux, 1 h; (f)AlCl$_3$, 120-160 °C, 3-24 h.

Scheme 9

435

In section 2.2 we have shown that TMSOTf catalyses the tandem Mannich-Michael cyclisation of unsubstituted *N-Boc* derivative **4** (X, R=H) with silyloxydienes **7** (R^1=R^2=H) and **6** (R^1=OMe, R^2=H) providing benzo[c]quinolizin-3-ones **8** and **1** in moderate yields. To extend this methodology to *N*-Boc-α-ethoxy derivatives such as **10a, 10b, 10c**, bearing a chlorine atom on the benzene ring we studied the use of different Lewis acids to promote the formation of *N*-(acyloxy)iminium ions, finding out that TiCl$_4$ promoted the iminium ion formation and the succeeding cyclization step better than TMSOTf. The new procedure applied to *N*-Boc-α-ethoxy derivative **10a** (X=Cl, R=H) and diene **7** afforded **9b** (X=Cl, R=H) in 34% yield, whereas the same compound was obtained in 17% yield employing TMSOTf as a Lewis acid.

The TiCl$_4$ methodology was applied to the synthesis of benzo[c]quinolizinones 8-H, 8-Cl and 8-CH$_3$ substituted **8, 15, 9b, 9g, 16, 9c, 2-9e** (Scheme 10), bearing no substituents on the A ring and therefore deriving from cyclisation of the corresponding α-ethoxycarbamates with Danishefsky diene **6** and 2-trimethylsilyloxy-1,3-butadiene **7**. Only in the case of compound **15** the TMSOTf procedure furnished higher yield (44%) than TiCl$_4$.

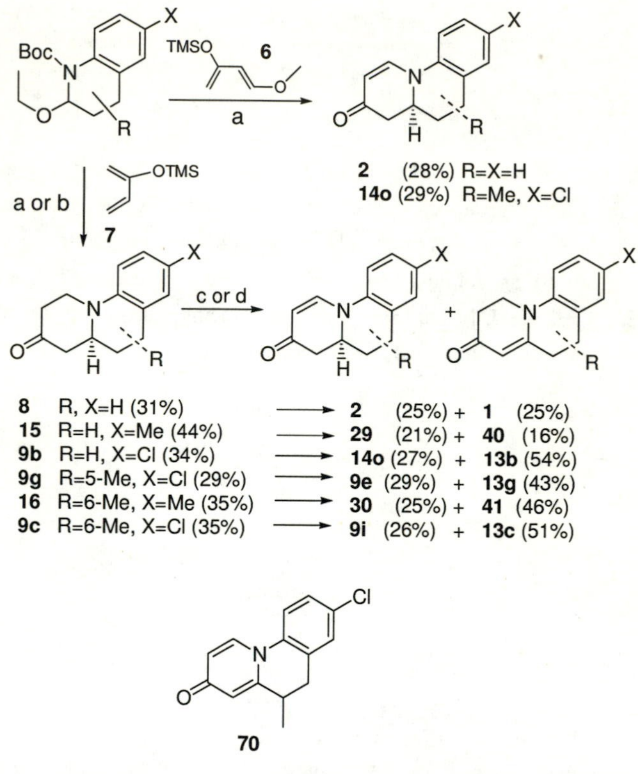

2 (28%) R=X=H
14o (29%) R=Me, X=Cl

8 R, X=H (31%)
15 R=H, X=Me (44%)
9b R=H, X=Cl (34%)
9g R=5-Me, X=Cl (29%)
16 R=6-Me, X=Me (35%)
9c R=6-Me, X=Cl (35%)

⟶ **2** (25%) + **1** (25%)
⟶ **29** (21%) + **40** (16%)
⟶ **14o** (27%) + **13b** (54%)
⟶ **9e** (29%) + **13g** (43%)
⟶ **30** (25%) + **41** (46%)
⟶ **9i** (26%) + **13c** (51%)

70

Reagents and conditions: (a) 1 M TiCl$_4$ in CH$_2$Cl$_2$, -30 °C, 10 min, then **7** or **6**, -30 → 25 °C, 30 min; then NaHCO$_3$ (satd), 45 min; (b) methyl vinyl ketone, TMSOTf, Et$_3$N, CH$_2$Cl$_2$, 0 °C, 30 min; then α-ethoxycarbamate, TMSOTf, 0 → 25 °C, 45 min; then NaHCO$_3$, 25 °C, 36 h; (c) Hg(OAc)$_2$, EDTA tetrasodium salt, 5-50% CH$_3$COOH (aq), 90 °C, 2 h; (d) LDA, Me$_3$SiCl, THF, -78 → 25 °C; then DDQ, 25 °C, 18 h.

Scheme 10

The reaction of 3- and 4-substituted iminium ions from **10b, 69, 10c** with diene **7** was stereoselective, providing the isomer with the methyl group and the bridgehead 4a proton in cis relative position as the major product. These (and the following) stereochemical assignments were possible by inspection of the coupling constants in the ^{1}H NMR spectrum of the saturated and, then, the corresponding Δ^1- or Δ^4-oxidized derivatives. Compounds **9g**, **16**, **9c** were isolated in 29-35% yield after chromatographic purification, while their trans isomers were only detected (< 3%) by ^{1}H NMR analysis of the crude reaction mixtures. The reactions of **10b** and **69** with Danishefsky's diene **6**, which afforded Δ^1-unsaturated compounds **9e** (29%) and **8** (28%), occurred with similar stereoselectivity.

The oxidation of compounds **8**, **15**, **9b** and **9g**, **16**, **9c** to the 1,2- or 4,4a-unsaturated analogues (Scheme 10) was performed by treatment with Hg(AcO)$_2$. As an alternative, after formation of the corresponding silyl enol ethers by treatment with LDA and TMSCl in THF at –78 °C, oxidation can be achieved by employing DDQ at room temperature.[52] Usually, the latter conditions afforded almost equimolar amounts of Δ^1 and Δ^4 isomers (as in the oxidation of **8** and **15**) and lower yields than Hg(OAc)$_2$, which instead provided always mixtures containing compounds of the 1H series (**1**, **13b**, **13g**, **41**, **13c**) as the major isomers. On the other hand, complete oxidation of the A ring was in some cases observed (oxidation of **9g** and the **17-18** mixture) under the more drastic conditions required with mercuric acetate and compounds **70** (Scheme 10) and **71** (Scheme 12) were obtained in 16 and 12% yield.

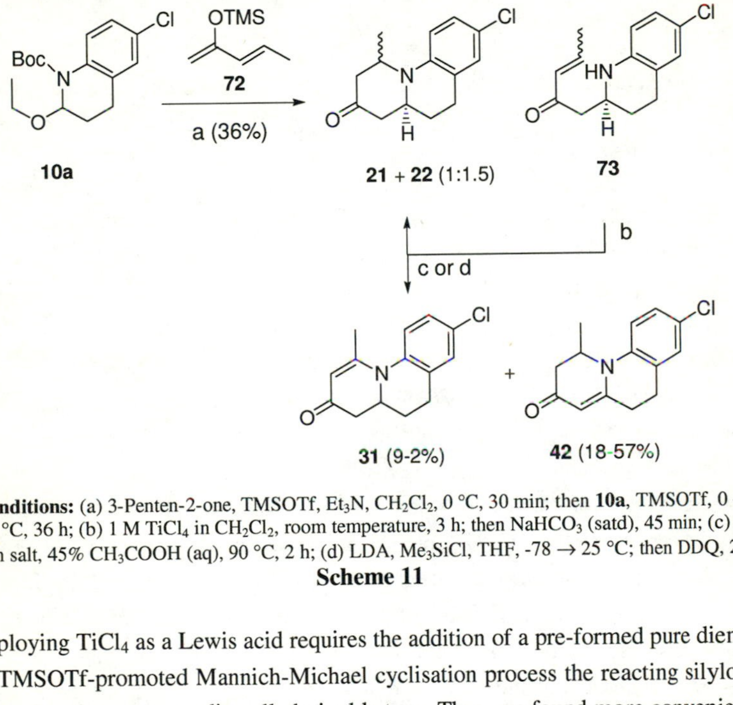

Reagents and conditions: (a) 3-Penten-2-one, TMSOTf, Et$_3$N, CH$_2$Cl$_2$, 0 °C, 30 min; then **10a**, TMSOTf, 0 → 25 °C, 45 min; then NaHCO$_3$, 25 °C, 36 h; (b) 1 M TiCl$_4$ in CH$_2$Cl$_2$, room temperature, 3 h; then NaHCO$_3$ (satd), 45 min; (c) Hg(OAc)$_2$, EDTA tetrasodium salt, 45% CH$_3$COOH (aq), 90 °C, 2 h; (d) LDA, Me$_3$SiCl, THF, -78 → 25 °C; then DDQ, 25 °C, 18 h.

Scheme 11

While employing TiCl$_4$ as a Lewis acid requires the addition of a pre-formed pure diene to the reaction mixture, in the TMSOTf-promoted Mannich-Michael cyclisation process the reacting silyloxydiene may be generated *in situ* using the corresponding alkyl vinyl ketone. Thus, we found more convenient to employ the TMSOTf procedure in the preparation of 1- and 4-methyl substituted benzo[c]quinolizin-3-ones (Schemes 11 and 12) by the *in-situ* generation of 4-methyl-2-[(trimethyl)silyloxy]-1,3-butadiene **72** from 3-penten-2-one with the TMSOTf/Et$_3$N system in dichloromethane. The addition of **10a** and a further amount of TMSOTf (Scheme 11) to the solution of **72**, afforded an approximatively 1:1 mixture of 1-methyl substituted compounds **21** and **22**, together with a smaller amount of open chain product **73**. Treatment of this mixture with TiCl$_4$ caused complete cyclisation of **22** to a mixture of **21** and **22**. Oxidation of the mixture of **21** and **22** was carried out by Hg(OAc)$_2$, providing Δ^4 compound **42** in 96% yield and its Δ^1 isomer **31** in 4% yield (18 and 9% yield, respectively, by the DDQ oxidation).

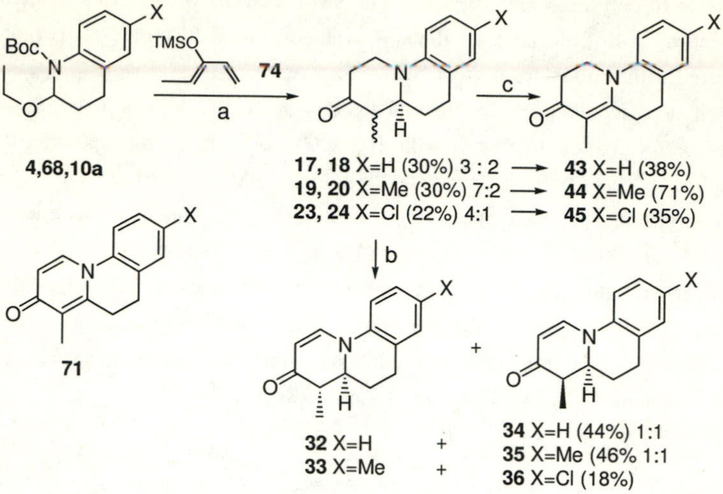

Reagents and conditions: (a) 1-Penten-3-one, TMSOTf, Et$_3$N, CH$_2$Cl$_2$, 0 °C, 30 min; then **4, 68, 10a**, TMSOTf, 0 → 25 °C, 45 min; then NaHCO$_3$, 25 °C, 36 h; (b) LDA, Me$_3$SiCl, THF, -78 → 25 °C; then DDQ, 25 °C, 18 h; (c) Hg(OAc)$_2$, EDTA tetrasodium salt, 5-8% CH$_3$COOH (aq), 90 °C, 2 h.

Scheme 12

With a similar procedure, α-ethoxycarbamates **4, 68, 10a,** were reacted with silyloxydiene **74** derived from 1-penten-3-one (Scheme 12), yielding 4-methyl substituted compounds **17-18, 19-20,** and **23-24** in 22-30% yield as 4α/β variable mixtures. Only in the case of the **19-20** mixture were we able to obtain, after chromatography, pure β isomer **20**. A portion of each mixtures was then oxidised by Hg(OAc)$_2$, affording Δ^4-unsaturated compounds **43-45** (35-71%), whereas another portion was subjected to the DDQ procedure, obtaining only Δ^1-unsaturated compounds **32-35** in 44-46% yield as epimeric 4α/β mixtures. Compound **36** was obtained as a single 4β isomer, but taking into account the very low final yield (18%), the other isomer might have been lost during the chromatographic purification.

Diene **74** was employed for the synthesis of 4,5- and 4,6-dimethyl substituted benzo[c]quinolizinones **25-28** (Scheme 13) according to the TMSOTf procedure. However, very complex isomeric mixtures were obtained after reaction of **74** with 3- and 4-methyl substituted carbamates **10b, 10c, 69**. A careful chromatographic separation allowed the recovery of the 1:2 mixture of epimers **25** and **26**, having the methyl at C-5 in cis relative position with the 4a-H proton, but in low yield (20%). Both 4,6-dimethyl derivatives **27** (X=Me) and **28** (X=Cl) were obtained in 14% yield as mixtures in which the diastereoisomer having both the methyl groups cis relatively to the 4a-H was prevailing. Usual oxidation of the **25-26** mixture by Hg(OAc)$_2$ yielded Δ^4 compound **46**(14%) besides the 1:2 mixture of Δ^1 isomers **37** and **38** (11%) which were separated after repeated chromatographies. Similarly, oxidation of **27** and **28** gave rise to Δ^4 compounds **47** (27%) and **48** (34%), respectively, together with Δ^1 isomers **39** and **40**.

The stereochemical outcome of the reaction of **74** with the iminium ions generated from **10b** seems to be in accordance with the initial *anti* approach of the diene as already discussed, as well as the formation of

438

compounds **27** and **28**, that apparently derive from the *anti* approach of **74** to the iminium ions from **10c** and **69**.

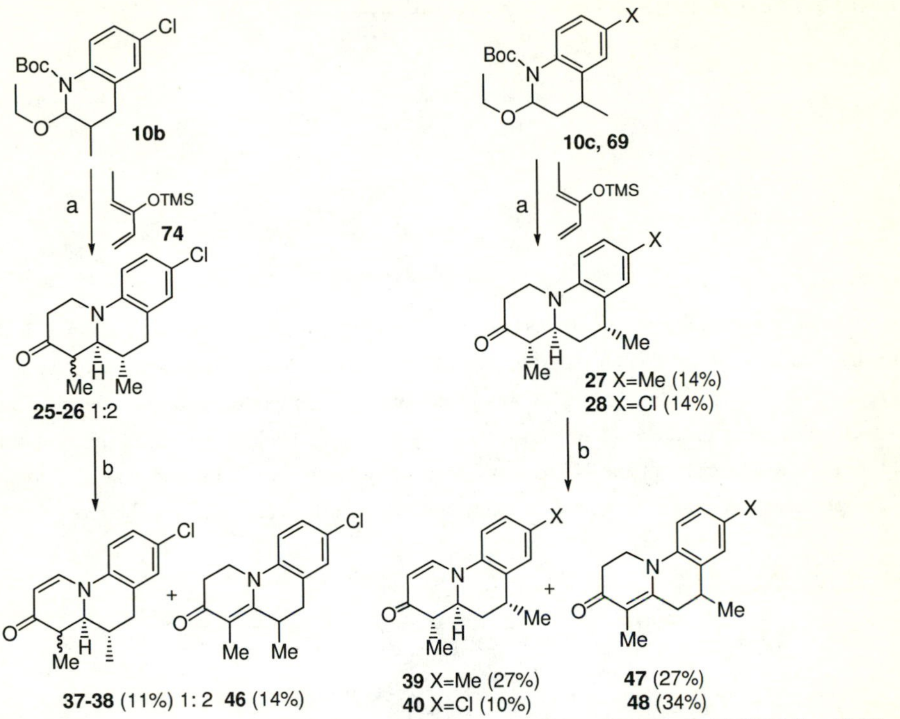

Reagents and conditios: (a) 1-Penten-3-one, TMSOTf, Et$_3$N, CH$_2$Cl$_2$, 0 °C, 30 min; then **10b, 10c, 69**, TMSOTf, 0 → 25 °C, 45 min; then NaHCO$_3$, 25 °C, 36 h; (b) Hg(OAc)$_2$, EDTA tetrasodium salt, 50% CH$_3$COOH (aq), 90 °C, 2 h.

Scheme 13

2.5. Modified aza Robinson annulation

As described in the previous paragraph, we based the synthesis of the benzo[c]quinolizidine skeleton on the Lewis-acid promoted Mannich-Michael tandem reaction of *N*-Boc iminium ions with 2-silyloxybutadiene derivatives. Because 4,4a-unsaturated compounds (1H-series), with a methyl group at 4 position, revealed to be the most potent inhibitors, we envisioned that a synthetic strategy based on an aza-Robinson annulation-type reaction could be more useful for the preparation of 4-substituted benzo[c]quinolizine in large scale as an alternative to the previously reported methodology. The aza-Robinson methodology has not been widely used for the synthesis of heterocyclic compounds. Due to the low reactivity of the lactam C=O toward C-nucleophiles, the oxygen must in fact be replaced by a sulfur atom to have a more reactive thiocarbonyl group which can undergo, under specific conditions, the cyclization reaction. There are only a few examples of application of this methodology to the synthesis of N-bridgehead heterocycles: indolizinones and quinolizinones have been prepared by Danishefsky and other authors,[53,54,55] who made use of a Rh(II)-catalysed reaction of α-diazoketones with a thiolactam for the ring closure step [Scheme 14, eq. (1)].

However, the use of diazomethane for the preparation of the α-diazoketone could be a limitation for the application of the methodology to a large scale synthesis.

439

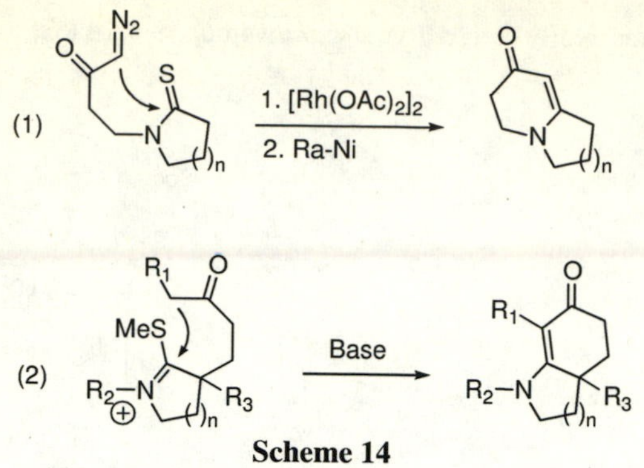

(1)

1. [Rh(OAc)$_2$]$_2$
2. Ra-Ni

(2)

Base

Scheme 14

Another possibility is the conversion of the thiolactam into an iminium ion which then undergoes attack by the C-nucleophile on the side chain [Scheme 14, eq. (2)]. This methodology has been already described by Heathcock[56] and other authors;[38] however, in the reported examples, the side chain bearing the C-nucleophile was on the α-carbon to the C=S group, thus leading to indole and quinoline derivatives. Tethering the side chain at the N atom and then performing the cyclisation onto the adjacent carbon by nucleophilic addition to the thioiminium ion would instead afford N-bridgehead heterocycles.

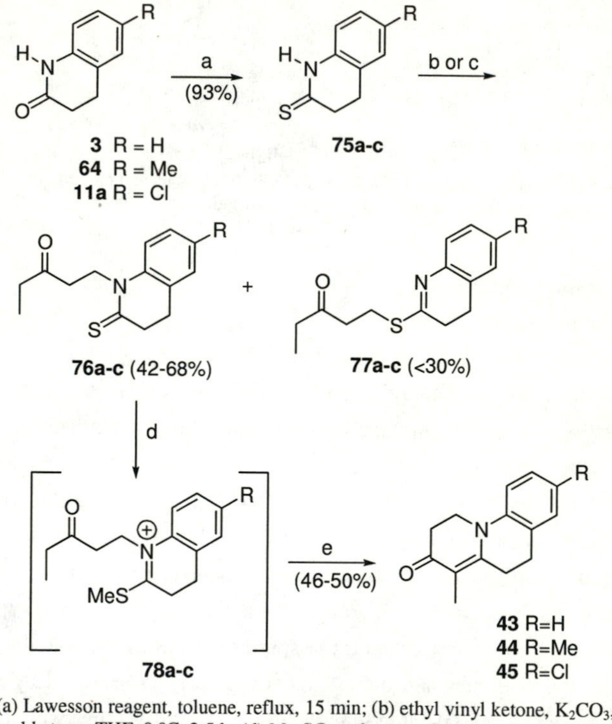

3 R = H
64 R = Me
11a R = Cl

a
(93%)

75a-c

b or c

76a-c (42-68%)   +   **77a-c** (<30%)

d

78a-c

e
(46-50%)

43 R=H
44 R=Me
45 R=Cl

Reagents and conditions: (a) Lawesson reagent, toluene, reflux, 15 min; (b) ethyl vinyl ketone, K$_2$CO$_3$, 18-C-6, THF, 0 to 25 °C; (c) DBU, ethyl vinyl ketone, THF, 0 °C, 2.5 h; (d) Me$_2$SO$_4$, toluene, reflux, 15 min; (e) DBU, reflux, 20 min.
Scheme 15

We have demonstrated that our modification of the aza-Robinson annulation is applicable to the synthesis of *N*-bridgehead heterocyclic compounds. In particular this procedure is suitable for the large-scale preparation (up to 0.5 kg was prepared using this strategy) of the potent 5αR-1 inhibitors 4-methyl-1H-benzo[c]quinolizin-3-ones **44-45** by using inexpensive starting materials and reagents and in only three steps (Scheme 15).

For the synthesis of **43-44** (wherein R=H, Me,Cl, and R'=H) we started from lactams **3, 11a, 64** prepared from the corresponding anilines.[57]

The conversion of **3, 11a, 64** to the corresponding thiolactams **75a-c** (93%) was achieved by the Lawesson reagent in boiling toluene.[58] N-Alkylation of thiolactams **75a** with ethyl vinyl ketone to give **76a** was quite troublesome, since it always gave in a variable extent also S-alkylated compounds **77a** as by-product. There are only few examples of N-alkylation of thiolactam with Michael acceptors.[59,54] In all cases the alkylating agents were α,β-unsaturated esters and S-alkylation has not been reported. However, in our case, the formation of **77a** could be favoured by the conjugation of the thioimine moiety with the aromatic ring. We studied the reaction of **75a** with ethyl vinyl ketone under different conditions in order to reduce the extent of S-alkylation: the use of DBU as a base,[38] in THF at 0 °C, after 2 h brought to the formation of both **76a** and **77a** in 42 and 27% yield, respectively, after chromatographic purification. The use of NaOH[54] or NaH[59] as bases for N-deprotonation in THF, instead caused a decrease of the yields of the N-alkylated product if compared with the reaction with DBU. Better results were obtained with K_2CO_3 in anhydrous THF and in the presence of a catalytic amount of 18-crown-6. Although also under these conditions a certain amount of the S-alkylated by-product was formed, it was possible to reduce completely the S-alkylation by slow addition of an excess of ethyl vinyl ketone at 0 °C during the period of 2 h. In these conditions N-alkylated compounds **76a-c** were obtained in 53-68% yield after chromatographic purification.

The last step was performed by treating **76a-c** with Me_2SO_4 (1.7 equiv) in boiling toluene in order to generate thioiminium ions **78a-c**. The formation of **78a-c** (a dark-red oil that separates from toluene) was very fast if compared with other cases (15 min vs. 16 h) reported in literature.[38] Ring closure was finally attained by addition to the boiling suspension containing **78a-c** of DBU (1.7 equiv), which resulted in the rapid (20 min) dissolution of the red oil and formation of **43-44** (obtained in 46-50% yield after chromatography).

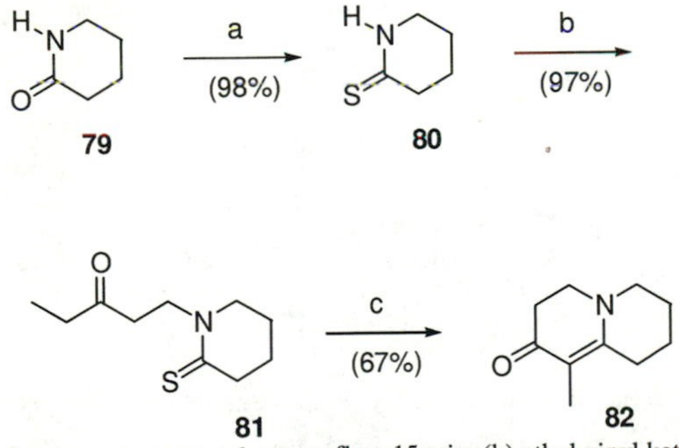

Reagents and conditions: (a) Lawesson reagent, toluene, reflux, 15 min; (b) ethyl vinyl ketone, K_2CO_3, 18-crown-6, THF, 0 to 25 °C; (c) Me_2SO_4, toluene, reflux, 15 min; then DBU, reflux, 20 min.

Scheme 16

A major difference of this procedure from the strategy depicted in Scheme 14 (reaction 2) is that, in the reported examples,[56,38] the cyclisation was done using iminium ions that could not isomerise to the corresponding enamines because the adjacent carbon atom was quaternary. In our case this isomerisation is possible, however we did not find the enamine in the crude reaction mixtures (by ^{1}H NMR analysis) or after chromatography. Presumably, decomposition of the thioiminium ion has instead occurred, which lowered the yields to a certain extent. In order to extend the scope of this work to thioiminium ions not conjugated to an aromatic ring, we applied the procedure to thiolactam **80**[54] (Scheme 16) which was *N*-alkylated with ethyl vinyl ketone to give compound **81** by the usual protocol. The reaction was faster (30 min) and afforded **81** in very good yield (97%). No traces of the S-alkylated compound were detected in this case. Ring closure finally gave quinolizinone **82** in 67% yield after chromatographic purification.

2.6. Synthesis of 8-substituted 4-methyl-benzo[c]quinolizin-3-ones

The best strategy for the preparation of a large number of 8-substituted 4-methyl-benzo[c]quinolizin-3-one inhibitors is to have a bulk quantity of a common intermediate to be easily, and possibly in a single step, transformed into the target compounds.

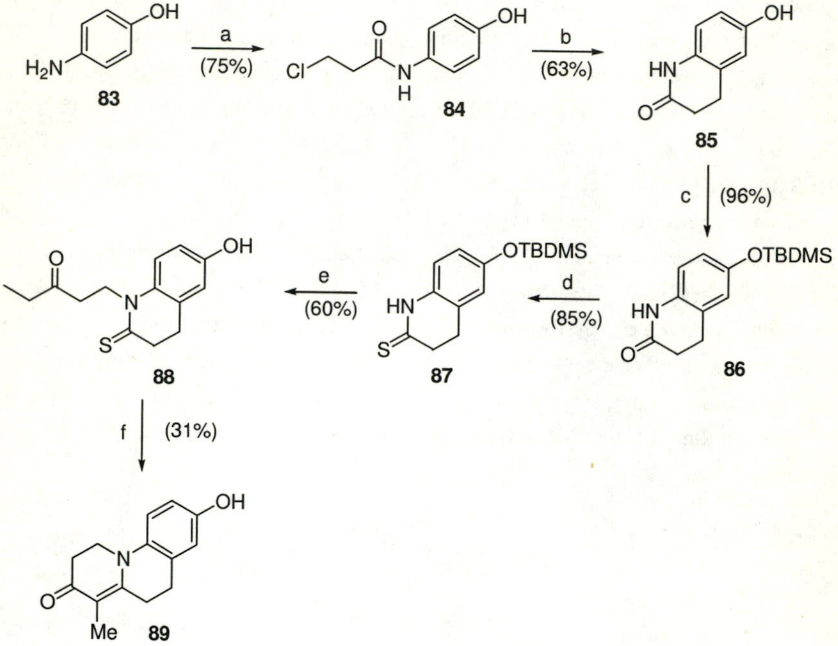

Reagents and conditons: (a) ClCH$_2$CH$_2$COCl, acetone, reflux, 1 h; (b) AlCl$_3$, fusion, 24 h; (c) TBDMSCl, imidazole, DMF, 50° C, 2 h, then r.t., 14 h; (d) Lawesson's reagent, toluene, reflux, 15 min; (e) ethylvinylketone, K$_2$CO$_3$, 18-crown-6, THF, r.t., 3 h; (f) Me$_2$SO$_4$, DBU, toluene, reflux, 40 min.

Scheme 17

To this aim, we focused our attention on the synthesis of two derivatives to be used as common intermediates, for instance 8-hydroxy-4-methyl-benzo[c]quinolizin-3-one **89** (Scheme 17), whose 8-OH group could be functionalized in order to obtain ethers, and 8-bromo-4-methyl-benzo[c]quinolizin-3-one **93** (Scheme 18), in which the aryl halide moiety could be used in Pd-catalysed cross-coupling reactions for the preparation of 8-alkenyl- and alkynyl-derivatives, and carbopalladation reactions for the introduction of ester moieties. Our modification of the aza-Robinson proved suitable for the preparation of both compounds **89** and **93** as depicted in Schemes 17 and 18.

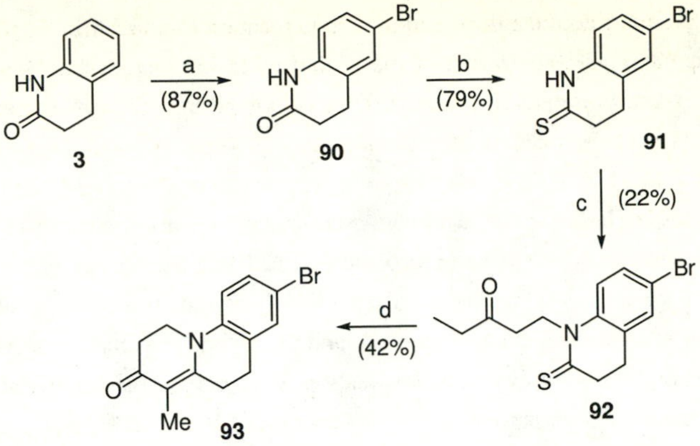

Reagents and conditions: (a) NBS, DMF, 0° C, 2 h; (b) Lawesson's reagent, toluene, reflux, 15 min; (c) ethylvinylketone, K$_2$CO$_3$, 18-crown-6, THF, r.t., 3 h; (d) Me$_2$SO$_4$, DBU, toluene, reflux, 40 min.

Scheme 18

Regarding the synthesis of compound **89** (Scheme 17), 4-hydroxy-aniline **83** was initially converted by treatment with β-chloropropionyl chloride to amide **84** which underwent an intramolecular Friedel-Craft alkylation by fusion at high temperature in the presence of AlCl$_3$ to give lactam **85** in 63% yield. After protection of the hydroxy group as a TBDMS ether, Lawesson's reagent was used to obtain thiolactam **87** (85% yield), which was subsequently N-alkylated with ethyl vinyl ketone in THF and in the presence of potassium carbonate as a base, to give **88** in 60% yield, with the TBDMS protection being lost in the latter reaction. The final cyclization was achieved through generation of the thioiminium ion by treatment first with Me$_2$SO$_4$ in refluxing toluene and then with DBU as a base, to give **89** in 31% yield after chromatography.

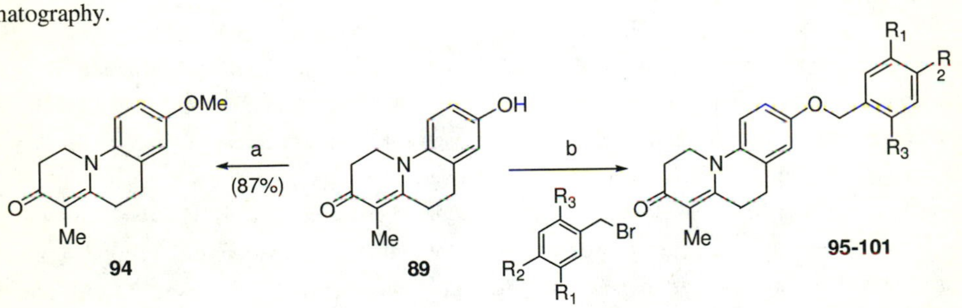

95 R$_1$=H, R$_2$=H, R$_3$=H (70%)
96 R$_1$=H, R$_2$=H, R$_3$=Cl (65%)
97 R$_1$=Cl, R$_2$=H, R$_3$=H (32%)
98 R$_1$=H, R$_2$=Cl, R$_3$=H (38%)
99 R$_1$=H, R$_2$=CF$_3$, R$_3$=H (72%)
100 R$_1$=H, R$_2$=CF$_3$, R$_3$=CF$_3$ (32%)
101 R$_1$=CF$_3$, R$_2$=H, R$_3$=CF$_3$ (75%)

Reagents and conditions: (a) MeI, K$_2$CO$_3$, acetone, reflux, 8 h; (b) K$_2$CO$_3$, acetone, reflux, 6 h.

Scheme 19

The synthesis of **93** was carried out (Scheme 18) starting from commercially available lactam **3**. This was easily converted to bromo derivative **90** by using *N*-bromosuccinimide in DMF. Applying the same

443

strategy as above, after formation of the thiolactam **91**, in this case a low yield (22%) was obtained in the N-alkylation step to give **92**. This result was essentially due to competitive S-alkylation of the thiolactam. Despite this, after the final cyclisation compound **93** was obtained in sufficient amount for all subsequent functionalizations.

3. Further functionalisation by Pd-catalysed cross-coupling processes. Synthesis of Dual Inhibitors

As mentioned earlier, the synthesis and use of dual 5αR-1 and 5αR-2 inhibitors is one of the current therapeutic model to completely reducing the circulating DHT and thus to treat BPH. Aimed at discovering new non-steroidal, dual inhibitors of 5α-reductases 1 and 2, having possible therapeutic applications, we started a program based on the synthesis and the evaluation of a series of benzo[c]quinolizin-3-ones bearing diverse substituents at position 8. In particular the introduction of F and CF_3 groups in these inhibitors has been dictated by the knowledge that H–C(α)–C=O fragments in an enzyme active site provide a pronounced fluorophilic environment due to occurrence of C–F···C=O contacts that are best described in terms of multipolar interactions between the intrinsically polar C–F and C=O units. Such F-interactions could be effectively exploited for enhancing ligand affinity or selectivity in structure based design. Dutasteride, a known azasteroidal dual inhibitor, indeed possesses two CF_3 groups on the phenyl ring of the 17-amide moiety. Also, PNU 157706, the most potent dual inhibitor (IC_{50} 3.9 nM and IC_{50} 1.8 nM, for type 1 and type 2 isozyme, respectively), has two geminal CF_3 groups in the 17-amide moiety.[60]

As mentioned above, compound **89** was used as the starting material for the preparation of ethers (Scheme 19): 8-methoxy-quinolizinone **94** was obtained in 87% yield by treatment with MeI in refluxing acetone and in the presence of K_2CO_3. Also, a wide variety of substituted benzyl ethers (**95-101**) were prepared with the same procedure using substituted benzylbromides in yields ranging from 32 to 75% after chromatographic purification.

8-Bromo-4-methylbenzo[c]quinolizin-3-one **93** was a more versatile precursor for the introduction of new substituents on position 8 because of the possibility of performing a series of different Pd-catalysed coupling reactions, including Sonogashira,[61] Stille,[62] Suzuki-Miyaura,[63] and palladium-catalysed carbonylation reactions. The first three processes were used for the introduction of small to medium-sized lipophilic groups and heterocyclic rings on position 8. Under the conditions of the Sonogashira reactions, **93** reacted with phenylacetylene (Scheme 20) to give alkynyl derivative **102** in 31% yield. In this case, conversion to **102** was not complete, but we were lucky to witness precipitation of the coupling product in Et_3N (used as a base and solvent) mixed with triethylammonium salts. Derivative **102** was afterwards hydrogenated to **103** over Lindlar catalyst, in order to obtain the Z alkenyl derivative **103** (89% yield after chromatography). These conditions ensured regioselectivity, since only partial reduction of the triple bond occurred. Suzuki-Miyaura cross-couplings were employed in order to insert heterocyclic substituents on the tricyclic structure (Scheme 20): 3-pyridyl, 2-furanyl and 2-thiophenyl derivatives **104**, **105** and **106**, respectively, were easily prepared performing couplings with the suitably substituted boronic acids at 80 °C under $(Ph_3P)_2PdCl_2$ (5%) catalysis in THF with 2 M Na_2CO_3(aq) as a base.

The couplings were complete in 4-24 h and afforded products in 67-77% yield after chromatography. Synthesis of E styryl derivative **107** was achieved in a similar way: the Suzuki reaction was carried out under the same conditions as above with 2-styryl-benzo[1,3,2]dioxaborole prepared by addition of catecholborane to phenyl acetylene and gave **107** in 33% yield after chromatographic purification. A Stille cross-coupling was finally exploited to obtain vinyl derivative **108** (Scheme 20), employing tributylvinylstannane as the nucleophilic reagent; selective hydrogenation of the vinyl moiety, leaving the enaminone moiety unaltered, was achieved by using the Wilkinson catalyst. This last reaction was carried out in benzene at 40 °C and was complete in 6 h, affording **109** in 75% yield.

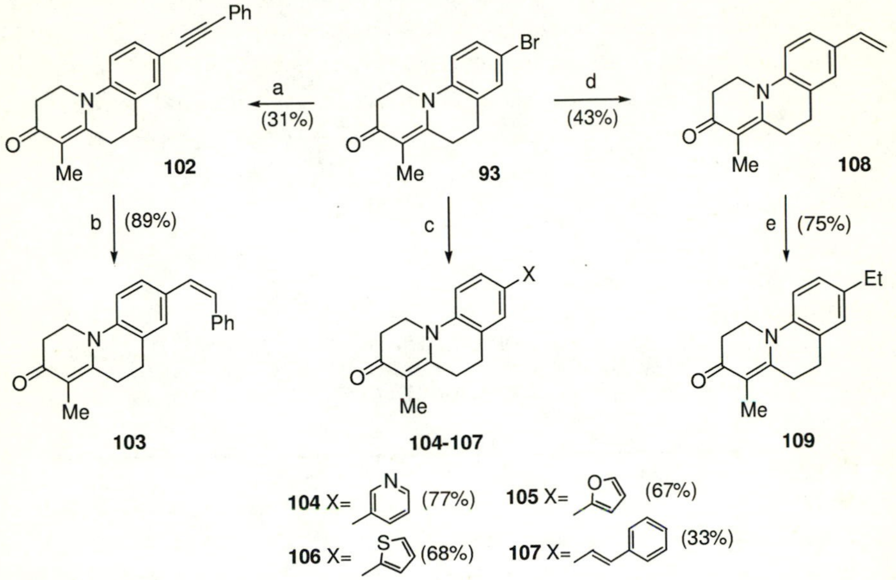

104 X= (pyridyl) (77%)   **105** X= (furyl) (67%)

106 X= (thienyl) (68%)   **107** X= (styryl) (33%)

Reagents and conditions: (a) PhCCH, PdCl₂(PPh₃)₂, CuI, Et₃N, reflux, 6 h; (b) H₂, Pd/BaSO₄, pyridine, r.t., 16 h; (c) R₂BX, PdCl₂(PPh₃)₂, Na₂CO₃ aq. 2 M, THF, 80° C, 4-24 h; (d) Bu₃SnCH=CH₂, Pd(OAc)₂, PPh₃, Et₃N, 95° C, 24 h; (e) H₂, (PPh₃)₃RhCl, C₆H₆, 40° C, 6 h.

Scheme 20

The synthesis of 8-phenyl substituted compound **115** (Scheme 21), which is structurally related to the heterocyclic derivatives **104-107**, could be realized through a Suzuki reaction using phenylboronic acid.

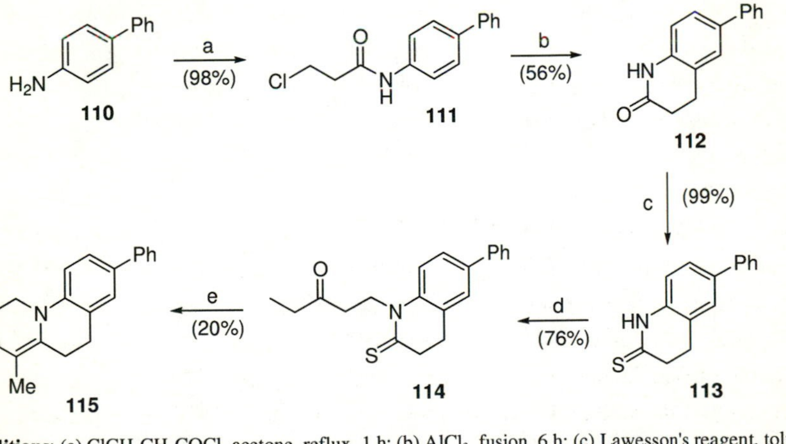

Reagents and conditions: (a) ClCH₂CH₂COCl, acetone, reflux, 1 h; (b) AlCl₃, fusion, 6 h; (c) Lawesson's reagent, toluene, reflux, 15 min; (d) ethylvinylketone, K₂CO₃, 18-crown-6, THF, r.t., 3 h; (e) Me₂SO₄, DBU, toluene, reflux, 40 min.

Scheme 21

445

However we did not evaluate this possibility, since this compound had previously been obtained starting from commercially available biphenyl-4-ylamine **110** by employing the thiolactam-based strategy already described for **89** and **93**, as depicted in Scheme 21. Benzo[c]quinolizin-3-ones, bearing ester groups on the position 8, were obtained by palladium-catalysed carbonylation of 8-Br-derivative **93** in the presence of a nucleophile (alcohols and phenols were used) under high pressure (50 bar) of CO (Scheme 22).

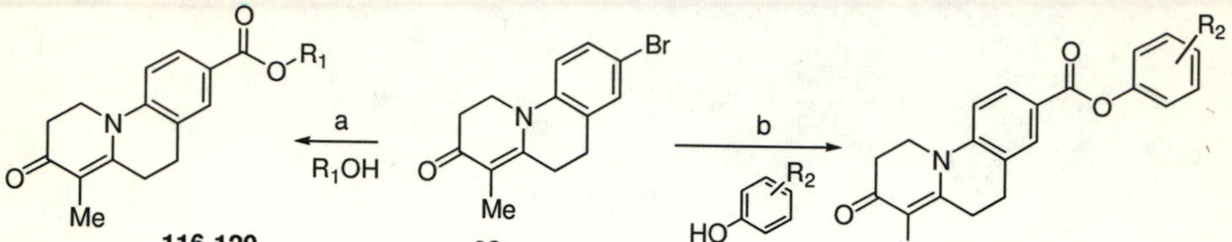

116 R$_1$=Me (80%)   **119** R$_1$= ^{t}Bu (11%)
117 R$_1$=Et (60%)   **120** R$_1$=Bn (71%)
118 R$_1$= ^{i}Pr (59%)

Reagents and conditions: (a) R$_1$OH, CO, PdCl$_2$, Ph$_3$P, Et$_3$N, C$_6$H$_6$, 115 °C, 24 h;
(b) R$_4$C$_6$H$_4$OH, CO, PdCl$_2$, Ph$_3$P, Et$_3$N, C$_6$H$_6$, 115 °C, 24 h

Scheme 22

Table 1.

Entry	Compound	R$_2$	5αR-1 [a] IC$_{50}$ (nM)[c]	5αR-2 [b] IC$_{50}$ (nM)[c]
1	**121**	H	942	295
2	**122**	p-(t-Bu)	560	low activity[d]
3	**123**	p-CO$_2$Me	129	584
4	**124**	m-CO$_2$Me	965	467
5	**125**	p-Me	102	553
6	**126**	p-OMe	116	1200
7	**127**	p-OEt	209	3300
8	**128**	p-Cl	1400	low activity[d]
9	**129**	p-F	93	119
10	**130**	m-F	160	134
11	**131**	o-F	138	166
12	**132**	p-CF$_3$	149	1200
13	**133**	m-CF$_3$	271	1000
14	**134**	o-CF$_3$	42	368
15	**135**	p-NH$_2$	272	801
16	**136**	p-NHAc	707	255
17	**137**	p-CONH$_2$	361	308
18	**138**	p-CONHPr	268	665
19	**139**	p-CONEt$_2$	463	372
20	**140**	p-CONH-t-Bu	225	low activity[d]

[a]Isozyme expressed by CHO 1827 cells [b]Isozyme expressed by CHO 1829 cells.
[c]Error in the 8-20% range. [d]% Inhibition < 50% at 10 μM.

We employed the same reaction conditions in all cases, only increasing the amount of the catalyst in case of poorer nucleophiles such as phenols bearing an electron withdrawing group on the ring. The reactions were conducted in a steel autoclave, using 5% $PdCl_2$ as a catalyst in the presence of 10% Ph_3P, Et_3N as a base and in benzene as a solvent, for 24 h at 120 °C. Different equivalents of nucleophile were used. Yields of carbopalladation ranged from 46 to 88% after chromatography, with the exception of t-butyl ester derivative **119** which underwent acid-catalysed hydrolysis during purification on silica gel, and p-tolyl ester derivative **125** which required an additional chromatographic purification.

By this methodology a series of alkyl esters **116-120** with increasing bulkiness in the alcoholic moiety R_1 was prepared, as well as a series of aryl esters **121-140** with polar and lipophilic substituents R_2 on the aromatic moiety. All of esters **121-140** were tested toward 5αR-1 and 2 expressed by CHO 1827 and CHO 1829, respectively (Table 1).[39] They maintained activity toward 5αR-1, with IC_{50} values always below 1 μM (one exception only) some of them being potent inhibitors of this isozyme. In particular, among the most potent were those having on the phenol moiety a small lipophilic group among which compound **129** (p-F substituted), **132** (p-CF$_3$ substituted) and o-CF$_3$ derivative **134** with IC_{50} value in the 42-149 nM range, *i.e.* close to the inhibition value of the most potent 5αR-1 inhibitor belonging to the benzo[c]quinolizinone series, *i.e.* 8-Cl-benzo[c]quinolizin-3-one (IC_{50}=7.6 nM). As for the 5αR-2 inhibition, we were pleased to find that the most potent inhibitor was again a compound with the F atom on the phenol moiety, *i.e.* p-F substituted compound **129** (entry 9) with an IC_{50} value of 119 nM. This was in particular the best dual inhibitor of the series, with IC_{50} values of about 100 nM for both isozyme. Also *meta*- and *ortho*-F-substituted derivatives **130** and **131** showed good activity toward 5αR-2.

4. SAR (Structure activity relationship) and a 3D-QSAR model for the inhibition of 5α-Reductase Type 1

The design of benzo[c]quinolizin-3-ones structure as non-steroidal inhibitors of 5α-reductase came from the observation that an increase of structural planarity in the corresponding 19-nor-10-azasteroidal 5α-reductase inhibitors[24,25] (obtained by introducing a double bond in the C ring) afforded molecules with higher inhibition potency toward 5αR-1. Benzo[c]quinolizin-3-ones, while maintaining the A ring enaminone moiety as an essential feature of the 19-nor-10-azasteroids, lack the D ring and incorporate a benzene ring in place of the C ring, in order to have a more planar overall structure. In general, with the exceptions reported in the above chapter, we found that benzo[c]quinolizin-3-one derivatives are selective and some of them very potent, competitive inhibitors of 5αR-1. The potency of these compounds depends on the presence and position of double bonds on the A ring, and the type and number of substituents introduced at positions 1, 4, 5, 6, and 8. We have observed that the activity is strongly affected by the substituent at the position 8 (for example compounds with a lipophilic group such as a methyl or a chlorine atom at position 8 are very active against 5αR-1).[39,40,41] Moreover the 4-methyl group is also of noteworthy importance, since its presence seems always associated with an increase of activity, irrespective of the group at position 8. In general, the compounds of the 1*H*-series resulted significantly more active than those of the 4a*H*-series, the IC_{50} values of the latter being approximately 10-fold higher. In fact the inhibition values ranged from 137 to 9100 nM for the 4a*H* compounds and from 7.6 to 376 nM for the 1*H* compounds.

More in particular (Table 2) the presence of a substituent at position 8, for instance a chlorine or a methyl group, generally increased the potency of the inhibitors in both series. Thus, in the 4a*H*-series compounds **29** (IC$_{50}$ 176 nM) and **14o** (IC$_{50}$ 459 nM), bearing a 8-methyl and a 8-chlorine, respectively, were significantly more active than unsubstituted compound **2** (IC$_{50}$ 5130 nM). Analogously, in the 1*H*-series the chlorine atom at position 8, either alone or in presence of other substituents on the two aliphatic rings, increased noticeably the potency toward 5αR-1. Thus 8-Cl substituted compounds **13b** (IC$_{50}$ 49 nM) and **45** (IC$_{50}$ 7.6 nM) were significantly more active than unsubstituted compound **1** (IC$_{50}$ 298 nM) and **43** (IC$_{50}$ 185 nM), respectively.

Table 2.

Entry	Compound		5αR-1 [a] IC$_{50}$ (nM)	Entry	Compound		5αR-1 [a] IC$_{50}$ (nM)
1	**1**		298	9	**41**		14.3
2	**2**		5130	10	**42**		204
3	**13b**		49	11	**43**		185
4	**13c**		346	12	**44**		20
5	**9i**		14.4	13	**45**		7.6
6	**14o**		459	14	**46**		15.6
7	**29**		176	15	**47**		15.8
8	**40**		376		**48**		8.5

[a]Isozyme expressed by CHO 1827 cells

The introduction of a methyl group at position 4 in both series was effective in increasing the inhibitory potency. The extent of the increase is slighter in the 4a*H* series and higher in the 1*H*-series. In the

1*H*-series the introduction of a methyl at 4 position determines a strong increase of potency, in particular when the 8-position is substituted with a chlorine or methyl. Thus, whereas 4-methyl derivative **43** displayed an inhibition activity (IC_{50} 185 nM) not too significantly different from the unsubstituted compound **1** (IC_{50} 298 nM), a very strong increase of potency is observed in 8-chloro-4-methyl derivative **45** (IC_{50} 7.6 nM) and 4,8-dimethyl derivative **44** (IC_{50} 20 nM).

The substitution with a methyl group at position 6 positively affected the potency of the inhibitors, although more markedly in the 1*H*- than in the 4a*H*-series. So compound **13c** (IC_{50} 14.4 nM) was significantly more active than not substituted at position 6 and compound **41** displayed high activity (14.3 nM). The further substitution with a methyl group at position 4 in trisubstituted compounds **47** (IC_{50} 15.8 nM) and **48** (IC_{50} 8.5 nM) maintained the inhibitory activity compared to 6,8-disubstituted compounds **41** and **13c**.

The results discussed so far are consistent with those reported for structurally related compounds,[33] whose potency toward 5αR-1 increased noticeably by introducing a methyl group at the position 4 of the skeleton and a chlorine at the position 8. Also, the beneficial effect of the methyl at position 6 seems consistent with the observation that the introduction of the same group on the corresponding position 7 in 4-azasteroids increased their 5αR-1 selectivity.[64,65]

By contrast, the presence of a methyl group at position 5 in general reduced the potency unless it was associated to another methyl group at the position 4. Finally, the introduction of a methyl at the position 1 in 8-Cl substituted compound **42** (IC_{50} 204 nM) slightly decreased the activity toward 5αR-1 in comparison with the homologous derivative **13b** (IC_{50} 49 nM) . This qualitative SAR analysis is summarised in Figure 2.

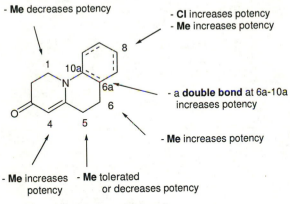

Figure 2

The construction of this homogeneous set of molecules has allowed a 3D QSAR study which, due to the lack of models for the active site of 5αR-1, could be very helpful in interpreting the biological data as well as in providing new insight for the design of new inhibitors. The model was obtained by taking into account the magnitude of dipole moment and the logP of the molecules along with the classical electrostatic and steric CoMFA fields. Moreover, to enhance the quality of the description of the electrostatic contribution, Density Functional Theory-based computations were carried out to evaluate both the atomic partial charges and the dipole moment values of each inhibitor. This CoMFA model has also been validated by predicting the biological activities of a set of molecules purposely synthesized and not used in deriving the 3D QSAR equations.

The CoMFA model accounting for the 3D QSAR is illustrated by the contour maps shown in Figure 3. Sterically favorable regions (Figure 3a, red contours) are quite split with a main appearance around the 4-methyl group, which was already pointed out as a fundamental substituent for an optimal 5αR-1 inhibitory profile.[39] Sterically unfavorable regions (Figure 3a, blue contours) are located around the substituent at position 8. In particular, the phenyl group of the styryl-substituted derivative **107**, along with all inhibitors aligned onto it, protrudes inside the sterically unfavorable region, whereas the diastereoisomer Z does not make any contact with that region. Electrostatic positive and negative regions (Figure 3b) are located around the substituent at position 8, as well. In particular, the negative CoMFA contours (yellow) are close to the C8 atom of the benzo[c]quinolizinone moiety, where the chlorine (compound **13c**) and the bromine (compound **93**) atoms, as well as the triple bond of **102**, and the oxygen of **95-100, 116, 117, 120,** and **123** are located. This means that in such a region, where relatively more potent compounds orient their 8-substituent, an increase of negative electrostatic potential should provide more potent 5αR-1 inhibitors. Concerning the positive electrostatic CoMFA contours (Figure 3b, green), two main regions were identified. One (bottom) was generated around both the carbonyl oxygen of the esters **116, 117, 120,** and **123**, as well as the sp³ oxygen of the ethers and esters **101, 118, 119, 121,** and **122**.

a)

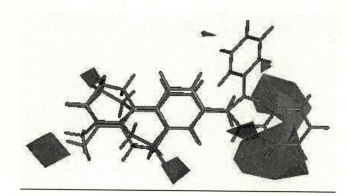

b)

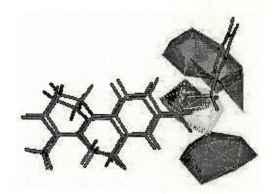

Figure 3

The other (top) was corresponding to the carbonyl oxygen of **118**, **119**, **121**, and **122**, as well as to the phenyl ring of the inhibitors aligned onto **103** (**95-100**, **116**, **117**, **120**, **123**), the latter being clearly embedded into the electrostatic positive region. All these molecules bear relatively lower potency than **93**, **102**, and **13c** and for this reason, the graphical model shows some apparently contradictory features. However, the meaning is that an increase of the positive electrostatic potential in that region should be required in order to obtain more potent compounds.

5. Conclusion

In conclusion we have reported on the synthesis and biological evaluation of a series of potent and selective inhibitors of 5α-reductase type 1, the isozyme involved in the transformation of testosterone to dihydrotestosterone mainly in human skin and scalp tissues and thus possibly causing the development and maintenance of disorders such as acne and androgenic alopecia in men, and hirsutism in women. The synthesized inhibitors are benzo[c]quinolizin-3-ones derivatives bearing at position 1, 4, 5, and/or 6 a methyl group and at position 8, *i.e.* on the aromatic ring, an hydrogen, a methyl group or a chlorine atom. Depending on the position of the double bond on the A ring, two classes of inhibitors can be identified, *i.e.* those having the double bond at position 1,2 (4a*H*-series) and those having the double bond at position 4,4a (1*H*-series). All these compounds were tested toward 5αR-1 and 5αR-2 expressed in CHO cells (CHO 1827 and CHO 1829, respectively) resulting selective inhibitors of the type 1 isoenzyme, with inhibitory potencies (IC_{50}) ranging from 7.6 to 9100 nM. The most potent inhibitors had a chlorine atom (or a methyl group) at position 8, although it was the presence of a methyl group at position 4, associated to the substitution at position 8, that determined the highest inhibition potency (IC_{50} from 7.6 to 20 nM). This suggests the presence of a small hydrophobic pocket in the enzyme active site accommodating the methyl group at position 4. A 3D QSAR model has been also developed, due to the lack of models for the active site of 5αR-1, which could be very helpful in interpreting the biological data as well as in providing new insight for the design of new inhibitors. Two new potent and selective 5αR-1 inhibitors were found which are the best candidates for the development of a drug for the treatment of acne and androgenic alopecia in men, hirsutism and polycystic ovarian syndrome in woman. Moreover, by the appropriate substitution at the position 8, also some potent dual inhibitors of 5α-reductase, with IC_{50} ranging between 93 and 166 nM for both isozymes were found. These dual inhibitors could be developed as drugs in the treatment of 5α-reductase related diseases where total suppression of circulating DHT is necessary, as for example in the treatment of BPH.

References
1. Geller, J. "Benign Prostatic Hyperplasia: Pathogenesis and Medical Therapy"; *J. Am. Geriatr. Soc.* **1991**, *39*, 1208-1266.
2. (a) Gormley, G. J. "Role of 5α-Reductase Inhibitors in the Treatment of Advanced Prostatic Carcinoma"; *Urol. Clin. North Am.* **1991**, *18*, 93-98. (b) Brawley, O. W.; Ford, G. L.; Thompson, I.; Perlman, J. A.; Kramer, B. S. "5α-Reductase Inhibition and Prostate Cancer Prevention"; *Cancer Epidemiol. Biomarkers Prev.* **1994**, *3*, 177-182.
3. Dallob, A. L.; Sadick, N. S.; Unger, W.; Lipert, S.; Geissler, L. A.; Gregoire, S. L.; Nguyen, H. H.; Moore, E. C.; Tanaka, W. K. "The Effect of Finasteride, a 5α-Reductase Inhibitor, on Scalp Skin Testosterone and Dihydrotestosterone Concentrations in Patients with Male Pattern Baldness"; *J. Clin. Endocrinol. Metab.* **1994**, *79*, 703-706.
4. Smith, L. S.; Tegler, J. J. "Advances in Dermatology"; *Annu. Rep. Med. Chem.* **1989**, *24*, 177-186.

5. (a) Price, V. H. "Treatment of Hair Loss"; *N. Engl. J. Med.* **1999**, *341*, 964-973, and references therein. (b) Gadwood, R. C.; Fiedler, V. C. "Pathogenesis and Treatment of Alopecia"; *Annu. Rep. Med. Chem.* **1989**, *24*, 187-196.

6. Brooks, J. R. "Treatment of Hirsutism with 5α-Reductase Inhibitors"; *Clin. Endocrinol. Metabol.* **1986**, *15*, 391-405.

7. Metcalf, B. W.; Levy, M. A.; Holt, D. A. "Inhibitors of Steroid 5α-Reductase in Benign Prostatic Hyperplasia, Male Pattern Baldness and Acne"; *Trends Pharm. Sci.* **1989**, *10*, 491-495.

8. (a) Holt, D. A.; Levy, M. A.; Metcalf, B. W. "Inhibition of Steroid 5α-Reductase" In *Advances in Medicinal Chemistry*; Maryanoff, B. E.; Maryanoff, C. A., Eds.; JAI Press Inc: Greenwich USA, 1993; Vol. 2, p. 1-29. (b) Abell, A. D.; Henderson, B. R. "Steroidal and Non-Steroidal Inhibitors of Steroid 5α-Reductase"; *Curr. Med. Chem.* **1995**, *2*, 583-597. (c) Li, X.; Chen, C.; Singh, S.; Labrie, F. "The Enzyme and Inhibitors of 4-Ene-3-oxosteroid 5α-Oxidoreductase"; *Steroids*, **1995**, *60*, 430-441. (d) Frye, S. V. "Inhibitors of 5α-Reductase"; *Curr. Pharm. Des.* **1996**, *2*, 59-84.

9. (a) Russell, D. W.; Wilson, J. D. "Steroid 5α-Reductases: Two Genes/Two Enzymes"; *Annu. Rev. Biochem.* **1994**, *63*, 25-61. (b) Wilson, J. D.; Griffin, J. E.; Russell, D. W. "Steroid 5α-Reductase 2 Deficiency"; *Endocrine Rev.* **1993**, *14*, 577-593.

10. Rasmusson, G. H. "Chemical Control of Androgen Action"; *Annu. Rep. Med. Chem.* **1986**, *22*, 179-188.

11. Thigpen, A. E.; Silver, R. I.; Guileyardo, J. M.; Casey, M. L.; McConnell, J. D.; Russell, D. W. "Tissue distribution and ontogeny of steroid 5alpha-reductase isozyme expression"; *J. Clin. Invest.* **1993**, *92*, 903-910.

12. Mahony, M. C.; Swanlund, D. J.; Billeter, M.; Roberts, K. P.; Pryor, J. L. "Regional distribution of 5alpha-reductase type 1 and type 2 mRNA along the human epididymis"; *Fertil. Steril.* **1998**, *69*, 1116-1121.

13. Bayne, E. K.; Flanagan, J.; Einstein, M.; Ayala, J.; Chang, B.; Azzolina, B.; Whiting, D. A.; Mumford, R. A.; Thiboutot, D.; Singer, I. I.; Harris, G. "Immunohistochemical localization of types 1 and 2 5alpha-reductase in human scalp"; *Br. J. Dermatol.* **1999**, *141*, 481-491.

14. Ando, Y.; Yamaguchi, Y.; Hamada, K.; Yoshikawa, K.; Itami, S. "Expression of mRNA for androgen receptor, 5alpha-reductase and 17beta-hydroxysteroid dehydrogenase in human dermal papilla cells"; *Br. J. Dermatol.* **1999**, *141*(5), 840-845.

15. Melcangi, R. C.; Poletti, A.; Cavarretta, I.; Celotti, F.; Colciago, A.; Magnaghi, V.; Motta, M.; Negri-Cesi, P.; Martini, L. "The 5alpha-reductase in the central nervous system: expression and modes of control"; *J. Steroid. Biochem. Mol. Biol.* **1998**, *65*, 295-299.

16. Span, P. N.; Benraad, Th. J;. Sweep, C. G.; Smals, A. G. "Kinetic analysis of steroid 5alpha-reductase activity at neutral pH in benign prostatic hyperplastic tissue: evidence for type I isozyme activity in the human prostate"; *J. Steroid. Biochem. Mol. Biol.* **1996**, *57*, 103-108.

17. (a) Kenny, B.; Ballard, S.; Blagg, J.; Fox, D. "Pharmacological Options in the Treatment of Benign Prostatic Hyperplasia"; *J. Med. Chem.* 1997, *40*, 1293-1314. (b) Harris, G. S.; Kozarich, J. W. "Steroid 5α-Reductase Inhibitors in Androgen-dependent Disorders"; *Curr. Opin. Chem. Biol.* 1997, *1*, 254-259. (c) Occhiato, E. G.; Guarna, A.; Danza, G.; Serio, M. "Selective non-steroidal inhibitors of 5α-reductase type 1. *J. Steroid. Biochem. Mol. Biol.* **2004**, *88*(1), 1-16. (d) Machetti, F.; Guarna, A. Novel Inhibitors of 5α-Reductase"; *Expert Opin. Ther. Patents* **2002**, *12*, 201-215.

18. Thiboutot, D.; Harris, G.; Iles, V.; Cimis, G.; Gilliland, K.; Hagari, S. "Activity of the Type 1 5α-Reductase Exhibits Regional Differences in Isolated Sebaceous Glands and Whole Skin"; *J. Invest. Dermatol.* **1995**, *105*, 209-214.

19. Rasmusson, G. H.; Reynolds, G. F.; Utne, T.; Jobson, R. B.; Primka, R. L.; Berman, C.; Brooks, J. R. "Azasteroids as inhibitors of rat prostatic 5 alpha-reductase"; *J. Med. Chem.* **1984**, *27*, 1690-1701.

20. Rasmusson, G. H.; Reynolds, G. F.; Steinberg, N. G.; Walton, E.; Patel, G. F.; Liang, T.; Cascieri, M. A.; Cheung, A. H.; Brooks, J. R.; Berman, C. "Azasteroids: structure-activity relationships for inhibition of 5 alpha-reductase and of androgen receptor binding"; *J. Med. Chem.* **1986**, *29*, 2298-2315.

21. Bull, H. G. "Mechanism-Based Inhibition of Human Steroid 5α-Reductase by Finasteride: Enzyme-Catalysed Formation of NADP-Dihydrofinasteride, a Potent Bisubstrate Analog Inhibitor"; *J. Am. Chem. Soc.* **1996**, *118*, 2359-2365.

22. Frye, S. V.; Haffner, C. D.; Maloney, P. R.; Mook, R. A., Jr.; Dorsey, G. F., Jr.; Hiner, R. N.; Batchelor, K. W.; Bramson, H. N.; Stuart, J. D.; Schweiker, S. L.; Van Arnold, J.; Bickett, D. M.; Moss, M. L.; Tian, G.; Unwalla, R. J.; Lee, F. W.; Tippin, T, K.; James, M. K.; Grizzle, M. K.; Long, J. E.; Schuster, S. V. "6-Azasteroids: potent dual inhibitors of human type 1 and 2 steroid 5 alpha-reductase"; *J. Med. Chem.* **1993**, *36*, 4313-4315.

23. Frye, S. V.; Haffner, C. D.; Maloney, P. R.; Mook, R. A., Jr.; Dorsey, G. F., Jr.; Hiner, R. N.; Cribbs, C. M.; Wheleer, T. N.; Ray, J. A.; Andrews, R. C.; Batchelor, K. W.; Bramson, H. N.; Stuart, J. D.; Schweiker, S. L.; Van Arnold, J.; Croom, S.; Bickett, D. M.; Moss, M. L.; Tian, G.; Unwalla, R. J.; Lee, F. W.; Tippin, T, K.; James, M. K.; Grizzle, M. K.; Long, J. E.; Schuster, S. V. "6-Azasteroids: structure-activity relationships for inhibition of type 1 and 2 human 5 alpha-reductase and human adrenal 3 beta-hydroxy-delta 5-steroid dehydrogenase/3-keto-delta 5-steroid isomerase"; *J. Med. Chem.* **1994**, *37*, 2352-2360.

24. Guarna, A.; Belle, C.; Machetti, F.; Occhiato, E. G.; Payne, H. P.; Cassiani, C.; Comerci, A.; Danza, G.; De Bellis, A.; Dini, S.; Marrucci, A.; Serio, M. "19-Nor-10-azasteroids: A Novel Class of Inhibitors for Human Steroid 5α-Reductases 1 and 2"; *J. Med. Chem.* **1997**, *40*, 1112-1129.

25. Guarna, A.; Occhiato, E. G.; Machetti, F.; Marrucci, A.; Danza, G.; Serio, M.; Paoli, P. "19-Nor-10-azasteroids, a New Class of Steroid 5α-Reductase Inhibitors. 2. X-ray Structure, Molecular Modeling, Conformational Analysis of 19-Nor-10-azasteroids and Comparison with 4-Azasteroids and 6-Azasteroids"; *J. Med. Chem.* **1997**, *40*, 3466-3477.

26. Guarna, A.; Occhiato, E. G.; Machetti, F.; Scarpi, D. A "Concise Route to 19-Nor-10-azasteroids, a New Class of Steroid 5α-Reductase Inhibitors. 3. Synthesis of (+)-19-Nor-10-azatestosterone and (+)-17β-Acetyloxy-(5β)-10-azaestr-1-en-3-one"; *J. Org. Chem.* **1998**, *63*, 4111-4115.

27. Occhiato, E. G.; Scarpi, D.; Machetti, F.; Guarna, A. "Stereoselectivity in the TiCl$_4$-catalysed Reaction of Danishefsky's Diene with a *N*-(acyloxy)iminium ion: Synthesis of 5α versus 5β Δ$^{1(2)}$-19-Nor-10-azasteroids. Part 4"; *Tetrahedron* **1998**, *54*, 11589-11596.

28. Guarna, A.; Occhiato, E. G.; Machetti, F.; Giacomelli, V. "19-Nor-10-azasteroids. 5.[1] A Synthetic Strategy for the Preparation of (+)-17-(3-Pyridyl)-(5β)-10-azaestra-1,16-dien-3-one, a Novel Potential Inhibitor for Human Cytochrome P450$_{17α}$ (17α-Hydroxylase/C$_{17,20}$-lyase)"; *J. Org. Chem.* **1999**, *64*, 4985-4989.

29. Levy, M. A.; Brandt, M.; Sheedy, K. M.; Dinh, J. T.; Holt, D. A.; Garrison, L. M.; Bergsma, D. J.; Metcalf, B. W. "Epristeride is a Selective and Specific Uncompetitive Inhibitor of Human Steroid 5α-Reductase Isoform 2"; *J. Steroid. Biochem. Molec. Biol.* **1994**, *48*, 197-206.

30. Baine, N. H.; Owings, F. F.; Kline, D. N.; Resnick, T.; Ping, L.; Fox, M.; Mewshaw, R. E.; Tickner, A. M.; Kowalski, C. J. "Improved Syntheses of Epristeride, a Potent Human 5.alpha.-Reductase Inhibitor"; *J. Org. Chem.* **1994**, *59*, 5987-5989.

31. Holt, D. A.; Levy, M. A.; Oh, H.; Erb, J. M.; Heaslip, J. I.; Brandt, M.; Lan-Hargest, H.; Metcalf, B. W. "Inhibition of steroid 5 alpha-reductase by unsaturated 3-carboxysteroids"; *J. Med. Chem.* **1990**, *33*, 943-950.

32. Levy, M. A.; Brandt, M.; Heys, J. R.; Holt, D. A.; Metcalf, B. W. "Inhibition of rat liver steroid 5 alpha-reductase by 3-androstene-3-carboxylic acids: mechanism of enzyme-inhibitor interaction"; *Biochemistry* **1990**, *29*, 2815-2824.

33. Jones, C. D.; Audia, J. E.; Lawhorn, D. E.; McQuaid, L. A.; Neubauer, B. L.; Pike, A. J.; Pennington, P. A.; Stamm, N. A.; Toomey, R. E.; Hirsch, K. R."Nonsteroidal inhibitors of human type I steroid 5-alpha-reductase" *J. Med. Chem.* **1993**, *36*, 421-423.

34. Wikel, J. H.; Bemis, K. G.; Audia, J. E.; McQuaid, L. A.; Jones, C. D.; Pennington, P. A.; Lawhorn, D. E.; Hirsch, K. R.; Stamm, N. B. "QSAR study of benzoquinolinones as inhibitors of human type 1 5-α-reductase"; *Bioorg. Med. Chem. Lett.* **1993**, 1157-1162.

35. Abell, A. D.; Erhard, K. F.; Yen, H.-K.; Yamashita, D. S.; Brandt, M.; Mohammed, H.; Levy, M. A.; Holt, D. "Preparative chiral HPLC separation of all possible stereoisomers of LY191704 and

LY266111 and their *in vitro* inhibition of human types 1 and 2 steroid 5α-reductases"; *Bioorg. Med. Chem. Lett.* **1994**, *4*, 1365-1368.

36. Smith, E. C. R.; McQuaid, L. A.; Goode, R. L.; McNulty, A. M.; Neubauer, B. L.; Rocco, V. P.; Audia, J. E. "Synthesis and 5α-reductase inhibitory activity of 8-substituted benzo[f]quinolinones derived from palladium mediated coupling reactions"; *Bioorg. Med. Chem. Lett.* **1998**, *8*, 395-398.

37. (a) Abell, A. D.; Prince, M. J.; McNulty, A. M.; Neubauer, B. L. "Simple Bi- and tricyclic inhibitors of human steroid 5α-reductase"; *Bioorg. Med. Chem. Lett.* **2000**, *10*, 1909-1911. b) Hartmann, R. W.; Reichert, M. "New nonsteroidal steroid 5 alpha-reductase inhibitors. Syntheses and structure-activity studies on carboxamide phenylalkyl-substituted pyridones and piperidones"; *Arch. Pharm.* (Weinheim) **2000**, *333*, 145-153. (c) Hartmann, R. W.; Reichert, M.; Göring, S. "Novel 5 alpha reductase inhibitors. Synthesis and structure–activity studies of 5-substituted 1-methyl-2-pyridones and 1-methyl-2-piperidones"; *Eur. J. Med. Chem.* **1994**, *29*, 807-817. (d) Göring, S.; Reichert, M.; Hartmann, R. W. "FC2 5α-reductase inhibitors: Synthesis and structure-activity studies on new non-steroidal compounds"; *Eur. J. Pharm. Sci.* **1994**, *2*, 101. (e) Baston, E.; Palusczak, A.; Hartmann, R. W. "6-Substituted 1H-quinolin-2-ones and 2-methoxy-quinolines: synthesis and evaluation as inhibitors of steroid 5alpha reductases types 1 and 2"; *Eur. J. Med. Chem.* **2000**, *35*, 931-940. (f) Abell, A. D.; Phillips, A. J.; Budhia, S.; McNulty, A. M.; Neubauer, B. L. "The Preparation and Biological Activity of Lactam-Based, Non-Steroidal, Inhibitors of Human Type-1 Steroid 5α-Reductase"; *Aust. J. Chem.* **1998**, *51*, 389-396.

38. Mook, R. A., Jr.; Lackey, K.; Bennet, C. "Synthesis of phenanthridin-3-one derivatives: non-steroidal inhibitors of steroid 5-α-reductase"; *Tetrahedron Lett.* **1995**, *36*, 3969-3972.

39. Guarna, A.; Machetti, F.; Occhiato, E. G.; Scarpi, D.; Comerci, A.; Danza, G.; Mancina, R.; Serio, M.; Hardy, K. "Benzo[c]quinolizin-3-ones: a novel class of potent and selective nonsteroidal inhibitors of human steroid 5alpha-reductase 1"; *J. Med. Chem.* **2000**, *43*, 3718-3735.

40. Guarna, A.; Occhiato, E. G.; Scarpi, D.; Tsai, R.; Danza, G.; Comerci, A.; Mancina, R.; Serio, M. "Synthesis of benzo[c]quinolizin-3-ones: selective non-steroidal inhibitors of steroid 5 alpha-reductase 1"; *Bioorg. Med. Chem. Lett.* **1998**, *8*, 2871-2876.

41. Guarna, A.; Occhiato, E. G.; Scarpi, D.; Zorn, C.; Danza, G.; Comerci, A.; Mancina, R.; Serio, M. "Synthesis of 8-chloro-benzo[c]quinolizin-3-ones as potent and selective inhibitors of human steroid 5alpha-reductase 1"; *Bioorg. Med. Chem. Lett.* **2000**, *10*, 353-356.

42. (a) Abell, A. D.; Brandt, M.; Levy, M. A.; Holt, D. A. "A comparison of steroidal and non-steroidal inhibitors of human steroid 5α-reductase: new tricyclic aryl acid inhibitors of the type-1 isozyme"; *Bioorg. Med. Chem. Lett.* **1996**, *6*, 481-484. (b) Reichert, W.; Jose, J.; Hartmann, R. W. "5 alpha-reductase in intact DU145 cells: evidence for isozyme I and evaluation of novel inhibitors"; *Arch. Pharm.* (Weinheim) **2000**, *333*, 201-204. (c) Baston, E.; Hartmann, R. W. "N-substituted 4-(5-indolyl)benzoic acids. Synthesis and evaluation of steroid 5α-reductase type I and II inhibitory activity"; *Bioorg. Med. Chem. Lett.* **1999**, *9*, 1601-1606. (d) Igarashi, S.; Inami, H.; Hara, H.; Fujii, M.; Koutoku, H.; Oritani, H.; Mase, T. "A novel class of inhibitors for human steroid 5alpha-reductase: synthesis and biological evaluation of indole derivatives. II"; *Chem. Pharm. Bull.* (Tokyo) **2000**, *48*, 382-388. (e) Ishibashi, K.; Nakajima, K.; Sugioka, Y.; Sugiyama, M.; Hamada, T.; Horikoshi, H.; Nishi, T. "Synthesis of 2-phenylbenzofuran derivatives as testosterone 5 alpha-reductase inhibitor"; *Chem. Pharm. Bull.* (Tokyo) **1999**, *47*, 226-240. (f) Ishibashi, K.; Nakajima, K.; Sugioka, Y.; Sugiyama, M.; Hamada, T.; Horikoshi, H.; Nishi, T. "Synthesis and 5α-reductase inhibitory activities of benzofuran derivatives with a carbamoyl group"; *Bioorg. Med. Chem. Lett.* **1998**, *8*, 561-566. (g) Blagg, J.; Ballard, S. A.; Cooper, K.; Finn, P. W.; Johnson, P. S.; MacIntyre, F.; Maw, G. N.; Spargo, P. L. "The development of non-steroidal dual inhibitors of both human 5α-reductase isozymes"; *Bioorg. Med. Chem. Lett.* **1996**, *6*, 1517-1522.

43. Occhiato, E. G; Guarna, A.; Brandi, A.; Goti, A.; De Sarlo, F. "N-Bridgehead Polycyclic Compounds by Sequential Rearrangement-Annulation of Isoxazoline-5-Spirocyclopropanes. 6. A General Synthetic Method for 5,6-Dihydro-7(8H)- and 2,3,5,6-Tetrahydro-7(1H)-indolizinones"; *J. Org. Chem.* **1992**, *57*, 4206-4211.

44. Ochoa, E.; Mann, M.; Sperling, D.; Fabian, J. "A Combined Density Functional and *ab initio* Quantum Chemical Study of the Brandi Reaction"; *Eur. J. Org. Chem.* **2001**, *22*, 4223-4231.

45. Brandi-Guarna Rearrangement *Tetrahedron Organic Chemistry Series*, Vol. 22 in *Organic Synthesis Based on Name Reactions*, A Hassner , C. Stumer , pag. 42, Second Edition, Pergamon.

46. Belle, C.; Cardelli, A.; Guarna; A. "Sequential Rearrangement-Annulation of Isoxazoline-5-Spirocyclopropanes. Total Synthesis of (±)-$\Delta^{9(11)}$-19-Nor-10-Aza-Testosterone"; *Tetrahedron Lett.* 1991, *32*, 6395-6398.

47. Pilli, R. A.; Dias, L. C.; Maldaner, A. O. "One-Pot Preparation of Quinolizidin-2-one and Indolizidin-7-one Ring Systems. Concise Total Syntheses of (±)-Myrtine, (±)-Lasubine II, and (-)-Indolizidine 223AB"; *J. Org. Chem.* 1995, *60*, 717-722.

48. Mali, R. S.; Yadav, V. J. A. "Useful Synthesis of Ethyl Indole-2-carboxylates and 3,4-Dihydrocarbostyrils"; *Synthesis* 1984, 862-865.

49. Mayer, F.; van Zütphen, L.; Philipps, H. "Eine neue Darstellungweise von Hydro-carbostyril und seinen Abkömmlingen. (A new method of preparation of hydrocarbostyril and its derivatives)"; *Chem. Ber.* 1927, *60*, 858-864.

50. Greene, T. W.; Wutz, P. G. M. *Protective Groups in Organic Synthesis*; John Wiley & Sons, Inc.: New York, 1991.

51. Von Isler, O.; Gutmann, H.; Montavon, M.; Rüegg, R.; Ryser, G.; Zeller, P. "Use of the Wittig Reaction for the Synthesis of Bixins and Crocetins Esters"; *Helv. Chim. Acta* 1957, *40*, 1242-1249.

52. Fleming, I.; Paterson, I. "A simple Synthesis of Carvone Using Silyl Enol Ethers"; *Synthesis* 1979, 736-738.

53. Kim, G.; Chu-Moyer, M. Y.; Danishefsky, S. J.; Schulte, G. K. "The total synthesis of indolizomycin"; *J. Am. Chem. Soc.* 1993, *115*, 30-39.

54. Fang, F. G.; Prato, M.; Kim, G.; Danishefsky, S. J. "The aza-Robinson annulation: an application to the synthesis of iso-A58365A"; *Tetrahedron Lett.* 1989, *30*, 3625-3628.

55. Maggini, M.; Prato, M.; Ranelli, M.; Scorrano, G. "Synthesis of (-)-8-deoxy-7-hydroxyswainsonine and (±)-6,8-dideoxycastanospermine"; *Tetrahedron Lett.* 1992, *33*, 6537-6540.

56. Heathcock, C. H.; Davidsen, S. K.; Mills, S. G.; Sanner, M. A. "Daphniphyllum alkaloids. 10. Classical total synthesis of methyl homodaphniphyllate"; *J. Org. Chem.* 1992, *57*, 2531-2544.

57. Guarna, A.; Lombardi, E.; Machetti, F.; Occhiato, E. G.; Scarpi, D. "Modification of the Aza-Robinson Annulation for the Synthesis of 4-Methyl-Benzo[c]quinolizin-3-ones as Potent and Selective Inhibitors of Steroid 5a-Reductase 1"; *J. Org. Chem.* 2000, *65*, 8093-8095.

58. Pederson, B. S.; Sheibye, S.; Nilsson, N. H.; Lawesson, S. "Studies on organophosphorus compounds. XX. Syntheses of thioketones"; *Bull. Soc. Chim. Belg.* 1978, *87*, 223-228.

59. Michael, J. P.; Jungmann, C. "New syntheses of (±)-lamprolobine and (±)-epilamprolobine"; *Tetrahedron* 1992, *48*, 10211-10220.

60. di Salle, E.; Giudici, D.; Radice, A.; Zaccheo, T.; Ornati, G.; Nesi, M.; Panzeri, A.; Délos, S.; Martin, P. M. "PNU 157706, a Novel Dual Type I and II 5α reductase Inhibitor"; *J. Steroid Biochem. Molec. Biol.* 1998, *64*, 179-186.

61. Elangovan, A.; Wang, Y.-H.; Ho, T.-I. "Sonogashira Coupling Reaction with Diminished Homocoupling"; *Org. Lett.* 2003, *5*, 1841-1844.

62. Milstein, D.; Stille, J. K. "A general, selective, and facile method for ketone synthesis from acid chlorides and organotin compounds catalyzed by palladium"; *J. Am. Chem. Soc.* 1978, *100*, 3636-3638.

63. Miyaura, N.; Suzuki, A. "Palladium-Catalyzed Cross-Coupling Reactions of Organoboron Compounds"; *Chem. Rev.* 1995, *95*, 2457-2483.

64. A full discussion of the SAR data put in the context of other classes of non-steroidal inhibitors is reported in ref. 17c.

65. Bakshi, R. K.; Patel, G. F.; Rasmusson, G. H.; Baginsky, W. F.; Cimis, G.; Ellsworth, K.; Chang, B.; Bull, H.; Tolman, R. L.; Harris, G. S. "4,7b-Dimethyl-4-azacholestan-3-one (MK386) and related 4-Azasteroids as Selective Inhibitors of Human Type 1 5α-Reductase"; *J. Med. Chem.* 1994, *37*, 3871-3874.